Encyclopedia

of the

OPERA

Encyclopedia

of the

OPERA

NEW ENLARGED EDITION

by

DAVID EWEN

 HILL AND WANG · NEW YORK

TO MY BROTHER

Louis

*in gratitude for a lifetime
of faith and encouragement*

Preface

~~~~~~

The ENCYCLOPEDIA OF THE OPERA is the first book of its kind in any language. The Editor has attempted to make it a comprehensive source book about opera and opera performance. It is not exclusively a book of opera plots; or a book of biographies of composers and performers; or a history of opera.

The opera lover will find within this volume the following materials:

STORIES OF OPERAS. The most significant operas (over a hundred) are treated in detail, act by act, scene by scene, with descriptions of the principal arias and other important excerpts and information about the composition and the performance history. Several hundred other operas are treated more succinctly with brief capsule plots and citations of the musical highlights.

The operas included in the encyclopedia embrace all works that are frequently performed and most of the works that make an intermittent appearance on the opera stages of the world; operas that have special historic importance or interest; operas with unusual technical approaches (atonal operas, quartertone operas, monodramas, and so on); operas that were once highly successful and have gone into discard, and operas that have been discarded but which have recently been revived. The attempt, in short, has been to provide basic information about any opera in which the operagoer might be interested for one reason or another.

Some of the operas are listed in their original foreign titles; others in titles translated into English. The general practice, in order to make the encyclopedia more functional, has been to use the names most familiar to Americans. Since *The Jewels of the Madonna* is better known to Americans than *I gioielli della Madonna* that opera is listed under the English title; and since *Il trovatore* is more familiar than *The Troubadour,* the original Italian title is used. The capitalization of foreign titles has been followed in the foreign form. The music lover consulting this encyclopedia should, however, have no difficulty in locating the opera he wants since there are ample cross references to guide him.

CHARACTERS OF OPERAS. There will be found here an alphabetical listing of all the characters of a hundred or so major operas, and of the principal characters of several hundred other operas. This, the Editor believes, is the most complete "who's who" of opera characters to be found anywhere. Characters with titles are listed not under the title but under their name: Count Ceprano or Don Basilio appears not under "Count" or "Don" but under "Ceprano" and "Basilio."

EXCERPTS FROM OPERAS. The most important excerpts from particular operas—arias, duets, ensemble numbers, choruses, orchestral passages—are not only found in the discussion of the operas themselves, but are listed separately, in the proper alphabetical order, together with the source. An opera lover curious about the origin of a favorite aria or duet can now easily trace its source.

BIOGRAPHIES. The major figures in every facet of operatic composition and performance are given succinct biographical treatment. These figures embrace: composers; librettists; singers; conductors; stage directors; impresarios; teachers; critics; musicologists; and so on.

HISTORY OF OPERA. The history of opera—its origin, evolution, and development in all the major countries of the world—is discussed in a special article. But there are also additional listings for, and brief explanations of, every form of opera (Masque, Ballad Opera, Singspiel, Opera Buffa, Music Drama, and so forth) together with explanations of every major trend and style in opera.

OPERA IN PERFORMANCE. The history of every major opera house of the world, and of every important festival emphasizing opera, is to be found here, along with general articles on the history of opera performances in Europe and America, and articles on opera over radio, television, and in the recordings.

LITERARY SOURCES. Listings are made of all the major authors ot the world, from Aeschylus to Eugene O'Neill, whose writings have been a source for opera texts; there are also listings of many of the great classics of literature and the operas derived from them.

SPECIAL ARTICLES. Specific subjects of operatic interest are discussed in special articles. The varied material thus treated includes the Aria, the Overture, Marionette Opera, Children's Opera, Ballet in Opera, Claque, Castrato, etc.

TERMS IN OPERA. All the important technical terms found in opera are defined.

PRONUNCIATION GUIDE. A special section in the back of the volume is devoted to pronunciation of the names of operas, arias, composers, and performers. Pronunciations are given for all of the foreign names and titles which appear as main entries in the book, and for English names as well when their pronunciation is difficult or doubtful.

It is hoped that the veteran operagoer and the musician will find this volume a convenient source for any information they may seek about the opera. But this volume has also been designed with an eye on the new audience of opera lovers springing up throughout America. Through radio, television, motion pictures, and recordings, opera is now available to millions. Besides, grassroot opera companies are springing up in all parts of the country, bringing live performances to places where formerly opera had been only a word. It seemed to this Editor that these new audiences, whose ranks are continually swelling, required a reference book providing them with the basic information needed for a maturer understanding of this great art form. It is hoped that this book will answer all the questions which will inevitably arise as these new audiences penetrate ever deeper into the fabulous world of the opera.

The Editor would like to express his indebtedness to several persons, the importance of whose co-operation would be difficult to overestimate: to Leon Wilson, for his painstaking copy editing; to Hedy D. Jellinek, for her meticulous proofreading; to Esther Gelatt for rewarding advice, criticism, and suggestions; to Reason A. Goodwin, for the preparation of the Pronunciation Guide; and finally to

George Jellinek and Nicolas Slonimsky, whose scholarship combed the pages of this book for inaccuracies, inconsistencies, and misstatements. Nicolas Slonimsky often provided the Editor with carefully documented information which enabled him to correct some of the errors in dates and place names that appear in many other reference books.

DAVID EWEN

Supplement

A

Aniara, opera by Karl-Birger Blom-dahl. Libretto by Erika Lindegren, based on a text by Harry Martinson. Première: Stockholm, May 31, 1959. "Aniara" is a spaceship carrying eight thousand passengers from earth to Mars. The ship has been traveling for twenty years when those aboard receive news that it is off course and must travel in space forever. Much of the score's interest lies in its use of electronic sounds to describe interstellar atmosphere and, at one point, to simulate the voices of an incalculable multitude.

Antheil, George, composer. Died New York, February 12, 1959.

Assassinio nella cattedrale (Murder in the Cathedral), opera by Ildebrando Pizzetti. Libretto by Alberto Castelli, based on a poetical drama by T. S. Eliot. Première: La Scala, Milan, March 1, 1958. The action revolves around the assassination of Thomas à Becket, the twelfth-century Arch-bishop, in the Canterbury Cathedral.

Auric, Georges, composer and opera manager. Born Lodève, France, February 15, 1899. He was a pupil of Vincent d'Indy and Albert Roussel. He came to prominence in the early 1920s, when his name was linked with the names of five other young French composers (including Milhaud, Honegger, and Poulenc) in an *avant-garde* group identified as "The Six." Auric later achieved success as a composer with scores for ballets and motion pictures. In 1962 he was appointed head of both the Paris Opéra and the Opéra-Comique.

B

Badings, Henk, composer. Born Bandoeng, Java, January 17, 1907. After completing his music study with Willem Pijper, he became in 1937 professor at the Rotterdam Conservatory and in 1941 director of the Royal Conservatory at The Hague. Since World War II, he has achieved recognition as one of Holland's leading composers. He has written several important operas, including *The Night Watch* and *Martin Korda, D.P.* The latter was introduced at the Holland Festival on June 15, 1960.

Ballad of Baby Doe, The, opera by Douglas Moore. Libretto by John La Touche. Première: Central City, Colorado, July 7, 1956. The setting is Leadville, Colorado, during the gold rush in the nineteenth century. The opera traces the career of Horatio Tabor, a prospector who becomes wealthy and politically powerful. He deserts his wife for Baby Doe, a blond beauty from Wisconsin. Tabor eventually loses his wealth and power, but Baby Doe remains true to him until his death.

Barber, Samuel, composer. Born West Chester, Pennsylvania, March 9, 1910. His aunt was the celebrated opera singer Louise Homer. He attended the Curtis Institute in Philadelphia, following which he won several important prizes for instrumental compositions and a Pulitzer Traveling Scholarship. By virtue of numerous distinguished works for orchestra and chamber-music groups, he won an eminent position in American music. His first opera came after he had achieved full maturity as a composer, with *Vanessa* (1958), one of the most significant American operas of our time. It was introduced by the Metropolitan Opera on January 15, 1958.

Bayreuth Festival. In 1960 Wagner's *Ring* cycle was presented with new staging under the direction of Rudolf Kempe. A year later Grace Bumbry, mezzo-soprano, became the first Negro artist to appear at the Festival, scoring as Venus in *Tannhäuser.* Tradition was shattered in the summer of 1962, when the silence customary during a performance was broken by an audience ovation for Irene Dalis, American mezzo-soprano, following her rendition

of Ortrud's invocation in *Lohengrin.*

Beecham, Sir Thomas, conductor. Died London, March 8, 1961.

Berlin Opera (Städtische Oper). A new auditorium was built in West Berlin, on the site of the old opera house destroyed during World War II. It opened in the fall of 1961 with a gala performance of *Don Giovanni,* Ferenc Fricsay conducting.

Bernstein, Leonard, composer and conductor. He made his Covent Garden debut in 1961 in *Falstaff.*

Bing, Rudolf, opera manager. In 1959 he was made Commander of the Order of Merit by the Italian government and received the grand Silver Medal of Honor from Austria.

Bjoerling, Jussi, tenor. Died Siaro, Sweden, September 9, 1960.

Blacher, Boris, composer. His opera *Rosamunde Floris,* libretto by Gerhart von Westerman based on a play by Georg Kaiser, was introduced in Berlin in the fall of 1960.

Blech, Leo, conductor. Died Berlin, August 24, 1958.

Blitzstein, Marc, composer. His opera *Juno,* based on Sean O'Casey's *Juno and the Paycock,* was introduced in New York, for a Broadway theater run, in 1959.

Blomdahl, Karl-Birger, composer. Born Växjö, Sweden, October 19, 1916. His musical studies took place with Hilding Rosenberg in Sweden, as well as in Paris and Rome. He subsequently attracted notice with various works in a progressive idiom—including the opera *Aniara,* which employs electronic sounds and whose world première took place in Sweden in 1959.

Böhm, Karl, conductor. He resigned as director of the Vienna State Opera in 1956, but continued appearing there as conductor. His American debut took place with the Chicago Symphony on February 9, 1956. Since then he has made distinguished appearances with the Metropolitan Opera.

Bori, Lucrezia, soprano. Died New York, May 14, 1960.

Borkh, Inge, soprano. Born Mannheim, Germany, May 26, 1921. She made appearances as an actress in Austria before turning to opera. After a period of vocal study in Italy and at the Mozarteum in Salzburg, she made her debut in Lucerne, Switzerland, in Johann Strauss's *Zigeunerbaron* during 1940–41. She scored an outstanding success in Berlin in 1951 in Menotti's *The Consul.* Further successes came in Paris (where she sang opposite Alexander Welitsch, baritone, whom she later married), Bayreuth, and Edinburgh. Her American debut took place with the San Francisco Opera in the fall of 1953, when she appeared as Elektra in Richard Strauss's opera. Later appearances at Covent Garden, the Stuttgart Opera, La Scala, and the Metropolitan Opera added significantly to her fame. In 1955 she created the role of Cathleen in Egk's *Irische Legende* at the Salzburg Festival.

Brecht, Bertolt, poet and dramatist. Died Berlin, East Germany, August 14, 1956.

Britten, Benjamin, composer. His opera *A Midsummer Night's Dream*—libretto by the composer in collaboration with Peter Pears, based on Shakespeare—was introduced at the Aldeburgh Festival in England on June 11, 1960.

C

Callas, Maria Meneghini, soprano. She made a sensational debut at the Metropolitan Opera on October 29, 1956, as Norma.

Castelnuovo-Tedesco, Mario, composer. Born Florence, Italy, April 3, 1895. He studied music first at the Cherubini Institute in Florence, then privately with Ildebrando Pizzetti. His first important work was *Fioretti,* a setting of three verses of St. Francis of Assisi, in 1920. This was followed by *La Mandragola,* an opera that won the Italian Prize and was introduced in Venice in 1926. He originally became known throughout the music world, however, as a composer of orchestral music. In 1939 he came to America, and later he established permanent residence in Beverly Hills as a teacher of composition and creator of motion picture scores. In 1956 he completed two operas based on Shakespeare, *All's Well That Ends Well* and *The Merchant of Venice.* The latter received first prize in a 1958 international competition sponsored by La Scala. Its première took place at the Florence May Music Festival in 1961.

Cervantes, Miguel de, novelist, dramatist, poet. Goffredo Petrassi's one-act opera *Il Cordovani* was based on Cervantes' *El Viejo gedoso.*

Charpentier, Gustave, composer. Died Paris, February 18, 1956.

Chaucer, Geoffrey, poet. Two of Erik Chisholm's operas were derived from Chaucer—*The Pardoner's Tale* came from several Chaucer verses and *The Canterbury Tales* from the narrative poem of the same name.

Christoff, Boris, bass-baritone. Born Sofia, Bulgaria, May 18, 1918. After studying law, he decided upon a career in music. He took singing lessons in Rome with Riccardo Stracciari and then spent several years in Salzburg studying the German repertory. His debut, at a concert, took place in Rome, 1945. A year later he appeared successfully at La Scala, and was first heard at Covent Garden in 1949. His American debut took place with the San Francisco Opera on September 10, 1956, in one of his most celebrated roles, Boris Godunov.

Cocteau, Jean, novelist, playwright, poet. Poulenc's one-act opera with a single character, *La Voix humaine,* was based on Cocteau's text.

Cooper, Emil, conductor. Died New York, November 16, 1960.

Corelli, Franco, tenor. Born Ancona, Italy, April 8, 1923. Self-taught in singing, he made his opera debut at Spoleto in 1952 in *Carmen.* His success there brought him an invitation to appear in Rome, in the same opera, later that year. In 1953 he was heard at the Florence May Music Festival in the first performance of Prokofiev's *War and Peace* outside the Soviet Union. Corelli made his debut at La Scala on the opening night of the 1953–54 season in *La Vestale.* Since then he has been a leading tenor of that company and has made notable

guest appearances in other leading European opera houses. He made a highly successful debut at the Metropolitan Opera as Manrico in *Il Trovatore* on January 27, 1961.

Covent Garden. Georg Solti, who had made his Covent Garden debut as conductor in December 1959 in *Der Rosenkavalier,* was appointed musical director in September 1961. His first performance in that capacity took place on September 14, 1961, in *Iphigénie en Tauride.* Among the most significant new operas heard at Covent Garden since 1956 were Britten's *A Midsummer Night's Dream* on February 2, 1961, and Michael Tippett's *King Priam* on June 8, 1962; neither, however, was a world première.

Crucible, The, opera by Robert Ward. Libretto by Bernard Stambler based on the play by Arthur Miller. Première: New York City Opera, October 26, 1961. Commissioned by and presented under the auspices of the Ford Foundation, *The Crucible* received the Pulitzer Prize in music in 1962. Its setting is Salem, Massachusetts, during the era of the witch trials.

D

Del Monaco, Mario, tenor. Born Florence, Italy, May 27, 1915. He attended the Pesaro Conservatory, but studied music mostly by listening to records. His professional debut, as Pinkerton, took place at the Teatro Puccini in Milan on January 1, 1941. His career acquired a new dimension after World War II, with successes at Covent Garden, La Scala, and throughout South America. He appeared first in the United States with the San Francisco Opera in 1950, and his Metropolitan

Opera debut was on November 27, 1950, in *Manon Lescaut.*

Dialogues des Carmélites, Les (The Dialogues of the Carmelites), opera by Francis Poulenc. Libretto by George Bernanos, based on a novel by Gertrude von Le Fort and a motion-picture scenario by Philippe Agostini and the Rev. R. V. Bruckberger. Première: La Scala, January 26, 1957. In Paris, during the French Revolution, sixteen nuns accept death at the guillotine rather than permit their order to be broken up. The score is deeply religious in content and feeling, achieving a state of exaltation in such moments as the second-act "Ave Maria" and the closing-scene "Salve Regina." The opera received its American première at the San Francisco Opera on September 22, 1957. It was telecast over the NBC network in 1957 and received the New York Music Critics Circle Award as the best opera of the season.

Ditson Fund. Since 1955 the Ditson Fund of Columbia University has provided funds to help mount or record American operas. Operas whose premières were partly financed by the Fund are: Jack Beeson's *The Sweet Bye and Bye* (1957) and Harry Partch's dance opera *Bewitched* (1959), both produced by the Juilliard School of Music. The Fund also helped to finance a complete recording of Robert Ward's Pulitzer Prize opera, *The Crucible.*

Dzerzhinsky, Ivan, composer. In 1961 he completed a new opera, *The Fate of Man.*

E

Easton, Florence, soprano. Died New York, August 13, 1955.

Ebert, Carl, opera manager. In 1959 he resigned as artistic director of the Glyndebourne Opera.

Edinburgh International Festival of Music and Drama. These are some important recent events: in 1959, the first performance outside Sweden of Karl-Birger Blomdahl's *Aniara;* in 1961, a presentation of Britten's *A Midsummer Night's Dream;* in 1962, revivals of Prokofiev's *The Love for Three Oranges* and *The Gambler.*

In 1959 Robert Posonby resigned as artistic director, a post he had held since 1955; he was succeeded by the Earl of Harewood.

Egk, Werner, composer. His opera *Der Revisor,* based on Gogol, was introduced in Schwetzingen, Germany, in 1957.

Elegie für Junge Liebende (Elegy for Young Lovers), opera by Hans Werner Henze. Libretto by W. H. Auden and Chester Kallmann. Première: Schwetzingen, Germany, May 20, 1961. The central character has been described as a "genius who feeds on others and who in a fit of rage sends two young lovers, one of them his mistress who has just left him, to their death." The musical style is a modification of the twelve-tone technique. After a highly successful première, the opera was seen in its first year at Glyndebourne (in an English-language presentation) and in Munich. In October 1962 it was produced at the twelfth annual Berlin Cultural Festival, when it was singled out as the Festival's "best new opera production."

F

Falla, Manuel de, composer. His last work, the oratorio *L'Atlàntida,* received its first stage performance at La Scala on June 18, 1962, Thomas Schippers conducting.

Farrell, Eileen, dramatic soprano. Born

Willimantic, Connecticut, February 13, 1920. She studied singing with Merle Alcock and Eleanor McLellan. She then was heard over the radio and she toured the United States in song recitals. She appeared in several modern operas given in concert version by the New York Philharmonic Orchestra, including Berg's *Wozzeck* and Milhaud's *Les Choëphores*. In March 1956 she made her opera debut in Tampa, Florida, in *Cavalleria rusticana*. After that came significant appearances with the San Francisco Opera and the Lyric Opera of Chicago. Her debut at the Metropolitan Opera took place on December 6, 1960, in the title role of Gluck's *Alceste*. She was starred in *Andrea Chénier* on opening night of the Metropolitan Opera's 1962–63 season.

Festival of Two Worlds, The. An annual summer festival of the arts held in Spoleto, Italy. It was founded in 1958 by Gian-Carlo Menotti, who has since served as its artistic director. The program of the first festival season included Verdi's *Macbeth*. Since then the more significant events in the field of opera have included such novelties as Henze's *Der Prinz von Homburg*, Prokofiev's *The Love for Three Oranges*, and Rossini's *Le Comte Ory*.

Fevrier, Henri, composer. Died Paris, July 8, 1957.

Flagstad, Kirsten, soprano. She served as the director of the Norwegian Opera in Oslo during the 1959–60 season. She died in Oslo, Norway, on December 7, 1962.

Florence May Music Festival (Maggio Musicale Fiorentino). Here are the most significant operatic events to take place recently at this annual spring festival: the world premières of Malipiero's *Il figliuol prodigo* and *Venere prigioniera* in 1957; a revival of Verdi's *La Battaglia di Legnano,* 1959; the revival of Peri's *Euridice* and Cherubini's *Elisa* in 1960; the world première

of Castelnuovo-Tedesco's *The Merchant of Venice* and presentations of Leonard Bernstein's *West Side Story* and Pizzetti's *Assassinio nella cattedrale* in 1961; revivals of Traetta's *Antigona* and Paisiello's *Lo Molinara* in 1962.

Floyd, Carlisle, composer. Born Latta, South Carolina, June 11, 1926. He studied piano with private teachers, then received a piano scholarship at Converse College. Later he studied with Ernst Bacon at Syracuse University and with Rudolf Firkušny. In 1947 he was appointed to the music faculty of Florida State University, in Tallahassee. He wrote two operas— *Slow Dusk* (1949) and *Fugitives* (1951)—before achieving success with *Susannah,* introduced in Tallahassee, Florida, on February 24, 1955. In 1956 *Susannah* was produced by the New York City Opera and received the New York Music Critics Circle Award; in 1958 it was given at the Brussels Exposition in Belgium. *Wuthering Heights,* an opera based on the novel of Emily Brontë, was seen in Santa Fe, New Mexico, in 1958 and was produced a year later by the New York City Opera. *The Passion of Jonathan Wade,* written under the auspices of the Ford Foundation, received its world première at the New York City Opera on October 11, 1962.

Ford Foundation Grants for Opera. The Ford Foundation has supported an ambitious program to aid American opera by providing grants to American composers for the creation of new operas. These are some of the operas subsidized by this program: Norman Dello Joio's *Blood Moon;* Douglas Moore's *The Wings of the Dove;* Vittorio Giannini's *The Harvest;* Abraham Ellstein's *The Golem;* Robert Ward's *The Crucible;* and Carlisle Floyd's *The Passion of Jonathan Wade.* Other composers commissioned to write new operas include Marc Blitzstein, Lee Hoiby, Marvin David Levy, Ned Rorem, and Hugo Weisgall.

The Ford Foundation has also provided great financial assistance to opera companies (the Metropolitan Opera, Chicago Lyric Opera, San Francisco Opera, and the New York City Opera) for the promotion of American opera.

Fortner, Wolfgang, composer. Born Leipzig, Germany, October 12, 1907. After attending the Leipzig Conservatory, he joined the faculty of the Evangelical Church Music Institute in Heidelberg in 1931. In 1954 he was appointed professor of composition at Northwestern Music Academy in Detmold, and in 1957 he became Dozent at the Musikhochschule in Freiburg-im-Breisgau. His numerous compositions in all forms include several notable operas: *Die Witwe von Ephesus* (1952); *Der Wald* (1953); *Die Bluthochzeit* (1957); and *In seinem Garten* (1962).

Foss, Lukas, composer. *Griffelkin,* an opera commissioned for television, received its world première over the NBC network on November 6, 1955. Foss subsequently wrote a nine-min-

ute opera, *Introductions and Goodbyes,* presented in a concert version by the New York Philharmonic on May 7, 1960.

Françaix, Jean, composer. His later works include a one-act chamber opera for two singers and fourteen instrumentalists, *Le Diable boiteux.*

Fricsay, Ferenc, conductor. Born Budapest, August 9, 1914. After completing his music study with Béla Bartók and Zoltan Kodály, he made his conducting debut in Szeged in 1936. In 1945 he became a principal conductor of the Budapest Opera, and between 1945 and 1949 he made distinguished appearances in opera performances in Vienna, Salzburg, and Holland. In 1949 he was appointed principal conductor of the RIAS Orchestra in the American sector of Berlin. His American debut took place with the Boston Symphony on November 13, 1953. For a brief period Fricsay was music director of the Houston Symphony; he then returned to Europe, where he became principal conductor of the West Berlin Opera.

G

Gedda, Nicolai, tenor. Born Stockholm, Sweden, July 11, 1925. He is the son of the choirmaster of the Russian Orthodox Church in Stockholm, who for a number of years was also a member of the famed Don Cossack Chorus. Nicolai Gedda received his academic schooling at the Soedra Latin in Stockholm and his training in voice from private teachers. In 1952 he made his opera debut with the Stockholm Opera in Adam's *Le Postillon de Longjumeau.* One year later he ap-

peared for the first time at La Scala, in the role of Don Ottavio in *Don Giovanni.* He then made his debut at the Paris Opéra in *Faust,* at Covent Garden in *Rigoletto,* and at the Vienna Opera in *Carmen.* His first appearance at the Metropolitan Opera took place on November 1, 1957, in *Faust.* In 1958 he created there the role of Anatol in Samuel Barber's *Vanessa.*

Ghedini, Giorgio Federico, composer. His later operas include *Le Baccanti* and *L'Ipocrita felice,* the latter derived from Max Beerbohm's *The Happy Hypocrite.*

Giannini, Vittorio, composer. *The Harvest,* written under a Ford Foundation grant, was introduced by the Lyric Opera of Chicago on November 25, 1961. *Rehearsal Call,* described as a "horseplay opera," was given in New York on February 15, 1962.

Gigli, Beniamino, tenor. Died Rome, November 30, 1957.

Glyndebourne Opera. In 1956, in celebration of Mozart's bicentenary, the Glyndebourne Opera devoted its entire repertory to that master, featuring six operas, two in new productions. In 1959 there were new productions of *Fidelio* and *Der Rosenkavalier;* in 1961, an English-language presentation of Henze's *Elegie für Junge Liebende (Elegy for Young Lovers);* and in 1962, a distinguished revival of Monteverdi's *L'Incoronazione di Poppea.*

Gogol, Nikolai, author. *The Inspector General* of Gogol was the source of Werner Egk's opera *Der Revisor.*

Goossens, Eugène, (III), composer and conductor. Died London, June 13, 1962.

Graf, Herbert, stage director. Between 1960 and 1963 he was the artistic director of the Zurich Opera.

Gretchaninov, Alexander, composer. Died New York, January 3, 1956.

H

Harshaw, Margaret, contralto and soprano. Born Narbeth, Pennsylvania, May 12, 1912. After winning the Metropolitan Opera Auditions in 1942, she made her debut at the Metropolitan Opera on November 25, 1942, as a contralto, singing the part of the Second Norn in *Götterdämmerung.* She continued to sing contralto parts for about eight years. In 1950 she began to sing soprano roles, specializing in the Wagnerian repertory. Her first soprano appearance, as Senta in *The Flying Dutchman,* took place on November 22, 1950. Since then she has won world-wide acclaim in all the leading soprano roles of Wagner's music dramas.

Haug, Hans, composer. He completed the opera *Der Spiegel der Agrippina* in 1954.

Haydn, Franz Joseph, composer. His opera buffa *Il Mondo della luna,* forgotten for almost two centuries, was revived at the Holland Music Festival on June 24, 1959. The manuscript of a lost Haydn Singspiel, *Die Feuerbrunst,* was found in the Yale University library, and this opera received its first modern performance in Sweden during the summer of 1961. In June 1962, Haydn's long neglected opera *L'Infedeltà delusa,* edited by H. C. Robbins Landon, was revived at the Stockholm Festival.

Hempel, Frieda, soprano. Died Berlin, October 7, 1955.

Henze, Hans Werner, composer. The world première of his opera *König Hirsch* stirred a tempest when introduced in Berlin during the Festival Weeks in the fall of 1956. Considerable antagonism and controversy were aroused by its advanced musical style, which is a modification and a personalization of the twelve-tone technique. The same idiom characterizes his later operas: *Der Prinz von Homburg,* introduced in Hamburg, Germany, in the spring of 1960; *Elegie für Junge Liebende,* a huge success when introduced in Schwetzingen, Germany, on May 20, 1961; and *The Emperor's Nightingale* (1962).

Hindemith, Paul, composer. In 1953 Hindemith settled permanently in Zurich, Switzerland, where he became a member of the music faculty at the University. He received the Sibelius Award of $35,000 in 1954 for outstanding creative achievement in music. His opera *The Long Christmas Dinner,* on a libretto by Thornton Wilder, was introduced in Mannheim on December 17, 1961.

Holland Music Festival. Recent significant events in opera at the Holland Festival include: in 1959, revivals of Janáček's *Katya Kabanova* and Haydn's *Il Mondo della luna,* the latter after two hundred years; in 1960, the world première of Henk Badings' *Martin Korda, D.P.* and the Holland première of Britten's *A Midsummer Night's Dream;* in 1961, a revival of Berlioz's *Benvenuto Cellini;* and in 1962, revivals of Hendrik Andriessen's *Philomela,* Monteverdi's *Il ritorno di Ulisse,* and Busoni's *Doktor Faust.*

Honegger, Arthur, composer. Died Paris, November 27, 1955.

I

Ibert, Jacques, composer. He was the artistic director of the Paris Opéra and Opéra-Comique between 1955 and 1957. He died in Paris on February 5, 1962.

J

James, Henry, novelist and playwright. Born New York, April 15, 1843; died England, February 28, 1916. Britten's *The Turn of the Screw* and Douglas Moore's *The Wings of the Dove* were based on novels of the same names by James.

Jochum, Eugen, conductor. He made his American debut in 1961 with the visiting Concertgebouw Orchestra of Holland.

Johnson, Edward, tenor and opera manager. Died Guelph, Ontario, Canada, April 20, 1959.

K

Kaiser, Georg, dramatist. Blacher's opera *Rosamunde Floris* was derived from a play by Kaiser.

Karajan, Herbert von, conductor. In 1956 he became artistic director of the Vienna State Opera and the Salzburg Festival. He subsequently also assumed the posts of principal conductor and member of the board of directors of La Scala.

Keilberth, Joseph, conductor. His American debut took place with the touring Bamberg Symphony in Carnegie Hall, New York, on April 4, 1954.

Khrennikov, Tikhon, composer. His opera *Mother* was produced in Moscow on October 26, 1957.

Kirsten, Dorothy, soprano. She made a triumphant tour of the Soviet Union in 1962, becoming the first American to sing with the Tiflis Opera.

Klebe, Giselher, composer. Born Mannheim, Germany, June 28, 1925. After completing his studies with Josef Rufer and Boris Blacher, he attracted attention in 1954 with a *Rhapsody,* for orchestra, which won first prize in the International Festival in Rome. In 1957 his opera *Die Räuber,* based on Schiller, was introduced in Düsseldorf. He subsequently completed a second opera, *Alkmene.*

Kleiber, Erich, conductor. Died Zurich, January 27, 1956.

Kleist, Heinrich Wilhelm von, author and dramatist. His tragedy *Der Prinz von Homburg* was the source of an important opera by Henze.

Korngold, Erich Wolfgang, composer. Died Hollywood, California, November 29, 1957.

Křenek, Ernest, composer. His one-act opera *The Bell Tower,* based on a story by Herman Melville, was produced at Urbana, Illinois, on March 17, 1957.

Krips, Josef, conductor. Born Vienna, August 8, 1902. He attended the Vienna Academy, where he studied conducting with Felix Weingartner. At nineteen he was chorus master of the Volksoper in Vienna. In 1926 he became one of the youngest musicians

ever to fill the post of musical director of the city of Karlsruhe. He stayed there for seven years and was then appointed conductor of the Vienna State Opera. When the Nazis came to power in Austria, Krips was forced to leave his post, but he returned to the Vienna State Opera in 1945. During the next few years he developed into one of Europe's most significant opera conductors. In 1947 he led an important cycle of Mozart operas in London, and in 1949 directed a one-week Mozart cycle at the Florence May Music Festival. His American debut took place with the Buffalo Philharmonic on November 7, 1954. His first opera performance in America, *Don Giovanni,* also took place in Buffalo. Since then Krips has conducted opera in Vienna, Salzburg, Bayreuth, and other music centers. In the fall of 1962 he was named musical director of the San Francisco Symphony.

Kubelik, Rafael, conductor and opera manager. Born Býchory, Czechoslovakia, June 29, 1914. He is the son of Jan Kubelik, celebrated violin virtuoso. His music study took place at the Prague Conservatory. From 1936 to 1939 he was musical director of the Czech Philharmonic; between 1939 and 1941 he was conductor of the Brünn (Brno) Opera; and he then returned to the Czech Philharmonic, as its director, for a six-year period. In 1948 he led distinguished performances of opera at Glyndebourne and the Edinburg Festival. His American debut took place in November 1949 with the Chicago Symphony. Between 1950 and 1953 he was the principal conductor of the Chicago Symphony. In 1955 he became musical director of Covent Garden. His own opera, *Veronika,* was produced in Brünn (Brno) in 1947.

L

Leinsdorf, Erich, conductor. In 1956–57 he served as musical director of the New York City Opera. He returned to the Metropolitan Opera in 1957 as "musical consultant" and conductor. In September 1957 he led the American première of Poulenc's *Les Dialogues des Carmélites* with the San Francisco Opera, and in 1959 he led performances of *Die Meistersinger* at Bayreuth. In the fall of 1962 he assumed the post of music director of the Boston Symphony.

Liebermann, Rolf, composer. His opera *The School for Wives* received its world première in Louisville, Kentucky, on December 3, 1955. It was performed in Europe for the first time at the Salzburg Festival in 1957.

Long Christmas Dinner, The, opera in one act by Paul Hindemith. Libretto by the composer and Thornton Wilder, based on the play of the same name by Wilder. Première: Mannheim, Germany, December 17, 1961. This is an intimate work, requiring only eleven singers, a small orchestra, and no chorus. The critic of the *Neue Zürcher Zeitung* said of it: "There is perhaps no other recent opera in which the spirit and style of text and music are so completely integrated."

M

Maazel, Lorin, conductor. Born Paris, France, March 5, 1930. The son of Americans living abroad, he was brought as a child to the United States, where he studied conducting with Vladimir Bakaleinikoff. He gave remarkable performances as a prodigy conductor with leading orchestras in America—including the NBC Symphony, to which he was invited by Toscanini. He then continued his academic education by attending the University of Pittsburgh for three years. From 1948 to 1951 he was second violinist and apprentice conductor of the Pittsburgh Symphony. He then went to Europe on a Fulbright Fellowship for research in early Italian music. A year later he began his European conducting career through guest appearances at several major festivals and with some of the leading symphony orchestras. Between 1952 and 1960 he averaged fifty performances a year in all parts of Europe. In 1960 he became the youngest conductor—and the first American—to appear at the Bayreuth Festival, where he conducted *Lohengrin.* After that he directed performances in most of the important European opera houses. Maazel made his New York debut as a mature conductor on October 1, 1962, with the visiting L'Orchestre National Français. A month later, on November 1, he made his debut at the Metropolitan Opera in *Don Giovanni.*

Maison, René, tenor. Between 1950 and 1957 Maison taught singing at the Julius Hart School in Boston, and from 1957 to 1962 at the Chalof School of Music in the same city. He became an American citizen in 1959. He died in Mont-Doré, France, in July 1962.

Malipiero, Gian Francesco, composer.

His one-act operas *Il figliuol prodigo* and *Venere prigioniera* received their world premières at the Florence May Festival on May 11, 1957.
Maria Golovin, musical drama by Gian-Carlo Menotti. Libretto by the composer. Première: Brussels, August 20, 1958. This opera was commissioned by the National Broadcasting Company for the Brussels Exposition. Its American première took place in New York on November 5, 1958, and early in 1959 it was telecast by NBC. The opera is set near a European frontier, a few years after "a recent war." The heroine, Maria Golovin, becomes involved romantically with a blind maker of bird cages, while her husband is serving time as a prisoner of war. When Maria's husband is finally freed, she is forced to leave her blind lover. In despair, the blind man shoots her but unknowingly misses aim. Thus he is left with the delusion that Maria is dead and that nobody will ever possess her again.
Martin, Frank, composer. Born Geneva, Switzerland, September 15, 1890. His music studies consisted mainly of private lessons with Joseph Lauber. Between 1923 and 1925 he lived in Paris, where his early style became crystallized, influenced by the French Impressionists. He later adopted a contrapuntal idiom into which a modern harmonic language was integrated. In this style he scored a major success with *Le Vin herbé,* an oratorio based on the Tristan saga, which has received stage presentations in Europe's leading opera houses. *The Tempest,* an opera based on Shakespeare, was introduced in Vienna on June 17, 1956, and *Le Mystère de la Nativité* was produced at the Salzburg Festival in 1960.
Martinu, Bohuslav, composer. His last operas were *Greek Passion* (1956) and *Ariane* (1958). He died in Liestal, Switzerland, on August 28, 1959.

Melton, James, tenor. Died New York, April 21, 1961.
Melville, Herman, author. Ernest Křenek's opera *The Bell Tower* was based on a story by Melville.
Menotti, Gian-Carlo, composer. In 1958 he organized an annual summer festival of the arts, "The Festival of Two Worlds," in Spoleto, Italy, serving as its artistic director. His later operas are: *The Unicorn, The Gorgon and the Manticore* (1956); *Maria Golovin* (1958); and *The Last Superman* (1961).
Metropolitan Auditions of the Air. Winners of the annual Auditions since 1955 include: Carlotta Ordassy, Ezio Flagello, Martina Arroyo, Roald Reitan, Teresa Stratas, Mary MacKenzie, Francesco Roberto, George Shirley, and Janice Martin.
Metropolitan Opera House. Samuel Barber's *Vanessa,* on January 15, 1958, provided the Metropolitan with its first world première and its first première of an American opera from the time when Rudolf Bing became general manager. Here are some of the more significant premières and revivals of recent seasons: Offenbach's *La Périchole* in 1956 and Berg's *Wozzeck* in 1959, both produced at the Metropolitan Opera for the first time; revivals of Johann Strauss's *The Gypsy Baron* in 1959, Verdi's *Nabucco* in 1960 (with which the season opened), Flotow's *Martha* in 1960–61, and Giordano's *Andrea Chénier* (with which the 1962–63 season opened).
Milhaud, Darius, composer. His one-act opera *Fiesque* was introduced in a concert version by the New York Philharmonic on April 6, 1960.
Miller, Arthur, playwright. Born New York, October 17, 1915. Robert Ward's opera *The Crucible,* which received the Pulitzer Prize in 1962, and Renzo Rossellini's opera *A View from the Bridge* were derived from Miller's plays of the same names.

Mitropoulos, Dimitri, conductor. In 1958 he resigned as musical director of the New York Philharmonic Orchestra. He subsequently made guest appearances with leading opera companies in Europe and the United Sates. He died of a heart attack on November 2, 1960, while directing a symphonic rehearsal at La Scala.

Molière (born Jean Baptiste Poquelin), playwright. His play *L'École des femmes* was the source of operas of the same name by Rolf Liebermann and Virgilio Mortari.

Monteux, Pierre, conductor. In 1961, at the age of eighty-six, he was appointed principal conductor of the London Symphony.

Moore, Douglas, composer. His opera *The Ballad of Baby Doe* was introduced in Central City, Colorado, on July 7, 1956, after which it was presented with outstanding success by the New York City Opera. His later operas were *Gallantry*, a one-act satire presented at Columbia University, New York, in March 1958 and introduced over television on August 30, 1962 (CBS), and *The Wings of the Dove*, presented by the New York City Opera on October 12, 1961.

Müller, Maria, soprano. Died Bayreuth, Germany, March 13, 1958.

N

Newman, Ernest, musicologist. Died Tadworth, Surrey, England, July 7, 1959.

New York City Opera Company. In 1956–57 Erich Leinsdorf became musical director. During his regime, there took place the American premières of Frank Martin's *The Tempest* and Carl Orff's *Der Mond*, the New York première of Carlisle Floyd's *Susannah,* and a revival of Offenbach's *Orpheus in the Underworld.* Julius Rudel succeeded Leinsdorf. The New York City Opera has produced many American operas, financed by grants from the Ford Foundation. These are some of the significant American operas performed by the New York City Opera since 1955: Marc Blitzstein's *The Cradle Will Rock* (in its first performance by an opera company); Norman Dello Joio's *The Triumph of St. Joan;* Abraham Ellstein's *The Golem;* Carlisle Floyd's *Susannah, Wuthering Heights,* and *The Passion ˙of Jonathan Wade;* Douglas Moore's *The Ballad of Baby Doe* and *The Wings of the Dove;* Robert Ward's *He Who Gets Slapped* and *The Crucible;* and Hugo Weisgall's *Six Characters in Search of an Author.* In the spring of 1962 the New York City Opera presented the American première of Britten's *The Turn of the Screw.*

Nilsson, Birgit, soprano. Born Karup, Sweden, May 17, 1922. She studied singing at the Royal Academy in Stockholm. In 1946–47 she sang the role of Agathe in *Der Freischütz* at the Stockholm Royal Opera. The following season she scored a major success there as Lady Macbeth in Verdi's opera. From 1947 to 1951 she made guest appearances at the Glyndebourne Opera, and in 1954–55 her performances as Brünnhilde in the Wagner *Ring* cycle were acclaimed in Munich. She made her American debut in 1957 with the San Francisco Opera. During the same summer she was heard as Isolde in Bayreuth. She was most enthusiastically received as Isolde in her

Metropolitan Opera debut in 1959. Since then she has been acknowledged as the most significant Wagnerian soprano since Kirsten Flagstad.

O

Opera—Television. On November 6, 1955, Lukas Foss's *Griffelkin,* commissioned by NBC-TV, received its world première over that network. An operatic event of the first importance took place on television in 1957, when NBC presented Poulenc's *Les Dialogues des Carmélites.* On January 1, 1961, NBC-TV gave the world première of Leonard Kastle's *Deseret,* and on August 30, 1962, CBS presented the première of Douglas Moore's *Gallantry,* a one-act operatic satire on soap operas.

Opéra, L' (Paris). Jacques Ibert served as artistic director of both the Paris Opéra and the Opéra-Comique from 1955 to 1957. The Opéra entered upon a new regime in 1959, when A. M. Julien became administrator and Gabriel Dussurget was appointed artistic counselor. Its first presentation, a newly realized production of *Carmen,* was directed by Roberto Benzi, the first twenty-one-year-old conductor ever to appear in that institution. In 1962 Georges Auric replaced A. M. Julien, becoming artistic director of both the Opéra and Opéra-Comique.

Orff, Carl, composer. His opera *Ludus de natio Infante Mirificus* received its world première at the Württemberg Opera on December 11, 1960. During the summer of 1962 Orff visited Canada, to serve as chairman of a conference on music education at the University of Toronto's Royal Conservatory of Music.

P

Périchole, La, opéra-bouffe by Jacques Offenbach. Libretto by Henri Meilhac and Ludovic Halévy, based on *La carrosse du Saint-Sacrement* by Mérimée. Première: Théâtre des Variétés, Paris, October 6, 1868. American première: Pike's Opera House, New York, January 4, 1869. The setting is eighteenth-century Peru. "La Périchole" is a gypsy street singer in love with her singing partner, Paquillo. The Viceroy becomes interested in her and takes her to his palace as lady-in-waiting. When Paquillo discovers that his sweetheart has become the Viceroy's favorite, he denounces her; by the final curtain, however, the lovers are reconciled and have received the Viceroy's blessings. One of Offenbach's most celebrated melodies is found in this opera: La Périchole's letter song to Paquillo ("O mon cher amant, je te jure"). *La Périchole* was revived brilliantly by the Metropolitan Opera on December 21, 1956, in a new English translation.

Peters, Roberta, soprano. Born New York City, May 4, 1930. She received her vocal training from William Pierce Herman, her only teacher. Her opera debut was an unscheduled appearance, when she was called in to substitute for Nadine Conner in a performance of *Don Giovanni* at the Metropolitan

Opera on November 17, 1950. During the next few years she established her reputation by appearing in leading soprano roles in the Italian and French repertory, and in several Mozart operas, at the Metropolitan Opera. Her first European opera engagement was in 1951, at Covent Garden, in a gala performance of Balfe's *The Bohemian Girl*.

Pinza, Ezio, basso. Died Stamford, Connecticut, May 9, 1957.

Pizzetti, Ildebrando, composer. His opera *Assassinio nella cattedrale* received its world première at La Scala on March 1, 1958. In 1960 he completed the opera *Lo Stivale d'argento*.

Polacco, Giorgio, conductor. Died New York, April 30, 1960.

Poulenc, Francis, composer. He received an honorary doctorate from Oxford University in 1958. His opera *Les Dialogues des Carmélites* scored a major success when it was introduced at La Scala on January 26, 1957, after which it was performed by major opera companies in Europe and the United States and, in 1957, was televised in the United States over the NBC network. His one-act opera for a single character, *La Voix humaine*, text by Jean Cocteau, was first produced in 1959.

Price, Leontyne, soprano. Born Laurel, Mississippi, February 10, 1927. She studied singing at the Juilliard School of Music for four years, mainly with Florence Page Kimball. For two years she played Bess in Gershwin's *Porgy and Bess,* during the opera's tour of Europe and the United States. In 1954 she initiated a successful career on the concert stage, and in 1955 she was starred in a performance of *Tosca* telecast over the NBC network. Her grand opera debut took place in 1957, with the San Franciso Opera, in Poulenc's *Les Dialogues des Carmélites*. In 1958 she made an outstandingly successful opera debut in Europe by appearing in *Aïda* at the Vienna State Opera. On January 27, 1961, she was acclaimed for her performance of Leonora in *Il Trovatore* at her Metropolitan Opera debut. On October 22, 1961, she became the first Negro singer to appear in a starring role on opening night of the Metropolitan Opera season. The opera was Puccini's *The Girl of the Golden West*.

Pulitzer Prize (for music). Pulitzer Prizes had been instituted in 1918 in various fields of literary and journalistic endeavor. The prize for music was created in 1943, its first recipient being William Schuman, for his cantata *A Free Song*. The first opera to win the award was *The Consul,* by Gian-Carlo Menotti, in 1950. A year later Douglas Moore received the Pulitzer Prize for his opera *Giants in the Earth*. Later operas to win the Pulitzer Prize were: *The Saint of Bleecker Street*, by Menotti, in 1955; *Vanessa*, by Samuel Barber, in 1958; and *The Crucible*, by Robert Ward, in 1962.

R

Rennert, Günther, stage director. Born Essen, Germany, April 1, 1911. He served his apprenticeship in various German theaters, as stage director for both dramatic presentations and operas. In 1942 he became stage director of the Berlin State Opera. In 1945 he directed the performance of *Fidelio* which reopened the Munich Opera following the end of World War II. From 1946 to 1956 he was stage director of the Hamburg Opera.

Since then he has been a free lance, working for most of the leading European opera houses and at the Salzburg and Edinburgh festivals as well. He directed his first opera at the Metropolitan Opera, Verdi's *Nabucco,* on opening night of the 1960–61 season.

Resnik, Regina, soprano and mezzo-soprano. Born New York, August 20, 1922. She studied voice in New York with Rosalie Miller. In 1944 she won the Metropolitan Auditions of the Air, and on December 6, 1944, made her Metropolitan Opera debut in *Il Trovatore.* During the next decade she sang principal soprano roles at the Metropolitan. In July 1953 she made her first appearance at Bayreuth, as Sieglinde. In 1955 she gave up soprano roles and became a mezzo-soprano, making her debut in this new range at the Cincinnati Zoo Opera as *Aïda*'s Amneris. She made her first Metropolitan Opera appearance as mezzo-soprano in the role of Marina in *Boris Godunov* on February 15, 1956.

Rosenberg, Hilding, composer. In 1956 he completed an opera, *The Portrait.*

Rosenstock, Joseph, conductor. He resigned as artistic director of the New York City Opera in 1955, but remained one of the conductors of that company. In 1958 he became music director of the Cologne Opera. He returned as conductor to the Metropolitan Opera on January 31, 1961, in *Tristan und Isolde.*

Rudolf, Max, conductor. He became principal conductor of the Cincinnati Symphony in 1958.

Rysanek, Leonie, soprano. Born Vienna, November 12, 1926. After studying voice at the Vienna Conservatory, mainly with Alfred Jerger, she made her opera debut in Innsbruck in 1949 as Agathe in *Der Freischütz.* She continued her vocal studies at Innsbruck with Rudolf Grossman, whom she subsequently married. From 1950 to 1952 she appeared in opera performances at Saarbrücken, and in 1951 she was heard as Sieglinde in the first post-war festival at Bayreuth. In 1952 she became a member of the Munich Opera, where her fame was established. She scored a major success during her American debut with the San Francisco Opera in 1957. In 1958 and 1959 she was acclaimed at Bayreuth, and in 1959 she made a remarkable debut at the Metropolitan Opera.

S

Sachse, Leopold, stage director. Died Englewood Cliffs, New Jersey, April 4, 1961.

Salzburg Festival. In 1956, Herbert von Karajan became the artistic director of the festival, and in 1959 Bernhard Paumgartner assumed the office of President. The new Festspielhaus was inaugurated on July 26, 1960, with *Der Rosenkavalier.* Among the contemporary operas heard at the Salzburg Festival since 1955 were: in 1958, Barber's *Vanessa,* the first American opera ever given there; in 1959, the world première of Erbse's *Julietta;* and in 1960, Frank Martin's *Le Mystère de la Nativité.*

San Francisco Opera Company. Since 1955 this company has offered the following significant American premières: Poulenc's *Les Dialogues des Carmélites* in 1957; Richard Strauss's *Die Frau ohne Schatten* in 1959; and Britten's *A Midsummer Night's Dream* in 1961. In 1961 the company gave the world première of an American opera, Norman Dello Joio's *Blood Moon*.

Sauguet, Henri, composer. His opera *La Dame aux camélias* was introduced in Berlin, during the Berlin Festival, on September 29, 1957.

Sawallisch, Wolfgang, conductor. Born Munich, Germany, August 26, 1923. During World War II he studied piano with Wolfgang Ruoff, theory with Hans Sachsse, and conducting at the Munich Conservatory. In 1947 he became a coach at the Augsburg Opera. In 1950–51 he conducted operettas, and in 1953 operas, with that company. Between 1953 and 1957 he served as general music director in Aachen. From 1957 to 1960 he was the principal conductor of the Wiesbaden Opera, and in 1960 he became general music director in Cologne. Meanwhile, in 1957, he made a triumphant appearance at the Bayreuth Festival conducting *Tristan und Isolde*. Sawallisch now occupies a place of first importance among German conductors of opera and symphonic music.

Scala, La, *see* TEATRO ALLA SCALA.

Schiller, Friedrich von, poet and dramatist. Schiller's drama *Die Räuber* was the source of an opera of the same name by Giselher Klebe.

Schippers, Thomas, conductor. Born Kalamazoo, Michigan, March 9, 1930. His musical training took place at the Curtis Institute, Yale University, and the Juilliard School of Music. He made his debut in New York in 1950, conducting Menotti's *The Consul*. On April 9, 1952, he made his New York City Opera debut in Menotti's *Amahl and the Night Visitors*. He conducted performances at the New York City Opera for the next three years, during which time he led the world première of Copland's *The Tender Land*. In 1954 he directed the world première of Menotti's *The Saint of Bleecker Street*

in New York. In May 1955 he made a successful debut at La Scala. On December 24, 1955, he made his first appearance at the Metropolitan Opera, conducting *Don Pasquale*. Since then he has directed distinguished opera performances at La Scala, the Metropolitan Opera, and at the Festival of Two Worlds in Spoleto. In June 1962 he conducted the first staged presentation of Manuel de Falla's *L'Atlantida,* at La Scala.

Schoeck, Othmar, composer. Died Zurich, March 8, 1957.

Schoenberg, Arnold, composer. The world première of his unfinished Biblical opera, *Moses und Aron,* took place in Zurich on June 6, 1957.

Schwarzkopf, Elisabeth, soprano. Her American debut took place in October 1955, with the San Francisco Opera, as the Marschallin in *Der Rosenkavalier.*

Shakespeare, William, poet and dramatist. His comedies and tragedies provided material for four recent operas: *All's Well That Ends Well,* Mario Castelnuovo-Tedesco; *Hamlet,* Mario Zafred; *The Merchant of Venice,* Mario Castelnuovo-Tedesco; and *A Midsummer Night's Dream,* Benjamin Britten.

Shostakovich, Dmitri, composer. He completed *Tcheremuski,* a light opera, in 1958.

Siepi, Cesare, basso. He made his Broadway debut in 1962, in the musical comedy *Bravo, Giovanni!*

Simionato, Giulietta, mezzo-soprano. Born Forli, Italy, December 15, 1910. After studying voice with Ettore Lucatello in Rovigo, she received first prize in a Bel Canto competition in Florence in 1933. She appeared in minor roles with opera companies in Florence, Milan, and Padua before joining the La Scala company, where she assumed major roles. When Toscanini directed a La Scala concert performance of Boïto's *Nerone* in 1947, he selected Simionato for the part of Asteria. In the same year her performance of Cherubino in *The Marriage of Figaro* was acclaimed at the Edinburgh Festival. In Palermo, in 1954, Bellini's rarely heard *I Capuletti ed i Montecchi* was revived for her. By the time she made her American debut, she had become known to American opera lovers through her many fine recordings. That debut took place in 1954 with the Chicago Lyric Opera. In 1955 she was heard with the San Francisco Opera and in 1957 she made her New York debut in a concert performance of Donizetti's *Anna Bolena*. Her debut at the Metropolitan Opera took place on October 26, 1959, as Azucena in *Il Trovatore*.

Solti, Georg, conductor. Born Budapest, Hungary, October 21, 1912. He was graduated from the High School for Music in Budapest in 1930 with diplomas in conducting, piano, and composition. He made his conducting debut in *The Marriage of Figaro* at the Budapest Opera. During World War II he lived in Switzerland, where he directed performances of the Swiss Radio Orchestra. From 1946 to 1952, while serving as director-general of music in Munich, Germany, he gave distinguished performances with the Munich Opera. In 1952 he became the director-general of music in Frankfurt-am-Main, musical director of the Frankfurt Opera, and conductor of the Museum concerts in that city. His American debut took place with the San Francisco Opera in September of 1953, in Strauss's *Elektra*. Solti made his debut at the Metropolitan Opera on December 17, 1960, in *Tannhäuser*. Georg Solti made his Covent Garden debut in December 1959 in *Der Rosenkavalier*. In September 1961 he was appointed musical director of Covent Garden.

Spoleto Festival. *See* FESTIVAL OF TWO WORLDS, THE.

Städtische Oper, *see* BERLIN OPERA.

Stich-Randall, Teresa, soprano. Born West Hartford, Connecticut. She was a scholarship student at the Hartford School of Music in Connecticut. While attending Columbia University, where she specialized in music, she appeared in the world premières of Virgil Thomson's *The Mother of Us All* and Otto Luening's *Evangeline,* as well as in other operas. Toscanini selected her to sing in his NBC Symphony performances of *Aïda* and *Falstaff*. In 1951 she went to Europe on a Fulbright Fellowship. There she received first prize in the Concours International for opera singers in Lausanne, Switzerland. In August 1952 she made her debut with a professional opera company as Violetta in *La Traviata* at the Vienna State Opera. Her performances as a permanent member of that company were supplemented by many significant guest appearances at leading European festivals and major European opera houses. On November 13, 1955, she made her debut at the Chicago Lyric Opera in *Rigoletto,* and on October 24, 1961, she appeared for the first time at the Metropolitan Opera in *Così fan tutte*.

Stokowski, Leopold, conductor. Born London, April 18, 1882. Although his career has been identified primarily with symphonic music—particularly by virtue of his brilliant career as music director of the Philadelphia Orchestra—Stokowski has periodically distinguished himself in opera. In 1929 he led the American première of *Boris Godunov* in its original version as Mussorgsky wrote it; this was a concert performance. On March 19, 1931, he directed the American première of Berg's *Wozzeck* in a staged presentation. In recent years Stokowski has identified himself more closely than ever with opera. Between 1959 and 1961 he was a conductor at the New

York City Opera, where he led memorable performances of such novelties as Monteverdi's *Orfeo,* Luigi Dallapiccola's *Il prigioniero,* and Orff's *Carmina Burana.* During 1961–62 he made his debut with the Metropolitan Opera in *Turandot.*

Susannah, musical drama by Carlisle Floyd. Libretto by the composer. Première: Tallahassee, Florida, February 24, 1955. In the composer's description, this opera's central theme is "persecution and the concomitant psychological ramifications." The time is the present, the place a farm in New Hope Valley in the Tennessee mountains. Susannah is victimized by the gossip of her townspeople. When Reverend Blitch calls on her, to help her, he is so moved by her agony—and so sympathetic to her beauty—that he makes advances to her. He is discovered by Susannah's brother, who kills him in cold blood. The crowd now descends on Susannah, to vent on her their fury, but is kept at bay by her menacing shotgun. As the curtain descends, Susannah is standing in the doorway of her farm, a forlorn and lonely creature. This opera scored a major success at the New York City Opera in 1956, when it received the New York Music Critics Circle Award. In the summer of 1958 it was produced at the Brussels Exposition in Belgium.

Sutherland, Joan, soprano. Born Sydney, Australia, November 7, 1929. She was a student at the Sydney Conservatory, where she made her opera debut in 1950 in Goossens' *Judith.* In 1952 she joined the company of Covent Garden, where during her first season she was heard in operas by Mozart and Verdi, and in the world première of Britten's *Gloriana.* In 1955 she created the leading female role in Michael Tippett's *The Midsummer Marriage.* In 1956 she became the first British artist to sing the role of the Countess in *The Marriage of Figaro* at Glyndebourne. Her American debut took place in Dallas, Texas, on November 17, 1960, in Handel's *Alcina.* On February 21, 1961, she made her first appearance in New York, in Bellini's *Beatrice di Tenda.* She was received with the greatest enthusiasm at her first appearance with the Metropolitan Opera. This took place on November 26, 1961, when she sang Lucia in *Lucia di Lammermoor.* In 1961 she was made Commander of the British Empire by Queen Elizabeth.

T

Teatro alla Scala (La Scala). The significant performances at La Scala since 1955 included: world premières of Poulenc's *Les Dialogues des Carmélites* in 1957, Pizzetti's *Assassinio nella cattedrale* in 1958, and the stage version of Manuel de Falla's *L'Atlantida* in 1962; revivals of Handel's *Hercules* in 1958, Glinka's *A Life for the Czar* in 1959, and Bloch's *Macbeth* in 1960.

Telva, Marion, contralto. Died Norwalk, Connecticut, October 23, 1962.

Theater-an-der-Wien. The Viennese theater, reconstructed following damages in a bombing attack during World War II, was reopened on May 28, 1962, with a gala performance of *Fidelio.*

Thomas, John Charles, baritone. Died Apple Valley, California, December 13, 1960.

Tibbett, Lawrence, baritone. Died New York, July 15, 1960.

Tippett, Michael, composer. Born London, January 2, 1905. He attended the Royal College of Music, then came to the fore with various instrumental works that were Romantic in style and strongly influenced by Sibelius. He later developed his own idiom, in which Romanticism was combined with polyphony. His first opera, *The Midsummer Marriage,* was produced in London on January 27, 1955. In the spring of 1962 his opera *King Priam* was introduced at Covent Garden.

Tokatyan, Armand, tenor. Died Pasadena, California, June 12, 1960.

Toscanini, Arturo, conductor. Died New York, January 16, 1957. His body was brought to Milan for burial on February 18.

V

Vanessa, opera by Samuel Barber. Libretto by Gian-Carlo Menotti. Première: Metropolitan Opera, January 15, 1958. In a Scandinavian city, in 1905, Vanessa has waited twenty years in her grim baronial manor for her lover to return. He is dead, but his son, Anatol, has come in his stead to marry Vanessa and to take her away. Vanessa's dreary manor is now taken over by her niece, Erika, who has been seduced and deserted by Anatol. *Vanessa,* Barber's first opera, became the first American opera performed at the Salzburg Festival; this took place in 1958. *Vanessa* was the recipient of the Pulitzer Prize in music.

Vaughan Williams, Ralph, composer. Died London, August 26, 1958.

Vienna State Opera (Staatsoper). Since 1956, Herbert von Karajan has served as its artistic director.

W

Walter, Bruno, conductor. Died Beverly Hills, California, February 17, 1962.

Ward, Robert, composer. Born Cleveland, Ohio, September 13, 1917. He attended the Eastman School of Music, the Juilliard School, and the Berkshire Music Center. He first attracted attention as a composer with several major orchestral works, including two symphonies. Between 1949 and 1951 he received Guggenheim Fellowships. In 1955 he became president of the American Composers Alliance, and a year later he was appointed editor of the Galaxy Music Corporation. His first opera, *He Who Gets Slapped,* was introduced in New York on May 17, 1956. In 1962 he received the Pulitzer Prize in music for *The Crucible,* an opera based on the play of the same name by Arthur Miller; its première took place at the New York City Opera on October 26, 1961.

Warren, Leonard, baritone. Died New York, March 4, 1960 (on the stage of the Metropolitan Opera during a performance of *La Forza del destino*).

Wilder, Thornton, novelist and playwright. Born Madison, Wisconsin, April 17, 1897. His celebrated 1928 Pulitzer Prize novel, *The Bridge of San Luis Rey,* was made into an opera by Hermann Reutter *(Die Brücke von San Luis Rey).* Wilder wrote the texts for *Alcestiad,* an opera by Louise Talma, and *The Long Christmas Dinner,* a one-act opera by Hindemith.

Encyclopedia
of the
OPERA

A

Abbey, Henry Eugene, impresario. Born Akron, Ohio, June 27, 1846; died New York City, October 17, 1896. He began his career as impresario in Buffalo in 1876 when, with Edward Schoeffel, he acquired and managed the local Academy of Music. In 1880 he produced theatrical presentations in New York, managing such celebrated stage personalities as E. H. Sothern, Henry Irving, Edwin Booth, and Sarah Bernhardt. In 1883 he helped organize the Metropolitan Opera Company in New York, assembling a brilliant roster of singers. He lost close to half a million dollars during the first season, and was replaced by the Metropolitan's second manager, Leopold Damrosch. Some years later, from 1891 to 1896, Abbey shared the management of the Metropolitan with Edward Schoeffel and Maurice Grau, this time with better results. He died a month before the opening performance of the 1896–1897 season.

Abbott, Emma, dramatic soprano. Born Chicago, December 9, 1850; died Salt Lake City, January 5, 1891. After studying with Achille Errani in New York, and Mathilde Marchesi, Antonio Sangiovanni and Enrico Delle Sedie in Europe, she made her opera debut at Covent Garden in 1876 in *The Daughter of the Regiment*. A year later she made her first appearance in America, in the same opera, at the Academy of Music in New York. She married Eugene Wetherell in 1878 and helped him organize the Emma Abbott English Grand Opera Company, which he managed and in which she appeared in principal soprano roles. Despite a limited voice, she had an immense audience appeal. She did not hesitate to interpolate popular ballads into operas or to couple a serious opera with a Gilbert and Sullivan comic opera on the same evening.

Abdallah, a corsair (baritone) in Weber's *Oberon.*

Abduction from the Seraglio, The (Die Entführung aus dem Serail), comic opera in three acts by Wolfgang Amadeus Mozart. Libretto by Gottlieb Stephanie, adapted from the text by Christoph Friedrich Bretzner for an earlier opera, *Belmont und Constanze* (the music of which was by Johann André). Première: Burgtheater, Vienna, July 16, 1782. American première: German Opera House, Brooklyn, New York, February 16, 1860.

Characters: Constanza (soprano); Blonde, her maid (soprano); Belmonte, Spanish nobleman (tenor); Pedrillo, his servant (tenor); Selim Pasha (speaking role); Osmin, his overseer (bass); a mute; slaves; guards.

The action takes place in Turkey in the sixteenth century.

Act I. A square before the Pasha's palace. There is a brief overture in gay Turkish style. Constanza and Blonde have been kidnaped by pirates, brought to Turkey, and sold to the Pasha. Belmonte, Constanza's lover, has followed her. He expresses the hope of a reunion ("Hier soll ich dich denn sehen"). He comes upon Osmin, the fat overseer, who expresses a cynical attitude toward all women ("Wer ein Liebchen hat gefunden") and refuses to help Belmonte. The imminent arrival of the Pasha and Constanza fills Belmonte

3

with joy, since he is about to catch a glimpse of his beloved ("Constanze! dich wiederzusehen"). In hiding, Belmonte hears Constanza tell the Pasha she can never be his, since her heart belongs elsewhere ("Ach, ich liebte"). Enraged, the Pasha demands that she change her mind by the morrow.

Act II. The palace garden. Blonde repulses the amorous Osmin ("Ich gehe, doch rathe ich dir"). Constanza bewails her unfortunate lot ("Traurigkeit ward mir zum Lose"), and remains defiant even as the Pasha threatens with torture if she refuses him ("Martern aller Arten"). Blonde now learns that Belmonte has a plan to rescue her mistress and herself. Her joy is unbounded ("Welche Wonne, welche Lust"). The first part of the plan consists of getting Osmin helplessly drunk. After this is accomplished, Belmonte and Constanza exchange words of love ("Wenn der Freude Thränen fliessen"), and Blonde and Pedrillo, Belmonte's helpful servant, do likewise.

Act III. Before the Pasha's palace. Coming to effect the escape of their women, Belmonte serenades Constanza ("Ich baue ganz auf deine Stärke"), after which Pedrillo sings to Blonde ("Im Mohrenland gefangen war"). The four flee, are apprehended, and brought before the Pasha. Osmin is triumphant ("Ha! wie will ich triumphieren"). The Pasha at first upbraids Constanza for trying to run away. Then he magnanimously forgives her and her lover and the two servants. Belmonte sings a hymn of praise to the generous Pasha ("Nie werd' ich deine Huld verkennen") and everybody takes up the joyous refrain.

The Abduction from the Seraglio was Mozart's first stage work after he had settled in Vienna in 1781. It was written for the Burgtheater, where Emperor Joseph II encouraged domestic productions. Mozart followed the form of the singspiel, a kind of musical comedy then in vogue in Germany and Austria. His subtle musical characterization, poignant lyricism, and advanced harmonic and orchestral techniques transformed a popular medium into a vehicle for great art and made this work the first important opera in the German language.

Abends, will ich schlafen geh'n, the prayer of Hansel and Gretel in Act II of Humperdinck's *Hansel and Gretel.*

Abigaille, Nebuchadnezzar's daughter (soprano) in Verdi's *Nabucco.*

Abimelech, Satrap of Gaza (bass) in Saint-Saëns' *Samson et Dalila.*

Abscheulicher! wo eilst du him Recitative and aria of Leonore in Act I of Beethoven's *Fidelio.*

Abstrakte Oper No. 1, opera by Boris Blacher. Libretto by Werner Egk. Première: Frankfurt, Germany, June 28, 1953. The libretto of this provocative abstract opera consists almost entirely of arbitrary sounds; where real words are used, their meaning is unintelligible. The opera created a scandal: laughter, catcalls, whistles, and shouts of disapproval disturbed the performance. One critic described it as "the worst opera ever written." Egk has stated that this opera was written for radio, but could also be performed in concert or on the stage. The scoring is for three solo voices, two narrators, chorus and orchestra.

Abul Hassan, the barber (bass) in Cornelius' *The Barber of Bagdad.*

Academy of Music, for over thirty years the principal opera house of New York City. Successor to the Astor Place Opera House, it was situated on Irving Place and Fourteenth Street. Its first manager, Max Marctzek, opened the theater on October 2, 1854, with *Norma,* starring Grisi and Mario. The Academy of Music continued to present outstanding opera performances until the Metropolitan Opera was founded and became a successful rival. The Academy was demolished by fire in 1866. It was rebuilt the following year and continued to house opera until the end of the century. Adelina Patti made her opera debut at the Academy in 1859. Under various managers—John Henry Mapelson, Ole Bull, Maurice Strakosch, Theodore Thomas—the Academy of Music presented the American premières of many outstanding operas, including: *L'Africaine; Aïda; Andrea Chénier; Un ballo in maschera; Don Carlos; Mefistofele; Otello; Rigoletto; Roméo et Juliette; The Sicilian Vespers; La traviata; Il trovatore.*

Accompagnato, Italian for "accompanied." The term refers to a recitative the instrumental accompaniment of which (more or less elaborate) has been written out by the composer. A recitativo accompagnato is distinguished from a recitativo secco (literally, "dry recitative"), in which the declamation is supported by the simplest sort of chords provided by a harpsichord or piano. The earliest example of accompagnato is found in the fourth act of Monteverdi's *Orfeo* (1607). Monteverdi's successors made much use of accompanied recitatives.

Ach, das Leid hab' ich getragen, Nureddin's aria in Act I of Cornelius' *The Barber of Bagdad.*

Ach, ich fühl's, Pamina's aria in Act II, Scene 4, of Mozart's *The Magic Flute.*

Ach, ich liebte, Constanza's aria in Act I of Mozart's *The Abduction from the Seraglio.*

Achilles, Iphigenia's betrothed (tenor) in Gluck's *Iphigénie en Aulide.*

Acis and Galatea, (1) a masque by Handel. Libretto by John Gay. Première: Cannons, the country home of the Duke of Chandos, 1720. This work contains several famous songs: Acis' "Love in her eyes sits playing," Galatea's "Heart, the seat of soft delight," and Polyphemus' "O ruddier than the cherry!" The story is based on the Greek myth in which the shepherd Acis, crushed beneath a rock by the giant Polyphemus, is transformed by his beloved, the sea nymph Galatea, into a bubbling fountain. In Italy, in 1708, Handel produced a "serenata" entitled *Aci, Galatea e Polifemo.* In 1732 he added some lines from Alexander Pope, John Hughes, and John Dryden to Gay's 1720 libretto, combined music of the early Italian serenata and the 1720 masque, and produced the resulting work at the King's Theatre on June 10. This, the *Acis and Galatea* we know today, was called a serenata; it was staged with scenery and costumes but no action.

(2) Opera by Jean-Baptiste Lully. Libretto by Jean Galbert de Campristron. Première: Anent, France, September 6, 1686; Paris Opéra, September 17, 1686.

Ackté, Aïno, soprano and impresario. Born Helsinki, Finland, April 23, 1876; died Nummela, August 8, 1944. After completing her music study in Paris, she made her opera debut at the Paris Opéra in 1897 in *Faust.* She was a leading soprano of the Paris Opéra for seven years. On November 25, 1904, she made her American debut at the Metropolitan Opera as Micaëla in *Carmen.* She remained a member of this company until 1905, thereafter, until 1910, singing at Covent Garden, where she scored one of her greatest triumphs as Salome in the first London performance of Richard Strauss's opera. Strauss later invited her to appear in this role in Paris and Dresden. In 1911 she founded the Finnish Opera Company at Savonhima, and in 1938 she became manager of the National Opera in Finland—a governmental post in which she supervised all opera within the nation—holding this post only one year.

Act Tune, a term found in English operas of the seventeenth and early eighteenth centuries, denoting music played between the acts, with curtain drawn. Also called curtain tune. Modern equivalents are intermezzo and en-tr'acte.

Adalgisa, Virgin of the Temple of Esus (mezzo-soprano) in Bellini's *Norma.*

Adam, Adolphe Charles, composer. Born Paris, July 24, 1803; died there May 3, 1856. He helped establish the form of the opéra comique. His father, Jean Louis Adam, was a well-known pianist, composer, and teacher. Adolphe attended the Paris Conservatory, where he came under the influence of Fran-çois Boieldieu, then at the height of his success. Adolphe's first dramatic work, the one-act opéra comique *Pierre et Catherine,* was successfully produced at the Opéra-Comique in 1829. A year later came the three-act *Danilowa* and, in 1836, *Le Postillon de Longjumeau,* Adam's best work and one still per-formed in Europe. Largely due to the success of this opera, Adam was made a member of the Legion of Honor in 1836, and a member of the Institut de France in 1844. In 1847, he organized his own opera company, the Théâtre National, devoted to presenting works by young and unrecognized composers. This venture put Adam heavily in debt. He was able to meet his obligations by writing a number of highly successful comic operas which made him one of the most popular composers of France. He also wrote several ballets, *Giselle* being the most famous. In 1849 he be-came a professor of composition at the Paris Conservatory. The best of his nearly fifty operas include: *Le châlet* (1834); *Le postillon de Longjumeau* (1836); *Le fidèle berger* (1838); *Le brasseur de Preston* (1838); *La rose de Péronne* (1840); *Le roi d'Yvetot* (1842); *Le toréador* (1849); *Giralda* (1850).

Adamastor, roi des vagues profondes, Nelusko's aria in Act III of Meyer-beer's *L'Africaine.*

Adams, Suzanne, soprano. Born Cam-bridge, Massachusetts, November 28, 1872; died London, England, February 5, 1953. She made her debut at the Paris Opéra in 1895 as Juliet, coached by Gounod himself. After successful appearances at both the Opéra and Covent Garden, she made her Ameri-can debut at the Metropolitan Opera on January 4, 1899, again as Juliet. She remained at the Metropolitan until 1903, and from 1903 to 1906 was seen at Covent Garden where she created the role of Hero in Stanford's *Much Ado About Nothing.* After 1903 she made her home in London. In 1898 she married the celebrated cellist Leopold Stern. His premature death in 1904 led to his wife's retirement from the opera stage only a few years later.

Addio alla madre, Turiddu's farewell to his mother in Mascagni's *Cavalleria rusticana.*

Addio del passato, Violetta's farewell to the world in Act IV of Verdi's *La traviata.*

Addio dolce svegliare, duet of Rodolfo and Mimi in Act III of Puccini's *La Bohème.*

Addio, fiorito asil, Pinkerton's farewell to his Nagasaki home in Act III of Puccini's *Madama Butterfly.*

Addio, senza rancor, Mimi's farewell to Rodolfo in Act III of Puccini's *La Bohème.*

Adele, Rosalinde's maid (soprano) in Johann Strauss's *Die Fledermaus.*

Adieu donc, vains objets, the prayer of John the Baptist in Act IV, Scene 1, of Massenet's *Hérodiade.*

Adieu, Mignon, courage! Wilhelm's farewell to Mignon in Act II, Scene 1, of Thomas's *Mignon.*

Adieu, mon doux rivage, Inez' aria in Act I of Meyerbeer's *L'Africaine.*

Adieu, notre petite table, Manon's fare-well in Act II of Massenet's *Manon.*

Adina, wealthy young woman (so-prano) in Donizetti's *L'elisir d'amore.*

Adina credimi, Nemorino's aria in Act I, Scene 2, of Donizetti's *L'elisir d'amore.*

Adler, Herman Peter, conductor. Born

Jablonec, Bohemia, December 2, 1899. He attended the Prague Conservatory, where one of his teachers was Alexander Zemlinsky. At age twenty-two he became musical director of the city of Jablonec. Subsequently, he held various posts as opera and symphonic conductor, mostly in Czechoslovakia. He came to New York in 1939, making his debut on January 24, 1940, in a special Philharmonic concert for Czech relief. In 1944 he became director of the Columbia Concerts Opera Company. Since 1949, when the NBC-TV Opera Company was organized, Adler has been its artistic and musical director.

Adler, Kurt Herbert, conductor and opera manager (not to be confused with Kurt Adler, chorus master of the Metropolitan Opera). Born Vienna, April 2, 1905. He conducted in various theaters in central Europe, was one of Toscanini's assistants at the Salzburg Festival of 1936, then conducted opera and radio concerts in Czechoslovakia. He settled in the United States in 1938 and is now an American citizen. After conducting at various opera houses, including the Chicago Opera (1941), he became associated with the San Francisco Opera in 1943. In 1953, upon the death of Gaetano Merola, Adler became the San Francisco Opera's artistic director.

Admetos, King of Pharae (tenor), in Gluck's *Alceste.*

Adolar, Count of Nevers (tenor), in Weber's *Euryanthe.*

Adorno, Gabriele, Genoese nobleman (tenor) in Verdi's *Simon Boccanegra.*

Adriana Lecouvreur, opera by Francesco Cilèa. Libretto is the play of the same name by Eugène Scribe and Gabriel Jean Baptiste Legouvé. Première: Teatro Lirico, Milan, November 26, 1902. In Paris, in the early eighteenth century, Adriana Lecouvreur is an actress of the Comédie Française. She is a rival of Princesse de Bouillon for the love of Maurice de Saxe. The Princess kills Adriana by means of poisoned flowers, and the actress dies in Maurice's arms after he reveals that he loves her alone.

The finest arias include Adriana's "Io sono l'umile ancella" in Act I; Maurice's "L'anima ho stanca" in Act II; and Adriana's "Poveri fiori" in Act IV.

Adriano, Stefano Colonna's son (mezzo-soprano) in Wagner's *Rienzi.*

Aegisthus, Klytemnestra's beloved (tenor) in Richard Strauss's *Elektra.*

Aegyptische Helena, Die (The Egyptian Helen), opera by Richard Strauss. Libretto by Hugo von Hofmannsthal, based on an episode from Homer's *Odyssey.* Première: Dresden, Germany, June 6, 1928. Menelaus, back from Troy, plans to kill Helen, his unfaithful wife whose beauty has brought so many to their death. However, the magic potion of the Egyptian sorceress Aithra causes him to forgive her.

Aelfrida, daughter of the Thane of Devon (soprano) in Deems Taylor's *The King's Henchman.*

Aeneas, (1) Trojan hero (tenor) in Berlioz' *Les Troyens.*

(2) Trojan hero (baritone) in Purcell's *Dido and Aeneas.*

Aeneid, The, epic poem by Virgil, describing the adventures of Aeneas after Troy's fall. Material from this poem is found in Berlioz' cycle of two operas, *Les Troyens,* and in Purcell's *Dido and Aeneas.*

Aeschylus, tragic dramatist. Born Eleusis, Greece, 525 B.C.; died Gela, Sicily, 456 B.C. Some of his dramas have been made into operas. The Orestes trilogy (*Agamemnon, Choëphoroi, and Eumenides*) was utilized for Darius Milhaud's *Oreste* trilogy, Sergei Taneyev's *Oresteia,* and Felix Weingartner's *Orestes. Prometheus Bound* was used by Maurice Emmanuel (*Prométhée enchaîné*) and Gabriel Fauré (*Prométhée*), *The Suppliants* by Maurice Emmanuel (*Salamine*).

Aethelwold, Earl of East Anglia (tenor) in Deems Taylor's *The King's Henchman.*

Africaine, L' (The African Maid), grand opera in five acts by Giacomo Meyerbeer. Libretto by Eugène Scribe. Première: Paris Opéra, April 28, 1865. American première: Academy of Music, New York, December 1, 1865.

Characters: Vasco da Gama, officer in the Portuguese Navy (tenor); Don Pedro, King's Councilor (basso); Don Alvar, King's Councilor (tenor); Don Diego, King's Councilor (tenor); Inez, Don Diego's daughter (soprano); Anna, her attendant (contralto); Selika, an African queen (soprano); Nelusko, her slave (baritone); Grand Inquisitor (bass); priests; soldiers; councilors; tribesmen. The settings are Lisbon and Madagascar; the time is the fifteenth century.

Act I. Council chamber in the palace of the King of Portugal. Inez is concerned over the fate of Vasco da Gama, gone over two years on one of his expeditions. Tenderly she recalls his beautiful song of farewell ("Adieu, mon doux rivage"). Her concern mounts when her father, insisting that Vasco must be dead, urges her to accept Don Pedro as her husband. The councilors now enter the chamber and acclaim the Grand Inquisitor ("Dieu, que le monde révère"). Unexpectedly, Vasco da Gama arrives to inform the councilors about his expedition to a strange new land ("J'ai vu, nobles seigneurs"). He introduces two slaves—Selika and Nelusko—whom he has brought back with him and asks for another ship so that he may return to this land and claim its riches for Portugal. When the councilors refuse, Vasco denounces them violently and is imprisoned.

Act II. The prison. Vasco has been in prison a month, looked after solicitously by Selika, who is in love with him. She sings him a lullaby ("Sur mes genoux, fils du soleil"). Nelusko, jeal-

ous, tries to kill Vasco in his sleep, but is stopped by Selika. Passionately, Nelusko declares that his only allegiance is to Selika, the queen of her realm ("Fille des rois, à toi l'hommage"). After Nelusko departs, Selika tells Vasco of a new secret route to her land. Vasco is jubilant ("Combien tu m'es chère"). Don Pedro and Inez arrive, and find Selika and Vasco in each other's arms. Don Pedro accuses Vasco of being unfaithful to Inez. Vasco denies this vehemently; he even gives Selika to Inez as her personal slave. Vasco then learns that Inez has bought his freedom by offering to marry Don Pedro who, with the help of the councilors, will undertake an expedition to the land discovered by Vasco.

Act III. Don Pedro's Ship. Don Pedro, Inez, Selika, and Nelusko are sailing for Selika's land. The sailors sing a rousing chantey as they attend to their tasks ("Debout! matelots!"). Treacherously, Nelusko steers the ship toward a reef, singing of Adamastor, monarch of the seas, who brings ships to their doom ("Adamastor, roi des vagues profondes"). Vasco, meanwhile, overtakes Don Pedro's ship and warns him of impending disaster. The distrustful Pedro orders him seized and executed. Before his command could be carried out, the ship is wrecked. Selika's tribesmen swarm aboard, killing or capturing all the Portuguese. Vasco is protected by Selika.

Act IV. The temple of Brahma on the Island of Madagascar. Selika is back on her throne, honored by her people. Vasco is enraptured by the beauty of the island ("O Paradis"). The High Priest comes to insist that the remainder of the Portuguese be executed. In order to save Vasco, Selika says she is secretly married to him. Enraptured, Vasco expresses his great love for her and she responds with equal ardor ("O transport, O douce extase").

But as the marriage of Vasco and Selika is to be solemnized by native rites, Vasco hears Inez' distant song of farewell. He had believed her dead. Knowing now that she yet lives, he hopes to save her.

Act V, Scene 1. The Queen's garden. (This scene is sometimes omitted.) When Inez is brought before the queen, Selika is so moved by the intensity of Inez' love that she frees both her and Vasco and orders Nelusko to put them on a ship bound for Portugal.

Scene 2. A promontory above the Sea. Watching Vasco's ship recede from sight, Selika addresses the sea, which, she says, is as boundless as her own misery ("D'ici je vois la mer immense"). Then she turns to a poisonous manchineel tree, hoping for peace ("O temple magnifique"). As she lies dying, Nelusko comes to her side to breathe the tree's deadly blossoms and join his beloved. An invisible chorus comments that in death all are equal ("C'est ici le séjour").

L'Africaine was Meyerbeer's last opera. It absorbed him for many years, for he was conscious that he was producing his finest work. He started the first sketches in 1838 and completed the opera twenty-two years later. Even then, he continued to make revisions to the last days of his life. He did not live to see the première. Like Meyerbeer's other grand operas, L'Africaine shows his predilection for spectacular scenes, but the work shows a more highly developed musical style and more refinement of detail than its predecessors. The best pages of L'Africaine reveal Meyerbeer at the peak of his melodic inspiration.

Afron (or Aphron), Prince, son of King Dodon (baritone) in Rimsky-Korsakov's *Le coq d'or.*

Agamemnon, King of the Greeks (baritone) in Gluck's *Iphigénie en Aulide.* *See* also AESCHYLUS.

Agathe, Kuno's daughter (soprano) in Weber's *Der Freischütz.*

Agathe's Prayer, *see* LEISE, LEISE.

Agnes, Micha's wife (soprano) in Smetana's *The Bartered Bride.*

Ah! bello, a me ritorna, Norma's aria in Act I of Bellini's *Norma.*

Ah, chacun le sait (Ah, ciascun lo dice), Marie's aria in Act I of Donizetti's *The Daughter of the Regiment.*

Ah! che la morte ognora (Miserere), the celebrated ensemble scene of Manrico, Leonora, and chorus in Act IV of Verdi's *Il trovatore.*

Ah! che tutta in un momento, duet of Dorabella and Fiordiligi in Act I, Scene 4, of Mozart's *Così fan tutte.*

Ah, ciascun lo dice, *see above,* AH, CHACUN LE SAIT.

Ah! dite alla giovine, duet between Violetta and Germont in Act II, Scene 1, of Verdi's *La traviata.*

Ah! du wolltest mich nicht deinen Mund küssen lassen! Salome's apostrophe to the head of John the Baptist, in Richard Strauss's *Salome.*

Ah! fuyez, douce image, Des Grieux's aria in Act III of Massenet's *Manon.*

Ah guarda, sorella, duet of Dorabella and Fiordiligi in Act I, Scene 2, of Mozart's *Così fan tutte.*

Ah! io veggio quell' anima bella, Ferrando's aria in Act II, Scene 2, of Mozart's *Così fan tutte.*

Ah! je respire enfin! Pelléas' expression of relief on leaving the castle vaults, in Act III, Scene 3, of Debussy's *Pelléas et Mélisande.*

Ah! je suis seule, Thaïs' monologue on her world weariness, in Act II, Scene 1, of Massenet's *Thaïs.*

Ah! lève-toi, soleil, Romeo's serenade in Act II of Gounod's *Roméo et Juliette.*

Ah! Louise, si tu m'aimes, Julien's plea to Louise, in Act II, Scene 1, of Charpentier's *Louise.*

Ah! Manon, mi tradisce, Des Grieux's aria in Act II of Puccini's *Manon Lescaut.*

Ah! mon fils!, Fidès' address to her son in Act II of Meyerbeer's *Le prophète*.

Ah! non credea mirarti, Amina's aria in Act III of Bellini's *La sonnambula*.

Ah! non giunge, Amina's final aria in Act III of Bellini's *La sonnambula*.

Ah, qual colpo, ensemble scene (trio) in Act III of Rossini's *The Barber of Seville*.

Ah! ritrovarla nella sua capanna, Flammen's aria in Act III of Mascagni's *Lodoletta*.

Ah! se intorno a quest' urna funesta, elegy of shepherds, shepherdesses, and nymphs at Euridice's grave in Act I of Gluck's *Orfeo ed Euridice*.

Ah! sì, ben mio, Manrico's aria in Act III, Scene 2, of Verdi's *Il trovatore*.

Ah! si, fa core e abbraccia, Norma's aria in Act II (originally Act I) of Bellini's *Norma*.

Ah, un foco insolito, Don Pasquale's aria in Act I, Scene 1, of Donizetti's *Don Pasquale*.

Aïda, opera in four acts by Giuseppe Verdi. Libretto by Antonio Ghislanzoni, based on a plot by Mariette Bey. Première: Cairo, Egypt, December 24, 1871. American première: Academy of Music, New York, November 26, 1873.

Characters: The King of Egypt (bass); Amneris, his daughter (mezzosoprano); Aïda, an Ethiopian slave (soprano); Radames, captain of the Egyptian guard (tenor); Ramfis, High Priest (bass); Amonasro, King of Ethiopia, Aïda's father (baritone); priests; priestesses; soldiers; slaves; prisoners; Egyptians. The action takes place in Memphis and Thebes during the reign of the Pharaohs.

Act I, Scene 1. A hall in the palace at Memphis. Ramfis informs Radames that the Ethiopian enemy is advancing upon the Nile valley and that the goddess Isis has chosen an Egyptian warrior to lead the defenders. Radames dreams that he will be this warrior, and that he will place the fruits of his victory at Aïda's feet ("Celeste Aïda").

Amneris, who wants to marry Radames, discovers that he and Aïda are secretly in love. The King of Egypt now enters with his entourage, followed by a messenger announcing the enemy's approach. The King appoints Radames commander of the Egyptian army, calling on his people to rally under their new leader ("Su! del Nilo al sacro lido!"). Aïda prays for Radames' victorious return ("Ritorna vincitor!"). Remembering that his victory must spell defeat for her own people, she is torn by inner conflict.

Scene 2. The temple of Vulcan. Ramfis is at the altar, praying with his priests for victory. Priestesses perform a ritual dance. Radames enters to be blessed with sword and armor. The High Priest now prays for divine protection ("Nume, custode e vindici") and the people join in the prayer.

Act II, Scene 1. A hall in Amneris' apartments. Amneris' servants are garbing her in silks and jewels for the imminent victory celebration. Slave girls sing of the joys of love ("Chi mai fra gli inni e i plausi"). Moorish boys perform a dance. Reclining on her couch, Amneris thinks of her beloved Radames and pines for his return ("Vieni, amor mio"). When Aïda arrives, Amneris dismisses her servants and other slaves, hoping to learn the true state of Aïda's feelings for Radames. By suggesting that he has been killed in action, then quickly admitting that she had lied, she uncovers Aïda's love. Proudly, Amneris insists that only she, the daughter of the Egyptian King, can win Radames. She threatens her slave with death and the latter, though a princess in her own right, is obliged to beg for mercy.

Scene 2. Outside the city walls. The King, Amneris, their courtiers and slaves, the priests and the people welcome the victorious Egyptian army. The people raise a hymn to victory ("Gloria all' Egitto"), after which the priests thank the gods ("Della vittoria agli arbitri"). Dancing girls perform a ballet symbolizing the victory. The celebration reaches a climax as the Egyptian army arrives, headed by Radames whom Amneris honors with a laurel wreath. The captive Ethiopians are brought in, among them King Amonasro who entreats the King to be lenient ("Ma tu, o Re, tu possente"). Aïda and Radames join in the plea. At the advice of the High Priest, the King frees the captives, but holds Amonasro as a hostage. Then, to reward Radames for his victory, he gives him the hand of Amneris. The populace sings the praise of Egypt and Isis.

Act III. The banks of the Nile. In a near-by temple priests sing to Isis. Ramfis brings Amneris to invoke divine blessings for her imminent marriage to Radames. After they enter the temple, Aïda appears. The slave recalls her beloved homeland ("O patria mia"). Amonasro enters suddenly, enraged over his daughter's love for Radames. He confides to her that the Ethiopians are about to strike back at the Egyptians and that victory would be assured if Radames will name the pass his men are soon to march through. Brokenheartedly, Aïda promises to wrest the secret from Radames. At his approach, Amonasro hides. Radames is overjoyed to see Aïda again, and soon resolves to flee with her to her own land. By asking what route they should take, Aïda tricks Radames into revealing what her father wants to know. Amneris, who has come from the temple in time to overhear, denounces Radames as a traitor. Aïda and her father flee. Aware of the enormity of his crime, Radames surrenders his sword to Ramfis.

Act IV, Scene 1. A room in the palace. Torn between love and jealousy, Amneris sends for Radames. She begs him to confess his guilt before the priests and plead for mercy, promising

to intercede on his behalf if he will give up Aïda ("Già i sacerdoti adunansi"). Radames refuses. In the judgment chamber the priests gather to pass sentence on him. Three times the High Priest calls on Radames to defend himself, three times Radames remains silent. The priests pronounce sentence: Radames is to be buried alive. ("Radames, è deciso il tuo fato").

Scene 2. The temple of Vulcan. Within his tomb Radames bewails the fact that he will not see Aïda again ("La fatal pietra"). But now Aïda emerges from the shadows—she has entered the tomb ahead of Radames to share his fate ("Presago il core della tua condanna"). Above the tomb the chanting priests perform a ritual dance. Radames and Aïda bid the world farewell ("O terra, addio"), while above, in the temple, Amneris prays to Isis for peace ("Pace t'imploro").

Verdi wrote Aïda on a commission from the Khedive of Egypt. The opera was meant for performance in a new theater in Cairo, as part of a celebration attending the opening of the Suez Canal. The theater and canal were opened, as planned, in 1869, but Verdi did not finish writing Aïda until 1871. The costumes and scenery, designed and executed in Paris, had to remain there until after the end of the Franco-Prussian War (1871). The première was an occasion worth the long delay. The audience included the cream of European and Egyptian society and royalty. Arabian trumpeters, a Cairo military band, and a cast of three hundred participated in the triumphal march in the second act. Verdi himself was absent. He detested publicity and was afraid of ocean travel. His opera did not disappoint. Filled with pageantry, ballets, and dramatic situations, Aïda was the concluding work of his second creative period. It was not only Verdi's most ambitious work so far, but his most successful attempt at achieving a natural fusion of drama and music. Because of his rich treatment of the orchestra and his avoidance of florid vocal writing, Verdi was accused by some critics of imitating Wagner. Yet, for all its advanced techniques, *Aïda* is pure Verdi throughout —in the beauty of its Italian melodies, in the happy way in which these melodies lie for the voice, in the emotional intensity of the score and in the felicitous musical characterizations.

Aiglon, L' (The Eaglet), opera by Arthur Honegger and Jacques Ibert. Libretto by Henri Cain, based on the play of the same name by Edmond Rostand. Première: Monte Carlo, March 11, 1937. The central character is the tubercular son of Napoleon, known as the "eaglet." Honegger wrote Acts II, III, and IV; Ibert, Acts I and V. In addition, an orchestral piece by Ibert was interpolated in the third act as ballet music.

Ai nostri monti, duet of Azucena and Manrico in Act IV, Scene 2, of Verdi's *Il trovatore.*

Ainsi que la brise légère, waltz with chorus in Act II of Gounod's *Faust.*

Air de la poupée, *see* OISEAUX DANS LA CHARMILLE, LES.

Air du miroir, *see* DIS-MOI QUE JE SUIS BELLE.

Aithra, an Egyptian sorceress (soprano) in Richard Strauss's *Die aegyptische Helena.*

Alain, a shepherd (tenor) in Massenet's *Grisélidis.*

Albanese, Licia, soprano. Born Bari, Italy, July 22, 1913. She studied singing with Giusepinna Baldassare-Tedeschi, and made an unscheduled debut at the Teatro Lirico in Milan in 1934, substituting for an indisposed prima donna in *Madama Butterfly*. In 1935 she won first prize in a national singing contest in Italy. Her formal debut took place at the Teatro Reale in Parma, on December 10, 1935, again in *Madama Butterfly*. Engagements in Milan, London (during the Coronation festivities in 1936), and Paris preceded her Metropolitan Opera debut on February 9, 1940, in *Madama Butterfly*. Here, as elsewhere, she was an immediate success. During World War II she lived in Italy, then returned to New York. In 1945 she married a New York stockbroker and became an American citizen. Ever since her return she has been one of the Metropolitan's principal singers. She was one of the few women singers to perform at the Vatican for Pope Pius XI, and was the first singer of her sex to broadcast over the Vatican radio station. Before World War II she gave command performances at the Italian court. In 1946 she was selected by Arturo Toscanini to sing the role of Mimi in his fiftieth anniversary performance of *La Bohème* with the NBC Symphony.

Albani, Emma (born MARIE-LOUISE-CÉCILE-EMMA LAJEUNESSE), soprano. Born Chambly, Canada, November 1, 1847; died London, England, April 3, 1930. In her eighth year she entered a Montreal convent school and here she first attracted attention to her unusual musical talent. In 1868 she went to Europe where she studied with Gilbert Duprez in Paris and Francesco Lamperti in Milan.

She made her opera debut at Messina in 1870, singing the role of Amina in *La sonnambula,* and assuming at this time her stage name of Albani. Additional appearances and further study with Lamperti preceded her Covent Garden debut in 1872. This marked the beginning of her twenty-four-year association with Covent Garden, where she was the toast of London's opera public. She gave the first of several command performances for Queen Victoria in 1874. On October 21, 1874, she made her American debut at the Academy of Music in *La sonnambula.* She returned to America in 1883 under Mapleson's direction, and again in

1889 under the wing of Maurice Grau. On March 24, 1890, she made her debut at the Metropolitan Opera in *Otello.* A year later she became the leading soprano of the company. Her last American appearance in opera took place on March 31, 1892, in *Der fliegende Holländer.* In July, 1896, she sang at Covent Garden for the last time. She retired from opera in 1906, but continued to sing in concerts. Financial reverses compelled her later to start a new career in music halls. In 1925 she was made a Dame of the British Empire.

Alberich, a Nibelung (baritone or bass) in Wagner's *Das Rheingold, Siegfried,* and *Die Götterdämmerung.*

Albert, (1) officer of the imperial guard (bass) in Halévy's *La Juive.*

(2) Husband (baritone) of Charlotte in Massenet's *Werther.*

Albert, Eugène d', *see* D'ALBERT, EUGÈNE.

Albert Herring, comic opera by Benjamin Britten. Libretto by Eric Crozier, based on a short story by Guy de Maupassant, *Le rosier de Madame Husson.* Première: Glyndebourne, England, June 20, 1947. Lady Billows of Suffolk, producing a May Day festival, offers a prize for a May Queen whose virtue is beyond doubt. Since no such young woman can be found, Lady Billows determines to have a May King, in the person of Albert Herring. When a practical joker fills Herring's lemonade glass with rum, the May King becomes inebriated and disappears into the night, searching for disreputable pleasures. He returns to the festival the following morning, disheveled and haggard, but proud of his new-won emancipation.

Albine, an abbess (mezzo-soprano) in Massenet's *Thaïs.*

Alboni, Marietta, contralto. Born Cesena, Italy, March 6, 1823; died Ville d'Avray, France, June 23, 1894. While she was studying in Bologna, Rossini heard her and was so impressed that he coached her in the leading contralto roles of some of his operas. Her formal debut took place in Bologna on October 3, 1842 in Pacini's *Saffo.* After a successful appearance at La Scala in Rossini's *Assedio di Corinto* on December 30, 1842, she made an extensive tour in Germany, Austria, and Russia. She was brought to London in 1847 as a rival attraction to Jenny Lind; her appearances at Covent Garden were sensational. She repeated her triumphs in Paris in 1849, and in 1853 toured North America. She retired in 1867 after the death of her husband, Count Pepoli, but appeared to sing at Rossini's funeral in 1868.

Alceste, opera in three acts by Christoph Willibald Gluck. Libretto by Ranieri da Calzabigi, based on the tragedy of Euripides. Première: Burgtheater, Vienna, December 26, 1767. American première: Metropolitan Opera, New York, January 24, 1941.

Characters: Admetos, King of Pharae (tenor); Alceste, his wife (soprano); High Priest (bass); Apollo (baritone); Evander, a messenger (tenor); Hercules (bass); Thanatos, God of Death (bass); priests; priestesses; people of Pharae. The setting is Pharae, in ancient Thessaly.

Act I. The temple of Apollo. King Admetos is dying. Alceste begs the gods to spare his life, and the people of Pharae join in the prayers. Apollo responds that Admetos can be saved only if someone dies in his place. The people shrink back in terror. Alceste offers herself as the sacrifice ("Divinités du Styx").

Act II. The King's palace. The King has recovered. When he discovers that Alceste has saved his life at the expense of her own, he entreats the gods to allow him to accompany her to the other world. Hercules promises him he will save Alceste.

Act III. The entrance to Hades. Alceste and Admetos have come to share death. When Thanatos tries to claim Alceste, Hercules rescues her. Appeased, Apollo allows both Alceste and Admetos to live. The people rejoice.

Alceste was written five years after *Orfeo ed Euridice*. Gluck's operatic reforms, first realized in *Orfeo*, are achieved even more boldly in the later work. *Alceste* is, indeed, a complete realization of the composer's ideal to make music and drama a single entity, to endow both with human values, and to arrive at simplicity. When *Alceste* was published in 1769 it contained a preface in which the composer explained his ideas (*see* GLUCK). *Alceste* was so far ahead of its time that its première was a failure. Only a handful of Gluck's friends realized what he had accomplished. The story goes that when the disheartened composer left the theater, he told a friend: *"Alceste* has fallen." The friend replied: "Yes— fallen from heaven."

See also ALKESTIS.

Alcindoro, Musetta's rich admirer (bass) in Puccini's *La Bohème.*

Alda, Frances (born FRANCES DAVIS), soprano. Born Christchurch, New Zealand, May 31, 1883; died Venice, September 18, 1952. After studying with Mathilde Marchesi in Paris she made her debut at the Opéra-Comique on April 15, 1904, in *Manon,* assuming the stage name of Alda. Her first appearance at La Scala, in 1908, in the Italian première of *Louise,* was a triumph and she was forthwith engaged by the Metropolitan Opera. Her first appearance there took place on December 7, 1908, in *Rigoletto.* On April 3, 1910, she married Giulio Gatti-Casazza, the Metropolitan's general manager. Fearing that her further appearances at the Metropolitan might cause criticism of her husband, she resigned from the company. She returned in 1911 and remained for twenty years, a luminous figure in a brilliant operatic

era. She appeared in over thirty roles, mostly in the French and Italian repertory, and created the roles of Roxanne in Damrosch's *Cyrano de Bergerac,* Cleopatra in Hadley's *Cleopatra's Night,* and Madeleine in Victor Herbert's opera of the same name. She sang in such significant Metropolitan premières and revivals as *La cena delle beffe, Francesca da Rimini, Mârouf,* and *Prince Igor.* When she appeared in *Le Roi d'Ys,* Lalo introduced a new aria in the third act at her request. Alda was divorced from Gatti-Casazza in 1928, and on December 28, 1929, she made a gala last appearance at the Metropolitan in *Manon Lescaut.* Subsequently, she taught singing and was the first important operatic singer to make radio appearances. She remarried in 1941. During World War II she was active in war-relief work. Her autobiography, *Men, Women and Tenors,* appeared in 1937.

Aleko, opera by Sergei Rachmaninoff. Libretto by Nemirovich-Danchenko, based on Pushkin's *The Gypsies.* Première: Moscow, May 9, 1893. The text avoids the philosophical implications of Pushkin's poem and concentrates on the unhappy love affair of a young gypsy and Aleko's wife. Rachmaninoff's score is in the style of Tchaikovsky and Borodin.

Alerte! Alerte! Trio of Faust, Marguerite, and Méphistophélès in Act V, Scene 2, of Gounod's *Faust.*

Alessandro Stradella, opera by Friedrich von Flotow. Libretto by Wilhelm Friedrich. Première: Hamburg, December 30, 1844. Stradella was a seventeenth century composer (*see* STRADELLA). An incident in his life provides the story for this opera. Stradella, here a singer, elopes with Leonora, whose old guardian, Bassi, wants her for himself. Twice, Bassi hires assassins to murder Stradella, but his singing moves them so deeply that they cannot raise a hand against him. Bassi finally approves of the marriage.

Alessio, a peasant (bass) in love with Lisa in Bellini's *La sonnambula*.

Alexis, Prince, Stephana's lover (tenor) in Giordano's *Siberia*.

Alfano, Franco, composer. Born Naples, Italy, March 8, 1876; died San Remo, October 27, 1954. His music study took place in Naples and Leipzig. His first operas, *Miranda* (1896) and *La fonte di Enschir* (1898), were failures. But *Risurrezione*, given in Turin in 1904, was a major success and made Alfano famous. Later operas established his position in Italian opera: *L'ombra di Don Giovanni* (1914); *La leggenda di Sacùntala* (1921); *Madonna Imperia* (1927); *Cyrano de Bergerac* (1937); *Il Dottor Antonio* (1949). Alfano at first combined Italian lyricism and sentiment with the elaborate harmonic and instrumental textures of the Wagnerian school; his later writing owes much to Ravel. In 1924, when Puccini's death left his *Turandot* unfinished, Alfano was chosen to write the closing pages. Alfano held several important directional positions, including those of the Bologna Liceo Musicale (1919–1926), the Turin Liceo Musicale (1926–1937), and the Rossini Conservatory in Pesaro (after 1937).

Al fato dan legge, duet of Ferrando and Guglielmo in Act I, Scene 2, of Mozart's *Così fan tutte*.

Alfio, a teamster (baritone) in Mascagni's *Cavalleria rusticana*.

Alfonso, son (tenor) of the Viceroy of Naples in Auber's *La muette di Portici*.

Alfonso, Don, (1) Duke of Ferrara (baritone) in Donizetti's *Lucrezia Borgia*.

(2) An old bachelor (basso buffo) in Mozart's *Così fan tutte*.

Alfonso XI, King of Castille (baritone) in Donizetti's *La favorita*.

Alfred, Rosalinde's admirer (tenor) in Johann Strauss's *Die Fledermaus*.

Alfredo, see GERMONT, ALFREDO.

Alice, Robert's foster sister (soprano) in Meyerbeer's *Robert le Diable*.

Alice Ford, see FORD, ALICE.

Alice M. Ditson Fund, see DITSON FUND.

Alidoro, the Prince's friend (bass), in Rossini's *La cenerentola*.

Alim, King of Lahore (tenor) in Massenet's *Le Roi de Lahore*.

Alisa, Lucia's companion (mezzo-soprano) in Donizetti's *Lucia di Lammermoor*.

Alkestis, (1) opera by Rutland Boughton. Libretto is Gilbert Murray's English translation of the Euripides drama. Première: Glastonbury, England, August 26, 1922.

(2) Opera by Egon Wellesz. Libretto by Hugo von Hofmannsthal, based on Euripides. Première: Mannheim, Germany, March 20, 1924.

See also ALCESTE.

Alla Cà d'Oro, chorus in Act III, Scene 2, of Ponchielli's *La Gioconda*.

All' idea di quel metallo, duet between Figaro and Almaviva in Act I of Rossini's *The Barber of Seville*.

Allmächt'ge Jungfrau, Elisabeth's Prayer in Act III of Wagner's *Tannhäuser*.

Allmächt'ger Vater, blick' herab, Rienzi's prayer in Act V of Wagner's *Rienzi*.

Almanzor, Emir of Tunis (baritone), in Weber's *Oberon*.

Almaviva, Count, (1) a nobleman of Andalusia (baritone) in Mozart's *The Marriage of Figaro*.

(2) The same (tenor) in Rossini's *The Barber of Seville*.

Alphonse, Camille's beloved (tenor) in Herold's *Zampa*.

Altair, desert chieftain (baritone) in Richard Strauss's *Die aegyptische Helena*.

Althouse, Paul, tenor. Born Reading, Pennsylvania, December 2, 1889; died New York City, February 6, 1954. He combined an exhaustive academic education with music study with Perley Dunn Aldrich in Philadelphia, and Oscar Saenger and Percy Richor Stephens in New York. His debut took place with the Hammerstein Opera Company in Philadelphia. On March 19, 1913, he made his Metropolitan Opera debut as Dmitri in the first American performance of *Boris Godunov*. During the next decade he appeared in the French and Italian repertory at the Metropolitan and sang in the world premières of four American operas: De Koven's *The Canterbury Pilgrims,* Joseph Breil's *The Legend,* Herbert's *Madeleine,* and Cadman's *Shanewis.* During a visit to Bayreuth, he became fired with the ambition of becoming a Wagnerian tenor. For nine years he studied the leading Wagnerian roles, then sang them in Berlin, Stuttgart, and Stockholm. In 1932 he scored a major success in an all-Wagner program conducted by Toscanini in New York. He made his second "debut" at the Metropolitan Opera—this time as a Wagnerian tenor—on February 3, 1934, as Siegmund. Six weeks later he became the first American-born singer to appear there as Tristan. Althouse remained at the Metropolitan through the 1939–1940 season, singing all the major Wagnerian tenor roles. Thereafter, he devoted himself to teaching and coaching. His students included Eleanor Steber and Richard Tucker.

Alto (or Contralto), the lowest range of the female voice, reaching approximately two octaves upward from E or F below middle C. The term is also applicable to boys' voices of this range.

Altoum, Emperor, Turandot's father (tenor) in Puccini's *Turandot.*

Alvar, Don, member of the King's council (tenor), in Meyerbeer's *L'Africaine.*

Alvarez, Albert Raymond (born GOURRON), tenor. Born Bordeaux, France, 1861; died Nice, February 26, 1933. After studies with A. de Martini in

Paris, a debut in Ghent, Belgium, and successful appearances in several French cities, he was engaged by the Paris Opéra in 1892. He became a great favorite there, particularly in the French repertory. He created the leading tenor parts in many new French operas, including *Thaïs.* On December 18, 1899, he made his Metropolitan Opera debut, on opening night of the season, as Roméo. He remained with the Metropolitan several seasons, then appeared with the Manhattan Opera Company.

Alvarez, Marguerite d', *see* D'ALVAREZ, MARGUERITE.

Alvaro, Don, the tragic hero (tenor) of Verdi's *La forza del destino.*

Alvary, Max (born ACHENBACH), tenor. Born Düsseldorf, Germany, May 3, 1856; died Gross-Tabarz, Germany, November 7, 1898. He studied principally with Julius Stockhausen and Francesco Lamperti. After appearances in Europe he made his American debut on November 25, 1885, as Don José, a performance in which Lilli Lehmann also appeared at the Metropolitan Opera for the first time. He remained a principal tenor of the Metropolitan Opera for four years, singing chiefly in the Wagnerian repertory. He was the first Loge and the first Siegfried in America. In 1891 he appeared as Tristan at the Bayreuth Festival.

Alvise, official of the Inquisition (bass) in Ponchielli's *La Gioconda.*

Amahl and the Night Visitors, one-act opera by Gian-Carlo Menotti. Libretto by the composer. Première: NBC Television Network, December 24, 1951. Menotti derived his theme from the painting *The Adoration of the Magi,* by the Flemish artist Hieronymus Bosch. His story concerns the Three Wise Men who, on their way to the Manger in Bethlehem, stop at the hovel of a crippled boy, Amahl. When Amahl offers them his crutches as a gift to the

Holy Child, he is miraculously healed. This was the first opera written expressly for television broadcast. It was subsequently produced by several opera companies in America, the first time at Indiana University, February 21, 1952. It was also given at the Florence Music Festival in 1953.

Amantio di Nicolao, a lawyer (bass), in Puccini's *Gianni Schicchi.*

Amato, Pasquale, baritone. Born Naples, March 21, 1878; died New York City, August 12, 1942. After attending the Naples Conservatory he made his debut at the Teatro Bellini in Naples in 1900 in *La traviata.* He sang in leading opera houses in Europe and visited Buenos Aires and Egypt before being contracted by Gatti-Casazza for the Metropolitan. His American debut took place on November 20, 1908, in *La traviata;* his role was the elder Germont. He remained at the Metropolitan over a decade and was acclaimed for his interpretations of Scarpia, Valentin, Rigoletto, Figaro (in *The Barber of Seville*), Iago, and Barnaba. He created the title role in Damrosch's *Cyrano de Bergerac,* that of Jack Rance in *The Girl of the Golden West,* and Napoleon in *Madame Sans-Gêne.* He also appeared in the American premières of *L'amore dei tre re, Francesca da Rimini, Germania,* and *Lodoletta.*

Because of illness he went into retirement in 1924, spending most of his time in Italy. On February 26, 1933, he returned to the Metropolitan to help celebrate Gatti-Casazza's silver jubilee, and on November 20 of the same year he celebrated his own twenty-fifth anniversary of his first appearance in New York by appearing at the Hippodrome Theater in the role of the elder Germont, the role of his American debut. On April 1, 1934, he became director of the Hippodrome Opera Company, but he held this post only briefly. In 1935 he became head of the voice and opera departments at the Louisiana State University School of Music, a position he held until his death.

Amelia, Renato's wife (soprano) in Verdi's *Un ballo in maschera.*

Amelia Goes to the Ball, one-act comic opera by Gian-Carlo Menotti. Libretto by the composer (English translation by George Meade). Première: Academy of Music, Philadelphia, April 1, 1937. This was Menotti's first opera as a mature composer, and his first success. The setting is Milan, the time 1910. The tongue-in-cheek text concerns Amelia's frantic and seemingly frustrated efforts to go to a ball. Her husband discovers a letter from her lover, her lover arrives through the window, the two men quarrel. Amelia lands her husband in the hospital by smashing a vase over his head, and her lover goes to jail when she puts a false charge against him of having attacked her husband. With the rivals out of the way, Amelia goes to the ball—escorted by the police officer.

Amelia Grimaldi, the assumed name of Maria (soprano) in Verdi's *Simon Boccanegra.*

American operas, *see* OPERA, section OPERA IN AMERICA.

Amfiparnaso, L', a madrigal comedy composed by Orazio Vecchi and published in 1597. The work consists of fourteen madrigals, all but one for five voices, the whole to be sung without instrumental support. The theatrical element in this work of Vecchi's, and similar works by other Italian composers of his time, is so pronounced that the madrigal comedy is considered an important forerunner of opera. There are two parallel plots in *L'Amfiparnaso.* One, "lirico tragica," involves the love of Isabella and Lucio; the other, "grottesco comica," portrays the doings of the buffoons of the commedia dell' arte. The work has occasionally been given a stage presentation, an instance being at the 1938 Florence Music Festival.

Amfortas, keeper of the Holy Grail (baritone) in Wagner's *Parsifal.*

Amico Fritz, L', opera by Pietro Mascagni. Libretto by P. Suardon (pseudonym for N. Daspuro), based on a novel by Erckmann-Chatrian. Première: Teatro Costanzi, Rome, October 31, 1891. Fritz Kobus, a rich bachelor, eventually marries a farmer's daughter, Suzel, as a result of the machinations of Rabbi David. The opera contains a fine duet in Act II, "Tutto tace," sometimes known as the "Duet of the Cherries." An orchestral intermezzo between the second and third acts; Fritz's aria "O amore, o bella luce," and Suzel's aria "Non mi resta che il pianto," both in Act III, are also noteworthy.

Amina, sleepwalking village maiden (soprano) in Bellini's *La sonnambula.*

Amis, l'amour tendre et rêveur, Hoffmann's "Couplets Bachiques" in Act II of Offenbach's *The Tales of Hoffmann.*

Am Jordan Sankt Johannes stand, David's hymn to St. John in Act III of Wagner's *Die Meistersinger.*

Amleto (Hamlet), opera by Franco Faccio. Libretto by Arrigo Boïto, based on Shakespeare's *Hamlet.* Première: Teatro Carlo Felice, Genoa, May 20, 1865. One of its soprano arias, "Sortita d'Ofelia," has become a famous concert number.

Amneris, the King of Egypt's daughter (mezzo-soprano) in Verdi's *Aïda.*

Amonasro, King of Ethiopia (baritone) Aïda's father, in Verdi's *Aïda.*

Amor, god of love (soprano) in Gluck's *Orfeo ed Euridice.*

amore dei tre re, L' (The Love of Three Kings), opera in three acts by Italo Montemezzi. Libretto by Sem Benelli, adapted from his own verse tragedy of the same name. Première: La Scala, Milan, April 10, 1913. American première: Metropolitan Opera, New York, January 2, 1914.

Characters: Archibaldo, King of Altura (bass); Manfredo, his son (baritone); Avito, a former prince of Altura (tenor); Fiora, Manfredo's wife (soprano; Flaminio, Archibaldo's servant (tenor). The action takes place in an Italian castle; the time is the Middle Ages.

Act I. A hall in Archibaldo's castle. Unable to sleep, Archibaldo recalls the time when he led the barbarian invasion of Italy ("Italia! Italia! e tutto il mio ricordo!). Flaminio reminds him how Fiora gave up her beloved Avito to marry Manfredo in order to insure peace in the land. When Archibaldo leaves for his chambers, Avito appears, followed by Fiora. They are still in love and they embrace passionately, unmindful of approaching dawn. Archibaldo returns as Avito departs. He wishes the name of the person to whom Fiora has been speaking. She insists that she has been talking to herself. The blind King knows that she is lying and he cries out against the affliction which prevents him from reading her face. A flourish of trumpets announces the homecoming of Manfredo from battle. Manfredo is overjoyed with his victory and his return ("Oh, padre mio"). Fiora welcomes her husband frigidly, a circumstance which does not escape the blind King.

Act II. A terrace atop the castle walls. Manfredo, about to return to war, bids his wife a sorrowful farewell ("Dimmi, Fiora, perchè ti veggo ancora"). He asks her mount the battlement and wave her scarf as he rides down the valley. After Manfredo has left, Avito enters, disguised as a castle guard. Fiora begs him to leave, but cannot resist Avito's last plea for a kiss ("Ho sete! Ho sete!"). They succumb passionately to one another. Archibaldo interrupts this idyll. After Avito manages to escape, Archibaldo bitterly denounces Fiora for her unfaithfulness and demands to know her lover's name.

Fiora admits her infidelity but refuses to name Avito. In a fit of rage Archibaldo strangles her. Manfredo unexpectedly returns, worried over Fiora. Brokenheartedly, Archibaldo reveals that he has murdered her.

Act III. The palace crypt. Mourners sing a dirge before Fiora's bier. After they depart, Avito comes to bid his beloved a desperate farewell ("Fiora . . . E silenzio!"). He bends to kiss her lips for the last time. Archibaldo, attempting to trap Fiora's lover, has poisoned her lips. As Avito begins to die, Manfredo, coming to pay his last respects, discovers him. Unable to hate Avito and unable to live on without his wife, he kisses Fiora in order to die. When Archibaldo discovers that his trick has not only destroyed Fiora's lover but also his son, he succumbs to despair.

L'amore dei tre re was not only Montemezzi's most successful opera, it has remained one of the crowning works of twentieth century Italian opera. It forged no new trails, but within familiar patterns showing the influences of Wagner, Verdi, and Debussy, it combined impressive craftsmanship with a powerful story.

amore medico, L' (Doctor Love), opera buffa by Ermanno Wolf-Ferrari. Libretto by Enrico Golisciani, based on Molière's *L'amour médecin*. Première: Dresden, Germany, December 4, 1913 (under the German title of *Der Liebhaber als Arzt*). Lucinda, suffering from a prolonged malady (which after all proves to be only love), is cured after her lover, Clitandro, disguised as a doctor, prescribes a mock marriage which turns out to be real.

Amore o grillo, Pinkerton's avowal of love for Cio-Cio-San in Act I of Puccini's *Madama Butterfly*.

Amor ti vieta, Count Loris' avowal of love in Act II of Giordano's *Fedora*.

amour des Trois Oranges, L', see LOVE FOR THREE ORANGES.

amour est un oiseau rebelle, L', Car-

men's Habanera in Act I of Bizet's *Carmen*.

amour médecin, L', see AMORE MEDICO, L'; MOLIÈRE.

Amour! viens aider ma faiblesse! Dalila's appeal to the god of love in Act II of Saint-Saëns' *Samson et Dalila*.

Am stillen Herd, Walther's narrative in Act I of Wagner's *Die Meistersinger*.

Andersen, Hans Christian, poet and writer of fairy tales. Born Odense, Denmark, April 2, 1805; died Copenhagen, August 4, 1875. Among the operas based on tales by Andersen are: Rudolf Wagner-Régény's *Der nackte König (The Emperor's New Clothes);* Ernst Toch's *The Princess on the Pea;* Stravinsky's *Le rossignol (The Nightingale);* Charles Stanford's *The Travelling Companion;* Bernard Rogers' *The Nightingale;* Alfred Bruneau's *Le Jardin du Paradis;* August Enna's *The Princess on the Pea;* Niels-Eric Fougstedt's radio opera, *The Tinderbox;* Margaret More's *The Mermaid;* Hermann Reutter's *Die Prinzessin und der Schweinehirt;* Bernhard Sekles' *Die zehn Küsse.* Andersen also wrote *Bruden fra Lammermoor,* a libretto for the Danish composer Frederik Bredal, and librettos for operas by Franz Gläser and Johann Peter Hartmann.

Anderson, Marian, contralto. Born Philadelphia, Pennsylvania, February 17, 1902. She was the first Negro to appear at the Metropolitan Opera in a major role: Ulrica in *Un ballo in maschera.* Her first appearance there on January 7, 1955 also marked the opera debut of this world-famed concert artist.

Andrea Chénier, opera in four acts by Umberto Giordano. Libretto by Luigi Illica. Première: La Scala, Milan, March 28, 1896. American première: Academy of Music, New York, November 13, 1896.

Characters: Andrea Chénier, a poet (tenor); Charles Gérard, revolutionary leader (baritone); Countess de Coigny (mezzo-soprano); Madeleine, her daughter (soprano); Bersi, Madeleine's maid (mezzo-soprano); Roucher, Chénier's friend (bass); Fouquier-Tinville, public prosecutor (bass); Fléville, a writer (baritone); Schmidt, a jailer (baritone); Madelon, an old blind woman (mezzo-soprano); Mathieu, a waiter (baritone); a spy; an abbé; soldiers; revolutionaries; prisoners; servants. The action takes place in Paris before and during the French Revolution.

Act I. Ballroom in the Château de Coigny. Gérard, a servant, bitter at social injustice, predicts imminent doom for the aristocracy ("T' odio casa dorata"). When Madeleine appears, Gérard remarks on her beauty, for he loves her. Madeleine complains of the tortures suffered by a young lady who must always be dressed fashionably ("Si! io penso alla tortura"). Guests now arrive, one of them Andrea Chénier. A ballet pantomime is performed. Then Chénier recites one of his love poems ("Un di all' azzurro spazio"), a thinly disguised attack on the rich which horrifies the guests. The embarrassment is relieved by dancing. But suddenly beggars, headed by Gérard, burst in and beg for charity ("La notte e giorno"). When Gérard is ordered to send them away, he tears off his livery and announces his sympathy with the poor ("Si, me ne vo, Contessa"). The footmen eject Gérard and the beggars, and the dancing continues.

Act II. The Café Hottot. The Revolution is in full swing. Chénier, having denounced Robespierre, is held in suspicion by the revolutionaries. He is sitting at a table when Bersi slips him a note in which an unnamed friend seeks help. Roucher urges Chénier to flee but Chénier refuses ("Credo a una possanza arcana"). Besides, he is eager to help the unknown writer of the note. The writer, who soon comes disguised to the café, is Madeleine ("Eravate possente"). Ché-

nier and Madeleine now realize how much they love each other. They conspire to flee ("Ora soave, sublime ora d'amore"). But before they can escape, Gérard, now an important revolutionary leader, comes for Madeleine. A duel ensues in which Gérard is wounded. Remorsefully, Gérard urges Chénier to escape. When Gérard's friends arrive, he pretends not to know who wounded him.

Act III. The revolutionary tribunal. Gérard learns that Chénier has been caught and arrested. He must now denounce the poet formally. As he does so, he recalls the poet's nobility and loyalty ("Nemico della patria?"). Despite his inner conflicts, Gérard signs the paper dooming Chénier. Madeleine appears to plead for the poet. Gérard tells her that fate has decreed that they belong to one another ("Perchè ciò volle il mio voler possente"). Madeleine tells Gérard of her mother's death ("La mamma morta"), and offers herself to Gérard in return for Chénier's freedom. Gérard promises help. During the trial Chénier defends himself ("Si, fui soldato"), but the mob demands his death and Gérard cannot help him.

Act IV. Prison of Saint-Lazare. Awaiting execution, Chénier writes a farewell poem ("Come un bel di di maggio"). Madeleine comes to die with the man she loves; she has bribed a jailer to substitute her name on the death list for that of a victim. Chénier and Madeleine embrace, repeating their devotion for each other ("Vicino a te s'aqueta"). Together, they walk to the guillotine.

Giordano became famous with *Andrea Chénier,* an opera dealing freely with a historical figure, the poet André de Chénier. While he wrote several fine operas after this one, none equaled *Chénier* in power of inspiration, beauty of melody, and sustained dramatic interest. *Chénier* is filled with passionate arias and recitatives, stirring emotional situations and climaxes. It has stirred audiences in every country.

Andrei, son (tenor) of Prince Ivan Khovansky in Mussorgsky's *Khovantchina.*

Andrès, Stella's servant (tenor) in Offenbach's *The Tales of Hoffmann.*

Andreyev, Leonid, author and playwright. Born Orel, Russia, June 18, 1871; died Helsinki, Finland, September 12, 1919. A number of his works have been the source of operas: *The Abyss* (Rebikov); *He Who Gets Slapped* (Robert Ward); *Ol-Ol* (Alexander Tcherepnin).

Ange adorable, love duet of Roméo and Juliette in Act I of Gounod's *Roméo et Juliette.*

Angeles, Victoria de los, *see* DE LOS ANGELES, VICTORIA.

Angelica, a nun (soprano) in Puccini's *Suor Angelica.*

Angelotti, Cesare, a political plotter (bass) in Puccini's *Tosca.*

Anges du paradis, Vincent's cavatina in Act III of Gounod's *Mireille.*

Anges purs, anges radieux, Marguerite's aria in Act V, Scene 2, of Gounod's *Faust.*

Anita, (1) an opera singer (soprano) in Křenek's *Jonny spielt auf.*

(2) A girl from Navarre (soprano) in Massenet's *La Navarraise.*

An jenem Tag, Hans Heiling's aria in Act I of Marschner's *Hans Heiling.*

Anna, Inez's maid (mezzo-soprano) in Meyerbeer's *L'Africaine.*

Anna, Donna, Don Pedro's daughter (soprano) in Mozart's *Don Giovanni.*

Annchen, Agathe's friend (soprano) in Weber's *Der Freischütz.*

Annina, (1) Valzacchi's accomplice (contralto) in Richard Strauss's *Der Rosenkavalier.*

(2) Violetta's maid (soprano or mezzo-soprano) in Verdi's *La traviata.*

Annunzio, Gabriele d', *see* D'ANNUNZIO, GABRIELE.

Antheil, George, composer. Born Trenton, New Jersey, July 8, 1900. He studied with Constantine von Sternberg, Ernest Bloch, and at the Settlement School in Philadelphia. In 1922 he toured Europe as a pianist. His determination to become a composer made him give up concert work and settle in Paris. There he began writing music in a revolutionary and provocative vein, and for some years he was known as an *enfant terrible* of music. The most celebrated of his early works was the *Ballet mécanique,* first given in Paris in 1927. He also wrote a jazz opera, *Transatlantic,* introduced by the Frankfort Opera in May, 1930, with outstanding success. His second opera, *Helen Retires,* written in 1932, was performed at the Juilliard School of Music in February, 1934. In 1933 Antheil settled in Hollywood and for a time wrote music for motion pictures. He now entered a new creative period, abandoning his former iconoclasm and writing music that was classical in style and approach. In this vein he completed a third opera, *Volpone,* first produced at the University of Southern California on January 9, 1953. A later opera, *The Brothers,* was introduced in Denver in 1954, and in 1955 *The Wish* —commissioned by a Rockefeller Foundation grant—was first given in Louisville, Kentucky.

Antigone, (1) drama by Sophocles. Daughter of Oedipus, Antigone buries her brother Polynices contrary to command, and for this she is put to death by the tyrant Creon.

(2) Opera by Arthur Honegger. Libretto by Jean Cocteau, based on Sophocles. Première: Théâtre de la Monnaie, December 28, 1927.

(3) Opera by Carl Orff. Libretto is Johann Christian Friedrich Hölderlin's translation of Sophocles' tragedy. Première: Salzburg Festival, 1949.

Antigono, opera by Gluck. Libretto by Metastasio, based on Sophocles' *Anti-*

gone. Première: Teatro Argentina, Rome, February 9, 1756.

Antonia, Hoffmann's third beloved (soprano) in Offenbach's *The Tales of Hoffmann.*

Antonida, Ivan Susanin's daughter (soprano) in Glinka's *A Life for the Czar.*

Antonio, (1) Lodoletta's foster father (bass) in Mascagni's *Lodoletta.*

(2) a gardener (bass) in Mozart's *The Marriage of Figaro.*

(3) a servant (bass) in Thomas's *Mignon.*

Antonio e Cleopatra, opera by Francesco Malipiero. Libretto by the composer, based on Shakespeare's *Antony and Cleopatra.* Première: May Music Festival, Florence, May 4, 1938. Malipiero explained: "Rather than the typical atmosphere of the play I have stressed the human drama, reduced to a few essential characters. Even the chorus has a secondary function in the opera; it culminates in the banquet scene on Pompey's galley where a few brief dances take place. *Antonio e Cleopatra* is really the tragedy of two human beings. Their actions form the central nucleus about which is built the music drama."

Anvil Chorus, *see* CHI DEL GITANO I GIORNI ABBELLA?

Aphrodité, music drama by Camille Erlanger. Libretto by Louis de Gramont, based on the novel of the same name by Pierre Louÿs. Première: Paris Opéra-Comique, March 23, 1906. In Alexandria, the rich sculptor Demetrios tries to gain the love of Chrysis. She will be his on condition that he acquire for her three things: the courtesan Bacchis' mirror, a comb of the High Priest's wife, and a necklace from the statue of Aphrodité. He steals all three and wins Chrysis, but remorse at his crimes turns his love to hatred. He asks Chrysis to appear in public, wearing the three items. The aroused people seize her and put her in prison where she drinks poison and dies.

Aphron, *see* AFRON.

Apollo, a god (baritone) in Gluck's *Alceste.*

Apostrophe, L', one-act opera by Jean Françaix. Libretto by the composer, based on one of Balzac's *Droll Tales.* Première: Holland Music Festival, Amsterdam, July 1, 1951. The opera concerns the futile love of the hunchback Darnadas for the coquettish Tascherette, and his death at the hands of her jealous husband.

Aprila, o bella, Rafaele's serenade in Act II of Wolf-Ferrari's *The Jewels of the Madonna.*

Arabella, opera by Richard Strauss. Libretto by Hugo von Hofmannsthal. Première: Dresden, July 1, 1933. American première: Metropolitan Opera, February 10, 1955. This was Hofmannsthal's last libretto for Strauss. Count Waldner, an impoverished Viennese aristocrat, must make a favorable marriage for his lovely daughter Arabella who, however, has other ideas. Waldner sends her photograph to an old crony, a landowner in Slavonia. Meantime, this old fellow has died, and his nephew and heir Mandryka opens the letter, falls in love with Arabella on sight, sweeps into Vienna with the pomp and circumstance of a wealthy country squire, supplies Count Waldner with cash, and proposes. Complications are provided by Zdenka, Arabella's younger sister, who has been raised as a boy for reasons of economy. In love with Arabella's ardent admirer Matteo, Zdenka meets him in the dark, pretending that she is Arabella. Mandryka thus becomes suspicious of his intended bride. After all misunderstandings are cleared up, Waldner's insolvency is alleviated by Arabella's marriage, and Zdenka wins Matteo.

With some justification, *Arabella* is generally regarded as a weaker *Der Rosenkavalier.* But if *Arabella* has not the earlier opera's inspiration, it certainly has the same atmosphere of waltz-happy Vienna, the same element of implausible disguise, the same light touch. Added is the intensely emotional manner with which Strauss uses musical lore: one of the most moving numbers, for example, the love duet in the ball scene of Act II, is based on an old Dalmatian melody.

Arabian Nights, a cycle of stories of Persian or Indian origin. Among the operas based on this tenth century classic are: Benno Bardi's *Fatme;* Peter Cornelius' *The Barber of Bagdad;* Issai Dobrowen's *A Thousand and One Nights;* Henri Rabaud's *Mârouf;* Ernest Reyer's *La Statue;* Bernhard Sekles' *Schaharazade;* Victor de Sabata's *Mille e una notte;* Julia Weissberg's *Gulnara.*

Araquil, a soldier (tenor) in love with Anita, in Massenet's *La Navarraise.*

Archibaldo, blind King of Altura (bass) in Montemezzi's *L'amore dei tre re.*

Archy and Mehitabel, concert opera by George Kleinsinger. Libretto by Joe Darion, based on tales by Don Marquis. Première: New York, December 6, 1954, by the Little Orchestra Society. The leading characters of this jazzy comedy are a philosophic cockroach, Archy, and an alley cat, Mehitabel. The story concerns Archy's attempts to shape Mehitabel's destiny—a task made thankless by Mehitabel's inborn waywardness.

Ardon gl'incensi, Lucia's "Mad Scene" in Act III, Scene 1, of Donizetti's *Lucia di Lammermoor.*

Aria, an extended solo for voice in an opera (or oratorio). The earliest operas of Peri and Caccini consisted entirely of recitatives. One of the earliest true arias was the famous "Lament" in Monteverdi's *Arianna,* performed in 1608. This single aria is the only portion of the music that has been preserved. A two-part aria has two contrasting sections (A-B); the three-part adds a repetition of the first part after the contrasting middle section (A-B-A).

In the early eighteenth century the Italians standardized a number of different classes of aria. The most important were: (1) the aria cantabile, a free-flowing emotional melody in which the singer was permitted to introduce displays of vocal virtuosity through embellishments; (2) the aria di portamento, a dignified air characterized by long notes and smooth delivery; (3) the aria parlante, a declamatory kind of song; (4) the aria di bravura, a highly florid air intended to display the singer's technique; (5) the air d'imitazione, in which the voice and accompanying instruments imitated the sounds of nature. In eighteenth century opera it was the practice not to have two arias of the same variety follow in succession.

The Italian aria which dominated the operatic writing of the eighteenth and nineteenth centuries was characterized by warmth of lyricism and florid passages and achieved an advanced stage of development in the writing of Bellini, Rossini, Donizetti, and the earlier Verdi. A more dramatic sort of aria, simpler and more emotional, was evolved by Gluck and developed further by Weber. The formal aria was abandoned by Wagner in his music dramas and by Verdi in *Otello.* The opera score now became a coherent and indivisible whole, without division between recitative and aria. This departure from the formal aria structure is found to an even more marked degree in Mussorgsky's *Boris Godunov* and Debussy's *Pelléas et Mélisande.* In Alban Berg's *Wozzeck* and other operas of the Schoenbergian school the aria assumes the inflections of speech (*see* SPRECHSTIMME).

Ariadne auf Naxos, opera by Richard Strauss. Libretto by Hugo von Hofmannsthal. Première: Stuttgart, October 24, 1912. This is a play within a play. In the prologue, an opera company prepares to perform for a select audience in the private theater of a wealthy eighteenth century patron. At his request, a serious and a comic opera are now performed simultaneously, the story revolving around the lovesick Ariadne, whom the frivolous Zerbinetta and her three helpers vainly try to cheer, whose joy of life is ultimately reawakened by Bacchus. In an earlier version of the work, produced by Max Reinhardt in Stuttgart in 1912, the prologue was a condensation of Molière's *Le bourgeois gentilhomme,* with incidental music by Strauss. (This music was preserved as an orchestra suite.) Notable numbers in the opera are Zerbinetta's recitative and aria, "Grossmächtigste Prinzessin," and the Composer's aria, "Seien wir wieder gut."

Ariane et Barbe-Bleue (Ariadne and Bluebeard), opera by Paul Dukas. Libretto is Maurice Maeterlinck's play of the same name. Première: Paris Opéra, May 10, 1907. Though this has never been a popular work, artistically is one of the finest French operas after Debussy's *Pelléas et Mélisande.* The score is marked both by subtlety of detail and massive brilliance. The story concerns Bluebeard's sixth wife to whom he gives seven keys. Six, of silver, open vaults of precious jewels, and these she may use. But the seventh, of gold, intended for a strange door, is prohibited her. She opens the door and finds Bluebeard's earlier wives. As Ariane frees them, she decides to leave her husband, but the other wives prefer to stay with him.

Arianna, opera by Claudio Monteverdi. Libretto by Ottavio Rinuccini. Première: Mantua, Italy, May 28, 1608. Only a single number survives from this opera, the "Lament of Arianna" ("Lasciatemi morire"), one of the most poignant lyric pages in the early history of opera. Monteverdi subsequently arranged the Lament as a madrigal for five voices and published it in his sixth book of madrigals (1614).

Aricie, Hippolytus' beloved (soprano) in Rameau's *Hippolyte et Aricie.*

Aristophanes, writer of comedies. Born Athens, Greece, about 450 B.C.; died about 380 B.C. Some of his brilliantly satirical, often poetically moving, comedies have been used for operas. Among them: *The Birds* (George Auric; Walter Braunfels); *The Frogs* (Granville Bantock); *Lysistrata* (Reinhold Glière; Raoul Gunsbourg; Engelbert Humperdinck; Victor de Sabata; Franz Schubert's *Der häusliche Krieg*); *Peace* (Marcel Delannoy); and *The Wasps* (Ralph Vaughan Williams).

Arkas, captain of the guards (bass) in Gluck's *Iphigénie en Aulide.*

Arkel, King of Allemonde (bass) in Debussy's *Pelléas et Mélisande.*

Arlecchino, (1) one-act opera by Ferruccio Busoni. Libretto by the composer. Première: Zurich, May 11, 1917. Busoni's story is a typical product of the commedia dell' arte. In the four sections, Arlecchino appears in turn as a rogue, warrior, husband, and conqueror. In the first, he makes love to Matteo's wife; in the second, he is a recruiting officer who sends Matteo off to war; in the third, he discovers his wife with a rival, and slays him with a wooden sword; in the fourth, the rival comes back to life, Matteo returns from the wars, and all the characters take their final bow.

(2) Pantolone's servant (baritone) in love with Colombina, in Wolf-Ferrari's *Le donne curiose.*

Arlesiana, L' (The Girl from Arles), opera by Francesco Cilèa. Libretto by Leopoldo Marenco, based on Daudet's play of the same name. Première: Teatro Lirico, Milan, November 27, 1897. The story concerns the love of Federico for a young woman of questionable reputation. His family stands in the way of their affair and arranges for him to marry a childhood sweetheart. On the eve of the wedding,

Federico commits suicide by throwing himself out of the farmhouse loft. Cilèa's score is noteworthy for its effective use of French folk songs. One of the most famous arias is Federico's lament, "E la solita storia."

Armida, (1) the central character in a dramatic poem by Tasso, *La Gerusalemme liberata.* The poem has been the source of several operas. Its action takes place during the first Crusade. Armida is a beautiful queen endowed with supernatural powers. Rinaldo, leader of the Crusaders, at first resists her charm. Armida tries to overcome her passion for Rinaldo, but then changes her mind. Eventually, Rinaldo falls under her spell. When duty calls him away, she sets her palace on fire and disappears into the air.

(2) Opera by Dvořák. Libretto by Jaroslav Vrchlicky, based on a Czech translation from Tasso. Première: Czech Theater, Prague, March 25, 1904. This was Dvořák's last opera.

(3) Opera by Haydn. Libretto by Jacopo Durandi. Première: Esterház, Hungary, February 26, 1784.

(4) Opera by Rossini. Libretto by Giovanni Schmidt. Première: San Carlo, Naples, November 11, 1817.

See also RINALDO.

Armide, opera by Gluck. Libretto by Philippe Quinault, based on Tasso's poem *La Gerusalemme liberata.* Première: Paris, Opéra, September 23, 1777. The charming ballet music in Act V, accompanying the dances and tableaux with which Armide entertains Rinaldo at her palace, is still occasionally heard, particularly in the concert suites arranged by Felix Mottl and by François Gevaert.

Arne, Thomas, composer. Born London, March 12, 1710; died there March 5, 1778. Originally directed to law, he studied the violin and spinet in secret, and acquired such proficiency that his father finally removed all objections to a musical career. In 1733 his first opera,

Rosamond (libretto by Joseph Addison), was produced and well received. Other operas and masques followed until 1738, when he achieved a triumph with his music for Milton's *Comus.* In 1740 he wrote the masque *Alfred,* in which the celebrated anthem "Rule Britannia" appears. Five years later he became the official composer for Vauxhall Gardens. During this period he wrote incidental music for many of Shakespeare's plays; some of his settings of Shakespeare's lyrics are among his finest creations. In 1759 he received an honorary doctorate in music from Oxford. His last years were darkened by domestic troubles, poor health, and financial problems. His best operas and masques: *Rosamond* (1733); *Dido and Aeneas* (1734); *Comus* (1738); *The Judgment of Paris* (1740); *Alfred* (1740); *Britannia* (1755); *Artaxerxes* (1762); *L'Olimpiade* (1764); *The Fairy Prince* (1771).

Arnold, Swiss patriot (tenor) in Rossini's *William Tell.*

Arnolfo, a rich landowner (baritone) in Wolf-Ferrari's *L'amore medico.*

Arnould, Sophie, soprano. Born Paris, February 13, 1740; died there October 22, 1802. After studying with Marie Fel and Mlle. Hippolyte Clairon she became a member of the Chapelle Royale. She made her debut at the Paris Opéra on December 15, 1757. For more than twenty years one of the stars of the Opéra, she created the leading soprano role in Gluck's *Iphigénie en Aulide* in 1774. She went into retirement in 1778.

Arrêtez, ô mes frères, Samson's aria to the Hebrews in Act I of Saint-Saëns' *Samson et Dalila.*

Arrigo, a commoner (tenor) in love with Elena in Verdi's *The Sicilian Vespers.*

Artemidor, a crusader (tenor) in Gluck's *Armide.*

Artemis (or Diana), goddess (soprano) in Gluck's *Iphigénie en Aulide.*

Artôt, Désirée (MARGUERITE JOSEPHINE DESIREE MONTAGNEY ARTOT), mezzo-soprano. Born Paris, July 21, 1835; died Berlin, April 3, 1907. The daughter of a horn professor at the Brussels Conservatory, she studied singing with Pauline Viardot-García, after which she toured Belgium, Holland, and England in concert appearances. Meyerbeer having engaged her for the Paris Opéra, she made a notable debut as Fidès in his *Le prophète*, February 5, 1858. Despite her success in Paris, she soon embarked on an extensive tour of Italy. Late in 1859 she scored a major success in Berlin, and soon after was a sensation in London, particularly in several Rossini operas. During the next decade she sang in Germany and England with outstanding success, though now as a soprano rather than a mezzo-soprano. She visited Russia in 1868. Tchaikovsky proposed marriage to her but she chose, instead, the Spanish baritone Mariano Padilla y Ramos. For the next few years she and her husband appeared in opera performances in Germany, Austria, and Russia. She retired from the stage in 1887, thereafter teaching singing in Berlin and later (after 1889) in Paris.

Arvino, Pagano's brother (tenor), husband of Viclinda, in Verdi's *I Lombardi*.

Ase, Aelfrida's servant (mezzo-soprano) in Deems Taylor's *The King's Henchman*.

A Serpina penserete, Serpina's aria in Act II of Pergolesi's *La serva padrona*.

Ashby, Wells-Fargo agent (bass) in Puccini's *The Girl of the Golden West*.

Ashton, Lord Enrico (or Henry), head of the house of Lammermoor (baritone), Lucia's brother, in Donizetti's *Lucia di Lammermoor*.

Asrael, opera by Alberto Franchetti. Libretto by Ferdinand Fontana, based on an old Flemish legend. Première: Municipal Theater, Reggio Emilia, Italy, February 11, 1888. Asrael and

Nefta are angels. Losing Nefta, Asrael searches for her in hell and on earth. After many vicissitudes he is reunited with her in heaven.

Assad, Sulamith's betrothed (tenor) in Goldmark's *The Queen of Sheba*.

Assoluta, Italian for "absolute." A "prima donna assoluta" is the leading female singer of an opera company.

Assur, a prince (baritone) in Rossini's *Semiramide*.

Astaroth, slave (soprano) in Goldmark's *The Queen of Sheba*.

A terra! si nel livido, Desdemona's lament over the loss of her husband's love in Act III of Verdi's *Otello*.

Athanaël, a Cenobite monk (baritone) in Massenet's *Thaïs*.

Atonality, a term often applied to the music of Arnold Schoenberg in which there is a calculated avoidance of anything suggestive of a normal tonal center and key relationships. Schoenberg himself objected to the term as a misnomer, explaining that the music in question gave each of the twelve tones of the octave an equal importance, with no interrelationship except that of one to another. The principles and rules for composing atonal music became elaborate, and some of the resulting works show an extraordinary complexity. Operas written in this system include those of Schoenberg and his best-known disciple, Alban Berg.

A travers le désert, Mârouf's aria in Act II of Rabaud's *Mârouf*.

Atterberg, Kurt, composer. Born Göteborg, Sweden, December 12, 1887. He combined a training in engineering with musical studies at the Stockholm Conservatory and private lessons with Max von Schillings. For many years he was employed at the Royal Patent Bureau, while pursuing the careers of music critic, conductor, and composer. A government subsidy finally enabled him to give up his extramusical occupation. In 1940 he became Secretary of the Royal Academy of Music in Stockholm. His music has drawn melodic and rhythmic ideas from Swedish folk sources. His operas: *Härward der Harfner* (1918); *Bäckahästen* (1924); *Fanal* (1932); *Aladdin* (1941); *The Tempest* (1947).

At the Boar's Head, opera by Gustav Holst. Libretto by the composer, based on Shakespeare's *Henry IV*. Première: Manchester, England, April 3, 1925.

Aubade, a French term (derived from *aube*, "dawn") originally applied to music suitable for performance in the morning, as distinct from evening music —nocturnes and serenades. In operatic usage, an aubade is an aria of light, often sentimental nature.

Auber, Daniel François, composer. Born Caen, France, January 29, 1782; died Paris, May 12, 1871. One of the earliest masters of opéra comique, Auber was trained for business but his passion for music led him to study three years with Luigi Cherubini and then undertake composition. His early operas had a lukewarm reception, but in 1820 *La bergère châtelaine* was a major success in Paris. From then until 1869 Auber wrote over forty operas, many to librettos by Eugène Scribe. His best ones were outstandingly successful, and there were few seasons in which Paris did not see at least one new work by Auber. *La muette de Portici* (1828) was his greatest success, though *Fra Diavolo* (1830) and *Le domino noir* (1837) were close rivals. In 1829 Auber became a member of the French Academy. In 1842 he was appointed director of the Paris Conservatory, a post he held until his death. In 1857 he was made maître de chapelle to Napoleon III. Auber died soon after witnessing the first riots of the Paris Commune in 1871. His best operas after those mentioned above: *Emma* (1821); *Le concert à la cour* (1824); *Fiorella* (1826); *Le philtre* (1831); *Le cheval de bronze* (1835); *Les diamants de la couronne* (1841); *La sirène* (1844);

La Barcarolle (1845); *Zerline* (1851); *Manon Lescaut* (1856); *La fiancée du Roi de Gerbe* (1864); *Le premier jour de bonheur* (1868); *Le rêve d'amour* (1869).

Au bruit de la guerre (Io vidi la luce nel camp guerrier), duet of Sulpizio and Maria in Act I of Donizetti's *The Daughter of the Regiment.*

Au bruit des lourds marteaux, Vulcan's song in Act I of Gounod's *Philemon et Baucis.*

Aucassin et Nicolette, a French romance of the thirteenth century about Aucassin's love for Nicolette, a captured Saracen girl who turns out to be a princess. Eventually the lovers are united. This story was made into operas by Mario Castelnuovo-Tedesco, August Enna, André Grétry, John Knowles Paine, Clifton Parker, Johann von Piszl, and Renzo Rosselini, among others.

Au fond du temple, duet of Nadir and Zurga in Act I of Bizet's *Les pêcheurs de perles.*

A un dottor della mia sorte, Bartolo's aria in Act I, Scene 2, of Rossini's *The Barber of Seville.*

Au secours de notre fille (Ti rincora, amata figlia), chorus of the French soldiers in Act II of Donizetti's *The Daughter of the Regiment.*

Aus einem Totenhaus (From a House of the Dead), opera by Leoš Janáček (completed by Břetislav Bakala). Libretto by the composer, based on Dostoyevsky's *Memoirs from a House of the Dead.* Première: Brünn, April 12, 1930. This was Janáček's last opera, left unfinished at his death. The setting is a Siberian penal labor colony, and the story tells how some of the prisoners came there. The opera is unconventional in that there are no female characters, no arias, and no real plot.

Austral, Florence (born WILSON), dramatic soprano. Born Melbourne, Australia, April 26, 1894. She attended the

Melbourne Conservatory, the London School of Opera, and in 1918 studied privately with Sibella in New York. At this point she was offered a contract with the Metropolitan Opera but refused it, feeling she was not yet ready for such an appearance. Her debut took place at Covent Garden in 1922 when she sang Brünnhilde in *Die Walküre*. She was an immediate success. Subsequently she appeared in the entire Ring cycle at Covent Garden. In 1925 she made her American debut at the Evanston and Cincinnati festivals, thereafter appearing five successive seasons in song recitals and guest opera appearances. After 1930 she was a leading soprano of the Berlin State Opera. Her last American tour took place in 1935–1936. She returned to the Berlin State Opera and appeared there until the outbreak of World War II, at which time she went into retirement in London.

Avant de quitter ces lieux, Valentin's aria in Act II of Gounod's *Faust*.

Ave Maria, Desdemona's prayer in Act IV of Verdi's *Otello*.

Avis de clochettes, *see* OU VA LA JEUNE HINDOUE?

Avito, a former prince of Altura (tenor), lover of Fiora, in Montemezzi's *L'amore dei tre re*.

Azora, opera by Henry Hadley. Libretto by David Stevens. Première: Chicago Opera, December 26, 1917. The setting is Mexico in the fifteenth century. Azora is to marry an Aztec general, but she loves Xalca. When Xalca asks for Azora's hand, Montezuma orders both killed. But a beam of light falling on the victims and on a white cross is interpreted as a divine message, and the lovers are freed and allowed to marry.

Azucena, a gypsy (contralto), Manrico's mother, in Verdi's *Il trovatore*.

B

Baba Mustapha, a Cadi (tenor) in Cornelius' *The Barber of Bagdad*.

Babekan, a Persian prince (baritone) in Weber's *Oberon*.

Babinsky, a robber (baritone) in Weinberger's *Schwanda*.

Baccaloni, Salvatore, basso-buffo. Born Rome, April 14, 1900. He studied architecture, receiving his degree at twenty-one. Giuseppe Kaschmann directed him to music and taught him singing. Baccaloni made his debut in Rome in 1921 in *The Barber of Seville*. In 1926 Arturo Toscanini engaged him for La Scala and advised him to concentrate on buffo roles. While a member of La Scala, Baccaloni also sang at other leading European opera houses. In

1934 he was made a Knight of the Crown of Italy. Meantime his American debut had taken place with the Chicago Opera during the 1930–1931 season. He first appeared at the Metropolitan Opera on December 3, 1940, in *The Barber of Seville;* three weeks later he was acclaimed in *Don Pasquale*, revived for him. Baccaloni's repertory includes nearly a hundred and fifty roles in five languages. He is best known for his Don Pasquale, Dr. Bartolo, Dr. Dulcamara, Falstaff, and Leporello.

Bacchanale, an orgiastic dance of no special musical form, performed by a corps de ballet. Notable operatic bacchanales are those in *Samson et Dalila,*

Tannhäuser, and Szymanowski's *King Roger.*

Bacchus, a god (tenor) in Richard Strauss's *Ariadne auf Naxos.*

Bach, Johann Christian, composer. Born Leipzig, Germany, September 5, 1735; died London, January 1, 1782. This youngest son of Johann Sebastian Bach studied with his brother Carl Philipp Emanuel in Berlin and Padre Martini in Italy. His first opera, *Artaserse,* was well received when introduced in Turin in 1761. With two succeeding operas his fame grew so great that he was invited to write an opera for the King's Theatre in London. This was *Orione,* a gratifying success when presented in 1763. The same year Bach was appointed Music Master to Queen Charlotte, a post he held until his death. In 1772 Bach was called to Mannheim to produce a new opera for the Elector Karl Theodor. The work, brilliantly produced, was *Temistocle.* It was a triumphant success. Bach's later operas fared less well. Fashions were changing —the new operas of Gluck were winning favor—and Bach, though a sensitive orchestrator and a fine melodist, continued to write essentially undramatic works in the older Italian style. In this respect Bach's dramatic works resemble those of another man who was not endowed to be a major figure in opera—Joseph Haydn. Bach's best operas, besides those already mentioned: *Alessandro nell' Indie* (1762); *Lucio Silla* (1776); *La clemenza di Scipione* (1778); *Amadis des Gaules* (1779).

Bacon, Ernst, composer. Born Chicago, Illinois, May 26, 1898. His teachers included Ernest Bloch and Eugene Goossens. He has held teaching and directorial posts with various conservatories and colleges, including Hamilton College and the School of Music at Syracuse University. His first opera, *A Tree on the Plains,* premièred in New York in 1943, received the David Bispham medal. *A Drumlin Legend,* commissioned by the Alice M. Ditson Fund, was introduced in New York in 1949. Bacon received a Pulitzer Traveling Fellowship in 1932, and Guggenheim Fellowships in 1939 and 1942.

Balducci, the Pope's treasurer (bass) in Berlioz' *Benvenuto Cellini.*

Balfe, Michael William, composer and baritone. Born Dublin, Ireland, May 15, 1808; died Rowney Abbey, England, October 20, 1870. He went to London in his sixteenth year, earning his living as a violinist and singer. Sponsored by Count Mazzara, he went to Italy to study with Vicenzo Federici and Filippo Galli. Balfe wrote his first ambitious score, a ballet, in Milan (1826). Through his friendship with Rossini he became principal baritone of the Italian Opera in Paris, appearing in the French première of *The Barber of Seville* in 1829. In the same year his first opera, *I rivali di se stesso,* was performed in Palermo. In 1835 *The Siege of Rochelle* was given successfully in London. On November 27, 1843, the comic opera by which he is chiefly remembered, *The Bohemian Girl,* started a sensational run in London and was soon given on the Continent in several different languages. Between 1846 and 1856 Balfe traveled widely in connection with various productions of his operas. In 1864 he retired to his estate. After his death his statue was placed in the vestibule of Drury Lane, the theater in which he had often sung and in which most of his works had first been seen. His principal operas: *The Siege of Rochelle* (1835); *Maid of Artois* (1836); *Joan of Arc* (1837); *Falstaff* (1838); *The Bohemian Girl* (1843); *The Maid of Honour* (1847); *The Armourer of Nantes* (1863); *Blanche de Nevers* (1863); *The Knight of the Leopard* (*Il talismano*) (1874).

Ballad of Queen Mab, Mercutio's aria in Act I of Gounod's *Roméo et Juliette.*

Ballad of the King of Thule, *see* IL
ETAIT UN ROI DE THULE.

Ballad Opera, an English theatrical
form popular in the eighteenth century, it consisted of a spoken comedy,
often satirical, with songs and incidental music. The song texts were written
for the occasion, the airs they were set
to were chiefly folk and popular songs.
The ballad opera was sometimes
known as the people's opera, chiefly because of its simple form, its everyday
subject matter, and its colloquial
speech. The first ballad opera was Allan
Ramsay's *The Gentle Shepherd* (1725).
The form became famous in 1728 with
the production of *The Beggar's Opera,*
text by John Gay, the airs arranged by
John Christopher Pepusch. The phenomenal success of this work started a
wave of ballad operas. The new craze
was largely responsible for the decline
of interest in serious opera and the
final failure of Handel's Royal Academy of Music. The first opera to be performed in the American colonies was
the ballad opera *Flora (Hob-in-the-
Well),* presented in Charleston, South
Carolina, in 1735. The first ballad
opera performed in New York was
The Mock Doctor (1750).

Ballata del fischio, Mephistopheles'
aria in Act I of Boïto's *Mefistofele.*

Ballatella, a small ballad—a song of no
particular form. A noted *Ballatella* in
opera is Nedda's aria "Che volo d'augelli" in Act I of Leoncavallo's *Pagliacci.*

Ballet in opera. From the beginnings of
opera, ballets have provided diversion
from the dramatic action. In Peri's
Dafne and *Euridice,* the two earliest
operas, the ballet offered a change of
interest from the succession of recitatives. As Italian opera developed, composers often introduced ballet sequences even when they were not essential for the action. Typical examples of
ballet in Italian opera are the finale of
La Vestale, the "Passo a sei" in *William*

Tell, the minuet in Act I of *Rigoletto,*
the oriental dances in Act II of *Aïda.*
While the ballet was an entertaining
element in Italian opera, it was rarely
an integral feature. In France, however, where ballet was born, the case
was otherwise. The earliest opera composers—Lully and Rameau—placed
considerable importance on the ballet.
Even Gluck who tried to free opera of
nonessentials, remained faithful to the
French tradition by introducing dances
in *Orfeo ed Euridice, Alceste, Iphigénie en Aulide,* and *Iphigénie en Tauride.* In 1767 Jean Jacques Rousseau
deplored the irrelevant use of dancing
in French opera, but he could not stop
the development. Meyerbeer, by whom
French grand opera was established,
made extensive use of ballet sequences
in *Les Huguenots, Robert le Diable,*
and *L'Africaine.* For the remainder of
the nineteenth century, ballet continued
prominent in French opera. Notable examples are the Indian ballet in *Lakmé,*
the waltzes in *Faust* and *Roméo et Juliette,* the minuets in *Mignon,* and *Manon,* the bacchanale in *Samson et Dalila.* Even operas which did not originally contain dancing have received
ballet scenes to suit the French taste.
At the Paris Opéra, music from Bizet's
Arlésienne Suite has been used for a
ballet in the last act of *Carmen.* When
Der Freischütz was given in Paris in
1841, Weber's *Invitation to the Dance*
was orchestrated by Berlioz and introduced into the opera. When *Tannhäuser* was first given in Paris, Wagner
had to extend his Venusberg music into
a bacchanale for the opening of the
first act. The German composers generally liked dancing in opera only when
it was essential to the dramatic action.
The "Dance of the Apprentices" in *Die
Meistersinger,* the "Dance of the
Flower Maidens" in *Parsifal,* the
"Dance of the Seven Veils" in *Salome*
evolve naturally from respective plots.
In Russian opera, ballets have been

used prominently, often to contribute pageantry, as in the "Polovtsian Dances" in *Prince Igor* and the polonaise in *Eugene Onegin*. Sometimes the dance is an integral element in the play, as is the polonaise in *Boris Godunov* and the ballet in the third act of *Sadko*. Russian folk dances are basic to such early national Russian operas as *A Life for the Czar* and Dargomizhsky's *Rusalka*.

Ballet of the Seasons, ballet in Act III of Verdi's *The Sicilian Vespers*.

Ballet-Opera, a form of opera created in France in the eighteenth century, with restricted dramatic content and almost continuous dancing. Lully and Rameau wrote numerous ballet-operas.

ballo in maschera, Un (A Masked Ball), opera in five acts by Giuseppe Verdi. Libretto by Antonio Somma, based on Eugène Scribe's libretto for Auber's *Gustavus III*. Première: Teatro Apollo, Rome, February 17, 1859. American première: Academy of Music, New York, February 11, 1889.

Characters: Riccardo, King of Sweden (tenor); Renato, his secretary (baritone); Amelia, Renato's wife (soprano); Ulrica, a fortuneteller (contralto); Oscar, a page (soprano); Silvano, a sailor (baritone); Samuel and Tommaso, conspirators (basses); courtiers; dancers. The action takes place in Sweden in the eighteenth century.

Act I. A hall in Riccardo's palace. The King is hailed by his subjects and given a list of guests invited for a ball. Among the names is that of Amelia. The sight of it causes Riccardo to sing of his love for her ("La rivedrà nell' estasì"). The courtiers leave, Renato enters and warns Riccardo of rebellion brewing ("Alla vita che t'arride") but Riccardo scorns fear. Next comes a judge with a decree of exile for Ulrica, a woman charged with witchcraft. Oscar, the page, pleads for the woman ("Volta la terrea"). Riccardo becomes

curious and decides to visit Ulrica's den in disguise.

Act II. Ulrica's hut. Ulrica is at her cauldron performing an incantation ("Re dell' abisso affretati"). Riccardo appears. He hides as a servant announces Amelia. In love with Riccardo, Amelia has come seeking a remedy for her desire. Ulrica tells her about a magic herb that must be plucked near a gallows at midnight ("Della città all' occaso"). With Amelia gone, Riccardo, now joined by his courtiers, asks to have his fortune told (Barcarolle: "Di' tu se fedele"). Ulrica predicts that he will be murdered by the first man to shake hands with him. The spectators are horrified, but Riccardo laughs ("E scherzo od e follia"). Renato bursts in, and, happy to find Riccardo unharmed, grasps his hand. Riccardo is reassured, doffs his disguise, and throws Ulrica a purse.

Act III. A deserted heath near a gallows. Amelia has come searching for the magic herb that will destroy her great love ("Ma dall' arido"). A clock strikes twelve. She nearly faints as she sees a figure approaching: Riccardo. She entreats him to leave, but their overpowering love throws them into each other's arms ("O qual soave brivido"). They are disturbed by Renato, who has followed Riccardo to warn him of assassins. Amelia veils her face; Renato persuades Riccardo to flee and offers to escort the unknown lady. Samuele, Tommaso, and their accomplices appear and attack Renato. As Amelia rushes to protect him, her veil drops. That a man has a rendezvous with his own wife is a source of infinite merriment to the conspirators. Renato is humiliated and outraged, and he asks Samuele and Tommaso to come to his house.

Act IV. A study in Renato's house. Blinded by jealousy and rage, Renato threatens Amelia with death. Unable to convince him of her innocence, she begs for a chance to bid her son farewell ("Morrò, ma prima in grazia"). Alone before Riccardo's portrait, Renato decides to punish not his wife but instead his disloyal friend and sovereign ("Eri tu che macchiavi"). When the conspirators arrive he joins them. By lot, he is chosen to assassinate Riccardo at the masked ball.

Act V. The palace. Riccardo, at his desk, signs a document sending Renato and his family abroad, and laments about his renunciation ("Ma se m'è forza perderti"). Curtains part, revealing a huge ballroom, and Riccardo joins his guests despite an anonymous warning. He and Amelia bid each other a tender farewell ("T'amo, si, t'amo, e in lagrime"), which Renato interrupts by fatally wounding his erstwhile friend. Dying, Riccardo attests Amelia's innocence and bids his courtiers spare her husband's life.

The locale of *Un ballo in maschera* has repeatedly been shifted. Originally, the setting was Sweden and the opera dealt with the assassination of King Gustavus III at a court ball in 1792. But while Verdi was en route to Naples to supervise the première, an attempt on the life of Napoleon III was made in Paris and the Naples government banished the portrayal of regicide from the stage. Rather than fit his music to a new libretto provided by the censors, Verdi withdrew the opera and was threatened with a fine and arrest. Neopolitans passionately sided with him and demonstrated in front of his hotel. He became a symbol of independence to patriots striving for the unification of Italy under the House of Savoy. Eventually, Verdi left Naples unmolested and produced the opera in Rome after agreeing to shift the locale overseas. So the King of Sweden became a governor of Boston, Massachusetts. Oscar (patterned after Gainsborough's "Blue Boy") and Amelia were the only characters left unchanged. However,

since the atmosphere of a brilliant eighteenth century court hardly fits the austerity of colonial New England, the Swedish setting is restored in most modern productions, all incongruities being taken for granted. In Paris, in 1862, when the tenor Mario refused to don Puritan costumes, the opera's action was moved to Naples.

Balstrode, Captain, a retired merchant-skipper (baritone) in Britten's *Peter Grimes.*

Baltasar (or Balthazar), prior (bass) in Donizetti's *La favorita.*

Balzac, Honoré de, novelist and short-story writer. Born Tours, France, May 20, 1799; died Paris, August 18, 1850. One of the founders of French realism in fiction, Balzac wrote numerous novels and short stories which were used for operas. Among these operas are: Franco Alfano's *Madonna Imperia;* Jean Françaix's *L'apostrophe;* Boris Koutzen's *The Fatal Oath;* Charles Levadé's *La peau de chagrin;* Jean Nouguès' *L'auberge rouge;* Othmar Schoeck's *Massimilia Doni;* and Herrmann Waltershausen's *Oberst Chabert.*

Bampton, Rose, soprano. Born Cleveland, Ohio, November 28, 1909. For five years she was a pupil of Queena Mario at the Curtis Institute. In the summer of 1929 she appeared with the Chautauqua Opera. After three years as member of the Philadelphia Opera Company she made her Metropolitan Opera debut on November 28, 1932, as Laura in *La Gioconda.* She continued in contralto or mezzo-soprano roles until 1936, when she retrained her voice; her debut as a soprano took place at the Metropolitan Opera on May 29, 1937, as Leonora in *Il trovatore.* In 1942 she made the first of several appearances at the Teatro Colón in Buenos Aires. In 1943 she sang the Wagnerian roles of Elisabeth, Kundry, and Elsa for the first time. She remained at the Metropolitan until 1950, when she sang with the New York City Opera. In the years since, she has made opera appearances in South America, television appearances in Canada, and concert appearances in the United States. Miss Bampton is the wife of Wilfred Pelletier, who was for many years a conductor at the Metropolitan.

Barbaja, Domenico, (or BARBAIA), impresario. Born Milan, Italy, 1778; died Posilipo, Italy, October 16, 1841. Before becoming interested in opera he earned his living as a coffee-house waiter and became famous by concocting a delicacy henceforth popular in Vienna and Italy: whipped cream on coffee or hot chocolate. He became wealthy through speculations in army contracts during the Napoleonic wars and investments in gambling rooms at La Scala. He then turned opera impresario and managed two theaters in Naples and Milan. In 1815 he engaged Rossini to write for him two operas a year and assist in their productions; one of these works was *Otello,* produced in 1816. In 1821 Barbaja went to Vienna where for seven years he directed the Kärntnerthor Theater and the Theater an der Wien without relinquishing his directorial activities in Italy. In 1821 he brought Rossini to Vienna and then and later introduced many of Rossini's operas there. He commissioned Weber to write *Euryanthe,* produced under his direction in Vienna in 1823. He also commissioned Bellini to write his first successful operas, and introduced many of Donizetti's operas. In 1828 he went into retirement. Some of the greatest singers of the day were in Barbaja's various companies: Lablanche, Grisi, Rubini, and Sontag.

Barbarina, the gardener's daughter (soprano) in Mozart's *The Marriage of Figaro.*

Barbarino, an assassin (tenor) in Flotow's *Alessandro Stradella.*

Barbe-Bleue, Bluebeard (bass) in Dukas' *Ariane et Barbe-Bleue.*

Barber of Bagdad, The (Der Barbier von Bagdad), comic opera by Peter Cornelius. Libretto by the composer, based on the story "The Tale of the Tailor" from the *Arabian Nights.* Première: Weimar, Germany, December 15, 1858. Nureddin is in love with Margiana, the Caliph's daughter. His barber friend Abul Hassan assists in arranging a rendezvous. Upon the intrusion of the Caliph, Nureddin hides in a chest and almost suffocates until discovered. The Caliph finally consents to Nureddin's marriage. Outstanding excerpts include Nureddin's aria in Act I, "Ach, das Leid hab' ich getragen," and the love duet or "dove duo" in Act II, "O holdes Bild."

Barber of Seville, The (Il barbiere di Siviglia), opera buffa in three acts by Gioacchino Rossini. Libretto by Cesare Sterbini, based on *Le barbier de Séville* and *Le mariage de Figaro,* both by Beaumarchais. Première: Teatro Argentina, Rome, February 20, 1816. American première: Park Theater, New York, May 3, 1819.

Characters: Count Almaviva (tenor); Fiorello, his servant (tenor); Dr. Bartolo, a physician (bass); Rosina, his ward (soprano); Don Basilio, a music teacher (bass); Figaro, a barber (baritone); Berta, a maid (mezzo-soprano). The setting is Seville in the seventeenth century.

The overture, which Rossini had already used for two other operas, consists of two sprightly themes preceded by a slow introduction.

Act I. A square in Seville. Count Almaviva is serenading Rosina ("Ecco ridente in cielo"). Figaro appears, describing his vigorous activities as jack-of-all-trades ("Largo al factotum"). Count Almaviva decides to woo Rosina under the name of Lindoro, since he does not want her to be influenced by his high station. As the humble Lindoro, he sings her a second serenade in which he regrets he can give her only love in place of wealth ("Se il mio nome"). The Count offers to pay Figaro well if he will help him meet Rosina. Figaro explains lightly that nothing in the world is so stimulating as gold ("All' idea di quel metallo") and unfolds a plan of action. First, the Count is to pose as a drunken soldier and get quarters in Bartolo's house.

Act II. Drawing room in Bartolo's house. Alone, Rosina is reading a love letter from "Lindoro" ("Una voce poco fa"). Figaro enters but hides as Dr. Bartolo approaches with his friend Don Basilio. Bartolo confides that he intends to marry his ward. Basilio says that Almaviva is often seen in the vicinity, evidently trying to court Rosina, and that his reputation can readily be demolished by slander ("La Calunnia"). Bartolo prefers his own scheme and urges Basilio to draw up a marriage contract without delay. After Basilio and Bartolo leave, Figaro and Rosina reappear. Figaro bears the happy news of Lindoro's love and promises a meeting with him. As soon as Figaro leaves, Bartolo returns and scolds Rosina for trying to deceive a man of his high station ("A un dottor della mia sorte"). No sooner has he finished his tirade than Almaviva enters in his soldier's disguise. Drunkenly, he demands to be quartered. Bartolo objects and there is an uproar which draws an officer and a squad of soldiers from the street. The officer wants to arrest "Lindoro" but when Almaviva whispers his true name, the officer snaps to attention and salutes —to the amazement of Rosina and Bartolo.

Act III. Again, Bartolo's drawing room. Almaviva returns, this time disguised as a music teacher in order to substitute for the supposedly ailing Don Basilio. He greets Bartolo and his ward unctuously ("Pace e gioia sia con voi"). Bartolo insists on remaining during the singing lesson, yet Almaviva and Rosina manage to exchange hasty words of endearment. When the unsuspecting Basilio arrives, he is bribed by Almaviva and soon leaves. At this point, Figaro insists on shaving Bartolo, making it easy for the lovers to plot their elopement. At last the deception becomes clear to Bartolo. He sends for a notary to draw up a marriage contract and wins Rosina's consent by creating the impression that her supposedly devoted "Lindoro" is planning to turn her over to the notorious Count Almaviva. During Bartolo's absence Almaviva returns, clears up the misunderstanding, and the lovers express their devotion ("Ah, qual colpo"). Figaro appears and urges haste and silence ("Zitti, zitti, piano, piano"). Basilio returns once more, bringing a marriage contract. A little pressure induces him to alter the husband's name from Bartolo to Almaviva, and when Bartolo enters, his ward has become Almaviva's wife. Bartolo accepts his fate philosophically (particularly when he learns that Almaviva does not want Rosina's dowry).

The Barber of Seville is probably the best loved Italian comic opera, and it is opera buffa at its best, turning easily from sentimentality to laughter, from drama to burlesque. The sardonic mockery, sophistication, gallantry, and intrigues of Beaumarchais's plays find their happy equivalent in Rossini's nimble melodies, mercurial rhythms, subtle dynamics, and fleeting patter tunes. Strangely enough, *The Barber* had a disastrous première. The Rome audience resented the fact that young Rossini should use a subject previously used by Paisiello, and still popular. Organized malice was manifested in whistles, laughter, catcalls. A rather shoddy performance combined with several unforeseeable accidents provided sufficient excuse for the demonstration. The derision grew so great that the second act could not be heard.

At the end of the performance Rossini (who had accompanied the recitatives at the piano) was hissed. The second performance (which Rossini did not attend) was another story. The organized hostility was absent, the singers performed better, and some minor changes (including the addition of the lovely serenade "Ecco ridente") found favor. Unfortunately, the opera season ended soon after this performance and the success was easily forgotten. But the opera enjoyed a triumphant tour through Italy five years later, and soon established its enduring reputation. The opera found strong support outside Italy, even among such musical anti-Italians as Berlioz, Beethoven, and the young Wagner.

Barbier, Jules, librettist. Born Paris, March 8, 1825; died there, January 16, 1901. He collaborated with Michel Carré on many French opera librettos, most notably those for Gounod's *Faust, Polyeucte, Philémon et Baucis, Roméo et Juliette.* They also wrote the librettos for Massé's *Galatée* and *Les noces de Jeannette,* Meyerbeer's *Le pardon de Ploërmel,* Napoléon-Henri Reber's *Les papillotes de M. Benoît;* and Thomas's *Francesca da Rimini* and *Hamlet.*

Barbieri, Fedora, mezzo-soprano. Born Trieste, Italy, June 4, 1920. She studied at the Opera School of the Teatro Communale in Florence, making her debut in that city in 1940. After successful appearances at La Scala and other major European opera houses she made her American debut at the Metropolitan Opera on November 4, 1950, in *Don Carlos,* the opening performance of Rudolf Bing's regime as manager of the Metropolitan. The singer had previously acquired a reputation in this country through her phonograph records. After her appearances in this country she returned to Italy, where she has sung chiefly at La Scala, though making guest appearances at other Italian opera houses.

Barcarolle, a boat song, probably originating with the Venetian gondoliers. The most celebrated example in opera is "Belle nuit, ô nuit d'amour" in *The Tales of Hoffmann.* Others are "Di' tu se fedele" in *Un ballo in maschera,* and "Pescator, affonda l'esca" in *La Gioconda.*

Bardi, Count Giovanni, scholar and music patron. Born Florence, 1534; died Rome, 1612. At his palace in Florence assembled the Camerata, the group responsible for the rebirth of ancient lyric drama and the birth of opera (*see* CAMERATA; OPERA). The seventeenth century scholar Giovanni Doni says that Count Bardi made the suggestions that prompted the rest of the group to develop the new musical form. Bardi is thought to have written the texts for some of the first operas by members of the Camerata, and it is also believed that some of these works were first performed in his home.

Bardolph, one of Falstaff's followers (tenor) in Verdi's *Falstaff.*

Baritone, the male voice between tenor and bass. Its normal range is approximately two octaves upward from the A a tenth below middle C.

Barnaba, a spy (baritone) of the Inquisition in Ponchielli's *La Gioconda.*

Barrientos, Maria, coloratura soprano. Born Barcelona, March 10, 1884; died Ciboure, France, August 8, 1946. She studied piano and composition at the Barcelona Conservatory. After only six months of vocal lessons she made her opera debut in Barcelona as Selika. Additional study in Milan was followed by a successful debut at La Scala as Lakmé. For fifteen years she toured Europe and South America. In 1913 she began a three-year retirement, then made her North American debut at the Metropolitan Opera on January 30, 1916 as Lucia. She remained at the Metropolitan Opera until 1920, during which time *I Puritani, Lakmé,* and *La* *sonnambula* were revived for her. She went into retirement in 1939.

Bartered Bride, The (Prodaná Nevěsta), comic opera in three acts by Bedřich Smetana. Libretto by Karel Sabina. Première: National Theater, Prague, May 30, 1866. American première: Metropolitan Opera, New York, February 19, 1909.

Characters: Kruschina, a Bohemian peasant (baritone); Kathinka, his wife (soprano); Marie, their daughter (soprano); Micha, a wealthy landowner (bass); Agnes, his wife (mezzo-soprano); Wenzel, their son (tenor); Hans, Micha's son by a previous marriage (tenor); Kezal, a marriage broker (bass); Springer, manager of a circus troupe (bass); Esmeralda, a dancer (soprano); Muff, a comedian (tenor); circus performers; villagers. The setting is a Bohemian village; the time, the late nineteenth century.

Act I. A square before an inn. It is the annual church festival, and people are singing and dancing to celebrate spring ("See the buds burst on the bush"). Only Hans and Marie are not gay: Marie has been ordered by her father to marry the bumpkin Wenzel, a match arranged by Kezal. Marie tells Hans where her love truly lies ("Gladly do I trust you"). After a tender farewell the lovers part and Marie's parents appear. They are followed by Kezal who soon falls to praising their prospective son-in-law ("A proper young man"). Marie reappears and protests that she loves another. Her father angrily insists that she will marry Micha's son, as arranged. The act closes as villagers throng into the square to dance a spirited polka.

Act II. Inside the inn. Hans tells his companions of the joys of true love. Kezal scornfully upholds the view that money is more important. Villagers enter the inn, dance a rousing furiant (peasant dance), and leave. Wenzel appears—a timid, well-meaning stut-terer ("Ma—Ma—Mamma so dear"). Marie enters and craftily warns Wenzel against the bride Kezal has picked for him: she is a terrible shrew, Marie reveals. After Marie and Wenzel leave the inn, Kezal meets Hans and tries to convince him to give up Marie. Hans talks vaguely of his faraway home ("Far from here do I love"), and Kezal begs him to return there. Hans insists that he will marry Marie. Kezal now describes another attractive girl ("One I know who has money galore"), but Hans is not interested. Hans finally allows himself to be persuaded by a bribe, insisting, however, that Marie's contract specify that she marry only Micha's son. The announcement of this barter causes amazement among the villagers.

Act III. The square before the Inn. Wenzel is bemoaning the fact that love has cost him anguish. His cares vanish with the appearance of a circus troupe which performs the famous "Dance of the Comedians." One of its members is Esmeralda, a tight rope dancer with whom Wenzel instantly falls in love. But his parents soon drag him away to his intended bride. By this time, Marie has learned that Hans has given her up for a bribe ("How strange and dead"). She announces her willingness to marry Wenzel. When Hans tries to explain, Marie turns a deaf ear ("My dearest love, just listen"). Micha finally recognizes Hans as his long-absent son, at which point Hans explains his trick: since the contract specifies that Marie must marry the son of Micha, he—the son of Micha—can be Marie's husband. Hans and Marie are reconciled. Only Kezal is heartbroken at the strange turn of events which has cost him his bride.

The Bartered Bride is the first important Bohemian folk opera; it is the foundation on which Bohemian national music rests. To this day it is one of the finest folk operas ever written anywhere, a colorful and spirited picture of village life, filled with catchy songs and dances. The opera originated as a play with incidental music. Smetana soon sensed its operatic potentialities and revised the score, substituted recitatives for the spoken dialogue, and wrote some new numbers. Transformed into an opera, *The Bartered Bride* was reintroduced in Vienna in 1892 and was a huge success.

Bartolo, an old physician (bass) who appears in Mozart's *The Marriage of Figaro* and Rossini's *The Barber of Seville.*

Basilio, Don, a music master who appears in Mozart's *The Marriage of Figaro* (where he is a tenor) and in Rossini's *The Barber of Seville* (in which he is a bass).

Bass (or basso), the lowest male voice, ordinarily ranging two octaves upward from E an octave and a sixth below middle C. The lowest variety of bass is known as basso profundo (Italian: basso profondo). A range somewhat higher than the normal bass is called bass-baritone.

Bassi, Leonora's guardian (baritone) in Flotow's *Alessandro Stradella.*

Bastien und Bastienne, singspiel by Mozart. Libretto by Andreas Schachtner, based on a French parody of Jean Jacques Rousseau's *Le devin du village.* First performed in the garden theater of Anton Mesmer, the celebrated hypnotist, in Vienna, 1768. Mesmer commissioned the work.

Bat, The, see FLEDERMAUS, DIE.

Batti, batti, o bel Masetto, Zerlina's aria in Act I, Scene 4, of Mozart's *Don Giovanni.*

Battistini, Mattia, baritone. Born Rome, February 27, 1856; died Collebaccaro, near Rome, November 7, 1928. He made his debut at the Teatro Argentina in Rome in 1878 in *La favorita.* His first season at Covent Garden (1883) was not particularly successful. He established his reputation in Italy during the

next four years and was acclaimed now an outstanding exponent of bel canto. He was a noted Don Giovanni. Returning to London in 1887, he was well received. Subsequently, he gathered laurels in Spain and Russia. Because he had a horror of ocean travel, he turned down all offers to appear in the United States. He appeared in song recitals nearly until his death, his voice as agile and beautiful as ever.

Baucis, Philémon's wife (soprano) in Gounod's *Philémon et Baucis.*

Bayreuth, a city in Franconia, Germany, the home of the Wagnerian Festival Theater and festivals of Wagnerian music dramas. It was in Bayreuth that Wagner spent the last years of his life, at Villa Wahnfried. Originally, King Ludwig II of Bavaria, Wagner's patron, expressed interest in, and promised support for, a Wagner theater in Munich. When he became aware of the tremendous scope of Wagner's ideas through the plans drawn by Gottfried Semper, he withdrew his support. It was then that the town of Bayreuth provided Wagner with free land for both his theater and his home. Funds for both structures were raised throughout the world by Wagner societies. Additional sums came from concerts conducted by Wagner. When the cornerstone of the theater was laid on May 22, 1872—an event celebrated with a performance of Beethoven's Ninth Symphony under Wagner—only one third of the required $250,000 was in hand. The construction proceeded as more funds were raised. The theater opened on August 13, 1876, with *Das Rheingold. Die Walküre* was given the following day; *Siegfried* (the world première) on the 16th; and *Götterdämmerung* (also the world première) on the 17th. The most notable Wagnerian singers of the day were in the casts, including Lilli Lehmann, Amalia Materna, Albert Niemann, and Georg Unger. The conductors included Felix Mottl, Hans

Richter, and Anton Seidl. The event attracted world attention. Composers who attended included Saint-Saëns, Grieg, Anton Rubinstein, Gounod, and Tchaikovsky. Newspapers from all parts of the world sent correspondents. A great deal of excitement and controversy was generated. But for all the attention and interest, the first festival was a financial failure, suffering a deficit of about $30,000.

The theater closed until 1882 when it reopened for the world première of *Parsifal*. Hermann Levi conducted and the cast included Herman Winkelmann as Parsifal, Amalia Materna as Kundry, Theodore Reichmann as Amfortas, and Emil Scaria as Gurnemanz. In 1883 and 1884 *Parsifal* was once again the only opera given, while in 1886 *Parsifal* alternated with *Tristan und Isolde*. Between 1888 and 1893 there were five festivals in which *Tannhäuser* and *Lohengrin* were given with *Parsifal, Die Meistersinger,* and *Tristan und Isolde*. The *Ring* returned in 1896, given five times that season. From then on, with intermissions in 1898, 1900, 1903, 1905, 1907, 1911, and 1913, the festivals were repeated until the outbreak of World War I. They were resumed in 1924 with performances of the *Ring, Parsifal, Tristan,* and *Die Meistersinger* and continued (except in 1926, 1929, and 1932) until World War II. During the summer of 1930 Arturo Toscanini, the first Italian conductor to direct at the festival, conducted *Tannhäuser* and *Tristan*. He returned in 1931 for *Parsifal* and *Tannhäuser*. In 1933 he refused to appear in protest against the Nazi regime which had come into power a few months before the festival season. Richard Strauss substituted for him.

From the time of Richard Wagner's death in 1883 up to 1909, the Bayreuth productions were under the artistic direction of his widow Cosima. In 1909 Siegfried Wagner—son of Richard and Cosima—took over the managerial responsibilities. In 1923–1924 Siegfried Wagner toured the United States as conductor to raise funds for the reopening of the festival theater in 1924. When he died in 1930 the direction of the festivals passed to his wife Winifred.

When the festivals resumed after World War II (in 1951) they were under the direction of Wagner's grandsons Wieland and Wolfgang. Considerable controversy was raised over the unorthodox procedures in production and the new designs instituted by Wieland. There was a revolutionary modernization of scenery and stage techniques with emphasis on economy and simplicity. Lighting played a major part in emphasizing mood. The general opinion, however, was that a revitalization of the performances had taken place through these innovations, and they continued in succeeding seasons. Since World War II the principal conductors at Bayreuth have been Josef Keilberth, Herbert von Karajan, Eugen Jochum, and Clemens Krauss. Outstanding recent singers have included these Americans in leading roles: George London (Amfortas); Astrid Varnay (Brünnhilde, Isolde, Ortrud); Regina Resnik (Sieglinde); Eleanor Steber (Elsa).

Beatrice, Ottavio's wife (mezzo-soprano) in Wolf-Ferrari's *Le donne curiose.*

Béatrice et Bénédict, opera by Berlioz. Libretto by the composer, based on Shakespeare's *Much Ado about Nothing.* Première: Baden-Baden, Germany, August 9, 1862.

Beaumarchais, Pierre Augustin Caron de, dramatist. Born Paris, January 24, 1732; died there May 18, 1799. His writings did much to precipitate the French Revolution. His most famous plays are the two comedies *Le barbier de Séville* and *Le mariage de Figaro,* which have received significant operatic treatment. Both plays are centered around the colorful character of the barber Figaro, who serves as a symbol of middle-class revolt against autocracy. *Le barbier de Séville* was banned for two years before it was finally given in 1775. The first performance was a failure, but major revisions resulted in success after the second presentation. *Le mariage de Figaro,* given in 1784, was such a triumph that it ran for eighty-six consecutive performances. Napoleon said of it that it was the "revolution already in action." The most important operas derived from these comedies are Rossini's *The Barber of Seville* and Mozart's *The Marriage of Figaro.* Others include: Friedrich Ludwig Benda's *Der Barbier von Sevilla;* Dittersdorf's *The Marriage of Figaro;* Paër's *Il nuovo Figaro;* and Paisiello's *The Barber of Seville.*

Beckmesser, town clerk (bass) in Wagner's *Die Meistersinger.*

Beecham, Sir Thomas, conductor. Born St. Helens, Lancashire, England, April 29, 1879. Beecham is the son of the prosperous manufacturer of Beecham's Pills. His musical education was haphazard. After leaving Oxford (without a degree) he helped organize an amateur orchestra in Huyton. During a visit of the Hallé Orchestra to that town, Beecham substituted for the regular conductor, Hans Richter, who was elsewhere detained. For a while Beecham wandered aimlessly throughout Europe absorbing musical experiences. In 1902 he became conductor of the Kelson Truman Opera Company that toured the English provinces. Three years later he made his London debut by directing a concert of the Queen's Hall Orchestra. This appearance encouraged him to found the New Symphony Orchestra in 1905. Three years later Beecham created still another orchestra, the Beecham Symphony, which consistently featured English music. In 1910 Beecham organized his own opera company and gave the English première of Delius' *A Village Romeo and Juliet.* A year later he took over the management of Covent Garden. Under his direction it became one of the most dynamic and progressive opera companies in Europe. He helped present more than sixty novelties, including a season of Russian operas with Chaliapin, a season of opéras comique in English, cycles of the Wagnerian music dramas, and English premières of many Richard Strauss operas. The important revivals and premières included works by Eugen d'Albert, Frederick Delius, Joseph Holbrooke, Nikolai Rimsky-Korsakov, Ethel Smyth, Charles Stanford, and Arthur Sullivan.

Due to financial reverses, Beecham gave up the direction of Covent Garden in 1919. Four years later he came out of his temporary retirement to conduct symphonic music, appearing with most of the major English orchestras and also becoming a principal conductor at Covent Garden. In 1932 he founded the London Symphony and became artistic director of Covent Garden. Guest appearances took him throughout the world of music. He made his American debut conducting the New York Philharmonic-Symphony in 1928, thus beginning a long and fruitful association with the United States. He first appeared at the Metropolitan Opera on January 15, 1942, conducting a dual bill consisting of *Le coq d'or* and Johann Sebastian Bach's cantata *Phoebus and Pan,* staged as an opera. He remained at the Metropolitan through the 1943–1944 season. Before returning to England in 1944 he married the concert pianist Betty Humby. Since World War II he has directed notable opera performances at Covent Garden. In 1947 he founded the Royal Philharmonic Orchestra, bringing it to the United States for a tour in 1949. In 1948 and 1949 Beecham conducted operas at Glyndebourne, and on May

4, 1953, in Oxford, England, he led the world première of Frederick Delius' first opera, *Irmelin*.

Beethoven, Ludwig van, composer. Born Bonn, Germany, December 16, 1770; died Vienna, Austria, March 26, 1827. The titan of the symphony, sonata, string quartet, concerto, and the *Missa Solemnis* wrote only a single opera, *Fidelio*. It is not difficult to understand why Beethoven waited until age thirty-five to write an opera in an age when Viennese composers naturally gravitated to the theater as a major source of revenue. And it is also not difficult to comprehend why, having written *Fidelio,* Beethoven did not essay a second opera. While Beethoven had a pronounced dramatic gift in his symphonic music, it was "for the interior psychological drama that is alien to footlights and backdrops," as Robert Haven Schauffler has pointed out. "Beethoven's [thought] was usually too deep for words." Following the same vein, J. W. N. Sullivan remarked that Beethoven's "most important states of consciousness, what he would have called his 'thoughts', were not of the kind that can be expressed in language." It can further be noted that Beethoven, while writing some lovely songs, was none too happy in creating for the voice; his thinking was essentially instrumental. We know what an immense struggle it was for Beethoven to set his libretto to music, how much anguish it cost him to impose on himself the restrictions of stage action and the stylized traditions of the singspiel. Beethoven himself once said that *Fidelio* earned him "the martyr's crown." His inspiration needed the wings of freedom in order to soar. After the harrowing experience of writing one opera, Beethoven was not likely to undertake another, particularly since he knew only too well his own limitations as an opera composer.

Beggar's Opera, The, ballad opera with dialogue and verses by John Gay, music arranged by John Christopher Pepusch. Première: Lincoln's Inn Fields, London, January 29, 1728. This was the first successful ballad opera, responsible for the tremendous vogue of the ballad opera in London in the first half of the eighteenth century. It had the unprecedented run of sixty-three nights, earning a profit of four thousand pounds for the producer and seven hundred for Gay. It became the model for all future works in this form. It is believed that Gay got the idea for his opera from Jonathan Swift. He called it "The Beggar's Opera" because in the prologue a beggar (representing the author) explains why the work was written; he does not appear in the rest of the play. The hero of the work is Captain Macheath, a highwayman who loves Polly Peachum. His love affair and betrayal of Polly, his incarceration, and his reprieve on the day of his execution, provide the author with an excuse to embark on a travesty of Italian opera, to poke malicious fun at English politicians, political corruption, English mores, and the pretensions of high society. Pepusch's contribution consisted of an original overture and the basses to the songs, the airs of which were English and Scotch ballad tunes and other popular music of the day, including a march from Handel's opera *Rinaldo* and a song by Henry Purcell. In 1729 Gay and Pepusch turned out a sequel, *Polly,* which was not allowed to be staged, though it was promptly published. Both these operas have been successfully revived in the twentieth century. Benjamin Britten is one of the numerous composers who have arranged the songs of *The Beggar's Opera* in recent times. In 1953 *The Beggar's Opera* was made as a motion picture with Laurence Olivier acting and singing the part of Macheath. *See also* THREEPENNY OPERA, THE.

Behüt' dich Gott, Werner's aria in Act

II of Nessler's *Der Trompeter von Säkkingen*.

Bei Männern, welche Liebe fühlen, duet of Papageno and Pamina in Act I, Scene 2, of Mozart's *The Magic Flute*.

Bekker, Paul, writer on music, impresario. Born Berlin, September 11, 1882; died New York City, March 7, 1937. He was a music critic of the *Berliner Allgemeine Zeitung* and the *Frankfurter Zeitung* before becoming intendant of the Prussian State Theater in Kassel in 1925. Between 1927 and 1932 he was director of the Wiesbaden State Theater. The rise of Hitler made him leave Germany for good. He settled in the United States and became the music critic of the *New York Staatszeitung und Herold*. He wrote several books of interest to opera lovers, including biographies of Offenbach (1909) and Franz Schreker (1919); *The Changing Opera* (1935); and a biography of Wagner (1931). The last two were published in English in the United States.

Belasco, David, playwright and theatrical manager. Born San Francisco, July 25, 1859; died New York City, May 14, 1931. He was one of the foremost New York theatrical managers in the last two decades of the nineteenth century and the first two of the twentieth. He was also the author and adapter of more than two hundred plays. Two of these—*The Girl of the Golden West* and *Madama Butterfly*—were the sources of operas by Puccini.

Bel canto, Italian for "beautiful song." The term is used to distinguish an Italian manner of singing which emphasizes beauty of tone, purity of texture, facility of voice production, agility in ornamental passages, and the lyrical quality of song. This style is opposed to a more declamatory kind of singing in which the emotional or dramatic element is pronounced.

Belcore, sergeant of the garrison (bass) in Donizetti's *L'elisir d'amore*.

Belinda, Dido's maid (soprano) in Purcell's *Dido and Aeneas*.

Bella figlia dell' amore, the celebrated quartet of the Duke, Maddalena, Gilda, and Rigoletto in Act IV of Verdi's *Rigoletto*.

Bella siccome un angelo, Dr. Malatesta's aria in Act I, Scene 1, of Donizetti's *Don Pasquale*.

Belle, ayez pitié de nous, Laërtes' madrigal in Act II, Scene 1, of Thomas's *Mignon*.

Belle nuit, ô nuit d'amour, the famous Barcarolle (duet of Giulietta and Nicklausse) in Act II of Offenbach's *The Tales of Hoffmann*.

Bellincioni, Gemma, dramatic soprano. Born Como, Italy, August 18, 1864; died Naples, April 23, 1950. She created the roles of Fedora, Santuzza, and Sapho. Her debut took place in Naples in 1881 in Carlo Pedrotti's *Tutti in maschera*. She then toured Europe extensively and in 1899 appeared in South America and the United States. Her repertory included over thirty French and Italian roles. She achieved her greatest successes as Violetta, Manon, and Carmen. In 1890 she appeared in the world première of *Cavalleria Rusticana,* and in 1897 and 1898 she appeared in the world premières of Massenet's *Sapho* and Giordano's *Fedora*. She also had a notable career as teacher of singing, first in Berlin and Vienna and after 1932 at the Naples Conservatory.

Bellini, Vincenzo, composer. Born Catania, Sicily, November 3, 1801; died Puteaux, France, September 23, 1835. The descendant of a long line of musicians, Bellini was given an early musical training by his father. A Sicilian nobleman became impressed with his promise and provided the funds for a comprehensive musical education. Bellini now entered the Naples Conservatory; while there he wrote a cantata, *Ismene,* and his first opera, *Adelson e Salvina.* On the strength of this work, Bellini now wrote a second opera, *Bianca e Fernando,* for the impresario Domenico Barbaja, performed at the San Carlo in Naples in 1826. This brought a second commission from Barbaja, intended for the tenor Rubini. The new opera, *Il pirata,* introduced at La Scala in 1827, was a huge success. Bellini achieved true greatness with *La sonnambula,* introduced in Milan on March 6, 1831. It was soon heard throughout Europe, and was introduced to the English-speaking world in an English version featuring Maria Malibran. An even more distinguished work followed, Bellini's masterpiece *Norma,* introduced at La Scala on December 26, 1831.

In 1833 Bellini visited London to attend performances of several of his operas. Wherever he went he was the object of adulation, particularly in fashionable salons where he assisted Giuditta Pasta (creator of *Norma*) in performances of his best-loved arias. His next destination was Paris. Here, encouraged by Rossini, he wrote his last opera, *I Puritani,* for the Théâtre des Italiens. After its successful première, on January 25, 1835, he withdrew to a secluded villa to work on two new operas. Here he was stricken by intestinal fever which proved fatal. In his delirium he saw before him the great singers of his day—Pasta, Rubini, Tamburini—who had appeared in his operas; just before he died he imagined that a performance of *I Puritani* was taking place in his bedroom.

Bellini was one of the masters of Italian opera. His art differed sharply from that of his celebrated contemporaries, Rossini and Donizetti. Rossini was essentially the genius of the comic; Donizetti, a master of tragedy as well as comedy. Bellini was primarily the apostle of beautiful lyricism. He did not have a pronounced dramatic feeling, and his skill at harmony and instrumentation was limited. But his gift of song was unrivaled. His melodies were perfect in design and structure, aristocratic in style, varied in expression, and endowed with genuine feelings. Lyricism served his every emotional and dramatic need.

His principal operas: *Il pirata* (1827); *La straniera* (1829); *Zaira* (1829); *I Capuletti ed i Montecchi* (1830); *La sonnambula* (1831); *Norma* (1831); *I Puritani* (1835).

Bell Song, *see* OU VA LA JEUNE HINDOUE?

Belmonte, Spanish nobleman (tenor) in Mozart's *The Abduction from the Seraglio.*

Ben, Lucy's lover (baritone) in Menotti's *The Telephone.*

Benda, Georg (or Jiři), composer. Born Jungbunzlau, Bohemia, June 30, 1772; died Köstritz, November 6, 1795. An early and highly successful composer of singspiels, Benda is believed to have originated a form of musical melodrama popular in his day, consisting entirely of spoken dialogue with the music merely an accompaniment.

Benda came from a long line of professional musicians, served as a chamber musician in Berlin and Gotha, and became court kapellmeister in Gotha in 1748. He returned to Gotha after a visit to Italy and wrote his first musical melodrama, *Ariadne auf Naxos,* in 1774. It aroused considerable excitement. Benda now wrote and produced in Gotha a series of melodramas and singspiels that enjoyed considerable vogue. He resigned his post as kapellmeister in 1788 and withdrew from professional life. His best operas and melodramas were: *Ariadne auf Naxos* (1774); *Medea* (1774); *Der Dorfjahrmarkt* (1776); *Romeo und Julie* (1776); *Der Holzhauer* (1778).

Benedict, Sir Julius, composer and conductor. Born Stuttgart, November 27, 1804; died London, June 5, 1885. In his youth he knew such German and Austrian musicians as Hummel, Weber,

Beethoven, and Mendelssohn. On Weber's recommendation, the young man became a conductor at the Kärntnerthortheater in Vienna in 1823. Two years later he went to Italy where he became a conductor at the San Carlo in Naples. There he wrote and produced his first opera, *Giacinta ed Ernesto* (1829). A second opera, *I Portoghesi in Goa,* written for performance in Stuttgart, was not successful there but fared well in Naples. In 1835 Benedict settled in England, his home for the rest of his life. In 1838 he wrote his first English opera, *The Gypsy's Warning.* Soon after this he became a conductor at the Drury Lane Theatre, where several of his operas were performed. He accompanied Jenny Lind on her tour of the United States in 1850, directing many of her concerts. Back in England, he became a conductor at Her Majesty's Theatre. He was knighted in 1871, and four years later was made Knight Commander by the Emperor of Austria. His principal operas: *The Gypsy's Warning* (1838); *The Brides of Venice* (1843); *The Crusaders* (1846); *The Lily of Killarney* (1862); *The Bride of Song* (1864).

Benelli, Sem, dramatist and poet. Born Prato, Italy, about 1875; died Genoa, December 18, 1949. One of the most successful Italian dramatists of his time, his works were the source of two important operas, Montemezzi's *L'amore dei tre re* and Giordano's *La cena delle beffe.*

Bennett, Arnold, novelist. Born Staffordshire, England, May 27, 1867; died England, March 27, 1931. One of the most popular of early twentieth century novelists, Bennett wrote the librettos for two operas by Eugene Goossens: *Don Juan de Mañara* and *Judith.*

Benoît, a landlord (bass) in Puccini's *La Bohème.*

Benoît, Pierre Léonard, composer. Born Harlebeke, Belgium, August 17, 1834; died Antwerp, March 8, 1901.

He created the Belgian school of national music. His three operas—the most successful being *Het dorp in't Gebergte* (The Village in the Mountains)—realized his national aims. In line with his ideals, he founded the Flemish School of Music in 1867 (Antwerp), remaining its director until his death, when he was succeeded by Jan Blockx. His operas: *Het dorp in't Gebergte* (1857); *Isa* (1867); *Pompeja* (1896).

Benson, Mrs., governess (mezzo-soprano) in Delibes's *Lakmé.*

Benvenuto Cellini, opera by Hector Berlioz. Libretto by Léon de Wailly and Auguste Barbier. Première: Paris Opéra, September 10, 1838. Berlioz' first full-length opera is built around Cellini's creation of the statue of Perseus.

Benvolio, Roméo's friend (tenor) in Gounod's *Roméo et Juliette.*

Beppe, (1) a clown (tenor) in Leoncavallo's *Pagliacci.*

(2) a gypsy (soprano) in Mascagni's *L'Amico Fritz.*

Berceuse, a song or instrumental piece in which the melody has the character of a lullaby. The most celebrated berceuse in opera is found in Godard's *Jocelyn:* "Cachés dans cet asile." The aria of Louise's father in *Louise,* "Reste, repose-toi," is also a berceuse.

Berg, Alban, composer. Born: Vienna, February 9, 1885; died Vienna, December 24, 1935. A disciple of Arnold Schoenberg, Berg wrote one of the most provocative operas of our time—*Wozzeck.* He did not begin formal study of music until his nineteenth year. After becoming a government official in 1905 he devoted his free time to musical interests. His meeting with Schoenberg was a turning point in his life: Berg was profoundly affected by the older man's esthetics and revolutionary ideas of musical composition. After some preliminary creative experiments in which he imitated the styles of Wagner and the French Impressionists, Berg wrote, in the atonal style, five songs with orchestral accompaniment. He started work on the atonal *Wozzeck* before World War I. During the war he served in the Austrian army. Afterward, he returned to his opera, completing it in 1920. On December 14, 1925, the Berlin State Opera introduced it. It created a sensation and was soon seen throughout Europe. Leopold Stokowski introduced the work to the United States with performances in Philadelphia and New York in 1931.

Berg wrote only one more opera, *Lulu*, left unfinished at his untimely death. Though a leading exponent of the Schoenbergian school, Berg often arranged the twelve tones of the atonal octave in such a way that they conveyed a certain feeling of tonal structure. This was one of the features that has made his music more readily comprehensible than that of his teacher. Other qualities that Berg's music possesses in marked degree are dramatic power, intensity of emotion, and poignant beauty.

Berger, Erna, soprano. Born Dresden, Germany, October 19, 1900. Much of her childhood was spent in Paraguay. In Dresden she worked as a governess for a French family in order to finance her musical education. She applied for a scholarship at the Dresden Opera School, but when Fritz Busch heard her sing he immediately engaged her for the regular company. She subsequently became the principal soprano of the Berlin State Opera, where she was recognized as one of the leading Mozart interpreters of our time. She has made guest appearances in numerous European opera houses and has performed at the Bayreuth and Salzburg festivals. In 1949 she made her American debut at the Metropolitan Opera, on the opening night of the season, in *Der Rosenkavalier*. She has distinguished herself in the French and Italian repertory as well as in Mozart.

Bergmann, Carl, conductor. Born Ebersbach, Germany, April 11, 1821; died New York City, August 16, 1876. A pioneer in promoting Wagner's music in America, he came to this country in 1850. After playing the cello in the Germania Orchestra in New York, he was appointed conductor of the New York Philharmonic Orchestra in 1855, a post he held for over two decades. Soon after taking over the baton, he led the first performance in America of a Wagnerian excerpt: the *Lohengrin* Prelude. A year later, he played the *Tannhäuser* Overture for the first time in America, and in 1859 he conducted at the New York Stadttheater the first American performance of a complete Wagner opera: *Tannhäuser*. Though he incurred the wrath of critics and audiences for his continued espousal of Wagner's music he continued to perform it as long as he was a conductor.

Berkenfeld, Countess of, Marie's mother (mezzo-soprano) in Donizetti's *The Daughter of the Regiment*.

Berkshire Symphonic Festival, a music festival performed by the Boston Symphony Orchestra each summer at Tanglewood, in Lenox, Massachusetts. Since the organization of the Berkshire Music Center on the festival grounds by Serge Koussevitzky, significant opera performances have been given during the festival season by members of the Opera School, under the direction of Boris Goldovsky. Among the events have been the American premières of Benjamin Britten's operas *Peter Grimes* and *Albert Herring,* Jacques Ibert's *Le Roi d'Yvetot,* and Mozart's *Idomeneo.* Mozart's *La clemenza di Tito* received its first stage performance in America at Tanglewood. Rossini's *Il Turco in Italia,* performed in 1948, received its first American hearing since 1826. Tchaikovsky's *Pique Dame* was revived in 1951,

Grétry's *Richard Coeur-de-Lion* in 1953, and Mozart's *Zaide* in 1955.

In 1953 the Opera School announced that it would encourage the writing of operas by American composers by commissioning two or three each season to write one-act operas for production at Tanglewood; these commissions would be provided through the Koussevitzky Music Foundation.

Berlin Opera (Staatsoper), the most important operatic theater in Berlin. Its history goes back two centuries. Before World War I it was called the Berlin Royal Opera (Hofoper).

It originated as a private theater built by Frederick the Great and inaugurated on December 7, 1742, with a gala performance of Karl Heinrich Graun's *Cleopatra e Cesare*. In 1756 the opera house closed because of the Seven Years' War. For almost two decades after the war it was in a comparatively somnolent state. While operatic activity was renewed in 1775 with the appointment of Johann Friedrich Reinhardt as intendant, or artistic director, it failed to realize artistic importance, largely because of the general apathy of the Emperor to opera. For a long time the major operatic activity in Berlin took place at the National Theater in Gendarmenplatz, where Mozart's greatest operas were introduced to the city.

After the turn of the nineteenth century the Royal Opera assumed increasing importance. Count Karl von Brühl became intendant, and he aimed to develop a significant theater emphasizing the highest artistic values. Since the Emperor Friedrich Wilhelm preferred more meretricious and spectacular productions, Count von Brühl's efforts were greatly hampered. He had to indulge the Emperor's passion for the more ornate Italian operas of the period. In line with this policy, Gasparo Spontini became musical director in 1819. Under his supervision many of his operas were given lavish presentations. Occasionally Von Brühl had his way. Despite Spontini's aversion to, and envy of, Weber, the world première of *Der Freischütz* took place at the Royal Opera, and was a success of formidable proportions. (This was one of the few important operas to be introduced by this opera house.) But the Italian vogue continued until the Emperor's death in 1840, when Spontini resigned.

In 1842, Giacomo Meyerbeer became musical director, Otto Nicolai one of the principal conductors. A fire demolished the opera house in 1843. A new building was erected on the same site, opening with Meyerbeer's *Das Feldlager in Schlesien*, written especially for the new singer Jenny Lind. Meyerbeer not only directed his own operas but produced and conducted *Euryanthe* and *Rienzi* and was responsible for the production of *Der fliegende Holländer*. In 1849 Nicolai presented the world première of his comic opera *The Merry Wives of Windsor*.

After Meyerbeer's resignation and the appointment of Count Botho von Hülsen as intendant in 1850, an increasingly high level of performances was realized. The level was maintained by Von Hülsen's successor, Count Bolke von Hochberg. With these two intendants the so-called "classic age of the Royal Opera" unfolded. If the house was comparatively delinquent in presenting new operas of importance—it was particularly negligent in the case of Wagner—it nevertheless touched a new standard of operatic presentation. Under the later artistic directions of Felix Weingartner (1891–1898), Karl Muck (1908–1912), and Richard Strauss ((1918–1919) the Wagnerian dramas came into their own at the Royal Opera.

Meanwhile, in 1896, Emperor Wilhelm II ordered the erection of a new opera house, the old one having outlived its usefulness. The Prussian State Theater bought the Kroll Theater, intending to tear it down and build on the site. But plans were long delayed, and were completely disrupted by the outbreak of World War I. After the war the old opera house was remodeled, reopening in 1924 with a gala performance of *Die Meistersinger* conducted by Erich Kleiber, who served as musical director until 1931 (holding the post again, briefly, a few years later). The artistic directors (intendants) during this period were Max von Schillings (1919–1925) and Heinrich Tietjens (1927–1945). The years of Kleiber's leadership were marked by interest in new operas, notable premières being those of Alban Berg's *Wozzeck* and Darius Milhaud's *Cristophe Colomb.* Wilhelm Furtwaengler, Kleiber's successor, lost his post as a result of his differences with the Nazis (see *Mathis der Maler*), but was later reinstated, remaining musical director until shortly after World War II.

During the war the opera house was demolished. Musical activities were afterward transferred to the Admiralspalast, in Berlin's Russian zone. Ernest Legal became artistic director in 1945, with Joseph Keilberth and Karl Fischer the principal conductors. In 1953 Carl Ebert became artistic director with a long-term contract. The rebuilt opera house was scheduled to oe reopened in 1955.

Berlioz, Hector, composer. Born La Côte-Saint-André, France, December 11, 1803; died Paris, March 8, 1869. The first of the great Romantics in music, Berlioz achieved significance in symphonic and choral music. His operas, for all their many fine pages, are lesser creations; attempts to revive them have failed to arouse sustained interest outside France. By temperament and natural gifts, Berlioz was seemingly an ideal composer for the theater. In his concert music he displayed a pronounced dramatic gift; his music, like his personality, was given to theatricalism. Yet in writing for the stage, he lost much of the ardor, boldness, power, and occasional sublimity we find in his best orchestral and choral works. There are moments in the opera *Benvenuto Cellini* (1837) that are theatrically effective. There are pages in *Béatrice et Bénédict* (1862) that have an ingratiating light touch. There are passages in *Les Troyens* (1856–1858) that possess classic beauty. But the defects in all these operas outweigh the merits. The characterizations are often effete, the dramatic action is too often static, the climaxes frequently lack the necessary impact. It is perhaps significant that Berlioz' best "opera" was not written for the stage but for the concert hall: the cantata *La damnation de Faust.* Because of its pronounced theatrical qualities this cantata was given a stage presentation in Monte Carlo in 1893, revealing enough vitality and dramatic interest to justify repeated performances at the Paris Opéra up to the present time.

Berlioz attended the Paris Conservatory where, after seven years, he won the Prix de Rome. His first major work was the *Symphonie fantastique,* written in 1830. After a tempestuous courtship, he married the actress Henrietta Smithson in 1833, but the marriage was unhappy and ended after a few years in separation. Berlioz composed prolifically for thirty years, but the failure of the second part of his *Les Troyens* in 1863 put an end to his creative days. His last years were marked by poor health, disappointments, and an unhappy second marriage.

Bernauerin, Die, opera by Carl Orff. Libretto (in old Bavarian dialect) by the composer, based on an old legend. Première: Stuttgart, June 15, 1947. A young nobleman marries Agnes, daughter of a public bathhouse owner;

during her husband's absence she is accused of being a witch and is drowned.

Bernstein, Leonard, composer and conductor. Born Lawrence, Massachusetts, August 25, 1918. While he is essentially a symphonic conductor, having appeared with outstanding success with most of the major orchestras of the world, he has also led opera performances, including the American premières of Britten's *Peter Grimes* and Poulenc's *Les mamelles de Tirésias*. In 1953 he became the first American-born conductor to direct a performance at La Scala, that of Cherubini's *Medea*. He wrote a one-act opera, *Trouble in Tahiti,* introduced at the Festival of Creative Arts at Brandeis University, Waltham, Massachusetts, on June 12, 1952, and subsequently performed over the NBC television network.

Bersi, Madeleine's maid (mezzo-soprano) in Giordano's *Andrea Chénier.*

Berta, a maid (mezzo-soprano) in Rossini's *The Barber of Seville.*

Bertha, John of Leyden's fiancée (soprano) in Meyerbeer's *Le prophète.*

Bertram, the devil in human form (bass) in Meyerbeer's *Robert le Diable.*

Bervoix, Flora, Violetta's friend (mezzo-soprano) in Verdi's *La traviata.*

Bess, Porgy's sweetheart (soprano) in Gershwin's *Porgy and Bess.*

Bess, You Is My Woman Now, love duet of Porgy and Bess in Act II of Gershwin's *Porgy and Bess.*

Betz, Franz, dramatic baritone. Born Mainz, Germany, March 19, 1835; died Berlin, August 11, 1900. He created the role of Hans Sachs (1868) and, at the first Bayreuth Festival (1876), the role of Wotan in *Siegfried.* He made his debut in Berlin in 1859 in *Ernani,* making such a good impression that he was engaged as a permanent member of the Berlin Royal Opera company. He remained with it until his retirement in 1897. He distinguished himself particularly in Wagnerian roles,

but he was also acclaimed as Falstaff, Don Giovanni, and William Tell.

Biaiso, a public letter writer (tenor) in Wolf-Ferrari's *The Jewels of the Madonna.*

Bianca al par hermine, *see* PLUS BLANCHE QUE LA BLANCHE HERMINE.

Bildnis Aria, *see* DIES BILDNIS IST BEZAUBERND SCHÖN.

Billy Budd, opera by Benjamin Britten. Libretto by E. M. Forster and Eric Crozier, based on the story of the same name by Herman Melville. Première: Covent Garden, December 1, 1951. The central theme (like that of another Britten opera, *Peter Grimes*) is man's inhumanity to man. Billy Budd is forcibly mustered into the British Navy (the period is the eighteenth century). He is hated by the master-at-arms, John Claggert, who builds up a false charge of treason against him. Overwhelmed and enraged by this unjust accusation, Billy Budd kills the tyrant, for which act he is court-martialed and hung. Captain Vere realizes that, while naval justice has been done, Budd had sufficient justification to commit murder. There are several unusual points about this opera. It is written exclusively for male voices (consequently there is no love interest). There are few arias and ensemble numbers; with the exception of a few chanteys and one or two lyrical pages, the opera consists entirely of recitatives, with the principal musical interest concentrated in the orchestra. Britten's emphasis is on the drama, and he recruits every musical means at his command to point up the personal tragedy of his protagonist.

An opera on the same subject was written in 1948 by the Italian composer Giorgio Federico Ghedini.

Bing, Rudolf, opera manager. Born Vienna, Austria, January 9, 1902. Since 1950 he has been the general manager of the Metropolitan Opera. He studied music and art at the University of Vienna. In 1923 he became a manager of concert artists in Vienna. Four years later he went to Germany as head of a unit supplying artists to more than eighty state and municipal opera houses. In 1929 he was appointed musical secretary of the Darmstadt Municipal Theater, and in 1931 he held a similar post with the Charlottenburg Municipal Opera. He left Germany when Hitler came to power. In 1934 Bing helped organize the first season of the Glyndebourne Opera Company (England), subsequently becoming general manager; he held this post until 1939. After becoming a British subject in 1946, he helped create the Edinburgh Festival in 1947, becoming its general manager. His success in organizing these annual festivals was responsible for bringing him an appointment as general manager of the Metropolitan Opera when Edward Johnson resigned in 1949. He assumed this new post on June 1, 1950. For his achievements at the Metropolitan Opera *see* METROPOLITAN OPERA HOUSE.

Birds, The, *see* ARISTOPHANES.

Bis, French for "twice," a call equivalent to "encore" and used by audiences desiring repetition of a number.

Bispham, David, baritone. Born Philadelphia, Pennsylvania, January 5, 1857; died New York City, October 2, 1921. He was the first American-born opera baritone to win international acclaim. Without any preliminary musical education he appeared during boyhood in amateur opera performances. When he was twenty-eight he went to Europe and studied singing with Luigi Vannuccini in Florence and Francesco Lamperti in Milan. In 1891 he was selected from fifty applicants for the role of the Duc de Longueville in Messager's *La Basoche,* in London. This was his first professional stage appearance. His first Wagnerian role was that of Kurwenal, in which he appeared under Gustav Mahler's direction at Drury Lane in 1892. His success brought him a con-

tract for Covent Garden, where he appeared for the next few years, primarily in Wagnerian roles. On November 18, 1896, he made his American debut as Beckmesser at the Metropolitan Opera. He appeared there until 1903, singing not only in the Wagner dramas but also in the American premières of Ignace Jan Paderewski's *Manru* and Ethel Smyth's *Der Wald.* After this period Bispham's appearances in opera were few, but he enjoyed great success as a recitalist. He made a point of singing English versions of songs by Beethoven, Schubert, and Schumann. In 1916 he appeared in an English-language performance of Mozart's *The Impresario* in New York. This was such a success that it led to the formation of the Society of American Singers, which, with Bispham's inspiration, gave three seasons of light operas in English.

Bispham Memorial Medal Award, an award created by the Opera Society of America, in Chicago, soon after the death of David Bispham, for opera in English by American composers. Recipients of the award have been: George Antheil, Ernst Bacon, Charles Wakefield Cadman, Walter Damrosch, George Gershwin, Louis Gruenberg, Henry Hadley, Howard Hanson, Victor Herbert, Otto Luening, and Deems Taylor.

Biterolf, a minstrel-knight (bass) in Wagner's *Tannhäuser.*

Bittner, Julius, composer. Born Vienna, April 9, 1874; died there January 19, 1939. For many years he divided his activities between the law (which he practiced successfully up to 1920) and composition. His first opera, *Die rote Gret,* was introduced in Frankfurt in 1907 and was well received. His following operas were successfully performed in Austria and Germany. Bittner arrived at a popular style without sacrificing an original approach; his finest works are graced by an engaging sense of humor. His principal operas and musical plays: *Die rote Gret* (1907); *Der Musikant* (1910); *Der Bergsee* (1911, revised 1922); *Der Abenteurer* (1913); *Der liebe Augustin* (1917); *Die Kohlhaymerin* (1921); *Das Rosengärtlein* (1923, revised 1928); *Mondnacht* (1928); *Der unsterbliche Franz* (1930); *Das Veilchen* (1934).

Bizet, Georges, composer. Born Paris, October 25, 1838; died Bougival, June 3, 1875. The son of a singing teacher, he entered the Paris Conservatory when he was nine years old; his teachers included Antoine Marmontel, François Benoist, and Jacques Halévy. In 1857 he won the Prix de Rome, and in the same year he completed a one-act *opéra comique*, *Le Docteur Miracle.* It won first prize in a contest sponsored by Jacques Offenbach and was introduced at the Bouffes-Parisiens. Returning to Paris in 1860 after his years in Rome, Bizet embarked on a career as opera composer. His *Les pêcheurs de perles,* given at the Théâtre Lyrique on September 30, 1863, was only moderately successful. *La jolie fille de Perth,* in 1867, was a failure; so was a one-act opera given by the Opéra-Comique in 1872, *Djamileh.* Meanwhile, in 1869, Bizet married Geneviève Halévy, the daughter of his teacher. He lived the humble existence of an unrecognized composer until 1872, when he was acclaimed for his incidental music to Alphonse Daudet's *L'Arlésienne,* to this day his most popular orchestral work. His last opera, *Carmen,* was completed in 1875 and was introduced at the Opéra-Comique on March 3 of the same year. It received thirty-seven performances that season, an indication that it was no failure, though its great popularity began later. Bizet died exactly three months after the première of his greatest work. He brought to French opera a fine feeling for colorful background and exotic atmospheres. His sensuous melodies, vivid harmonies and orchestration, and captivating rhythms were ideally suited to such subjects. He had a keen dramatic sense. His use of recurrent musical themes prompted the criticism that he was being too Wagnerian. But Bizet, for all the Wagnerian influences and oriental subjects, remains a typically French composer in the refinement and sensitivity of his style and the purity of his lyricism.

Bjoerling, Jussi, tenor. Born Stockholm, Sweden, February 2, 1911. He attended the Stockholm Conservatory. As a boy he joined his father and two brothers in the Bjoerling Male Quartet which toured Scandinavia in native dress, and then appeared in the United States in 1920–1921. In his eighteenth year he entered the Royal Opera School, where his teachers included John Forsell and Tullio Voghera. Immediately after graduation (1929) he made his debut as Don Ottavio at the Stockholm Royal Opera. Between 1931 and 1934 he appeared in major European opera houses, making his Covent Garden debut in the spring of 1936, and in 1937 appearing at the Salzburg Festival in performances of *Don Giovanni* which were highly acclaimed. He made his American debut over a radio network in 1937. His opera debut took place a month later with the Chicago Opera. On November 24, 1938, he made his bow at the Metropolitan Opera as Rodolfo. He has since appeared in over fifty leading roles in Italian and French operas at the Metropolitan Opera and in other leading operatic institutions.

Blacher, Boris, composer. Born China, January 6, 1903. He received his early academic and musical schooling in the Far East. When he was nineteen he went to Berlin (from then on his permanent home) where he attended the Technische Hochschule as a student of architecture, and the Hochschule für Musik. He received his first musical assignment in 1926: to collaborate with another composer in writing a two-hour score for a silent film. He now began to earn his living through music: as copyist, orchestrator, and performer in motion-picture theaters. A ballet, *Festival in the South*, introduced in Kassel in 1937, was his first outstanding success. It was given in about fifty German theaters. He wrote several more ballets before undertaking his first opera, *The Princess Tarakanova*, first given in Wuppertal in 1941. This was followed by a chamber opera, *Romeo and Juliet*, which, after being introduced in Berlin in 1946, was successfully performed in New York and at the Salzburg Festival. After this came a chamber opera written primarily for radio, *The Flood* (1946); a ballet-opera, *Prussian Fairy Tales* (1952); and a highly provocative opera made up of abstractions, *Abstrakte Oper No. 1* (1953). In 1953 Blacher became director of the Berlin Hochschule für Musik.

Blanche Dourge, Lakmé's aria in Act I of Delibes's *Lakmé*.

Blaze, François-Henri-Joseph (better known as CASTIL-BLAZE), writer on music. Born Cavaillon, France, December 1, 1784; died Paris, December 11, 1857. He is frequently described as the father of French music criticism. After completing music study at the Paris Conservatory in 1820, he wrote his first book, the one by which he is most often remembered: *De l'opéra en France*, a definitive study of the operatic activity of his day. For a decade he was the music critic of the *Journal des Débats*. His books include a history of the Paris Opéra (1855), one on Italian opera (1856), and another on French opera (1856). He translated many librettos into French, including those of *Don Giovanni, Fidelio, Der Freischütz* and *The Marriage of Figaro*. His son, Baron Henri Blaze de Bury (1813–1888), wrote biographies of Rossini and Meyerbeer.

Blech, Leo, conductor and composer.

Born Aix-la-Chapelle, Germany, April 21, 1871. He considered making business a career, but in 1890 began music study at the Berlin Hochschule für Musik. From 1893 to 1898 he was conductor at the Aix-la-Chapelle Stadttheater, and from 1899 to 1906 at the Deutsches Landestheater in Prague. In 1906 he was appointed principal conductor at the Berlin Royal Opera. During the next fifteen years he distinguished himself as one of the leading opera conductors in Germany; at the request of Richard Strauss, he led the première of *Elektra* in 1909. In 1923 he visited the United States at the head of a German company that performed the Wagnerian repertory, making his American bow on February 12 with *Die Meistersinger*. He now became the artistic director of the Berlin Volksoper, and two years later he returned to the Berlin State Opera as principal conductor. He left Germany in 1937 because of differences with the Nazi regime. He conducted opera in Riga in 1939 and, after World War II, in Stockholm. In 1949 he returned to Berlin and became principal conductor of the Municipal Theater. He wrote several operas early in his career, the most famous being *Versiegelt* (*Sealed*), introduced in Hamburg in 1908 and produced by the Metropolitan Opera in 1912. His other operas: *Das war ich* (1902); *Alpenkönig und Menschenfeind* (1903, revised 1917); *Aschenbrödl* (1905).

Blick' ich umher, Wolfram's hymn to pure love in Act II of Wagner's *Tannhäuser*.

Blitzstein, Marc, composer. Born Philadelphia, Pennsylvania, March 2, 1905. His formal music study took place at the Curtis Institute, with Alexander Siloti in New York, Nadie Boulanger in Paris, and Arnold Schoenberg in Berlin. His earliest works revealed an interest in advanced techniques and unorthodox approaches, but his increasing social consciousness made him abandon this style for a more popular kind of music. He first became famous with an opera in this new vein, *The Cradle Will Rock,* introduced in New York in 1937. After receiving a Guggenheim Fellowship in 1940 he completed a second social opera, *No for An Answer.* During World War II he served in the air force where his duties were mostly of a musical nature. After the war a third opera, *Regina,* was introduced on Broadway, in 1949. He subsequently made an adaptation of Kurt Weill's *The Threepenny Opera,* introduced in Waltham, Massachusetts, in 1952, and later performed in New York.

Blockx, Jan, composer. Born Antwerp, Belgium, January 25, 1851; died there May 26, 1912. He carried on the traditions of the Belgian national school, based on Flemish elements, first established by Pierre Leonard Benoît. He studied with Benoît at the Antwerp School of Music and later with Reinecke at the Leipzig Conservatory. In 1886 he was appointed professor of harmony at the Antwerp School and director of the Cercle Artistique. When Benoît died, Blockx succeeded him as director of the Antwerp School of Music. His operas were distinguished by their vivid recreation of Flemish life. They are: *Iets Vergeten* (1877); *Maître Martin* (1892); *De Herbergprinses* (1896); *Thyl Uylenspiegel* (1900); *De Bruid der Zee* (1901); *De Capel* (1903); *Baldie* (1904).

Blonde (or Blonda), Constanza's maid soprano) in Mozart's *The Abduction rom the Seraglio.*

Bluebeard, *see* BARBE-BLEUE.

Bluebeard's Castle, one-act opera by Béla Bartók. Libretto by Béla Balázs. Première: Budapest, May 24, 1918. The story of Bluebeard and his last wife, Judith (Bluebeard and Judith are the only singing characters in the opera), is the basis of a psychological text emphasizing the eternal conflict between man and woman.

Bluebird, The (L'oiseau bleu), fairy opera by Albert Wolff. Libretto is Maurice Maeterlinck's poetic fantasy of the same name. Première: Metropolitan Opera, December 27, 1919. The familiar Maeterlinck tale concerns the search of the children Tyltyl and Mytyl for the bluebird, symbol of happiness. The search carries them to many strange places: the Land of Memory, the Palace of Night, the Palace of Happiness. When they find the bird at last, it is in their humble home.

Bob, a tramp (baritone) in Menotti's *The Old Maid and the Thief.*

Boccaccio, Giovanni, author. Born Florence (?), Italy, 1313; died Certaldo, Italy, December 21, 1375. His classic *The Decameron* was the source of several operas including: Marcel Delannoy's *Ginevra;* Claude Delvincourt's *Boccacerie;* and Rodolphe Kreutzer's *Imogène.* Boccaccio is the central character in a popular comic opera by Franz von Suppé, *Boccaccio.*

Boccanegra, *see* SIMON BOCCANEGRA.

Bodanzky, Artur, conductor. Born Vienna, December 16, 1877; died New York City, November 23, 1939. From 1915 to the time of his death he was the principal conductor of German operas at the Metropolitan Opera. After graduating from the Vienna Conservatory he became a violinist in the Imperial Opera orchestra. In 1902 he became Gustav Mahler's assistant at the Vienna Opera, and in 1904 he was chosen to lead a performance of *Die Fledermaus* in Vienna that was outstandingly successful. After conducting at the Prague Opera for two years, he became musical director of the Mannheim Opera in 1909. In 1914 he led a performance of *Parsifal* at Covent Garden that so impressed Gatti-Casazza that the latter engaged him for the Metropolitan to succeed Alfred Hertz.

Bodanzky's American debut took place on November 18, 1915, with *Die Götterdämmerung.* Except for a brief hiatus in 1928, Bodanzky remained at the Metropolitan for the remainder of his life, distinguishing himself in the fidelity and painstaking thoroughness of his performances. Bodanzky also conducted symphonic and choral music in New York. He prepared new editions of *Fidelio, Oberon,* and Von Suppé's *Boccaccio,* all of them given at the Metropolitan.

Böhm, Karl, conductor. Born Graz, Austria, August 28, 1894. He studied to be a lawyer, but his passion for music led him to attend the Graz Conservatory and to study privately with Eusebius Mandyczewski. He was engaged by the Graz Opera as prompter in 1917, and three years later he became first conductor. On the recommendation of Karl Muck, he was engaged by the Munich State Opera as conductor in 1920, where he remained several years. In 1927 he conducted in Darmstadt, in 1931 became musical director at the Hamburg Opera, and in 1933 musical director of the Dresden State Opera. In Dresden he led the world première of Richard Strauss's *Daphne* and Heinrich Sutermeister's *The Magic Isle,* both of which are dedicated to him. After World War II, Böhm conducted extensively in Europe and South America and was a frequent participant in the Salzburg Festivals. In 1954 he received a five-year contract as musical director of the Vienna State Opera.

Bohème, La (1) opera by Ruggiero Leoncavallo. Libretto by the composer, based on Henri Murger's novel *Scènes de la vie de Bohème.* Première, Teatro la Fenice, May 6, 1897. Leoncavallo's opera has been thrown completely into the shade by the more popular and more significant opera on the same subject by Puccini (see below). Both operas were written in the same period,

and each composer was aware that the other was setting the Murger novel. A spirited contest developed as to which opera would be performed first. Puccini won. Curiously enough, Leoncavallo's opera was far better received than Puccini's at their premières. The growing popularity of Puccini's opera, however, spelled doom for Leoncavallo's. Two of Leoncavallo's arias have survived: "Io non ho che una povera stanzetta" and "Testa adorata."

(2) Opera in four acts by Giacomo Puccini. Libretto by Giuseppe Giacosa and Luigi Illica, based on Murger's novel *Scènes de la vie de Bohème.* Première: Teatro Regio, Turin, February 1, 1896. American première: Los Angeles, October 14, 1897.

Characters: Rodolfo, a poet (tenor); Marcello, a painter (baritone); Colline, a philosopher (bass); Schaunard, a musician (baritone); Mimi, a seamstress (soprano); Benoît, a landlord (bass); Parpignol, vendor of toys (tenor); Alcindoro, a state councilor (bass); Musetta, a girl from the Latin Quarter (soprano); customhouse sergeant; students; girls; shopkeepers; soldiers; waiters; vendors. The setting is Paris, in the middle of the nineteenth century.

Act I. An attic. In this chilly home of four Bohemians, Marcello is about to make a fire by burning a chair; the poet Rodolfo prefers using one of his unpublished manuscripts. Suddenly, Schaunard appears, his arms overflowing with food, drink, and fuel bought with money just acquired from a patron. The friends celebrate; then they decide to continue their merrymaking in a Latin Quarter café; only Rodolfo stays behind. Mimi, a neighbor, comes seeking a light for her candle. Before she leaves she is seized by a coughing fit and begins to faint. Rodolfo revives her and she is able to leave. But a moment later she returns for her key, lost in Rodolfo's room. As Rodolfo and

Mimi go on their knees to hunt for it, Rodolfo touches Mimi's cold hand. He takes it in his and begins to tell her about himself ("Che gelida manina"). Mimi now reveals to him her hunger for beauty of flowers and the warmth of springtime ("Mi chiamano Mimi"). From below, in the street, come the voices of Rodolfo's friends urging him to join them. Rodolfo opens the window and moonlight streams into the room. He turns to Mimi and rhapsodizes over her beauty. The two voices join in an ecstatic outpouring of love ("O soave fanciulla"), after which Rodolfo and Mimi go off to join his friends.

Act II. The Café Momus in the Latin Quarter. On Christmas Eve, Rodolfo and Mimi stop off at a milliner's shop to buy her a hat. Then they join their friends at the café. Musetta, one-time sweetheart of Marcello, appears with the wealthy councilor, Alcindoro. Coquettishly she reveals how men are attracted to her (Musetta's Waltz: "Quando me'n vo' soletta"). It is obvious that Marcello, as he listens to her, is seized by his old feeling of love, and that Musetta is still responsive to him. She sends Alcindoro to a cobbler's shop, feigning that her shoe is too tight. Then she rushes to Marcello. Mixing in a passing parade, the Bohemians escape. The returning Alcindoro finds not only that he has been jilted but left with a large café bill.

Act III. At one of the city's gates. From the adjoining tavern comes the sound of gay voices. Mimi appears, coughing and shivering. She inquires from a policeman where she can find Marcello, and is informed that he is now employed as a sign painter in the inn. When she finds Marcello, she confides how difficult life has become with Rodolfo, since he is insanely jealous of her. The appearance of Rodolfo sends Mimi hiding behind a tree, where she overhears his complaints about her.

When Rodolfo announces his intention of giving her up for good, she emerges from hiding. Seeing her revives the poet's ardor. He tenderly takes her in his arms. But Mimi insists that they must separate for his own good. The lovers bid one another farewell ("Addio, dolce svegliare"), but even now they realize they cannot separate. In a renewed wave of tenderness they depart together.

Act IV. Again the attic. Once more Rodolfo and Marcello have quarreled with their sweethearts. Nostalgically, they recall how happy they used to be with Mimi and Musetta. Marcello takes Musetta's ribbons from his pocket and kisses them. Rodolfo lives over bygone days with Mimi ("Ah, Mimi, tu più"). Their reveries are punctuated with food and drink brought by Colline and Schaunard. The Bohemians' spirits lift, and a quadrille and a mock duel ensue. The revelry is at its height when Musetta bursts in to say that Mimi, who is outside, is deathly sick. She is brought into the attic and tenderly placed on Rodolfo's bed. Once again the lovers are reconciled. When Rodolfo's friends go to buy medicine for Mimi, the lovers are alone. They repeat their true feelings. When the friends return, Mimi closes her eyes wearily. Rodolfo goes to the window to cover it and obscure the light. But Schaunard notices that Mimi is not asleep—but dead. One glance at his friends, another at Mimi, and Rodolfo knows the tragic truth. He cries out Mimi's name, rushes to her bed, and sobs over her body.

La Bohème is Puccini's most down-to-earth opera. The central interest is in the everyday problems, the little joys and sorrows, of several Parisian artists. The opera has no big scenes, the action never gets involved, there are no breathtaking climaxes. Puccini's concern is not so much his story as his characters, and it is the characters who dominate music as well as libretto. Frequently, the main arias serve to throw a light on the characters who sing them; throughout the opera these recurring melodies are subtly changed to produce new insights into the characters' personalities. The naturalism of the story—combined with the restraint and tenderness of Puccini's music—makes for a poignant human drama. It is undoubtedly for this reason that *La Bohème* has through the years remained Puccini's best-loved work. For the same reason—since its effect on an audience is subtle rather than overpowering—*La Bohème* was not at first successful. At its première in Turin the audience was apathetic, the critics outrightly hostile. When *La Bohème* was given in Rome, soon afterward, it was still received coldly. To outward appearances it seemed that Puccini had produced a failure. But its third presentation the same year, in Palermo, was a triumph. There was such an ovation that the entire death scene had to be repeated. From this performance on, the opera passed from one triumph to another and to presentations on all the opera stages of the world.

Bohemian Girl, The, comic opera by Michael William Balfe. Libretto by Alfred Bunn, based on the ballet *The Gypsy* by Vernoy Saint-Georges. Première: Drury Lane Theatre, London, November 27, 1843. The setting is eighteenth century Hungary, where Arline, daughter of Count Arnheim, has been kidnaped and raised by gypsies. As a beautiful young woman, she is falsely accused of stealing a medallion and is imprisoned by the Count's men. When she appears before the Count to plead for clemency, he recognizes her by her scar. The opera is noteworthy for such songs as "I dreamt I dwelt in marble halls," "The heart bowed down," and "Then you'll remember," and for its gypsy songs and dances.

Bohnen, Michael, bass-baritone. Born Keulen, Germany, January 23, 1888. He made his debut at Düsseldorf in 1910 in the part of Caspar in Weber's *Der Freischütz,* a part in which he later attained universal success. After appearances in Europe, he made his Metropolitan Opera debut in the American première of Schilling's *Mona Lisa* on March 1, 1923. Bohnen remained at the Metropolitan until 1932, excelling in baritone as well as bass roles. He went into temporary retirement during World War II in Germany, but emerged for a guest appearance at the Berlin State Opera during the 1950–1951 season, appearing in the roles of Scarpia and Baron Ochs.

Boieldieu, François Adrien, composer. Born Rouen, France, December 16, 1775; died Jarcy, France, October 8, 1834. He was one of that triumvirate of early opéra comique composers that includes Auber and Adam. After studying with a Rouen organist, he wrote his first opera, *La fille coupable,* in his eighteenth year. Two years later he went to Paris. A meeting with Cherubini was a decisive event in his life. Cherubini accepted him as a pupil and in 1798 appointed him professor of the piano at the Conservatory. In 1798 Boieldieu achieved his first success as a theatrical composer with *Zoraïme et Zulnare.* Two years later his *Le Calife de Bagdad* was so well received that it ran for seven hundred performances. In 1803 Boieldieu visited Russia, where he was showered with honors. He was appointed kapellmeister by the Czar and given a contract to write three operas a year. He returned to Paris in 1811, having written numerous works he felt unworthy of presenting in Paris. In 1817 he succeeded Méhul as professor of composition at the Conservatory, and a year later he became a member of the Institut de France. His masterwork, and one of the finest works in the opéra comique repertory, was *La* *dame blanche,* introduced in 1825. It was a sensational success and earned its composer a government pension. After this, Boieldieu produced a failure or two and wrote no more. The fall of the monarchy in 1830 deprived him of his pension; soon after, poor health compelled him to give up his Conservatory post. The new government finally came to his help with an annual grant. Boieldieu's principal operas: *La fille coupable* (1793); *Zoraïme et Zulnare* (1798); *Le Calife de Bagdad* (1800); *Ma tante Aurore* (1803); *Calypso* (*Télémaque*) (1806); *Jean de Paris* (1812); *Le petit chaperon rouge* (1818); *La dame blanche* (1825).

Boisfleury, Marquis de, brother (bass) of the Marchionesse de Serval in Donizetti's *Linda di Chamounix.*

Boïto, Arrigo, librettist and composer. Born Padua, Italy, February 24, 1842; died Milan, June 10, 1918. He achieved importance as the composer of *Mefistofele* and as the librettist of Ponchielli's *La Gioconda* and Verdi's *Otello* and *Falstaff.*

Entering the Milan Conservatory in his fourteenth year, Boïto remained there six years. In 1861 he collaborated with Franco Faccio in writing a cantata which won for both composers a two-year traveling scholarship. Back in Italy after a fruitful stay in Paris, Boïto led a movement to reform Italian music by encouraging composers to write symphonic music; to clarify and propagandize his ideas, Boïto wrote many brilliant theoretical essays.

While trying to get others to write symphonic music, he himself preferred working in opera. He began his masterwork, *Mefistofele,* in 1866. The outbreak of war between Austria and Germany delayed its composition, Boïto joining the Garibaldian volunteers and seeing active service. He completed his opera early in 1868 and on March 5 it was introduced at La Scala, the composer conducting. The conflicting opinions on the opera, and the violent demonstrations during its three performances, made it seem that Boïto had written a failure. Seven years later, after revisions by the composer, it was produced in Bologna, and this time it found favor. Ever since, it has remained a favorite opera in Italy. In 1916 Boïto finished his second opera, *Nerone,* after many years of meticulous work. It was not performed during his lifetime. After its presentation at La Scala in 1924 it was thought to be a less fine work than *Mefistofele.*

Boïto's first important effort at writing librettos was in 1865 when he provided his friend Faccio with the book for *Amleto;* eleven years later he did a similar service for Ponchielli with *La Gioconda.* His major achievements in this direction came in 1886 and 1893 with his librettos for Verdi's last two operas, *Otello* and *Falstaff.* Boïto's texts for these works are considered two of the finest librettos in all Italian opera.

Bolero, originally a vigorous Spanish dance in triple time, later a dancelike song for solo voice. Boleros were introduced in a number of nineteenth century operas. Examples occur in Méhul's *Les deux aveugles,* Auber's *La muette de Portici;* and Verdi's *The Sicilian Vespers* (Elena's aria "Mercè, dilette amiche").

Bonci, Alessandro, tenor. Born Cesena, Italy, February 10, 1870; died Vitterba, August 8, 1940. He was a master of bel canto and one of the most celebrated tenors of his generation. After five years of study at the Rossini Conservatory in Pesaro with Carlo Pedrotti and Felice Coen he made his debut at the Teatro Regio in Parma in 1896 in *Falstaff.* He was an immediate success; before the end of his first season he was engaged by La Scala, where he made his debut in *I Puritani.* Appearances throughout Europe followed. On December 3, 1906, he made his American debut with the Hammerstein company at the Manhattan Opera House in *I Puritani*. He stayed with this company two seasons, a competitive attraction to Enrico Caruso, then scoring immense successes at the Metropolitan Opera. In 1908 Bonci joined the Metropolitan Opera, and in 1914 he became a member of the Chicago Opera. Meanwhile, in 1910–1911, he made an extensive transcontinental tour in song recitals. During World War I he served in the Italian army. After the war he toured America for three seasons, appearing at the Metropolitan and with the Chicago Opera during the 1920–1921 season. In 1922–1923 he was the principal tenor of the Teatro Costanzi in Rome. A year later he taught master classes in singing in the United States. After 1925 he went into partial retirement and devoted himself primarily to teaching in Milan. Though his voice lacked volume, it had exceptional beauty of texture and lyrical sweetness. He was a master of phrasing and expression.

Boniface, monastery cook (baritone) in Massenet's *Le jongleur de Notre Dame*.

Bononcini (or Buononcini), Giovanni Battista, composer. Born Modena, Italy, July 18, 1670; died Vienna, July 9, 1747. He is principally remembered for his association and rivalry with Handel in London. It was this rivalry that contributed to our language the phrase "tweedledum and tweedledee." It is first found in a satirical poem by John Byrom:

Some say, compar'd to Bononcini,
That Mynheer Handel's but a ninnny;
Others aver that he to Handel
Is scarcely fit to hold a candle.
Strange all this difference should be
'Twixt Tweedledum and Tweedledee.

Bononcini came from a long line of professional musicians. He studied first with his father and afterward in Bologna, where he published a volume of masses. In 1690 he was appointed maestro di cappella at the Church of San Giovanni in Monte. Four years later he was in Rome, where his first two operas were produced. After some years of success in Berlin and Vienna, he went to London in 1716 to be joint director (with Handel) of the newly organized Royal Academy of Music, with which he was associated for over half a dozen years. He also wrote many operas for the English stage. London was for a long time divided between two factions: those favoring Bononcini (this group was headed by the powerful Duke of Marlborough, who paid Bononcini an annual stipend of five hundred pounds); and those on Handel's side. In 1731 Bononcini became discredited when a madrigal, submitted by him to the Academy of Ancient Music, was said to be by another composer. A year later he left England in disgrace. After living for a time in France he spent several years as court composer in Venice, and died in Vienna. He wrote about seventy-five operas, the most successful being: *Endimione* (1706); *Turno Aricino* (1707); *Mario fuggitivo* (1708); *Abdolonimo* (1709); *Astarto* (1714); *Ciro* (1722); *Crispo* (1722); *Griselda* (1722); *Erminia* (1723); *Calpurnia* (1724); *Astianatte* (1727); *Alessandro in Sidone* (1737).

Bonze, The (or Bonzo), Cio-Cio-San's uncle (bass) in Puccini's *Madama Butterfly*.

Bori, Lucrezia (born BORJA), soprano. Born Valencia, Spain, December 24, 1888. Until her eighteenth year she was educated in a convent. After deciding to become a singer, she went to Milan for coaching. She made her debut at the Teatro Costanzi on October 31, 1908, as Micaëla. After auditioning for Gatti-Casazza and Toscanini she was engaged by La Scala, where she appeared for the first time in 1910 in *Il matrimonio segreto*. Her first American appearance took place on the opening night of the Metropolitan Opera's 1912–1913 season. The opera was *Manon Lescaut;* she was acclaimed. In 1915 a growth in her throat necessitated a delicate operation. It seemed that her career might be over, but she refused to lose faith, and kept on working assiduously. She returned to the opera stage in Monte Carlo in 1918 and to the Metropolitan Opera on January 29, 1921. She remained at the Metropolitan for the next fourteen years, starred in French and Italian operas. Among the important premières and revivals in which she appeared were: *L'amore dei tre re, L'amore medico, L'oracolo, Pelléas et Mélisande, Peter Ibbetson,* and *La rondine.* In 1925 she became one of the first important opera artists to sing on the radio; she was heard in a nationwide hookup with John McCormack. In 1933, when the Metropolitan Opera faced an economic crisis, she became chairman of a committee "to save the Metropolitan." She helped raise a considerable amount of money, and in 1934 once again served as chairman to raise funds.

Her final opera performance at the Metropolitan took place on March 21, 1936, in *La rondine.* A week later the Metropolitan Opera gave a gala concert in her honor, in which she sang excerpts from her favorite operas, and was given a twenty-minute ovation. After her retirement, Bori sang occasionally over the radio, and became a member of the board of directors of the Metropolitan Opera Association.

Boris Godunov, opera in three acts (also given in four acts with a different distribution of scenes) by Modest Mussorgsky. Libretto by the composer, based on the drama of the same name by Alexander Pushkin. Première: Maryinsky Theater, St. Petersburg, February 8, 1874. American première: Metropolitan Opera, New York, March 19, 1913.

Characters: Boris Godunov (bass); Xenia, his daughter (soprano); Feodor, his son (mezzo-soprano); Marina, a Polish landowner's daughter (mezzo-soprano); Prince Shuisky, Boris' advisor (tenor); Gregory, a novice, later Dmitri the Pretender (tenor); Varlaam, a monk (bass); Missail, a monk (tenor); Pimen, a monk (bass); Stchelkalov, secretary of the Duma (baritone); Jesuits; monks; boyars; an innkeeper's wife; a police official; a nurse; an idiot. The settings are Russia and Poland; the years 1598 to 1605.

Act I, Scene 1. Moscow—a square before the Novodievich Monastery. A crowd is kneeling in prayer that Boris Godunov accept the Russian crown ("Why hast thou abandoned us?") Stchelkalov comes out of the monastery to inform the people that Boris has not yet accepted. He urges them to continue their prayers. From the distance is heard the chant of pilgrims approaching to join the people in prayer.

Scene 2. A cell in the Monastery of the Miracles. Pimen is chronicling the recent events in Russia (Pimen's Narrative: "Still one page more"). He informs the novice Gregory of the murder of Czarevich Dmitri by Boris' men. When Pimen further informs Gregory that the heir to the Russian throne was his age, Gregory is fired with the ambition to appear as Dmitri and avenge his murder.

Scene 3. The square between the Cathedral of the Assumption and the Cathedral of the Archangels. A rejoicing crowd fills the square—Boris Godunov is to be crowned ("Coronation Scene"). The pealing of cathedral bells heralds the approach of a procession of boyars. Boris appears with his two children and promises the people that he will work for their good and for that of Russia. Calling for the help of God, he entreats the people to join him in prayer.

Act II, Scene 1. An inn on the Lithuanian border. Gregory and the monks Varlaam and Missail have come here in their flight from the monastery. Gregory is disguised as a peasant, since he is wanted by the police for spreading the false rumor that Dmitri is alive. After a few drinks, Varlaam sings an earthy song ("In the town of Kazaan"), then falls into a drunken sleep. Soldiers appear searching for Gregory. Skillfully, Gregory directs their suspicions to the drunken Varlaam; while they are arresting him, Gregory escapes.

Scene 2. The apartment of Czar Boris in the Kremlin. Boris' children, Xenia and Feodor, are with their nurse. She sings them an amusing little ditty ("Song of the gnat"). Boris appears, praises his son for the way he has been learning his lessons, and reminds him that some day he will rule Russia. He soliloquizes on the torment of his rule, surrounded as he is by conspirators, and blamed by the people for all the evils in the land ("I have attained the highest power"). Boris' anguish is intensified when Prince Shuisky arrives with tidings of the false Dmitri, and how the people are rallying under him. Boris is now obsessed with the belief that the dead can arise from their graves, and that the false Dmitri is really the true one. After Shuisky departs, panic seizes Boris. He sees the ghost of the murdered Dmitri. Falling on his knees, he prays to God for mercy ("Ah, I am suffocating").

Act III, Scene 1. The garden of a Polish palace. Gregory has come for a rendezvous with Marina, whom he expects to make his queen. Marina and her guests emerge into the garden where a brilliant polonaise is danced. After the guests go back into the palace, Marina meets Gregory, and an ardent love scene follows ("Oh! Czarevitch").

Scene 2. The forest of Kromy. Peasants are dragging a captured boyar. They mock him and mock his Czar. A simpleton appears singing a ditty. From a distance the voices of Varlaam and Missail are heard denouncing Boris. Two Jesuits sing the praises of the new Czar, Dmitri. The people attack them, for they do not want the help of the clergy. The emotional climate gets stormy. With a blare of trumpets, Gregory appears, accompanied by his soldiers. The people acclaim him as Dmitri and follow him as he starts for Moscow. Only the simpleton remains to sing mournfully of the coming doom of Russia.

Scene 3. The Kremlin. Stchelkalov reads a message from Boris to the Duma (state council) informing it that a traitor is leading a revolt against the Czar. Shuisky expresses concern over the mental state of Boris, and the boyars are shocked. They are further horrified when Boris appears. He is out of his mind and seems to be fleeing from someone. He soon takes hold of himself, however, ascends his throne, and gives an audience to Pimen. The old monk relates a strange story about a blind shepherd sent by the voice of Dmitri to his tomb in Uglich Cathedral where, by prayer, the shepherd recovered his vision. The story overwhelms Boris. He cries out for help. Then he summons Feodor. Taking his son in his arms, the Czar bids him farewell ("Farewell my son, I am dying"). He counsels the boy against traitors and blesses him as the new Czar. Bells begin tolling and the sounds of people in prayer are heard. Boyars and monks fill the room. Boris designates his son as his successor, prays to God for mercy, collapses, and dies.

If any single work can be said to realize the artistic goals of the Russian national school, it is *Boris Godunov*. The "Five" (Balakirev, Borodin, Cui, Rimsky-Korsakov, and Mussorgsky) aspired to produce a great musical art by deriving inspiration and subject matter from Russian culture and history; at the same time they aimed for a musical art derived from Russian folk songs and dances. *Boris Godunov* fulfilled these specifications completely. It is a mighty drama of the Russian people taken from Russian history. It is a drama about the inner torment and anguish of a Czar; it also is a drama about the shifting forces of the Russian people. Mussorgsky produced a score in which the Russian soul speaks out with force and conviction. For his lyricism, Mussorgsky went to Russian folk songs and liturgical music, adapting their individual harmonic and rhythmic traits for his own purposes. He also devised a melody that followed the inflections of the Russian language. To his harmonic and rhythmic language, and to his orchestral colors, he brought a strength well suited to the personal drama of Boris and the even greater drama of the Russian people.

There exist several different versions of this opera. The first is Mussorgsky's original concept. When *Boris Godunov* received its première, however, changes were made to please the opera-house directors; a few pleasing arias and some love interest were interpolated, and the order of some of the scenes was shifted. When the opera was revived in 1904, still a third version appeared, prepared by Rimsky-Korsakov, who refined away much of Mussorgsky's characteristic harmonic and orchestral styles. It was in this version that *Boris Godunov* (with its greatest interpreter, Feodor Chaliapin, in the title role) achieved its first major success. In 1908 Rimsky-Korsakov decided to revise his version to conform a bit more to Mussorgsky's original intentions. It is this fourth version that is now most often performed in this country and abroad. From time to time there have been attempts to revive the opera as Mussorgsky originally wrote it, since it is generally felt that much of the primitive force and cogency of the opera are lost in Rimsky-Korsakov's cultured adaptation. On February 26, 1928, Mussorgsky's first version was introduced at the Bolshoi Theater in Moscow. A year and a half later Leopold Stokowski and the Philadelphia Orchestra gave the American première of the original *Boris* in a concert version. Later adapters—Dmitri Shostakovich and Karol Rathaus to name two—have tried to make a suitable compromise between Mussorgsky's original and the Rimsky-Korsakov edition.

Borodin, Alexander, composer. Born St. Petersburg, Russia, November 11, 1833; died St. Petersburg, February 27, 1887. A member of the Russian nationalist school, Borodin wrote an opera, *Prince Igor,* that realized the ideals and principles of Russian nationalism. All his life Borodin divided his energies and interests between medicine and music. He was educated at the Academy of Medicine and Surgery in St. Petersburg, becoming there assistant professor of pathology and therapeutics in 1856, receiving his degree in medicine in 1858. After 1859, and up to the end of his life, he did significant research in chemistry, and for many years was professor at the Academy. Music, his passion since childhood, he followed seriously for the first time in 1862, when he met and became a pupil of Balakirev. Balakirev inflamed Borodin with his own national ideals, so much so that Borodin eagerly joined him—and Mussorgsky, Cui, and Rimsky-Korsakov—in spreading the cult of Russian national music. He completed his first symphony, some songs, and a farcical opera *The Bogatyrs* in 1867. Later works—notably the Second Symphony, the Second String Quartet, the tone poem *In the Steppes of Central Asia,* and the monumental opera that absorbed him on and off for twenty years, *Prince Igor*—made him one of the most significant of the nationalists.

Borodin's health was seriously affected in 1884 by an illness believed to be cholera. Afterward he was frequently in poor health, and a victim of mental depressions. He died from a ruptured aneurism while enjoying a party with musical friends. He did not live to finish *Prince Igor;* it was completed by Rimsky-Korsakov and Alexander Glazunov. Besides the earlier opera, *The Bogatyrs,* Borodin also wrote a portion of *Mlada,* a composite opera (never finished) whose other composers were Cui, Moussorgsky, and Rimsky-Korsakov.

Borov, a doctor (baritone) in Giordano's *Fedora.*

Borromeo, Carlo, a Cardinal (baritone) in Pfitzner's *Palestrina.*

Borsa, a courtier (tenor) in Verdi's *Rigoletto.*

Bostana, Margiana's attendant (mezzo-soprano) in Cornelius' *The Barber of Bagdad.*

Boston Opera House, a theater built in 1909 as the home for the then newly organized Boston Opera Company, directed by Henry Russell. It opened with a splendid performance of *La Gioconda,* with Lillian Nordica and Louise Homer in the cast. The opera company flourished for some years. In 1912 Felix Weingartner was engaged to direct some of the Wagner dramas, and it was here that Weingartner made his American debut, conducting *Tristan und Isolde.* The company presented the American première of Raoul Laparra's *La Habanera* in 1910, and the Boston première of *Pelléas et Mélisande.* In 1914, the year the company expired from lack of support, it visited Paris for a two-month season at the Théâtre des Champs Elysées. Subsequently, an effort was made to create another resident company at the Boston Opera House. This time it was under the direction of Max Rabinoff. The venture was short-lived. Since then the theater has been the home for visiting opera companies.

Boughton, Rutland, composer. Born Aylesbury, England, January 23, 1878. Before World War I, Boughton nursed the ambition to create an English equivalent of Bayreuth where Wagner-like music dramas glorifying English traditions would be performed. To realize this mission, Boughton settled in Glastonbury, where, in a small hall and with semiprofessional casts, he performed his own musico-dramatic works based on the Arthurian legends. One of these works is his finest opera, *The Immortal Hour.* The Glastonbury performances were interrupted by the outbreak of World War I, but resumed afterward. In August, 1920, there took place performances of Boughton's *The Birth of Arthur* and *The Round Table,* the first two parts of a projected cycle of Arthurian legends. Boughton now sought to build a special theater with up-to-date equipment. But the needed financial support was not forthcoming and the entire project collapsed. Boughton's operas: *The Immortal Hour* (1913); *The Round Table* (1916); *Agincourt* (1918); *Alkestis* (1922); *The Queen of Cornwall* (1924); *The Ever Young* (1928); *The Lily Maid* (1934).

Bouillon, Princesse de, Adriana's rival (mezzo-soprano) for Maurice in Cilèa's *Adriana Lecouvreur.*

Boulevard Solitude, opera by Hans Werner Henze. Libretto by Grete Weil, based on Abbé Prévost's story *L'histoire du chevalier des Grieux et de Manon Lescaut.* Première: Hanover, February 17, 1952. This is a new adaptation of the celebrated Manon Lescaut story previously used by Massenet and Puccini. In this version, Prévost's original treatment is adhered to by making Des Grieux, and not Manon, the principal character, and tracing his disintegration as a result of his relations with her. The time of the opera has been moved up to 1950. *See also* MANON; MANON LESCAUT.

bourgeois gentilhomme, Le, *see* MO-
LIERE.

Bradford, Puritan clergyman (baritone), "Wrestling Bradford," in Hanson's *Merry Mount.*

Brander, Faust's friend (bass) in Berlioz' *The Damnation of Faust.*

Brangäne, Isolde's attendant (mezzo-
soprano) in Wagner's *Tristan und
Isolde.*

Brangäne's Warning, *see* HABET ACHT.

Braslau, Sophie, contralto. Born New
York City, August 16, 1892; died there
December 22, 1935. Her music study
took place at the Institute of Musical
Art, where she specialized in the piano.
After her voice was discovered, she
studied singing with Herbert Witherspoon, Marcella Sembrich, and Mario
Marafioti. Her opera debut took place
at the Metropolitan Opera on November 27, 1914, as the offstage Voice in
Parsifal. She did not assume major roles
for several years; the first time she did
so was on March 23, 1918, when she
created the title role in Cadman's
Shanewis. For the next two years she
continued to appear in important contralto roles. She created for America
the parts of Amelfa in *Le coq d'or* and
Hua-Quee in *L'oracolo.* Her last opera
appearance took place in 1920, at the
Metropolitan, after which she devoted
herself to concert appearances.

Braut von Messina, Die, *see* SCHILLER,
FRIEDRICH.

Bravour Aria, German term for an aria
with bravura passages.

Bravura, an Italian term (literally:
"bravery") applied to a song or passage requiring brilliance and technical
adroitness on the part of the singer.

Brecht, Bertolt, poet and dramatist.
Born Augsburg, Germany, February
10, 1898. Brecht provided Kurt Weill
with the book for their modern adaptation of *The Beggar's Opera: The Threepenny Opera.* Other Weill operas with
librettos by Brecht are *Der Jasager* and
The Rise and Fall of Mahagonny.

Brecht's one-act play *The Trial of Lu-
cullus* provided the texts for operas by
Paul Dessau and Roger Sessions.
The Informer was made into an opera
of the same name by Daniel Sables.
Brecht was one of the founders of an
esthetic cult popular in Germany in
the 1920's—that of *Gebrauchsmusik*
(which see).

Breisach, Paul, conductor. Born Vienna,
June 3, 1896; died New York City,
December 26, 1952. He attended the
Vienna State Academy, after which he
served as Richard Strauss's assistant at
the Vienna State Opera. In 1921 he
became conductor of the Mannheim
National Theater, and in 1924 of the
Deutsches Opernhaus in Berlin. He
made his American debut at the Metropolitan Opera on December 12, 1941,
conducting *Aïda.* He stayed at the
Metropolitan through the 1945–1946
season, when he became principal conductor of the San Francisco Opera.

Breitkopf und Härtel, one of the world's
great music-publishing institutions. It
was founded in Leipzig as a general
printing establishment by Bernhardt
Christoph Breitkopf in 1719. In the
middle of the eighteenth century,
Johann Gottfried Breitkopf devised the
use of movable musical type. The first
significant musical achievement of the
firm was the publication of an opera
score in 1756. After Gottfried Cristoph
Härtel took over the firm from Breitkopf's grandson, it became the leading
music-publishing organization in Germany. It published many works of Mozart and Haydn and, subsequently,
monumental complete editions of most
of the great German composers.

Bréval, Lucienne (born BERTHE SCHIL-
LING), soprano. Born Berlin, November 4, 1869; died Paris, August 15,
1935. For almost thirty years she was
the leading soprano of the Paris Opéra,
specializing in Wagnerian roles. Her
studies took place at the Conservatories of Geneva and Paris. Her debut

took place at the Paris Opéra, on January 20, 1892, when she was acclaimed as Selika. During the three decades she remained with that company she created the principal soprano roles in many French operas, including *Ariane et Barbe-Bleue, Le Cid, Grisélidis, Pénélope,* and *Salammbô.* In 1899 she made guest appearances at Covent Garden. She made her American debut at the Metropolitan Opera on January 16, 1901, as Chimène in *Le Cid.* She returned to the Metropolitan for the season of 1901–1902.

Bridal Chorus, see TREULICH GEFÜHRT.

Bridal Procession, procession in Act III of Rimsky-Korsakov's *Le coq d'or.*

Bride of Abydos, The, see BYRON, GEORGE NOEL GORDON, LORD.

Bride of Lammermoor, The, see SCOTT, SIR WALTER.

Brindisi, a drinking or toasting song. Operatic examples include: "Il segreto per essere felice" in *Lucrezia Borgia;* "Viva il vino" in *Cavalleria rusticana;* "O vin dissipe la tristesse" in *Hamlet;* "Inaffia l'ugola" in *Otello;* and "Libiamo, libiamo," in *La traviata.*

British National Opera Company, an important English opera company founded in 1922 under the artistic direction of Percy Pitt, and including many of the leading singers and instrumentalists who had previously been associated with Sir Thomas Beecham. Its first performance (*Aïda*) took place in Bradford in 1923. Later the same year the company appeared at Covent Garden. Subsequently, it toured England extensively, sometimes with ambitious performances. The company was responsible for the premières of several English operas, including Boughton's *Alkestis,* Holst's *The Perfect Fool,* and Vaughan Williams' *Hugh the Drover.* Frederick Austin succeeded Pitt as artistic director in 1924, and remained in the post until the company was dissolved five years later.

Britten, Benjamin, composer. Born Lowestoft, Suffolk, England, November 22, 1913. He demonstrated extraordinary creative talent in childhood, writing his first string quartet when he was nine. At sixteen he had written half a dozen quartets, ten piano sonatas, and a symphony. His music teachers included Frank Bridge, John Ireland, and Arthur Benjamin; between 1930 and 1933 he attended the Royal College of Music. His first major success came in 1938 with the orchestral *Variations on a Theme of Frank Bridge,* introduced at a festival of the International Society for Contemporary Music. An avowed pacifist, Britten came to the United States in 1939 and remained in this country during the early years of World War II. He completed several important orchestral works in this country, and with them his first opera, *Paul Bunyan.* Given at Columbia University in 1941, the work was severely criticized. While in America, Britten received a commission from the Koussevitzky Foundation to write a second opera. He returned to England in 1942. Exempt from military duty because of his convictions, he helped in the war effort by giving concerts in hospitals and shelters. His commissioned opera, *Peter Grimes,* was so successful when introduced in London on June 7, 1945, that it established Britten as one of the major opera composers of our time. He now went on to write *The Rape of Lucretia, Albert Herring, Billy Budd, Gloriana* (written on a commission from the British government for the Coronation ceremonies in June, 1953 —the first time an opera was ordered for such an event), and *The Turn of the Screw.* What distinguishes Britten's operas are his natural gift for theatrical effect, his ability in finding the proper musical equivalent for every demand of the stage, and his projection of atmosphere. His gamut is a wide one: he can be passionate and intense, as in *Peter Grimes* and *Billy Budd;* satirical and witty, as in *Albert Herring;* spacious and grand, as in *Gloriana.* He can write equally well for large forces, in a complex style, and for the simple means of his children's work, *Let's Make an Opera.*

Brod, Max, author. Born Prague, May 27, 1884. Primarily a literary man, Brod has interested himself in contemporary opera. He translated many of Janáček's operas into English and wrote this composer's biography. He was also the man who discovered Weinberger's *Schwanda* and helped get it performed. Celebrated as the literary executor of Franz Kafka, Brod preserved and brought to publication Kafka's novel *The Trial,* later the source of an opera by Gottfried von Einem.

Brogny, Cardinal, head of the Council of Constance (bass) in Halévy's *La Juive.*

Brothers Karamazov, The, *see* DOSTOYEVSKY, FEODOR.

Brownlee, John, baritone. Born Geelong, Australia, January 7, 1901. As a boy he became a junior naval cadet in the Australian navy, serving during World War I. Following service, he studied accounting. Engaged in the latter profession, he entered a singing competition in Ballarat, and though he had never had a lesson, won first prize. Several singing engagements followed. One of these, a performance of the *Messiah,* was attended by Nellie Melba, who convinced him to go to Paris for serious study with Dinh Gilly. His debut took place at Covent Garden on June 8, 1926, in the performance of *La Bohème* in which Melba made her farewell appearance. That fall he was engaged by the Paris Opéra, the first time a British subject was made a permanent member of that company; his Paris debut was in *Thaïs* in 1927. On February 17, 1937, he appeared for the first time at the Metropolitan Opera. The opera was *Rigoletto.* Since then, Brownlee has sung at Covent Garden,

the Paris Opéra, and the Metropolitan, besides making important guest appearances elsewhere. His greatest successes have been in the Mozart repertory, particularly at the Glyndebourne Festival; he has also been acclaimed in *Salome* and *Pelléas et Mélisande.*

Bruch, Max, composer. Born Cologne, Germany, January 6, 1838; died Friedenau, October 2, 1920. His boyhood compositions won him a scholarship at the Mozart foundation in Frankfort. His other teachers were Ferdinand Hiller, Carl Reinecke, and Ferdinand Breuning. For a while he taught music in his native city, there completing his first opera, *Scherz, List und Rache,* performed in 1858. In 1861 he settled temporarily in Munich, where he completed a new opera, *Die Loreley* (whose libretto had been intended for Mendelssohn); so successful was it when introduced in Mannheim on June 14, 1863, that it was soon repeated in Leipzig. Bruch's third and last opera, *Hermione,* based on Shakespeare's *A Winter's Tale,* was written in 1871 and given in Berlin in 1872. During his long career Bruch held various conducting posts with orchestras in Germany and England. From 1892 to 1910 he was head of the master school in composition at the Berlin Royal High School. After 1910 he lived in retirement. Among the honors he received were a membership in the French Academy and the Prussian Order of Merit. Bruch is remembered not for his operas but for his G minor violin concerto and his *Kol Nidrei* for cello and orchestra.

Brüderchen, komm tanz' mit mir, Gretel's song and dance in Act I of Humperdinck's *Hansel and Gretel.*

Brüll, Ignaz, composer. Born Prossnitz, Moravia, November 7, 1846; died Vienna, September 17, 1907. He studied in Vienna with Julius Epstein and Felix Otto Dessoff. In 1861, Epstein performed Brüll's first piano concerto in Vienna. Brüll wrote his first opera,

Der Bettler von Samarkand, in 1864. His second opera appeared eleven years later: *Das goldene Kreuz.* Introduced at the Berlin Royal Opera on December 22, 1875, it was such an outstanding success that it was soon performed throughout Europe. It was given at the Metropolitan Opera in 1886. Later operas: *Der Landfriede* (1877); *Bianca* (1879); *Königin Mariette* (1883); *Gloria* (1886); *Das steinerne Herz* (1888); *Gringoire* (1892); *Schach dem König* (1893); *Der Husar* (1898).

Brünnhilde, a Valkyrie (soprano), daughter of Wotan, in *Die Walküre, Siegfried* and *Die Götterdämmerung.*

Brünnhilde! Heilige Braut! Siegfried's farewell to Brünnhilde in Act III of Wagner's *Die Götterdämmerung.*

Bruneau, Alfred, composer. Born Paris, March 3, 1857; died there June 15, 1934. After attending the Paris Conservatory, where he won the Prix de Rome, he had one of his orchestral works performed by the Pasdeloup Orchestra in 1884. Turning to the stage, he completed his first opera, *Kérim,* given in Paris in 1887. From then on he specialized in opera. His friendship with Emile Zola was a decisive influence. Impressed by Zola's naturalism, he made "Naturalisme" the backbone of his own esthetic principles. In this vein he wrote his first important work, *Le rêve,* based on a Zola story and introduced at the Opéra-Comique in 1891. Bruneau now adapted other works by Zola, including *L'attaque du moulin* (1893) and *La faute de l'abbé Mouret* (1907). His musical style matched the naturalism of his subjects with its vigor and realism. When the Dreyfus affair placed a temporary stigma on Zola, Bruneau's operas lost favor. Not until 1905 was interest revived in his work, after the première of *L'enfant roi,* his first success in a decade. This was followed by a revival of some of his earlier operas. After World War I, Bruneau's work lost

favor, the new generation considering it old-fashioned and occasionally crude. For many years, beginning in 1904, Bruneau was the music critic of *Le Matin.* He was made Officer of the Legion of Honor in 1904. Besides the operas mentioned, he wrote: *Messidor* (1897); *L'ouragan* (1901); *Nais Micoulin* (1907); *Les quatre journées* (1916); *Le Roi Candaule* (1920); *Le jardin du paradis* (1921); *Virginie* (1931).

Bucklaw, Lord Arthur (Arturo), Lucia's husband (tenor) in Donizetti's *Lucia di Lammermoor.*

Büchner, Georg, playwright. Born Goddelau, Germany, October 17, 1813; died Zurich, February 19, 1873. His play *Wozzeck* was the source of Alban Berg's opera of the same name. (Manfred Gurlitt, a German composer, also made an opera of *Wozzeck.*) Another of Büchner's plays, *Dantons Tod,* was made into an opera by Gottfried von Einem.

Buffa (or Buffo), Italian for "comic," as in opera buffa and basso buffo.

Bülow, Hans von, conductor and pianist. Born Dresden, January 8, 1830; died Cairo, Egypt, February 12, 1894. World-famous as conductor and pianist, Von Bülow is of importance in the history of opera through his intimate associations with Wagner. As a conductor, he promoted Wagner's music. It was a hearing of *Lohengrin* in Weimar in 1850 that convinced Von Bülow to become a professional musician. He joined Wagner in Zurich, and for a year was Wagner's apprentice in conducting. Subsequently, he studied the piano with Franz Liszt, after which he toured extensively as a virtuoso. In 1864 he became principal conductor of the Munich Royal Opera. Later he became famous as the conductor of the Meiningen Orchestra. In 1857 he married Liszt's daughter Cosima. She later became Wagner's mistress and bore him two children before she left Von Bülow to join Wagner in Switzerland. After her divorce, in 1869, she married Wagner. Von Bülow made adaptations of Wagner's works for the piano, including the entire score of *Tristan und Isolde*.

Bulwer-Lytton, Edward George Earle, novelist and dramatist. Born London, May 25, 1803; died Torquay, England, January 18, 1873. Several of Bulwer-Lytton's romances were made into operas. Wagner's *Rienzi* came from his novel of the same name. (Another composer, Vladimir Kashperov, also wrote a *Rienzi*.) *The Last Days of Pompeii* was the source of *Le dernier jour de Pompéi*, an opera by Victorin de Joncières; *Pompeji*, an opera by Marziano Perosi, and Enrico Petrella's *Ione*. An early American opera, William Fry's *Leonora*, was based on Bulwer-Lytton's *The Lady of Lyons*. Frederick H. Cowen's *Pauline* was also drawn from this novel.

Bunyan, John, religious writer. Born Elstow, England, November 1628; died London, August 31, 1688. His classic *The Pilgrim's Progress* was the source for an early opera by Ralph Vaughan Williams, *The Shepherds of the Delectable Mountain,* and also for Vaughan Williams' later opera, *The Pilgrim's Progress*.

Buona Figliuola, La, *see* CECCHINA, LA.

Buononcini, *see* BONONCINI.

Bürgschaft, Die (The Pledge), opera by Kurt Weill. Libretto by Caspar Neher, based on a fable by Johann Herder. Première: Städtische Oper, Berlin, March 10, 1932. A judge in Africa is called upon to decide the ownership of a bag of money found in a sack of chaff. He solves the problem by ordering the children of the litigants to marry one another, giving them the money as a dowry. Weill's music is primarily in a jazz vein, and his unusual instrumentation calls for two pianos with electric amplifiers.

Burgtheater, the Vienna court theater where most of the important opera productions took place before the opening of the Vienna Royal Opera in 1869. The Burgtheater was opened on February 5, 1742, with a performance of an Italian opera: Giuseppe Carcano's *Amleto*. Support of opera was so poor that the theater had to close in 1747. It reopened on May 14, 1748, with Gluck's *Semiramide riconosciuta*. Gluck's career was intimately connected with the theater. First, his operas in the Italian vein were successfully performed there, then his first works with which he opened new directions for opera: *Orfeo ed Euridice, Alceste,* and *Paride e Elena*. Gluck also served as a conductor at the Burgtheater in 1754.

Mozart was the next important composer connected with the Burgtheater. His first opera written in Vienna—and the first important opera in the German language—was introduced there in 1782: *The Abduction from the Seraglio*. *The Marriage of Figaro* was given its première at the Burgtheater four years later, and *Così fan tutte* in 1790. In 1792 the Burgtheater gave the première of one of the most delightful comic operas in the repertory: Cimarosa's *Il matrimonio segreto*.

Ten years after Mozart's death the Burgtheater gave the first performances in Vienna of Cherubini's *The Water Carrier* and *Medea*. In addition, the theater commissioned Cherubini to write a new opera. The work, *Faniska,* was finally given not at the Burgtheater but at the Kärntnerthortheater.

In 1857 the Burgtheater gave the Viennese première of *Lohengrin,* under Wagner's supervision. It was so successful that the theater was emboldened in 1861 to attempt the first performance of *Tristan und Isolde*. But delay followed delay. A full year passed, and though there had been seventy-two rehearsals, no performance was in view. When Luise Dustmann, scheduled to sing Isolde, fell ill in the fall of 1863, the project was abandoned.

In 1869, the Vienna Royal Opera was built on the Ring. Henceforth, the most important operatic productions took place there, with the Burgtheater becoming mainly a place for spoken drama.

Burr, Aaron, chief conspirator (baritone) in Damrosch's *The Man Without a Country.*

Busch, Fritz, conductor. Born Siegen, Germany, March 13, 1890; died London, September 14, 1951. He entered the Cologne Conservatory in 1906, and only a year later became musical director of the Riga Stadttheater. After holding various other conductorial posts he became musical director of the Stuttgart Opera in 1918. Three years later he was engaged for a similar post with the Dresden Opera. There he gave outstanding performances of Wagner and Mozart and works of twentieth century composers. Many of Richard Strauss's later operas received their world premières, as did Hindemith's *Cardillac* and Weill's *Der Protagonist,* among other works. When the Nazis came to power, Busch (though not a Jew) had to leave Germany because of his antipathy to the new regime. For the next few years he conducted in Scandinavia and served as the musical director of the Glyndebourne Festival in England. His first visit to the United States had taken place during the 1927–1928 season, when he served as guest conductor of the New York Symphony Society. He returned in 1941 and 1942 to lead various American orchestras. From 1942 to 1945 he conducted extensively in South America. In 1945 he became a conductor of the Metropolitan Opera, making his debut there on the opening night of the season (November 26, 1945) with *Lohengrin.* He remained a principal conductor of the Metropolitan until his death.

Busch, Hans, stage director. Born Aachen, Germany, April 4, 1914. The son of Fritz Busch (see above), his education took place in Dresden, at the Geneva University, and at the Reinhardt School in Vienna. In 1937 he assisted Toscanini at the Salzburg Festival. He came to the United States in 1940 and directed two productions for the New York City Opera Company. While serving in the United States Army in 1942, he acted in an advisory capacity in the rebuilding of La Scala. After the war he served as stage director for the Metropolitan Opera. In 1948 he began serving in a similar capacity with the University of Indiana's Opera Workshop, a position he still holds. Here he assisted with the American première of Britten's *Billy Budd,* and the world premières of Weill's *Down in the Valley* and Foss's *The Jumping Frog of Calaveras County.*

Busoni, Ferruccio, composer and pianist. Born Empoli, Italy, April 1, 1866; died Berlin, July 27, 1924. He toured Europe as a child pianist, then studied composition with Wilhelm Mayer-Remy. He began his career as teacher by joining the piano faculty of the Helsingfors Conservatory. In 1890 he taught at the Moscow Conservatory, and from 1891 to 1894 at the New England Conservatory in Boston. Returning to Europe, he achieved tremendous success as a piano virtuoso. His first opera, *Die Brautwahl* (The Bridal Choice), was introduced in Hamburg in 1912. Two one-act operas followed during World War I, *Turandot* and *Arlecchino,* both introduced on the same program in Zurich on May 11, 1917. After the war Busoni became a professor of composition in Berlin. There he worked on his magnum opus, the opera *Doktor Faust.* He did not live to complete it; it was finished by his pupil Philip Jarnach and introduced in Dresden in 1925.

Büsser, Paul Henri, composer and conductor. Born Toulouse, France, January 16, 1872. He attended the Paris Conservatory, where his teachers included Charles Widor, Charles Gounod, and César Franck, and where he won the Prix de Rome in 1893. After returning to France, he held various posts as organist and chorus master until, in 1902, he was appointed conductor at the Paris Opéra. He also became a professor at the Paris Conservatory. In 1938 he succeeded Pierné as member of the Institut de France, and in 1947 he was elected president of the Académie des Beaux Arts. His operas: *Daphnis et Chloé* (1897); *Colomba* (1921); *Les noces Corinthiennes* (1922); *Le carrosse du Saint Sacrement* (1948).

Butterfly, *see* MADAMA BUTTERFLY.

Buzzy, a journalist (baritone) in Leoncavallo's *Zaza.*

Byron, George Noel Gordon, Lord, poet. Born London, January 22, 1788; died Missolonghi, Greece, 1824. The epic poems and poetic dramas of Byron have been used for a number of opera librettos. Examples are: Paul Lebrun's *La fiancée d'Abydos* (The Bride of Abydos); Hans von Bronsart's *Der Corsair* and Verdi's *Il corsaro* (*The Corsair*); Zdeněk Fibich's *Hedy* (*Don Juan*); Enrico Petrella's *Manfredo* and Carl Reinecke's *König Manfred* (Manfred); Donizetti's *Marino Faliero* and *Parisana;* Anatol Bogatirev's *The Two Foscari* and Verdi's *I due Foscari* (*The Two Foscari*); Natanael Berg's *Leila* (Giaour); and Aimé Maillart's *Lara.*

C

Cabaletta, a type of brief aria with several repeats, found in the operas of Rossini and some of his contemporaries. In the later part of the nineteenth century, and particularly in the works of Verdi, the term refers to the final part of an extended aria or duet.

Caccini, Giulio, composer. Born Rome, about 1546; died Florence, December 10, 1618. He was one of the earliest composers to write operas. In Florence, where he spent most of his life, he served as court singer and lutist to the Grand Duke of Tuscany. He joined the Camerata (which see), participating in their discussions on music and art. A direct result of these discussions was his volume of accompanied arias and madrigals for a single voice, called *Le nuove musiche,* published in 1601. This was an epoch-making work, for it marked the first break with the then-existing style of polyphony. There is good reason to believe that the first opera ever written—Peri's *Dafne*—contained some music by Caccini. Caccini contributed a few arias to Peri's second opera, *Euridice* (1600). He then wrote an opera of his own on the same text. Later he wrote two operas in collaboration with Peri. Caccini's daughter, Francesca, was a professional singer. She appeared in the role of Euridice in Peri's opera when that work was first performed. She was also the composer of an opera, *La liberazione di Ruggiero* (1625).

Cachés dans cet asile, the celebrated Berceuse in Godard's *Jocelyn.*

Cadman, Charles Wakefield, composer. Born Johnstown, Pennsylvania, December 24, 1881; died Los Angeles,

California, December 30, 1946. He was the first composer to make successful use of American Indian themes and rhythms as a basis for songs and larger work. His interest in Indian music was first aroused by Nelle Richmond Eberhart, who wrote the lyrics for Cadman's song cycle, "Four American Indian Songs," in which is found "From the Land of the Sky-Blue Water." In 1909 Cadman spent a summer with the Omaha Indians, studying their ceremonials, love calls, dances and songs. His most ambitious work in this idiom was *Shanewis,* an opera first produced by the Metropolitan Opera in 1918. Cadman later wrote another opera, *A Witch of Salem,* first performed by the Chicago Opera Company in 1926. American in subject, this opera has no suggestion of Indian influences. Cadman's other dramatic works include the three-act *The Land of Misty Water* and the one-act *The Garden of Mystery.*

Caffarelli (born GAETANO MAJORANO), male soprano. Born Bitonto, Italy, April 12, 1710; died Naples, January 31, 1783. One of the most famous castrati (see CASTRATO) of the eighteenth century, he studied singing with Niccolò Antonio Porpora, who pronounced him to be the greatest singer in Europe. Assuming his stage name, he made his debut in Rome in 1724 in a female role and was an immediate success. In additional appearances he became the idol of the Italian opera public. He appeared in London for the first time in 1738, singing in Handel's *Faramondo.* Later triumphs came in France, Spain, and Austria. He amassed a considerable fortune, enabling him to purchase the dukedom of San Donato.

Caius, Dr., intended husband (tenor) of Nannetta Ford in Verdi's *Falstaff.*

Calaf, the Unknown Prince (tenor) in Puccini's *Turandot.*

Calatrava, Marquis of, Leonora's father (bass) in Verdi's *La forza del destino.*

Calchas, high priest (bass) in Gluck's *Iphigénie en Aulide.*

Caldara, Antonio, composer. Born Venice, about 1670; died Vienna, December 28, 1736. In Vienna, in 1714, he filled the post of imperial chamber composer to Emperor Charles VI, and from 1716 on he was Johann Joseph Fux's assistant as court kapellmeister. He wrote over sixty operas, none of which has survived, though some of these were performed with outstanding success in his lifetime. He clung to the traditions of the Italian school of his day, but achieved greater simplicity of structure and greater melodic sobriety than his Italian contemporaries.

Calderón de la Barca, Pedro, poet and dramatist. Born Madrid, January 17, 1600; died there May 25, 1681. One of the most notable of Spanish writers, Calderón wrote several plays that were later adapted into operas. These include: Johann Georg Conradi's *Der königliche Prinz* (one of the earliest operatic adaptations from Calderón, 1693), Werner Egk's *Circe,* Benjamin Godard's *Pédro de Zalaméa,* Malipiero's *La vita è sogno,* Raff's *Dame Kobold,* Schubert's *Fierrabras,* Richard Strauss's *Der Friedenstag,* and Weingartner's *Dame Kobold.*

Caliph, The, a character (baritone) in Cornelius' *The Barber of Bagdad.*

Callas, Maria Meneghini (born KALLAS), soprano. Born New York City, December 3, 1923. Of Greek parentage, she was taken to Greece as a child. She attended the Athens Conservatory of Music. In her fifteenth year she made her debut at the Athens Royal Opera House. She made her Italian debut in 1947 in *La Gioconda,* so impressing the conductor, Tullio Serafin, that he coached her for several months. She joined La Scala in 1947, scoring triumphs as Elvira in *I Puritani,* and as Tosca, Norma, and Lucia. She also sang the roles of Isolde and Brünnhilde. When she appeared at Covent Garden, one critic said she was the "greatest singer, male or female, since Nordica." Between 1951 and 1953 she appeared at the Florence May Music Festivals in revivals of Rossini's *Armida* and Cherubini's *Médée* and the first stage representation of Joseph Haydn's *Orfeo ed Euridice*. By this time she had become a familiar name in the United States through recordings of her La Scala performances in *Tosca* and *I Puritani*. She turned down an offer from the Metropolitan Opera for "trifling reasons," as she put it, and made her American debut with the Lyric Theatre in Chicago on November 1, 1954, as Norma, creating a sensation. The same season she appeared in *Lucia di Lammermoor* and *La traviata*.

Calvé, Emma (born DE ROQUER), dramatic soprano. Born Décazeville, France, August 15, 1858; died Millau, France, January 6, 1942. She is most often thought of in the role of Carmen, a portrayal that gave her her greatest triumphs. For several years she attended a convent in Montpellier, after which she studied singing for two years with Paul Puget. She made her opera debut at the Théâtre de la Monnaie, on September 29, 1882, as Marguerite. After an additional year of study with Mathilde Marchesi, she made a successful first appearance in Paris in 1884. For three seasons she was the principal soprano of the Opéra-Comique. In 1887 she appeared at La Scala where she made such a poor impression as Ophelia that she was hissed off the stage. After another period of study, she returned to La Scala, once again as Ophelia, this time to be acclaimed. In 1890 she created the role of Santuzza in Rome. She achieved heights in the closing years of the century. Her first appearance at Covent Garden, in 1892, was a success of the first magnitude. On November 29, 1893, she was acclaimed at her American debut at the Metropolitan Opera as Santuzza. Three weeks later she appeared as Carmen and was a sensation. From this time on she was one of the most celebrated interpreters of that role.

She had a comparatively limited repertory, but her voice had such richness of texture and color, and her interpretations were so electric, that she maintained a position of first importance among the singers of her time. She created the title role in Massenet's *Sapho* (Paris, 1897), and that of Anita in his *La Navarraise* (1894), a work written for her.

After 1910, and until 1924, she devoted herself primarily to concert appearances. In 1940, she emerged from retirement to appear in a motion picture. She wrote her autobiography, *My Life* (1922), and a volume of reminiscences, *Sous tous les ciels j'ai chanté* (1940).

Calzabigi, Ranieri da, poet and librettist. Born Leghorn, Italy, December 23, 1714; died Naples, July, 1795. He wrote the librettos for three of Gluck's operas.

Before settling in Vienna, in 1761, he had distinguished himself as a historian, literary man, and adventurer. In Vienna he became chamber councilor to the exchequer, and allied himself with Gluck and Count Durazzo (director of court theaters) in bringing about a reform in opera based on French dramatic principles, ideas, and culture. Their first collaboration was a ballet, *Don Juan,* given in 1761. A year later they produced *Orfeo ed Euridice,* with which their revolution of opera was finally realized. After writing the librettos for Gluck's *Alceste* and *Paride ed Elena,* Galzabigi left Vienna, having become involved in a theatrical scandal.

Cambert, Robert, composer. Born Paris about 1628; died London, 1677. He wrote the first French opera: *Pomone.* After completing harpsichord study with Chambonnières, he wrote, in 1659, the music for a comedy, *La Pastorale d'Issy.* In 1666 he became director of music to the Queen Dowager, Anne of Austria. Three years later he obtained a patent with Abbé Pierre Perrin to establish the Académie Royale de Musique for the purpose of presenting operas; this Académie subsequently became the Paris Opéra. For this theater he wrote, in 1671, a work now credited as being the first French opera: *Pomone*. A second opera, *Les peines et les plaisirs de l'amour*, was written in the same year and later produced in London. In 1671, through the machinations of Lully, Cambert lost his patent for the Académie. His disappointment driving him out of Paris, he settled in England where he served as a military bandmaster.

Camerata, the group of musicians and poets who gathered in Florence in the sixteenth century to discuss music, poetry, and the theater, and who were responsible for the birth of opera. Leaders of the Camerata (literally: the group that meets in a room) were two noblemen, Giovanni Bardi and Jacopo Corsi, and in their homes the fruitful meetings took place. Other members included the composers Vincenzo Galilei (father of the astronomer), Giulio Caccini, Jacopo Peri, and Emilio de' Cavalieri, and a poet, Ottavio Rinuccini. For a more detailed account of the Camerata's accomplishments see OPERA.

Camille, opera by Hamilton Forrest. Libretto by the composer based on Alexander Dumas's novel *La dame aux camélias*. Première: Chicago Civic Opera, December 10, 1930. *See also* LA TRAVIATA.

Cammarano, Salvatore, librettist. Born Naples, March 19, 1801; died there July 17, 1852. After writing several prose dramas, he began writing opera librettos in 1834, and a year later completed *Lucia di Lammermoor* for Donizetti. He subsequently wrote librettos for Verdi (including *Alzira, La battaglia di Legnano, Luisa Miller* and *Il Trovatore*) as well as for other composers, including Giovanni Pacini and Giuseppe Mercadante.

Campanari, Giuseppe, dramatic baritone. Born Venice, November 17, 1855; died Milan, May 31, 1927. Originally, he was a cellist in the orchestra of La Scala, and later with the Boston Symphony Orchestra. While still in Milan he began to study singing, and in 1893 made his debut in New York with the Gustav Hinrichs Opera Company; during the same year he created for America the role of Tonio. He was immediately acclaimed both for the beauty of his voice and the vitality of his characterizations. On January 7, 1895, he made his debut at the Metropolitan Opera as Valentin. He remained at the Metropolitan for three seasons, scoring great successes as Figaro in *The Barber of Seville,* and as Count di Luna, Escamillo, and Falstaff. After abandoning opera he concentrated on concert appearances and teaching.

campana sommersa, La (The Sunken Bell), opera by Ottorino Respighi. Libretto by Claudio Guastalla, based on the play of the same name by Gerhart Hauptmann. Première: Hamburg Opera, November 18, 1927. Heinrich, the bell founder, comes under the spell of a fairy. He deserts his people to follow her into the mountains. The death of his wife, Magda, brings him home, but he cannot forget his sweetheart. On his deathbed, Heinrich calls for the fairy, and she returns to him.

Campanini, Cleofonte, conductor. Born Parma, Italy, September 1, 1860; died Chicago, December 19, 1919. His musical studies took place at the Parma Conservatory. He made his debut as conductor in Parma in 1883 with *Carmen.* In the same year he was appointed assistant conductor of the Metropolitan Opera during its inaugural season. Five years later he was brought back to the United States expressly to direct the American première of Verdi's *Otello* at the Academy of Music.

The Desdemona of that performance was Eva Tetrazzini, who had become Campanini's wife in 1887. After conducting at La Scala for three years, he was appointed artistic director and principal conductor of the newly organized Oscar Hammerstein Opera Company at the Manhattan Opera House. He stayed there three years, finally resigning because of differences with Hammerstein over artistic policies. In 1910 he became first conductor of the Chicago Opera, a post he retained to the time of his death. In 1918 he brought the Chicago company to New York for a four-week season, when the then sensational coloratura soprano, Galli-Curci, made her first New York appearance. Campanini was one of the outstanding conductors of his generation, particularly of the French repertory. Among the operas he introduced to America were *Hérodiade, The Jewels of the Madonna, Louise, Pelléas et Mélisande,* Massenet's *Sapho* and *Thaïs.*

Campanini, Italo, tenor. Born Parma, Italy, June 30, 1845; died Villa Vigatto, Italy, November 14, 1896. He was the brother of the conductor discussed above, and one of the most successful opera tenors in the United States before Caruso. After study in Parma he made his opera debut in Odessa in *Il trovatore* (1869). After an additional period of study with Francesco Lamperti in Milan, he returned to the stage and scored his first major success in Bologna in 1871 in the Italian première of *Lohengrin.* After touring the United States in 1873 and 1879–1880, he appeared at the Metropolitan Opera as Faust in the company's first performance (October 22, 1883). After 1883 he lived principally in New York, and was the leading tenor of the Metropolitan Opera. The beautiful texture of his voice and his flawless delivery made him a favorite.

Campiello, Il (The Piazza), opera buffa

by Wolf-Ferrari. Libretto by the composer based on Goldoni's comedy of the same name. Première: La Scala, February 12, 1936. The action takes place in Venice about 1750 and concerns an eventful day in the lives of four families inhabiting a Venetian piazza. A climax comes with a street brawl in which two elderly mothers (represented by tenors!) participate.

Campra, André, composer. Born Aix, Provence, December 4, 1660; died Versailles, June 29, 1744. He was one of the early masters to establish the foundations of French opera. After studying with Guillaume Poitevin, he was appointed maître de musique of the Toulon cathedral. In 1694 he went to Paris and soon became maître de chapelle at Notre Dame. At about this time he began writing operas. For fear of losing his ecclesiastical post he presented them as his brother's. The first was *L'Europe galante,* given in Paris in 1697. Two years later he wrote *Le carnaval de Venise.* The success of these works encouraged him to give up his church post and devote himself exclusively to secular music. From this time on he produced many operas which made him the logical successor to Lully as the foremost French composer for the theater. His opera *Tancrède,* given in 1702, held the stage in Paris for half a century. In 1718 he received a life pension, and in 1722 he was appointed musical director of the Chapelle Royale, and to the Prince de Conti. His principal operas: *L'Europe galante* (1697); *Le carnaval de Venise* (1699); *Aréthuse* (1701); *Tancrède* (1702); *Iphigénie en Tauride* (1704); *Télémaque* (1704); *Le Triomphe de l'amour* (1705); *Idoménée* (1712); *Les Amours de Mars et Vénus* (1712); *Camille* (1717); *Achille et Déidamie* (1735).

Canio, head of a theatrical troupe (tenor) in Leoncavallo's *Pagliacci.*

Canterbury Pilgrims, The, opera by

Reginald De Koven. Libretto by Percy MacKaye, based on a tale of Chaucer. Première: Metropolitan Opera, March 8, 1917. The tale concerns the assembling of the Canterbury pilgrims in the courtyard of the Tabard Inn at Southwark, near London, on April 16, 1387. By losing a wager, Chaucer must decide to marry Alisoun; he is finally rescued by legal technicalities. Charles Villiers Stanford wrote an opera with the same title.

Canzonetta, a short song, generally of light, cheerful character.

Caponsacchi, opera by Richard Hageman. Libretto is Arthur Goodrich's play of the same name, adapted from Robert Browning's poem *The Ring and the Book*. Première: Freiburg, Germany, February 18, 1932 (in a German translation). Caponsacchi is accused of having been Pompilia's lover. Pompilia is subsequently murdered by her husband, Count Guido Francheschini. Brought before a papal hearing, Caponsacchi reveals how he has been falsely accused by Guido. The papal court pronounces Caponsacchi innocent and sentences Guido to death.

Capriccio, one-act opera by Richard Strauss, libretto by Clemens Krauss. Première: Munich Opera, October 28, 1942. An opera about an opera, *Capriccio* was described by its authors as "a conversation piece for music." In eighteenth century Paris—the period of Gluck's reforms in opera—the question of the relative importance of music and text is argued in the salon of a Countess Madeleine. The Countess has two lovers: a musician, Flamand; a poet, Olivier. Each tries to win her to his art. The Countess is just as incapable of choosing sides in the operatic debate as she is in deciding between the lovers. The final verdict is that poetry and music are equally important.

Capulet, a nobleman (bass), head of the house of Capulet, in Gounod's *Roméo et Juliette*.

Cardillac, opera by Paul Hindemith. Libretto by Ferdinand Lion, based on E. T. A. Hoffmann's *Das Fräulein von Scuderi*. Première: Dresden Opera, November 9, 1926 (original version). Première of revised version (completely new libretto by the composer but with the score unchanged): Zurich State Opera, June 20, 1952. The opera revolves around the character of Cardillac, a master jeweler, who murders the purchasers of his art works rather than have to part with his creations.

Card Song, *see* EN VAIN POUR EVITER.

Carestini, Giovanni, male contralto. Born Filottrano, Italy, 1705; died there about 1760. He made his debut in Rome in 1721 in the principal female role in Bononcini's *Griselda*. From then to 1733 he established his reputation with appearances throughout Italy. Though originally a soprano, his voice deepened and he acquired what was described by his contemporaries as the finest contralto voice ever heard. In 1723 he was invited to sing at the coronation of Charles VI as King of Bohemia. A decade later he went to London, making his debut in a pasticcio. He soon distinguished himself as a member of Handel's opera company where he proved to be a successful rival to Farinelli, then acknowledged to be the greatest of the castrati. Among Handel's operas in which Carestini appeared in London were *Alcina, Ariodante,* and *Il Pastor Fido*. Carestini returned to Italy in 1735. For the next twenty years he was a reigning favorite in Italian, German, and Russian opera houses.

Carlo, Don (baritone), son of the Marquis of Calatrava in Verdi's *La forza del destino*.

Carlos, Don, (1) king of Castile (baritone) in Verdi's *Ernani*.

(2) *See* DON CARLOS.

Carl Rosa Opera Company, The Royal, English company which has been producing operas in English for over three-quarters of a century. It was organized in 1873 by Carl Rosa, giving its first performance in Manchester: William Vincent Wallace's *Maritana*. During the first season the company gave operas by Balfe, Bellini, Gounod, Mozart and Verdi, all in English. The company was reorganized in 1875, and it first appeared in London with *The Marriage of Figaro*. For the next fourteen years, the company was outstandingly successful, primarily in the provinces, but for several seasons also in London. The personnel included many fine American, English, and French artists, including Minnie Hauk and Alwina Valleria from America, Sir John Bentley, and Marie Roze from France. Seven new British operas were commissioned. Carl Rosa proved the popularity of opera by amassing a fortune. After his death, in 1889, the management passed to Augustus Harris, while the ownership remained with Rosa's widow. For the next decade the company continued to prosper artistically and financially. In 1891 its performance of *Carmen* was so popular that a special company was formed to tour the provinces, with Marie Roze in the title role. In 1893 the company gave a command performance of *Fra Diavolo* at Balmoral Castle, after which Queen Victoria conferred on it the honorary title of "royal." In 1897 Puccini helped supervise the first English performances of *La Bohème*. Operas by English composers were not neglected: during the 1890's seven native works were introduced at Covent Garden, and several others in the provinces.

In 1900 the company was taken over by Alfred van Noorden, who became its manager, while his brother Walter was principal conductor. Under this regime, the company gave the first English performances of Karl Goldmark's *The Cricket on the Hearth* and *The Queen of Sheba*. In 1918 Sir Joseph

Beecham's Quinlan Opera Company merged with the Carl Rosa company, and five years later the manager of the Quinlan Opera, H. B. Phillips, became full owner. He retained control up to the time of his death in 1950, when the ownership passed to his widow who, for several years previously, had been artistic director. Under Phillips, the company continued its policy of presenting new operas and interesting revivals.

This company has occupied a position of major importance in English musical life by virtue of its sponsorship of opera in English, and its contribution in arousing and keeping alive an interest in opera in the provinces. It was also significant in helping train, and giving experience to, several generations of singers.

Carmela, Gennaro's mother (mezzo-soprano) in Wolf-Ferrari's *The Jewels of the Madonna*.

Carmen, opera in four acts by Georges Bizet. Libretto by Henri Meilhac and Ludovic Halévy, based on Prosper Mérimée's story of the same name. Première: Opéra-Comique, March 3, 1875. American première: Academy of Music, New York, October 23, 1878.

Characters: Don José, a guardsman (tenor); Carmen, a gypsy (mezzo-soprano); Escamillo, a toreador (baritone); Micaëla, a peasant girl (soprano); Frasquita, a gypsy friend of Carmen (soprano); Mercédès, another gypsy friend (mezzo-soprano); Zuniga, captain of the guards (bass); Moralès, an officer (bass); Le Remendado, a smuggler (tenor); Le Dancaïre, another smuggler (baritone); cigarette girls; gypsies; smugglers; dragoons. The setting is in and near Seville, about 1820.

Act I. A square in Seville. A vigorous prelude alternates between gaiety and foreboding. The principal themes are Escamillo's "Toreador Song" and an ominous theme for cellos suggesting impending doom. With the rise of the curtain a girl timidly approaches a guardsman to inquire of the whereabouts of Don José. She is informed he will appear with the change of the guards. The guards finally arrive. From a near-by cigarette factory, girls emerge for the noonday respite. One of these, Carmen, makes flirtatious overtures to Don José as she mockingly sings of love (Habanera: "L'amour est un oiseau rebelle"). Her song ended, she flings a flower at him and rushes back into the factory. Don José picks up the flower and conceals it in a pocket near his heart. The timid girl now reappears. She is Don José's sweetheart Micaëla, come with news from home. Tenderly they recall their childhood happiness ("Ma mère, je la vois"). Don José then sends her back with a message for his mother. When she has gone, he takes the flower from his pocket and is about to throw it away when he is attracted by noises from the factory. When the women come rushing out, he learns that Carmen has stabbed one of the girls. Carmen is seized by a dragoon who ties her hands and leaves her in Don José's custody while he goes off to seek a warrant for her arrest. Coyly, with light heart, Carmen insists that she and Don José will soon meet again in the tavern of Lillas Pastia, outside the city walls (Seguidilla: "Près des remparts de Seville"). Don José is now under her spell. He unties Carmen's hands. When the dragoons come to conduct her to prison, she pushes them aside and escapes.

Act II. The tavern of Lillas Pastia. Gypsies are dancing and singing. Carmen receives the news that Don José, who had been arrested for complicity in her escape, has been released from prison. A moment later the famous toreador, Escamillo, arrives. Proudly he tells his admirers of the excitement of bullfighting ("Toreador Song"), then departs, trailed by the crowd. Carmen and two of her friends are ap-

proached by smugglers who want to employ them as lures for the coast guard. Carmen is sympathetic, but before she can accept the offer she wants to await Don José's arrival. Already his voice is heard in the distance. When he appears, Carmen welcomes him passionately. She plays on his emotions; she dances for him. The sudden sound of a bugle call reminds Don José that he must return to his barracks, but he is now so hopelessly in love that he cannot leave Carmen. He removes from his pocket the flower Carmen had once thrown him (Flower Song: "La fleur que tu m'avais jetée"). However, he is also torn by pangs of conscience. He is about to return to duty when Zuniga appears and mocks Carmen for taking up with a mere soldier when he, an officer, wants her. Don José attacks Zuniga. This act of insubordination makes it impossible for him to return to military duty. He now joins Carmen in her association with the smugglers.

Act III. A mountain pass. A short orchestral prelude sets the mood with a night song for the flute.

The smugglers tell of their dangerous trade, and how they continually must be on the watch ("Ecoute, écoute, compagnon"). Don José is sad and reflective because, as he tells Carmen, he is thinking of his mother. Bitterly, Carmen tells him to go home, a suggestion so upsetting to Don José that he threatens to kill her if she repeats it. Near-by, the gypsies Frasquita and Mercédès are reading fortunes with cards. When Carmen begins to read her own, the cards tell of impending disaster. (Card Song: "En vain pour éviter les réponses amères"). Her tensions are relieved with the announcement by the smugglers that the time has come for them to carry their contraband through the mountain pass. Don José is left behind to keep guard. When the smugglers are gone, Micaëla comes seeking her lover. Poignantly, she prays to heaven for protection ("Je dis que rien ne m'épouvante"). When a shot rings out, she hides. The shot has come from Don José's gun, fired at the approach of a stranger. Escamillo has come seeking Carmen. Recognizing each other as rivals, Don José and Escamillo fall upon each other with drawn daggers. Only the sudden return of the smugglers prevents a tragedy. Placated, Escamillo departs with his customary swagger, inviting all the smugglers to be his guests at his next bullfight. Micaëla is now discovered by the smugglers. She reveals to Don José that his mother is dying. Don José must now leave Carmen, but he warns her that they will meet again.

Act IV. Another square in Seville. The festive prelude is alive with Spanish rhythms and melodies. It is the day of the bullfight. Escamillo appears, Carmen with him. When the toreador enters the arena, Carmen's gypsy friends warn her that José is lurking near-by to avenge himself for her desertion. The crowd surges into the arena. When Carmen is left alone, José comes to plead for her love. Icily, the gypsy tells him she loves him no longer. Don José continues to plead; Carmen remains deaf. Suddenly, shouts from the arena hail the victorious Escamillo. Carmen is about to join her hero when José stops her and kills her with a dagger. When the toreador emerges to inform Carmen of his triumph he finds her stretched out dead, Don José sobbing at her side.

Before *Carmen* was universally accepted as one of the finest products of the French lyric theater, and before it became one of the best loved operas in the repertory, it had to live down two devastating accusations made against it. When first produced, *Carmen* was criticized by several French writers as being a feeble imitation of the Wagner dramas. While it is true that Bizet had recourse to the *leitmotiv* technique—

though not in the Wagnerian way—and that he had Wagner's respect for the stage and his inventiveness of harmonic and orchestral language, *Carmen* is not the stereotype of a Wagnerian drama by any stretch of the imagination. The essence of Bizet's musical style and dramatic approach was French. Greater familiarity with *Carmen* proved how wrong these Parisian critics were, but hardly had this controversy died down when a new attack was directed against the opera. Several musicologists, particularly Spanish ones, took the opera to task for its supposedly pseudo-Spanish style. But Bizet had no intention of writing national Spanish music. What he wanted to do, and what he succeeded in doing, was capture the spirit of Spanish song and dance in music essentially his own. Bizet's borrowing from popular melodies and rhythms (as when he expropriated a melody by Sebastian Yradier for his famous Habanera) were used as the Mediterranean spice for his dish; but his dish was still representative of the French cuisine.

While *Carmen* was at first the victim of unjust criticisms, it was by no means the failure that sentimental historians and biographers led us to believe. Once audiences got over the shock of the sensual story and the lurid characterizations, they began responding enthusiastically to Bizet's wonderful music. Though the première took place late in the season, the opera was seen thirty-seven times in its first year. It remained in the repertory the following season, which most certainly would not have been the case had it been a failure. Ludovic Halévy, one of the librettists, recorded that the box-office receipts were "respectable, and generally in excess of those for other works in the repertory."

Originally, *Carmen* was an opéra-comique, not a grand opera. In other words, it had spoken dialogue between its musical numbers. However, as seen in the United States and in most European opera houses outside France, *Carmen* is sung throughout, the spoken words having been replaced by recitatives composed by Ernest Guiraud.

Carmen, the volatile gypsy girl, is one of the most colorful characters in all opera, and many famous sopranos have distinguished themselves with their vivid, at times highly individual portrayals of the part. Célestine Galli-Marié, the first Carmen, became widely famous in the part. The second celebrated Carmen was Minnie Hauk, who introduced the opera in England (June 22, 1878) and America (October 23, 1878). Other notable Carmens have been Emma Calvé, Geraldine Farrar, Mary Garden, Rosa Ponselle, and Bruna Castagna.

Carmina Burana, opera or "scenic cantata" by Carl Orff. Libretto by the composer, based on poems of anonymous 13th century monks. Première: Frankfort on the Main, 1937. This is the first opera of a trilogy entitled *Trionfi,* of which the subsequent two parts are: *Catulli Carmina* and *Trionfo d'Aphrodite.* The libretto (in medieval Latin, with parts in medieval German) is built around the activities of medieval students who wandered about appearing as jesters. The sections of *Carmina Burana* are entitled "Springtime," "In the Tavern," and "The Court of Love."

Carmosine, see MUSSET, ALFRED DE.

Carolina, Paolina's wife (soprano) in Cimarosa's *Il matrimonio segreto.*

Caro nome, Gilda's aria in Act II of Verdi's *Rigoletto.*

Carré, Albert, impresario. Born Strasbourg, France, June 22, 1852; died Paris, December 12, 1938. He was the director of the Opéra-Comique for over a decade. His initiation into the theater was as an actor in Paris. After directing various theaters in France, including the Comédie Française, he was appointed, in 1898, the director of the

Opéra-Comique in succession to Léon Carvalho. He introduced so many fine operas, including *Le Juif polonais, Louise* and *Pelléas et Mélisande,* that the Opéra-Comique became a formidable rival of the Opéra. Carré remained director until 1912. After World War I, he was co-director with the Isola brothers until 1925, after which he was made honorary director. He wrote several opera librettos, including that for Messager's *La Basoche.* His uncle, Michel Carré, collaborated with Jules Barbier in writing the librettos for eight of Gounod's operas including *Faust* and *Roméo et Juliette,* and for Massé's *Paul et Virginie,* Meyerbeer's *Le pardon de Ploërmel,* Offenbach's *The Tales of Hoffmann,* Thomas's *Hamlet* and *Mignon,* among other operas.

Carrosse du Saint—Sacrement, La, see MERIMEE, PROSPER.

Caruso, Enrico (tenor). Born Naples, February 25, 1873; died there August 2, 1921. One of the greatest operatic tenors of all time, he was the idol of the opera world for over two decades. A musical child, in his ninth year he joined the choir of his parish church. Formal musical training came comparatively late. When he was eighteen he began a three-year period of study with Guglielmo Vergine, completing his training with Vincenzo Lombardi. After several appearances in Naples, he made what he regarded as his official debut there on November 16, 1894, in *L'Amico Fritz.* In 1898 he was engaged by the Teatro Lirico in Milan, where he created the principal tenor roles in *Adriana Lecouvreur* and *Fedora.* In 1901 he became a member of the La Scala company, where he was featured in leading roles ot the Italian and French repertory. Here he created the principal tenor roles in Franchetti's *Germania* and Mascagni's *Le Maschere.* Arturo Toscanini said, after hearing him in *L'elisir d'amore:* "If this Neapolitan continues singing like this, he will make the whole world talk about him." Caruso's international fame began in Monte Carlo in 1902, where he was so highly acclaimed that he was engaged for three additional seasons and received contracts from Covent Garden and the Metropolitan Opera. His American debut took place at the Metropolitan Opera on November 23, 1903, in *Rigoletto*. This was the opening night of the Metropolitan season. Largely due to his nervousness, Caruso did not make a good first impression. The critics pointed to his "tiresome Italian mannerisms" (particularly his excessive use of the so-called "Rubini sob"). But before the end of the season he became an outstanding favorite. He achieved a personal triumph as Nemorino in *L'elisir d'amore*. The following season he again appeared at the Metropolitan's opening-night performance maintaining a tradition which lasted seventeen years in all, with only one year skipped. He was now the shining light of the company. When he sang, the box office prospered; his performances brought in close to a hundred thousand dollars a season. He duplicated his Metropolitan Opera triumphs in all the major opera houses, becoming the highest-paid singer, and the most adulated, in the world. When he sang in Germany and Austria, seats for his performances were often sold at auction. His income from his phonograph recordings totaled almost two million dollars during his lifetime. He appeared successfully in almost fifty roles. His voice was admired for its range, tone, and shading. It was powerful yet supple, exquisite in upper range, sensuous in middle tones, extraordinarily expressive in lower registers. "I have never heard a more beautiful voice," Edouard de Reszke once wrote to him.

His career ended in 1920. During a performance of *L'elisir d'amore* at the Brooklyn Academy on December 11, he coughed blood. The diagnosis at this time was "intercostal neuralgia" and he continued to sing. His 607th and last appearance at the Metropolitan (and his last in opera anywhere) took place on Christmas Eve, when he sang in *La Juive* while suffering acute pain. Pleurisy developed into bronchial pneumonia. An operation removed the fluid from his pleural cavity and it seemed certain that Caruso would recover and sing again. However, in February, 1921, he developed complications. After more treatment, Caruso returned to Italy for a long rest. During the summer he recovered sufficiently to work with his voice again, but a relapse proved swiftly fatal. (It is believed that an Italian doctor made an examination of Caruso with an unsterilized instrument, causing the final infection.) His death was mourned throughout the world, which paid a tribute to him such as few other opera singers before or since have received.

Carvalho, Léon (born Carvaille), impresario. Born Port-Louis, France, January 18, 1825; died Paris, December 29, 1897. For many years he was the manager of the Opéra-Comique. After attending the Paris Conservatory he sang at the Opéra-Comique, where he met and, in 1835, married the prima donna Marie-Caroline Miolan. For a period he was the director of the Théâtre Lyrique in Paris, and after that was stage manager of the Opéra. In 1876 he became manager of the Opéra-Comique, holding this post for a decade. In 1887 he was found guilty of negligence in a fire that destroyed the opera house and killed 131 people. He was fined and spent some time in prison, being released on appeal. In 1891, restored to good graces, he resumed his management of the Opéra-Comique. His wife appeared alternately at the Opéra and the Opéra-Comique between 1868 and 1883. She was seen in the premières of several

Gounod operas, including *Faust, Mireille,* and *Roméo et Juliette.*

Cary, Annie Louise, contralto. Born Wayne, Maine, October 22, 1841; died Norwalk, Connecticut, April 3, 1921. She was the first American woman to appear in a Wagnerian role in the United States. After preliminary training in America, she went to Europe in 1866 and studied with Giovanni Corsi in Milan. Her debut took place in Copenhagen in *Il trovatore.* During the next few years she combined opera appearances with further study with Pauline Viardot-García. In 1868 she became a principal soprano of the Hamburg Opera, and in 1870 made her bow at Covent Garden. After making her American debut in concert, she joined the company at the Academy of Music where, in 1877, she created for America the role of Amneris. Her first Wagnerian role was Ortrud, sung in 1877. A serious throat ailment compelled her to abandon opera in 1881 at the height of her popularity.

Casella, Alfredo, composer. Born Turin, Italy, July 25, 1883; died Rome, March 5, 1947. He attended the Paris Conservatory and first achieved success as a composer with an orchestral rhapsody, *Italia,* in 1909. Though he was in the vanguard of a movement among twentieth century Italian composers to bring about a renascence of instrumental music, he did not neglect the theater. His first opera, *La donna serpente,* introduced in Rome in 1932, was in the style of the commedia dell' arte. His second was *La favola di Orfeo* (1932). A one-act opera, *Il deserto tentato,* given at the Florence Music Festival in 1937, was severely criticized by the outside world for its open espousal of fascism and its glorification of the Italian conquest of Ethiopia. His autobiography, *Music in My Time,* was published in the United States in 1954.

Caspar, a huntsman (bass) in Weber's *Der Freischütz.*

Cassandra, opera by Vittorio Gnecchi. Libretto by the composer. Première: Bologna, December 5, 1905. This opera became a storm center in 1909 when Giovanni Tebaldini wrote an article in the *Rivista musicale italiana* in which he quoted examples from Richard Strauss's *Elektra* and tried to prove that they had been plagiarized from *Cassandra.* Strauss's friends in Germany insisted that the similarities were remote, but one of them confessed that Strauss had seen the score of *Cassandra* before writing his own opera. In a letter to Romain Rolland, Strauss ridiculed the idea that any similarity existed between his opera and Gnecchi's or that he was a plagiarist. Soon after this, Rolland came to Strauss's defense in an article in the *Bulletin français de la S. I. M.*

Cassio, Otello's lieutenant (tenor) in Verdi's *Otello.*

Casta Diva, Norma's aria with chorus in Act I of Bellini's *Norma.*

Castagna, Bruna, contralto. Born Bari, Italy, October 15, 1908. She had only three months of vocal instruction before she made her opera debut in Mantua, in her seventeenth year, in *Boris Godunov.* Tullio Serafin thereupon engaged her for the Teatro Colón in Buenos Aires. After singing three years in South America, she was engaged by Arturo Toscanini for La Scala, where for five years she was a favorite; it was to provide a proper opportunity for her voice that Rossini's *L'Italiana in Algeri* was revived there. She first came to the United States in 1934, making her American debut in *Carmen* at the Hippodrome Theater in New York. On March 2, 1936, she made her debut at the Metropolitan Opera as Amneris. For the next decade she made many successful appearances at the Metropolitan and other major opera houses in such roles as Adalgisa, Amneris, Azucena, Delilah, Laura, and Santuzza. Since 1946 she appeared ex tensively in South America and Europe.

Castil-Blaze, *see* BLAZE, FRANÇOIS-HENRI-JOSEPH.

Castor et Pollux, opera by Jean Philippe Rameau. Libretto by Pierre Joseph Bernard. Première: Paris Opera, October 24, 1737. The work was overwhelmingly successful and was long regarded in France as Rameau's masterpiece. The story is based on Greek and Roman mythology. When Castor dies in a quarrel, his twin brother, Pollux, begs Zeus to allow him to die in his brother's place. Unable to grant this, since Pollux is immortal, Zeus allows him to spend alternately one day with the gods and another with his brother in Hades. The ballet and funeral music, and Télaïre's aria in the first act, "Tristes apprêts," are noteworthy.

Castrato, an Italian term for a eunuch with a high-pitched voice. Such singers were extensively employed for opera performances in the 18th century. Boys with exceptional voices were castrated to prevent a change of voice at puberty. These eunuchs came to have voices delicate in texture, voluptuous in tone, and flexible in range. Many of the famous opera composers of the 18th century wrote their most florid arias for the castrati, who sometimes contributed additional embellishments of their own. Of these operas, one that has remained vital in the repertory is Gluck's *Orfeo ed Euridice,* in which the male role of Orfeo —written for Gaetano Guadagni— is today sung by a contralto. Among the most celebrated castrati were Caffarelli, Carestini, Farinelli, Guadagni, and Senesino. They were all highly paid and adulated. A journalist reported in 1720 that "women from every grade of society—peeresses incognito, melancholy wives of city merchants, wretched little streetwalkers—all jostled each other . . . hungry for a look or a word" from one of these singers. One woman

is reported to have said: "There is only one God, and one Farinelli."

Catalani, Alfredo, composer. Born Lucca, Italy, June 19, 1854; died Milan, August 7, 1893. His studies took place with his father, with Fortunato Magi, and at the Conservatories of Paris and Milan. A one-act opera, *La Falce,* was performed in Milan in 1876. Four years later he wrote his first full-length opera, *L'Elda,* presented in Turin in 1880. His first major success was a revision of *L'Elda,* renamed *Loreley,* performed in Turin in 1890. His most popular opera, *La Wally,* was given at La Scala in 1892. His other operas were *Dejanire* (1883), and *Edmea* (1886). In 1886 Catalani succeeded Ponchielli as professor of composition at the Milan Conservatory.

Catalani, Angelica, dramatic soprano. Born Sinigaglia, Italy, May 10, 1780; died Paris, June 12, 1849. After receiving her education at the Convent Santa Lucia di Gubbio in Rome, she made her opera debut at the Teatro la Fenice in 1795. Other appearances in Italy, climaxed by performances at La Scala in 1801, brought her immense success. In 1804 she became a member of the Italian Opera in Lisbon. Her debut in London in 1806 was brilliant; so great was her drawing power in England that she earned unprecedented sums. She remained in that country seven years (appearing in the first performance in England of *The Marriage of Figaro* in 1812), then stayed in Paris three years, where she managed, without success, the Théâtre des Italians. Having sung in Russia, Poland, and Germany, and again in Italy and England, she retired (after 1828) and directed a singing school near Florence.

Catalogue Song, *see* MADAMINA, IL CATALOGO E QUESTO.

Catherine, (1) a laundress known as Madame Sans-Gêne (soprano) in Giordano's *Madame Sans-Gêne.*

(2) Petruchio's terrible-tempered

wife (soprano) in Goetz's *The Taming of the Shrew.*

Catulli Carmina, opera by Carl Orff. Libretto by the composer, based on the poems of Catullus. Première: Stuttgart Opera, 1952. The composer describes this work as a "dramatic ballet." Its music consists of unaccompanied solos and consists entirely of choruses; only the prelude and postlude require an instrumental background of four pianos and percussion. The libretto is in classic Latin. This is the second opera in a trilogy called *Trionfi,* the first of which is *Carmina Burana,* and the last *Trionfo d'Aphrodite.*

Cavalieri, Emilio de', composer. Born Rome, about 1550; died there, March 11, 1602. He was a member of the Camerata (which see) and wrote several operas, all of them with texts by Laura Guidiccioni. He was one of the first composers to employ the technique of basso continuo—a continuous bass part provided with figures and signs indicating the harmonies to be employed in the instrumental accompaniment. His most celebrated work was a sacred opera, *La rappresentazione dell' anima e del corpo,* performed and published in 1600. His other operas: *Il Satiro* (1590); *La disperazione di Fileno* (1590); *Il Giuoco della cieca* (1595).

Cavalieri, Lina, dramatic soprano. Born Viterbo, December 25, 1874; died Florence, February 8, 1944. She was trained in Paris by Madame Mariani-Masi. In 1901 she made her debut in Lisbon in *Pagliacci.* This was followed by appearances in opera houses in Italy, Poland, Russia, France, and England. On December 5, 1906, she made her American debut at the Metropolitan Opera in the title role of *Fedora.* She stayed with the Metropolitan until 1908, after which she joined the Manhattan Opera Company. In 1915–1916 she sang with the Chicago Opera. She scored her greatest successes in French opera, particularly in such roles as

Hérodiade, Manon, and Thaïs. She retired soon after World War I, living first in Paris, then in Florence. She married the celebrated tenor, Lucien Muratore. She was killed during an air raid in World War II.

Cavalleria rusticana (Rustic Chivalry), one-act opera by Pietro Mascagni. Libretto by Giovanni Targioni-Tozzetti and Guido Menasci, based on a short story of the same name by Giovanni Verga. Première: Teatro Costanzi, Rome, May 17, 1890. American première: Grand Opera House, Philadelphia, September 9, 1891.

Characters: Santuzza, a village girl (soprano); Turiddu, a soldier (tenor); Mamma Lucia, his mother (contralto); Alfio, a teamster (baritone); Lola, his wife (mezzo-soprano); peasants; villagers. The setting is a Sicilian village square; the time, the latter part of the nineteenth century.

The voice of Turiddu is heard off stage praising his one-time sweetheart, Lola, who is now Alfio's wife (Siciliana: "O Lola"). It is Easter morning. Villagers are entering the church. Santuzza meets Mamma Lucia and pleads with her to reveal where Turiddu is. Lucia inquires why she is so inquisitive. Santuzza is about to confess that she loves him when Alfio appears, a lusty tune on his lips about his profession as teamster ("Il cavallo scalpita"). When from inside the church the music of the "Regina Coeli" is heard, the villagers assume reverent attitudes and join in the singing. All of them then enter the church, leaving behind only Santuzza and Lucia. It is now that Santuzza can tell Lucia of her love affair with Turiddu ("Voi lo sapete"). Shocked, Lucia rushes into the church to pray for Santuzza. Turiddu now appears, and Santuzza confronts him with his infidelity. They become involved in a bitter quarrel as Turiddu accuses Santuzza of unwarranted jealousy. Their bitter tirades are interrupted by the arrival of Lola. She sings a gay song ("Fior di giaggiolo"), after which the two women exchange harsh words. Lola lightly shrugs off Santuzza's veiled accusations and enters the church. Enraged by this scene, Turiddu curses Santuzza, throws her angrily to the ground, and enters the church. At this critical moment Alfio reappears and Santuzza reveals to him that his wife has been unfaithful. Alfio swears to seek revenge. The Easter services come to an end. The strains of the "Intermezzo" are heard, contrasting the peace of the holiday with the stormy emotions of the principal characters. The villagers file out of the church. Some of the men fill wine glasses as Turiddu sings a rousing drinking song (Brindisi: "Viva il vino spumeggiante"). When Turiddu offers a glass of wine to Alfio the latter turns it down. Insulted, Turiddu challenges Alfio and is accepted. Sensing the approach of doom, Turiddu bids his mother farewell ("Mamma, quel vino è generoso"). His mother tries to follow as he leaves to meet Alfio. Santuzza stops her. Suddenly villagers rush into the square with the dreadful news that Turiddu has been killed.

When *Cavalleria rusticana* won first prize in a contest for one-act operas sponsored by the publishing house of Edoardo Sonzogno, it started a trend in Italian opera known as "verismo" (naturalism). Later examples of verismo operas are *Pagliacci* and *La Bohème*. The première of *Cavalleria rusticana* created a sensation equaled by few other operas. An obscure, impoverished composer, Mascagni suddenly became famous. He took forty curtain calls; outside the theater thousands waited to acclaim him. Before many months passed, parades were held in his honor and medals were sold with his picture. By 1892 *Cavalleria* had been seen throughout Europe and in New York. This single opera made

Mascagni famous and rich. Unfortunately, he was never able to duplicate his first success, though he wrote some creditable operas. His own comment was: "It is a pity I wrote *Cavalleria* first. I was crowned before I became king."

Cavalli, Francesco (born CALETTI-BRUNI), composer. Born Crema, Italy, February 14, 1602; died Venice, January 14, 1676. He was the immediate successor of Monteverdi, and a leader in the Venetian school of opera. He assumed the name of his patron, Federico Cavalli, the Podesta of Crema, who was attracted to the boy and took him to Venice in 1616 where he became a singer in the choir of St. Marks and a pupil of Monteverdi. In 1640 Cavalli was appointed organist at St. Marks, and in 1668, maestro di cappella. Meanwhile, in 1639, he wrote his first opera, *Le nozze di Teti e di Peleo,* a work of historical importance since it was the first to be designated by its composer an opera (specifically, "opera scenica") instead of a "dramma per musica." Cavalli wrote over forty operas, the most famous being *Giasone* (1649) and *Serse* (1654). In 1660 he was invited to Paris to help produce *Serse* in conjunction with the marriage ceremonies of Louis XIV. He returned to Paris two years later to present another of his operas, *Ercole amante.* Cavalli's significance rests in the tunefulness of his arias and his use of the recitativo secco (*see* RECITATIVE).

Cavaradossi, Mario, artist (tenor) in Puccini's *Tosca.*

Cavatina, Italian term (French: **cavatine**) for a melody simpler in style and more songlike than an aria.

Cavatine du page, *see* NOBLES SEIGNEURS, SALUT!

Cebotari, Maria, soprano. Born Kishinev, Bessarabia, February 10, 1910; died Vienna, June 9, 1949. As a child she sang in school and church choirs, and at fourteen appeared as a singer and dancer with a traveling company. After a period of study with Oskar Daniel in Berlin, she joined the Dresden Opera where she made her debut as Mimi in 1930. One year later she scored a major success at the Salzburg Festival, and after that in major opera houses in Germany, at the Vienna State Opera, and at Covent Garden, excelling in such lyric soprano parts as Violetta, Mimi, Butterfly, and in several leading soprano roles in Mozart operas. She died at the peak of her career, a victim of cancer. She made numerous appearances in foreign motion pictures. In 1942 she was heard in the première of Sutermeister's *Romeo und Julia,* and in 1947 in Einem's *Dantons Tod.*

Ce bruit de l'or . . . ce rire, Manon's hymn to gold in Act IV of Massenet's *Manon.*

cecchina, La or **La buona figliuola (The Good Girl),** opera buffa by Nicola Piccinni. Libretto by Carlo Goldoni, based on Samuel Richardson's novel, *Pamela.* Première: Teatro delle Dame, Rome, February 6, 1760. Richardson's story of a servant girl who resists the advances of her master and finally becomes his wife was the source of Piccinni's most famous opera, one of the finest and most successful opera buffas before those of Rossini.

Celeste Aïda, Radames' hymn to Aïda's beauty in Act I, Scene 1, of Verdi's *Aïda.*

Cellini, Benvenuto, Florentine sculptor, goldsmith, author. Born Florence, November 1, 1500; died there, February 14, 1571. He is the central figure in several operas, the most notable being Berlioz' *Benvenuto Cellini.* Others include: *Ascanio* by Saint-Saëns; *Benvenuto* by Eugene Diaz; *Benvenuto Cellini* by Franz Lachner; *Benvenuto Cellini a Parigi* by Lauro Rossi.

cena delle beffe, La (The Feast of the Jest), opera by Umberto Giordano. Libretto is the play of the same name by Sem Benelli. Première: La Scala,

December 20, 1924. In fifteenth century Florence, Gianetto—a physically weak poet and painter—is the victim of cruel jests at the hands of Neri, a captain of mercenaries. When Neri prevents Gianetto's marriage to Ginevra, Gianetto succeeds in triumphing over Neri by means of trickery until the latter goes mad.

Cendrillon (Cinderella), opera by Massenet. Libretto by Henri Cain, based on the fairy tale by Charles Perrault. Première: Opéra-Comique, May 24, 1899. *See also* CENERENTOLA, LA; CINDERELLA.

cenerentola, La (Cinderella), (1) opera by Rossini. Libretto by Jacopo Ferretti, based on the fairy tale by Charles Perrault. Première: Teatro Valle, Rome, January 25, 1817. The magical elements of the Cinderella tale have been omitted in this version. The fairy godmother becomes Alidoro, a practical philosopher employed by the Prince who—disguised as a beggar—receives help from Cinderella after having been rudely turned down by her stepsisters. Though the father of the stepsisters plans to have one of them marry the Prince, Alidoro contrives to have Cinderella become his fortunate bride.

(2) Opera by Wolf-Ferrari. Libretto by Pezze-Pescolato, based on the fairy tale by Charles Perrault. Première: Teatro la Fenice, Venice, February 22, 1900.

See also CENDRILLON; CINDERELLA.

C'en est donc fait et mon coeur va changer (Me sedur han creduto), Marie's lament in Act II of Donizetti's *The Daughter of the Regiment.*

Ceprano, a nobleman (bass) in Verdi's *Rigoletto.*

Cervantes, Miguel de, novelist, dramatist, poet. Born Alcalá de Henares, Spain, 1547; died Madrid, April 23, 1616. He was the author of *Don Quixote.* For operas based on this story, *see* DON QUIXOTE. His story *La Gitanella* was the source of Pius Alexander

Wolff's *Preciosa,* for which Carl Maria von Weber wrote an overture, choruses, melodramas, dances, and a song. Carl Orff's *Astutuli* and Hans Henze's *Das Wundertheater* were derived from Cervantes' *El Teatro Magico.* Other operas drawn from various works of this author include Henri Barraud's *Numance* and Grétry's *Le trompeur.*

Cesti, Marc' Antonio, composer. Born Arezzo, Italy, August 5, 1623; died Florence, October 14, 1669. A major figure in the Venetian school of opera, he studied with Carissimi in Florence and in 1646 became maestro di cappella to Ferdinand II de' Medici. Fourteen years later he was appointed tenor to the chapel choir. During the last three years of his life he was assistant Kapellmeister to Emperor Leopold I of Vienna. Cesti returned to Venice just before his death. He wrote eight operas, the first, *L'Orontea* (1649), being so successful that it was performed in several Italian cities besides Venice. Later works were: *Cesare amante* (1651), *La Dori* (1663), and *Semiramide* (1667). Cesti placed great emphasis on lyricism, filling his operas with gracious, flowing melodies charged with feeling.

C'est ici le séjour (questo sol è il soggiorno), concluding chorus in Meyerbeer's *L'Africaine.*

C'est l'histoire amoureuse, an aria known as the "Laughing Song," found in Auber's *Manon Lescaut.* It is frequently interpolated in the Lesson Scene of Rossini's *The Barber of Seville.*

Chabrier, Emmanuel, composer. Born Ambert, France, January 18, 1841; died Paris, September 13, 1894. He was employed in the Ministry of Interior, following music as an amateur. Influenced by the vogue for Offenbach he wrote his first stage work, a comic opera, *L'Étoile,* successfully performed at the Bouffes Parisiens in 1877. In March, 1880, he took a three-day leave of absence from his government post to attend a performance of *Tristan und Isolde* in Munich. Then and there Chabrier decided to dedicate himself completely to music. A confirmed Wagnerian, he completed an opera, *Gwendoline,* in the Wagnerian style; it was introduced in Brussels in 1886. *Le roi malgré lui,* another comic opera, was introduced by the Opéra-Comique in 1887. His last opera, *Briséis,* was left unfinished.

Chaconne, an old dance in moderately slow ¾ time, presumably of Spanish origin and so similar to another old dance, the passacaglia, that the terms were often used interchangeably. Chaconnes appear frequently in seventeenth and eighteenth century operas. There is a chaconne in Monteverdi's *Orfeo;* Lully, Rameau, and Gluck frequently ended their operas with one. The chaconne in Gluck's *Paride ed Elena* reappears as a passacaglia in *Iphigénie en Aulide.* One of the most affecting chaconnes in opera is Dido's song "When I am laid in earth," in Purcell's *Dido and Aeneas.* Other examples of operatic chaconnes are those in Handel's *Rodrigo,* Lully's *Cadmus et Hermione,* and Mozart's *Idomeneo.*

Chaliapin, Feodor (sometimes **Shaliapin**), bass. Born Kazan, Russia, February 13, 1873; died Paris, April 12, 1938. One of the most celebrated Russian basses and singing actors. The son of a peasant, he was given few opportunities to acquire an education. Without benefit of formal musical training he joined a provincial opera company in 1890. Two years later he studied with Usatov in Tiflis. An engagement followed with the St. Petersburg Opera. In 1896 he joined the company of S. I. Mamontov in Moscow, where he was assigned leading bass roles in Russian operas. He made powerful impressions as Boris Godunov, as Ivan the Terrible in Rimsky-Korsakov's *Maid of Pskov,* and as the miller in Dargomizhsky's

Russalka. His first appearance outside Russia took place in 1901 at La Scala in *Mefistofele.* He made his American debut in the same opera at the Metropolitan Opera on November 20, 1907. While certain facets of his art were appreciated, he was not acclaimed at this time. Indeed, Henry E. Krehbiel found elements of "vulgarity" in his performance. After further successes in Russia and London, Chaliapin returned to the Metropolitan Opera in *Boris Godunov,* on December 9, 1921. This time his success was unqualified; Krehbiel could now say that only the actor Salvini was Chaliapin's equal. Chaliapin remained at the Metropolitan Opera eight seasons. Roles for which he was famous included Boris, Don Basilio, Don Quixote (in Massenet's *Don Quichotte*), Leporello, and Mephistopheles (in Boïto's *Mefistofele*). Toward the end of his life he appeared as Don Quixote in a motion picture. He wrote two books of memoirs: *Pages from My Life* (1926), and *Man and Mask* (1932).

Chamber opera, an opera of modest proportions and intimate character, calling for limited forces and stage paraphernalia. Richard Strauss's *Ariadne auf Naxos* and Pergolesi's *La serva padrona* are examples from different periods.

Champs paternels, Joseph's aria in Act I of Méhul's *Joseph.*

Chanson bachique, *see* O VIN DISSIPE LA TRISTESSE.

Chanson de la puce (Song of the Flea), Mephistopheles' aria in Act II of Berlioz' *The Damnation of Faust.*

Chanson huguenote, *see* POUR LES COUNVENTS C'EST FINI.

Chanson hindoue, *see* SONG OF INDIA.

Charfreitagszauber, *see* GOOD FRIDAY SPELL.

Charlotte, young woman (soprano) in love with Werther in Massenet's *Werther.*

Charlottenburg Opera (Deutsches Opernhaus), a leading Berlin opera house, built by the municipal council of the Charlottenburg district in 1911–1912. It opened on November 9, 1912, with *Fidelio,* conducted by Ignaz Waghalter. During World War I the opera house went into an artistic decline and after the war it was temporarily closed. In 1925 the reorganized company opened as the Städtische Oper with a performance of *Die Meistersinger* conducted by its new musical director, Bruno Walter. Heinz Tietjen was intendant until 1930, Carl Ebert between 1931 and 1933, and Wilhelm Rode from 1934 until the theater was finally closed. Bruno Walter left in 1929; succeeding musical directors were Robert Denzler, Fritz Stiedry, Arthur Rother, and Leopold Ludwig. The building was destroyed by a bomb during World War II. Notable premières given here included Julius Bittner's *Mondnacht,* Franz Schreker's *Der Schmied von Ghent,* and Kurt Weill's *Die Bürgschaft.*

Charmant oiseau, Mysoli's aria in David's *La perle du Brésil.*

Charpentier, Gustave, composer. Born Dieuze, France, June 25, 1860. The librettist and composer of *Louise,* an opera which provoked controversy and won popularity by reason of its novel working-class atmosphere. Charpentier attended the Lille Conservatory, where he won several prizes. Entering the Paris Conservatory in 1881, he won the Prix de Rome in 1887. During his stay in Rome he wrote his first successful work for orchestra, *Impressions of Italy.* After his return to Paris he became interested in socialism and wrote songs with a political viewpoint. The writing of *Louise* took him ten years. It was introduced at the Opéra-Comique on February 2, 1900. Charpentier's only subsequent major work was a sequel to *Louise* entitled *Julien.* Produced at the Opéra-Comique on June 3, 1913, it proved a failure.

Charton-Demeur, Anne, *see* DEMEUR, ANNE ARSÈNE.

Chartreuse de Parme, La (The Carthusian Monastery), opera by Henri Sauget. Libretto by Armand Lunel, based on the novel of the same name by Stendhal. Première: Paris Opéra, March 16, 1939. The opera traces the career of Fabrice del Dongo through wars, political intrigues, and imprisonment until he becomes a priest and enters a monastery.

Chaucer, Geoffrey, (1) poet. Born England about 1340; died there, 1400. Material from his narrative, the *Canterbury Tales,* was used in two operas named *The Canterbury Pilgrims,* one by Reginald de Koven, the other by Charles Villiers Stanford. His poem, *Troilus and Cressida,* was the source of an opera of the same name by William Walton. Voltaire's *La Fée Urgèle* —made into operas of the same name by Egidio Duni and Ignaz Pleyel—was derived from Chaucer.

(2) The poet (baritone), principal character in Reginald De Koven's *The Canterbury Pilgrims.*

Che farò senza Euridice? Orfeo's lament at the loss of his wife, Euridice, in Act III (Act IV in some versions) of Gluck's *Orfeo ed Euridice.*

Che gelida manina, Rodolfo's narrative in Act I of Puccini's *La Bohème.*

Chekhov, Anton, story writer and dramatist. Born Taganrog, Russia, January 17, 1860; died Badenweiler, Germany, July 2, 1904. One of the most distinguished authors of his time, Chekhov is represented in the world of opera by the following works drawn from his writings: Pierre-Octave Ferroud's *Le Chirugie;* Constantine Nottara's *Over the Highway;* and Henri Sauget's *La Contrebasse.*

Ch'ella mi creda libero, Dick Johnson's aria in Act III of Puccini's *The Girl of the Golden West.*

Chénier, *see* ANDREA CHENIER.

Che puro ciel! Orfeo's description of the beauties of Elysium in Act II of Gluck's *Orfeo ed Euridice.*

Cherubini, Maria Luigi, composer. Born Florence, September 14, 1760; died Paris, March 15, 1842. Though trained in the Italian school, he was a dominant figure in the development of French opera. He began studying music with his father, and between 1773 and 1777 wrote several masses and similar works. The Duke of Tuscany became interested in him and supported his study with Giuseppe Sarti. Cherubini wrote his first opera, *Il Quinto Fabio,* in 1780, performed three years later. His first success came with *Armida,* in 1782. After writing five more operas, all in the strict Italian pattern of the time, Cherubini went to London where he remained two years, producing two more operas without success and serving as composer to the king. In 1780 he settled permanently in Paris. Becoming dissatisfied with the Italian traditions, he aspired to write operas in Gluck's style, in which drama and musical resources were emphasized. His first work in this new vein, *Démophon,* was a failure in 1788. But in *Lodoïska,* performed in Paris in 1791, his new style became effective. Between 1794 and 1801 he wrote half a dozen operas, including the work generally deemed his finest: *Les deux journées* (known in German as *Der Wasserträger,* in English as *The Water Carrier*), first heard in Paris in 1800. Cherubini suffered severe mental depressions after the French Revolution, due primarily to his unhappy marriage and the failures of some of his operas. In 1805 he was invited to Vienna in conjunction with the Viennese première of *Les deux journées.* For production here, Cherubini wrote a new opera, *Faniska,* an outstanding success. Back in Paris, and out of favor with Napoleon, Cherubini devoted himself more to church music than to opera and concentrated on his teaching duties at the

Conservatory. From 1822 until just before his death he was one of the Conservatory's directors.

Cherubini's most famous operas: *Armida* (1782); *Adriano in Siria* (1782); *Alessandro nell' Indie* (1784); *Ifigenia in Aulide* (1787); *Lodoïska* (1791); *Elisa* (1794); *Médée* (1797); *La punition* (1799); *La prisonnière* (1799); *Les deux journées* (1800); *Anacréon* (1803); *Faniska* (1806); *Les Abencérages* (1813); *Bayard à Mézières* (1814)—a collaboration with François Boiedieu, Charles Catel, and Nicolo Isouard.

Cherubino, Count Almaviva's page (soprano) in Mozart's *The Marriage of Figaro.*

Che soave zeffiretto, the letter duet of Susanna and Countess Almaviva in Act III of Mozart's *The Marriage of Figaro.*

Che vita maledetta, Despina's aria in Act I, Scene 3, of Mozart's *Così fan tutte.*

Che volo d'augelli! Nedda's ballatella in Act I of Leoncavallo's *Pagliacci.*

Chicago Opera Company, The, a company organized in Chicago, Illinois, in 1910 with members of the then recently disbanded Oscar Hammerstein Opera Company of New York. Under the artistic direction of Andreas Dippel, and with Cleofonte Campanini as prinicpal conductor, the company opened on November 3, 1910, with *Aïda.* The following evening Mary Garden appeared in *Pelléas et Mélisande,* and within a few weeks as Louise and Salome. For the next twenty years the personality of Mary Garden dominated the company. With funds provided by social leaders of Chicago, it was able to maintain a standard of artistic excellence found otherwise only at the Metropolitan Opera. The repertory included revivals and premières. When *The Jewels of the Madonna* received its first American performance in 1912, its composer, Ermanno Wolf-Ferrari,

was invited to supervise the production. In 1922, Serge Prokofiev led the world première of his *The Love for Three Oranges.* Several American operas received their first performances, including Henry Hadley's *Azora* in 1917 (the composer conducting) and Charles Wakefield Cadman's *A Witch of Salem* in 1926. Karl Goldmark's *The Cricket on the Hearth* and Alfredo Catalani's *Loreley* were other operas given American premières in 1912 and 1918, respectively. When Campanini died in 1919, the musical direction passed to Gino Marinuzzi. In January, 1921, Mary Garden was appointed artistic director (the first time a woman was made head of a major opera house). She spent money with a lavish hand and under her regime the company suffered a deficit of over a million dollars. Two of the most important sponsors, Mr. and Mrs. Harold McCormick, withdrew their support. In the reorganization that followed the name of the organization was changed to the Chicago Civic Opera Company. The Chicago industrialist, Samuel Insull, became head of a group of businessmen guaranteeing an annual fund of $500,-000. Mary Garden returned to her original status as prima donna and the direction of the company passed to Giorgio Polacco. Under Polacco's regime, the company maintained a high position. Singers bound to it by exclusive contract now included Mary Garden, Lotte Lehmann, Frida Leider, Claudia Muzio, Rosa Raisa, Alexander Kipnis, and Tito Schipa. In 1929 a new house was built, the magnificent thirty-million-dollar Chicago Opera House on Wacker Drive. The company failed in 1932 because of diminished financial support. Thereafter, other companies tried to fill the gap. The most recent, the Chicago Lyric Theatre, opened November 1, 1954, with *Norma.*

Chi del gitano i giorni abbella? The

second part of the Anvil Chorus in Act II, Scene 1, of Verdi's *Il trovatore*.

Children's operas. A number of operas have been written for audiences of children, many of them intended for performance by children: Wheeler Beckett's *The Magic Mirror* (based on *Snow White*); Nicolai Berezowsky's *Babar;* Benjamin Britten's *Let's Make an Opera;* Aaron Copland's *The Second Hurricane;* Arnold Franchetti's *The Lion;* Engelbert Humperdinck's *Hänsel und Gretel;* Eduard Poldini's *Aschenbrödl, Dornröschen,* and *Die Knuspershexe;* Francesco Pratella's *La Ninna nanna della Bambola;* Vladimir Rebikov's *The Christmas Tree;* Alfred Soffredini's *Il piccolo Haydn;* Kurt Weill's *Der Jasager;* Alec Wilder's *Chicken Little.*

Children's Prayer, *see* ABENDS, WILL ICH SCHLAFEN GEH'N.

Chillingworth, Roger, Roger Prynne's assumed name in Damrosch's *The Scarlet Letter.*

Chi mai fra gli inni e i plausi, chorus of the slave girls in Act II, Scene 1, of Verdi's *Aïda.*

Chimène, Count de Gormas' daughter (soprano) in Massenet's *Le Cid.*

Chi mi frena? The celebrated sextet of Lucia, Edgardo, Alisa, Arturo, Raimondo, and Enrico in Act II of Donizetti's *Lucia di Lammermoor.*

Chi vide mai a bimbo, Cio-Cio-San's aria in Act II of Puccini's *Madama Butterfly.*

Choéphores, Les (The Libation Bearers), opera by Darius Milhaud. Libretto by Paul Claudel based on the tragedy of Aeschylus. Première: Théâtre de la Monnaie, March 27, 1935. This is the second opera in a trilogy entitled *Oreste,* the first of which is *Agamemnon* and the last, *Les Euménides.* In this drama, Orestes avenges the death of his father by returning to Argos and killing his mother, Klytemnestra, and her lover, Aegisthus. *See* AESCHYLUS; ORESTE.

Chorus, a body of singers singing ensemble music. In operas of the sixteenth and seventeenth centuries the chorus was utilized extensively to render musical numbers, but it was not required to be an actual part of the drama. Gluck was one of the first composers to make the chorus an indispensable part of the dramatic action.

Chorus of the bells, *see* DIN, DON, SUONA VESPERO.

Chorus of the Levites, chorus in Act I of Verdi's *Nabucco.*

Chorus of the Swords, *see* DE L'ENFER QUI VIENT EMOUSSER.

Christmas Eve, *see* GOGOL, NIKOLAI.

Christophe Colomb, opera by Darius Milhaud. Libretto by Paul Claudel. Première: Berlin State Opera, May 5, 1930. The life of Columbus is told in a philosophical allegory, mystical and religious in tone. An innovation here is the use of motion pictures to supplement the stage action. *See also* COLUMBUS.

Chrysis, Demetrios' beloved (soprano) in Camille Erlanger's *Aphrodité.*

Chrysothemis, Elektra's sister (soprano) in Richard Strauss's *Elektra.*

Cicillo, a Camorrist (tenor) in Wolf-Ferrari's *The Jewels of the Madonna.*

Cid, Le, opera by Massenet. Libretto by Adolphe d'Ennery, Louis Gallet, and Edouard Blau, based on the historical drama by Pierre Corneille. Première: Paris Opéra, November 30, 1885. The central characters are Rodrigo, called Le Cid (the Conqueror), and Chimène. Chimène's father, a Spanish nobleman, is killed in a duel by the Cid. She demands vengeance, but when King Ferdinand, elated at the news of Rodrigo's victory over the Moors, directs Chimène to pronounce the death sentence, she loses heart and instead embraces the forgiven conqueror. The ballet music in Act II is famous.

Cieco, Iris' blind father (bass) in Mascagni's *Iris.*

Cielo e mar! Enzo's song of praise to sky and sea in Act II of Ponchielli's *La Gioconda.*

Cilèa, Francesco, composer. Born Palmi, Italy, July 26, 1866; died Verazza, Italy, November 20, 1950. He wrote and produced his first opera, *Gina,* in 1889, while still a student at the Naples Conservatory. It brought him a commission from Edoardo Sonzogno, the publisher, to write *La Tilda,* produced in Florence in 1892. Four years later Cilèa's *L'Arlesiana* was introduced in Milan. His next work, *Adriana Lecouvreur,* was first performed in 1902. His last opera, *Gloria,* appeared in 1907. From 1896 to 1904 he was professor of composition at the Musical Institute in Florence. From 1913 to 1916 he was director of the Palermo Conservatory, and after 1916 of the Majella Conservatory. About 1930 Cilèa rewrote his opera *Gloria;* this revised version was performed at La Scala in 1932.

Cimarosa, Domenico, composer. Born Aversa, Italy, December 17, 1749; died Venice, January 11, 1801. He wrote one of the most celebrated opera buffas before those of Rossini, *Il matrimonio segreto.* For eleven years he attended the Conservatorio Santa Maria di Loreto in Naples, where his teachers included Antonio Sacchini and Nicola Piccinni. His first opera, *Le stravaganze del conte,* was produced in Naples in 1772. His next, *La finta Parigina,* was a major success. During the next two decades he lived alternately in Rome and Naples, writing operas for both cities. In 1787 he was invited to Russia by Catherine II where he served as her court composer. He stayed there three years and wrote three operas. Replaced by Giuseppe Sarti in 1792, he went to Vienna where he succeeded Antonio Salieri as court kapellmeister. It was here that he wrote the work for which he is known today, *Il matrimonio segreto,* introduced Feb-

ruary 7, 1792, with formidable success. When Emperor Leopold II died, Cimarosa left Vienna and returned to Naples in 1793 to become maestro di cappella to the king and teacher of the royal children. When the French Republican army entered Naples in 1799 Cimarosa openly expressed his sympathy for the invaders. For this he was imprisoned and sentenced to death; only his great popularity saved his life. He was finally pardoned by King Ferdinand, on the condition that he leave Naples for good. Broken in health and spirit, he collapsed and died in Venice while en route to St. Petersburg. After *Il matrimonio segreto* his most popular operas were: *La finta Parigina* (1773); *L'Italiana in Londra* (1778); *Artaserse* (1784); *La ballerina amante* (1782); *Cleopatra* (1791); *L'amante disperato* (1795); *Penelope* (1794); *Semiramide* (1799).

Cinderella, a fairy tale by Charles Perrault (1628–1703), the subject of many operas. A few are: Leo Blech's *Aschenbrödl;* Massenet's *Cendrillon;* Eduard Poldini's *Aschenbrödl;* Rossini's *La cenerentola;* and Wolf-Ferrari's *La cenerentola.*

Cio-Cio-San, a geisha girl (soprano) in love with Pinkerton in Puccini's *Madama Butterfly.*

Claggert, John, master-at-arms (bass) in Britten's *Billy Budd.*

Claque, a group of people engaged either to applaud a performer or a performance, and thereby stimulate the audience into audible signs of appreciation, or to voice disapproval and thus create a disturbance. A claque is most often engaged by an individual singer who instructs it as to when an ovation is to be encouraged. Sometimes opera houses employ claques. The claque originated in France in 1820 when two enterprising Frenchmen organized the *Assurance des succès dramatiques.* The idea took hold immediately; it flourished during the age of Meyerbeer. In the middle of the 19th century claques sprouted in most Italian opera houses. They were frequently employed not only to arouse interest in an opera or a singer, but for political purposes, since many operas had texts that could be interpreted as political propaganda. In 1919 the London *Musical Times* reported the fees paid to members of an Italian claque: five lire for each "interruption of bravo" to fifty lire for a "bis [encore] at any cost"; and a special sum for "wild enthusiasm." Claques have also been employed in English and American opera houses.

Clara, a young mother (soprano) in Gershwin's *Porgy and Bess.*

Claudel, Paul, poet, dramatist, and diplomat. Born Villeneuve-sur-Fère, France, August 6, 1868; died Paris, February 23, 1955. He wrote the texts for several French operas, including Honegger's *Jeanne d'Arc au bûcher,* Milhaud's trilogy *Oreste,* and Milhaud's *Christophe Colomb.*

Claudius, king of Denmark (baritone) in Thomas's *Hamlet.*

Claussen, Julia, mezzo-soprano. Born Stockholm, June 11, 1879; died there May 1, 1941. Her musical education took place at the Royal Academy of Music in Stockholm and the Royal Academy of Music in Berlin. She made her opera debut on January 19, 1903, in Stockholm in *La favorita.* She remained with the Stockholm Opera Company for nine years. In 1913 she appeared with the Chicago Opera, after which she made successful appearances at Covent Garden and in Paris. On November 23, 1917, she made her debut at the Metropolitan Opera as Delilah. She remained with the Metropolitan until 1932, then went into retirement.

Clément, Edmond, tenor. Born Paris, March 28, 1867; died Nice, February 23, 1928. He attended the Paris Conservatory, after which he made a successful debut at the Opéra-Comique, on November 29, 1889, in Gounod's *Mireille.* For the next twenty years he was the principal tenor of the Opéra-Comique, where he created the leading tenor roles in many French operas including Bruneau's *L'attaque du moulin,* Erlanger's *Le Juif polonais,* Godard's *La Vivandière,* Hahn's *L'ile de rêve,* and Saint-Saëns' *Phryné* and *Hélène.* On December 6, 1909, he made his first appearance at the Metropolitan Opera in *Manon.* Between 1911 and 1913 he was a member of the Boston Opera Company. His art, both as actor and singer, was marked by restraint and understatement.

clemenza di Tito, La (The Clemency of Titus), opera by Mozart. Libretto by Caterino Mazzola, adapted from a libretto by Metastasio. Première: National Theater, Prague, September 6, 1791. This was Mozart's last opera, written in the year of his death for the coronation of Emperor Leopold II as King of Bohemia. The central character is a former tyrant grown benevolent. Nevertheless, Vitellius and Sextus conspire to overthrow him. When Sextus sets fire to the palace, it is believed Titus is killed in the flames. However, he has been saved, and he magnanimously forgives the traitors. Gluck also wrote an opera of the same title, using the Metastasio libretto.

Cleopatra, Egyptian queen celebrated for her beauty, the central character in several operas including: Cimarosa's *Cleopatra;* Graun's *Cleopatra e Cesare;* Hadley's *Cleopatra's Night;* Massé's *Une Nuit de Cléopâtre;* Massenet's *Cléopâtre;* and Mattheson's *Cleopatra.*

Cleopatra's Night, opera by Henry Hadley. Libretto by Alice Leal Pollock, adapted from Gautier's story *Une Nuit de Cléopâtre.* Première: Metropolitan Opera, January 31, 1920. The story concerns the surrender of Cleopatra to Meïamoun for one night in return for his willing death the following morning.

Cléophas, the name assumed by Joseph in Egypt in Mehul's *Joseph.*

Clitandro, Lisetta's lover (tenor) in Wolf-Ferrari's *L'amore medico.*

Cloak, The, *see* IL TABARRO.

Clotilda, Norma's confidante (soprano) in Bellini's *Norma.*

Clytemnestra, *see* KLYTEMNESTRA.

Coates, Albert, conductor and composer. Born St. Petersburg, Russia, April 23, 1882; died Capetown, South Africa, December 11, 1953. In 1902 he entered the Leipzig Conservatory where he studied with Nikisch, whose assistant he later became at the Leipzig Opera. Coates conducted his first opera, *The Tales of Hoffmann,* when Nikisch fell ill. In 1906, Coates became the principal conductor of the Elberfeld Opera where he directed over forty operas. After conducting operas in Dresden, Mannheim, St. Petersburg, and London, he was engaged by Sir Thomas Beecham as codirector and conductor at Covent Garden for the first postwar season. He was the first British conductor to conduct at the Paris Opéra. In 1920 he visited the United States for the first time, and three years later he was appointed the musical director of the Rochester Philharmonic Orchestra. In 1928 and 1929 he led opera performances in Italy, including at La Scala. For a five-year period he was in charge of a two-month opera season in Barcelona. After that he appeared as guest conductor in most of the leading opera houses of Europe, gave successful opera performances in the Soviet Union, and was director of the British Opera season at Covent Garden. He wrote several operas including: *Sardanapalus* (1916), *Samuel Pepys* (1929), and *Pickwick* (1936).

Cobblers Song (Schusterlied), *see* JERUM! JERUM!

Cochenille, Spalanzani's servant (tenor) in Offenbach's *The Tales of Hoffmann.*

Cocteau, Jean, novelist, playwright, and poet. Born Maisons-Lafitte, France, July 5, 1891. A friend of the French composers known as the "Six," he frequently prepared texts and poems for their works. These include the librettos for Honegger's *Antigone* and Milhaud's *Le pauvre Matelot.* He also wrote the text for Stravinsky's *Oedipus Rex.* His play *Les mariés de la Tour d'Eiffel (The Wedded Pair of the Eiffel Tower)* has been made into an opera by Lou Harrison.

Colas Breugnon, opera by Dimitri Kabalevsky. Libretto by V. Bragin, based on the novel of the same name by Romain Rolland. Première: Leningrad State Opera, February 22, 1938. Colas Breugnon is a Burgundian craftsman of the 16th century who approaches every problem of life with laughter. In Kabalevsky's opera, Breugnon is used as a symbol criticizing the social customs and economic problems of the sixteenth century. The score is filled with French folksongs, particularly those originating in Burgundy. The overture is frequently performed.

Colbran, Isabella (soprano). Born Madrid, February 2, 1785; died Bologna, Italy, October 7, 1845. She was the first wife of Rossini and a celebrated prima donna. After studying with Girolano Crescentini in Italy she made her bow in opera in Paris in 1801. Six years later she made a successful debut at La Scala. In 1811, Domenico Barbaja engaged her for his opera company in Naples where she had a triumph in Paisiello's *Nina.* Four years later Rossini wrote *Elisabetta* for her. He fell in love with her and she soon deserted Barbaja (whose mistress she had been) to live with the composer; they were married on March 15, 1822. Eventually they were estranged, and Rossini deserted her for Olympe Pélissier, whom he married two years after Colbran's death.

Colline, philosopher (bass) in Puccini's *La Bohème.*

Colomba, *see* MERIMEE, PROSPER.

Coloratura, Italian for "colored." The term is used in vocal music to denote passages highly ornamented with runs and figures. A coloratura soprano is one who specializes in such music.

Columbus, Christopher, the central character in operas by Ramón Carnicer, Werner Egk, Alberto Franchetti, Darius Milhaud, and Sergei Vassilenko.

Combien tu m'es chère, Vasco da Gama's aria in Act II of Meyerbeer's *L'Africaine.*

Com' è gentil, Ernesto's serenade in Act III, Scene 2, of Donizetti's *Don Pasquale.*

Come in quest' ora bruna, Amelia's aria in Act I, Scene 1, of Verdi's *Simon Boccanegra.*

Come scoglio, Fiordiligi's aria in Act I, Scene 3, of Mozart's *Così fan tutte.*

Come un bel dì di maggio, Chénier's aria in Act IV of Giordano's *Andrea Chénier*

Comic opera, a general term for any musico-dramatic work of nonserious nature. For particular forms of comic opera, *see* OPERA BOUFFE and OPERA BUFFA. The term opéra comique (which see) means something quite different.

Comme autrefois dans la nuit sombre, Leila's cavatina in Act II of Bizet's *Les pêcheurs de perles.*

Commedia per musica, "comedy through music," a term prevailing in 18th century Naples to designate comic operas.

Comme une pâle fleur, Hamlet's arioso in Act V of Thomas's *Hamlet.*

Comte Ory, Le (Count Ory), opera buffa by Rossini. Libretto by Eugène Scribe and Delestre-Poirson. Première: Paris Opéra, August 20, 1828. This is the last but one of Rossini's operas, and one of the two he wrote for production in Paris. The central character is a licentious count who does not hesitate to employ any means at his command to win young ladies. In his attempt to seduce Countess Adele he assumes various disguises, including those of a hermit and a nun, and is finally exposed for the fraud that he is.

Comus, *see* MASQUE.

Concetta, a Camorrist (soprano) in Wolf-Ferrari's *The Jewels of the Madonna.*

Concitato, Italian for "agitated." In opera the term has been used for a kind of recitative employed by Monteverdi and some of his contemporaries.

Connais-tu le pays? Mignon's aria in Act I of Thomas's *Mignon.*

Conried, Heinrich (born Cohn), impresario. Born Bielitz, Austria, September 13, 1855; died Meran, Germany, April 26, 1909. He was the general manager of the Metropolitan Opera for five years. He began his career in the theater as an actor in the Vienna Burgtheater, and with various traveling troupes. In 1887 he was appointed director of the Bremen Stadttheater. In 1888 he came to the United States and became manager of the Germania Theater. He held other managerial posts in this country with various theatrical and opera companies, including the Thalia, Casino, and Irving Place theaters. In 1903 he succeeded Maurice Grau as manager of the Metropolitan Opera, organizing the Heinrich Conried Opera Company and directing opera performances there for five years. He proved to be an astute and farsighted impresario, combining a feeling for good showmanship with high ideals (*see* METROPOLITAN OPERA). He was the recipient of numerous honors from foreign countries, including the Order of the Crown from the German Emperor, the Cross of Knighthood from the Austrian Emperor, and the Order of the Crown from the King of Italy.

Constanza, Belmonte's beloved (soprano) in Mozart's *The Abduction from the Seraglio.*

Constanze! dich wiederzusehen! Belmonte's aria in Act I of Mozart's *The Abduction from the Seraglio.*

Consuelo, a novel about Chopin by George Sand which was adapted for operas by Vladimir Kashperov, Giacomo Orefice, and Alfonso Rendano, all three entitled *Consuelo.*

Consul, The, opera by Gian-Carlo Menotti. Libretto by the composer. Première: New York City, March 15, 1950 (after out-of-town tryouts). The action takes place in the present in an unspecified country. Magda tries desperately to get a visa out of a police state, is enmeshed in the red tape of a dictatorial regime, and finds escape in suicide. Originating in a Broadway theater, rather than in an opera house, *The Consul* became a triumph of the 1950 season, receiving the Pulitzer Prize, the Drama Critics' Award, and enjoying prosperity at the box office.

contes d'Hoffmann, Les, *see* TALES OF HOFFMANN, THE.

Contratador dos diamantes (The Diamond Merchant), opera by Francisco Mignone. Libretto by the composer. Première: Teatro Municipal, Rio de Janeiro, September 20, 1924. This opera, whose setting is eighteenth century Brazil and whose theme is the exploitation in Brazil's diamond mines, was written in the Italian style and tradition. It contains a brilliant Afro-Brazilian dance, a Congada, in Act II. This dance has become one of the composer's best known pieces.

Contralto, *see* ALTO.

Converse, Frederick Shepherd, composer. Born Newton, Massachusetts, January 5, 1871; died Boston, June 8, 1940. He received his musical training at Harvard University, with George Chadwick in Boston, and at the Munich Conservatory in Germany. After returning to the United States he taught harmony at the New England Conservatory. In 1904 he became a professor of composition at Harvard. After 1907 he devoted himself exclusively to composition. His opera *The Pipe of Desire,* produced at the Metropolitan

Opera in 1910, was the first American opera ever performed there. Later operas: *The Sacrifice* (1911); *Sinbad the Sailor* (1913); *The Immigrants* (1914).

Convien partir, o miei compagni d'arme, *see* IL FAUT PARTIR, MES BONS COMPAGNONS.

Cooper, Emil, conductor. Born Kherson, Russia, December 20, 1877. Of Russian parentage, he attended the Vienna Conservatory and made his conducting debut in 1896 in Odessa. In 1900 he directed opera performances at the Kiev Opera. Four years later he was given a similar post with the Zimin Opera in Moscow. When Feodor Chaliapin and a Russian company visited London and Paris with *Boris Godunov,* Cooper was the conductor. After the revolution in Russia, Cooper helped found the Leningrad Philharmonic Orchestra. In 1923 he made a world tour as conductor of opera and symphonic performances. In 1929 he was the principal conductor of the Chicago Civic Opera. In 1934 he conducted special performances of Russian operas at La Scala, and in 1944 he joined the conductorial staff of the Metropolitan Opera, where he made his debut on January 26 with *Pelléas et Mélisande.* Leaving the Metropolitan in 1950, Cooper became principal conductor of the Montreal Opera Guild.

Copland, Aaron, composer. Born Brooklyn, New York, November 14, 1900. His music study took place with Victor Wittgenstein and Rubin Goldmark, at the American Conservatory at Fontainebleau, and privately with Nadia Boulanger in Paris. After returning to America in 1924 he came to prominence with several orchestral works, performed by major organizations. He wrote his first opera, *The Second Hurricane,* in 1937, as a "play opera" for performance by high school children. Much more ambitious was his second opera, *The Tender Land,* introduced by the New York City Opera Company in 1954. Copland's later works are strongly influenced by American folk music.

Coppelius, a magician (his other personalities are Dr. Miracle and Dapertutto) (baritone) in Offenbach's *The Tales of Hoffmann.*

coq d'or, Le (The Golden Cockerel), opera by Nikolai Rimsky-Korsakov. Libretto by Vladimir Bielsky, based on a tale by Alexander Pushkin. Première: Moscow, October 7, 1909. The golden cockerel has a gift for prophecy. It is given as a gift to King Dodon by his astrologer. When the cockerel crows, it is a sign of imminent danger. Danger comes to the astrologer, who is killed by Dodon when he insists upon payment for his cockerel; and it comes to Dodon himself, who is killed by the avenging cockerel. The "Hymn to the Sun" of the Queen of Shemakha in Act II, "Salut à toi, soleil," is famous.

Corneille, Pierre, dramatist and poet. Born Rouen, France, June 6, 1606; died Paris, October 1, 1684. His dramas *Le Cid, Polyeucte,* and *Robert Devereux* were adapted into operas by many composers. These operas include: Cornelius' *Der Cid;* Donizetti's *Poliuto,* and *Roberto Devereux;* Gounod's *Polyeucte;* Handel's *Flavio* (based partly on *Le Cid*); Massenet's *Le Cid;* Mercadante's *Roberto Devereux;* Sacchini's *Il Gran Cid;* and Johan Wagenaar's *The Cid.* Pierre Corneille's younger brother, Thomas, was the author of the drama *Médée,* used for operas of the same name by Marc-Antoine Charpentier and Luigi Cherubini, among others.

Cornelius, Peter, composer. Born Mainz, Germany, December 24, 1824; died there October 26, 1874. His early musical education was haphazard, since he aspired to be an actor. An unsuccessful stage debut turned him to music. He went to Berlin in 1845 where he studied counterpoint with Siegfried Wilhelm Dehn and wrote his first major works. During a visit to Weimar in 1852 he met Franz Liszt, who became interested in him. Henceforth Cornelius was a Liszt disciple as well as a champion of Wagner. It was in Weimar that he wrote his best-known work, the comic opera, *The Barber of Bagdad.* Its first performance was directed by Liszt on December 15, 1858. The opera was a failure, primarily because there was an organized cabal in the city against Liszt. It was due to the failure of this opera that Liszt decided to resign his post as kapellmeister. Cornelius became a close friend of Wagner in 1859. His creative achievements at this time included a second opera, *Der Cid,* performed in Weimar in 1865. A third opera, *Gunlöd,* was left unfinished at his death; it was completed by Eduard Lassen and produced in Weimar in 1891.

Coro delle campane, *see* DIN, DON, SUONA VESPERO.

Coronation March, the march in Act IV of Meyerbeer's *Le Prophète.*

Coronation of Poppea, The, *see* L' IN-CORONAZIONE DI POPPEA.

Coronation Scene, the coronation of Boris Godunov in Act I, Scene 3, of Mussorgsky's *Boris Godunov.*

Corps de Ballet, a group of dancers, or a ballet company, attached to an opera house.

Corregidor, Der (The Magistrate), opera by Hugo Wolf. Libretto by Rosa Mayreder-Obermayer based on Pedro Antonio de Alarcón's novel, *The Three-cornered Hat.* Première: Mannheim, June 7, 1896. The magistrate, Don Eugenio di Zuniga, pursues the lovely Frasquita. On one occasion he comes to her door soaked to the skin, having just fallen into a pond. When Frasquita threatens him with a rifle, he falls into a faint and he must be put to bed. Lucas, husband of Frasquita, finds the magistrate in his bed and becomes convinced that his wife has been unfaithful. He puts on the magistrate's clothes.

now dry, and goes forth to make advances to the magistrate's wife. Both Lucas and the magistrate receive sound thrashings in the confusion that follows, with the result that each wisely decides to confine his lovemaking to his own home. The same story is the basis of Manuel de Falla's ballet, *The Three-cornered Hat.*

Corsair, The, see BYRON.

Corsi, Jacopo, member of the Florentine group that created opera. Born Celano, Italy, about 1560; died Florence about 1604. It was at Corsi's palace that the first operas, by a member of the Camerata (which see), were performed: Peri's *Dafne* and *Euridice* (1597 and 1600 respectively). Corsi not only assisted in the performances of these first operas, playing the harpsichord, but also contributed some songs to *Dafne;* these are the only numbers to survive from that opera.

Così fan tutte (So Do They All), comic opera in two acts by Mozart. Libretto by Lorenzo da Ponte. Première: Burgtheater, January 26, 1790. American première: Metropolitan Opera House, March 24, 1922.

Characters: Fiordiligi, a lady of Ferrara (soprano); Dorabella, her sister (soprano or mezzo-soprano); Despina, their maid (soprano); Ferrando, an officer in love with Dorabella (tenor); Guglielmo, officer in love with Fiordiligi (baritone); Don Alfonso, an old bachelor (bass); soldiers; servants; musicians; boatmen; guests. The setting is Naples in the 18th century.

Act I, Scene 1. A café. While discussing women in general, Ferrando and Guglielmo express the conviction that their respective sweethearts, Dorabella and Fiordiligi, are devoted to them. Their friend Don Alfonso is a cynic. He wagers the soldiers that if their sweethearts were courted by other men they would be unfaithful. To win the wager, the soldiers must follow his instructions to the letter for twenty-four hours. The wager is accepted. Confident of the outcome, the soldiers drink a toast with Don Alfonso.

Scene 2. The garden of Fiordiligi's and Dorabella's villa. Dorabella and Fiordiligi ecstatically sing of the men they love ("Ah guarda, sorella"). Their idyllic mood is shattered when Don Alfonso arrives with sad news: Ferrando and Guglielmo have been recalled to their troops. The two soldiers come to say farewell ("Al fato dan legge"). After the soldiers depart, the sisters withdraw and Don Alfonso expresses his cynicism regarding the fidelity of women.

Scene 3. An anteroom in the house of Fiordiligi and Dorabella. Despina complains about the lot of a lady's maid ("Che vita maledetta"). Dorabella now enters and bewails her unhappy state ("Smanie implacibili"). Despina is unsympathetic; to her, all men are philanderers ("In uomini, in soldati"). She becomes a willing ally when Don Alfonso seeks her aid in duping her mistresses. Ferrando and Guglielmo appear, disguised as Albanian noblemen. At first, the ladies are cold; Fiordiligi protests her devotion to Guglielmo ("Come scoglio"). Guglielmo continues to woo her ardently ("Non siate ritrosi") only to be again rejected. When the ladies depart, the men express their delight, for their women have proved true. Ferrando even grows sentimental about the course of true love ("Un' aura amorosa"). After he and Guglielmo leave, Don Alfonso and Despina confer about the next move.

Scene 4. Once again in their garden, the two ladies are lamenting the absence of their lovers ("Ah! che tutta in un momento") when Ferrando and Guglielmo, still disguised, burst in upon them. They say they are ready to die for love and have taken poison. As they go through "death pangs," the women attend them solicitously. A doctor appears—none other than Despina in disguise. Muttering incantations and flourishing a magnet ("Questo e quel pezzo"), the "doctor" revives the stricken men who then proceed to make love to the women more ardently than ever.

Act II, Scene 1. Within the sisters' villa. Despina tries to convince her mistresses that there is much to be gained by being sympathetic to the attentions of the Albanians, and she describes the art of love ("Una donna a quindici anni"). After Despina leaves, the two women begin to agree that, with their lovers away, a flirtation might prove diverting. Dorabella decides to be receptive to Guglielmo while Fiordiligi expresses preference for the disguised Ferrando ("Prenderò quel brunettino"). Having made their decision, they accompany Don Alfonso to the garden where, as he has informed them, they are to be pleasantly surprised.

Scene 2. The garden. The "Albanians" are in a flower-bedecked boat, surrounded by musicians and guests. Upon the arrival of the women, the two men sing a serenade ("Secondate, aurette amiche"). The four lovers then pair off. Guglielmo and Dorabella exchange pendants and tender words ("Il core vi dono"). Ferrando is less successful. Rejected, he reiterates his passionate feelings for Fiordiligi ("Ah! io veggio quell' anima bella") and then departs. Fiordiligi is upset, for she is not altogether immune to temptation ("Per pietà, ben mio perdona"). With the women gone, the soldiers meet and compare experiences. Ferrando, understandably, becomes furious. Learning that only one of the sisters has proved fickle, Don Alfonso reminds his friends that the test is not yet over.

Scene 3. In the house Fiordiligi expresses disapproval of the way her sister has behaved, while Dorabella, in a more practical vein, insists that it is wisest to follow the dictates of love ("E' amore un ladroncello"). Fiordiligi is still unconvinced. She decides to pursue an honorable course: to put on an officer's uniform and go off to fight at the side of her lover. Her good intentions vanish with the appearance of Ferrando, still dressed as an Albanian. Determined to avenge himself on Guglielmo and Dorabella, Ferrando intensifies his advances, then threatens to kill himself. Helplessly, Fiordiligi succumbs. This turn of affairs arouses the fury of both soldiers, since it is now apparent that neither of their sweethearts has remained faithful. But Don Alfonso is more philosophic: he advises the soldiers to marry their sweethearts as they originally planned, since in the matters of the heart all women are unpredictable.

Scene 4. A banquet room. The weddings of the "Albanians" and the sisters are about to be performed. Despina, now disguised as a notary, reads the terms of the marriage contracts. At the last moment a drum roll announces the return of the two soldiers from the war. In the confusion that ensues, the "Albanians" disappear and replace their disguises with their uniforms. They feign surprise at the coldness with which their sweethearts greet them and amazement at the signed marriage contracts. Finally, they reveal that they were the "Albanians." Humiliated, the sisters blame Don Alfonso for their troubles. Don Alfonso convinces them that what has happened has been for the best and advises the lovers to patch up their differences.

Così fan tutte was written in 1790, after *Don Giovanni* and before *The Magic Flute*. It represents Mozart at his fullest mastery as an opera composer. In some respects, *Così fan tutte* is the most remarkable of Mozart's operas. Using as his point of departure an inconsequential comedy of love and infidelity—a text which for all its occasional wit is hardly calculated to make exacting demands on a composer— Mozart produced a miraculous score, subtle in characterization, profound in psychological insight, and traversing a wide gamut of feelings. In his other operas he is at times nobler, more passionate, and more eloquent; but he is never nimbler, nor is his touch ever surer. With amazing dexterity he maintains in *Così fan tutte* a subtle balance between comedy and burlesque, sentimentality and mockery, tenderness and broad satire. The music continually catches the nuances of the play, points them up, brings artistic value to trivialities of stage business. This opera is much more than a succession of wonderful arias and ensemble numbers: it is operatic comedy at its best, with music and libretto equal collaborators in a gay adventure.

Costa, Michael, conductor and composer. Born Naples, February 4, 1808; died Hove, England, April 29, 1884. He went to England in 1829 after having composed a number of operas in Italy. When Covent Garden became an opera house in 1847, and was called the Royal Italian Opera, Costa became its principal conductor. In 1871 he was appointed conductor at Her Majesty's Opera. He wrote several now forgotten operas: *Il sospetto funesto* (1826); *Il delitto punito* (1827); *Il carcere d'Ildegonda* (1828); *Malvina* (1829); *Don Carlos* (1844).

Costanzi, see TEATRO COSTANZI.

Couplets Bachiques, see AMIS, L'AMOUR TENDRE ET REVEUR.

Covent Garden, the leading opera house in England. It is situated in London in the heart of a produce market, on the site of what once was a convent garden, and afterwards two theaters destroyed by fire. The first of these theaters opened in 1732. From 1734 to 1737, Handel was associated with it, presenting there many of his operas including *Ariodante, Alcina,* and *Ata-*

lanta. He directed the first London performance of the *Messiah* there in 1743. This theater also witnessed the première of *The Beggar's Opera.* After being destroyed by fire in 1808, it was replaced by a new theater on whose stage took place the English première of *Der Freischütz* and the world première of *Oberon.*

Converted into a luxurious and well-equipped opera house, Covent Garden was opened on April 6, 1847, with Rossini's *Semiramide.* For the first time a formal opera company was established at Covent Garden. This was the Royal Italian Opera, directed by Frederick Beale, with Michael Costa as principal conductor. After 1851, under the direction of Frederick Gye—and with such stars as Grisi, Lucca, Patti, and Viardot—this company achieved international significance. However, the venture collapsed in 1884. In 1888 a new opera company was organized under the management of Augustus Harris. Four years later this company became known as the Royal Opera, and in the same year it presented the first performance in England of the entire *Ring* cycle, under the direction of Mahler. Under Harris, and after him Maurice Grau, Covent Garden became the home of some of the most brilliant singing of that day, since the company included Lilli Lehmann, Battistini, Melba, Calvé, Ternina, the De Reszkes, and Bispham. During this period, in 1894, Massenet's *La Navarraise* received its world première.

A marked decline of artistic standards took place between 1900 and 1914. Except for a memorable cycle of Wagner under Richter's direction in 1908, the world première of Leoni's *L'Oracolo* in 1905, and some stimulating performances under Sir Thomas Beecham in 1910 (particularly of *Elektra* and *A Village Romeo and Juliet*), presentations at Covent Garden were substandard. During World War I the opera house was closed. It reopened in 1919 for the first of two summer seasons of opera under Beecham. Several different opera companies then occupied Covent Garden for winter seasons, among them the Beecham Opera Company, the Carl Rosa Company, and the British Opera Company. In 1924, the operatic activity at Covent Garden assumed an international character. The German Opera Syndicate presented a season of German opera with Bruno Walter directing some of the foremost artists from German and Austrian opera houses; this was followed by a season of Italian and French operas with artists from Italy and France. The international character of the performances continued from 1933 to 1939 under the artistic direction of Sir Thomas Beecham.

Once again war closed Covent Garden in 1940. After World War II the Covent Garden Opera Trust was created under the chairmanship of Lord Keynes. The theater reopened on February 20, 1946, with a series of ballet performances by the Sadler's Wells Company. Opera returned with visits of foreign opera companies, and with performances by a newly formed native company under the musical direction of Karl Rankl, and with the financial support of the publishing house of Boosey and Hawkes. Among the notable world premières at Covent Garden since World War II have been those of Benjamin Britten's *Billy Budd* and *Gloriana,* Arthur Bliss's *The Olympians,* and Ralph Vaughan Williams' *The Pilgrim's Progress.*

Cradle Will Rock, The, a musical play by Marc Blitzstein. Libretto by the composer. Première: Venice Theater, New York, June 16, 1937. The story takes place in a night court and concerns the efforts of steel workers to form a union, and the attempt of employers to smash it. Mr. Mister, who controls Steeltown, forms a committee of leading citizens to attack the workers' effort, but the union defies these attacks and ultimately proves triumphant.

With *The Cradle Will Rock,* American proletarian opera emerges. The early history of this opera was both dramatic and turbulent. It was written for the Federal Theater, a unit of the WPA. The radical theme of the play impelled several government officials to demand its withdrawal. Just before the opening, and with the first-night audience gathering at the theater, the Federal Theater announced that the production was canceled. A frantic last-minute maneuver transferred both the production and audience to a near-by theater. Denied the support of the Federal Theater, the troupe could not avail itself of costumes, scenery or orchestra. Consequently the play was performed on a bare stage, with performers dressed in street clothes. The composer performed his score on a piano. Between scenes, he explained to the audience what was about to happen on the stage. This simple, straightforward way of presenting the work added, rather than detracted, from its force and emotional impact.

Credo a una possanza arcana, Chénier's aria in Act II of Giordano's *Andrea Chénier.*

Credo in un Dio crudel, Iago's aria in Act II of Verdi's *Otello.*

Crespel, Antonia's father (bass) in Offenbach's *The Tales of Hoffmann.*

Cricket on the Hearth, The (Das Heimchen am Herd), opera by Karl Goldmark. Libretto by A. M. Willner, based on the story of the same name by Charles Dickens. Première: Vienna Opera, March 21, 1896. The cricket is the guiding spirit of an English household in the early nineteenth century and extricates its members from various personal difficulties. *See also* DICKENS, CHARLES.

Crime and Punishment, see DOSTOYEV-SKY; RASKOLNIKOFF.

Cristoforo Colombo, opera by Alberto Franchetti. Libretto by Luigi Illica. Première: Genoa, October 6, 1892. Queen Isabella provides Columbus with royal jewels to finance his expedition to the new world. Aboard the "Santa Maria," Columbus' enemy, Roldano, incites the sailors to mutiny. The situation is saved by the sight of land. Roldano continues to intrigue against Columbus until he succeeds in discrediting him and having him sent back to Spain in chains. In the epilogue, Columbus commits suicide at the tomb of Isabella.

Crobyle, a slave girl (soprano) in Massenet's *Thaïs.*

Crooks, Richard Alexander, tenor. Born Trenton, New Jersey, June 26, 1900. He made appearances as a boy singer and continued to appear publicly up to the time of World War I. After the war he studied with Frank La Forge and Sydney H. Bourne. In 1922 he made ten appearances with the New York Symphony Society. His opera debut took place in Hamburg, in 1927, in *Tosca.* His American debut took place in Philadelphia, in the same opera, on November 27, 1930. On February 25, 1933, he made his first appearance at the Metropolitan Opera in *Manon* and was acclaimed. He remained with the Metropolitan for the next decade; after leaving, he specialized in concert appearances, radio concerts, and performances in oratorios. He retired in 1946.

Cross, Milton, radio announcer. Born New York City, April 16, 1897. He attended the Institute of Musical Art, after which he embarked on a career as singer. In 1922 he became a radio announcer. He has been the announcer for all the broadcasts of the Metropolitan Opera for over two decades. He is the author of *Milton Cross' Complete Stories of the Great Operas* (1947), and co-author (with David Ewen) of *Encyclopedia of the Great Composers and Their Music* (1953).

Crown, a stevedore (bass) in Gershwin's *Porgy and Bess.*

Crudel! perchè finora, duet of Susanna and Almaviva in Act III of Mozart's *The Marriage of Figaro.*

Csárdás, see KLÄNGE DER HEIMAT.

Cui, César, composer. Born Vilna, Russia, January 18, 1835; died St. Petersburg, March 24, 1918. A member of the Russian nationalist school, the "Five," he wrote ten operas in some of which he tried to realize the artistic goals set by the nationalists. His first important work in any form was the opera *The Captive in the Caucasus* (1859). An earlier work of the same year, *The Mandarin's Son,* was a failure. His third opera, *William Ratcliffe,* was a major success when introduced in St. Petersburg in 1869. Later operas: *Angelo* (1876); *Le Filibustier* (1894); *The Saracen* (1899); *A Feast in Time of Plague* (1901): *Mlle. Fifi* (1903); *Matteo Falcone* (1908); *The Captain's Daughter* (1911). Cui had a pronounced melodic gift, but he was too derivative to provide sustained interest; all his operas have lapsed into oblivion.

Curra, Leonora's maid (mezzo-soprano) in Verdi's *La forza del destino.*

Curtain Tune, see ACT TUNE.

Cyrano de Bergerac, (1) a poetic play by Edmond Rostand centering around the poet-soldier Cyrano, who is in love with the beautiful Roxanne. Disfigured by a huge nose which makes him unattractive to women, he makes love to Roxanne through the presentable person of Christian, writing his love letters and making his love speeches. Only when Cyrano is dying does Roxanne realize that Christian has been a front for the poet and that she is really in love with Cyrano.

(2) Opera by Franco Alfano. Libretto by Henri Cain, based on the

Rostand play. Première: Teatro Reale, Rome, January 22, 1936.

(3) Opera by Walter Damrosch. Libretto by W. J. Henderson, based on the Rostand play. Première: Metropolitan Opera, February 27, 1913.

Czar's Bride, The, opera by Nikolai Rimsky-Korsakov. Libretto by I. F. Tyumenev, based on the play of the same name by Lev Alexandrovich Mey. Première: Moscow Opera, November 3, 1899. Martha, who has been selected by Czar Ivan for his bride, is loved by two other men. One of them, whose love she returns, is the boyar Lykov. The other, Griaznoy, contrives to win her love by having her drink a love potion; but his mistress substitutes poison. When Martha lies dying in the Kremlin she learns that her beloved, Lykov, has been beheaded by the Czar on suspicion of having poisoned her. Martha goes mad. Griaznoy kills his mistress.

D

Da capo aria, a three-part aria in which the third section, after a contrasting second section is a repetition of the first. The form was developed by the Venetian and Neapolitan schools of opera composers.

Dafne, (1) opera by Jacopo Peri. Libretto by Ottavio Rinuccini. Première: Corsi Palace, Florence, 1597. This is the first opera ever written. Peri's music has not survived. The story is based on Greek mythology. Pursued by the god Apollo, Dafne is transformed by her mother into a laurel tree. This story was used extensively by the early composers of opera.

(2) Opera by Heinrich Schütz. Libretto by Martin Opitz with material from the Rinuccini libretto. Première: Torgau, Germany, April 23, 1627. This work is regarded as the first German opera, since it is the first known opera set to a German text.

(3 Bucolic tragedy by Richard Strauss. Libretto by Joseph Gregor. Première: Dresden Opera, October 15, 1938.

da Gama, Vasco, officer in the Portuguese navy (tenor) in Meyerbeer's *L'Africaine.*

Dai campi, dai prati, Faust's aria in Act I of Boïto's *Mefistofele.*

Daland, a Norwegian sea captain (bass) in Wagner's *Der fliegende Holländer.*

d'Albert, Eugène, composer and pianist. Born Glasgow, April 10, 1864; died Riga, Latvia, March 3, 1932. He entered the National School of Music in London in his twelfth year. Two years later he made his debut as pianist. In 1881 he won the Mendelssohn Prize, which enabled him to study with Hans Richter in Vienna and Liszt in Weimar. He lived mostly in Germany, but toured extensively. He was a prolific composer. His most successful opera was *Tiefland,* introduced in Prague on November 15, 1903. Another major success was *Die toten Augen,* first heard in Dresden in 1916. He held several important musical posts, including that of director of the Hochschule für Musik, in Berlin. His most important operas besides those mentioned were: *Der Rubin* (1893); *Die Abreise* (1898); *Kain* (1900); *Flauto solo* (1905); *Der Golem* (1926); *Mister Wu* (1932).

Dalibor, opera by Bedřich Smetana.

Libretto by Joseph Wenzig. Première: National Theater, Prague, May 16, 1868. Dalibor, captain of the guards to the king of Bohemia, is accused of being an insurgent and is imprisoned. Milada, in love with him, heads a band which storms the prison and frees him. In the flight, Milada is fatally wounded and dies in her lover's arms.

Dalila, the High Priest's daughter (mezzo-soprano) in Saint-Saëns' *Samson et Dalila. See also* DELILAH.

Dal labbro il canto estasïato, the song of Oberon (Fenton) in Act III, Scene 2, of Verdi's *Falstaff.*

Dalla sua pace, Ottavio's aria in Act I, Scene 3, of Mozart's *Don Giovanni.*

Dalle stanze, ove Lucia, Raimondo's aria in Act III of Donizetti's *Lucia di Lammermoor.*

Dalmorès, Charles, tenor. Born Nancy, France, January 1, 1871; died Hollywood, California, December 6, 1939. He made his opera debut at the Théâtre des Arts in Rouen on October 6, 1899. He subsequently appeared for six seasons at the Théâtre de la Monnaie, and for seven at Covent Garden. His American debut took place at the Manhattan Opera House on December 7, 1906, in *Faust.* He remained with that company four years, specializing in French roles, and creating for America the roles of Julien in *Louise* and Jean Gaussin in Massenet's *Sapho.* In 1910 he joined the Chicago-Philadelphia Opera, and in 1917 the Chicago Opera, where he essayed for the first time the roles of Parsifal and Tristan.

d'Alvarez, Marguerite, contralto. Born Liverpool, England, about 1886; died Alassio, Italy, October 18, 19ʳ3. She attended the Brussels Conservatory, after which she made her debut in Rouen, France, as Dalila. She made her American debut with the Manhattan Opera on August 30, 1909, as Fidès. After a season in New York, she helped inaugurate Hammerstein's London Opera in 1911, scoring great success

in *Hérodiade* and *Louise.* She subsequently appeared at Covent Garden and other leading European opera houses, and in America with the Chicago Opera and the Boston Opera. In 1944 she was seen in a motion picture, *Till We Meet Again.* She wrote an autobiography, *Forsaken Altars* (1954).

Dame aux camélias, La, novel by the younger Alexandre Dumas, the source of Verdi's *La traviata.* Also based on this novel is Hamilton Forrest's *Camille.*

Damian, Werner Kirchhofer's rival (tenor) for Marie in Nessler's *Der Trompeter von Säkkingen.*

Damnation of Faust, The, dramatic cantata or "legend" in four parts by Berlioz. Libretto by the composer and Almire Gandonnière, based on Gérard de Nerval's version of Goethe's drama. Première: Opéra-Comique, December 6, 1846 (as an oratorio); Monte Carlo, May 18, 1893 (as an opera). The most celebrated portions of this work are staples in the symphonic repertory: the "Rákóczy March," the "Dance of the Sylphs," and the "Minuet of the Will-o'-the-Wisps." Among the more popular vocal excerpts are: "Chanson de la puce" and "Sérénade de Méphisto," both sung by Mephistopheles in the second and third parts respectively; Marguerite's romance, "D'amour l'ardente flamme" and Faust's invocation to Nature, "Nature immense," both in Part Four.

D'amor sull' ali rosee, Leonora's aria in Act IV, Scene 1, of Verdi's *Il trovatore.*

D'amour l'ardente flamme, Marguerite's romance in Part Four of Berlioz' *The Damnation of Faust.*

Damrosch, Leopold, conductor. Born Posen, Germany, October 22, 1832; died New York City, February 15, 1885. He was the first conductor of German opera at the Metropolitan Opera. In Germany he led the Breslau Philharmonic Society and was a friend of Wagner. In 1862 he organized and led the Breslau Orchesterverein. He came to New York in 1871, where he founded the Oratorio Society of New York and the New York Symphony Society, conducting both organizations up to the time of his death. As a guest conductor of the New York Philharmonic Orchestra during 1876–1877 he gave the American première of the third act of *Siegfried*. With the New York Symphony he led the American premières of Berlioz' *The Damnation of Faust* and the first act of *Die Walküre*. In 1884 Damrosch was chosen principal conductor of the Metropolitan Opera, which was then emphasizing German opera, and operas presented in the German language. Damrosch led all the performances—operas by Beethoven, Weber, Wagner, and Mozart as well as representative French and Italian operas—in German. Henry Krehbiel spoke of the Damrosch regime as "the beginning of an effort to establish grand opera in New York on the lines which obtain in Continental Europe." On February 10, 1885, Damrosch was stricken with pneumonia. His young son Walter, called upon to assume his father's post, conducted the performance on February 11. Leopold Damrosch died four days later.

Damrosch, Walter, conductor and composer. Born Breslau, Germany, January 30, 1862; died New York City, December 22, 1950. He was a major influence in the development of musical culture in America, and he played a significant role in the early history of Wagnerian performances in America. The son of Leopold Damrosch, he was five years old when his father came with his family to America. After preliminary music study with his father and several other teachers, Walter Damrosch returned to Germany to study with Felix Draeske and Hans von Bülow. Back in New York, Walter Damrosch helped his father prepare the rehearsals of the New York Symphony, the Oratorio Society, and the German season at the Metropolitan Opera. When Leopold died in 1885, Walter took over his father's various conductorial activities, including the direction of operas at the Metropolitan Opera. In 1885, he enlisted the services of Anton Seidl to help him carry on his labors at the Metropolitan. He also recruited leading European singers, including Lilli Lehmann, Emil Fischer, and Max Alvary. During the 1885–1886 season—Damrosch's first as a full-fledged conductor at the Metropolitan—nine operas were performed, including the American premières of *Rienzi, Die Meistersinger,* and *The Queen of Sheba*. Damrosch remained principal conductor of German opera at the Metropolitan for the next five seasons, during which period he led the American premières of Brüll's *The Golden Cross,* Karl Goldmark's *Merlin,* and Cornelius' *The Barber of Bagdad*. The vogue for German opera at the Metropolitan went into a decline after 1890. In 1895 Damrosch founded his own company for the purpose of presenting German operas (*see* DAMROSCH OPERA COMPANY). After the dissolution of this company he once again became a conductor at the Metropolitan Opera. Serving there in the seasons of 1900–1901 and 1901–1902, he led all the German operas, including the complete *Ring* cycle. In 1903 Damrosch reorganized the New York Symphony Society, establishing it on a permanent basis. When he retired from the concert field in 1926, he devoted himself to radio work. He was a pioneer in broadcasting symphonic music and music-education programs for children. He appeared as himself in two motion pictures: *The Star Maker* and *Carnegie Hall*. He wrote four operas: *The Scarlet Letter* (1896); *Cyrano de Bergerac* (1913, revised 1939); *The Man Without a Country* (1937); *The Opera Cloak* (1942).

Damrosch Opera Company, The, a company organized by Walter Damrosch in 1894 to present German operas. It was an important force in developing a consciousness for the Wagnerian music dramas in the United States. Damrosch organized it to combat the growing apathy of the opera public in general, and the Metropolitan Opera management in particular, to German opera after 1890. The immediate impetus was Damrosch's success in 1894 with concert versions of *Die Götterdämmerung* and *Die Walküre*. He raised the money for his company by selling his New York house. In the summer of 1894 he toured Europe to select his leading singers, returning with contracts with Johanna Gadski and Rosa Sucher (both appearing in America for the first time), and Max Alvary and Emil Fischer among others. The company opened an eight-week season in New York at the Metropolitan Opera on February 25, 1895, with *Tristan und Isolde*. An extensive tour followed, and the first season netted a profit of $53,000. The company continued five years. In the second season, Milka Ternina, Katharina Klafsky, and David Bispham were added to the company, the last making his American debut. In 1896 the company presented the world première of Damrosch's first opera, *The Scarlet Letter*. One season later the company changed its name to the Damrosch-Ellis Company and expanded its activities to embrace Italian and French operas as well as German; at the same time, Melba and Calvé joined the company. The deficits, which started to accumulate after the second season, mounted rapidly, and proved so burdensome that the company had to suspend operations in 1900.

Dancaïre, Le, a smuggler (baritone) in Bizet's *Carmen.*

Dance of the Apprentices, dance in Act III, Scene 2, of Wagner's *Die Meistersinger.*

Dance of the Blessed Spirits, dance in Elysium in Act III of Gluck's *Orfeo ed Euridice.*

Dance of the Buffoons (or Tumblers), dance in Act III of Rimsky-Korsakov's *The Snow Maiden.*

Dance of the Camorristi, dance in Act III of Wolf-Ferrari's *The Jewels of the Madonna.*

Dance of the Comedians, dance in Act III of Smetana's *The Bartered Bride.*

Dance of the Furies, dance in Hades in Act II of Gluck's *Orfeo ed Euridice.*

Dance of the Hours, dance symbolizing the victory of right over wrong, in Act III, Scene 2, of Ponchielli's *La Gioconda.*

Dance of the Moorish Slaves, dance in Act II, Scene 1, of Verdi's *Aïda.*

Dance of the Seven Veils, Salome's dance before Herod in Richard Strauss's *Salome.*

Dance of the Sylphs, dance as Faust dreams of Marguerite in Part Two of Berlioz' *The Damnation of Faust.*

Dandini, a valet (baritone) in Rossini's *La cenerentola.*

Daniello, a violinist (baritone) in Křenek's *Jonny spielt auf.*

D'Annunzio, Gabriele, novelist, poet, dramatist. Born Pescara, Italy, March 12, 1863; died Vittoriale, Italy, March 1, 1938. Several of D'Annunzio's dramas have provided effective material for opera texts. One of Pizzetti's finest operas, *Fedra,* is based on d'Annunzio's play of the same name. Others are: *Francesca da Rimini* (Zandonai); *La figlia di Jorio* (Franchetti); *La nave* (Montemezzi); *Parisina* (Mascagni); *Il sogno d'una mattina di primavera* (Malipiero). D'Annunzio's *Flame of Life* is of interest to opera goers be· cause it describes an imaginary meeting with Wagner on a Venetian canal steamer while the master was suffering his last illness. The façade of the Palazzo Vendramin, where Wagner died, bears a plaque with an inscription in which D'Annunzio poetically linked

Wagner's dying breath with the waves that touch the building's walls.

Dans la cité lointaine, Julien's serenade in Act II, Scene 2, of Charpentier's *Louise.*

Dante Alighieri, poet. Born Florence, May 1265; died Ravenna, September 14, 1321. The author of the *Divine Comedy.* The tragic love of Paolo and Francesca (fifth canto of the "Inferno") has provided the material for several operas (*see* FRANCESCA DA RIMINI; PAOLO E FRANCESCA). Dante appears as the central character in Benjamin Godard's opera *Dante.*

Dantons Tod (Danton's Death), opera by Gottfried von Einem. Libretto by Boris Blacher and the composer, based on a play of the same name by Georg Büchner. Première: Salzburg Festival, August 6, 1947. The opera, based on the tragic fall of the revolutionist Danton during the French Revolution, is written almost exclusively in a recitative style approximating speech.

Dapertutto, the magician (baritone) whose other personalities are Coppelius and Dr. Miracle in Offenbach's *The Tales of Hoffmann.*

Daphne, see DAFNE.

Da Ponte, Lorenzo (born EMMANUEL CONEGLIANO), librettist and poet. Born Ceneda, Italy, March 10, 1749; died New York City, August 17, 1838. On the occasion of his baptism he acquired the name of Da Ponte. He became involved in intrigues and scandals in Venice, resulting in his banishment in 1780. In Vienna he was eventually able to get an appointment as court poet, and he became a favorite of Joseph II. After writing librettos for various composers, he collaborated with Mozart in writing *The Marriage of Figaro* in 1785. In the next few years he also wrote for Mozart the librettos of *Don Giovanni* and *Così fan tutte.* After the death of Joseph II, Da Ponte lost favor with the court. He went to London in 1793 and for a while worked at the Drury Lane

Theatre, then a home for opera. Financial difficulties with the threat of imprisonment forced him to leave England secretly, and he came to the United States in 1805. He taught Italian languages and literature for many years, occupying a chair in the Italian language at Columbia University from 1826 to 1837. (In 1929, Columbia established a Lorenzo da Ponte professorship.) During this period he wrote his autobiography. Da Ponte was also active as an opera impresario in America. With Manuel García he became, in 1825, one of the first to present Italian opera in the United States. In 1832 he again participated in opera performances by collaborating with a French tenor, Montresor, in bringing his opera troupe from Europe, and in 1833 he was responsible for the erection of the Italian Opera House in New York.

Dardanus, opera by Jean Philippe Rameau. Libretto by Charles Antoine Leclerc de la Bruère. Première: Paris Opéra, October 19, 1739. Dardanus, according to Greek legend, is the founder of the royal house of Troy. He kills his brother Iasus and flees to Samothrace, and from there to Phrygia where he marries Princess Batea and builds the city of Troy. The dances from this opera have been collected into two orchestral suites edited by Vincent d'Indy.

Dargomizhsky, Alexander (sometimes DARGOMIJSKY), composer. Born Tula, Russia, February 14, 1813; died St. Petersburg, January 17, 1869. He was Glinka's immediate successor in the writing of national operas. He entered civil service in 1831, remaining there for thirteen years. A meeting with Glinka in 1834 filled him with the ambition to become a composer. After completing his musical training, he wrote an opera, *Esmeralda,* in 1839, which, when produced in 1847, was a failure. He now concentrated on writing smaller pieces and songs, but in 1848 he began working on his most ambitious and most popular work, the opera *Russalka*. When introduced in St. Petersburg in 1856, *Russalka* was a mild success, but later performances established its popularity. Personal contact with members of the Russian "Five" fired Dargomizhsky with the ideal of writing national music. Incorporating the ideas of the nationalist group, he composed his last and most important work, *The Stone Guest,* an opera based on Pushkin's version of the Don Juan story. His death left the orchestration uncompleted; it was finished by César Cui and Rimsky-Korsakov.

Das schöne Fest, Pogner's address in Act I of Wagner's *Die Meistersinger.*

Das süsse Lied verhallt, love duet of Lohengrin and Elsa in Act III, Scene 1, of Wagner's *Lohengrin.*

Da-ud, Altair's son (tenor) in Richard Strauss's *Die aegyptische Helena.*

Daudet, Alphonse, novelist and playwright. Born Nîmes, France, May 13, 1840; died Paris, December 16, 1897. Daudet's plays and novels provided the material for several operas, among them Cilèa's *L'Arlesiana,* Massenet's *Sapho,* and Emile Pessard's *Tartarin sur les Alpes.*

Daughter of the Regiment, The (La fille du regiment; La figlia del reggimento), opera in two acts by Gaetano Donizetti. Libretto by Jean François Bayard and Vernoy Saint-Georges. Première: Opéra-Comique, February 11, 1840. American première: New Orleans, March 6, 1843.

Characters: Marie, vivandière (canteen-manager) of the French 21st Regiment (soprano); the Countess of Berkenfeld, her mother (mezzo-soprano); Ortensio, her servant (bass); Tonio, a peasant (tenor); Sulpizio, a sergeant in the 21st Regiment (bass); a peasant; a corporal; a notary; a duchess; soldiers;

peasants; servants; ladies in waiting. The setting is the Tyrol about 1815.

Act I. A mountain pass. The French, under Napoleon, have invaded the Tyrol. Marie, adopted in infancy by the 21st Regiment, has grown to young womanhood and become the regiment's vivandière (canteen-manager). She sings of her joy in camp life and battle ("Au bruit de la guerre"—"Io vidi la luce nel camp guerir"). She then discloses that she is in love with Tonio, who is soon afterwards dragged in by the French and accused of being a spy. When Marie explains that Tonio once saved her life, the French are kinder to him and hail him as a recruit to their regiment. This inspires Marie to sing a song of praise to the regiment ("Ah, chacun le sait, chacun le dit"—"Ah, ciascun lo dice, ciascun lo sà"), in which the soldiers join. After the soldiers leave, Tonio expresses his love for Marie and insists on his willingness to die for her; Marie, too, speaks of her love and exclaims that for her sake he must live, not die ("Depuis l'instant où dans mes bras"—"Perchè v'amo"). The Countess of Berkenfeld, who has been hiding during the fighting, now reappears and learns from Sulpizio that Marie is her long-lost niece. The Countess insists that Marie return with her to her castle. The French soldiers now appear, singing a stirring song to war and victory ("Rataplan"). Heartbroken, Marie bids farewell to her regiment ("Il faut partir, mes bons compagnons"—"Convien partir, o miei compagni d'arme"). Overwhelmed with grief that he can no longer be with Marie, Tonio angrily tears the regimental colors from his hat and stamps on them.

Act II. A room in the Berkenfeld Castle. Marie is being raised as a lady of nobility. The Countess gives her a singing lesson ("Le jour naissait dans le bocage"—"Sorgeva il dì del bosco in seno"). While Marie is singing the first strains of the song, Sulpizio—who, wounded, has been allowed to stay in the castle—reminds Marie of their regimental song. Almost helplessly, she bursts into the refrain ("Le voilà, le voilà"—"Egli è la"), to the horror of the Countess, who storms out of the room. Since Marie is being compelled to marry the Duke of Crackenthorp, and since she is still in love with Tonio, she laments her misfortune ("C'en est donc fait et mon coeur va changer"—"Me sedur han creduto"). French soldiers suddenly storm into the castle. Tonio is with them. Sulpizio, Marie, and Tonio are overjoyed at their reunion ("Tous les trois réunis"—"Stretti insiem tutti tre"). The Countess, returning, is shocked to find Tonio present; she now discloses that Marie is not her niece but her daughter, and she insists that it is Marie's duty to obey her mother and marry the Duke. The soldiers reappear from the garden and shout their disapproval of the mother's decision, insisting they will never permit Marie to marry any but the man she loves ("Au secours de notre fille"—"Ti rincora, amata figlia"). After Marie voices her own sentiments ("Quand le destin au milieu"—"Quando fanciulla ancor l'avverso"), the Countess relents and is ready to accept Tonio as a son-in-law. Rejoicing prevails, and the assemblage sings a stirring salute to France ("Salut à la France"—"Salvezza alla Francia").

Donizetti wrote this gay and martial opera for Naples, using his own Italian translation of a French libretto. But another of his operas had antagonized the Neapolitan authorities with its political implications and Donizetti had to leave Italy for France. He took with him his new opera, and it was in Paris that it was introduced in its original French version. The title role has attracted coloratura sopranos throughout the years, and many famous prima donnas are identified with it. Anna Thillon created the role at the Paris première; later sopranos who scored major successes in the part included Albani, Hempel, Jenny Lind, Pons, Patti, Sembrich, Sontag, and Luisa Tetrazzini. When the opera was revived by the Metropolitan for Frieda Hempel during the 1917–1918 season, America was at war; the martial character of the opera had special significance. In keeping with the times, Hempel interpolated the English ballad, "Keep the Home Fires Burning." During World War II—and soon after the Nazis occupied France—the opera again acquired special political interest; Lily Pons draped herself in the flag of the Free French Forces and sang the *Marseillaise*.

David, (1) opera by Darius Milhaud. Libretto by Armand Lunel. Première: Jerusalem, June 1, 1954 (concert version). Milhaud wrote this opera on a commission from the Koussevitzky Foundation. The Biblical story begins with the visit of prophet Samuel to the house of David's father and ends with the annointment of Solomon as king of Israel in succession to David. A chorus, dressed in present-day garb, draws parallels between the Biblical tale and modern Israeli history.

(2) A rabbi (baritone) in Mascagni's *L'amico Fritz*.

(3) An apprentice (tenor) to Hans Sachs in Wagner's *Die Meistersinger*.

David, Félicien, composer. Born Cadenet, France, April 13, 1810; died St. Germain-en-Laye, August 29, 1876. After four years of music study at the Jesuit College in Aix he became conductor of a theater orchestra, and after that maître de chapelle at the Cathedral. In 1830 he went to Paris for further music study at the Conservatory. A year later he joined a religious brotherhood, the Saint-Simonists, and lived in monastic seclusion. After the brotherhood was dissolved in 1833, David began an extended period of travel in the Near East. His impressions inspired his most famous work, a symphonic ode, *Le désert*, introduced in 1844 with such success that it created a vogue for compositions with exotic backgrounds. His first successful work for the stage was an opéra comique, *La perle du Brésil*, given in Paris in 1851. A second opera, *Herculanum*, won a national prize in 1859. His most celebrated opera, *Lalla-Roukh*, was acclaimed at its première at the Opéra-Comique in 1862. Later operas: *Le saphir* (1865); *La captive* (not performed). In 1869 David succeeded Berlioz as member of the Academy and as librarian at the Conservatory.

Da zu Dir der Heiland kam (Kirchenchor), chorale at beginning Act I of Wagner's *Die Meistersinger*.

Dead City, The, *see* TOTE STADT, DIE.

Deane, Mrs., a widow (mezzo-soprano) in Deems Taylor's *Peter Ibbetson*.

Death of Don Quixote, *see* ECOUTE, MON AMI.

Death of Thaïs, *see* SE SOUVIENT-IL DU LUMINEUX VOYAGE.

Debora e Jaele (Deborah and Jael), opera by Ildebrando Pizzetti. Libretto by the composer. Première: La Scala, December 16, 1922. This is the first of several Biblical works by Pizzetti, and one of his most powerful operas. The text was derived from the Book of Judges and concerns the victory of the Israelites over the Canaanites, and the part played by Deborah, the prophetess and judge, and the heroine Jael, in that victory.

Debout! matelots (Su, su, marinar), the sailors' chorus in Act III of Meyerbeer's *L'Africaine*.

De Brétigny, a nobleman (baritone) in Massenet's *Manon*.

Debussy, Claude Achille, composer. Born Saint-Germain-en-Laye, France, August 22, 1862; died Paris, March 25, 1918. The father of musical impressionism completed only a single opera; but that work, *Pelléas et Mélisande*, represents the quintessence of his art and is one of the monuments in postWagnerian opera. Debussy attended the Paris Conservatory for eleven years. His teachers included Antoine Marmontel, Emile Durand, and Ernest Guiraud. In 1884 he won the Prix de Rome for the cantata *L'enfant prodigue*. After his sojourn in Rome, Debussy returned to Paris where he was affected by the Symbolist movement in poetry and the Impressionist school of painting, then flourishing. These influences, combined with his conversations with the iconoclastic French musician, Erik Satie, crystallized Debussy's thinking about the kind of music he wanted to write and set him off in the direction of musical impressionism. By exploiting a number of characteristic technical devices he achieved a highly personal idiom. In this vein he produced his first songs, his single quartet for strings, pieces for piano, and *L'après-midi d'un faune* for orchestra. His most ambitious work, the opera *Pelléas et Mélisande*, was begun in 1892 and completed a decade later. The circumstances surrounding the writing and production of this work appear elsewhere (*see* PELLEAS ET MELISANDE). Debussy tried writing another opera even while he was engaged on *Pelléas: Rodrigue et Chimène*, begun in 1892, but abandoned a few years later after the completion of only two acts. He also intended adapting Edgar Allan Poe's *The Fall of the House of Usher* into an opera, but this project never progressed beyond the planning stage.

Decameron, The, *see* BOCCACCIO, GIOVANNI.

Declamation, music in which the voice approximates the inflections of speech and in which the text assumes greater significance than the melody. Declamation is used in opera for dramatic episodes.

de Falla, Manuel, *see* FALLA, MANUEL DE.

Deh! con te li prendi, Norma's aria in Act II (or, in some arrangements, Act III) of Bellini's *Norma.*

Deh non parlare al misero, Rigoletto's duet with Gilda in Act II (originally Act I, Scene 2) of Verdi's *Rigoletto.*

Deh placatevi con me! Orfeo's plea to the furies in Act II of Gluck's *Orfeo ed Euridice.*

Deh! proteggimi o Dio! Adalgisa's aria in Act I of Bellini's *Norma.*

Deh, vieni alla finestra, Don Giovanni's serenade in Act II, Scene 1, of Mozart's *Don Giovanni.*

Deh vieni, non tardar, Susanna's aria in Act IV of Mozart's *The Marriage of Figaro.*

Déjà les hirondelles, duet for two sopranos in Delibes's *Le Roi l'a dit.* It has become a celebrated concert number.

de Jouy, *see* JOUY, VICTOR DE.

De Koven, Reginald, composer. Born Middletown, Connecticut, April 3, 1859; died Chicago, January 16, 1920. His music study took place in Germany and Paris. After returning to the United States he wrote several operettas before achieving his first success with the work by which he is today remembered, *Robin Hood,* introduced in Chicago in 1890. In 1917 he was commissioned by the Metropolitan Opera to write the grand opera, *The Canterbury Pilgrims,* performed in 1917. A second grand opera, *Rip Van Winkle,* was introduced in Chicago in 1920. De Koven was the music critic for the *New York Herald* from 1898 to 1900 and again from 1907 to 1912.

de Lara, Isidore, composer. Born London, August 9, 1858; died Paris, September 2, 1935. After appearing as a child-prodigy pianist, he entered the Milan Conservatory in his fifteenth year. His first opera, *The Light of Asia,* an adaptation of a cantata, was given at Covent Garden in 1892. His most successful opera, *Messaline,* was introduced in Monte Carlo in 1899, and given at the Metropolitan Opera three years later. De Lara's style derived from that of Massenet. His other operas: *Amy Robsart* (1893); *Moïna* (1897); *Le réveil de Bouddha* (1904); *Sanga* (1908); *Soléa* (1907); *Naïl* (1911); *Les trois masques* (1912); *Les trois mousquetaires* (1921).

De l'enfer qui vient émousser, the chorus of the swords in Act II of Gounod's *Faust.*

Delibes, Léo, composer. Born St. German du Val, France, February 21, 1836; died Paris, January 16, 1891. While most famous for his ballets, Delibes is also the composer of *Lakmé,* a staple in the French operatic repertory. He attended the Paris Conservatory and in 1855 wrote an operetta, *Deux sous de charbon,* introduced in Paris. Between 1862 and 1871 he was organist at the church of St. Jean St. François. In 1865 he became second chorusmaster at the Opéra. His first ballet, *La source,* was a major success when introduced at the Opéra in 1866. His second ballet, *Coppélia,* performed in 1870, was an even greater triumph. In 1873 he wrote his first opera, *Le Roi l'a dit,* given at the Opéra-Comique. *Lakmé* came a decade later, bringing its composer to the front rank of French composers for the stage. In 1881, Delibes was appointed professor of composition at the Conservatory, and in 1884 member of the Institut de France. Besides the operas mentioned above, Delibes also wrote *Jean de Nivelle* in 1880, and *Kassya,* which was finished by Massenet after the composer's death.

Delilah, the High Priest's daughter in Bernard Rogers' *The Warrior. See also* DALILA.

Delius, Frederick, composer. Born Bradford, England, January 29, 1862; died Grez-sur-Loing, France, June 10, 1934. After working two years in his father's wool establishment, he came to the United States to superintend an orange plantation in Florida which his father bought for him. Here he began to study music intensively by himself, and to compose his first works, including *Appalachia,* for chorus and orchestra. In 1886 he returned to Europe to attend the Leipzig Conservatory. He next went to Paris where his first works were published. After marrying Jelka Rosen in 1899 he established his home in Grez-sur-Loing. Here he lived the rest of his life and here he wrote his most celebrated works, including his tone poems for orchestra, and his opera *A Village Romeo and Juliet* (1900–1902). Soon after World War I, Delius' health failed and he became a victim of paralysis and blindness. Composition was not abandoned. He dictated his last works, note by note, to an amanuensis. In 1929 he appeared in London to attend a festival of six concerts devoted to his music. His operas: *Irmelin* (1892); *The Magic Fountain* (1893); *Koanga* (1897); *A Village Romeo and Juliet* (1900–1902); *Margot la Rouge* (1902); *Fennimore and Gerda* (1910).

Dell' aura tua profetica, chorus of the Druid priests in Act I of Bellini's *Norma.*

Della città all' occaso, Ulrica's aria in Act I, Scene 2 of Verdi's *Un ballo in maschera.*

Della vittoria agli arbitri, chorus of the priests in Act II, Scene 2, of Verdi's *Aïda.*

delle Sedie, Enrico, baritone and teacher of singing. Born Leghorn, Italy, June 17, 1824; died Paris, November 28, 1907. Before embarking on his musical career he distinguished himself as a soldier, fighting in the war for Italian independence in 1848–1849. Taken prisoner by the Austrians during the Battle of Curtatone, he was released, later retiring from the army with the rank of lieutenant and several decorations. He now began the study of singing with various teachers, including Orazio Galeffi, and in 1851 he made his debut in Pistoia, Italy, in

Nabucco. Success came three years later in Florence, after which he appeared in leading baritone roles of the Italian and French repertories throughout Italy, and in Vienna, Paris, and London. In 1867 he abandoned the stage to become a professor of singing at the Paris Conservatory. He wrote a valuable treatise, *L'art lyrique* (1874), published in the United States as *A Complete Method of Singing.*

Delmas, Jean-François, dramatic basso. Born Lyons, France, April 14, 1861; died St. Alban de Monthel, September 27, 1933. After studying at the Paris Conservatory, where he won first prize in singing, he made his opera debut at the Opéra on September 13, 1886 in *Les Huguenots.* He became a regular member of the company and for the next four decades distinguished himself in both the French and German repertory. He was particularly significant in the Wagner music dramas, creating for France most of the leading Wagnerian baritone and bass roles. He also created the principal bass parts in *Ariane et Barbe-bleue, L'Etranger, Monna Vanna, Salammbô,* and *Thaïs.*

de los Angeles, Victoria, soprano. Born Barcelona, Spain, November 1, 1923. She attended the Conservatorio del Liceo for three years, after which, in her twentieth year, she made her debut in opera as the Countess in *The Marriage of Figaro* in Barcelona. In 1947 she won first prize in the International Singing Contest in Geneva, which was followed by concert and opera appearances throughout Europe. In March 1950 she made a successful debut at the Paris Opéra, and in June of that year at La Scala. In the fall of 1950 she made her American debut in a recital at Carnegie Hall after which she was described by Virgil Thomson as a "vocal delight unique in our time." In March, 1951, she made her American opera debut at the Metropolitan Opera in *Faust.* She has since appeared at the Metropolitan Opera with outstanding success, as well as in the foremost opera houses of Europe. She has also been acclaimed at the Edinburgh, Holland, and Florence music festivals.

del Puente, Giuseppe, baritone. Born Naples, January 30, 1841; died Philadelphia, Pennsylvania, May 25, 1900. As a child he entered the Naples Conservatory to study the cello, but eventually he turned to singing. After only a single year of vocal study he made his opera debut in Jassy. Appearances followed in leading Italian opera houses. In 1873 Del Puente made a successful debut at Covent Garden. He was a great favorite in England, where he created for that country the roles of Escamillo and Valentin. In 1873 he also made his American debut with the Strakosch Opera Company. He helped make operatic history in this country by appearing in the American première of *Carmen* and in the performance of *Faust* with which the Metropolitan Opera was opened. As a member of the Metropolitan, and subsequently of the Gustav Hinrich Company, he appeared in the American premières of *L'amico Fritz, La Gioconda, Manon Lescaut, Les pêcheurs de perles,* and *Roméo et Juliette.*

Del Tago sponde addio, *see* ADIEU MON, DOUX RIVAGE.

de Luca, Giuseppe, baritone. Born Rome, December 29, 1876; died New York City, August 27, 1950. For two decades he was the principal baritone of the Metropolitan Opera, appearing more than eight hundred times in about a hundred different roles. After completing his studies at the Schola Cantorum and the Santa Cecilia Academy, both in Rome, he made his opera debut on November 7, 1897, in Piacenza as Valentin. Other important opera appearances followed, notably in Genoa, Milan, and Buenos Aires. In the winter of 1903 he was engaged by La Scala where he remained for eight years and

where he created the role of Sharpless in 1904. He also appeared in the world premières of *Adriana Lecouvreur* and *Siberia.* After touring as star in leading opera houses of the world, he made his American debut at the Metropolitan Opera on November 25, 1915, in *The Barber of Seville.* During the next twenty years he appeared in virtually every important baritone role of the French and Italian repertory, and was heard in the world première of *Goyescas* and the American premières of *La campana sommersa, Gianni Schicchi,* and *Mârouf.* He was a towering figure in the golden age of opera in America, a master of bel canto.

He left the Metropolitan Opera after the 1934–1935 season. The next four years he spent singing in Europe after which he returned to the Metropolitan on February 7, 1940, in *La traviata.* Back in Italy during World War II, he refused to sing for five years because, as he explained, "I was not in good humor." The ejection of the Nazis from Italy restored his good humor and he sang for Allied troops at the Rome Opera. Persuaded to return to America, he appeared in a New York recital in March, 1946, when he received a tumultuous ovation. On November 7, 1947, his golden jubilee as a singer was also celebrated with a New York recital. In the last two years of his life he devoted himself to teaching voice in New York. He was the recipient of many honors from his native land, including a decoration of the Santa Cecilia Academy, and decorations of the Donatao First Class, Knights of Malta, Cavalier of the Great Cross, Crown of Italy, and Commendatore of S.S. Maurizio and Lazzaro.

de Maupassant, *see* MAUPASSANT, GUY DE.

Demetrios, a rich sculptor (tenor) in Camille Erlanger's *Aphrodité.*

Demeur, Anne Arsène (born CHARTON), soprano. Born Saujon, France,

March 5, 1824; died Paris, November 30, 1892. She distinguished herself in performances of Berlioz operas, creating the leading soprano roles in *Béatrice et Bénédict* and *Les Troyens à Carthage*. After studying voice with Bizot, she made her opera debut in Bordeaux in 1842 in *Lucia di Lammermoor*. After appearances in France, Belgium, and London she married (1847) a flutist, Jules Antoine Demeur, after which she appeared under the name of Charton-Demeur. In 1849 and again in 1853 she appeared at the Opéra-Comique, but was poorly received on both occasions. Success came in Italian opera with performances in St. Petersburg, Vienna, Paris, and South America. In 1862 her performance in the première of *Béatrice et Bénédict* so delighted the composer that he invited her to create the role of Dido in *Les Troyens à Carthage* in 1863. After she went into retirement she emerged in 1870 to appear in a Berlioz festival at the Paris Opéra, and again in 1879 to sing in Berlioz' *La Prise de Troie* in a performance given by the Pasdeloup Orchestra.

De' miei bollenti spiriti, Alfredo's aria in Act II, Scene 1, of Verdi's *La traviata*.

Demon, The, opera by Anton Rubinstein. Libretto by A. N. Maikov, revised by Viskotov, based on a poem by Mikhail Lermontov. Première: St. Petersburg Imperial Opera, January 25, 1875. The demon, eager to capture the beautiful Tamara for himself, provokes the Tartars to kill her betrothed. Tamara finally succumbs to the demon. After receiving his first kiss, she dies. Angels put the demon to flight, then carry Tamara off to heaven.

Demoni, fatale, chorus of the demons in Act III of Meyerbeer's *Robert le Diable*.

de Musset, Alfred, *see* MUSSET, ALFRED, DE.

de Nangis, Raoul, Huguenot nobleman (tenor) in Meyerbeer's *Les Huguenots*.

Dent, Edward Joseph, musicologist. Born Ribston, England, July 16, 1876. After being educated at Eton and Cambridge he became, in 1902, Fellow of King's College at Cambridge, and after 1926 professor of music. In 1922 he helped organize the International Society for Contemporary Music, serving as its president for many years. In 1931 he was elected president of the Société Internationale de Musicologie. He was particularly active in the field of opera, both as scholar and as impresario. He edited and produced many old English operas, particularly Purcell's *Dido and Aeneas*. He prepared new English translations of well-known operas. He also wrote many books about opera and opera composers, among them biographies of Alessandro Scarlatti (1905), Busoni (1933), and Handel (1934), and *Mozart's Operas: A Critical Study* (1913), *Foundations of English Opera* (1923), *Music of the Renaissance in Italy* (1934), and *Opera* (1940).

De Paris tout en fête, Julien's apostrophe to Paris in Act III of Charpentier's *Louise*.

de Puiset, Eglantine, Euryanthe's false friend (mezzo-soprano) in Weber's *Euryanthe*.

Depuis le jour, Louise's aria in Act III of Charpentier's *Louise*.

Depuis l'instant óu dans mes bras (Perchè v'amo), Marie's and Tonio's duet in Act I of Donizetti's *The Daughter of the Regiment*.

Depuis longtemps j'habitais cette chambre, Julien's aria in Act I of Charpentier's *Louise*.

de Ravoir, Géronte, Manon's elderly lover (bass) in Puccini's *Manon Lescaut*.

de Reszke, Edouard (brother of Jean), bass. Born Warsaw, Poland, December 22, 1853; died Garnek, Poland, May 25, 1917. One of the most celebrated singers of the late nineteenth century, he was the brother of one of the most adulated tenors of all time, Jean de Reszke, and of a famous soprano, Josephine de Reszke. Edouard's first singing lessons were given him by his brother Jean. He later studied in Italy with Filippo Coletti and another teacher named Steller. The director of the Paris Opéra was so impressed by his voice that he recommended him to Verdi for the French première of *Aïda*, and on April 22, 1876, De Reszke made his debut in Paris in the role of the King at the Théâtre des Italiens. During the next few seasons he appeared in Paris, Turin, Milan, Lisbon, and at Covent Garden. As a principal bass of the Paris Opéra he appeared in the five hundredth performance of *Faust* (his brother Jean sang the title role); he also was heard in *Roméo et Juliette* when that work entered the Paris Opéra repertory. His American debut took place during a visit of the Metropolitan Opera to Chicago on November 9, 1891, when he appeared as the King in *Lohengrin;* on this same occasion, Jean de Reszke made his American debut in the title role. A few weeks later, on December 14, Edouard was heard in New York in *Roméo et Juliette*. He remained at the Metropolitan for over a decade. It was here that he started to sing German roles in German (the *Lohengrin* performance in Chicago had been in Italian), beginning with a performance of *Tristan und Isolde,* in which Jean appeared as Tristan in a German-language performance for the first time. A giant figure, Edouard was a commanding personality on the stage; and his voice was like his figure, large and masterful. His last appearance at the Metropolitan took place on March 21, 1903, in *Faust*. For many years thereafter he lived in seclusion on his estate in Poland. The outbreak of war in 1914 proved a disaster to both his health and fortune. During the war years he lived in ex-

treme poverty. An improvement in his personal affairs in 1917 came too late to be appreciated, for by then he was broken in health.

de Reszke, Jean (born JAN MIECZYSLAW), tenor. Born Warsaw, Poland, January 14, 1850; died Nice, France, April 3, 1925. One of the foremost tenors of his century, he was the brother of Josephine and Edouard de Reszke, soprano and bass. His mother, also a singer, gave him his first music lessons, after which he studied in Warsaw with Ciaffei and in Milan with Antonio Cotogni. He was trained as a baritone, and it was as such, under the name of Giovanni di Reschi, that he made his debut in Venice (January, 1874) in *La favorita*. During the next few years he continued singing baritone roles without any special success, appearing in Italy, London, Dublin, and Paris. His brother Edouard and the singing teacher Giovanni Sbriglia convinced him to change from a baritone into a tenor. After a period of study with Sbriglia, he returned to the opera stage on November 9, 1879, this time as a tenor, in *Robert le Diable*. He was not well received. Regarding himself as a failure, De Reszke withdrew from opera completely for five years, devoting himself to concert appearances. Massenet induced him to return to opera for the Paris première of *Hérodiade* in 1884. De Reszke was a sensation as John the Baptist. From 1884 to 1889 he was the principal tenor of the Paris Opéra, where he created the leading tenor role of *Le Cid,* which Massenet wrote with him in mind. His career now established, it henceforth paralleled that of his brother. With Edouard, he appeared in *Roméo et Juliette* when it entered the Paris Opéra repertory, and he sang in the five hundredth performance of *Faust*. A visit to Bayreuth in 1888 turned him toward German opera. After an intensive period of study of the Wagnerian repertory, he appeared for the first time in a Wagner opera in London in 1898, singing the part of Tristan in Italian. He followed this with performances as Walther and Lohengrin (still in Italian) that brought him to the front rank of living Wagnerian tenors. Curiously, when Jean de Reszke was engaged by the Metropolitan Opera in 1891 it was to appear in French and Italian roles; and, more curious still, he, who was destined to become one of the most idolized interpreters of Wagner in America, had been imported by the Metropolitan to defeat the then growing vogue for Wagner in New York. Yet it was the immense personal appeal of Jean de Reszke that was finally instrumental in having the Metropolitan Opera restore the Wagner music dramas to their former prominence in the repertory.

Jean de Reszke's American debut took place in Chicago, during a visit to that city of the Metropolitan Opera, on November 9, 1891. The opera was *Lohengrin.* (Jean's brother Edouard made his American debut in the same performance.) On December 14, Jean de Reszke appeared for the first time in New York in *Roméo et Juliette,* once again with his brother in the cast, and was only moderately successful. It was only after Jean had begun appearing in the Wagnerian repertory that he became a matinee idol. He appeared for the first time in a German-language performance of *Tristan und Isolde* on November 27, 1895; for the next half dozen years he was considered the ideal Tristan, the standard by which all later Tristans were measured. His last appearance at the Metropolitan Opera took place on March 29, 1901, in *Lohengrin.* After only one other appearance on the opera stage, in *Pagliacci* at the Paris Opéra, he withdrew from an active career. For the next decade and a half he lived in Paris, teaching. Repeated attempts were made

to lure him out of his retirement, but they were futile. When his only son died during World War I, he completely lost interest in himself. After 1919 he lived in Nice. Jean de Reszke was one of the greatest tenors of all time. To artistic phrasing, perfect enunciation, and beauty of voice, he brought dramatic power and an arresting stage presence.

de Reszke, Josephine (Sister of Jean), soprano. Born Warsaw, Poland, June 4, 1855; died there February 22, 1891. She was the sister of the world-famous De Reszke brothers, Edouard and Jean. After preliminary studies with her mother and more formal instruction with Mme. Nissen-Salomon, she made her debut at the Paris Opéra on June 21, 1875, singing the role of Ophelia. She remained at the Opera for several years, scoring a triumph in the Italian and French repertories and creating the role of Sita in *Le Roi de Lahore.* She turned down a flattering offer to appear in the United States, preferring to remain in Europe. When her brother Jean made his debut as tenor she appeared with him; and in 1884 she appeared with both her brothers in the Paris première of *Hérodiade.* At the height of her fame she married Baron Leopold de Kronenberg and retired from opera. Her only further appearances were exclusively for charity, in recognition of which the city of Posen, Poland, presented her with a diamond.

Der kleine Sandmann bin ich, the Sandman's song in Act I, Scene 2 (originally Act II), of Humperdinck's *Hänsel und Gretel.*

Der kleine Taumann heiss' ich, the Dewman's song in Act II (originally Act III) of Humperdinck's *Hänsel und Gretel.*

Der Vogelfänger bin ich, Papageno's ditty in Act I, Scene 1, of Mozart's *The Magic Flute.*

de Sabata, *see* SABATA, VICTOR DE.

de Saxe, Maurice, young man (tenor) loved by both Adriana and the Princess de Bouillon in Cilèa's *Adriana Lecouvreur.*

Desdemona, Otello's wife (soprano) in Verdi's *Otello.*

Deserto sulla terra, the troubadour's serenade in Act I, Scene 2, of Verdi's *Il trovatore.*

des Grieux, Chevalier, young nobleman (tenor) in love with Manon in Massenet's *Manon* and Puccini's *Manon Lescaut.*

des Grieux, Comte, father (bass) of Chevalier des Grieux in Massenet's *Manon.*

de Silva, see SILVA, DON RUY GOMEZ DE.

Désiré, an attendant (tenor) in Giordano's *Fedora.*

De Siriex, a diplomat (baritone) in Giordano's *Fedora.*

de Sirval, Arthur, a nobleman (tenor) posing as a painter in Donizetti's *Linda di Chamounix.*

De son coeur j'ai calmé la fièvre, Lothario's berceuse in Act III of Thomas's *Mignon.*

Despina, a maid (soprano) in Mozart's *Così fan tutte.*

Destinn, Emmy (born KITTL), dramatic soprano. Born Prague, February 26, 1878; died Budejovice, Czechoslovakia, January 28, 1930. Her music study began with the violin, but the discovery that she had a beautiful voice led to vocal study in Prague with Marie Loewe-Destinn (whose name she assumed). In 1898 Destinn made her opera debut at the Kroll Opera House in Berlin in *Cavalleria rusticana.* A month later she was engaged by the Berlin Royal Opera, where she remained for a decade. In 1901 she was chosen by Cosima Wagner to appear as Senta in the first Bayreuth production of *Der fliegende Holländer;* a few years later Richard Strauss selected her for the Berlin première of *Salome.* Meanwhile, on May 2, 1904, she made her debut at Covent Garden as Donna Anna and was such a success that she returned there for the next ten years, appearing in the first performance in England of *Madama Butterfly.* Her American debut took place at the Metropolitan Opera on November 16, 1908 (the opening night of the season) in *Aïda;* it was a performance in which Toscanini was also making his American debut. Destinn was an instantaneous success. In the *New York Times,* Richard Aldrich described her voice as "of great power, body, and a vibrant quality, dramatic in expression, flexible and wholly subservient to her intentions." She remained a principal soprano of the Metropolitan for the next decade, appearing in the world première of *The Girl of the Golden West* and the American premières of *Germania, Pique Dame,* and *Tiefland.* She was a versatile artist. Her eighty or so roles included French, German, Italian, Russian, and Bohemian operas. She was at ease in every style. She had a pronounced histrionic ability as well as a remarkable voice; she was a tragedienne in the grand style.

During World War I she was interned at her Bohemian estate. She returned to opera in 1919 with appearances in Europe and America, retiring in 1921. After opera, her major interest was writing. She wrote a play that was produced, a novel, and a considerable amount of poetry.

Deutsches Opernhaus, see CHARLOTTENBURG OPERA.

deux journées, Les (Der Wasserträger), see WATER CARRIER, THE.

de Valois, Marguerite, Henry IV of Navarre's betrothed (soprano) in Meyerbeer's *Les Huguenots.*

Devil and Daniel Webster, The, opera by Douglas Moore. Libretto by the composer, based on a short story by Stephen Vincent Benét. Première: American Lyric Theatre, New York, May 18, 1939. Jabez Stone makes a pact with the devil to yield his soul in return for money. When the time comes for Jabez to keep his bargain, Daniel Webster defends him so eloquently before a jury comprising the famous villains of history that Jabez is released from his contract.

devin du village, Le (The Village Soothsayer), comic opera by Jean Jacques Rousseau. Libretto by the composer. Première: Fontainebleau, October 18, 1752. The celebrated French philosopher was inspired to write this opera by his enthusiasm for Pergolesi's *La serva padrona.* While the work is obviously imitative, it enjoyed a tremendous success in Paris, holding the stage for over half a century and achieving over four hundred performances. It is filled with lilting, though frequently slight, tunes, such as "Si des galants de la ville" and "Avec l'objet de mes amours." Mozart's little opera, *Bastien und Bastienne,* written in 1768, was based on *Le devin du village.* The text was a German adaptation of a French parody of the Rousseau opera written by Charles Simon Favart.

Dewman, The, a character (soprano) in Humperdinck's *Hansel and Gretel.*

Dewman's song, The, *see* DER KLEINE TAUMANN HEISS' ICH.

Dich, teure Halle, Elisabeth's apostrophe to the Hall of Minstrels in Act II of Wagner's *Tannhäuser.*

D'ici je vois la mer immense (Di qui io vedo il mar), Selika's aria in Act V of Meyerbeer's *L'Africaine.*

Dickens, Charles, novelist. Born Landport, England, February 7, 1812; died Gadshill, England, June 9, 1870. Three of Dickens' novels have provided texts for operas. *The Cricket on the Hearth* was adapted by Karl Goldmark (*Das Heimchen am Herd*), Mackenzie, and Zandonai (*Il grillo sul focolare*). *The Pickwick Papers* was the source of Albert Coates's *Pickwick* and Charles Wood's *Pickwick Papers. A Tale of Two Cities* was made into an opera by Arthur Benjamin.

Dido and Aeneas, opera in three acts by Henry Purcell. Libretto by Nahum Tate, based on the fourth book of Virgil's *Aeneid.* Première: Josias Priest's Boarding School for Girls, Chelsea, about 1689.

Characters: Dido, Queen of Carthage (contralto); Belinda, her maid (soprano); Aeneas, Trojan hero, legendary founder of Rome (baritone); sorceress (mezzo-soprano); attendant; witches; courtiers; sailors. The setting is ancient Carthage.

Act I, Scene 1. The royal palace in Carthage. Dido is tormented because she is in love with Aeneas and senses that disaster awaits her. When Aeneas arrives and reassures her of his great love, her doubts are dispelled. A dance of triumph follows.

Scene 2. A cave. The sorceress and two witches, who hate Dido, plot to destroy her by robbing her of her love. One of them is to assume the form of Mercury and command Aeneas to leave Carthage. The sound of hunting horns reveals that Aeneas and his party are at a hunt. The witches laugh demoniacally as they set forth to put their plan into operation. The scene ends with the Echo Dance of the Furies.

Act II. A grove. Aeneas and his party are on a hunt, and Dido is with them. At Diana's fountain there are festivities, interrupted by a sudden storm. After Belinda has led Dido to shelter, the false Mercury comes to Aeneas to bring him Jove's comand to fulfill his destiny by leaving Carthage immediately. Anguished, Aeneas promises to obey. The sorceress and witches sing and dance at the success of their maneuver.

Act III. The harbor. Aeneas' ship is ready to sail. The sailors sing a nautical tune as they prepare to weigh anchor. The sorceress and witches watch with delight, prophesying Dido's death and the destruction of Carthage. Dido arrives and learns for the first time that her lover is about to leave her. After Aeneas has departed, she voices a poignant lament ("When I am laid in earth") and dies in Belinda's arms while courtiers sing an elegy ("With drooping wings").

Of all seventeenth century operas *Dido and Aeneas* remains the one that comes closest to our present-day concept of what a musical drama should be. Surely nowhere else in the operas before Gluck do we find such integration of music and text, song and dance, into a single artistic entity. One can say more: Few operas of any era accomplish so much so economically. There are only three principal characters. The plot is bare, the dramatic action simple to the point of ingenuousness, the conflicts and climaxes direct. The play never lags, never lacks interest, and the famous lament of Dido, "When I am laid in earth," is surely one of the most affecting songs in all opera.

See also TROYENS A CARTHAGE, LES.

Di due figli vivea, Ferrando's narrative in Act I, Scene 1, of Verdi's *Il trovatore.*

Didur, Adamo, bass. Born Sanok, Poland, December 24, 1874; died Katowice, Poland, January 7, 1946. After studying with Kasper Wysocki in Lemberg and Carlo Emerich in Milan he made his concert debut in Milan as soloist in Beethoven's Ninth Symphony. His opera debut took place in Rio de Janeiro in 1894, after which he appeared in major opera houses in Russia, Poland, and London. His American debut took place at the Manhattan Opera House in 1907, and on November 14, 1908, he made his bow at the Metropolitan as Méphistophélès in *Faust.* In 1913 he created for America the role of Boris Godunov. He remained a principal bass of the Metropolitan for a quarter of a century, making his last appearance there on

February 11, 1932, in *The Tales of Hoffmann.* He then returned to Europe. Two months before the outbreak of World War II he was appointed director of the Lemberg Opera in Poland, but the war made musical activities impossible. After the war Didur taught singing for a brief time at the Katowice Conservatory.

Diego, Don, (1) Rodrigo's father (bass) in Massenet's *Le Cid.*

(2) Council member (bass) in Meyerbeer's *L'Africaine.*

Die Majestät wird anerkannt (Hymn to Champagne), finale of Act II of Johann Strauss's *Die Fledermaus.*

Dies Bildnis ist bezaubernd schön, Tamino's aria to the portrait of Pamina in Act I, Scene 1, of Mozart's *The Magic Flute.*

Dietsch, Pierre-Louis, composer and conductor. Born Dijon, France, March 17, 1808; died Paris, February 20, 1865. He is remembered for his unhappy part in Wagner's career. When Pillet, director of the Paris Opéra, turned down Wagner's *Der fliegende Holländer* on the basis of Wagner's sketches, he bought the libretto for Dietsch. Dietsch's opera, *Le vaisseau fantôme,* was produced in Paris in 1842 and was a failure. Two decades later, as conductor at the Paris Opéra, Dietsch conducted the Paris première of *Tannhäuser,* a fiasco.

Dieu, que le monde révère (Tu che la terra adora), the chorus of councilors in Act I of Meyerbeer's *L'Africaine.*

Dieu, que ma voix tremblante, Eléazar's prayer in the Passover scene in Act II of Halévy's *La Juive.*

Die Zukunft soll mein Herz bewahren, Max's aria in the finale of Weber's *Der Freischütz.*

di Luna, Count, a nobleman (baritone) in love with Leonora in Verdi's *Il trovatore.*

D'immenso giubilo, wedding chorus in Act III, Scene 1, of Donizetti's *Lucia di Lammermoor.*

Dimmesdale, Arthur, father (tenor) of Hester Prynne's child in Damrosch's *The Scarlet Letter.*

Dimmi, Fiora, perchè ti veggo ancora, Manfredo's aria in Act II of Montemezzi's *L'amore dei tre re.*

Din, don, suona vespero, the chorus of the bells in Act I of Leoncavallo's *Pagliacci.*

d'Indy, Vincent, composer. Born Paris, March 27, 1851; died there December 2, 1931. He studied the piano with Louis Diémer and Antoine Marmontel, harmony with Alexandre Lavignac. During the Franco-Prussian War he led a bayonet attack in the battle of Val-Fleuri. After the war he began studying with César Franck, whose influence on his work was profound. From 1875 to 1879 D'Indy was chorus master of the Colonne Orchestra (Les Concerts du Châtelet). He received recognition as a composer with a series of orchestral works and an opera, *Le Chant de la Cloche,* performed in 1886. His first important opera, *Fervaal,* introduced in Brussels on March 12, 1897, was acclaimed. Later works established him as one of the major figures in French music; these included the operas *L'Etranger* (1901), *La Légende de Saint-Cristophe* (1915), and *Le Rêve de Cynias* (1923). D'Indy helped found the Société Nationale de Musique in 1871 and in 1890 succeeded Franck as its president. In 1894 he was one of the founders of the Schola Cantorum, soon to become one of France's most distinguished music schools. D'Indy taught at the Schola for many years. He was also a professor of conducting at the Paris Conservatory. In 1905 and 1921 he visited the United States, conducting performances of his works in New York and Boston.

Dinorah (Le pardon de Ploërmel), opera by Meyerbeer. Libretto by Jules Barbier and Michel Carré. Première: Opéra-Comique, April 4, 1859. A sorcerer reveals to Hoël, a goatherd, the location of a buried treasure, but warns him that the first to touch it must die. Dinorah, in love with Hoël, goes mad in the belief that Hoël has left her for good. She is almost drowned, but is rescued by her lover. On seeing Hoël, Dinorah recovers her sanity, while Hoël promises to abandon his quest. This opera is nowadays rarely performed. It is remembered for its fine overture and one aria, the "Shadow Song" ("Ombre légère"), one of Meyerbeer's most brilliant coloratura arias.

Dio! mi potevi scagliar, Otello's bitter lament that his illusions are shattered, in Act III of Verdi's *Otello.*

Dio ti giocondi, Desdemona's protestation of innocence in Act III of Verdi's *Otello.*

Di pescatore ignobile, Gennaro's aria in Act II of Donizetti's *Lucrezia Borgia.*

Dippel, Andreas, dramatic tenor and impresario. Born Kassel, Germany, November 30, 1866; died Hollywood, California, May 12, 1932. He studied singing with Julius Hey, Franco Leoni, and Johann Ress. His opera debut took place in Bremen in 1887. While a member of the Bremen Opera he was given leave to appear at the Metropolitan Opera, where he made his American debut on November 26, 1890, in the American première of Franchetti's *Asrael.* Subsequently he sang with the Breslau Opera, Munich Opera, Vienna Opera, Covent Garden, and Bayreuth; from 1898 to 1908 he was a principal tenor of the Metropolitan Opera. His extensive repertory of about a hundred and fifty roles made him one of the most valuable members of the company, since he could always be counted on to make a last-minute substitute appearance for an indisposed tenor. For the 1908–1909 season of the Metropolitan he was appointed joint manager (with Gatti-Casazza) in charge of the German repertory. He left the post in the spring of 1910, and from then until 1913 he was general manager of the Philadelphia-Chicago Opera Company. After that he managed his own light-opera company.

Di Provenza il mar, the aria of the elder Germont recalling his Provençal home, in Act II of Verdi's *La traviata.*

Di quella pira, Manrico's dramatic aria in Act III, Scene 2, of Verdi's *Il trovatore.*

Di qui io vedo il mar, *see* D'ICI JE VOIS LA MER IMMENSE.

Di rigori armato, an Italian serenade in Act I of Richard Strauss's *Der Rosenkavalier.*

Dir töne Lob, Tannhäuser's hymn to Venus in Act I of Wagner's *Tannhäuser.*

di Signa, Betto, Buoso Donati's brother-in-law (baritone) in Puccini's *Gianni Schicchi.*

Dis-moi que je suis belle, Thaïs' mirror song in Act II, Scene 1, of Massenet's *Thaïs.*

Di tanti palpiti, one of Rossini's most beautiful love songs, an aria in Act I of *Tancredi.*

Ditson Fund, a fund established at Columbia University in 1944 to commission and produce American operas. Among the works sponsored by this fund have been Ernst Bacon's *A Drumlin Legend* and *A Tree on the Plains;* Normand Lockwood's *The Scarecrow;* Otto Luening's *Evangeline;* Gian-Carlo Menotti's *The Medium;* Virgil Thomson's *The Mother of Us All;* and Bernard Rogers' *The Warrior.*

Dittersdorf, Karl Ditters, von, composer and violinist. Born Vienna, November 2, 1739; died Neuhof, Bohemia, October 24, 1799. In 1761 he toured Italy with Gluck, appearing as a violinist. Between 1764 and 1769 he was Kapellmeister for the Bishop of Grosswardein, in Hungary, for whom he wrote his first comic opera. In 1769 he was employed in a similar capacity by Count Schaffgotsch, who subsequently appointed him Overseer of Forests, and after that Chief Magis-

trate, a post which carried with it a patent of nobility. He visited Vienna in 1773, where his oratorios *Esther* and *Job* were produced successfully. Emperor Joseph II offered him the post of Kapellmeister but Dittersdorf declined it. The death of his patron and employer in 1795 brought about a sharp reversal of fortune. Dittersdorf's pension was so meager that he was now frequently in want. Only when Count von Stillfried took his family into his own household was the composer's plight alleviated. Dittersdorf continued composing up to the end of his life. His voluminous output included numerous symphonies, concertos, string quartets, and piano sonatas. He wrote some twenty-eight operas, the best being: *Doctor and Apothecary* (1786); *Orpheus the Second* (1787); *Hieronymous Knicker* (1787); *The Merry Wives of Windsor* (1797).

Di' tu se fedele, Riccardo's barcarolle in Act I, Scene 2 (or, in a later arrangement, Act II) of Verdi's *Un ballo in maschera.*

Divine Comedy, The, *see* DANTE ALIGHIERI; FRANCESCA DA RIMINI.

Divinités du Styx, Alceste's aria in Act I of Gluck's *Alceste.*

Dmitri, (1) a groom (contralto) in Giordano's *Fedora.*

(2) The assumed personality of the novice Gregory (tenor) in *Boris Godunov.*

d'Obigny, Marquis, a nobleman (bass) in Verdi's *La traviata.*

Dobrowen, Issai, conductor. Born Nishni-Novgorod, Russia, February 27, 1893; died Oslo, Norway, December 9, 1953. He attended the Conservatories in Moscow and Vienna, after which, in 1917, he was appointed professor at the Moscow Conservatory. In 1919 he made his debut as opera conductor at the Moscow Grand Theater. Three years later he became a principal conductor of the Dresden Opera, in charge of Russian operas, scoring a major success with his performance of *Boris Godunov.* From 1924 to 1927 he was first conductor of the Vienna Volksoper, and in 1927–1928 he was general music director of the Royal Opera in Sofia, Bulgaria. In 1932 he settled in Norway and became conductor of the Oslo Philharmonic Orchestra, holding this post until the Nazi occupation. He then went to Sweden and became conductor of the Stockholm Opera. Between 1949 and 1951 he led performances of Russian operas at La Scala. In the United States he was known exclusively as a symphony conductor. He made his debut with the San Francisco Symphony in 1930, subsequently appearing as guest conductor of major American orchestras. Dobrowen wrote one opera, *A Thousand and One Nights,* produced in Moscow in 1921.

Doctor and Apothecary (Der Doktor und der Apotheker), comic opera by Karl von Dittersdorf. Libretto by Gottlieb Stephanie. Première: Kärntnerthortheater, Vienna, July 11, 1786. Despite the enmity of Dr. Krautmann for the apothecary Stöszel, their children, Gotthold and Leonore, are in love and want to get married. The doctor's wife becomes an ally of the lovers and effects not only their marriage but a reconciliation between the two households.

Dodon, the king (bass) in Rimsky-Korsakov's *Le coq d'or.*

Doktor Faust, opera by Ferruccio Busoni, completed by Philip Jarnach. Libretto by the composer, based on an old German folk tale. Première: Dresden Opera, May 21, 1925. This "epic of disillusionment and disenchantment" has no relation to the Goethe drama. In place of stressing theatrical values, love interest, or characterization, Busoni moves on a spiritual plane in which ideas become all important. In *Doktor Faust* we have the essence of Busoni's creative art, the synthesis of his varied approaches to music; as such it has in-

terest and significance, but it will probably never have popular appeal.

Two orchestral excerpts are sometimes performed at symphony concerts: "Sarabande," which gives a premonition of Faust's death, and "Cortège," describing the procession of guests at the wedding of the Duke and Duchess of Parma.

Doll Song, *see* OISEAUX DANS LA CHARMILLE, LES.

Dolores, La, opera by Tomás Bretón. Libretto by the composer, based on a tale of Salares. Première: Madrid Opera, March 16, 1895. This is one of the most prominent and successful operas of the Spanish nationalist school. The plot revolves around a waitress, the seductive Dolores, who captures the hearts of many men in the small town of Calatayud.

Don Carlos, opera in four acts by Verdi. Libretto by François Joseph Méry and Camille du Locle, based on the Schiller tragedy of the same name. Première: Paris Opéra, March 11, 1867. American première: Academy of Music, New York, April 12, 1877.

Characters: Philip II of Spain (bass); Don Carlos, Infante of Spain (tenor); Rodrigo, Marquis of Posa (baritone); Elizabeth of Valois (soprano); Grand Inquisitor (bass); Princess Eboli (mezzo-soprano); a friar; Countess of Aremberg; Count of Lerma; Theobald, a page; a royal herald; ladies; gentlemen; inquisitors; courtiers; pages; guards; soldiers; magistrates. The setting is France and Spain in the sixteenth century.

Act I. Forest of Fontainebleau. Don Carlos is in love with Elizabeth of Valois. He comes to her with a gift, pretending he is only a messenger from Don Carlos. When Elizabeth recognizes that the messenger is really Don Carlos she responds to him passionately. But their love is complicated by the fact that, for reasons of state, she is compelled to marry Don Carlos' father, Philip II.

Act II. The Convent of St. Just. Elizabeth and Philip II are married; but Don Carlos is unable to forget his feelings for her. His friend Rodrigo advises him to leave the country and forget her. When Don Carlos visits Elizabeth to tell her of his departure, and to ask the king's permission to go, the old flame bursts hot again, and they fall into each other's arms.

Act III, Scene 1. The Queen's garden. At carnival time, Don Carlos mistakes the masked Princess Eboli for Elizabeth. The Princess is herself in love with Don Carlos. When she realizes that his ardent words of love are intended not for herself but for Elizabeth, she furiously denounces Don Carlos and threatens to expose him to the king.

Scene 2. A square before Nostra Donna D'Atocha. Don Carlos appears at the head of a delegation of Flemings to ask the king for mercy for these people. When the king turns a deaf ear, Don Carlos draws his sword and proclaims his willingness to fight for the freedom of the Flemings. On the king's order, Rodrigo takes away Don Carlos' sword.

Act IV, Scene 1. The King's library. Advised by Princess Eboli of Elizabeth's infidelity, the king ponders his unfortunate state as he is unable to sleep ("Dormirò sol nel manto"). The Grand Inquisitor comes to advise the king that Don Carlos must be imprisoned, and the king consents. The Princess by now repents of having been an informer and speaks of her anguish in being the instrument of Don Carlos' destruction ("O don fatale").

Scene 2. The cloisters of the Convent of St. Just. Don Carlos, freed from prison, rushes to the tomb of Charles V in the monastery to meet Elizabeth secretly. The king discovers them there. He turns Don Carlos over to the officers of the Inquisition. Suddenly, a monk, dressed in the clothes of the Emperor, steps out of the tomb. The officers, in dread of what they believe to be an apparition, free Don Carlos, who is led by the monk into safety.

Don Carlos belongs to Verdi's productive middle period in which he wrote some of his most famous and best-loved works. Possibly because he wrote this opera for performance in Paris, it has an almost Meyerbeer character, with big ensemble scenes, elaborate settings, spectacular climaxes, and overpowering dramatics. Thus, *Don Carlos* is larger in design and of greater emotional power and visual impact than operas like *Rigoletto* or *La traviata,* which preceded it; in its emphasis on spectacle, it anticipates its immediate successor, *Aïda.*

Don Giovanni (Don Juan), "dramma giocoso" in two acts by Mozart. Libretto by Lorenzo da Ponte, based on Giuseppe Bertati's play of the same name. Première: National Theater, Prague, October 29, 1787. American première: Park Theater, New York, May 23, 1826.

Characters: Don Giovanni, nobleman (baritone); Leporello, his servant (bass); Don Pedro, the Commandant (bass); Donna Anna, his daughter (soprano); Don Ottavio, betrothed to Donna Anna (tenor); Masetto, a peasant (baritone or bass); Zerlina, his fiancée (soprano); Donna Elvira, lady of Burgos (soprano); peasants; musicians. The setting is Seville, Spain, in the middle of the seventeenth century.

Act I, Scene 1. Courtyard of the Commandant's palace. Don Giovanni emerges from the palace followed by Donna Anna. They exchange angry words. Donna Anna's father appears, and before long he and Don Giovanni are engaged in a duel in which the Commandant is killed. After the culprit makes his escape, Donna Anna—who has run for help—comes back with her betrothed, Don Ottavio. She is grief-stricken to find her father dead. Together with her lover she vows vengeance.

Scene 2. A square outside Seville. Don Giovanni, followed by Leporello, comes upon a weeping woman. She turns out to be Donna Elvira, one of Giovanni's former loves. After Giovanni makes a discreet departure, Leporello enumerates to Donna Elvira his Master's conquests (Catalogue Song: "Madamina, il catalogo"). Now it is Donna Elvira who swears she will some day destroy the fickle Don.

Scene 3. A country place near Don Giovanni's castle. The marriage of Masetto and Zerlina is about to be celebrated. Since Don Giovanni finds Zerlina attractive, he has Leporello invite the townspeople into the castle so that he can be alone with her. Zerlina tries her best to resist his charm ("Là ci darem la mano"). Donna Elvira now makes a timely appearance and denounces Don Giovanni before Zerlina, Donna Anna, and Don Ottavio. After Don Ottavio repeats his intent to avenge the Commandant's death, he sings about his beloved, who is the only one able to bring him peace of mind ("Dalla sua pace").

Scene 4. The garden of Don Giovanni's palace. A feast has been arranged. Masetto, jealous of Giovanni's flirtation with Zerlina, quarrels with her but has a change of heart after Zerlina's coy plea for forgiveness ("Batti, batti, o bel Masetto"). Among the guests are Donna Anna, Donna Elvira, and Don Ottavio, all three masked: Don Giovanni invites them, with his other guests, into the palace.

Scene 5. Inside the palace. While the guests are dancing the famous Minuet, Don Giovanni draws Zerlina into another room. Her outcries attract the attention of the guests. Suddenly, Don Giovanni enters, dragging Leporello after him and accusing his servant of having made advances to Zerlina. But the guests are not fooled, particularly not Donna Anna, Donna Elvira, and Don Ottavio. Drawing his sword, Don Giovanni forces his way through the crowd and escapes.

Act II, Scene 1. Before Donna Elvira's house. Don Giovanni comes here because he learned that Zerlina is in the house. He and Leporello exchange cloaks. Disguised as Leporello, Don Giovanni serenades Zerlina ("Deh, vieni, alla finestra"). Masetto comes on the scene determined to give Giovanni a thrashing, but fails to recognize Giovanni in Leporello's cloak. Being off guard, he is thrashed by Giovanni. Zerlina comes to Masetto to soothe him as he lies stretched out in the street. Masetto confesses he is hurt more grievously in heart than in body, and Zerlina consoles him ("Vedrai, carino").

Scene 2. The garden of the Commandant's palace. When Leporello, still disguised as Don Giovanni, is confronted by Don Ottavio, Donna Anna, Donna Elvira, and Zerlina, he reveals his true identity to escape punishment. Don Ottavio once again speaks of his great love for Donna Anna ("Il mio tesoro").

Scene 3. In a graveyard, Don Giovanni and Leporello meet near a statue raised to the memory of the late Commandant. Don Giovanni mockingly orders Leporello to invite the statue to dinner. To Leporello's horror, the statue nods its head in acceptance.

Scene 4. In the Commandant's palace. When Don Ottavio entreats Donna Anna to marry him, she replies sorrowfully that while she loves him she cannot be his wife as long as she is filled with her terrible grief ("Non mi dir").

Scene 5. Banquet hall in Don Giovanni's palace. While Don Giovanni dines alone, his private orchestra (on stage) plays for him such delightful morsels as the "Non più andrai" aria from Mozart's *The Marriage of Figaro*.

Donna Elvira comes to prevail on Giovanni to reform. Her quest proving futile, she departs. She quickly returns, screaming, and rushes out another door. Leporello, too, is a victim of terror when he goes to investigate the source of Elvira's fear. For the statue of the Commandant has come to keep his dinner appointment. He is welcomed fearlessly by the proud Don Giovanni. Since Don Giovanni refuses to change his ways, the statue consigns him to the fiery world below. The opera ends with a brief epilogue (sometimes omitted) in which Donna Anna, Donna Elvira, Zerlina, and Ottavio learn from Leporello of Don Giovanni's fate and rejoice over it.

Don Giovanni was the second collaboration of Mozart and the librettist, Da Ponte, coming just a year after *The Marriage of Figaro. Don Giovanni* was commissioned from the Bondini Opera Company in Prague, which had scored a resounding success with *Figaro.* What that company wanted was another comic opera in like vein, and it is probable that this is what the authors set out to do when they started writing. But, recognizing Mozart's power and passion, Da Ponte emphasized the dramatic element in his text, though without abandoning comic and satiric nuances; and Mozart, ever sensitive, alternated the light and comic strokes with darker colors of genuine tragedy. Recognizing that he had produced neither an opera buffa nor an opera seria, but a combination of both, Mozart applied to his new opera the designation of "dramma giocoso" or "gay drama." The gaiety is there: in the character of Leporello, in the playful quarrels and reconciliations of Masetto and Zerlina, in the fleet and witty music Mozart wrote for these opera buffa characters. The Italian grace of Don Giovanni's serenade and Don Ottavio's love songs are also in the opera-buffa tradition. But it is the tragic element, rather than

the comic, that is accentuated, particularly in the characterization of Don Giovanni. And it is with its tragedy that this opera arrives at its highest plane of eloquence: with Donna Anna's terrible grief in the first scene, or the cataclysmic music with which Don Giovanni meets his doom in the closing one. That this unconventional alternation of comedy and tragedy did not confuse its first audience speaks volumes for its discrimination. "Connoisseurs and artists say that nothing like this has been given in Prague," reported a contemporary journal. "Mozart himself conducted, and when he appeared in the orchestra, he was hailed by a triple acclamation." Since then, *Don Giovanni* has been generally accepted as Mozart's greatest opera; it is also one of the oldest operas that belongs in the permanent repertory of every major opera house.

See also DON JUAN.

Donizetti, Gaetano, composer. Born Bergamo, Italy, November 29, 1797; died there April 8, 1848. He began the study of music with Simon Mayr in Naples. The operas of Rossini, which he read in score, profoundly attracted him. Later, in Bologna, he studied with Rossini's teacher, Padre Stanislao Mattei, and aspired to become an opera composer. Since his father objected to this, the young man enlisted in the Austrian army. His military duties did not interfere with his music. In 1818 he completed his first opera, *Enrico di Borgogna,* and it was performed the same year in Venice with moderate success. His first major success came with his fourth opera, *Zoraïde di Granata,* introduced in Rome on January 28, 1822. This triumph brought the composer an official release from the army. Now able to concentrate on music, he produced operas with amazing rapidity —twenty-one between 1822 and 1828. In all these works the influence of Rossini is predominant. The first opera in which Donizetti's own personality asserted itself was *Anna Bolena,* given in Milan in 1830 and soon performed throughout Europe. Donizetti's increasing powers became even more evident in two later works, both of them still in the repertory. One was the comic opera, *L'elisir d'amore,* introduced in Milan in 1832; the other was *Lucia di Lammermoor,* first performed in Naples in 1835.

In 1837 Donizetti was appointed director of the Naples Conservatory. He held this post for two years. In 1839, aroused by a bitter feud with the censor over the performance of his opera *Poliuto,* Donizetti left Italy and went to Paris. Here he assisted in the performance of some of his operas. Several others were given their premières here, among them *The Daughter of the Regiment* and *La favorita,* both performed in 1840. Donizetti was now one of the most celebrated composers in Europe. He visited Vienna in 1842 for the première of *Linda di Chamounix* and was acclaimed; the Emperor conferred an honorary title on Donizetti. Back in Paris, the composer completed one of his best comic operas, *Don Pasquale,* and was present at its highly successful première on January 3, 1843. His last opera was *Caterina Cornaro,* given in Naples in 1844. Soon after this, Donizetti began suffering violent headaches and depressions. His mental deterioration necessitated his commitment to an asylum. He made enough improvement to be able to leave it in the custody of his brother at Bergamo. When he died a few months later, he was given a hero's funeral by his native city.

Donizetti was extraordinarily fertile and facile. He wrote sixty-seven operas, most of them now forgotten. Though he produced much that was trite and superficial, he was nevertheless a major figure in Italian opera in the era between Rossini and Verdi. At his best, Donizetti had a wonderful gift of melody and a sound instinct for effective theatricalism. The best pages of his serious operas have power and passion, and his best comic operas are marked by spontaneity, verve, and gaiety.

His finest operas were: *Anna Bolena* (1830); *L'elisir d'amore* (1832); *Parisina* (1833); *Lucrezia Borgia* (1833); *Gemma di Vergy* (1833); *Lucia di Lammermoor* (1835); *Il campanello di notte* (1836); *Roberto Devereaux* (1837); *Gianni di Parigi* (1839); *Poliuto* (1840); *La fille du régiment* (*The Daughter of the Regiment*) (1840); *La favorita* (1840); *Linda di Chamounix* (1842); *Maria di Rohan* (1843); *Don Pasquale* (1843).

Don José, brigadier (tenor) in love with Carmen in Bizet's *Carmen.*

Don Juan, libertine of legend, drama, and poetry. He is the central character of several notable operas. The most celebrated is Mozart's *Don Giovanni.* Others include Franco Alfano's *L'ombra di Don Giovanni;* Dargomizhsky's *The Stone Guest;* Eugène Goossens' *Don Juan de Mañara;* Paul Graener's *Don Juans letztes Abenteuer;* Henry Purcell's *The Libertine;* Vissarion Shebalin's *The Stone Guest;* and Felice Lattuada's *Don Giovanni.* See also BYRON, GEORGE NOEL GORDON, LORD.

Donna Diana, comic opera by Emil von Reznicek. Libretto by Julius Kapp based on a comedy of the same name by Moreto y Cavaña. Première: Prague Opera, December 16, 1894. The composer's most famous work is now seldom performed, but in its time it was extraordinarily popular throughout Germany. Its sprightly overture, however, has survived. The setting is Barcelona, during Catalonia's independence. Don Cesar is in love with Donna Diana, who is cold to him. He uses the ruse of appearing indifferent to her and thus is able to arouse her interest and finally her love. A waltz in the first act and the ballet music in the second are of particular interest.

Donna non vidi mai, Des Grieux's romance in Act I of Puccini's *Manon Lescaut.*

donna serpente, La (The Serpent Woman), opera by Alfredo Casella. Libretto by Cesare Lodovici, based on a comedy by Carlo Gozzi. Première: Teatro Reale, Rome, March 17, 1932. This is the same story that Wagner used for his first opera, *Die Feen.*

donne curiose, Le (The Inquisitive Women), comic opera by Wolf-Ferrari. Libretto by Luigi Sugana, based on a comedy by Carlo Goldoni. Première: Munich Opera, November 27, 1903. Three women, convinced that their husbands are carrying on an orgy at their men's club, manage to gain admission there and to secrete themselves. They discover to their delight that all their men are doing is eating a meal.

Donner, thunder god (bass) in Wagner's *Das Rheingold.*

Don Pasquale, opera buffa in three acts by Gaetano Donizetti. Libretto by Giacomo Ruffini and the composer, based on Angelo Anelli's libretto for an opera by Stefano Pavesi, *Ser Marcantonio.* Première: Théâtre des Italiens, Paris, January 3, 1843. American première: Park Theater, New York, March 9, 1846.

Characters: Don Pasquale, an old bachelor (bass); Ernesto, his nephew (tenor); Norina, a young widow (soprano); Dr. Malatesta, a physician (baritone); a notary; valets; chambermaids. The setting is Rome in the early nineteenth century.

Act I, Scene 1. A room in Don Pasquale's house. Don Pasquale is opposed to the love affair of his nephew Ernesto and the young widow Norina. The lovers, however, have found an ally in Dr. Malatesta, and they contrive a plot to win over the old bachelor. Dr. Malatesta rapturously describes to Don Pasquale his beautiful and wholly imaginary sister ("Bella siccome un angelo"), and tells the old bachelor that the girl is in love with him. Before long, Pasquale is convinced he loves the girl and expresses the wish to marry her ("Ah! un foco insolito"). He is even ready to cut Ernesto out of his will. Unaware that there is a plot afoot, Ernesto grows bitter at the betrayal at the hands of his good friend, Malatesta ("Sogno soave e casto").

Scene 2. In Norina's house. As she reads a novel, Norina insists she knows all the tricks of winning a man's love ("So anch' io la virtù magica"). Malatesta comes to Norina and reveals his plans for fooling Pasquale. He wants Norina to masquerade as his sister, wed the old man in a mock marriage, and then so torture him with her whims and caprices that Pasquale will eagerly seek a way out of his hasty and unhappy marriage.

Act II. Don Pasquale's house. Aware that without his uncle's money he will never be able to marry Norina, Ernesto once again laments his sad lot ("Cercherò lontana terra"). When he leaves, Malatesta arrives with the "bride" and presents her to the handsomely attired Pasquale. A mock marriage takes place without further delay during which Ernesto returns and realizes for the first time that a fraud is being enacted. The moment the ceremony ends, Norina becomes a hot-tempered shrew who harasses her husband with her vicious tongue and her extravagance.

Act III. Don Pasquale's house. When Norina, beautifully gowned, brazenly leaves the house and Pasquale discovers a love letter addressed to her, he realizes that he has come to the end of his rope. In a rage he calls Malatesta, who promises to set matters right. In Pasquale's garden Ernesto sings a serenade to his beloved ("Com è gentil") and Norina responds ardently. Trapped at last by old Pasquale, the conspirators explain their intrigue. Pasquale is so relieved at being freed from his distress that he forgives them and readily consents for Ernesto and Norina to marry.

Don Pasquale was Donizetti's last successful opera. He wrote four more, but they were failures. A good case can be made for the claim of some writers that *Don Pasquale* is Donizetti's finest work. Certainly it belongs with the greatest Italian opera buffas. Old traditions are adhered to. The characters belong to everyday existence and their problems are the little complications of everyday life. (As a matter of fact, in Donizetti's time it was customary to present *Don Pasquale* in contemporary dress.) This opera has the inevitable busybody of opera buffa who sets the intrigue into motion. There are the usual amatory complications, serious and comic, which are finally straightened out to the satisfaction of all concerned. But while the formula is a familiar one, the musical treatment gives the work its originality and distinction. The music never lacks sparkle and freshness, and there is also found here wonderful bel canto writing, together with passages of discreet sentimentality which bring a welcome change of pace.

When the opera was introduced it enjoyed one of the most notable casts ever assembled for an opera première. Pasquale was played by Lablache, one of the foremost interpreters of that role, about whom Ernest Newman has written "opera has perhaps never seen or heard his like before or since." Tamburini was Malatesta; Grisi, Norina; and Mario, Ernesto. Closer to our own day, one of the most notable of Pasquales has been Salvatore Baccaloni, for whom the Metropolitan Opera successfully revived the opera in 1940.

Don Pedro, (1) a three-act opera created in 1952 by Hans Erismann from heretofore neglected music by Mozart. A new libretto was written by Oskar Walterlin and Werner Galusser from material by Lorenzo da Ponte and Abbé G. B. Varesco. The music was derived from two uncompleted operas —*L'Oca del Cairo* and *Lo sposo deluso*

—and nineteen miscellaneous arias which Mozart wrote about 1783. These fragments were integrated by the Swiss musician Erismann, who added recitatives of his own. *Don Pedro* was introduced at the Zurich Municipal Theater under Erismann's direction in 1952; it was well received.

(2) Councilor (bass) in Meyerbeer's *L'Africaine*.

(3) The Commandant (bass) in Mozart's *Don Giovanni*.

Don Quichotte (Don Quixote), opera in five acts by Jules Massenet. Libretto by Henri Cain after a play by Jacques Le Lorrain, based on the novel of Cervantes. Première: Monte Carlo Opera, February 19, 1910. American première: New Orleans, January 27, 1912.

Characters: Don Quixote (baritone or bass); La Belle Dulcinée, courtesan (soprano); Sancho Panza (baritone); Pedro, a burlesquer (soprano); Garcias, another burlesquer (soprano); Rodriguez (tenor); Juan (tenor); valets. The setting is Spain.

Act I. A square in front of Dulcinea's house. Don Quixote arouses the jealousy of one of Dulcinea's admirers by singing her a serenade ("Quand apparaissent les étoiles"). Dulcinea intervenes to prevent a duel, then sends Don Quixote on a fool's errand to retrieve a necklace stolen by a brigand; she promises to marry him if he is successful.

Act II. Before the windmills. In search of the necklace, Don Quixote and his servant Sancho Panza approach some windmills. Mistaking them for giants, the Don attacks them.

Act III. The camp of the brigands. By his knightly manner, Don Quixote is able to win over the brigands to the point that they surrender Dulcinea's necklace.

Act IV. A salon in Dulcinea's house. The Don and Sancho come to Dulcinea's house to turn over the necklace. Dulcinea is overjoyed to receive it, but when Don Quixote demands her hand as his reward, she derides him and sends him away.

Act V. A forest path. Don Quixote is dying. He begs Sancho to pray for him (Mort de Don Quichotte: "Ecoute, mon ami"). In the distance, the Don hears the voice of Dulcinea singing to him of love. Transported, he dies, leaving the grief-stricken Sancho.

Don Quixote, the self-styled knight-errant of La Mancha, principal character in Cervantes' famous romance of the same name. He appears in several operas, including: Antonio Caldara's *Don Chisciotte* and *Sancio Panza;* Donizetti's *Il furioso;* Francesco Feo's *Don Chisciotte;* Manuel de Falla's *El Retablo de Maese Pedro;* Vito Frazzi's *Don Chisciotte;* Jacques Ibert's *Le Chevalier errant;* Wilhelm Kienzl's *Don Quixote;* Adolf Neuendorff's *Don Quixotte;* George Macfarren's *The Adventures of Don Quixote;* Massenet's *Don Quichotte* (see above); Felix Mendelssohn's *Die Hochzeit des Camacho;* Giovanni Paisiello's *Don Chisciotte;* Antonio Salieri's *Don Chisciotte.*

Dorabella, Fiordiligi's sister (soprano) in love with Ferrando, in Mozart's *Così fan tutte.*

Dorian Gray, see WILDE, OSCAR.

Dormirò sol nel manto, soliloquy of King Philip II in Act IV, Scene 1, of Verdi's *Don Carlos.*

Dorota, Schwanda's wife (soprano) in Weinberger's *Schwanda.*

Dorothea, Cinderella's stepsister (mezzo-soprano) in Massenet's *Cendrillon.*

Dositheus, leader of the Old Believers (bass) in Mussorgsky's *Khovantchina.*

Dostoyevsky, Feodor, novelist. Born Moscow, November 11, 1821; died St. Petersburg, February 9, 1881. His stories and novels have been used in the following operas: Leoš Janáček's *From a House of the Dead;* Otakar Jeremias' *The Brothers Karamazov;* Prokofiev's *The Gambler;* Heinrich Sutermeister's *Raskolnikoff;* and Arrigo Pedrollo's *Delitto e castigo*—the latter two based on *Crime and Punishment.*

Douphol, Baron, Alfredo's rival (baritone) for Violetta in Verdi's *La traviata.*

Dove Duo, see O HOLDES BILD.

Dove son? O qual gioia, see O TRANSPORT, O DOUCE EXTASE.

Dove sono, Countess Almaviva's aria in Act III of Mozart's *The Marriage of Figaro.*

Down in the Valley, folk opera by Kurt Weill. Libretto by Arnold Sundgaard. Première: Bloomington, Indiana, July 15, 1948. As the opening words of the "Leader" explain, this opera is about "Brack Weaver, who died on the gallows one morning in May; he died for the love of sweet Jennie Parsons, he died for the slaying of Thomas Bouche." Brack's story is then told in a series of flashbacks. Weill's score includes five authentic American folk songs: the title song, "The Lonesome Dove," "The Little Black Train," "Hop Up, My Ladies," and "Sourwood Mountain."

Dramma giocoso, an Italian term for "gay drama," or an opera buffa with interpolations of tragic situations. The term is applied to certain eighteenth-century operas; the most celebrated example is *Don Giovanni.*

Dramma per musica, Italian for "drama through music," the name by which opera was first designated when the form was created by the Camerata in Florence. The word "opera" first replaced "dramma per musica" in 1639 with Cavalli's *Le nozze di Teti e di Peleo.*

Dream Pantomime, the orchestral interlude in Act II of Humperdinck's *Hänsel und Gretel* accompanying the descent of the fairies to provide a protective ring for the children as they fall asleep in the forest.

Dreigroschenoper, Die, see THREE-PENNY OPERA, THE.

Dresden Amen, a famous choral amen,

believed to have been written by Johann Gottlieb Naumann (1741–1806). It was used by Wagner in *Parsifal*. Naumann, a brilliant figure in his day, wrote a number of operas (*Cora* was his most successful) and a great deal of church music.

Dresden Opera, one of the most important operatic institutions in Germany. Its predecessor was the Royal Opera of Saxony, which up to the middle of the nineteenth century was dominated by the Italian school. When Weber became director in 1816, a reorganization took place after which German operas were emphasized. A public opera house for the company was completed in 1841, and it was in this theater that *Rienzi* was first performed on October 20, 1842. On January 2 of the following year the Dresden Opera gave the première of *Der fliegende Holländer*. In February, 1843, Wagner became the kapellmeister, a post he held for six years; he was a vital factor in the artistic rehabilitation of the company. In 1869 the theater burned down. Rebuilt from the original plans (with modifications) it reopened in 1878. Under the artistic direction of Ernst von Schuch (from 1882 to 1914) and Fritz Busch (from 1926 to 1933) the Dresden Opera became one of the foremost opera houses of Germany, if not all of Europe. Most of Richard Strauss's operas were introduced in Dresden; other notable premières have included Hindemith's *Cardillac;* Paderewski's *Manru;* Kurt Weill's *Der Protagonist;* Wolf-Ferrari's *L'amore medico;* Ferruccio Busoni's *Doktor Faust;* Paul Graener's *Hanneles Himmelfahrt;* Othmar Schoeck's *Penthesilea;* Dohnányi's *Tante Simona.*

The opera house was partially destroyed during World War II. After being rebuilt, it opened on September 22, 1948 with *Fidelio*. Karl von Appen was the new artistic director; Joseph Keilberth the musical director. Keil-

berth was succeeded in 1950 by Rudolf Kempe. Important premières since the war have included two operas by Carl Orff: *Die Kluge* and *Antigone.*

Drinking Song, *see* TRINKE, LIEBCHEN, TRINKE SCHNELL.

Drumlin Legend, A, folk opera by Ernst Bacon. Libretto by Helene Carus. Première: New York City, May 4, 1949. This opera has an Irish-Scotch folk-music background. The story concerns the conflict of a former aviator between his desire to return to the cockpit and his wish to settle down with a country schoolteacher with whom he is in love. His dilemma is resolved when elves and woodsprites endow him with the power of flying in his imagination, while following his everyday humdrum existence.

Drury Lane Theatre, a theater in London which opened in 1696 had a long and notable history of operatic productions up to World War I. From 1738 to 1778 Thomas Arne was its official composer, and many of his operas were here first produced. It is interesting to note that the practice of providing analytical program notes was instituted by Arne at this theater in 1768. In 1833, under the managership of Alfred Bunn, Drury Lane produced Italian operas in English translations. Balfe's *The Bohemian Girl,* Wallace's *Maritana,* and Benedict's *The Crusaders* were some of the English works written for, and introduced at, Drury Lane during this period. From 1867 to 1877 James Henry Mapleson used Drury Lane for his annual season of summer operas. In 1870 the theater was under the direction of George Wood when the first performance of a Wagner opera took place in England: *Der fliegende Holländer.* In 1882 Hans Richter gave here the first performances in England of *Tristan und Isolde* and *Die Meistersinger.* A year later the Carl Rosa Company leased the house, giving the premières of several English operas, including Stanford's *The Canterbury Pilgrims*. In 1887 Augustus Harris began a long and distinguished career as opera impresario at Drury Lane. His company included the De Reszke brothers, then appearing in England for the first time. German opera was seen at Drury Lane in 1892–1893, light opera in 1895, and opera in English in 1896. In 1904 the Moody-Manners Company occupied the theater for a season of operas in English, and in 1913 and 1914 Sir Thomas Beecham directed highly successful seasons of Russian operas, including the English première of *Boris Godunov* with Chaliapin.

Dryad, a character (contralto) in Richard Strauss's *Ariadne auf Naxos.*

Dryden, John, poet and playwright. Born Aldwinkle, England, August 9, 1631; died London, May 1, 1700. Henry Purcell wrote theater music for many of Dryden's plays, among them *Amphitryon, Aurengzebe, The Indian Queen, Love Triumphant, King Arthur, Oedipus, The Spanish Friar* and *Tyrannic Love.* Thomas Arne's *Cymon and Iphigenia* and Handel's *Acis and Galctea* were both derived from Dryden's works.

Du bist der Lenz, Sieglinde's love song in Act I of Wagner's *Die Walküre.*

Duchess of Towers, Mary, character (soprano) in Deems Taylor's *Peter Ibbetson.*

due Foscari, I, (The Two Foscari), opera by Verdi. Libretto by Francesco Piave, based on Byron's drama *The Two Foscari.* Première: Teatro Argentina, Rome, November 3, 1844. Loredano, a member of the Venetian Council, vows to destroy the house of Foscari, believing it responsible for the death of both his father and uncle. He is instrumental in having the Council exile Jacopo Foscari for the crime, and finally sees Jacopo's father compelled to abdicate from the Council. Though the perpetrator of the murders is subsequently proved to be somebody other than Jacopo, Loredano's vengeance is complete: Jacopo dies on his way to exile, and his father collapses after his compulsory abdication.

Duet, a song for two voices.

Duet of the Cherries, *see* TUTTO TACE.

Dufresne, Milio, Zaza's lover (tenor) in Leoncavallo's *Zaza.*

Dukas, Paul, composer. Born Paris, October 1, 1865; died there, May 17, 1935. He attended the Paris Conservatory for eight years; his teachers included Théodore Dubois and Ernest Guiraud. Some recognition came to him in 1892 for an orchestral overture *Polyeucte,* but fame was realized in 1897 with his most celebrated composition, the orchestral scherzo, *The Sorcerer's Apprentice* (*L'Apprenti sorcier*). Subsequently, major works like the "danced poem" *La Péri* and the opera *Ariane et Barbe-Bleue* placed him in the front rank of contemporary French composers. The opera was introduced at the Opéra-Comique on May 10, 1907, and a few months later it entered the permanent repertory of the Paris Opéra. After 1910 Dukas wrote little, concentrating on teaching; he was a professor at the Conservatory for a time, and writing criticisms for the French journals. In the last year of his life he succeeded Alfred Bruneau at the Académie des Beaux-Arts.

Duke of Mantua, The, a nobleman (tenor) in Verdi's *Rigoletto.*

Dulcamara, a quack doctor (bass) in Donizetti's *L'elisir d'amore.*

Dulcinée, La belle, courtesan (soprano) in Massenet's *Don Quichotte.*

du Locle, Camille, librettist. Born Orange, France, July 16, 1832; died Capri, Italy, October 9, 1903. For many years he was secretary of the Paris Opéra, when Perrin was director, and afterwards he served as director of the Opéra-Comique. He helped write the texts for Verdi's *Aïda, Don Carlos* and *La forza del destino,* and he prepared the librettos for Ernest Reyer's *Salammbô* and *Sigurd*.

Dumas, Alexandre, (père), novelist and playwright. Born Villiers-Cotterets, France, July 24, 1802; died Puys, France, December 5, 1870. Dumas's works were used for the following operas: César Cui's *The Saracen;* Donizetti's *Gemma di Vergy;* Isidore de Lara's *Les trois mousquetaires;* Flotow's *La Duchesse de Guise;* Humperdinck's *Die Heirat wider Willen;* and Saint-Saëns' *Ascanio*.

Dumas, Alexandre, (fils), novelist and playwright. Born Paris, July 27, 1824; died there November 27, 1895. Son of Alexandre Dumas père, he was the author of the celebrated play *La dame aux camélias*, the source of Verdi's *La traviata*. Hamilton Forrest is another composer who made an opera (*Camille*) of this play.

Duo de la fontaine, (1) duet of Pelléas and Mélisande in Act II, Scene 1, of Debussy's *Pelléas et Mélisande*.

(2) Duet of Pelléas and Mélisande in Act IV, Scene 4, of the same opera.

Duo de la lettre, (1) *see* J'ECRIS A MON PERE.

(3) *See* VOICI CE QU'IL ECRIT.

Durand et Compagnie, the foremost music publishers of France. The company was founded in Paris in 1870 by Marie Auguste Durand (a professional music critic, organist, and composer) when he acquired the publishing house of Flaxland and altered the name to Durand et Schoenewerk. In 1891 Durand's son Jacques replaced Schoenewerk and the house became known as Durand et Fils. Still later, the name was changed to Durand et Compagnie. As the principal publishers for Massenet, Lalo, Saint-Saëns, Bizet, Debussy, Ravel, among others, the house of Durand issued the famous French operas of the late nineteenth and early twentieth centuries. Among its other significant contributions to opera were a monumental edition of the complete works of Rameau, edited by Saint-Saëns, and the first French editions of Wagner's *Lohengrin, Tannhäuser* and *Der fliegende Holländer*.

Durante, Francesco, composer and teacher. Born Frattamaggiore, Italy, March 31, 1684; died Naples, August 13, 1755. As the director of the Conservatorio San Onofrio from 1718 to 1742, and of the Conservatorio di Santa Maria di Loreto in Naples from 1742 to the time of his death, Durante taught an entire generation of Italian opera composers including Jommelli, Paisiello, Pergolesi, Piccinni, and Sacchini. His own creative output consisted primarily of religious and choral music.

Durch die Wälder, durch die Auen, Max's aria in Act I of Weber's *Der Freischütz*.

Dusk of the Gods, The, *see* RING DES NIBELUNGEN, DER.

Dutchman, The, (baritone) principal character in Wagner's *Der fliegende Holländer*.

Du trugest zu ihm meine Klage, Elsa's prayer in Act I of Wagner's *Lohengrin*.

Du und Du Waltzes, music in the finale of Act II of Johann Strauss's *Die Fledermaus;* known also in orchestral transcription.

Dvořák, Antonín, composer. Born Nelahozeves, Bohemia, September 8, 1841; died Prague, May 1, 1904. The Bohemian nationalist composer wrote nine operas (a tenth was abandoned).

Dvořák's music study began with a teacher in Zlonice, Antonin Liehmann, who was the first to recognize his talent. Encouraged by his teacher, Dvořák entered the Organ School in Prague when he was sixteen. His studies ended, he played in the orchestra of the National Opera for eleven years; during this period he came under the influence of Smetana, then conductor at the National Opera, who aroused in him the ambition to write Bohemian national music. Dvořák's first opera,

King Alfred, written in 1870, was, however, but a pale reflection of a Wagnerian music drama; the composer never allowed it to be published or performed. Smetana's increasing influence made it possible for Dvořák to free himself from Wagnerian influences. He now turned to Bohemian subjects for his operas, filling them with folk-like songs and dances. His first venture in this direction was a comic opera, *The King and the Collier,* a failure when produced by the National Opera in 1871. Another comic opera, *The Devil and Kate,* given by the National Opera on November 23, 1899, was such a triumph that it immediately entered the permanent Bohemian repertory; it was also performed in Germany and Austria. Another great success came with the national opera, *Rusalka,* performed in 1901. Dvořák's last opera, *Armida,* however, was a failure in 1904. It marked Dvořák's return from nationalism to mythology. The composer's disappointment in the reception given *Armida* is believed to have been contributory to the breakdown of his health and his premature death.

For three years, beginning in 1892, Dvořák lived in America, teaching in New York City as director of the National Conservatory, spending his summer vacations in Spillville, Iowa. The fruits of this period were several chamber and orchestral works (Opus 95, Symphony "From the New World") containing Negro or American Indian thematic material.

Dvořák's operas, in addition to those already mentioned: *The Pigheaded Peasants* (1874), *Vanda* (1875), *The Cunning Peasant* (1877), *Dimitrij* (1882), and *The Jacobin* (1888).

Dybbuk, The, (1) a famous Yiddish drama by S. Ansky, based on an old Hebrew belief that the spirit of a dead person may enter and obsess the body of a living one. In the play, Chanon, a Chassidic student, becomes absorbed in the mysteries and mysticism of the Kabala. He tries to acquire from the sacred book the power to gain wealth so that he may win Leah, with whom he is in love. He dies, and his spirit enters Leah's body. The play has been used as the basis of several operas.

(2) Opera by Lodovico Rocca. Libretto by R. Simoni, based on the Ansky play. Première: La Scala, March 7, 1934.

(3) Opera by David Tamkin. Libretto by Alex Tamkin (the composer's brother) based on the Ansky play. Première: New York City Opera, May 4, 1951.

Dyck, Ernest Van, see VAN DYCK, ERNEST.

Dzerzhinsky, Ivan, composer. Born Tambov, Russia, April 8, 1909. His musical education did not begin until his nineteenth year when he entered the Gnessin Music School. From 1930 to 1932 he attended the First State Musical School, and from 1932 to 1934 he was at the Leningrad Conservatory, where one of his teachers, Riazanov, was an important influence in his development. He wrote some piano music from 1932 to 1934 and in the latter year completed the opera that made him famous, *The Quiet Don,* produced at the Leningrad Little Opera Theater; within three years the work had two hundred performances in the Soviet Union. He subsequently completed several more operas: *Virgin Soil Upturned* (1937); *In the Days of Volochaiev; The Storm; The Blood of the People* (1942).

E

Eadgar of Wessex, king of England (baritone) in Deems Taylor's *The King's Henchman.*

Eames, Emma, soprano. Born Shanghai, China, August 13, 1865; died New York City, June 13, 1952. The daughter of American parents, she was brought to the United States when she was five. After study in Boston with Clara Munger, she was encouraged by Wilhelm Gericke, conductor of the Boston Symphony, to go to Paris. She arrived there in 1886 and for two years studied with Mathilde Marchesi. Her opera debut took place at the Paris Opéra on March 13, 1889, in *Roméo et Juliette* (the composer Gounod himself selected her for this performance). She was such a success that she was required to sing the role of Juliette ten times in a single month. She was also called upon to create the roles of Colombe in Saint-Saëns' *Ascanio* and Zaïre in Paul de la Nux's *Zaïre.* Intrigues and cabals by envious singers compelled her to leave the Paris Opéra after two seasons. On December 14, 1891, she made her American debut at the Metropolitan Opera, once again as Juliette. For the next eighteen years she was a favorite of the opera public in New York and London. Her voice was not large, but it was used with consummate artistry. Her greatest triumphs came in *Tosca* (a performance Puccini himself praised ecstatically), *Don Giovanni,* and *Aïda.* She also appeared in the American premières of *Falstaff* and *Werther.* Disagreeing with the artistic policies of the then new regime of Gatti-Casazza at the Metropolitan Opera, she resigned in 1909, her last appearance taking place in *Tosca* on February 15. She went into temporary retirement from which she emerged in 1911–1912 for two performances with the Boston Opera Company as Desdemona and Tosca. She subsequently appeared in recitals with her husband, Emilio de Gogorza. She was frequently decorated, her honors including the English Jubilee Medal (presented after a command performance for Queen Victoria in 1896) and the order of Les Palmes Académiques from the French Academy. She wrote her autobiography, *Some Memoirs and Reflections* (1927).

E Amore un ladroncello, Dorabella's aria in Act II, Scene 3, of Mozart's *Così fan tutte.*

Easton, Florence, soprano. Born Middlesbrough-on-Tees, England, October 25, 1884. As a child she made a public appearance as pianist in Canada. In her fourteenth year she entered the Royal Academy of Music in London. After voice coaching with Elliott Haslam in Paris she made her opera debut in 1903 with the Moody-Manners Opera Company as Cio-Cio-San. Two years later she appeared for the first time in America with the Henry Savage Opera Company, singing Kundry and Cio-Cio-San in the English versions of *Parsifal* and *Madama Butterfly.* Between 1907 and 1913 she was a member of the Berlin Royal Opera, where she achieved her first major successes. In 1910 she was heard at Covent Garden in the London première of *Elektra,* and from 1913 to 1915 she was the principal soprano of the Hamburg Opera. In 1915 she returned to the United States for two seasons with the Chicago Opera. On December 7, 1917, she made her debut at the Metropolitan Opera in *Cavalleria rusticana.* For twelve consecutive seasons she was one of the most highly esteemed members of the Metropolitan Opera company. She was extraordinarily versatile, her repertory including about a hundred and fifty roles in four languages. Her specialty was German opera, particularly the works of Wagner, Mozart, and Richard Strauss. She also appeared in major world and American premières, including *La cena delle beffe, Gianni Schicchi, Jonny spielt auf, The King's Henchman,* and Liszt's oratorio *Die Legende von der heiligen Elisabeth* (given a stage presentation in 1918).

Between 1930 and 1936 Easton sang in England, principally at Covent Garden. She returned to the Metropolitan for a single performance, on February 29, 1936, when she sang Brünnhilde in *Die Walküre.* Following this, she went into retirement. Died New York City, August 13, 1955.

Eastward in Eden, opera by Jan Meyerowitz. Libretto by Dorothy Gardner. Première: New York City, May 27, 1954. The central figure is the New England poetess Emily Dickinson, and the story concerns her frustrated love for Charles Wadsworth, a married minister.

Ebben? ne andrò lontana, Wally's aria in Act I of Catalani's *La Wally.*

Ebert, Carl, opera manager. Born Berlin, February 20, 1887. He was trained in the theater by Max Reinhardt, after which he had a distinguished career in Germany as actor, becoming the first German actor to receive the honorary title of "professor." He also helped found the first municipally subsidized drama school in Germany. In 1927 he became director of the Darmstadt Theater, where, for four years, he was in charge of operatic as well as dramatic productions. In 1931 he was appointed director of the Städtische Oper in Berlin. When the Nazis came to power he was offered the post of chief of all German opera houses. Unsympa-

thetic to the government, Ebert declined and voluntarily left his native land. In 1934 he helped found the Glydebourne Festival in England, serving as its artistic director from its inception. Two years later, the Turkish government called to him to help him establish a Turkish National Theater and Opera. In 1948 he formed an opera workshop at the University of Southern California and was general manager of the American Grand Opera Company. In 1953 he returned to Germany to become head of the Städtische Oper in Berlin on a long-term contract. He was also the director of the 1955 Berlin Music Festival.

Eboli, Princess, Don Carlos' admirer (mezzo-soprano) in Verdi's *Don Carlos.*

E casta al par di neve! Tonio's song in Act II of Leoncavallo's *Pagliacci*—Tonio acting the role of Taddeo.

Echo, a character (soprano) in Richard Strauss's *Ariadne auf Naxos.*

Ecoute, écoute, compagnon, the smugglers' chorus in Act III of Bizet's *Carmen.*

Ecoute, mon ami, the death of Don Quixote in Act V of Massenet's *Don Quichotte.*

Edgardo (or Edgar) of Ravenswood, Lucia's lover (tenor) in Donizetti's *Lucia di Lammermoor.*

Edinburgh International Festival of Music and Drama, one of the most significant European summer music festivals. It was organized in 1947 in Edinburgh, Scotland, by Mrs. John Christie and Rudolf Bing, the latter becoming artistic director. Bing's success in establishing the Edinburgh Festival was responsible for his appointment as general manager of the Metropolitan Opera in 1950; he was succeeded at Edinburgh by Ian Hunter. From the first festival on the Glyndebourne Opera has been an important participant in the varied programs, usually in performances of Mozart operas, but occasionally in the presentation of such novelties as the original version of Richard Strauss's *Ariadne auf Naxos,* a revival of Rossini's *Le Comte Ory,* and Stravinsky's *The Rake's Progress.* Occasionally, foreign opera companies give guest performances. In 1952 the Hamburg Opera gave a series of performances tracing the evolution of German opera from *The Magic Flute* to *Mathis der Maler.*

Edipo Re, opera in one act by Ruggiero Leoncavallo. Libretto by the composer, based on the drama of Sophocles. Première: Chicago Opera, December 13, 1920. This was Leoncavallo's last opera, and it was produced posthumously.

Edmondo, a young student (tenor) in Puccini's *Manon Lescaut.*

Egk, Werner, composer. Born Auchsesheim, Germany, May 17, 1901. After studying piano with Anna Hirzel-Langenhan and composition with Carl Orff, he started a career as conductor in Bavaria in 1929. He settled near Munich, and after the rise of the Hitler government succeeded Graener as head of the faculty of composition of the Reichsmusikkammer. In 1938 he was appointed conductor of the Berlin State Opera. His first opera, *Die Zaubergeige,* introduced at the Frankfurt Opera in 1935, was a major success, particularly after receiving the blessings of Nazi officials. Those officials first frowned on his opera *Peer Gynt,* performed in Berlin in 1938; but when Hitler himself expressed his enthusiasm, it immediately came into favor. Egk's later operas include: *Columbus* (1942); *Circe* (1948); *Irish Legend* (1954).

Eglantine de Puiset, see DE PUISET, EGLANTINE.

Egli è salvo! Don Carlos' aria in Act III, Scene 2, of Verdi's *La forza del destino.*

Egyptian Helen, The, see AEGYPTISCHE HELENA, DIE.

E il sol dell' anima, the Duke's love song to Gilda in Act II of Verdi's *Rigoletto.*

Einem, Gottfried von, composer. Born Bern, Switzerland, January 24, 1918. While he was still a child, his family settled in Austria, where he began his education. In 1938 he studied with Hindemith in Berlin, after which he worked for a time as coach at the Berlin State Opera. Suspected of harboring anti-Nazi sentiments, he and his mother were arrested by the Gestapo and confined for four months to a concentration camp. After his release, Einem studied for two years with Boris Blacher. He then wrote several orchestral works and a ballet, *Turandot.* These were well received when performed in Dresden and Berlin between 1943 and 1945. During this period Einem served as Karl Elmendorff's assistant at the Dresden Opera. In 1945 he settled in Salzburg, where he completed his first opera, *Dantons Tod,* a great success when introduced at the Salzburg Festival in 1947. His second opera, *Der Prozess,* based on Franz Kafka's novel *The Trial,* was introduced at the Salzburg Festival in 1953. Later the same year it was given in New York by the New York City Opera Company and was counted a failure.

Ein' feste Burg ist unser Gott, a setting by Martin Luther of the 46th Psalm. It became a battle song of the Reformation and has since entered the Protestant hymnbook. It is the basis of the overture to Meyerbeer's *Les Huguenots,* and is used prominently in the first and second acts. It was also quoted by Wagner in his *Kaisermarsch.*

Ein Mädchen oder Weibchen, Papageno's aria in Act II, Scene 5, of Mozart's *The Magic Flute.*

Ein Männlein steht im Walde, Gretel's folk song in Act II (or Act I, Scene 2) of Humperdinck's *Hansel and Gretel.*

Einsam in trüben Tagen, Elsa's aria,

known as "Elsa's Dream," in Act I of Wagner's *Lohengrin*.

Ein Schwert verhiess mir der Vater, Siegmund's aria in Act I of Wagner's *Die Walküre*.

Einstein, Alfred, musicologist. Born Munich, August 30, 1880; died El Cerrito, California, February 13, 1952. His music study took place in Munich, Italy, and England. In 1913 he became music critic of the *Münchner Post*, and in 1918 editor of the *Zeitschrift für Musikwissenschaft*. He settled in Berlin in 1927, where he became music critic of the *Berliner Tageblatt*, retaining this post until forced to leave Nazi Germany in 1933. For a while he lived in London. In 1939 he came to the United States and joined the music faculty of Smith College. In 1945 he became an American citizen. Einstein made outstanding contributions to musical scholarship. One of his major achievements was a revision (1937) of the Köchel catalogue of Mozart's works. This new edition is now a definitive source for the chronology of Mozart's operas. Other Einstein works of operatic interest include biographies of Gluck (1936) and Mozart (1945).

Eisenstein, Baron von, a rich banker (tenor) in Johann Strauss's *Die Fledermaus*.

Eleazar, Jewish goldsmith (tenor) in Halévy's *La Juive*.

Elektra, music drama in one act by Richard Strauss. Libretto by Hugo von Hofmannsthal, based on the tragedy of Sophocles. Première: Dresden Opera, January 25, 1909. American première: Manhattan Opera House, New York, February 1, 1910.

Characters: Klytemnestra, queen, and widow of Agamemnon (mezzo-soprano); Aegisthus, her lover (tenor); Orestes, her son (baritone); Elektra, her daughter (soprano); Chrysothemis, another daughter (soprano). The setting is Mycenae, in ancient Greece. The action takes place in a courtyard at the rear of King Agamemnon's palace.

Elektra mourns the death of her father, Agamemnon, murdered by his wife Klytemnestra and her lover Aegisthus. Her one consuming passion is to avenge this murder. When her mother comes to her with tales of terrible dreams, Elektra mockingly suggests a cure: the shedding of blood of someone near to her. In a sudden fit of rage, Elektra tells Klytemnestra that the murdered woman will be none other than Klytemnestra herself, and that her murderer will have the visage of the dead Agamemnon ("Was bluten muss?"). Klytemnestra's horror is relieved by a messenger's tidings that Orestes is dead. Since Orestes cannot now avenge his father's death, Elektra plans to do so herself with an ax. At this point the messenger reappears, and Elektra recognizes that he is none other than Orestes ("Orest! Es rührt sich niemand!"). Klytemnestra now meets her deserved doom. From within the palace come her shrieks as Orestes kills her. Aegisthus arrives, enters the palace, and is likewise slain. Delirious with joy that vengeance has come, Elektra sings a rapturous song, dances in triumph on her father's grave ("Schweig und tanze"), and sinks lifeless to the ground.

Elektra was the first opera in which Strauss collaborated with the Austrian poet, Hugo von Hofmannsthal. This artistic marriage of librettist and composer, destined to yield so many fine operas, was a success from the very beginning. In *Elektra,* as in the operas that succeeded it, we find the sensitivity with which dramatist and musician were attuned to each other's artistic demands. For Hofmannsthal's psychoneurotic play, in which the tragedy of Sophocles becomes filled with morbid lusts and even depravity, Strauss wrote one of his most realistic scores. Music and text are filled with emotional disturbances and demoniac frenzy. The discords are agonizing; the harmonies, oversensual; the flights of melody at times hysterical; the rhythms, nervous and high tensioned. When *Elektra* was first introduced, many critics condemned it as being the last word in decadence, using such adjectives as "blood-curdling" and "gruesome." Yet many of those who were shocked had to admit the opera's compelling impact. In the half century since its première, *Elektra* has lost little of its force—an effective performance is still about as overpowering an emotional experience as the opera stage has to offer. But today we know that *Elektra* is more than just a thunderbolt hurled at an audience by a musical Jove. We recognize its singular power and beauty, and find it an opera in which play and music are one in the projection of a mighty drama.

The first Elektra was Anna Krull. Florence Easton created the role in London, and Mariette Mazarin in New York. When *Elektra* was first given at the Metropolitan Opera (1932) the title role was assumed by Gertrude Kappel. A later revival at the Metropolitan (1938) had the services of one of the most compelling interpreters of the title role: Rosa Pauly.

Elena, Sicilian noblewoman in Verdi's *The Sicilian Vespers.*

Eleonora, Lelio's wife (soprano) in Wolf-Ferrari's *Le donne curiose.*

Elisa, Princess, a noblewoman (soprano) at Napoleon's court in Giordano's *Madame Sans-Gêne.*

Elisabeth, the Landgrave's niece (soprano) in Wagner's *Tannhäuser.* See also ELIZABETH.

Elisabeth's Prayer, *see* ALLMÄCHT'GE JUNGFRAU.

Elisetta, one of Geronimo's daughters (soprano) in Cimarosa's *Il matrimonio segreto.*

elisir d'amore, L' (The Elixir of Love), opera buffa by Gaetano Donizetti.

Libretto by Felice Romani based on Eugène Scribe's *Le philtre.* Première: Teatro della Canobbiana, Milan, May 12, 1832. American première: Park Theater, New York, May 22, 1844.

Characters: Nemorino, a young peasant (tenor); Adina, a wealthy woman (soprano); Belcore, a sergeant of the garrison (bass); Dr. Dulcamara, a quack (bass); Gianetta, a peasant girl (soprano); peasants; soldiers; villagers; a notary; a landlord. The setting is an Italian village in the nineteenth century.

Act I. The lawn of Adina's farm. As Adina is reading a book, Nemorino looks upon her longingly and speaks of his love ("Quanto è bella"). Sergeant Belcore comes to pay court to Adina. When he is rejected, Nemorino becomes encouraged to address her. But he, too, is rudely dismissed. The traveling quack, Dr. Dulcamara, appears with a trunkful of medicines. He proclaims his genius ("Udite, udite"). Desperate for Adina's love, Nemorino gives his last coin for a love elixir. No sooner does the young man drink it than he feels light of heart; he sings and dances. So confident is he of the powers of the elixir that he consciously slights Adina to arouse her jealousy. Piqued, Adina suddenly decides she will marry Belcore, and sets the wedding for that very day. Nemorino, having been told that the elixir must have time to work its magic, entreats Adina to delay the wedding ("Adina credimi"). She refuses to do so.

Act II. Interior of Adina's house. While preparations for the wedding are taking place, Nemorino complains bitterly to Dulcamara of the ineffectiveness of his elixir. The doctor suggests the purchase of a second bottle. Now, word comes of the death of Nemorino's uncle. Nemorino has been left a fortune. All the village girls quickly become attentive to the young man. The sight of Nemorino's sudden popularity brings tears to Adina's eyes. She is soon consoled by Nemorino ("Una furtiva lagrima"). With Adina and Nemorino in each other's arms, the quack doctor insists that his elixir has brought them together; and he finds many new customers for his products.

Auber's opera-comique *Le philtre* (1831) also makes use of Eugène Scribe's play.

Elizabeth, Philip II's wife (soprano) in Verdi's *Don Carlos.*

See also ELISABETH.

Elle a fui, la tourterelle, Antonia's romance in Act III of Offenbach's *The Tales of Hoffmann.*

Ellen, the British governor's daughter (soprano) in Delibes's *Lakmé.*

Elle ne croyait pas, Wilhelm Meister's romance in Act III of Thomas's *Mignon.*

Elmendorff, Karl, conductor. Born Düsseldorf, Germany, January 25, 1891. At first he studied philology, but in 1913 he entered the Hochschule für Musik in Cologne. His apprenticeship as an opera conductor took place in Wiesbaden and Mannheim, after which he became conductor of the Munich Opera, specializing in the Wagnerian repertory. Beginning in 1927 he conducted at the Bayreuth Festivals for several years. During the 1930's he conducted at the Berlin Opera and the Florence May Music Festival. After the rise of Hitler, Elmendorff fell into disfavor and had to relinquish his post with the Munich Opera. Leaving Germany, he appeared in other European opera houses. In 1946 he became musical director of the Dresden Opera. Since 1949 he has held a similar post with the Kassel Opera.

Elsa, noblewoman of Brabant (soprano) in Wagner's *Lohengrin.*

Elvira, Donna, (1) former sweetheart (soprano) of Giovanni in Mozart's *Don Giovanni.*

(2) Noblewoman (soprano) in Verdi's *Ernani.*

Emilia, Iago's wife (contralto) in Verdi's *Otello.*

Emma, young girl (contralto) in Mussorgsky's *Khovantchina.*

Emperor's New Clothes, The, see ANDERSEN, HANS CHRISTIAN.

Emperor Jones, The, opera by Louis Gruenberg. Libretto by Kathleen de Jaffa, based on the play of the same name by Eugene O'Neill. Première: Metropolitan Opera, January 7, 1933.

Characters: Brutus Jones, former Pullman porter (baritone); Henry Smithers, a cockney trader (tenor); native woman; Congo witch doctor; apparitions of soldiers, convicts, planters, slaves. The setting is a West Indian island; the time is the present.

Act I. The throne room of Jones's palace. An escaped convict, Jones has come to a West Indian island and made himself its emperor. He has looted the native people, and now he is making preparations to flee the island with his booty. Henry Smithers reveals to Jones that he is leaving none too soon: the tribesmen are about to revolt against his rule. When Jones calls his ministers into court nobody appears. Hurriedly, Jones sets about escaping. From the distance comes the sound of chanting and the beating of a tom-tom.

Act II. The forest. As Jones makes his way toward the coast he hears the continual throbbing of the tom-tom, as if pronouncing his doom. At first self-confident and full of bravado, Jones slowly begins to succumb to fright as the tom-tom sounds grow louder and the shadows of the forest lengthen. Hallucinations from his wicked past arise to haunt him, and he shouts at them wildly. Realizing that his pursuers are drawing closer, he falls on his knees and confesses his sins; he begs God for forgiveness and protection ("Standin' in de need of prayer"). But he knows he is doomed. When the witch doctor finds him and calls to his fellow tribesmen. Jones uses his last bullet—a silver one saved for this very purpose—to commit suicide. The tribesmen, finding him dead, dance joyfully around his body. The last word is Smithers'. "Yer died in the grand style, anyhow," he comments.

To convey in his music the ever-growing terror obsessing Brutus Jones, Gruenberg employs modern techniques. In place of arias there is song-speech; the harmonic writing is spiced with discords; the rhythmic patterns are complex; the choral writing is at times pierced with agonizing outcries. The opera progresses to the final orgy with relentless movement, and the tension mounts continually, making a great emotional experience in which the listener finds little contrast of mood and little relaxation.

Emperor of China, The, a character (bass) in Stravinsky's *Le rossignol.*

Encore, French for "again." The call is used by opera audiences to demand repetition of a number.

enfant et les sortilèges, L', (The Child and the Sorceries), opera by Maurice Ravel. Libretto by Colette. Première: Monte Carlo Opera, March 21, 1925. Variously designated a "fantasy," "ballet," and a "comedy of magic," *L'enfant et les sortilèges* is actually an opera requiring singing, dancing and pantomime. Colette's play concerns a mischievous boy who, scolded for failing to do his lessons, breaks up furniture and victimizes domestic animals. The furniture comes to life to taunt the boy. A princess emerges from a fairy-tale book he has torn to say she is through with him. When the boy flees out of the house he is harassed by the animals outside. In the confusion, a squirrel is hurt. The boy nurses him tenderly. This kind act induces the boy's tormentors to forgive him. Ravel's score is one of his wittiest and most satiric, including a provocative duet of cats (in cat language), a saucy dance of the cups and teapot, and an American fox trot.

En fermant les yeux (Le rêve), Des Grieux's aria in Act II of Massenet's *Manon.*

Enrico Ashton, *see* ASHTON, ENRICO (or HENRY).

En silence, pourquoi souffrir? Rozenn's and Margared's duet in Act I of Lalo's *Le Roi d'Ys.*

Entführung aus dem Serail, Die, *see* ABDUCTION FROM THE SERAGLIO, THE.

Entr'acte, French term meaning the interval between two acts. In opera usage it denotes music composed to fill this interval. The term is used interchangeably with "intermezzo."

Entrada, Spanish for "entrance," denoting music used as a prelude or introduction to an act of an opera.

Entrance of the Gods into Valhalla, The, the closing scene of Wagner's *Das Rheingold.*

Entrée, French for "entry." In eighteenth century opera an entrée was a piece of music in march time accompanying the entry of a procession. Subsequently, it came to mean a prelude accompanying the rise of the curtain.

En vain pour éviter, Carmen's card song in Act III of Bizet's *Carmen.*

Enzo Grimaldo, *see* GRIMALDO, ENZO.

E quest' asilo ameno e grato, Euridice's description of the peace and beauty of Elysium in Act III (or sometimes Act II, Scene 2) of Gluck's *Orfeo ed Euridice.*

Eravate possente, Madeleine's aria in Act II of Giordano's *Andrea Chénier.*

Erckmann-Chatrian, two French novelists who wrote as collaborators. They were: Emile Erckmann (1822–1899) and Alexandre Chatrian (1826–1890). Among the operas derived from their works are: *Le Juif polonais* by Camille Erlanger; *L'amico Fritz* and *I Rantzau* by Mascagni; *L'ami Fritz* by Roland-Manuel; and *The Polish Jew,* by Karel Weis.

Erda, spirit of the earth (contralto) in Wagner's *Das Rheingold* and *Siegfried.*

Erda's Warning, see WEICHE, WOTAN, WEICHE!

Erede, Alberto, conductor. Born Genoa, Italy, November 8, 1909. After attending the conservatories of Genoa, Milan, and Basel he became a pupil of Felix Weingartner in conducting. Between 1934 and 1939 he was assistant conductor of the Glyndebourne Festival. During much of this period he also conducted the Salzburg Opera Guild, with which organization he first came to the United States in 1937. Two years later he returned to the United States as guest conductor of the NBC Symphony, leading the world première of Gian Carlo Menotti's *The Old Maid and the Thief.* In 1945 he became artistic director and conductor of the Turin Radio Orchestra, in 1946 of the New London Opera Company (England). After this he conducted extensively throughout Europe. On November 11, 1950, he made his debut at the Metropolitan Opera with *La traviata.* Since then he has been a principal conductor of the Metropolitan.

Erik, Senta's beloved (tenor) in Wagner's *Der fliegende Holländer.*

Eri tu che macchiavi, Renato's aria in Act III, Scene 1, of Verdi's *Un ballo in maschera.*

Erkel, Franz (or FERENC), composer. Born Gyula, Hungary, November 7, 1810; died Budapest, June 15, 1893. He was the creator of Hungarian national opera. In 1838 he became conductor of the National Theater in Budapest, and it was here that his first opera, *Báthory Mária,* was produced in 1840. His most successful opera was *Hunyady László,* presented in 1844. For many years this was one of the most frequently performed of all Hungarian operas. In 1845 Erkel wrote the Hungarian national anthem. Subsequently, he helped direct the Philharmonic concerts in Budapest and was

the first professor of piano and instrumentation at the National Academy. His later operas: *Erzsébet* (1857); *Kúnok* (1858); *Bánk Bán* (1861); *Sarolta* (1862); *Dózsa György* (1867); *Brankovics* (1874); *King Stefan* (1874).

Erlanger, Camille, composer. Born Paris, May 25, 1863; died there April 24, 1919. He attended the Paris Conservatory, where he won the Prix de Rome in 1888. Success came in 1895 with the dramatic legend, *Saint-Julien l'Hospitalier,* based on a story by Flaubert, performed at the Concerts de l'Opéra. His most celebrated work was the opera, *Le Juif polonais,* introduced at the Opéra-Comique, April 9, 1900. His other operas: *Kermaria* (1897); *Le fils de l'etoile* (1904); *Aphrodité* (1906); *Bacchus triomphant* (1910); *L'Aube rouge* (1911); *La sorcière* (1912); *Le barbier de Deauville* (1917); *Forfaiture* (1919).

Ernani, opera in four acts by Verdi. Libretto by Francesco Piave, based on Victor Hugo's drama *Hernani.* Première: Teatro la Fenice, March 9, 1844. American première: Park Theater, New York, April 15, 1847.

Characters: Don Carlos, King of Castile (baritone); Don Ruy Gomez de Silva, a Spanish grandee (bass); Donna Elvira, his betrothed (soprano); Juana, her nurse (mezzo-soprano); Ernani, a bandit chief (tenor); nobles; ladies; followers of the King; followers of Don Silva; followers of Ernani. The action takes place in Aragon, Aix-le-Chapelle, and Saragossa, in 1519.

Act I, Scene 1. A mountain retreat in Aragon. Outlawed by the king, Ernani, son of a Spanish duke, has become a bandit. At his retreat, his followers sing a drinking song ("Evviva, beviam"). In love with Donna Elvira, who is about to marry Don Ruy Gomez de Silva, Ernani and his men set about abducting her.

Scene 2. Elvira's room in the castle.

Depressed by the necessity of having to marry a man she does not love, Elvira calls out to Ernani to save her ("Ernani, involami"). Don Carlos comes to her in disguise and tries to make love to her. She resists him and is saved by the sudden appearance of Ernani. The King and Ernani recognize one another, and a conflict seems inevitable. Suddenly, Don Silva appears. The King offers a lame excuse for being in Elvira's room, and Ernani makes his escape.

Act II. A hall in Don Silva's castle. Determined to prevent Elvira's marriage to Silva, Ernani comes to the grandee's castle disguised as a pilgrim, and is given shelter and protection. When Elvira appears, dressed as a bride, Ernani discloses his true identity. The laws of hospitality prevent Silva from doing Ernani harm; indeed, he must even protect him when the King comes to capture him. Failing to find Ernani, the King's men take Elvira as hostage. Silva challenges Ernani to a duel; Ernani refuses to accept, since his host has saved his life. Instead, he offers to join forces with Silva to rescue Elvira, promising to give her up for good. As a pledge, Ernani gives his host a hunting horn. When Silva shall blow on that horn, Ernani will kill himself.

Act III. A vault in the cemetery at Aix-le-Chapelle. Don Carlos has come to this solemn tomb because he has been informed that it is the hiding place of conspirators intending to kill him. Concealing himself, he hears Ernani and Silva plotting against him, with Ernani chosen to do the killing. From a secret door, electors and courtiers file in to announce that Don Carlos has just been proclaimed Emperor Charles V. Don Carlos now orders Ernani and Silva put to death. When Elvira pleads for their lives he rescinds the order and frees them. He even gives his blessings to the union of Ernani and Elvira. This generous gesture moves the assemblage to sing the praises of their Emperor ("O sommo Carlo"). Only Don Silva is bitter at the new turn of events.

Act IV. The terrace of Ernani's palace in Aragon. Elvira and Ernani are together at last, and they are happy. But Silva has not forgotten Ernani's promise. A hunting horn sounds—Don Silva appears and demands that Ernani keep his word. Ernani bids his beloved a tender farewell, then kills himself with a dagger.

Ernani was Verdi's fifth opera; it belongs in the first period of his creative development, when he was still uninhibited in his emotional responses, still lavish in his use of musical resources for passionate and melodramatic expression. There is vigor, theatricalism, and overwhelming feeling in this score, music which lends itself admirably to the ringing phrases of Hugo's eloquent poetry. While *Ernani* was Verdi's first opera to make him famous outside Italy, and is the earliest of his operas to survive in the repertory, it had a stormy career. When first produced, it ran into censorship trouble. Since Italy was then ruled by Austria, the authorities objected to the conspiratorial nature of the play, and the libretto had to be revised extensively before the opera could be presented. In Paris, the opera became a major issue between the young romanticists, who regarded it as a masterwork, and the classicists. Strange to say, Victor Hugo himself was opposed to the opera; he regarded both the adaptation of his play and Verdi's music as travesties.

Ernani, involami, Donna Elvira's aria in Act I, Scene 2, of Verdi's *Ernani*.

Ernesto, Don Pasquale's nephew (tenor) in Donizetti's *Don Pasquale*.

Ero e Leandro (Hero and Leander), opera by Luigi Mancinelli. Libretto by Arrigo Boïto and the composer. Première: Teatro Real, Madrid, November 30, 1897. The composer himself conducted the opera's Metropolitan Opera première on March 10, 1899. The opera is based on the Greek legend in which Leander, beloved of Hero, is by her crowned victor in festival games. When Ariofarno aspires to win Hero's love he is attacked by Leander. For this, Leander is exiled to Asia. He swims across the Hellespont to reach Hero, and both meet their doom in a storm.

Erwartung (Anticipation), monodrama by Arnold Schoenberg. Libretto by Marie Pappenheim. Première: Prague, June 6, 1924. This one-character opera, written in 1909 in an extremely dissonant style, is the composer's first attempt at writing for the stage. A woman searches for her lover in a dark forest. As she gropes through the darkness, her fears mount. At length, she stumbles over his dead body. Her fears and hallucinations are manifested in the long monologue she now sings.

Escamillo, a toreador (baritone) in Bizet's *Carmen.*

Eschenbach, Wolfram von, see WOLFRAM VON ESCHENBACH.

E scherzo od e follia, quintet of Edgardo, Ulrica, Riccardo, Samuele, and Tommaso in Act I, Scene 2, of Verdi's *Un ballo in maschera.*

Esmeralda, (1) opera by Dargomizhsky. Libretto by Victor Hugo, based on his novel *Notre Dame de Paris.* Première: Moscow Opera, December 17, 1847. This was the composer's first opera, and his first major work; it was not in the national style characterizing his later operas.

(2) Opera by Arthur Goring Thomas. Libretto by T. Marzials and A. Randegger, based on Hugo's *Notre Dame de Paris.* Première: Drury Lane Theater, London, March 26, 1883. This operatic adaptation differs from Hugo's well-known story. The gypsy girl Esmeralda saves the poet Gringoire from death by offering herself as his wife. But she is in love with Captain Phoebus. Frollo, an archdeacon, is

himself in love with Esmeralda, and when he finds her making love to Phoebus he stabs his rival. Esmeralda is jailed on a charge of attempted murder. In jail she is reunited with Phoebus. When Frollo attempts again to kill Phoebus, Quasimodo intercepts the thrust and is killed. Frollo is now arrested, and Esmeralda and Phoebus are able to share their love without interference.

See also HUGO, VICTOR.

E sogno? o realtà? Ford's monologue in Act II, Scene 1, of Verdi's *Falstaff.*

Esultate! Otello's first aria in Act I of Verdi's *Otello.*

etoile du nord, L' (The Star of the North), opera by Meyerbeer. Libretto by Eugène Scribe. Première: Opéra-Comique, February 16, 1854. Czar Peter is in love with the village girl Katherine, who disguises herself as a man and enters the Russian army in place of her brother. She brings the Czar a report about a conspiracy. The conspiracy is quickly rooted out. Disguised as a carpenter, the Czar woos and wins Katherine, who now becomes the Czarina.

Meyerbeer included in *L'etoile du nord* six numbers from his earlier opera, *Ein Feldlager in Schlesien.*

Euch Lüften, die mein Klagen, Elsa's aria in Act II of Wagner's *Lohengrin.*

Eudoxie, Princess, the Emperor's niece (soprano) in Halévy's *La Juive.*

Eugene Onegin, opera in three acts by Tchaikovsky. Libretto by Konstantin Shilovsky and the composer, based on the poem of the same name by Pushkin. Première: Moscow, March 29, 1879 (student performance); first public performance: Moscow Opera, January 23, 1881. American première: concert performance by the Symphony Society of New York, February 1, 1908.

Characters: Eugene Onegin, a young dandy (baritone); Lensky, his friend (tenor); Mme. Larina, a landowner (mezzo-soprano); Tatiana, her daughter (soprano); Olga, another daughter (contralto); Prince Gremin (bass); Triquet, a Frenchman (tenor); Filipievna, a nurse (mezzo-soprano); Petrovitch, a captain (baritone); Zaretski, Lensky's friend (baritone); Gillot, Onegin's servant. The setting is St. Petersburg and its environs, the time about 1815.

Act I, Scene 1. A garden adjoining Mme. Larina's home. Tatiana and Olga are singing to their mother, Mme. Larina, when their nurse announces the arrival of Eugene Onegin and Lensky. It is not long before Tatiana falls in love with Onegin.

Scene 2. Tatiana's room. Unable to sleep, Tatiana begs her nurse to tell her a story, which she does. Tatiana then reveals to her nurse how much she loves Onegin, and impetuously decides to write her beloved a letter telling him of her feelings ("Letter Scene").

Scene 3. The garden. Onegin responds to the letter by meeting Tatiana in her garden. He tells her that he is not the man for her and urges her to forget about him. Humiliated, Tatiana runs away.

Act II, Scene 1. A living room in Mme. Larina's home. A ball is taking place, celebrating Tatiana's birthday; it is here that the brilliant and well-known waltz occurs. Overhearing some women gossiping about Tatiana's love for him, Onegin decides to dispel their suspicions by paying attention to Olga. This arouses Lensky's jealousy, since he is in love with Olga, and he challenges Onegin to a duel.

Scene 2. A mill beside a stream. Waiting for Onegin, Lensky recalls his youth (Lensky's Air: "Faint Echo of Youth"). Onegin appears, the duel takes place, and Lensky is killed.

Act III, Scene 1. A hall in the palace of Prince Gremin. Six years have passed. Tatiana is married to Prince Gremin. Onegin is a guest at their palace during a gay reception. Here oc-

curs the familiar Polonaise. When Onegin meets Tatiana he is aware for the first time of how much he really loves her.

Scene 2. Tatiana's boudoir. Tatiana is awaiting Onegin—he has sent a message that he must talk with her. She is torn with conflicting feelings, for she is still in love with him, yet she wants to be true to her husband. When Onegin arrives, he pleads for her love. For a moment, Tatiana wavers and responds ardently. But she immediately assumes control of herself and sends her distraught lover forever away.

Because it lacks sustained dramatic interest and sharply defined characterizations—and because it does not exploit big emotional scenes and spectacles—*Eugene Onegin* has never been especially popular except in Russia. There it has long been recognized as the composer's finest opera. Tchaikovsky knew he was not writing a popular opera. "It is true that the work is deficient in theatrical opportunities; but the wealth of poetry, the humanity and the simplicity of the story . . . will compensate for what is lacking in other respects," he wrote. The opera was not at first successful, even in Russia. In describing its "weariness and monotony" César Cui was voicing the reaction of most Russian musicians. But repeated performances in Russia brought to the surface the deep humanity of the opera. Audiences began discovering the subtle inner conflicts of the characters, began to respond to the poignancy of Tatiana's unresolved love.

Eumenides, *see* AESCHYLUS; ORESTE.

Euridice, (1) opera by Jacopo Peri. Libretto by Ottavio Rinuccini. Première: Pitti Palace, Florence, October 6, 1600. Since the music of Peri's *Dafne* (the first opera ever written) has been lost, its immediate successor, *Euridice,* is sometimes spoken of as the first complete opera in history. Whereas *Dafne* was described by its librettist as a pastoral fable ("favola pastorale"), *Euridice* was designated a "tragedy." *Euridice* was written for, and performed in conjunction with, the festivities attending the marriage of Henry IV of France to Maria de' Medici. A few arias in this opera were written by Caccini. The story is a version of the myth concerning Orpheus and Euridice.

(2) The wife of Orfeo (Orpheus) in numerous operas, including Caccini's *Euridice,* Gluck's *Orfeo ed Euridice,* Haydn's *Orfeo ed Euridice,* Monteverdi's *Orfeo,* and Peri's *Euridice.*

Euripides, dramatist. Born Salamis, Greece, 480 B.C.; died Pella, Greece, 406 B.C. One of the greatest tragic dramatists of classical Greece, Euripides was the author of more than seventy-five plays. Among the operas based on works of Euripides are: Gluck's *Alceste, Iphigénie en Aulide,* and *Iphigénie en Tauride;* Ernst Křenek's *Das Leben des Orest;* Jean Philippe Rameau's *Hippolyte et Aricie;* Alessandro Scarlatti's *Mitridate Eupatore;* Domenico Scarlatti's *Ifigenia in Aulide* and *Ifigenia in Tauride;* Giorgio Ghedini's *La Baccanti;* Francesco Malipiero's *Ecuba;* Jean Martinon's *Hécube;* Darius Milhaud's *Médée;* André Campra's *Iphigénie en Tauride;* Egon Wellesz' *Alkestis* and *Die Bakchantinnen.*

Euryanthe, romantic opera in three acts by Carl Maria von Weber. Libretto by Helmina von Chézy, based on the thirteenth century romance, *L'histoire de Gérard de Nevers.* Première: Kärntnerthor Theater, Vienna, October 25, 1823. American première (reputed): Wallack's Theater, New York, 1863.

Characters: King Louis VI (bass); Adolar, Count of Nevers (tenor); Euryanthe of Savoy, his betrothed (soprano); Lysiart, Count of Forêt (baritone); Eglantine de Puiset, companion of Euryanthe (mezzo-soprano). The action takes place in the Castle of Premery, at the Palace of Nevers, and in a forest. The time is the twelfth century.

Act I, Scene 1. A hall in the Castle of Premery. The court sings the praises of Adolar's betrothed, Euryanthe. Lysiart, jealous of Adolar, insists that all women are not to be trusted; he even boasts that he can win Euryanthe's love. Adolar is willing to wager all his possessions on Euryanthe's constancy. The wager is accepted; Lysiart promises to show Adolar a token of Euryanthe's love.

Scene 2. The garden of the Palace of Nevers. Euryanthe confides to Eglantine a secret entrusted to her by Adolar. Adolar's sister, Emma (a suicide) has been buried with a particular ring. Eglantine, in love with Lysiart, realizes that the possession of this ring may be the means of his winning his wager.

Act II, Scene 1. Again the garden at Nevers. Eglantine, having removed the ring from Emma's grave, gives it to Lysiart in return for a promise of marriage.

Scene 2. The Castle of Premery. Adolar and Euryanthe are on the point of being married when Lysiart arrives and displays the ring as proof of Euryanthe's infidelity. Since Adolar realizes that nobody but Euryanthe could know of the ring, he is convinced that Lysiart is telling the truth. He gives up his possessions and goes into exile.

Act III, Scene 1. A forest. Euryanthe has followed Adolar. At first he is moved to kill her, but, remembering his love, he merely abandons her. The King's hunting party appears. Euryanthe reveals to the King how Eglantine betrayed her. Realizing that Euryanthe is innocent, the King promises to destroy both Eglantine and Lysiart.

Scene 2. The garden of the Palace of Nevers. Having learned of Euryanthe's innocence, Adolar comes to avenge himself. He arrives just as Lysiart and Eglantine are to be married. As Eglantine confesses her crime, Lysiart kills

her. Lysiart is condemned to be executed. Adolar has his property restored
to him and is reunited with Euryanthe.

Weber was commissioned to write
Euryanthe by the Vienna Kärntnerthor
Theater, where *Der Freischütz* had recently scored a resounding success. Unfortunately, the new opera was no
Freischütz. It lacked the original creative power and the strong folk elements
of the earlier opera; more important
still, it was burdened by a silly libretto.
Largely because of this, *Euryanthe* was
a failure in Vienna and since then has
passed out of the repertory. But it has
historical importance as an early example of German romantic opera. Its
music has many moments of grandeur
and beauty, with occasional anticipa-
tions of moods and themes later to be
heard in *Lohengrin*. But the parts are
finer than the whole; and it is only
through some of its parts (particularly
the wonderful overture) that the opera
is remembered.

Eva, Pogner's daughter (soprano),
loved by Walther, in Wagner's *Die
Meistersinger*.

Evander, a messenger (tenor) in
Gluck's *Alceste*.

Evangelimann, Der (The Evangelist),
opera by Wilhelm Kienzl. Libretto by
the composer, based on a story by L. F.
Meissner. Première: Berlin Opera, May
4, 1895. Mathis and Martha are in love,
but Martha is also loved by Mathis'
brother Johannes. The latter sets fire to
a monastery and accuses the innocent
Mathis of the crime. Mathis is sent to
prison for ten years. During this period,
Martha's grief leads her to suicide.
Many years later, Mathis becomes an
evangelist. Johannes, on his deathbed,
is tortured by his conscience. He confesses everything and dies peacefully in
the knowledge that his brother Mathis
has forgiven him.

Evangeline, opera by Otto Luening.
Libretto by the composer, based on the
poem by Longfellow. Première: Columbia University, New York, May 5,
1948. The opera follows Longfellow's
narrative poem closely, showing the
long-frustrated love and the prolonged
separation of the heroine and her beloved Gabriel. The music embodies
various American references of the
period, including Indian calls and
Lutheran hymns.

Evening Prayer, see ABENDS, WILL ICH
SCHLAFEN GEH'N.

Evenings on a Farm Near Dikanka, *see*
GOGOL, NIKOLAI.

Evening Star, *see* O DU MEIN HOLDER
ABENDSTERN.

Expressionism, a term borrowed from
the world of painting and used to describe a sort of music that attempts to
express subconscious states. Its use is
almost entirely confined to works of the
Schoenbergian school, particularly the
operas of Schoenberg and Berg.

F

Faccio, Franco, conductor and composer. Born Verona, Italy, March 8,
1840; died Monza, Italy, July 21, 1891.
He attended the Milan Conservatory,
where he collaborated with his fellow
student, Arrigo Boïto, in writing a cantata that won a government prize.
Faccio's first opera, *I profughi Fiam-
minghi,* was performed at La Scala in
1863. Two years later came *Amleto,*

libretto by Boïto, a major success following its première in Genoa. After seeing service in Garibaldi's army, Faccio was appointed professor of harmony at the Milan Conservatory in 1868. In 1871 he was appointed principal conductor of La Scala, where he led the world première of *Otello* and the Italian première of *Aïda*.

Fafner, a giant (bass) in Wagner's *Das Rheingold* and *Siegfried*. In *Siegfried* he assumes the form of a dragon.

Fair at Sorochinsk, The, unfinished opera by Mussorgsky. Libretto by the composer, based on the tale of Nikolai Gogol, *Evenings on a Farm Near Dikanka.* Première: St. Petersburg, December 30, 1911. Three versions exist of this opera, each completed by a different composer: I. Sakhnovsky, César Cui, and Nikolai Tcherepnin. The Tcherepnin version is the one most widely favored. The story concerns the efforts of the peasant Tcherevik to marry off his daughter to Pritzko. The peasant's wife, however, favors the pastor's son. When the wife is compromised with the pastor's son she becomes contrite and consents to the match between her daughter and Pritzko. The Hopak in Act III is frequently heard.

Faites-lui mes aveux, Siebel's serenade in Act III of Gounod's *Faust.*

Falcon, Marie-Cornélie (soprano) Born Paris, January 28, 1812; died there February 25, 1897. She attended the Paris Conservatory and made her debut at the Opéra in 1832 as Alice in *Robert le Diable.* She was a principal soprano of the Opéra for the next five years, after which she went into retirement, due to permanent loss of voice. The extraordinary dramatic quality of her voice became known as the "Falcon soprano," and sopranos assuming the roles in which she became famous were thus described.

Falla, Manuel de, composer. Born Cádiz, Spain, November 23, 1876; died

Alta Gracia, Argentina, November 14, 1946. He attended the Madrid Conservatory, where his teacher, Felipe Pedrell, was a decisive influence. Pedrell encouraged De Falla to become a composer and led him to write Spanish national music based on the foundations of Spanish folk songs. In this vein De Falla wrote his first opera, *La vida breve,* which won first prize in a national competition for Spanish operas in 1905. It was introduced in Nice in 1913, and was so successful that the Opéra-Comique gave it the following year. In 1915 De Falla completed his most famous work, the ballet *El amor brujo.* Seven years later came a charming opera for puppets, *El retablo de Maese Pedro.* When the Civil War broke out in Spain, De Falla at first sided with the Franco nationalist movement. For a while he served as president of the Institute of Spain. But he became disillusioned with the Franco regime and decided to leave his native land for good. In 1939 he settled in South America, and it was here that he died seven years later.

Falsetto, high-pitched notes of female quality produced by male singers singing above their normal voice range.

Falstaff, (1) the fat, jovial, unprincipled, and endearing knight of Shakespeare's *The Merry Wives of Windsor* and *Henry IV.* He appears in several operas, the most famous being Verdi's *Falstaff* (see below). Others include: Balfe's *Falstaff;* Holst's *At the Boar's Head;* Nicolai's *The Merry Wives of Windsor;* Salieri's *Falstaff;* Vaughan Williams' *Sir John in Love.*

(2) Lyric comedy in three acts by Verdi. Libretto by Arrigo Boïto, based on Shakespeare's *The Merry Wives of Windsor* and *Henry IV.* Première: La Scala, February 9, 1893. United States première: Metropolitan Opera, New York, February 4, 1895.

Characters: Sir John Falstaff (baritone); Ford, a wealthy burgher (baritone); Mistress Ford, his wife (soprano); Nannetta (Anne), their daughter (soprano); Fenton, Nannetta's suitor (tenor); Dr. Caius, a physician (tenor); Mistress Page (soprano); Dame Quickly (mezzo-soprano); Bardolph, one of Falstaff's followers (tenor); Pistol, another follower (bass). The action takes place in Windsor, England, in the fifteenth century.

Act I, Scene 1. A room in the Garter Inn. Falstaff is drinking wine. He has written love letters to two respectable married women—Mistress Page and Mistress Ford—hoping to begin a profitable liaison with one of them. He instructs his followers to deliver his proposals, but they refuse to do so on the grounds of honor. Falstaff sends his letters by page and upbraids Pistol and Bardolph for their cowardice.

Scene 2. In Ford's garden. Comparing the notes they have received, Mistress Page and Mistress Ford find them identical. They decide to avenge themselves by collaborating in a plot with Fenton, who is in love with Mistress Ford's daughter. Dame Quickly is dispatched to invite Falstaff to a rendezvous with Mistress Page. At the same time, an arrangement is made for Ford to meet Falstaff under an assumed name.

Act II, Scene 1. The Garter Inn. After Dame Quickly has arranged with Falstaff for him to meet Mistress Page, Ford appears. He tells Falstaff that his name is Signor Fontana and that his purpose is to bribe Falstaff to speak on his behalf to Mistress Ford, with whom he is very much in love. Falstaff says he will be delighted to do so, revealing that he himself has an appointment with the lady. Since Ford knows nothing of the projected meeting, he is suddenly led to suspect that his wife is unfaithful. While Falstaff retires to don his best clothes, Ford denounces all women for their faithlessness ("E sogno? o realtà?"). Falstaff reappears

and departs with Ford for the meeting with Mistress Ford.

Scene 2. A room in Ford's house. Falstaff, arriving, begins to make love to Mistress Ford. He tells her that though he is now old and fat, once, as a page to the Duke of Norfolk, he was handsome ("Quand'ero paggio del Duca di Norfolk"). Suddenly, Ford arrives, fuming because he suspects his wife is entertaining a lover. Confusion prevails as Falstaff hides behind a screen. Ford's feverish search proves futile. When he departs, Mistress Ford and Mistress Page conceal Falstaff in a basket of laundry. Ford returns, remembering the screen, behind which a lover could hide. The women dump the laundry basket out of the window and into the river below. As Falstaff scrambles out of the river, wet and unhappy, Ford sees him and joins in the laughter.

Act III, Scene 1. The Garter Inn. Falstaff is sitting outside the inn, sad and disillusioned, trying to find some comfort in wine. Dame Quickly revives his spirits by telling him that Mistress Ford regrets what has happened and would like to meet him at midnight in Windsor Park: he is to come disguised as the Black Huntsman.

Scene 2. Windsor Park. In the moonlight Fenton, disguised as Oberon, sings of his lady love ("Dal labbro il canto estasïato vola"). The other conspirators are also present, concealed by the darkness. When Falstaff comes for his rendezvous he hears eerie noises. Convinced that supernatural forces are around, he drops to the ground, terrified. The conspirators, dressed as fairies, emerge from hiding, and give him a sound thrashing. Merriment is now at its height, at the expense of poor Falstaff. After Ford consents to have his daughter marry Fenton (a marriage he has all the while opposed) the whole company—even Falstaff—joins in the remark that all the world is but a stage ("Tutto nel mondo è burla").

In writing this, his last opera, Verdi, though at the patriarchal age of eighty, was not afraid to venture in a new direction. *Falstaff* was a comedy, and for fifty years Verdi had written only tragedies. No longer was he concerned with large arias and big scenes. Instead he had to work in smaller dimensions and create a work whose appeal lay in delicacy of expression and wit rather than passion and intensity, in subtlety of detail rather than massive effects, in penetrating characterizations instead of overpowering emotions. Tone painting was required to help create mood and atmosphere. Delicate orchestral effects were employed to accentuate details of stage action. *Falstaff* is a score of the most consummate craftsmanship as well as artistry. Since Boïto's libretto is, at the same time, one of the finest in opera literature, the opera stands with *The Barber of Seville, The Marriage of Figaro,* and *Die Meistersinger* as an outstanding example of comic opera.

The world première of a new opera by the grand old man of Italian opera inevitably attracted pilgrims from all parts of the world. The brilliant audience to whom *Falstaff* was introduced at La Scala on February 9, 1893, was uninhibited in its enthusiasm; so were the critics. The greatest measure of this success belonged, of course, to the composer, who had produced another masterwork. But a part of the triumph belonged also to the interpreter of the title role—Victor Maurel, one of the world's great baritones, and after that evening one of the most celebrated interpreters of the role of Falstaff. Maurel also played the part when this opera was first introduced in Paris, London and New York.

fanciulla del West, La, *see* GIRL OF THE GOLDEN WEST, THE.

Fandango, a lively Spanish dance in triple time, accompanied by castanets, and frequently associated with songs. Mozart interpolated a fandango in the third act of *The Marriage of Figaro.*

Fanfare, a short flourish for trumpets, used in opera for special festivities, or to attract special attention. Beethoven used a fanfare in the second act of *Fidelio* to announce the arrival of the Prime Minister. The opening bars of the march in Wagner's *Tannhäuser* are a fanfare.

Fanget an! So rief der Lenz in den Wald, Walther's improvised love song in Act I of Wagner's *Die Meistersinger.*

Faninal, a rich merchant (baritone) in Richard Strauss's *Der Rosenkavalier.*

Fanny, a model (soprano) in Massenet's *Thaïs.*

Farandole, a lively dance of Provence in 6/8 time. Gounod wrote a choral farandole in the second act of his *Mireille.*

Farewell my son, I am dying, Boris Godunov's farewell to his son in Act III, Scene 2, of Mussorgsky's *Boris Godunov.*

Farfarello, a devil (bass) in Prokofiev's *The Love for Three Oranges.*

Farinelli (born CARLO BROSCHI), male soprano. Born Andria, Italy, January 24, 1705; died Bologna, July 15, 1782. He was one of the most celebrated castrati of the eighteenth century. After studying with Niccolò Porpora he made his stage debut in 1722 in his teacher's *Eumene,* on which occasion he assumed the stage name of Farinelli. In 1727 he engaged the celebrated castrato Antonio Bernacchi in a test of vocal skill and was defeated; but Bernacchi recognized the younger man's unusual ability and became his teacher. Farinelli now began to enjoy an adulation known by few other singers of his time. He was a sensation in Vienna in 1728 and 1731, and in London in 1734. Women swooned at his performances. It was Farinelli's two-year alliance with Handel's rivals in London that was largely responsible for bringing about that composer's eclipse as an opera composer. In 1737 the Queen of Spain offered Farinelli a lavish salary to become singer to the court. Accepting, Farinelli stayed in Spain twenty-five years, singing no more in public, and occupying a position of great honor and considerable political power. He was responsible for the long succession of opera productions that made Madrid famous during these years. In 1750 he received the Cross of Calatrava, one of the highest of Spanish orders. When Charles III ascended the throne in 1759, Farinelli was required to leave Spain, but he continued to collect his salary. Surrounded by paintings, harpsichords, and other valuable souvenirs of his court career, he spent the rest of his life in retirement in Bologna.

Farlaf, Ludmilla's suitor (bass) in Glinka's *Russlan and Ludmilla.*

Farlaf's Rondo, Farlaf's aria in Act II of Glinka's *Russlan and Ludmilla.*

Farrar, Geraldine, lyric soprano. Born Melrose, Massachusetts, February 28, 1882. She was the daughter of Syd Farrar, a professional baseball player. She began music study in her fifth year and started appearing publicly when she was fourteen. On the advice of Jean de Reszke she studied singing seriously with Emma Thursby in New York, Trabadello in Paris, and Francesco Graziani in Berlin. Later on, as an established opera star in Berlin, she received coaching from Lilli Lehmann. She made her debut with the Berlin Opera in *Faust* on October 15, 1901, and was a sensation. She remained with the company three years, appearing in eleven different roles, and giving a command performance for the Emperor. Successes in Monte Carlo and Paris preceded a triumphant return to her native country. On November 26, 1906 (the opening night of the season) she made her American debut at the Metropolitan Opera in *Roméo et*

Juliette. For fifteen years, Miss Farrar was a principal soprano of the Metropolitan Opera, making 493 appearances in 29 roles. She achieved her greatest successes as Carmen, Cio-Cio-San, the Goose Girl in *Königskinder* (a role which she created), Thaïs, Zaza, Juliet, Gilda, Manon, and Tosca. She sang in the first performance of Mascagni's *Amica* (Monte Carlo, 1905), and she appeared in the American premières of Dukas's *Ariane et Barbe-Bleu* (1911) and Charpentier's *Julien* (1914). She made her final appearance at the Metropolitan on April 22, 1922, in *Zaza.* Her striking beauty and personal magnetism were probably more outstanding than the quality of her voice; but when she was at her best she was an artist of compelling force both vocally and personally. While at the height of her fame, Miss Farrar appeared in many silent motion pictures including *Carmen, Joan the Woman,* and *The Riddle Woman.* For a period after her retirement from opera she appeared in song recitals. Some years later she served a single season as a commentator during the Metropolitan Opera broadcasts. In 1938 she published her autobiography, *Such Sweet Compulsion.*

Fasolt, a giant (bass) in Wagner's *Das Rheingold.*

Fata Morgana, a witch (soprano) in Prokofiev's *The Love for Three Oranges.*

Fatima, (1) Mârouf's wife (contralto) in Rabaud's *Mârouf.*

(2) Rezia's attendant (mezzo-soprano) in Weber's *Oberon.*

Fauré, Gabriel, composer. Born Pamiers, France, May 12, 1845; died Paris, November 4, 1924. He attended the École Niedermeyer, after which he served as organist for various Parisian churches, including the Madeleine. He first attracted attention as composer of songs. *Prométhée,* an opera, was introduced in Béziers in 1900. *Penelope,* another lyric work, was performed with moderate success in Monte Carlo in 1913. Fauré's operas, like his more familiar works, are characterized by beautiful balance and proportion, and purity of expression. In 1905, Fauré succeeded Théodore Dubois as director of the Paris Conservatory, holding this post fifteen years. In the last years of his life he was a victim of deafness. He was made member of the Académie des Beaux Arts in 1909, and in 1922 he received the highest class in the order of the Legion of Honor.

Faure, Jean-Baptiste, dramatic baritone. Born Moulins, France, January 15, 1830; died Paris, November 9, 1914. After attending the Paris Conservatory, he made his opera debut at the Opéra-Comique in Victor Massé's *Galathée* on October 20, 1852. After eight successful years at the Opéra-Comique, he became the first baritone of the Paris Opéra, remaining there fifteen years, a period during which he appeared in many notable first performances, including those of *L'Africaine, Don Carlos, Hamlet,* and *Faust.* His greatest successes were in *Faust, Don Giovanni, Les Huguenots, William Tell,* and *Le Prophète.* His last appearance at the Paris Opéra took place on May 13, 1876, in *Hamlet.* After making several appearances in London and Vienna, he retired from opera and was heard only in song recitals. Between 1857 and 1860 he taught singing at the Paris Conservatory. In 1859 he married Caroline Lefebvre, a singer of the Opéra-Comique. He wrote two volumes of songs; one of the songs, "Les Rameaux" ("The Palms"), is very popular.

Faust, (1) aged philosopher and alchemist, the central character of an old German legend treated by many German writers. The most celebrated version is that of Goethe. The theme of the Faust legend is his exchange of his soul for the return of youth and power. This subject has been treated in several operas, the most famous being that of Gounod (see below). Others include: Berlioz' *Le damnation de Faust;* Boïto's *Mefistofele;* Busoni's *Doktor Faust;* and Spohr's *Faust.*

(2) Opera in five acts by Charles Gounod. Libretto by Jules Barbier and Michel Carré, based on Goethe's drama of the same name. Première: Théâtre Lyrique, Paris, March 19, 1859. American première (reputed): Philadelphia, November 18, 1863.

Characters: Faust (tenor); Méphistophélès (bass); Marguerite (soprano); Valentin, her brother (baritone); Siebel, young man in love with Marguerite (soprano); Martha, Marguerite's friend (mezzo-soprano); Wagner, a student (baritone); soldiers; students; peasants; priests. The setting is Germany in the sixteenth century.

Act I. Faust's study. The venerable Faust, sick of life, is about to take poison when he hears young, gay voices outside his window. Envious of the gaiety of youth, he curses the young people and calls on Satan to help him. Méphistophélès appears and makes a bargain with Faust: he will restore Faust's youth in return for his soul. Faust hesitates until Méphistophélès evokes the image of the lovely Marguerite at her spinning wheel. Then he acquiesces, drinks a potion prepared by the devil, and is magically changed into a young and handsome man. Méphistophélès promises Faust he shall see Marguerite without delay.

Act II. A public square. Soldiers and villagers are celebrating the day of the fair ("Vin ou bière"). Valentin and Siebel appear, the former greatly concerned because he must join the army and leave his sister unprotected. When Siebel promises to watch over Marguerite, Valentin expresses his gratitude ("Avant de quitter ces lieux"). Wagner, a student, jumps on a table to sing a ditty about a rat, but is rudely interrupted by Méphistophélès, who provides a pleasanter song, a cynical comment on man's greed for gold ("Le veau d'or"). After that, the devil delights the crowd with feats of magic. He prophesies that any flower touched by Siebel will wither and die, particularly those he sends to Marguerite; the devil also produces wine by striking his sword on the sign of the near-by inn. When Méphistophélès proposes a toast to Marguerite, Valentin grows furious and rushes at him with his sword; the sword instantly snaps in half. Siebel, Wagner, and their friends now join Valentin. Realizing they are in the presence of the devil, they raise their swords in the form of a cross and confront him with it (Chorus of the Swords: "De l'enfer qui vient émousser"). Méphistophélès, terrified, withdraws. Faust comes seeking Marguerite. The villagers fill the square as they dance and sing (Waltz: "Ainsi que la brise légère"). Marguerite now passes, coming from church. Faust offers to escort her home. Marguerite rebuffs him. As she walks on, Faust sings of his great love for her ("O belle enfant! Je t'aime"). Méphistophélès is cynical about Faust's lack of success, and the townspeople continue their gay waltzing.

Act III. Marguerite's Garden. Siebel gathers flowers for Marguerite and asks them to carry his message of love ("Faites-lui mes aveux"). But the flowers die in his hands. Remembering Méphistophélès' prophecy, Siebel rushes to a near-by shrine and dips his hands in the holy water. The flowers he now picks go unharmed, and he places them tenderly at Marguerite's door. After Siebel leaves, Faust comes, thinking of Marguerite ("Salut! demeure chaste et pure"). His revery is disturbed by Méphistophélès, who places a casket of jewels near Siebel's flowers. Faust and Méphistophélès

hide as Marguerite comes from her house and sits at her spinning wheel. Musing on the handsome stranger who greeted her in the square, she sings a ballad about the King of Thulé ("Il était un roi de Thulé"). Suddenly she catches a glimpse of the flowers, and knows they are Siebel's. Finding the casket, she opens it and is overjoyed to find it filled with jewels. Putting them on while inspecting herself joyously in a mirror, she sings the Jewel Song ("Je ris de me voir"). As she is doing this, Martha enters and is amazed at the way the jewels enhance Marguerite's beauty. Faust and Méphistophélès come from hiding. The devil engages Martha by telling her that her husband is dead. Faust pursues Marguerite. Night begins to fall. Méphistophélès addresses the night, foretelling that lovers are about to be united ("O Nuit, étends sur eux ton ombre"); then he disappears. Faust and Marguerite reappear in the garden. Tearing herself away from Faust, Marguerite promises to meet him the following day. A moment later she appears at the window of her cottage, still thinking of her rapture. Faust rushes toward her; from a distance comes the sound of Méphistophélès' mocking laughter.

Act IV, Scene 1. Marguerite's room. (This scene is often omitted.) Marguerite, at her spinning wheel, is bemoaning her fate: she has been betrayed and deserted. Siebel comes to console her. Marguerite is grateful, but she cannot forget the man she loves.

Scene 2. The cathedral. Marguerite is praying for Faust and their unborn child when Méphistophélès comes to mock her for having yielded to temptation. The church choir sings of Judgment Day ("Quand du Seigneur le jour luira"). Overcome with terror, Marguerite falls into a faint.

Scene 3. The square before the cathedral. (In the original version, this scene came before the preceding one;

it is now customary to perform the scenes in the sequences here given.) Soldiers, returning from battle, jubilantly sing of their recent victory and their joy at being home (Soldiers' Chorus: "Gloire immortelle de nos aïeux"). Valentin is with them. Eagerly he questions Siebel about Marguerite. When Siebel is evasive, Valentin rushes into his sister's cottage. Faust and Méphistophélès appear, and the latter sings a mocking serenade beneath Marguerite's window ("Vous qui faites l'endormie"). Valentin emerges and challenges Faust and Méphistophélès to a duel in which, through the devil's magic, Valentin is fatally wounded by Faust. As the townspeople rush into the square, Faust and Méphistophélès disappear. Bitterly, Valentin condemns his sister and refuses to forgive her. Marguerite watches him die, denouncing her. The townspeople kneel at his side in prayer.

Act V, Scene 1. In the Harz Mountains. (This scene is frequently omitted.) Faust and Méphistophélès come to watch the revels of Walpurgis Night. At the height of the orgy, Faust sees a vision of Marguerite, apparently crushed by a blow from an axe. Shaken, he insists that the devil take him back to his beloved.

Scene 2. A prison. Marguerite has killed her child and is in prison awaiting execution for her crime. At dawn, hearing the voice of her lover, Marguerite becomes delirious with joy. Faust calls to her to follow him out of the prison, but Marguerite does not seem to hear what he is saying. Impatiently, Méphistophélès urges Faust and Marguerite to make their escape ("Alerte! Alerte!"), but both are deaf to his entreaty. Marguerite, on the threshold of death, prays to be borne heavenward ("Anges purs, anges radieux"), after which she voices her horror of Faust. As she dies, Méphis-

tophélès drags Faust to his doom. Voices of invisible angels sing of Marguerite's redemption ("Sauvée! Christ est ressuscité!").

The early career of Gounod's Faust was turbulent. Accepted by the Théâtre Lyrique, the opera was repeatedly delayed; when rehearsals finally began, there were new misfortunes: the censors objected to the cathedral scene, and the principal tenor had to withdraw because of ill health. The censors were placated, and a last-minute substitute was found for the tenor. When the opera was at last given, public and critics reacted unfavorably. Not until it was revived by the Paris Opéra in 1869 did Faust begin to enjoy the success it deserved. After that, the opera proceeded from triumph to triumph. In the next thirty-five years it was seen on the average of once every nine days at the Opéra; during the next forty years it was given an additional thousand performances. In England, where two rival companies competed to give its first performance (1863), the opera became one of the most popular items in the entire repertory. It was a particular favorite of Queen Victoria, who, just before her death, asked to have parts of it sung for her. In the United States, in 1883, Faust was the opera chosen to open the newly founded Metropolitan Opera. Only in Germany was Faust not popular. There it was considered a travesty of the Goethe drama. The success of Faust is as easy to explain as it is difficult to understand why it took a decade for this recognition to arrive. Faust overflows with wonderful melodies of all kinds: rousing choruses, mocking and satiric tunes, sentimental melodies, lilting and heart-warming pages. The music is direct in its emotional appeal, with no attempt at subtlety. It is the wonderful abundance of its lyricism that continues to place Faust in the front rank of the world's most popular operas even though we have an increasing awareness that its characterizations are stilted, the text is vulgar, and much of the music superficial.

Favart, Charles Simon, librettist and impresario. Born Paris, November 13, 1710; died there March 12, 1792. He was a theater manager in Paris and Brussels before becoming the manager of the Opéra-Comique in 1752. For the next twenty-seven years he presented over a hundred plays at both the Opéra-Comique and the Théâtre des Italiens; many of these works influenced the evolution of the opéra-comique form. Favart provided numerous composers with about a hundred and fifty librettos for operas, some of which were Grétry's *L'amitié à l'épreuve,* Gluck's *Cythère assiegée,* and Philidor's *Le jardinier supposé.* The Opéra-Comique was frequently called the "Salle Favart" after his death.

Favola del figlio cambiato, La (The Fable of the Changeling), opera by Francesco Malipiero. Libretto is the play of the same name by Luigi Pirandello. Première: Rome, March 24, 1934. Based on a Sicilian legend, the play concerns the son of a Sicilian peasant woman. In babyhood, the boy is exchanged for an idiot. The idiot turns out to be of royal birth. In the end, the peasant woman gets back her own son and the idiot rightfully becomes a prince. This opera had political repercussions when introduced in Fascist Italy. Mussolini, disturbed by the overtones of the play, ordered its removal from the stage two days after the première on the grounds that the story had "moral incongruity."

Favola per musica, Italian for "fable through music," one of the names given to opera in the sixteenth and seventeenth centuries. A favola per musica was usually an adaptation of a story on a legendary or mythological subject.

favorita, La, opera in four acts by Gaetano Donizetti. Libretto by Alphonse Royer and Gustave Vaëz, based on the drama *Le Comte de Comminges,* by Baculard-d'Arnaud. Première: Paris Opéra, December 2, 1840 (in French). American première: New Orleans, February 9, 1843.

Characters: Alfonso XI, King of Castile (baritone); Leonora de Guzman, his mistress (mezzo-soprano); Inez, her confidante (soprano); Fernando, a monastic novice (tenor); Baltasar, prior of the Monastery of St. James (bass); Don Gasparo, the King's officer (tenor). The action takes place in Castile, in 1340.

Act I, Scene 1. The cloister of St. James. Fernando confesses to Baltasar that he has fallen in love with an unknown woman who regularly passes the monastery window (Romanza: "Una vergine, un angiol di Dio"). The prior, shocked at the revelation, sends the novice away from the monastery. The latter is now free to search for the woman he loves.

Scene 2. The island of Leon. Leonora de Guzman, the King's favorite, has ordered that Fernando be brought to her. She is the unknown woman of whom he dreams; united, they rush to each other's arms. But Leonora knows that Fernando will desert her once he discovers that she is the King's mistress. She gives him a parchment and orders him to leave. Now convinced that Leonora is beyond his reach, Fernando finds that the parchment is his commission as an officer in the King's army.

Act II. The gardens of Alcázar. The King is beset by Baltasar, who threatens him and his court with excommunication for having abandoned his wife for a mistress. The King is torn between duty and love.

Act III. A hall of the palace. Now a victorious hero, Fernando is promised by the King any reward he desires. Still not knowing that Leonora is the King's mistress, Fernando asks for her hand in marriage. The King, eager to resolve his personal problems and thus avoid excommunication, is happy to grant this request. Knowing that Fernando will have nothing to do with her once he learns of her status, Leonora sings of her anguish and her determination to make any sacrifice for Fernando's happiness ("O mio Fernando"). She instructs Inez to deliver a letter which tells Fernando the truth, but the interference of the King prevents its delivery. After Fernando and Leonora are married, Baltasar returns to the palace and Fernando learns of the sordid deception.

Act IV. The Monastery. Fernando has come back to find peace. The monks comment on the heavenly splendor of religious life ("Splendon più belle"). Before taking part in the final rites that will make him a monk, Fernando recalls his lost love ("Spirito gentil"). Leonora, disguised as a novice, comes to the monastery to find him. Fernando and the stricken woman are briefly reunited before she falls dead in his arms.

Feast in Time of Plague, see PUSHKIN, ALEXANDER.

Federica, Duchess of Ostheim (contralto) in Verdi's *Luisa Miller.*

Federico, principal male character (tenor) in Cilèa's *L'Arlesiana.*

Fedora, lyric drama in three acts by Umberto Giordano. Libretto by Arturo Colautti, based on the drama of the same name by Victorien Sardou. Première: Teatro Lirico, Milan, November 17, 1898. American première: Metropolitan Opera House, December 5, 1906.

Characters: Princess Fedora Romazov (soprano); Countess Olga Sukarev (soprano); Count Loris Ipanov, Fedora's husband (tenor); De Siriex, a diplomat (baritone); Baron Rouvel (tenor); Grech, a police officer (bass); Borov, a doctor (baritone); Lorek, a surgeon (baritone); Dmitri, a young groom (contralto); Désiré, an attendant (tenor); Cyril, a cook (baritone). The action takes place in St. Petersburg, Paris, and Switzerland in the closing years of the nineteenth century.

Act I. The house of Count Vladimir Andreyevich. Princess Fedora, about to be married to Vladimir, is waiting for him. Seeing his picture, she kisses it, and addresses it tenderly ("O grandi occhi"). Suddenly, the Count is carried into his house fatally wounded. His assassin is believed to be Count Loris Ipanov, a suspected Nihilist. After Vladimir dies, Fedora vows to avenge his death.

Act II. Fedora's house in Paris. As part of her plot to get Count Loris to confess he killed Vladimir, Fedora contrives to have him fall in love with her. He is now a guest at a reception in her house. Finding her alone, Loris tells Fedora of his love for her ("Amor ti vieta") as Fedora leads him on coquettishly. At last, he confesses that he killed Vladimir. Fedora makes plans with the police officer to have Loris seized, when her guests have departed. But when the guests are gone, Loris reveals to Fedora the real reason for the murder: Vladimir had seduced Loris' wife and been responsible for her death. This information relieves Fedora of the necessity for revenge. Realizing now that she loves Loris, she helps him escape from the police.

Act III. Fedora's villa in Oberland. Fedora and Loris are married and living happily in Switzerland. A spy, however, has traced Loris' whereabouts. He comes with the news that both Loris' brother and mother have died as a consequence of his crime. It is only now that Loris learns that Fedora was the one who betrayed him to the police. Blinded by anger, he is about to kill her when she takes poison. As Loris forgives her and begs her to live for his sake, she dies in his arms.

Feen, Die (The Fairies), opera by Richard Wagner. Libretto by the composer, based on Carlo Gozzi's *La donna serpente*. Première: Munich, June 29, 1888. This was Wagner's first complete opera, written when he was twenty. It was never performed in the composer's lifetime. An opera on the same text, titled *La donna serpente* was written in 1932 by Alfredo Casella.

Feldlager in Schlesien, Ein, *see* ETOILE DU NORD, L'.

Feldmarschallin, The, *see* PRINCESS VON WERDENBERG.

Fenena, Nebuchadnezzar's sister (soprano) in Verdi's *Nabucco*.

Fenice, La, *see* TEATRO LA FENICE.

Fennimore and Gerda, opera by Frederick Delius. Libretto (in German) by the composer, based on Jens Peter Jacobsen's novel, *Niels Lyhne*. Première: Frankfort-on-the-Main, October 21, 1919. This was Delius' last opera. The love triangle in the story consists of the poet, Niels Lyhne, the painter Erik Refstrup, and Fennimore, a girl they both love. Niels and Fennimore marry. Later, the old love between Erik and Fennimore is revived. After Erik suffers accidental death, Niels finds consolation and happiness in the love of a childhood sweetheart, Gerda.

Fenton, (1) Anna's suitor (tenor) in Nicolai's *The Merry Wives of Windsor*.

(2) Nannetta's (Anne's) suitor (tenor) in Verdi's *Falstaff*.

Feodor, Boris Godunov's young son (mezzo-soprano) in Mussorgsky's *Boris Godunov*.

Fernando, a monastic novice (tenor) in love with Leonora in Donizetti's *La favorita*.

Fernando, Don, (1) Prime Minister of Spain (bass) in Beethoven's *Fidelio*.

(2) A young Spaniard (tenor) in Granados' *Goyescas*.

Ferrando, (1) Dorabella's suitor (tenor) in Mozart's *Così fan tutte*.

(2) Captain of the guards (bass) in Verdi's *Il trovatore*.

Feste! Pane! Opening chorus of Ponchielli's *La Gioconda*.

Festa Teatrale, Italian for "theatrical feast," a type of opera, generally on a mythological or allegorical subject, popular in the eighteenth century. It was written for such occasions as royal weddings, and was invariably highly festive in character.

Festspiel, German for "festival play." Wagner called his *Ring of the Nibelungen* a "stage festival play" (Bühnenfestspiel), and *Parsifal* a "stage consecration festival play" (Bühnenweihfestspiel).

Feuersnot (Fire Famine), one-act opera by Richard Strauss. Libretto by Ernst von Wolzogen. Première: Dresden Opera, November 21, 1901. In the distant past, during the St. John Eve's celebration in Munich, Diemut goes out collecting wood for the solstice fire. She comes to Kunrad's house, and the moment they meet they fall in love. But Kunrad is presumptuous in kissing her, and Diemut punishes him by making him ridiculous to his fellow villagers. This so enrages the young man that he calls on the Feursnot to take place: all lights are to be extinguished. Diemut repents her hasty act, wins back Kunrad, and the lights are restored. In the score Strauss uses quotations from Wagner's *Der fliegende Holländer* and themes from his own opera, *Guntram*.

Feuerzauber, *see* MAGIC FIRE SCENE, THE.

Février, Henri, composer. Born Paris, October 2, 1875. He attended the Paris Conservatory, his teachers including Massenet and Fauré. In 1906 his first opera, *Le Roi aveugle,* was performed at the Opéra-Comique. Three years later came the opera that made him internationally famous, *Monna Vanna,* introduced by the Paris Opéra. Later operas were: *Ghismonda* (1918); *La Damnation de Blanche-Fleur* (1920); *La Femme nue* (1932).

Fibich, Zděnek, composer. Born Seborsitz, Bohemia, December 21, 1850; died Prague, October 15, 1900. Influenced by Richard Wagner's theories regarding the fusion of music and poetry, Fibich developed an operatic form which had been popular in the eighteenth century—the melodrama—in which the spoken dramatic text is provided with an orchestral accompaniment. His teachers at the Leipzig Conservatory included Ignaz Moscheles, Carl Richter, and Salomon Jadassohn. After additional study in Paris and Mannheim, he became a teacher of music in Vilna in 1870. Four years later he returned to his native land and was appointed assistant conductor of the Prague National Theater. After 1881 he concentrated on composition, producing many works influenced by the German romanticists, particularly Schumann and Brahms. His first opera, *Bukovin,* was produced in 1874. He now began evolving the melodrama, producing his first work in this form in 1875, *Christmas Eve.* His most ambitious melodrama was a trilogy entitled *Hippodameia,* written between 1889 and 1891. Fibich then returned to writing operas in a more formal and traditional manner and achieved his greatest success with *Sarka,* a folk opera. His most famous operas and melodramas: *Bukovin* (1874); *Christmas Eve* (1875); *Blaník* (1877); *The Water Sprite* (1883); *Hakon* (1888); the trilogy *Hippodameia: Pelop's Wooing* (1890); *The Atonement of Tantalus* (1891); *The Death of Hippodameia* (1892); *The Tempest* (1894); *Hedy* (1895); *Sarka* (1897); *The Fall of Arcona* (1900).

Fidalma, the aunt (mezzo-soprano) of Carolina and Elisetta in Cimarosa's *Il matrimonio segreto.*

Fidelio, opera in two acts by Ludwig van Beethoven. Libretto by Josef Sonnleithner and George Treitschke, based on a play by Jean Nicolas Bouilly, *Lénore, ou l'amour conjugal.* Première:

Theater-an-der-Wien, Vienna, November 20, 1805. American première: Park Theater, New York, September 9, 1839.

Characters: Florestan, a nobleman (tenor); Leonore, his wife (soprano); Don Fernando, Prime Minister of Spain (bass); Pizarro, governor of the prison (bass); Rocco, chief jailer (bass); Marcellina, his daughter (soprano); Jacquino, Rocco's assistant (tenor); prisoners; soldiers; guards. The setting is a prison near Seville, Spain, in the eighteenth century.

The rise of the curtain is preceded by the playing of the so-called *Fidelio Overture,* made up of two principal themes, one for horn answered by clarinets, the other for strings.

Act I, Scene 1. Rocco's kitchen. (In Beethoven's original version, the entire first act takes place in the prison courtyard. Following an innovation of Gustav Mahler, it has been customary to divide the act into two scenes.)

Florestan, a fighter of despotism, has been thrown into prison by his enemy Pizarro, and he is slowly starving to death. Hoping to save him, Florestan's wife Leonore disguises herself as a young man, takes the name of Fidelio, and becomes Rocco's assistant at the jail. Rocco's daughter, Marcellina falls in love with the disguised woman, arousing the jealousy of her suitor Jacquino. The three of them, along with Rocco, express their reactions to these complications in a quartet ("Mir ist so wunderbar!").

Scene 2. The prison courtyard. Fidelio learns from Pizarro that the Prime Minister is about to inspect the prison. Afraid that the Prime Minister may discover Florestan, Pizarro decides to kill his enemy. The news overwhelms Leonore ("Abscheulicher! wo eilst du hin!"). She prevails on Rocco to allow the prisoners to leave their cells for a few minutes. The prisoners emerge haltingly, groping into the blinding sunlight, singing a paean to freedom (Prisoners' Chorus: "O welche Lust!"). Leonore is grief-stricken to find that Florestan is not with them, but is somewhat consoled when Rocco informs her that she will descend with him to Florestan's cell and help with the digging of his grave.

Act II, Scene 1. The dungeon. Florestan is chained to the wall. With anguish he recalls happier days with his beloved wife ("In des Lebens Frühlingstagen"). At this point, Rocco and Leonore enter. She is shocked to see her husband so emaciated, but controls herself lest Rocco become suspicious. Rocco and Leonore dig the grave. Pizarro arrives; with dagger in hand he rushes toward his victim. Leonore springs between them, declaring she will kill Pizarro with her pistol if he makes another move. A fanfare of trumpets announce the arrival of the Prime Minister. Florestan and Leonore are jubilant; Pizarro is apprehensive, Rocco relieved. When Pizarro departs, Florestan and Leonore rush into each other's arms ("O namenlose Freude!"), after which Leonore leads her husband out of the dungeon.

Scene 2. The courtyard. (It has been customary since Mahler's time to perform the *Leonore Overture No. 3* during this change of scene. This overture contains material from the opera: a theme of Florestan's aria "In des Lebens Frühlingstagen," given by clarinet and bassoon; and the trumpet fanfare announcing the arrival of the Prime Minister. The energetic coda expresses the joy of Leonore and Florestan at their reunion. By an edict of the Prime Minister, the prisoners are released. They emerge from their cells, headed by Leonore and Florestan. It is now that the first Prime Minister learns of Florestan's ordeal. The Minister orders Pizarro arrested, then gives Leonore the key to Florestan's chains and asks her to free her husband. Florestan sings a hymn of praise to his devoted wife, and the people join him in this tribute ("Wer ein holdes Weib errungen").

Beethoven's solitary venture in opera is a masterwork—a masterwork with flaws—which is not dwarfed by the stature of his symphonies, sonatas, and chamber music. It is true, as critics have noted, that Beethoven's writing in *Fidelio* was usually more instrumental than vocal, that he was never at his best in vocal music, and that the limitations of the stage constricted his thinking. But after these subtractions are made, *Fidelio* remains a work of outstanding inspiration. It is the only German opera between those of Mozart and Wagner to survive in the permanent repertory. It has a high-minded nobility, an all-encompassing humanity, and a profundity of feeling found in few other operas. To Beethoven, the story of Leonore and Florestan represented the eternal struggle of man for freedom. Leonore was the symbol of the liberator. This subject inspired the democratic Beethoven. He soared above his limitations as a dramatist. In music as well as text *Fidelio* is a proclamation of liberty, tolerance, and human dignity. There is probably nothing in German opera more filled with this feeling than Leonore's celebrated "Abscheulicher!" — Beethoven's outraged defiance of all tyrants; nor can we find many pages more moving than the Prisoners' Chorus, one of the most eloquent musical paeans to freedom ever written. The early performance history of *Fidelio* was stormy. When the opera was introduced, Vienna was experiencing economic depression and political confusion, the aftermath of the French invasion. A new opera was of little interest—even one by Beethoven—and *Fidelio* was a failure. Beethoven revised his opera, compressing three acts into two, simplifying his writing for the voices, and preparing a new overture. The new version was given in 1806 and was a huge success. The opera would now have enjoyed a long and successful run if Beethoven, as the result of a quarrel over money, had not withdrawn it. Not until 1814 was *Fidelio* given again, once more with outstanding success.

Beethoven wrote four different overtures to *Fidelio*. The most famous is the *Leonore No. 3*, one of Beethoven's mightiest orchestral epics, written for the 1806 revival. The *Leonore No. 2* was used at the première performance and subsequently discarded. This overture, simplified and concentrated, became the *Leonore No. 1*, intended for a Prague performance that never materialized. The *Fidelio Overture*, which now opens the opera, was written for the 1814 revival.

Ferdinando Paër's opera *Eleanora* (1804) is based on the same story as *Fidelio*.

Fidès, mother (mezzo-soprano) of John of Leyden in Meyerbeer's *Le Prophète*.

Fieramosca, sculptor (baritone) in Berlioz' *Benvenuto Cellini*.

Fiesco, Jacopo, a Genoese nobleman (bass) in Verdi's *Simon Boccanegra*.

Figaro, (1) A former barber, now Count Almaviva's valet (baritone), in Mozart's *The Marriage of Figaro*.

(2) A barber (baritone) in Rossini's *The Barber of Seville*.

Figlia che reggi il tremulo piè, trio of La Cieca, La Gioconda, and Barnaba in Act I of Ponchielli's *La Gioconda*.

figlia del reggimento, La, *see* DAUGHTER OF THE REGIMENT, THE.

Figlia di re, a te l'omaggio, *see* FILLE DES ROIS, A TOI L'HOMMAGE.

Filipievna, a nurse (mezzo-soprano) in Tchaikovsky's *Eugene Onegin*.

Fille des rois, à toi l'hommage (Figlia di re a te i'omaggio), Nelusko's aria in Act II of Meyerbeer's *L'Africaine*.

fille du régiment, La, *see* DAUGHTER OF THE REGIMENT, THE.

Finale, the concluding number of an opera or of an act of an opera. It is usually a large-scale affair and it may contain formal arias and ensemble numbers.

Finck, Henry Theophilus, music critic and author. Born Bethel, Missouri, September 22, 1854; died Rumford Falls, Maine, October 1, 1926. After graduating from Harvard College with highest honors he went to Bayreuth in 1876 to attend the first Wagner Festival there. During the next few years he lived in Europe, studying psychology and philosophy. Back in the United States, he joined the staff of the *New York Post* as music critic in 1881, a post he held more than forty years. In 1890 he became professor at the National Conservatory of Music in New York, then headed by Antonin Dvořák. Finck wrote many books that are of interest to opera lovers. They include: *Wagner and His Works* (1893); *Pictorial Wagner* (1899); *Anton Seidl* (1899); *Massenet and His Operas* (1910); *Richard Strauss* (1917). His autobiography, *My Adventures in the Golden Age of Music* (1926), is a rich source of information about operatic activity in the United States.

Finita è per frati, *see* POUR LES COUVENTS C'EST FINI.

Finn, a sorcerer (tenor) in Glinka's *Russlan and Ludmilla*.

Fiora, Manfredo's wife (soprano) in Montemezzi's *L'amore dei tre re*.

Fior di giaggiolo, Lola's song in Mascagni's *Cavalleria rusticana*.

Fiordiligi, a lady (soprano) in love with Guglielmo in Mozart's *Così fan tutte*.

Fioriture, Italian for "flourishes"— decorative tones, either written by the composer or improvised by the singer, ornamenting the basic notes of a melodic line. The use of fioriture was especially common in Italian operas of the 18th century.

Firenze è come un albero fiorito, Rinuccio's aria in Puccini's *Gianni Schicchi*.

Fischer-Dieskau, Dietrich, baritone. Born Berlin, May 28, 1925. After studying voice in Berlin with Georg Walter, he entered the Berlin Hochschule für Musik. He served in the German army during World War II and for two years was a prisoner of war. In 1947 he made his concert debut in Freiburg, Germany, as a soloist in Brahms's *Requiem*. A year later he was appointed a leading baritone of the Berlin Opera, where he distinguished himself in the German and Italian repertories. He has since appeared at several of the Bayreuth Festivals, and has achieved an international reputation as an interpreter of lieder. He made his American debut early in 1955, touring the country in recitals.

Five, The, a group of nineteenth century Russian composers whose ideal was music grounded in national backgrounds. The group was a leading force in establishing Russian folk opera. The members were César Cui, Alexander Borodin, Mili Balakirev, Modest Mussorgsky, and Nikolai Rimsky-Korsakov.

Flagstad, Kirsten, soprano. Born Oslo, Norway, July 12, 1895. One of the most distinguished Wagnerian sopranos of the twentieth century, she studied singing with her mother (a coach at the Oslo Opera) and Ellen Schytte-Jacobsen, after which she made her opera debut in Oslo on December 12, 1913, in a minor role in *Tiefland*. She then had additional training with Albert Westwang and Gillis Brant, and appeared for several years in operettas and musical comedies as well as opera. After she joined the Gothenburg State Opera in 1927 she concentrated on opera. Two years later she married a wealthy Oslo industrialist, Henry Johansen, following which she went into temporary retirement. But she was soon convinced to return to the opera stage. Until 1933 her opera appearances were confined exclusively to Scandinavia, but in the summer of 1933 she appeared at the Bayreuth Festival in minor roles. The following summer she was cast as Sieglinde. A scout from the Metropolitan Opera heard her and arranged for an audition. This proved successful, and on February 2, 1935, Flagstad made her American debut as Sieglinde. Virtually unknown before she stepped on the stage that day, she created a sensation. The critics were as enthusiastic as the audience. Lawrence Gilman wrote in the *Herald Tribune:* "Yesterday was one of those comparatively rare occasions when the exigent Richard might have witnessed with happiness an embodiment of his Sieglinde. For this was a beautiful and illusive recreation, poignant and sensitive throughout, and crowned in its greater moments with an authentic exaltation." Three days later Flagstad scored another triumph in her first American appearance as Isolde. During the same season she was also heard as Brünnhilde in both *Die Walküre* and *Götterdämmerung* and as Kundry. As a gesture of honor to its new star the Metropolitan Opera began its 1936–1937 season with a German opera (*Die Walküre*) for the first time in thirty-five years. The beauty of Flagstad's voice was matched by the dignity and magnetism of her characterizations. Her voice was extraordinary in power, flexibility, and expressiveness. These qualities were combined with a profound musical understanding and a penetrating insight into the characters she was portraying.

Flagstad's American triumphs were duplicated in Europe, first at Covent Garden and the Vienna State Opera in 1936, afterward in Zurich, Paris, and Prague. During World War II, she decided to leave the United States and return to her native land. The fact that Norway was at the time occupied by the Nazis, and that her husband was a Quisling, discredited Flagstad in the eyes of the freedom-loving world. Flagstad insisted after the war that her political conscience was clear: she had never been friendly or cooperative with the Nazis, and her return home was motivated exclusively by the wish of a mother and wife to be with her family in perilous times. The antagonism to Flagstad, however, persisted in America for a long time, and there was opposition to her return appearances in concerts. Eventually, the opposition died down. In 1949 she sang successfully with the San Francisco Opera. A year later she was back with the Metropolitan Opera. She appeared at the Metropolitan Opera for the last time in 1952, singing in *Alceste*. The previous season Flagstad had sung the role of Dido *in* a London revival of Purcell's *Dido and Aeneas*. The performance was extraordinarily successful, and during the Coronation season of 1953 she sang the role another twenty-seven times, later appearing in the same opera in Norway. She returned to New York in March, 1955, to make two appearances with the Symphony of the Air (formerly the NBC Symphony Orchestra) in all-Wagner programs.

Flamand, a musician (tenor) in love with the Countess in Richard Strauss's *Capriccio*.

Flaminio, King Archibaldo's servant (tenor) in Montemezzi's *L'amore dei tre re*.

Flammen, French painter (tenor) in love with Lodoletta in Mascagni's *Lodoletta*.

Flammen, perdonami, Lodoletta's aria in Act III of Mascagni's *Lodoletta*.

Flaubert, Gustave, novelist. Born Rouen, France, December 12, 1821; died Croisset, France, May 8, 1880. Flaubert's novels and stories have been used in operas, the most famous being Massenet's *Hérodiade*. *Saint Julien l'Hospitalier* is the source of operas of the same name by Camille Erlanger and Riccardo Zandonai; *Salammbô,* of

operas by Eugenius Morawski, Ernest Rayer, Carl Navrátil, and Joseph Hauer. Guido Pannain drew on *Madame Bovary* for an opera of the same name; *La Tentation de Saint Antoine* is the title of an opera by Cecil Gray.

Flavio, Pollione's centurion (tenor) in Bellini's *Norma*.

Fledermaus, Die (The Bat), operetta by Johann Strauss II. Libretto by Haffner and Genée, based on a German comedy by Roderich Bendix, *Das Gefängnis.* Première: Theater-an-der-Wien, Vienna, April 5, 1874. One of the most popular operettas of all time, *Die Fledermaus* has often been performed in the great opera houses of the world. In 1950–1951 the Metropolitan Opera gave it a highly successful revival in a new English version by Howard Dietz and under the direction of Garson Kanin.

Baron von Eisenstein is a rich banker who is sought by the police for a minor indiscretion. When the police arrive at his home, his wife Rosalinde is entertaining her lover, Alfred. The police mistake Alfred for the Baron and drag him off to jail. When the Baron learns that he is supposed to be in jail, he decides to make the most of his freedom by going to a masquerade ball at the palace of Prince Orlovsky. There he flirts with all the women. His wife is also a guest, and before long the Baron begins to flirt with his own wife. When the Baron's identity is finally discovered, he is compelled to fill out his time in prison.

The overture, drinking song ("Trinke, Liebchen, trinke schnell"), laughing song ("Mein Herr Marquis"), csárdás ("Klänge der Heimat") the hymn to champagne ("Die Majestät wird anerkannt") and the "Du und Du" waltzes are the operetta's most popular numbers.

Fleg, Edmond, novelist, dramatist, and poet. Born Geneva, Switzerland, November 26, 1874. Fleg wrote the librettos for Ernest Bloch's *Macbeth* and Georges Enesco's *Oedipe*.

fleur que tu m'avais jetée, La, Don José's Flower Song in Act II of Bizet's *Carmen*.

Fleurissait une sauge (Légende de la sauge), Boniface's aria in Act II of Massenet's *Le Jongleur de Notre Dame*.

fliegende Holländer, Der (The Flying Dutchman), opera in three acts by Richard Wagner. Libretto by the composer, based on an old legend adapted by Heine. Première: Dresden Opera, January 2, 1843. American première: Philadelphia Academy of Music, November 8, 1876.

Characters: Daland, a Norwegian sea captain (bass); the Dutchman (baritone); Erik, a huntsman (tenor); Senta, Daland's daughter (soprano); Mary, her nurse (contralto); Steersman (tenor); sailors; maidens; villagers; hunters. The setting is a Norwegian village in the eighteenth century.

The overture opens with the motive of the Flying Dutchman in horns and bassoons, and includes references to Senta and the Sailors' Chorus.

Act I. In return for his challenge of heaven and hell, the Dutchman is doomed to sail forever on his ship "The Flying Dutchman" until redeemed through the love of a faithful woman. Once every seven years he is permitted to go ashore to find that love. During one of these periods, he is driven by a storm to a Norwegian harbor. He moors his ship near that of Daland. Wearily, the Dutchman describes how he seeks escape from his doom, and how always he is driven from shore to shore ("Wie oft in Meeres tiefsten Schlund"). Daland is impressed by the Dutchman's wealth, for the latter exhibits a jewel-filled casket. The Dutchman first asks Daland to give him shelter in his house, then inquires if he has a daughter and whether he would consent to their marriage. Daland has a daughter and, thinking only of the Dutchman's wealth, he gives his consent. When the weather clears, Daland and the Dutchman set sail.

Act II. A room in Daland's house. Senta and her friends are busy spinning (Spinning Song: "Summ' und brumm' "). Senta grows impatient with the singing, and when her friends ask her for a better song she sings a ballad about the Flying Dutchman (Senta's Ballad: "Traft ihr das Schiff"). Her lover Erik arrives. He begs her to forget the Dutchman and to accept his love. But Senta only points to the picture of the Dutchman hanging on the wall. And now the Dutchman and Daland arrive. Astonished and confused, Senta learns from her father that the Dutchman has asked for her hand in marriage. The Dutchman confesses to Senta that she is the woman of his dreams.

Act III. The bay near Daland's home. Daland's sailors strike up a chantey (Sailors' Chorus: "Steuermann! Lass die Wacht!"). Erik pursues Senta to beg for her love. He rebukes her for her faithlessness ("Willst jenes Tag's du nicht dich mehr entsinnen"). The Dutchman overhears him. Reasoning that if Senta can be unfaithful to Erik she can also be unfaithful to him, he suspects that his hopes for redemption are again to be shattered. He returns to his ship and, though a storm is raging, sets sail. Senta climbs to the top of a cliff, shouting to the Dutchman that she has always been faithful to him and will continue to be until death. She then throws herself into the sea. The "Flying Dutchman" immediately vanishes beneath the waves. The Dutchman has, after all, been redeemed. Embracing, the forms of Senta and the Dutchman are seen rising heavenward.

Der fliegende Holländer is one of Wagner's early operas, coming a year after *Rienzi* and preceding *Tannhäuser* by four. The concept of the music drama had not yet been crystallized by the composer. *Rienzi* had been a work in the style of Meyerbeer. *Der fliegende Holländer* placed little emphasis on ornate scenes or spectacles, but concerned itself with subtle psychological insights into its two principal characters and the drama of their inner conflicts. The music pointed to Wagner's later development, since it was here that he made use for the first time of the leitmotiv technique, even though sparingly. The idea for the opera came to Wagner in 1839 when he crossed the North Sea during a storm in which his ship was almost wrecked. Recalling the legend of the "Flying Dutchman" (he had read it in Heine's *Memoirs of Herr von Schnabelewopski*), he now identified himself with the character of the Dutchman. He felt that he, too, was fated to wander in misery until he could find redemption through the love of a woman, and through the peace and security of a permanent home in his own land. Wagner began writing his opera in Paris. Unable to interest the directors of the Paris Opéra in his music, he sold his libretto to them. They turned it over to Pierre-Louis Dietsch, who used it for his opera *Le vaisseau fantôme*. Produced in 1842, this work was a failure. Meantime, notwithstanding his sale of the text, Wagner continued writing his own opera. The tremendous success enjoyed by *Rienzi* at the Dresden Opera encouraged the director of that company to produce Wagner's new opera. *Der fliegende Holländer* failed to repeat the triumph of its predecessor. The opera was too somber, too new in style, too subtle in its portrayal of character and atmosphere to be understood and appreciated at first hearing.

Flight of the Bumblebee, The, an orchestral interlude in Act III of Rimsky-Korsakov's *The Legend of Tsar Saltan*. It has become famous in various transcriptions and adaptations.

Flora, a medium (soprano) in Menotti's *The Medium*.

Florence May Music Festival (Maggio Musicale Fiorentino), a music festival held every May in Florence, Italy. Opera is prominently featured. The first festival took place in 1933, the second in 1935. In 1937, subsidized by the Italian government, it became an annual event. With the exception of the World War II period, the festival has functioned without interruption, attracting audiences from all parts of the world. There have been many revivals of forgotten operas and premières of new ones. The premières have included: Valentino Bucchi's *Il contrabasso;* Mario Castelnuovo-Tedesco's *Aucassin and Nicolette;* Luigi Dallapiccola's *Il prigionero;* Vito Frazzi's *Don Chisciotte* and *Re Lear;* Joseph Haydn's *Orfeo ed Euridice;* Adriano Lualdi's *Il Diavolo sul campanile* (a new version); Francesco Malipiero's *Antonio e Cleopatra;* Ildebrando Pizzetti's *Vanna Luppa.* Unusual revivals have included: Cavalli's *Didone;* Cimarosa's *Le astuzie feminili;* Donizetti's *Don Sebastiano;* Monteverdi's *Orfeo;* Rossini's *Armida, La pietra del paragone, L'assedio di Corinto,* and *Tancredi;* Schumann's *Genoveva;* Spontini's *Agnes von Hohenstaufen* and *Olympie.* The festival also presented the European première of Menotti's *Amahl and the Night Visitors* and the first performance outside the Soviet Union of Prokofiev's *War and Peace.*

Florestan, Leonore's husband (tenor) in Beethoven's *Fidelio.*

Floriana, a music-hall singer (contralto) in Leoncavallo's *Zaza.*

Floria Tosca, a celebrated prima donna (soprano) in Puccini's *Tosca.*

Florindo, young man (tenor) in love with Rosaura in Wolf-Ferrari's *Le donne curiose.*

Florville, Sofia's beloved (bass) in Rossini's *Il Signor Bruschino.*

Flosshilde, a Rhine maiden (contralto)

in Wagner's *Das Rheingold* and *Götterdämmerung.*

Flotow, Friedrich, Freiherr von, composer. Born Teutendorf, Mecklenburg, April 26, 1812; died Darmstadt, January 24, 1883. His father was a landed nobleman. He studied in Paris with Anton Reicha and Johann Pixis. The 1830 Revolution sent him back to Germany, where he completed his first opera, *Peter und Katharina.* Returning to Paris in 1831, Flotow moved with the most distinguished opera composers of the day. For a while, he concentrated on writing operettas but in 1838 a collaboration with a French musician, Albert Grisar, turned him to writing more ambitious works. Flotow became famous as an opera composer with *Alessandro Stradella,* introduced in Hamburg on December 30, 1844. Three years later the Vienna Royal Opera introduced the work by which he is best remembered, *Martha.* Between 1856 and 1863 Flotow served as intendant of the Schwerin Court Theater. The next decade of his life was spent in Paris, Vienna, and Italy. In his last years he settled in Darmstadt. His most successful operas, in addition to those mentioned, were: *Rübezahl* (1853); *La Veuve Grapin* (1856); *Zilda* (1866); *L'Ombre* (1870.

Flower Duet, *see* SCUOTI QUELLA FRONDA DI CILIEGIO.

Flower Maidens' Scene, the second scene in Act II of Wagner's *Parsifal.*

Flower Song, *see* FLEUR QUE TU M'AVAIS JETEE, LA.

Fluth, name for Ford in German-language version of Nicolai's *The Merry Wives of Windsor.* See FORD.

Flying Dutchman, The, *see* FLIEGENDE HOLLÄNDER.

Fontana, Signor, the name assumed by Ford to conceal his identity in Verdi's *Falstaff.*

Ford, a wealthy burger in Nicolai's *The Merry Wives of Windsor* and Verdi's *Falstaff.* In both operas the role is for a baritone.

Ford, Alice, Ford's wife (soprano) in Verdi's *Falstaff.*

Ford, Mistress, Ford's wife (mezzo-soprano) in Nicolai's *The Merry Wives of Windsor.*

Ford, Nannetta (or Anne), Ford's daughter (soprano) in Verdi's *Falstaff.*

Forest Murmurs (Waldweben), a scene in Act II of Wagner's *Siegfried.*

Forging Song, *see* NOTHUNG! NOTHUNG!

Forsell, John, baritone and opera manager. Born Stockholm, November 6, 1868; died there November 30, 1941. He completed his music study at the Conservatories of Stockholm and Paris. In 1896 he made his opera debut at the Stockholm Royal Opera, remaining a permanent member of that company from 1896 to 1901 and again from 1903 to 1909. On November 20, 1909, he made his American debut at the Metropolitan Opera as Telramund, but he stayed with the Metropolitan only one season. He subsequently appeared in leading European opera houses, acclaimed in the German repertory. One of his outstanding interpretations was that of Don Giovanni. In 1923 he became intendant of the Stockholm Royal Opera, holding this post until 1938. The list of his many pupils included Metropolitan Opera tenors Jussi Bjoerling and Set Svanholm. On his seventieth birthday his extensive career was reviewed in the book *Boken om John Forsell* (1938).

Forty Days of Musa Dagh, *see* WERFEL, FRANZ.

forza del destino, La (The Force of Destiny), opera in four acts by Giuseppe Verdi. Libretto by Francesco Piave, based on the play *Don Alvaro, o la fuerza del sino* by Angel de Saavedra. Piave's libretto was later revised by Antonio Ghislanzoni. Première: St. Petersburg, November 10, 1862. American première: Academy of Music, February 24, 1865.

Characters: The Marquis of Calatrava (bass); Leonora, his daughter (soprano); Curra, her maid (mezzo-soprano); Don Carlo, the Marquis' son (baritone); Don Alvaro, a nobleman of Inca origin (tenor); Preziosilla, a gypsy (mezzo-soprano); Padre Guardiano, an abbot (bass); Fra Melitone, a friar (baritone); the Alcalde of Hornachuelos (bass); Mastro Trabuco, a muleteer (tenor); a surgeon; peasants; soldiers; friars. The action takes place in Spain and Italy; the time is the end of the eighteenth century.

Act I. The home of the Marquis of Calatrava, Seville. Leonora, in love with Don Alvaro, laments that her proud family will never accept him ("Me pellegrine ed orfana"). She plans to elope with him. When Leonora's father discovers the plot, Don Alvaro insists that the proposed elopement was his doing. He demands that the Marquis punish him, not Leonora. As a gesture of submission, Alvaro tosses his pistol aside. It explodes, wounding the marquis. Cursing his daughter, he dies.

Act II, Scene 1. An inn at Hornachuelos. Leonora, having lost trace of Don Alvaro after her father's death, comes seeking him, disguised as a man. In the crowd she discovers her brother, Don Carlo, who has sworn to kill her and Alvaro. She flees.

Scene 2. The monastery at Hornachuelos. Leonora falls on her knees at the monastery door and prays to the Virgin for help ("Madre, pietosa Vergine"). Padre Guardiano offers her a penitent's haven in a mountain cave. The abbot and monks then pray that a curse befall whoever attempts to harm her. Leonora joins the monks in a prayer to the Virgin ("La Vergine degli angeli").

Act III, Scene 1. A battlefield in Italy. Don Alvaro, under an assumed name, is with the Spanish army, fighting the Germans. Believing Leonora dead, he recalls her nostalgically ("O, tu che in seno"). His thoughts are disturbed by the cries of a wounded man. He proves to be Don Carlo, but the two men do not recognize one another. After Alvaro saves Carlo's life, they swear eternal friendship. The sound of a bugle sends them off to battle.

Scene 2. Headquarters of the Spanish army. Don Alvaro has been seriouly wounded in battle. He begs Don Carlo to destroy a certain packet of letters, declaring that he can then die in peace ("Solenne in quest' ora"). While looking for the letters, Carlo comes upon his sister's picture. It is only now that he realizes who his friend is, and he swears to destroy him ("Egli è salvo").

Scene 3. A military camp. Don Alvaro, recovered from his wounds, finds that his friend has now become his enemy. He tries to convince Don Carlo that he is innocent of wrong. Carlo, however, insists that the matter can be settled only in a duel. Forced to fight, Alvaro wounds Carlo seriously. Horrified—for once again he may be responsible for the death of someone dear to Leonora—Don Alvaro decides to seek peace and salvation in holy vows at a monastery.

Act IV, Scene 1. The monastery at Hornachuelos. Five years have passed. As Father Raphael, Don Alvaro has found peace. Carlo, recovered from his wound, has located Alvaro's retreat and comes seeking vengeance ("Invano Alvaro"). Once again he demands that Alvaro fight until one of them is destroyed. Alvaro refuses and tries to convince Carlo that God alone can bring retribution. In the face of Carlo's bitter insults, however, Alvaro is aroused to a point where he can no longer be compassionate. Seizing a sword, he rushes out of the monastery to fight Carlo.

Scene 2. A wild place in the mountains. Leonora, in her hiding place, prays God to relieve her of her tortured dreams and memories ("Pace, pace mio Dio!"). She hears the clashing of swords. This time Don Alvaro wounds his opponent mortally. Carlo begs Alvaro for absolution, but the latter no longer considers himself a holy man and refuses. Instead, he summons the inhabitant of the mountain cave—and is overwhelmed to learn that the "hermit" is none other than Leonora. She rushes to her dying brother. He curses her and with his ebbing strength plunges a dagger into her. Leonora, dying, begs Alvaro to find salvation in religion. Alvaro rails against his destiny. Padre Guardiano commands him, rather, to ask forgiveness ("Non imprecare, umiliati").

La forza del destino belongs to the rich middle period of Verdi's creative life, a period that produced *La traviata, Il trovatore,* and *Aïda.* It preceded *Aïda,* the last opera of this period, by about a decade. With its pronounced dramatic content and enriched harmonic and orchestral writing, *La forza del destino* represents a gradual departure from the style of *La traviata* toward that of *Aïda.* There are many beautiful arias and effective ensemble numbers in *La forza del destino,* but the central point of interest is not in isolated excerpts but in the dramatic feeling that pervades the entire work.

Foss, Lukas, composer. Born Berlin, Germany, August 15, 1922. He came to the United States in 1937, and has since become an American citizen. For a number of years he served as official pianist of the Boston Symphony Orchestra. He achieved his first major success as a composer with a cantata, *The Prairie,* introduced in 1944. He subsequently received a Guggenheim Fellowship and the American Prix de Rome. His first opera, *The Jumping Frog of Calaveras County,* was successfully introduced in Bloomington, Indiana, in 1950, and was subsequently given at the Venice Music Festival, in

Italy. A second opera, *Griffelkin,* was commissioned by the Koussevitzky Foundation.

Fouché, a Revolutionary patriot (baritone) in Giordano's *Madame Sans-Gêne.*

Four Saints in Three Acts, opera by Virgil Thomson. Libretto by Gertrude Stein. Première: Hartford, Connecticut, February 8, 1934. Miss Stein's beguiling but enigmatic libretto, consisting of seemingly random statements and phrases, inspired Thomson to write a sensitive and melodious score full of delightful American and popular elements. It has been performed with an all-Negro cast. After the première the opera had a six-week run in New York. It was subsequently heard several times in a concert version and was given a stage revival in 1952 in New York and Paris.

Four Sea Interludes, the orchestral intermezzi ("Dawn," "Sunday Morning," "Moonlight," and "Storm") in Britten's *Peter Grimes.*

Fra Diavolo (Brother Devil), opéra-comique in three acts by Daniel Auber. Libretto by Eugène Scribe. Première: Opéra-Comique, Paris, January 28, 1830. Fra Diavolo is a notorious bandit who, masquerading as the Marquis of San Marco, compromises an innkeeper's daughter in the course of a jewel robbery. Later, betrayed by his henchmen, Fra Diavolo is shot. Facing death, he gallantly absolves the girl of wrongdoing and reunites her with her worthy young lover.

Fra Gherardo, opera by Ildebrando Pizzetti. Libretto by the composer, based on the 13th century Chronicles of Salimbene de Parma. Première: La Scala, May 16, 1928. The central character is a weaver who repents having had a love affair with Mariola and joins the Flagellant Order to become Fra Gherardo of the White Friars. He becomes the spearhead of an attack of the oppressed people of Parma against the nobility. For his part in this fight, he is burned at the stake. Mariola, true to him to the end, is killed by an insane woman.

Françaix, Jean, composer. Born Le Mans, France, May 23, 1912. He attended the Paris Conservatory and made his bow as a composer in 1932 with a work for string quartet, heard at the International Society for Contemporary Music Festival in Vienna. Later works for piano and orchestra added to his reputation. He visited the United States in 1938, making his American debut with the New York Philharmonic-Symphony as soloist in his own piano concerto. His opera *La main de gloire* had a successful première at the Bordeaux Festival in 1950. A second opera, *L'apostrophe,* was given at the Holland Music Festival in 1951.

France, Anatole (born JACQUES ANATOLE THIBAULT), author. Born Paris, April 16, 1844; died Tours, October 12, 1924. A dominant figure in French literature, France wrote novels and stories that have been used for operas. The most notable is Massenet's *Thaïs.* Others include Charles Levadé's *La rôtisserie de la reine Pédauque,* and Massenet's *Le Jongleur de Notre Dame,* derived from France's short story *L'etui de nacre.*

Francesca da Rimini, the wife (by proxy) of the Lord of Rimini. She falls in love with the proxy, the Lord's brother, Paolo. This tragic romance is described by Dante in the fifth canto of the "Inferno" in *The Divine Comedy.* It has been used in numerous operas, notably ones by Emil Abrányi, Robert Hernried, Luigi Mancinelli, Eduard Napravnik, Sergei Rachmaninoff, Ambroise Thomas, and Riccardo Zandonai.

Franchetti, Alberto (Baron), composer. Born Turin, Italy, September 18, 1860; died Viareggio, August 2, 1942. He was born into a wealthy, noble family.

He studied with Felix Draeseke in Dresden and at the Munich Conservatory, and became successful with his first and finest opera, *Asrael,* produced in Brescia in 1888. Having a fortune at his disposal, he produced his later operas under the best possible auspices, but it would be a mistake to say that his success and fame were purchased with coin; they were acquired by means of a fine lyrical gift and a talent for effective theatricalism. His most successful operas were: *Asrael* (1888); *Cristoforo Colombo* (1892); *Germania* (1902); *Notte di leggenda* (1914). He wrote two operas in collaboration with Giordano: *Giove a Pompei* (1921); and *Glauco* (1922).

Franck, César, composer. Born Liège, Belgium, December 10, 1822; died Paris, November 8, 1890. After attending the Paris Conservatory, he completed his first major work, the eclogue *Ruth,* for solo voices, chorus, and orchestra. This work was performed in 1846. For many years he lived in humble obscurity, serving as organist for various Parisian churches, including Ste. Clotilde, and writing works that met with little recognition during his lifetime. He is most famous today for his orchestral, chamber, and choral works. His three operas were lesser productions, and they are now forgotten. The first, *Le valet de ferme,* (1852), was not published. The others are *Hulda* (1885), and *Ghisèle* (1890).

Frantz, Crespel's servant (tenor) in Offenbach's *The Tales of Hoffmann.*

Fra poco a me ricovero, Edgardo's aria in Act III, Scene 3, of Donizetti's *Lucia di Lammermoor.*

Frasquita, gypsy friend (soprano) of Carmen in Bizet's *Carmen.*

Frau ohne Schatten, Die (The Woman Without a Shadow), opera by Richard Strauss. Libretto by Hugo von Hofmannsthal. Première: Vienna State Opera, October 10, 1919. The opera is a symbolical fairy tale. Because the Fairy Princess has married an Eastern Emperor she is no longer either human or fairy. The spirit world decrees that if she cannot find a shadow, the Emperor will be transformed into stone. Her search for a shadow brings her to the humble household of Barak, who is ready to sell his wife's shadow for treasure until he realizes the tragic consequences of such a barter. In a temple, the Empress is told by mysterious voices that she can secure the shadow of Barak's wife by drinking the water in a near-by fountain. But the Empress refuses to bring tragedy to Barak's wife even at the price of saving her husband. For this act of unselfishness, the spirit world awards her with a shadow.

Frazier, a lawyer (baritone) in Gershwin's *Porgy and Bess.*

Frazzi, Vito, composer. Born San Secondo, Italy, August 1, 1888. He attended the Parma Conservatory. His first opera, *Re Lear,* was introduced at the Florence May Music Festival in 1939 and was acclaimed. A second opera, *Don Chisciotte,* was also given at the Florence festival. For many years Frazzi was a member of the faculty of the Cherubini Institute in Florence.

Frédéric, Gerald's friend (baritone) in Delibes's *Lakmé.*

Frederick, a student (tenor) in Thomas's *Mignon.*

Freia, goddess of youth (soprano) in Wagner's *Das Rheingold.*

Freischütz, Der (The Free-Shooter), opera in three acts by Carl Maria von Weber. Libretto by Friedrich Kind, based on a tale in the *Gespensterbuch,* edited by Apel and Laun. Première: Schauspielhaus, Berlin, June 18, 1821. American première: Philadelphia, December, 1824.

Characters: Ottokar, a Prince of Bohemia (baritone); Kuno, head ranger to the Prince (bass); Agathe, his daughter (soprano); Ännchen, Agathe's friend (soprano); Caspar, a huntsman (bass); Max, another huntsman (tenor); Zamiel, the Black Huntsman (speaking part); Kilian, a peasant (tenor); a hermit; huntsmen; peasants; musicians; spirits; demons. The setting is Bohemia; the time, long ago.

The celebrated overture opens with a dignified, religious melody for horns, and its core, a mobile theme for violins, is a tonal picture of Agathe.

Act I. A forest shooting range. Kuno has arranged a marksmanship contest to choose a successor as head ranger. Max, in love with Kuno's daughter, is eager to win, but in a preliminary trial, he loses to the peasant Kilian. Disheartened by this setback, Max expresses his anguish ("Durch die Wälder, durch die Auen"). He is now responsive to Caspar, who has a method whereby Max can win. For Caspar has sold his soul to Zamiel, the Black Huntsman, in return for magic bullets which never miss their mark. Since Caspar must bring Zamiel a new victim or lose his life, he is eager to have Max cooperate with the Black Huntsman. Max agrees to go to the Wolf's Glen.

Act II, Scene 1. Agathe's house. Alone and apprehensive, Agathe looks out of the window and marvels at the beauty of the night ("Leise, leise"). When Max comes, he tells her he has shot a stag at the Wolf's Glen and must rush off to bring it back. Agathe knows that the Glen is a haunted place and she begs Max not to go there. Max, however, is deaf to her pleading.

Scene 2. At the Wolf's Glen. Max meets Caspar. Strange apparitions appear; weird incantations are sounded. There are flashes of lightning. Zamiel creates seven bullets for Max.

Act III, Scene 1. Agathe's room. Agathe is getting ready to marry Max. She begs heaven for protection ("Und ob die Wolke sie verhülle"), for she is apprehensive, having had a dream filled with dire omens. Her terror mounts when she receives bridal flowers that turn out to be a funeral wreath.

Scene 2. The shooting range. It is the day of the contest. Foresters sing a hymn to hunting (Huntsmen's Chorus: "Was gleicht wohl auf Erden"). The contest begins. Max amazes everyone with his six remarkable shots. The Prince commands Max to hit a passing dove with his seventh bullet. Agathe's voice is heard begging him not to shoot, for she is the dove. But Max fires. His bullet strikes Agathe, but her bridal wreath has caught the bullet and saved her life. Caspar, serving the devil's justice, dies in her place. It is then that Max reveals his pact with Zamiel. The Prince first orders Max banished. Eventually, he promises to forgive Max after a year's probation. Max thanks the Prince for his generosity ("Die Zukunft soll mein Herz bewahren"), after which the entire assemblage gives voice to a song of thanksgiving ("Ja! lasst uns zum Himmel die Blicke erheben").

While *Der Freischütz* is seldom performed today, being chiefly remembered through a few excerpts, including its dramatic overture, it is historically a most significant opera. With this work romanticism and nationalism came to German opera. From this time on, Italianisms discarded, German opera would have a physiognomy of its own. The text of *Der Freischütz* was derived from German legend and made use of German backgrounds and landscapes as well as the German love for superstition and supernatural and diabolical elements. The music employed folk elements with such success that some of the arias sound like authentic German folk songs. Weber said Richard Wagner, "was the most German of German composers." *Der Freischütz* is one of the few operas that shaped musical history and at the same time was acclaimed by its first audiences. Its première was a triumph. The German Romantics immediately saw in it the typification of their creed and they hailed it as a glorification of the German spirit and soul. In Vienna, Dresden, and London it gathered further triumphs; in London the opera had to be given simultaneously in three different theaters.

Fremstad, Olive, soprano. Born Stockholm, Sweden, March 14, 1871; died Irvington, New York, April 21, 1951. She was one of the greatest Wagnerian sopranos. Brought to the United States in her childhood, she first appeared as a prodigy pianist. She later turned to singing, became soloist in a church choir, and appeared in a production of *Patience,* by Gilbert and Sullivan. In 1890 she settled in New York, studying singing with F. E. Bristol, who trained her as a contralto; it was in this range that she made her concert debut, appearing as soloist with various orchestras. She was trained for opera by Lilli Lehmann in Berlin, and she made her opera debut with the Cologne Opera on May 21, 1895, in *Il trovatore.* She remained with the Cologne Opera for three years, appearing in principal soprano roles. She was also seen in Bayreuth in 1896, and London and Vienna a year after that. After an additional period of study in Italy, she joined the Munich Opera Company for three years, appearing in about seventy different roles. Her American opera debut took place at the Metropolitan Opera on November 25, 1903, in the role of Sieglinde (the conductor of the evening, Felix Mottl, was also making his American debut). The following season she appeared for the first time as Kundry, and on January 1, 1908 (an evening when Gustav Mahler made his American bow), as Isolde.

Now one of the most brilliant personalities of the Metropolitan, she appeared in all the major Wagnerian soprano roles, combining a remarkable voice with outstanding histrionic gifts.

Her interpretations were a standard by which later Wagnerian sopranos were measured. Besides her success in Wagner, she scored as Salomé in Strauss's opera, a role which she created for New York and Paris, and in the title role of Gluck's *Armide*. Her last appearance at the Metropolitan took place on April 23, 1914, as Elsa. She made a few more opera appearances in 1915 with the Boston Opera and the Chicago Opera before withdrawing completely from the stage. After appearing in song recitals until 1920, she went into complete retirement. Among the many honors given her were appointments as Officer of Public Instruction and Officer of the French Academy, both in Paris. The story of her life was told by Mary Watkins Cushing in *The Rainbow Bridge*, published in 1954.

French Opera House, one of the most important nineteenth century opera houses in the United States, situated in New Orleans. It was built in 1859 on a site were earlier opera houses had stood, notably the Theatre St. Pierre, the Theatre St. Philippe, and the Theatre d'Orleans. It opened on December 1, 1859, with *William Tell*. During the next few decades some of the most significant of French operas were here introduced to America, including: Berlioz' *Benvenuto Cellini;* Gounod's *The Queen of Sheba;* Lalo's *Le Roi d'Ys;* Massenet's *Don Quichotte* and *Hérodiade;* Reyer's *Salammbô* and *Sigurd;* and Saint-Saëns' *Samson et Dalila.*

French Six, The, *see* SIX, LES.

Freudig begrüssen wir die edle Halle, chorus of the knights, noblemen, and ladies in Act II of Wagner's *Tannhäuser.*

Friar Laurence, friar (bass) in Gounod's *Roméo et Juliette.*

Fricka, Wotan's wife (mezzo-soprano) in Wagner's *Das Rheingold.*

Friedenstag, Der (The Day of Peace), one-act opera by Richard Strauss. Libretto by Joseph Gregor, based on Calderón's play, *La Redención de Breda.* Première: Munich State Opera, July 24, 1938. This opera, whose action takes place on the day of the Westphalian Peace, October 24, 1648, had political repercussions. Hitler, who had previously spoken bitterly about the Westphalian Peace, was significantly absent from the première, and so were most of his henchmen. Despite this obvious official disapproval, the opera enjoyed a huge success, and the composer received many curtain calls.

Frogs, The, *see* ARISTOPHANES.

Froh, a god (tenor) in Wagner's *Das Rheingold.*

From a House of the Dead, *see* AUS EINEM TOTENHAUS.

From Boyhood Trained, Hüon's aria in Act I, Scene 1, of Weber's *Oberon.*

Fugitif et tremblant, Lothario's aria in Act I of Thomas's *Mignon.*

Fuoco di gioia, chorus in Act I of Verdi's *Otello.*

Furiant, a lively Bohemian dance. The most famous example in opera is found in Act II of Smetana's *The Bartered Bride.*

Furtwaengler, Wilhelm, conductor. Born Berlin, January 25, 1886; died Eberstein, Germany, November 30, 1954. After an intensive period of study with Joseph Rheinberger, Max Schillings and Anton Beer-Walbrunn, Furtwaengler received his conducting apprenticeship in Zurich, Strassburg, and Lübeck. In 1915 he succeeded Artur Bodanzky as principal conductor of the Mannheim Opera. During the next few years he gained stature with remarkable performances of symphonic music with the Berlin State Opera Orchestra and at the Museum Concerts in Frankfort-on-the-Main. When Arthur Nikisch died in 1922, Furtwaengler inherited two of the most important conducting posts in Europe: that of the Leipzig Gewandhaus Orchestra and the Berlin Philharmonic. He now advanced to the first position among German conductors. In 1924 he gave highly successful guest performances of the Wagner music dramas at the Berlin State Opera; during the following years he was to appear there frequently. On January 3, 1925, he made his American debut as a guest conductor of the New York Philharmonic Orchestra. He returned to America during the next three seasons, while filling his regular posts in Germany and making remarkably successful guest appearances in major European cities. His Wagner festivals at the Paris Opéra became a regular feature of the Paris spring music season. In 1931 he made his first appearance in Bayreuth, directing *Tristan und Isolde*. He subsequently achieved outstanding success in opera performances at the Vienna State Opera, the Bayreuth, and Salzburg festivals, and at Covent Garden. When the Nazis came to power Furtwaengler was appointed Deputy President of the Reich Chamber of Music, musical director of the Berlin State Opera, and principal conductor at Bayreuth. A year later he was at odds with the Nazi officials. He was plan-

ning the world première of Hindemith's opera *Mathis der Maler,* at the Berlin State Opera, when government leaders decided that neither the composer nor his opera were in harmony with the new spirit of Germany. The opera was not performed, and Furtwaengler was forced to resign his musical and political posts. Six months later he was again in favor, and he resumed some of his former posts. In 1939 he was made Director of Musical Life in Vienna in a Nazi effort to rehabilitate the musical life of that city. Because of his close association with the Nazi government, Furtwaengler became *persona non grata* in the United States. Efforts to give him conductorial assignments with major American orchestras had to be abandoned. In Europe his position as one of the world's great conductors of symphonic music and opera remained unchallenged. It was intended that Furtwaengler should lead the Berlin Philharmonic Orchestra during its 1955 tour of the United States, but when his death intervened Herbert von Karajan succeeded to the conductorship.

G

Gadski, Johanna, dramatic soprano. Born Anclam, Germany, June 15, 1872; died Berlin, February 22, 1932. She studied with Mme. Schroeder-Chaloupka in Stettin and made her opera debut in 1889 with the Kroll Opera in the title role of Lortzing's *Undine.* After appearances in various German opera houses she made her American debut with the Damrosch Opera Company on March 1, 1895, in

Lohengrin. During the next three seasons she was seen with that company in most of the principal Wagnerian soprano roles and in the world première of Walter Damrosch's *The Scarlet Letter.* Her first appearance at the Metropolitan Opera took place at a concert on December 11, 1898. Up to 1904 she appeared, and was acclaimed, in the Wagner dramas, particularly in the roles of Brünnhilde and Isolde. During this period she also appeared at Bayreuth and Covent Garden. In 1905 and 1906 she appeared at the Wagner festivals in Munich. Returning to the Metropolitan in 1907, she remained there for a decade. During World War I, when her husband Hans Tauscher (whom she had married in 1892) was deported as an undesirable alien, she left the United States and returned to Germany. During the years 1929–1931 she sang in America with the Wagnerian Opera Company, but she was then past her prime.

Gailhard, Pierre, bass and opera manager. Born Toulouse, France, August 1, 1848; died Paris, October 12, 1918. After study at the Paris Conservatory, he made his opera debut at the Opéra-Comique on December 4, 1867, in Thomas's *Le songe d'une nuit d'été.* As a member of that company for the next four years he created several roles, including the principal bass parts in Auber's *Rêve d'amour* and Offenbach's *Vert-Vert.* On November 3, 1871, he appeared for the first time at the Paris Opéra, as Méphistophélès in *Faust.* He remained a principal bass there for over a decade. In 1884 he retired as singer and joined Ritt as manager of the Opéra, becoming sole manager of the company from 1905 to 1908. He was responsible for the French premières of several of the Wagnerian music dramas and for the growing reputations of such singers as the De Reszke brothers, Eames, Melba, and Renaud.

Galatea, Acis' beloved (soprano) in Handel's pastoral opera, *Acis and Galatea.*

Galitsky, Prince, Prince Igor's brother-in-law (bass) in Borodin's *Prince Igor.*

Galitzin, Prince Vassily, a reformer (tenor), member of Young Russia, in Mussorgsky's *Khovantchina.*

Galli-Curci, Amelita, coloratura soprano. Born Milan, Italy, November 18, 1882. She studied the piano at the Milan Conservatory, but was encouraged by Mascagni to specialize in singing. She studied by herself and, without a single formal lesson, made her opera debut at the Teatro Costanzi, in 1909, as Gilda. After appearing in many opera houses in Europe and South America she sang in the United States for the first time on November 18, 1916, as Gilda, in a performance by the Chicago Opera. She was a sensation. On January 28, 1918, during a visit of the Chicago Opera, she appeared in New York for the first time, singing the role of Dinorah, and again was extravagantly acclaimed. Her Metropolitan Opera debut took place on November 14, 1921, as Violetta. For the next three seasons she was one of the few singers who were simultaneous permanent members of the Metropolitan and the Chicago Operas. In 1924 she withdrew from the Chicago company, remaining with the Metropolitan another half dozen years. While her coloratura singing frequently lacked finish, it had great beauty of tone, and her impersonations had warmth and charm. Her finest roles included Rosina, Lakmé, Dinorah, Gilda, Juliette, and Elvira in *I Puritani.* A throat ailment compelled her to bring her career to a temporary end. She was operated on in 1935, and on November 24, 1936, was able to make a return appearance in Chicago, as Mimi, but it was evident that her singing days were over. She then went into complete retirement, living in Beverly Hills, California.

Galli-Marié, Marie Célestine (born DE L'ISLE), mezzo-soprano. Born Paris, November, 1840; died Vence, France, September 22, 1905. She created the roles of Carmen and Mignon. Her only teacher was her father, a member of the Paris Opéra company. In 1859 she made her opera debut in Strasbourg, but she did not achieve success until 1862, when, in Rouen, she sang in *The*

Bohemian Girl. On August 12, 1862, she made her debut at the Opéra-Comique in *La serva padrona*, remaining a principal soprano of that company for over two decades, and appearing in many notable premières. Though her voice was not exceptional, she brought remarkable dramatic force to her characterizations.

Gallo, Fortune, impresario. Born Torremaggiore, Italy, May 9, 1878. He came to the United States in 1895 and soon after the turn of the century began his career as impresario by managing several traveling opera companies. In 1909 he founded the company with which his name is chiefly identified: the San Carlo Opera. This company made numerous tours of the United States and Canada and was a vital force in spreading the love of opera. Gallo also supervised many open-air opera performances in stadia, made a film version of *Pagliacci* in 1928, and subsequently was a pioneer in presenting opera in sound motion pictures.

Galuppi, Baldassare, composer. Born Burano, Italy, October 18, 1706; died Venice, January 3, 1785. After studying the violin with his father he earned his living in Venice playing the organ. There he wrote his first opera, *La fede nell' incostanza,* which was such a fiasco that for a while he gave up music. The composer Marcello convinced him to reconsider. After a period of study at the Conservatorio degli Incurabili, in Venice, Galuppi wrote a second opera, *Dorinda,* which was an outstanding success in 1729. Galuppi went on to write over a hundred operas, presented in the leading Italian opera houses. In 1741 he visited London, where two of his operas, *Scipione in Cartagine* and *Enrico,* were so successful that they influenced a whole school of English opera composers. He became particularly famous for his comic operas, so much so that he is sometimes described as the "father of opera

buffa." His most celebrated work in this vein was *Il filosofo di campagna,* produced in Venice in 1754. From 1748 to 1762 Galuppi was assistant maestro di cappella at St. Mark's in Venice, and from 1762 principal maestro. In 1766 he was invited to Russia by Catherine II, where for two years he assisted in producing his operas. He returned to Venice in 1768, where he remained for the rest of his life, active as maestro at St. Mark's, as director of the Conservatory, and as composer. His finest operas were: *Dorinda* (1729); *Issipile* (1738); *Adriano in Siria* (1740); *Scipione in Cartagine* (1742); *Enrico* (1743); *Didone abbandonata* (1752); *Il filosofo di campagna* (1754); *Il re pastore* (1762); *Ifigenia in Tauride* (1768).

Gama, *see* VASCO DA GAMA.

García, Manuel del Popolo Vicente, tenor, composer, and teacher. Born Seville, Spain, January 22, 1775; died Paris, France, June 2, 1832. He began singing in the Seville Cathedral choir when he was six, and by the time he was seventeen he was active as actor, singer, conductor, and composer. On February 11, 1808, he made his opera debut in Paris in Ferdinando Paër's *Griselda,* and was a great success. In 1811 he went to Italy, where he appeared for five years and profited through contact with Italian singing methods. It was with García in mind that Rossini created the principal male characters in *The Barber of Seville* and *Elisabetta.* From 1817 to 1819, in London, and from 1819 to 1824, in Paris, he was the idol of opera-goers; in Paris he created for France the role of Count Almaviva in *The Barber of Seville.* In 1823 he founded a highly successful school of singing in London. Two years later he came to the United States as an opera impresario. From November 29, 1825, to September 30, 1826, he gave seventy-nine performances at the Park and Bowery Theaters, including eleven operas new to America. From New York he went to Mexico. On his way back, after a successful year and a half, he was held up and robbed of all his earnings. Returning to Paris, he sang at the Théâtre des Italiens, after which he retired from the stage and devoted himself to teaching. His most celebrated pupils were his daughters Maria Malibran and Pauline Viardot-García, and the tenor Adolphe Nourrit. García wrote many Spanish, Italian, and French operas, all now forgotten.

His son, Manuel Rodriguez García (1805–1906), was a famous singing teacher, for many years a professor at the Parıs Conservatory and the Royal Academy of Music in London. His pupils included Jenny Lind, Henriette Nissen, Mathilde Marchesi, and his son Gustave (later a successful opera singer). Manuel García was a pioneer in investigating the physiology of voice production. He lived to the patriarchal age of a hundred and one.

Garcias, a burlesquer (soprano) in Massenet's *Don Quichotte.*

Garden, Mary, soprano. Born Aberdeen, Scotland, February 20, 1877. One of the most glamorous prima donnas of the twentieth century, she created the roles of Louise and Mélisande. While still a child she was brought to the United States, where she studied singing with Mrs. Robinson Duff. In her sixteenth year she appeared in an amateur performance of the Gilbert and Sullivan operetta *Trial by Jury.* Further vocal study took place in Paris with Mathilde Marchesi. The famous American soprano Sibyl Sanderson became interested in her and introduced her to Albert Carré, manager of the Opéra-Comique. Carré urged her to continue studying until he could find the proper role for her. It was found suddenly and unexpectedly. Some two months after the première of *Louise,* Marthe Rioton, creator of the

leading role, was taken ill during a performance. Garden, who had studied the role but had never yet sung before an audience, replaced the indisposed prima donna and was a sensation. She was subsequently invited to sing the role of *Louise* in most of the great opera houses of Europe and America and became identified with it. As a permanent member of the Opéra-Comique company, she was selected by Carré and Debussy in 1902 to create the role of Mélisande in *Pelléas et Mélisande,* this over the vociferous objections of Maeterlinck, author of the play, who wanted Georgette Leblanc, his common-law wife, to have the role. Mary Garden made the part so completely her own that for many years it was impossible to think of Mélisande without her. She created the principal soprano roles in several other French operas, including Camille Erlanger's *Aphrodité,* Massenet's *Sapho,* and Saint-Saëns' *Hélène.* Her American debut took place at the Manhattan Opera on November 25, 1907, in the American première of *Thaïs.* Some of the critics remarked that her singing had imperfections, but they could not deny that she was a brilliant personality and that her voice had natural beauty. She remained with the Manhattan Opera company until its dissolution in 1910, scoring a sensation in her first appearance as Salomé in the Richard Strauss opera, in which she dispensed with the services of the customary double and herself performed the Dance of the Seven Veils. In 1910, Garden joined the Chicago Opera, remaining with this company until the end of her career in opera. When Campanini, general manager of the company, died in 1919, Garden was selected as his successor, the first time a woman was called upon to direct a major opera company. She held this post a little less than three years. As a result of her lavish expenditures and resultant deficits, the company had to be reorganized. The direction passed to other hands, and Mary Garden returned to the company of singers, where she remained until 1931, appearing in several world and American premières, including those of Alfano's *Risurrezione,* Hamilton Forrest's *Camille,* and Honegger's *Judith.*

Oscar Thompson noted that her "work was disturbingly irregular," and that "there was a wide divergence between her best and her most commonplace performances of the same part." But when she was at her best as Mélisande, Louise, or Thaïs, she was an artist of incomparable magnetism and glamor. After her withdrawal from the opera stage, Mary Garden devoted herself to concert work, lectures, and teaching. For a period, she was vocal advisor to Metro-Goldwyn-Mayer, in Hollywood. She wrote her autobiography, *Mary Garden's Story,* in collaboration with Louis Biancolli (1951).

Garrick, David, actor and playwright. Born Hereford, England, February 19, 1717; died London, January 20, 1779. One of the great theatrical figures of his day, Garrick wrote several plays that were made into operas. One of these, written in collaboration with George Coleman, was *The Secret Marriage,* used by Cimarosa for his *Il matrimonio segreto,* and by Peter Gast for his less familiar opera, *Die heimliche Ehe.* Garrick also wrote the libretto for Thomas Arne's *Cymon.* He is the central character in a contemporary opera by Albert Stoessel, *David Garrick.*

Garrido, general in the Royalist army (bass) in Massenet's *La Navarraise.*

Gasparo, Don, the king's officer (tenor) in Donizetti's *La favorita.*

Gastone, Viscount of Letorières (tenor) in Verdi's *La traviata.*

Gatti, Guido Maria, musicologist. Born Chieti, Italy, May 30, 1892. In 1928 he founded *La Rassegna musicale,* one of Italy's most distinguished

music journals, and has edited it since then. In 1925 he founded and directed the Teatro di Torino, where performances of operas and symphonic music took place until 1931. He was also one of the founders of the Florence May Music Festival. His books include biographies of Bizet (1914) and Pizzetti (1952) and monographs on the operas *Debora e Jaele* (1922) and *The Barber of Seville* (1925).

Gatti-Casazza, Giulio, opera manager. Born Udine, Italy, February 3, 1869; died Ferrara, Italy, September 2, 1940. Though he expected to become an engineer, in 1893 he succeeded his father as director of the Municipal Theater in Ferrara. So brilliant was his work that in 1898 he was offered the most important post of this kind in Italy, the management of La Scala. He remained at La Scala for a decade. His regime did much to popularize the Wagnerian music dramas in Italy, many of them receiving their Italian premières, and also brought about the first Italian productions of such significant operas as *Boris Godunov* and *Pelléas et Mélisande.* In 1908 Gatti-Casazza came to the United States to become general manager of the Metropolitan Opera. He held this post with great distinction for a quarter of a century (*see* METROPOLITAN OPERA HOUSE). Retiring in 1935, he returned to Italy. He was married twice, first to Frances Alda, the famous prima donna of the Metropolitan Opera, and afterward to Rosina Galli, the Metropolitan Opera ballerina and ballet mistress. Just before his death he completed his autobiography, *Memories of the Opera.*

Gaubert, Philippe, flutist, conductor, and composer. Born Cahors, France, July 4, 1879; died Paris, July 10, 1941. He was the principal conductor of the Paris Opéra for over two decades. He attended the Paris Conservatory, where he won the Prix de Rome. Later he appeared throughout France as solo flutist. During World War I he served in the French army, receiving the Croix de Guerre for heroism in the Battle of Verdun. From 1919 to 1938 he was the principal conductor of the Paris Conservatory Concerts, and from 1920 to the time of his death was principal conductor of the Paris Opéra. He wrote two operas: *Sonia* (1913), and *Naïla* (1927).

Gaudenzio, Sofia's guardian (bass) in Rossini's *Il Signor Bruschino.*

Gaudio son al cuore queste pene dell' amor, the trio of Orfeo, Euridice, and Amor in the final scene of Gluck's *Orfeo ed Euridice.*

Gaussin, Jean, young man (tenor) in love with Sapho in Massenet's *Sapho.*

Gautier, Théophile, poet and novelist. Born Tarbes, France, August 31, 1811; died Neuilly, October 23, 1872. His story *Une nuit de Cléopâtre,* was used in two operas: Hadley's *Cleopatra's Night,* and Massé's *Une nuit de Cléopâtre.*

Gavotte, an old French dance in common time and beginning on the third beat. The dance became popular at the court of Louis XIV, and gavottes were used in operas by the early French composers (Lully and Rameau) and by Handel. There is a fine specimen in Gluck's *Iphigénie en Tauride,* another in Mozart's *Idomeneo.* Perhaps the most celebrated gavotte of all is that of Gossec, in his opera *Rosina.* Later instances occur in *Manon* (Manon's aria "Obéissons quand leur voix appelle") and *Mignon* (the orchestral entr'acte before Act II and Mignon's aria "Me voici dans son boudoir").

Gay, John, poet and playwright. Born Barnstaple, England, September 1685; died London, December 4, 1732. He wrote the texts for Handel's *Acis and Galathea* and for *The Beggar's Opera.*

gazza ladra, La (The Thieving Magpie), opera buffa by Rossini. Libretto by Giovanni Gherardini, based on *La*

pie voleuse, by D'Aubigny and Caigniez. Première: La Scala, May 31, 1817. A servant girl is accused of stealing a silver spoon and is condemned to the gallows. The spoon, however, is found in a magpie's nest, and the girl is exonerated. The overture is famous; an innovation in its orchestration is the use of two snare drums.

Gebrauchsmusik, a German term signifying "functional music"—describing a movement in twentieth century German music. While the term was usually applied to pieces for radio, screen, education, and so forth—particularly, such works composed early in his career by Paul Hindemith—it also described operas of that period utilizing a functional or popular style. Notable examples were Hindemith's *Neues vom Tage,* Ernst Křenek's *Jonny spielt auf,* and Kurt Weill's *The Threepenny Opera.* When the Third Reich came to power, this movement collapsed, since popular and functional elements in art were considered decadent.

Gellner, Wally's suitor (baritone), Hagenbach's rival, in Catalani's *La Wally.*

Geneviève (contralto), mother of Pelléas and Golaud in Debussy's *Pelléas et Mélisande.*

Gennaro, (1) Lucrezia Borgia's son (tenor) in Donizetti's *Lucrezia Borgia.*

(2) A blacksmith (tenor) in Wolf-Ferrari's *The Jewels of the Madonna.*

Genoveva, opera by Robert Schumann. Libretto by the composer, based on another libretto by Robert Reinick, in turn derived from dramas by Tieck and Hebbel. Première: Leipzig, June 25, 1850. This is Schumann's only opera, and it is rarely performed. It was revived at the Florence May Music Festival in 1951. The story concerns Siegfried, Count of the Palatinate, who, off to war, entrusts the care of his wife Genoveva to his friend Golo. Golo is madly in love with her. When she resists him, his love turns to hate and he desires her destruction. He brings Siegfried false news of his wife's infidelity. Siegfried orders Golo to kill Genoveva, but before the murder can take place, Siegfried learns of Golo's treachery and arrives in time to save his wife. Golo meets his doom by falling over a precipice.

Gérald, British officer (tenor), in love with Lakmé in Delibes's *Lakmé.*

Gérard, Charles, a revolutionary leader (baritone) in Giordano's *Andrea Chénier.*

Germania, lyric drama by Alberto Franchetti. Libretto by Luigi Illica. Première: La Scala, March 11, 1902. The setting is Germany in the early nineteenth century. A secret organization is created to fight Napoleon's rising power. Two of its members, Worms and Loewe, are rivals for Ricke's love. Loewe wins Ricke and marries her; he and Worms become bitter enemies. Later on, Ricke finds the bodies of both Loewe and Worms on the Leipzig battlefield.

Germont, Alfredo, Violetta's lover (tenor) in Verdi's *La traviata.*

Germont, Giorgio, Alfredo's father (baritone) in Verdi's *La traviata.*

Geronimo, a rich merchant (bass) in Cimarosa's *Il matrimonio segreto.*

Gershwin, George, composer. Born Brooklyn, New York, September 26, 1898; died Hollywood, California, July 11, 1937. He became popular as a composer of musical comedies, some of them of more than ordinary quality. One, *Of Thee I Sing,* was the first musical comedy to win a Pulitzer Prize. After the success of his *Rhapsody in Blue,* for piano and orchestra, in 1924, he devoted himself to both popular and serious music. His opera *Porgy and Bess* was the last and most important of his serious works. An earlier one-act Negro opera, in jazz style, *135th Street,* was performed one night only in the *George White scandals of 1922.* (It was then called *Blue Monday.*) It has been revived several times since, twice by Paul Whiteman (1925 and 1936), and on a television broadcast in 1953. Gershwin died, aged thirty-eight, after a brain operation.

Gerster, Etelka (coloratura soprano). Born Kaschau, Hungary, June 25, 1855; died Pontecchio, Italy, August 20, 1920. While studying with Mathilde Marchesi at the Vienna Conservatory she sang for Verdi an aria from *La traviata.* Verdi recommended her to the Teatro la Fenice, where she made her opera debut on January 8, 1876, as Gilda. She then toured Italy and Germany with a traveling opera company. In 1877 she married Carlo Gardini, an impresario, and the same year made a successful debut at Covent Garden. A year later she came to the United States with the Mapleson Opera Company, making her American debut in *La sonnambula.* She was so successful that she became a rival of Adelina Patti, then a New York favorite. After 1890 Gerster withdrew from opera and in Berlin founded a school of singing. She also taught for a year in New York. She wrote a treatise on singing, *Stimmführer* (1906).

Gertrude, (1) Juliet's nurse (mezzosoprano) in Gounod's *Roméo et Juliette.*

(2) Mother of Hansel and Gretel (mezzo-soprano) in Humperdinck's *Hansel and Gretel.*

(3) Hamlet's mother (soprano) in Thomas's *Hamlet.*

Gerusalemme liberata, La, epic poem by Tasso, the source of numerous operas. *See* ARMIDA; RINALDO; TASSO.

Gerville-Réache, Jeanne, dramatic contralto. Born Orthez, France, March 26, 1882; died New York City, January 5, 1915. She studied in Paris with Rosine Laborde, Pauline Viardot-García, and Jean Criticos. Her opera debut took place at the Opéra-Comique in 1900 in Gluck's *Orfeo ed Euridice.* During two seasons with that company she created the roles of Catherine in *Le Juif polonais* and Geneviève in *Pelléas et Mélisande*. From 1907 to 1910 she sang with the Manhattan Opera Company, where her electrifying interpretation of Dalila was largely responsible for making Saint-Saëns' *Samson et Dalila* popular in America. She married Dr. G. Gibier-Rambeaud, director of the Pasteur Institute in New York. After leaving the Manhattan Opera she appeared for one season with the Chicago Opera.

Gessler, tyrant governor (bass) of Schwitz and Uri in Rossini's *William Tell.*

Gezeichneten, Die (The Stigmatized Ones), opera by Franz Schreker. Libretto by the composer. Première: Frankfort-on-the-Main, April 25, 1918. This expressionist drama, set in 16th century Genoa, tells of the fruitless love of a cripple with a beautiful soul for a girl with a beautiful body.

Ghedini, Giorgio Federico, composer. Born Cuneo, Italy, July 11, 1892. He graduated from the Bologna Conservatory in 1911, after which he taught in the Conservatories of Bologna, Turin, and Milan. In 1951 he was appointed director of the Milan Conservatory. After achieving success with several orchestral works he wrote his first opera, *Maria d'Allesandria*, which was introduced in Bergamo in 1937. Later operas: *L'intrusa* (1938); *Re Hassan* (1939); *La pulce d'oro* (1940); *Le Baccanti* (1943); *Billy Budd* (1948).

Gherardino, Gherardo's son (mezzosoprano) in Puccini's *Gianni Schicchi.*

Gherardo, Donati's nephew (tenor) in Puccini's *Gianni Schicchi.*

Ghione, Francesco, conductor. Born Acqui, Italy, August 26, 1889. After studying at the Parma Conservatory he played the violin in various Italian opera houses. His debut as conductor took place at the Puglie Opera in 1913. Six years later he directed the Italian repertory at the Barcelona Opera, and

in 1922 was appointed Toscanini's assistant at La Scala. After serving as principal conductor of La Scala for several seasons and appearing with other Italian opera companies, he came to the United States in 1931 as guest conductor of the Detroit Civic Opera. From 1936 to 1939 he was the conductor of the Detroit Symphony. After 1940 he conducted the Italian repertory at the Teatro Colón in Buenos Aires and the Rio de Janeiro Municipal Theater.

Ghislanzoni, Antonio, dramatist and librettist. Born Lecco, Italy, November 25, 1824; died Bergamasco, July 16, 1893. He began his career as a singer but soon turned to writing and editing. For many years he was the editor of the *Gazzetta Musicale* in Milan. He wrote numerous librettos, the most important being the one for *Aïda.* Others were for Alfredo Catalani's *Edmea;* Vladimir Kashperov's *Maria Tudor;* Enrico Petrella's *I promessi sposi;* Ponchielli's *I Lituani* and *I Mori di Valenza.*

Già i sacerdoti adunansi, duet between Amneris and Radames in Act IV, Scene 1, of Verdi's *Aïda.*

Già nella notte densa, the love duet of Otello and Desdemona in Act I of Verdi's *Otello.*

Gianetta, a peasant girl (soprano) in Donizetti's *L'elisir d'amore.*

Gianetto, (1) Lodoletta's suitor (baritone), in Mascagni's *Lodoletta.*

(2) A poet (tenor), principal character in Giordano's *La cena delle beffe.*

Giannini, Dusolina (soprano). Born Philadelphia, Pennsylvania, December 19, 1902. She studied with Marcella Sembrich after which she made her concert debut in New York in 1923. Four years later her opera debut took place in Hamburg, Germany, in *Aïda.* Her success brought her engagements in Europe's leading opera houses. In 1934 she scored a major success as Donna Anna at the Salzburg Festival. On February 12, 1936, she made her debut at the Metropolitan Opera in *Aïda*. She remained at the Metropolitan through the 1940–1941 season. In Europe she created the role of Hester Prynne in *The Scarlet Letter,* an opera by her brother Vittorio. She is the daughter of Ferruccio Giannini, a noted operatic tenor.

Giannini, Vittorio, composer. Born Philadelphia, Pennsylvania, October 19, 1903. His sister is the prima donna Dusolina Giannini (see above). He attended the Milan Conservatory on a scholarship, then continued his music study at the Juilliard Graduate School. After receiving the grand prize of the American Academy of Rome in 1932, he wrote his first mature opera, *Lucedia,* introduced in Munich in 1934. Four years later the Hamburg Opera introduced *The Scarlet Letter,* his sister creating the role of Hester. In 1937, 1939, and 1940 he wrote three radio operas on a commission from the Columbia Broadcasting System: *Flora, Beauty and the Beast,* and *Blennerhasset.* In 1953 another opera, *The Taming of the Shrew,* was produced in Cincinnati. Giannini has been a member of the faculties of the Juilliard School and the Manhattan School of Music.

Gianni Schicchi, one-act opera buffa by Giacomo Puccini. Libretto by Gioachino Forzano, based on the history of a citizen of medieval Florence. World première: Metropolitan Opera, New York, December 14, 1918.

Characters: Zita, cousin of Buoso Donati (mezzo-soprano); Rinuccio, her nephew (tenor); Simone, cousin of Buoso (basso); Marco, his son (baritone); La Ciesca, Marco's wife (soprano); Gherardo, Buoso's nephew (tenor); Nella, his wife (soprano); Gherardino, their son (mezzo-soprano); Betto di Signa, Buoso's brother-in-law (baritone); Gianni Schicchi (baritone); Lauretta, his daughter (soprano); Spinelloccio, a doctor (bass); Amantio di Nicolao, a lawyer (bass); Pinellino, a

shoemaker (bass); Guccio, a dyer (bass). The setting is Florence in the thirteenth century.

Buoso Donati, a rich Florentine, has died, leaving his fortune to a monastery. The loss of an inheritance disturbs his family. Rinuccio, in love with Lauretta, tries to devise a scheme for acquiring Buoso's wealth. He suggests calling in the wily Gianni Schicchi. Schicchi at first appears reluctant to enter into discreditable maneuvers, but his daughter pleads with him ("O mio babbino caro") and effects a change of heart. Schicchi notes that Buoso's death has not yet been made public. This being the case, he offers to impersonate the dead man. The news of Buoso's imminent death is to be spread through Florence, after which Schicchi, as Buoso, will dictate a new will bequeathing his fortune to his family. The plot is set in motion. The family calls in a lawyer to draw up a new will. To the horror of the family, "Buoso Donati" bequeaths his fortune to one Gianni Schicchi. The relatives must remain silent, for to protest would be to incriminate themselves in the fraud. But when the lawyer leaves, they fall upon Schicchi. He drives them away with a stick. As Rinuccio and Lauretta tenderly express their love for one another ("Lauretta mia"), Schicchi remarks that no better use can be made of his new fortune than to help the lovers.

Gianni Schicchi is the third of a trilogy of one-act operas collectively entitled *Il Trittico (The Triptych). Il tabarro* and *Suor Angelica* are the other operas. Of the three, only *Gianni Schicchi* is a comedy. While tenderness and sentimentality were always Puccini's strongest points, he frequently showed flashes of wit even in his tragic operas. In *Gianni Schicchi* he fully indulged his flair for the comic, and succeeded in writing a distinguished opera buffa. The character of the crafty Gianni Schicchi is in the best traditions of opera buffa, as are some of the fluttering arias and ensemble numbers and the contrasts between touching sentiment and broad farce.

Giarno, leader of a gypsy band (bass) in Thomas's *Mignon.*

Gibichungs, the children of Gibich, Gunther and Gutrune, in Wagner's *Die Götterdämmerung.*

Gigli, Beniamino, tenor. Born Recanati, Italy, March 20, 1890. His formal study did not begin until his eighteenth year, when he entered the Santa Cecilia Academy in Rome. There he studied with Antonio Cotogni and Enrico Rosati, after which he made his opera debut in Rovigo, in 1914, in *La Gioconda.* He now made numerous appearances in leading Italian opera houses and became a great favorite. In 1918 Toscanini selected him to sing the role of Faust in a La Scala revival of Boïto's *Mefistofele* during a Boïto festival. On November 26, 1920, Gigli made his American debut at the Metropolitan Opera in *Mefistofele* and was acclaimed. He was considered the successor to Caruso after the latter's death in 1921. The beauty of his voice—its purity of tone, elegance of legato, and suppleness of range—made him an idol of opera-lovers everywhere.

When, during the economic crisis of 1931, Gigli was asked by the Metropolitan Opera management to accept a cut in fee, he refused and withdrew from the company. For the next few years he concentrated his activity in Europe, but returned to the United States in 1938–1939 for several concert and operatic appearances. On January 23, 1939, he was back on the stage of the Metropolitan Opera in *Aïda* and was given an enthusiastic welcome. When he returned to Europe that season he furnished the European press with some violent opinions about America in general and the Metropolitan Opera in particular. These remarks created a furor in the United States, and

Gigli lost favor with many Americans, who were further alienated when he identified himself completely with the Fascist regime in Italy and became an outspoken propagandist for the Nazi Axis during World War II. In Europe, however, Gigli's position as one of the world's foremost opera tenors remained unassailable. In 1955 the sixty-five-year-old tenor returned to the United States for a concert tour.

Gil, Count, Suzanne's husband (baritone) in Wolf-Ferrari's *The Secret of Suzanne.*

Gil Blas, *see* LE SAGE, ALAIN.

Gilda, Rigoletto's daughter (soprano) in Verdi's *Rigoletto.*

Gilgamesj, opera by Ture Rangström. Libretto by the composer, based on an ancient Babylonian epic. Première: Stockholm Opera, November 20, 1952. Gilgamesj, king of Uruk, is the personification of intellectual strength. He engages Engidu in a fight in which neither emerges victorious. They join forces and conquer the world. Defying the gods, Engidu meets his doom at the hands of Isjtar, goddess of love. Gilgamesj then renounces his earthly power and descends to the kingdom of death to search for his partner. The opera was left unfinished, and was completed and orchestrated by John Fernström. The première took place posthumously.

Gillot, Onegin's servant in Tchaikovsky's *Eugene Onegin.*

Gilman, Lawrence, music critic. Born Flushing, New York, July 5, 1878; died Franconia, New Hampshire, September 8, 1939. He was self-taught in music. After holding various posts as editor and music critic, he was appointed music critic of the New York *Herald Tribune* in 1923, a position he held until his death. He also served as program annotator for the New York Philharmonic-Symphony and the Philadelphia Orchestra. His books include several on operatic subjects: *Aspects* *of Modern Music* (1908); *Wagner's Operas* (1937); *Toscanini and Great Music* (1938). He also wrote guides to *Salomé* (1907) and *Pelléas et Mélisande* (1907).

Gingerbread Waltz (Juchhei, nun ist die Hexe todt!), duet of Hansel and Gretel in Act III of Humperdinck's *Hansel and Gretel.*

Gioconda, La, opera in four acts by Amilcare Ponchielli. Libretto by Arrigo Boïto, based on Victor Hugo's play *Angelo, tyran de Padoue.* Première: La Scala, April 8, 1876. American première: Metropolitan Opera, New York, December 20, 1883.

Characters: La Gioconda, a ballad singer (soprano); La Cieca, her blind mother (contralto); Alvise, official of the State Inquisition (bass); Laura, his wife (mezzo-soprano); Enzo Grimaldo, a nobleman (tenor); Barnaba, spy of the Inquisition (baritone); Isepo, a public letter writer (tenor); Zuane, a gondolier (bass); monks; sailors; senators; ladies; gentlemen. The setting is Venice in the seventeenth century.

Act I. "The Lion's Mouth." A street near the Adriatic shore. A brief prelude is built out of part of La Cieca's dramatic aria in Act I. Crowds are milling around and singing ("Feste! Pane!"). Barnaba, the spy, is contemptuous of the mob. La Gioconda arrives, leading her blind mother. The mother is grateful for her daughter's love; the daughter responds with devotion; and Barnaba, from a distance, expresses the hope that he can ensnare the daughter through the mother (Trio: "Figlia che reggi il tremulo piè"). When Barnaba accosts La Gioconda she brushes him rudely aside. Barnaba determines to destroy her. The crowd now returns from a regatta, bearing aloft the winner. Barnaba maliciously whispers to the loser that he lost because of the diabolical powers of La Cieca. The loser is aroused, and attacks the old woman as a witch. But the Genoese nobleman, Enzo Grimaldi,

whose ship is harbored in Venice, protects the old woman. The Grand Duke Alvise now emerges from the palace, accompanied by his wife Laura. Laura, once betrothed to Enzo, prevails on the Duke to save La Cieca from mob violence. In gratitude, La Cieca gives Laura a rosary ("Voce di donna"). After the crowd disperses, Barnaba—aware that Enzo is still in love with Laura—slyly tells him that Laura reciprocates and will come to see him that very night on his ship. Enzo is ecstatic ("O grido di quest' anima"). When he leaves, Barnaba dictates a letter to Laura's husband warning him of the projected rendezvous. La Gioconda overhears him; since she herself is in love with Enzo, she is overwhelmed to discover that he is having an affair with another woman. As he reads the letter he has dictated, Barnaba gloats over his powers as a spy ("O monumento!"). The crowd returns, and merriment is resumed.

Act II. "The Rosary." A lagoon near Venice. The sailors of Enzo's ship sing as they work (Marinaresca: "Ho! He! Ho!"). Disguised as a fisherman, Barnaba sings a fisherman's ballad (Barcarolle: "Pescator, affonda l'esca"). Since he is awaiting Laura, Enzo sends his sailors away. Alone, he is enraptured by the beauty of the night ("Cielo e mar!"). Barnaba now appears, leading Laura. Laura and Enzo embrace; the old passion is revived; they plan to flee on Enzo's ship. When Enzo orders his sailors to prepare to set sail, Laura falls on her knees and prays God to forgive her ("Stella del marinar"). At this point, La Gioconda appears in disguise and upbraids Laura for stealing her lover. She is about to stab Laura when a boat appears. Laura realizes with horror that her husband has followed her. She takes her rosary and prays again. Recognizing the rosary as the one her mother gave Laura, La Gioconda now decides to aid her rival, and arranges for her escape. When Enzo returns he finds not Laura but La Gioconda. Distraught at the loss of the woman he loves and faced with capture by the Duke's men, Enzo orders that his ship be set afire.

Act III. "The House of Gold." Scene 1. A room in Alvise's palace. Realizing that his wife loves another man, Alvise is planning her murder ("Sì, morir ella dé' "). Upon Laura's arrival, he denounces her for her faithlessness. He then gives her a vial of poison, demands that she drink it, and leaves. Laura is about to take the poison when La Gioconda appears and gives her instead a drug that will induce a deathlike sleep. When Alvise reappears, he is satisfied that his wife has killed herself.

Scene 2. Alvise's ballroom. The Duke is entertaining his guests at a ball. They sing their praises of his palace ("Alla cà d'oro"). There then takes place a spectacular ballet (Danza delle ore: "The Dance of the Hours"). When the dance ends, Barnaba enters, dragging La Cieca after him, accusing her of being a witch. The sound of tolling bells is heard. Barnaba informs one of the masked guests that they are being tolled for Laura. The guest is Enzo. When the Duke pulls aside a curtain in order to show his guests the dead body of his wife, Enzo, blinded by anger and sorrow, rushes to stab him. He is seized by the Duke's men and taken to prison.

Act IV. "The Orfano Canal." A ruined palace on the island of Giudeca. La Gioconda has brought Laura to a lonely island for safety. Feeling she has nothing more to live for, La Gioconda contemplates suicide ("Suicidio"). Enzo, released from prison, arrives at the island and is reunited with his beloved. The lovers bid La Gioconda a tender farewell ("Sulle tue mani l'anima") and depart. Barnaba comes to claim La Gioconda—for she has promised her body in return for Enzo's freedom. Before he can take her in his arms, La Gioconda kills herself with a dagger. Bitter with rage, Barnaba shouts into La Gioconda's unhearing ears that he has already had his revenge: he has come from strangling her mother.

La Gioconda is first of all a good show. The many big scenes, the stirring passions, the high tensions make for excellent theater. It is for this reason that the opera, despite an involved and often confusing libretto, remains a favorite. But *La Gioconda* is an important opera, too. Ponchielli, influenced by Wagner, brought to Italian opera an orchestral technique, an expressive lyricism, and a dramatic power that were to be influential upon the Verdi of *Otello* and *Falstaff* and upon Ponchielli's pupil Puccini.

gioielli della Madonna, I, *see* JEWELS OF THE MADONNA, THE.

Giordano, Umberto, composer. Born Foggia, Italy, August 27, 1867; died Milan, November 12, 1948. While attending the Naples Conservatory, where he was a student for nine years, he wrote his first opera, *Marina,* which he entered in the Sonzogno competition that was won by *Cavalleria rusticana.* The publisher Edoardo Sonzogno was so impressed with Giordano's opera that he commissioned him to write a new work, *Mala vita,* given in Rome on February 21, 1892. Recognition came to Giordano four years later with his finest and most celebrated work, *Andrea Chénier,* introduced at La Scala on March 28, 1896. Before the end of the century this opera was performed throughout Europe and in New York. The most significant of Giordano's later operas were *Fedora,* given at La Scala in 1897; *Siberia,* at La Scala in 1904; *Madame Sans-Gêne,* at the Metropolitan Opera in 1915; and *La cena delle beffe,* at La Scala in 1920. Among the many honors gathered by Giordano were those of Chevalier of the French Legion of Honor, Commander of the Crown of Italy, and a membership in the Italian Academy.

Giorgetta, Michele's wife (soprano) in Puccini's *Il tabarro*.

Giovanna, Gilda's nurse (mezzo-soprano) in Verdi's *Rigoletto*.

Giovanni, Paolo's deformed brother (baritone) in Zandonai's *Francesca da Rimini*.

Girl of the Golden West, The (La fanciulla del West), opera in three acts by Giacomo Puccini. Libretto by Guelfo Civinini and Carlo Zangarini, based on the David Belasco play of the same name. Première: Metropolitan Opera, New York, December 10, 1910.

Characters: Minnie, owner of the Polka Saloon (soprano); Jack Rance, sheriff (baritone); Nick, bartender (tenor); Dick Johnson, an outlaw, alias Ramerrez (tenor); Ashby, Wells-Fargo agent (bass); Billy Jackrabbit, an Indian (bass); Wowkle, his squaw (mezzo-soprano); Jake Wallace, traveling camp minstrel (baritone); José Castro, member of Ramerrez' gang (bass). The setting is the foot of Cloudy Mountain, California, during the Gold Rush days.

Act I. The Polka Saloon. Miners are singing and gambling. Ashby tells them he is hot on the trail of Ramerrez, the notorious outlaw. When Minnie appears, the atmosphere becomes charged, for all the men love her—particularly Jack Rance, who sings of his passion ("Minnie della mia casa"). When Rance tries to force his attentions on Minnie, she repels him with her revolver, reminding him that he is married and that she does not love him. At this point a stranger enters, introducing himself as Dick Johnson of Sacramento. Minnie and Dick recognize one another. They have met once before and felt a mutual attraction. Minnie does not know, however, that Johnson is really Ramerrez come to hold up the saloon. The sheriff, his men, and the miners depart, searching for

Ramerrez. They leave their gold with Minnie, who announces boldly that she will protect it with her life. But by now the outlaw is so much in love with Minnie that he is willing to give up the gold. Touched, Minnie invites him to her cabin.

Act II. Minnie's cabin. Minnie and Johnson are confiding to each other their love when shots are heard. Before a posse appears, Minnie hides her lover. Even when she discovers from the posse that Johnson is Ramerrez she refuses to disclose his hiding place. After the posse leaves, she bitterly attacks Johnson for having failed to tell her the truth. Johnson confesses all, explaining the circumstances that led him to become an outlaw, and promising to give up crime. Minnie, however, sends him away. No sooner does he step out of the cabin than he is shot. Minnie drags him back in and conceals him in the loft. A falling drop of blood reveals to the sheriff where the wounded man is hiding. Desperate, Minnie proposes a hand of poker—if the sheriff wins, she will marry him, and he can bring the outlaw to justice; but if he loses, he must give up his pursuit. Minnie cheats, winning the game and the life of her lover.

Act III. A forest. Johnson has recovered, has been captured, and is about to be strung up. He prays that Minnie be led to believe that he has gone away to find a new life ("Ch'ella mi creda libero"). Minnie arrives on horseback, gun in hand. She entreats the lynchers to remember their feelings for her, and for the sake of those feelings to spare the man she loves. The miners have a change of heart. Minnie and Johnson are free to begin a new life together.

Puccini was commissioned to write this opera by the Metropolitan Opera Association when he visited the United States in 1907 to assist in the first American performance of *Madama Butterfly*. Searching for suitable material, he came upon Belasco's play, then enjoying a huge success on Broadway. "I should think that something stunning could be made of the '49 period," he said at the time. Another thing about the play appealed to him: its author had been the source of his successful *Madama Butterfly,* and Puccini felt that a revival of this collaboration might promise another success of equal proportions. It took Puccini three years to write his opera. Its première was the most brilliant of the season, the cast including Caruso, Destinn, and Amato, with Toscanini conducting. Puccini himself was present. The critics and audiences were enthusiastic, but later audiences were less easily satisfied. The opera is now generally regarded as one of Puccini's lesser accomplishments. His Italian lyricism seemed unsuitable for a text so thoroughly American. The interpolated musical material from the American West lacked conviction and was incongruous with the Italian arias and with Puccini's intermittent excursions into impressionism.

Giulia, a vestal virgin (soprano) in love with Licino in Spontini's *La Vestale.*

Giulietta, one of Hoffmann's loves (soprano) in Offenbach's *The Tales of Hoffmann.*

Giulio Cesare (Julius Caesar), (1) opera by Francesco Malipiero. Libretto by the composer, based on the Shakespeare tragedy. Première: Teatro Carlo Felice, Genoa, February 8, 1936. This opera was criticized by freedom-loving people because the feeling prevailed that Malipiero wrote it to glorify Mussolini. Malipiero himself explained that he wrote it because "something in the air we breathe today urged me towards a Latin hero."

(2) Opera by Handel. Libretto by Nicolo Francesco Haym. Première: King's Theatre, London, February 20, 1724.

Giuseppe, Violetta's servant (tenor) in Verdi's *La traviata.*

Glaives pieux, saintes épées (Nobili acciar, nobili e santi), the "Benediction of the Swords," in Act IV of Meyerbeer's *Les Huguenots.*

Glaz, Herta, contralto. Born Vienna, Austria, September 16, 1914. After attending the Vienna State Academy she made her debut with the Breslau Opera, in her eighteenth year, as Erda in *Das Rheingold.* In 1935 she became the leading contralto of the Glyndebourne Festival, and in the same year she sang successfully with the Prague Opera and at the opera festival in Interlaken, Switzerland. She toured the United States with the Salzburg Opera Guild and made appearances with the San Francisco, St. Louis, and Chicago opera companies. On December 25, 1942, she made her debut at the Metropolitan Opera as Amneris. She has since appeared at the Metropolitan Opera in a great variety of roles, scoring her greatest successes in the German repertory.

Glinka, Michael, composer. Born Novospasskoye, Russia, June 1, 1804; died Berlin, Germany, February 15, 1857. His two operas—*A Life for the Czar* and *Russlan and Ludmilla*—laid the foundations for Russian national opera. The son of a prosperous landowner, Glinka became strongly conscious of music as a child, but formal study did not come for several years. In 1817 he was sent to St. Petersburg, where he attended the Pedagogic Institute for five years. At the same time he studied piano with John Field and Carl Meyer, and violin with Joseph Böhm. Poor health sent him to the Caucasus in 1823. There he studied harmony and orchestration from textbooks and became absorbed with the folk songs and dances of that region. Back in St. Petersburg, he entered civil service and for three years worked in the Ministry of Communications. But music was not neglected. In 1828 he resigned his post and went to Italy, where he developed a passing fondness for Italian opera.

Proceeding to Berlin, he studied theory with Siegfried Dehn. His homesickness led him to think of writing national music. "My most earnest desire," he wrote from Berlin, "is to compose music which would make all my beloved fellow countrymen feel quite at home, and lead no one to allege that I strutted about in borrowed plumes." Returning to Russia, he joined a literary and artistic group that included Pushkin and Gogol. He now knew he wanted to produce music of an unmistakable Russian identity, with intimate association with Russian history and the Russian people. With this aim in mind, he completed his first opera, *A Life for the Czar,* introduced at the Imperial Theater on December 9, 1836. The opera was a success, and the Czar presented Glinka with a valuable ring. He now began work on a second opera, *Russlan and Ludmilla,* the text derived from a poem by Pushkin. The new opera, introduced in St. Petersburg on December 9, 1842, was at first a failure. It was considered to lack the human interest and dramatic quality of the previous work. Not until a revival three years after Glinka's death did it receive the recognition it deserved. In the closing years of his life, Glinka lived in France and Spain, writing a considerable quantity of symphonic music. His death came suddenly in Berlin. He was buried there, but a few months later his body was taken back to St. Petersburg.

Glinka's significance in Russian music cannot be overestimated. He was the first Russian composer whose music made an impact on the world outside Russia, and whose best work has survived. More than this, he was the first Russian composer to realize a national art, setting the stage for Dargomizhsky and the composers grouped as the "Five," all of whom admired his operas profoundly.

Gloire immortelle de nos aïeux, the Soldiers' Chorus in Act IV, Scene 3, of Gounod's *Faust.*

Gloria all' Egitto, the hymn that introduces the triumphal scene of Act II, Scene 2, of Verdi's *Aïda.*

Gluck, Alma (born FIERSOHN, REBA), (soprano). Born Bucharest, Rumania, May 11, 1884; died New York City, October 27, 1938. She studied singing with Arturo Buzzi-Peccia, after which she made her opera debut at the Metropolitan Opera on November 16, 1909, in *Werther* and was a great success. She remained at the Metropolitan Opera for the next three seasons, appearing in about twenty roles. After an additional period of study with Marcella Sembrich, she left opera and became a concert singer. For many years she was immensely successful, particularly in joint recitals with her husband, the violinist Efrem Zimbalist, whom she married in 1914. Her daughter by a first marriage is Marcia Davenport, a successful novelist who wrote a biography of Mozart and a novel about a prima donna, *Of Lena Geyer.*

Gluck, Christoph Willibald, composer. Born Erasbach, Bavaria, Germany, July 2, 1714; died Vienna, November 15, 1787. A giant figure in the early history of opera, Gluck helped establish the musical drama, as distinguished from the more formal Italian opera, in works like *Orfeo ed Euridice* and *Alceste.* As a boy he attended Catholic schools were he studied the violin, organ, and harpsichord. In 1736 he went to Vienna where he was employed by Prince Ferdinand Lobkowitz. Another patron, Prince Melzi, took him to Italy, where he studied with G. B. Sammartini and wrote his first opera, *Artaserse,* produced in Milan in 1741. He now wrote six more operas, all of them in the Italian tradition, before proceeding to London, where two of his works were staged without success, and earning their composer the disapproval of Handel. Returning to Vienna by way of Paris, Hamburg, and Dresden, Gluck had his newest opera, *Semiramide riconosciuta,* performed at the Burgtheater when that theater reopened on May 14, 1748. The work, with text by Metastasio, was a success.

In 1754, four years after his marriage to wealthy Marianne Pergin, Gluck was appointed kapellmeister at the Vienna Court Theater. In this post he wrote numerous operas. But he was dissatisfied with them. Influenced by the French school, headed by Rameau, he felt the need for greater simplicity, naturalism, and dramatic truth; a departure from the mannerisms and pompous artificiality of the Italian tradition. Nor was he alone in these ideas. There were two other figures at the Vienna court who were also devotees of French art and culture. One was Count Giacomo Durazzo, director of the court theaters, the other was Raniere da Calzabigi, a poet who had come from Paris in 1761, aroused by the intellectual revolution then sweeping through England and France with Rousseau, Voltaire, and Diderot as spokesmen. The three kindred spirits combined forces to produce an art that once and for all would overthrow the Italian influence. Their first effort was a ballet, *Don Juan,* produced at the Burgtheater on October 17, 1761. The Calzabigi text was based on the play by Molière, the music was by Gluck. This was followed by an even bolder effort to establish their new ideas: the opera *Orfeo ed Euridice.* Given at the Burgtheater on October 5, 1762, the opera was a failure, too new and revolutionary for Viennese tastes. But the collaborators were not discouraged. They wrote a second opera in their new style, *Alceste,* seen on December 16, 1767, and after that, *Paride ed Elena,* performed in 1770. Both operas were failures.

Convinced of the truth of his message, and discouraged at his failures in Vienna, Gluck decided to go to Paris. There he wrote a new opera, *Iphigénie en Aulide.* In Paris, as in Vienna, Gluck encountered not appreciation and sympathy but envy, antagonism, and malice. A powerful clique upheld the Italian tradition, and so much obstruction was placed in Gluck's way that *Iphigénie en Aulide* might never have been produced had not Marie Antoinette herself intervened. It was finally given at the Opéra on April 19, 1774, and was a major triumph, the rage of the musical season. *Orfeo ed Euridice,* given its first French performance in 1775, was also an immense success. But the enemies of Gluck, whose leader was Jean François Marmontel, would not acknowledge defeat. They brought to Paris one of Italy's most celebrated composers, Niccolò Piccinni, and had the Opéra commission him to write an opera. An intense rivalry developed between the two composers and their supporters; the musical atmosphere in Paris became charged with bitterness and dissension. The directors of the Opéra stepped boldly into the controversy by commissioning both Piccinni and Gluck to write an opera on the same libretto, derived from Euripides: *Iphigénie en Tauride.* Gluck's opera was seen first, on May 18, 1779, and was so successful that Piccinni tried to withdraw his own work. Though Piccinni's opera had a run of seventeen consecutive performances, it was less well received than Gluck's. The opera war was over. Gluck's last opera, *Echo et Narcisse,* disappointed his Paris following. He now returned to Vienna, rich and honored, and died there of a stroke eight years later.

Gluck was the most significant opera composer before Mozart. At a time when opera was growing increasingly formal, stilted, and remote from truth and human experience, Gluck produced a new kind of musico-dramatic work in which music and text were integrated in a coherent whole. "My chief endeavor," as he himself explained, "should be to attain a grand simplicity and consequently I have avoided making a parade of difficulties at the cost of clearness. I have set no value on novelty as such unless it was naturally suggested by the situation and suited to the expression. In short, there was no rule which I did not consider myself bound to sacrifice for the sake of effect." No one before him brought to opera such touching sentiment, wealth of feeling, realism, and sympathetic understanding of text. Gluck's greatest operas represent a revolution which prepared the way for such later works as *Don Giovanni, Der Freischütz,* and *Tristan und Isolde.*

Glyndebourne Opera, an opera company organized in 1934 by John Christie and his wife, Audrey Mildmay, to present Mozart operas in a specially constructed theater on Christie's ancestral estate at Glyndebourne, in Lewes, Sussex, England. The aim was to achieve a "unity between sight and sound" and a synchronization of "stage time and musical time." Carl Ebert was chosen artistic director; Fritz Busch, musical director; and Rudolf Bing, general manager. The first festival opened on May 28, 1934, with *The Marriage of Figaro,* with Audrey Mildmay as Susanna, Aulikki Rautawaara as the Countess, Willy Domgraf-Fassbaender as Figaro, Lucy Mannen as Cherubino, and Fritz Busch conducting. For two weeks this opera alternated with *Così fan tutte,* whose principals included Heddle Nash as Ferrando, Ina Souez as Fiordiligi, Luise Helletsgrüber as Dorabella, and Vincenzo Bettoni as Alfonso. A year later the season was expanded to five weeks with *The Magic Flute* and *The Abduction from the Seraglio* added to the repertory. In 1938 operas by composers other than Mozart were heard when *Macbeth* (Verdi) and *Don Pasquale*

(Donizetti) were given. During World War II, the festival suspended activity, though the company performed one work, *The Beggar's Opera.* The festivals resumed after the war with premières of *The Rape of Lucretia* and *Albert Herring,* productions of the newly organized English Opera Company. The first postwar production of the Glyndebourne company was Gluck's *Orfeo ed Euridice,* with Kathleen Ferrier, in 1947. In the same year the company appeared for the first time at the Edinburgh Festival. Since then, the major new productions have included: *Alceste; Ariadne auf Naxos; Un ballo in maschera; La cenerentola; Le Comte Ory;* and *Idomeneo.* In 1952 the Glyndebourne Festival Society was organized to finance the annual festivals.

Among the singers who have made successful appearances are: Salvatore Baccaloni, Willy Domgraf-Fassbaender, John Brownlee, Kathleen Ferrier, Sena Jurinac, Erich Kunz, Marko Rothmüller, Mariano Stabile, and Italo Tajo. The following Americans have been acclaimed: Marita Farell, David Poleri, Eleanor Steber, and Blanche Thebom. Conductors have included: Fritz Busch, Vittorio Gui, Rafael Kubelik, Fritz Stiedry, and Walter Süsskind.

Godard, Benjamin, composer. Born Paris, August 18, 1849; died Cannes, January 10, 1895. He attended the Paris Conservatory. His first major work was an "operatic symphony," *Le Tasse,* which won first prize in a competition conducted by the city of Paris in 1878. In the same year he completed his first opera, *Les Bijoux de Jeannette.* Ten years later his opera *Jocelyn* was successfully introduced at the Théâtre de la Monnaie. Even more successful was *La Vivandière,* given by the Opéra-Comique in 1895, eleven months after the composer's death. Besides the operas already mentioned Godard wrote: *Pedro de Zalaméa* (1884); *Le Dante* (1890); *Jeanne d'Arc* (1891); *Les Guelfes; Ruy Blas.*

Godunov, Boris, see BORIS GODUNOV.

Goethe, Johann Wolfgang von, poet, dramatist, and novelist. Born Frankfort-on-the-Main, Germany, August 28, 1749; died Weimar, March 22, 1832. Many of Goethe's dramas and novels were adapted into operas. His masterwork, the epic drama *Faust,* was the source of Gounod's famous opera of the same name and of operas by Berlioz (*Le damnation de Faust*), Boïto (*Mefistofele*), and Spohr (*Faust*). Other of Goethe's works adapted for operas include: *Götz von Berlichingen* (Karl Goldmark); *Die Leiden des jungen Werthers* (Massenet's *Werther* and Alberto Randegger's *Werthers Schatten*); *Scherz, List und Rache* (Max Bruch and Egon Wellesz); *Wilhelm Meister* (Hans Gál's *Requiem für Mignon* and Thomas's *Mignon*).

Götterdämmerung, Die, see RING DES NIBELUNGEN, DER.

Goetz, Hermann, composer. Born Königsberg, Germany, December 17, 1840; died Hottingen, Switzerland, December 3, 1876. After attending the Stern Conservatory in Berlin he became organist in Winterthur. His first and most successful opera was *The Taming of the Shrew* (*Der Widerspänstigen Zähmung*), introduced in Mannheim on October 11, 1874, and soon after that produced in leading German and Austrian opera houses. A second opera, *Francesca da Rimini* was performed in Mannheim in 1877, after the composer's death.

Götz von Berlichingen. see GOETHE, JOHANN WOLFGANG VON.

Gogol, Nikolai, author. Born Poltava, Russia, March 21, 1809; died Moscow, March 4, 1852. One of the creators of the Russian novel, Gogol wrote many works that were adapted into operas. These include: *Christmas Eve* (Rimsky-Korsakov); *Evenings on a Farm*

Near Dikanka (Mussorgsky's *The Fair at Sorochinsk*); *The Inspector General* (Karel Weis's *The Revisor*); *The Marriage* (Gretchaninoff; Martinu; Mussorgsky); *May Night* (Rimsky-Korsakov); *The Nose* (Shostakovich); *Taras Bulba* (Vladimir Kashperov; Nikolai Lissenko).

Golaud, King Arkel's grandson (baritone) in Debussy's *Pelléas et Mélisande.*

Golden Cockerel, The, see COQ D'OR, LE.

goldene Kreuz, Das (The Golden Cross), opera by Ignaz Brüll. Libretto by Salomon Hermann Mosenthal, based on the play *Cathérine* by Mélesville and Brazier. Première: Berlin Opera, December 22, 1875. Christina, heartbroken that her brother Colas must go off to war, offers herself to any man serving as his substitute. A nobleman, Gontran de l'Ancre, acts as the substitute and receives from Christina a golden cross as token of her pledge. Three years later, Gontran returns from the war, wounded. The golden cross becomes the means through which he wins Christina as his wife.

Goldmark, Karl, composer. Born Keszthely, Hungary, May 18, 1830; died Vienna, January 2, 1915. After preliminary study of the violin and composition he attended the Vienna Conservatory. His studies there were aborted by the revolution of 1848, which for a time closed the school. For a while he earned his living playing the violin in Viennese theaters, teaching the piano, and writing music criticisms. Two concerts of his works—one in Vienna, the other in Budapest—failed to lift him out of poverty and obscurity. Success finally came with his finest work, the opera *The Queen of Sheba,* introduced at the Vienna Opera in 1875 to such acclaim that within a few years it was given almost three hundred times in Vienna alone. By 1885 it had been performed in most of Europe'ℰ leading opera houses as well as at the Metropolitan Opera. Goldmark subsequently wrote several other operas, the most famous being *The Cricket on the Hearth,* in 1896. The others: *Merlin* (1886); *Der Fremdling* (1897); *Die Kriegsgefangene* (1899); *Götz von Berlichingen* (1902); *A Winter's Tale* (*Ein Wintermärchen*) (1908).

Goldoni, Carlo, dramatist and librettist. Born Venice, February 25, 1707; died Versailles, January 6, 1793. Immensely popular in his day, Goldoni wrote some hundred and fifty plays and opera librettos. Some of the latter were *L'amore artigiano* (Florian Gassmann); *L'Arcadia in Brenta* (Galuppi); *Il Conte Caramella* (Galuppi); *Il filosofo di campagna* (Galuppi); *Il mondo alla roversa* (Galuppi); *Il mondo della luna* (Galuppi; Haydn); *Lo Speziale* (Haydn); *Il Tigrane* (Gluck). Comedies of Goldoni from which operas were made include: *Il Ciarlone* (Paisiello); *Le donne curiose* (Wolf-Ferrari); *I due litiganti* (Sarti); *Le gelosie villane* (Sarti); *Tre commedie Goldoniane: La bottega di caffè, Sior Todero Brontolon, Le baruffe chiozzotte* (Malipiero); *Vittorina* (Piccinni); *I quattro rusteghi* (Wolf-Ferrari).

Goldovsky, Boris, opera director and teacher, Born Moscow, Russia, June 7, 1908. He is the son of the violinist Lea Luboschutz. He attended the Moscow Conservatory and the Budapest Academy. In 1930 he came to the United States, where he studied conducting at the Curtis Institute and became an American citizen. He has served as head of the opera departments of the Cleveland Institute of Music, the New England Conservatory, and the Berkshire Music Center. Since 1946 he has been head of the New England Opera Theater and an annotator for the radio broadcasts of the Metropolitan Opera.

Goldsmith, Oliver, poet, playwright, novelist. Born Elphin, England, November 10, 1728; died London, April 4, 1774. Goldsmith's famous play, *She Stoops to Conquer,* was adapted into an opera of the same name by George Macfarren. Liza Lehmann's *The Vicar of Wakefield* and Victor Pelissier's *Edwin and Angelina* were also derived from Goldsmith's works.

Gonzalve, Concepcion's lover (tenor) in Ravel's *L'heure espagnole.*

Good Friday Spell (Charfreitagszauber), the music in Act III, Scene 1, of Wagner's *Parsifal,* as Parsifal is being bathed in preparation for his entrance into the Grail castle.

Good-Night Quartet, *see* SCHLAFE WOHL.

Goose Girl, The, principal female character (soprano) in Humperdinck's *Die Königskinder.*

Goossens, Eugène, (III), composer and conductor. Born London, May 26, 1893. The first Eugène Goossens (1845–1906) conducted opera performances in Belgium, France, Italy, and (after 1873) England; he was the first conductor of the Carl Rosa Opera Company. His son Eugène was also a conductor of the Carl Rosa Opera. Eugène Goossens III studied music at the Bruges Conservatory, the Liverpool College of Music, and the Royal College of Music in London. He received recognition as a conductor in 1916 when he led the première of Charles Stanford's opera *The Critic.* He now appeared as guest conductor with many leading English musical organizations, including the Carl Rosa Opera Company. In 1923 he came to the United States, where for eight years he was musical director of the Rochester Philharmonic, and for sixteen years of the Cincinnati Symphony. In 1947 he went to Australia to become conductor of the Sydney Symphony and director of the South Wales Conservatory. He was knighted in June 1955. He has written several operas: *Judith* (1929);

Don Juan de Mañara (1937); *Gainsborough.*

Gopak, see HOPAK.

Gorislava, Ratmir's beloved (soprano) in Glinka's *Russlan and Ludmilla.*

Gorky (or Gorki), Maxim, (born ALEKSEI PESHKOV), author. Born Nizhni-Novgorod (now Gorki), March 28, 1868; died Moscow, June 18, 1936. Raoul Gunsbourg's opera *Le vieil aigle,* Giacomo Orefice's *Radda,* and Valery Zhelebinsky's *The Mother* were derived from Gorky's works.

Gormas, Count de, Chimène's father (baritone) in Massenet's *Le Cid.*

Goro, marriage broker (tenor) in Puccini's *Madama Butterfly.*

Gossec, François-Joseph, composer. Born Vergnies, Belgium, January 17, 1734; died Passy, France, February 16, 1829. After serving as chorister at the Antwerp Cathedral and studying the violin, he went to Paris in 1751, where Rameau placed him in the employ of La Pouplinière as conductor of his orchestra. His first success as an opera composer came in 1766 with *Les pêcheurs,* successfully given at the Théâtre des Italiens. In 1770 he founded the Concerts des Amateurs, from 1773 to 1777 he conducted the Concerts Spirituels, and from 1780 to 1782 he was second conductor at the Opéra. He organized the Ecole Royale de Chant in 1784 and eleven years later he became the first director of the Paris Conservatory as well as a professor of composition. He was a prolific composer, his works including fifteen operas, none of which have survived. (He is remembered for a trifle, a gavotte, found in his opera *Rosine.*) He was, however, a major figure in the development of symphonic and chamber music and the art of orchestration. His finest operas were: *Les pêcheurs* (1766); *Sabinus* (1773); *Hylas et Sylvie* (1776); *La fête du village* (1778); *Thésée* (1782); *Rosine*

(1786); *Les sabots et le cerisier* (1803).

Gottfried, Elsa's brother, transformed into a swan, in Wagner's *Lohengrin.*

Gounod, Charles François, composer. Born Paris, June 17, 1818; died there, October 18, 1893. A major figure in French opera of the nineteenth century, he attended the Paris Conservatory, where his teachers included Jacques Halévy and Jean François Lesueur, and where, in 1839, he won the Prix de Rome. In Italy he became passionately interested in church music. After he returned to Paris he became a church organist and wrote choral music. During a fortuitous meeting with the opera singer Pauline Viardot he was asked to write an opera. That work was *Sapho,* introduced at the Opéra on April 16, 1851, with Viardot in the title role. Though it was a failure, Gounod did not lose interest in the stage. He kept on writing operas, and his fourth was the work that made him famous: *Faust,* given at the Théâtre Lyrique on March 19, 1859. At first only moderately successful, *Faust* became world famous after a triumphant revival at the Opéra in 1869. After *Faust,* Gounod wrote eight operas, only two of which were successful: *Mireille* (1864) and *Roméo et Juliette* (1867).

Between 1852 and 1860 Gounod directed a Parisian choral society, the Orphéon. This association revived his interest in religious and choral music and he wrote a great deal of diversified music in this field, including the celebrated "Ave Maria," and several masses and oratorios. At the outbreak of the Franco-Prussian War, Gounod settled in London where he lived for five years and appeared as conductor. He returned to Paris in 1875, increasingly absorbed in religious inspirations; he produced little secular music after 1881.

Gounod helped create a restrained

and sensitive operatic art that was filled with human values and parted company with the more ornate products of Meyerbeer and his imitators. Unlike his contemporaries Bizet and Halévy, he lacked an instinct for good theater: even his best works are dramatically weak and unconvincing. But he was a supreme melodist, the creator of a refined and expressive lyricism supported by a sensitive harmony and orchestration.

His operas: *Sapho* (1851); *La nonne sanglante* (1854); *Le médecin malgré lui* (1858); *Faust* (1859); *Philémon et Baucis* (1860); *La Reine de Saba* (1862); *Mireille* (1864); *La colombe* (1866); *Roméo et Juliette* (1867); *Cinq-Mars* (1877); *Polyeucte* (1878); *Le tribut de Zamora* (1881); and two posthumous works, *Maître Pierre* and *Georges Dandin*.

Goyescas, opera by Enrique Granados. Libretto by Fernando Periquet, suggested by setting and characters in Goya's paintings. Première: Metropolitan Opera, January 28, 1916. Don Fernando, captain of the guards, is in love with Rosario. When he learns that Rosario has been invited to a ball by a toreador, Paquiro, he is so overcome with jealousy that he decides to invade the ball. There Paquiro challenges him to a duel. Wounded fatally, Don Fernando dies in his beloved's arms. Granados' score makes skillful use of Spanish dances. One of the instrumental numbers is particularly famous: the Intermezzo, which the composer wrote after his score was completed, upon a suggestion from the directors of the Metropolitan.

Gozzi, Carlo, dramatist. Born Venice, December 13, 1720; died Italy, April 14, 1806. He was celebrated for his satires and fairy pieces in Venetian dialect, many of them written to ridicule rival writers. His plays have frequently been adapted as operas, notably: *La donna serpente* (opera by Casella; Wagner's *Die Feen*); *Fiaba dell' amore delle tre melarancie* (Prokofiev's *The Love for Three Oranges*); and *Turandot* (Antonio Bazzini, Busoni, Adolph Jensen, Puccini, and Karl Reissiger).

Graener, Paul, composer. Born Berlin, Germany, January 11, 1872; died Salzburg, November 13, 1944. After studying at Veit's Conservatory in Berlin, he earned his living conducting orchestras in small German theaters. In 1896 he settled in London, conducting the orchestra of the Haymarket Theatre and teaching at the Royal Academy of Music. In 1908 he became head of the New Conservatory in Vienna, and two years later, director of the Mozarteum in Salzburg. In 1920 he was appointed professor of composition at the Leipzig Conservatory, succeeding Max Reger. From 1930 to 1934 he was director of the Stern Conservatory in Berlin. His operas are representative of his style in which, as Hugo Leichtentritt once wrote, "the naive expression and the simple charm of the folksong has been deeply felt . . . and has been the source of some of his finest inspiration." His operas: *Das Narrengericht* (1913); *Don Juans letztes Abenteuer* (1914); *Theophano,* rewritten as *Byzanz* (1918); *Schirin und Getraude* (1920); *Hanneles Himmelfahrt* (1927); *Friedemann Bach* (1931); *Der Prinz von Homburg* (1935).

Graf, Herbert, stage director. Born Vienna, April 10, 1904. The son of a famous Austrian music critic, Max Graf, he was early directed to music. He attended the Vienna University where he received his doctorate in 1928. He then studied music and stage techniques at the Vienna Academy, following which he became stage director of the Münster Opera. For three years he served in a similar capacity in Breslau, and for three additional years in Frankfort-on-the-Main. In 1934 he came to the United States, where he joined the Philadelphia Opera

Company. In 1936 he became stage director of the Metropolitan Opera, where he has since served. He has also served as stage director for the San Francisco Opera, and for opera performances at the Salzburg Festival, the Florence May Music Festival, and in Vienna. He is the author of *The Opera and its Future in America* (1941).

Gralserzählung, see IN FERNEM LAND.

Granados, Enrique, pianist and composer. Born Lérida, Spain, July 29, 1867; died at sea, March 24, 1916. He early came under the influence of Felipe Pedrell, who directed him to the writing of Spanish national music. In his twentieth year he went to Paris where he stayed for two years, studying the piano. Back in Spain, he began writing piano music so strongly national in style and technique that he is considered by many the father of modern Spanish music. He wrote his first opera, *María del Carmen,* in 1898, successfully introduced in Madrid the same year. *Goyescas,* his more famous opera, was worked up from two volumes of piano pieces. Granados came to America to attend the world première of this work at the Metropolitan Opera in 1916. Invited to play for President Wilson at the White House, he delayed his return to Spain by one week. Returning aboard the *Sussex,* he drowned when the ship was torpedoed by a German submarine in the English Channel.

Grand opera, an opera with a serious theme and no spoken dialogue. The distinction is made between grand opera and such other types as light opera, opéra comique, and opera buffa.

Grane, Brünnhilde's horse in Wagner's *Die Götterdämmerung.*

Grassini, Josephina (or Giuseppina), contralto. Born Varese, Italy, 1773; died Milan, January 3, 1850. After studying at the Milan Conservatory, she made her opera debut at the Ducal Theater in Parma in 1789. For the next few years she appeared in Italian comic operas. In 1792 she turned to serious opera, and in 1794 was a triumphant success in Milan. Napoleon heard her in a special concert and engaged her for the Paris Opéra, where she created a furor at her debut on July 22, 1800. She then sang in London and again in Paris. For a number of years she was popular at the French court. She was nearly sixty when Paris heard her for the last time. A highly successful teacher, her pupils included Giuditta Pasta and the sisters Giuditta and Giulia Grisi. (The latter were Grassini's nieces.)

Grau, Maurice, opera manager. Born Brünn, Moravia, 1849; died Paris, March 14, 1907. He was brought to the United States as a child, and here he received his academic education. In 1872 he became an impresario, managing the first American tour of Anton Rubinstein. Subsequently, he managed the American tours of many notable European musical and dramatic personalities, including Offenbach, Wieniawski, Salvini, and Bernhardt. In 1890, in collaboration with Henry Abbey, he produced a season of opera at the Metropolitan Opera, giving twenty-one performances. The following year he joined Abbey and Edward Schoeffel in leasing the Metropolitan Opera House and presenting opera there for the next six years. In 1897 he became sole manager, holding this post with great distinction until 1903. He created such a high standard of performance that his era has since become known as the "golden age of opera." He brought to New York such outstanding personalities as the De Reszke brothers, Maurel, Plançon, Melba, Eames, and Calvé; he helped make stars of Nordica, Sembrich, and Campanari. He was a pioneer in giving American audiences operas in their original languages, was responsible for the first uncut performances of the Wagner music dramas in America, and he managed many notable American premières, one being *Tosca*. He retired in 1903 and spent the last years of his life in a small town near Paris.

Graun, Karl Heinrich, singer and composer. Born Wahrenbrück, Saxony, May 7, 1701; died Berlin, August 8, 1759. As a child he studied music at the Dresden Kreuzschule and sang publicly. In 1725 he became principal tenor at the Brunswick Opera, where he was soon elevated to the post of assistant kapellmeister and where he wrote his first opera, *Pollidoro,* a great success in 1726. In 1735 he became kapellmeister to Crown Prince Frederick at Rheinsberg; when Frederick became Emperor, Graun was appointed kapellmeister and opera director in Berlin. He secured the services of outstanding artists of Italy and Germany and maintained a high level of performance. Some thirty of his operas were produced in Berlin, making him one of the most celebrated composers of his generation. His most famous operas were: *Scipio Africanus* (1732); *Timareta* (1733); *Pharao Tubaetes* (1735); *Rodelinda* (1741); *Catone in Utica* (1744); *Adriano in Siria* (1745); *Mitridate* (1750); *Ezio* (1755); *Montezuma* (1755); *Merope* (1756).

Graupner, Christoph, composer. Born Hartmannsdorf, Saxony, January 13, 1683; died Darmstadt, Germany, May 10, 1760. The major part of his music study took place at the St. Thomasschule in Leipzig. He went to Hamburg in 1706, where he served as accompanist at the opera. Four years later he went to Darmstadt as kapellmeister, gaining considerable prestige there both as director of musical performances and as composer. He was extraordinarily fertile, producing over a thousand works of all kinds, including nine operas. In 1722 he was invited to take over the post of cantor at the Thomasschule. Too successful in Darmstadt to be interested in a change, he turned down the offer, which was then made available to Johann Sebastian Bach. His finest operas: *Dido* (1707); *Hercules und Theseus* (1708); *Bellerophon* (1709); *Simson* (1709); *Berenice und Lucio* (1710); *Telemach* (1711).

Grech, a police officer (bass) in Giordano's *Fedora.*

Greensleeves, a celebrated English folk song used by Ralph Vaughan Williams in his opera *Sir John in Love.*

Gregorio, a kinsman (baritone) of Capulet in Gounod's *Roméo et Juliette.*

Gregory, the Pretender Dmitri (tenor) in Mussorgsky's *Boris Godunov.*

Gremin, Prince, Tatiana's husband (bass) in Tchaikovsky's *Eugene Onegin.*

Grenvil, Dr., a physician (bass) in Verdi's *La traviata.*

Gretchaninov, Alexander, composer. Born Moscow, October 25, 1864. His musical education took place at the Conservatories of Moscow and St. Petersburg. He made his bow as a composer with a symphony, conducted by Rimsky-Korsakov in 1895. In 1901 he completed his first opera, *Dobrinya Nikitich,* produced in Moscow on October 17, 1903, with Chaliapin in the title role. It was well received. In 1925 Gretchaninov left Russia and for several years lived in Paris; during this period he made several concert tours of the United States. In 1939 he settled permanently in New York. He wrote an autobiography, *My Life* (1952). His operas: *Dobrinya Nikitich* (1901); *Sister Beatrice* (1910); *The Dream of a Little Christmas Tree,* children's opera (1911); *The Castle House,* children's opera (1911); *The Cat, the Fox, and the Rooster,* children's opera (1919); *The Marriage* (1946).

Gretel, Hansel's sister (soprano) in Humperdinck's *Hansel and Gretel.*

Grétry, André Ernest, composer. Born Liège, Belgium, February 8, 1741; died

Montmorency, France, September 24, 1813. He was one of the most significant composers of opéra comique in the late eighteenth century. While still a boy he heard a performance of *La serva padrona* which so enchanted him that he decided to dedicate his life to writing operas in the style of Pergolesi. After a period of study with Giovanni Battista Casali in Italy, Grétry wrote his first opera, *La Vendemmiatrice*, given successfully in Rome in 1765. Returning to Paris three years later, he acquired the patronage of the Swedish ambassador, with whose help his opera *Le Huron* was put on in 1768. Grétry now wrote operas prolifically: within twenty years more than fifty. When the Revolution broke out, Grétry allied himself with the proletariat and wrote several popular works reflecting the new ideology. The municipality of Paris named a street after him and later placed his statue in the Opéra-Comique. When the Paris Conservatory was founded, Grétry became an inspector; with the inauguration of the Institut de France, he was chosen a member. The monarchy restored, Grétry flexibly changed his allegiance and remained a powerful and popular figure in French music. Napoleon III honored him with a pension and with the Legion of Honor. Grétry lived the last years of his life in the home formerly inhabited by Jean Jacques Rousseau—*L'Eremitage* in Montmorency.

In his theoretical writings Grétry anticipated by a century some of Wagner's ideas about an opera theater: one with a curved stage so that the action might be visible from all parts of the house; with an orchestra completely concealed from view; with seats of only a single class, and no boxes. Grétry's most successful operas were: *Le tableau parlant* (1769); *Les deux avares* (1770); *Zémire et Azor* (1771); *Céphale et Procris* (1775); *La fausse magie* (1775); *L'amant jaloux* (1778); *L'epreuve*

villageoise (1784); *Richard Coeur de Lion* (1784); *La caravane du Caire* (1784); *Amphitryon* (1788); *La rosière républicaine* (1793).

Grieve my heart, Rezia's aria in Act III, Scene 2, of Weber's *Oberon*.

Grimaldo, Enzo, Genoese nobleman (tenor) in love with Laura in Ponchielli's *La Gioconda*.

Grimes, *see* PETER GRIMES.

Grimm, Friedrich Melchior, Baron von, writer. Born Ratisbon, Germany, December 26, 1723; died Gotha, December 18, 1807. Grimm fired the opening shot in the Parisian operatic war known as the "guerre des bouffons" in 1752. This consisted of a provocative letter published in the *Mercure de France* in which Grimm posed Pergolesi's *La serva padrona* as an example of true operatic art as opposed to the French operas of Lully and Rameau (*see* GUERRE DES BOUFFONS; RAMEAU). Two decades later, however, Grimm sided with Gluck in that composer's rivalry with Piccinni in Paris (*see* GLUCK). Grimm was also editor of the *Correspondence littéraire, philosophique et critique*. This circulated in manuscript during the years 1753–1790. Published in sixteen volumes a century later, the *Correspondence* provides valuable information about the history of opera in Paris.

Grimm, Jakob Ludwig, writer and philologist. Born Hanau, Germany, January 4, 1785; died Berlin, September 20, 1863. With his brother Wilhelm he collected, wrote, and published volumes of German fairy tales entitled *Kinder und Hausmärchen* (1812, 1815). These are known and loved throughout the world. Humperdinck's *Hänsel und Gretel*, Carl Orff's *Die Kluge* and *Der Mona*, Hugo Kaun's *Der Fremde*, Theodore Chanler's *The Pot of Fat*, and Leon Stein's *The Fisherman's Wife* are operas derived from these tales.

Grisélidis, opera by Jules Massenet. Libretto by Paul Armand Silvestre and

Eugène Morand. Première: Opéra-Comique, November 20, 1901. The setting is Provence in the fourteenth century, where the shepherd Alain loves Grisélidis, who is married to the Marquis. When the Marquis goes to war, the devil tries to convince Grisélidis of her husband's infidelity and suggests that she avenge herself by taking Alain as a lover. Grisélidis resists the devil, who then disappears with her child. Back from the war, the Marquis is at first suspicious about his wife's fidelity. But eventually he learns the truth and is able to retrieve their child from the devil. Among the familiar arias are Grisélidis's, "Il partit du printemps" in Act II, and Alain's, "Je suis l'oiseau" in Act III.

Grisi, Giulia, dramatic soprano. Born Milan, July 28, 1811; died Berlin, November 29, 1869. One of the most celebrated opera stars of her generation, she came from a family of celebrated artists. Her older sister Giuditta and her aunt Grassani were celebrated singers; her cousin, Carlotta, was a famous dancer. After studying singing with various teachers she made her debut in Milan in 1828 in Rossini's *Zelmira.* She was acclaimed, and for the next three seasons she continued to appear in Milan. Dissatisfied with conditions there, she broke her contract and went to Paris. Here she made her first appearance at the Théâtre des Italiens, on October 16, 1832, in Rossini's *Semiramide,* under the composer's direction. She remained at the Théâtre des Italiens for the next seventeen years, adored by Parisian opera-goers. In 1833 she visited Venice to appear in Bellini's *I Capuletti ed i Montecchi,* which the composer wrote for her. In 1834 she made her London debut in Rossini's *La gazza ladra;* with the exception of a single season she appeared regularly in London for the next twenty-seven years, a reigning favorite. In 1854 she toured the United States with the tenor

Mario, whom she had married a decade earlier. She went into temporary retirement in 1861, but six years later returned briefly to the opera stage in London. Following this, she sang at concerts. Donizetti wrote his *Don Pasquale* for a quartet of singers, one of whom was Grisi. Grisi was also one of four singers for whom Bellini wrote *I Puritani.*

Gritzko, a peasant (tenor) in Mussorgsky's *The Fair at Sorochinsk.*

Grossmächtigste Prinzessin, recitative and aria of Zerbinetta in Richard Strauss's *Ariadne auf Naxos.*

Gruenberg, Louis, composer. Born Brest-Litovsk, Russia, August 3, 1884. He came to the United States as an infant. After studying the piano with Adele Margulies he returned to Europe in 1903 for further music study with Ferrucio Busoni, and at the Vienna Conservatory. Between 1912 and 1913 he wrote two operas—*The Witch of Brocken* and *The Bridge of the Gods* —and a symphonic poem, *The Hill of Dreams,* that won a thousand dollar prize. In 1919 he returned to the United States and soon completed an opera based on a play by Anatole France, *The Man Who Married a Dumb Wife.* Failing to secure permission for the use of this play, Gruenberg was denied the rights to have his opera published or performed. In the 1920's he interested himself in the jazz style, producing several serious works in this vein. His most important work is his opera, *The Emperor Jones,* introduced at the Metropolitan Opera on January 7, 1933. He has also written a radio opera, *Green Mansions,* and a three-act opera, *Volpone.* Gruenberg has written extensively for motion pictures, three of his scores winning Academy Awards.

Guadagni, Gaetano, male soprano. Born Lodi, Italy, about 1725; died Padua, 1792. He created the role of Orfeo in Gluck's *Orfeo ed Euridice.* After making his debut in Padua in

1747, he went to London and scored a phenomenal success there between 1748 and 1754. During this period he was heard in Handel's oratorios the *Messiah* and *Samson*. After a period of study with Gioacchino Gizziello in Lisbon he made several sensational tours of Europe. He was in Vienna in 1762, and there appeared in the first performance of *Orfeo ed Euridice*. After 1770 he sang at the court of the Elector of Munich, and in 1774 he settled in Padua. Gluck rewrote *Telemaco* for him.

Guarany, Il, Brazilian national opera by Antonio Gomes. Libretto by the composer, based on a Brazilian novel by José Alencar. Première: La Scala, March 19, 1870. This is the only opera by a South American composer which has appeared in the repertory of several leading opera houses. The central figure of the opera is an Indian of the Guarany tribe; the music is rich with Amazon Indian melodies and rhythms.

Guard, William J., opera publicist. Born Limerick, Ireland, March 29, 1862; died New York City, March 3, 1932. He was brought to America as a child and was educated in this country. For many years he worked as journalist for various important newspapers. When the Manhattan Opera was organized in 1906, he was engaged as press representative. After the dissolution of that company he was engaged for similar duties with the Metropolitan Opera, remaining with that company until his death. He was decorated by the Italian government with the Order of the Crown of Italy in recognition of his services to Italian singers.

Guardiano, Padre, an abbot (bass) in Verdi's *La forza del destino*.

Gueden, Hilde (soprano). Born Vienna, September 15, 1917. After preliminary music study in Vienna, she received intensive training in singing in Milan, Rome, and Paris. Her debut took place in 1939 at the Zurich Opera, where she

appeared as Cherubino. After two years in Zurich, she assumed principal soprano roles at the Berlin Opera and the Munich Opera, where she scored major successes in operas of Mozart and Richard Strauss. In 1942 she was engaged by the Royal Opera in Rome, making her debut there in the role of Sophie in *Der Rosenkavalier*. During the German occupation of Italy, Gueden withdrew from public appearances, living in temporary retirement in Venice and near Milan. After the war, she was heard at La Scala, the Vienna State Opera, the Royal Opera of Rome, Covent Garden and at several Salzburg and Edinburgh Festivals. Her American opera debut took place at the Metropolitan Opera on November 15, 1951, when she sang Gilda; she appeared there in subsequent seasons not only in the traditional repertory but in the American première of *The Rake's Progress*.

Guerre des Bouffons (The Buffoons' War), a musical war waged in Paris in 1752, precipitated by a performance, by a visiting Italian company, of *La serva padrona*. Headed by Denis Diderot, Jean Jacques Rousseau, and Friedrich Grimm, among others, a cult arose proclaiming that Pergolesi's art was the only true one for opera. The chief target of the attack was Jean Philippe Rameau, who had broken with Italian traditions by writing operas that emphasized the dramatic element and utilized advanced harmonic and orchestral writing.

Guglielmo, an officer (baritone) in love with Fiordiligi in Mozart's *Così fan tutte*.

Guglielmo Ratcliff, *see* WILLIAM RATCLIFF.

Gui, Vittorio, conductor. Born Rome, September 14, 1885. He attended the Academy of Santa Cecilia, after which he made his baton debut in 1907 at the Teatro Adriano with *La Gioconda*. After appearances in various Italian opera houses, he succeeded Campanini as principal conductor of the San Carlo in Naples in 1910. In 1925 he was appointed music director of the Teatro Torino, where he helped inaugurate a vital repertory; under his direction, this opera house achieved national significance. In 1933 he became the principal conductor of the Florence May Music Festival, and in 1938 a principal conductor at Covent Garden. He has also led distinguished opera performances at Glyndebourne, where he was Fritz Busch's successor as musical director.

Guido, commander of the Pisan army (bass) and husband of Monna Vanna in Février's *Monna Vanna.*

Guidon, Prince, King Dodon's son (tenor) in Rimsky-Korsakov's *Le Coq d'or.*

Guillaume Tell, see WILLIAM TELL.

Guiraud, Ernest, composer. Born New Orleans, Louisiana, June 23, 1837; died Paris, May 6, 1892. He studied with his father, and in his fifteenth year wrote an opera, *Le Roi David,* performed in New Orleans. For several years he attended the Paris Conservatory, where he won the Prix de Rome. From 1876 to the time of his death he taught at the Conservatory; his many pupils included Debussy, André Gédalge, and Charles Martin Loeffler. In 1875 he was engaged to write accompanied recitatives for the dialogue passages of Bizet's *Carmen;* it is in this form, as an opera, that *Carmen* is chiefly known. In France, however, Guiraud's recitatives are not used, and *Carmen* is still performed as an opéra comique, with spoken dialogue. Guiraud wrote the following operas: *Sylvie* (1864); *En prison* (1869); *Le Kobold* (1870); *Madame Turlupin* (1872); *Piccolino* (1876); *La galante aventure* (1882); *Frédégonde* (completed by Saint-Saëns) (1895). In 1891 he succeeded Delibes as a member of the Institut de France.

Gunther, king of the Gibichungs (baritone) in Wagner's *Die Götterdämmerung.*

Guntram, opera by Richard Strauss. Libretto by the composer. Première: Weimar Opera, May 10, 1894. This was Strauss's first opera; it was modeled after the Wagnerian music dramas. The action takes place in a German duchy of the thirteenth century. The tyrant, Duke Robert, is murdered by the minstrel Guntram. For this act Guntram is imprisoned. When Freihild, who is in love with Guntram, becomes ruler of the duchy, Guntram is freed; but because of Freihild's high station he must renounce his love for her.

Gura, Eugen, dramatic baritone. Born Pressern, Bohemia, November 8, 1842; died Aufkirchen, Bavaria, August 26, 1906. After studies at the Munich Conservatory he made his debut in Munich in 1865 in Lortzing's *Der Waffenschmied.* After two years with the Munich Opera he appeared with the Breslau Opera, and from 1870 to 1876 with the Leipzig Opera, where he was acclaimed one of the foremost German baritones of the day. In 1876 he appeared in the first complete performance of *Der Ring der Nibelungen* at Bayreuth. From 1876 to 1883 he was principal baritone of the Hamburg Opera and from 1883 to 1895 of the Munich Opera. He wrote his autobiography *Erinnerungen aus meinem Leben* (1905). His son Hermann was also a noted Wagnerian baritone, and served as director of the Komische Oper in Berlin and the Helsingfors Opera.

Gurnemanz, Knight of the Grail (bass) in Wagner's *Parsifal.*

Gustav Hinrichs Opera Company, see HINRICHS, GUSTAV.

Gustavus III, see RICCARDO.

Gutheil-Schoder, Marie, dramatic mezzo-soprano. Born Weimar, Germany, February 16, 1874; died there, October 4, 1935. She studied singing with Virginia Gungl in Weimar, making her opera debut in that city in 1891. Acclaimed, she was immediately engaged by the Weimar Opera, where she remained for nine years. In 1900 she made her debut at the Vienna Opera. For the next twenty-six years she appeared in Vienna with outstanding success, particularly in such roles as Salomé, Elektra, and Carmen. In 1926 she retired as a singer, and for a brief period served as a stage director.

Gut'n Abend, Meister, Eva's address to Hans Sachs in Act II of Wagner's *Die Meistersinger*.

Gutrune, Gunther's sister (soprano) in Wagner's *Die Götterdämmerung*.

Guy Mannering, *see* SCOTT, SIR WALTER.

Gwendoline, opera by Emmanuel Chabrier. Libretto by Catulle Mendès, based on a medieval legend. Première: Théâtre de la Monnaie, April 10, 1886. Harald, king of the Vikings, falls in love with Gwendoline, daughter of his captive, the Saxon Arnel. Arnel seemingly consents to their marriage, but arranges for Harald's murder by his bride. At the last moment she refuses to kill her husband. Eventually, Arnel kills Harald and Gwendoline commits suicide.

Gypsy Baron, The (Der Zigeunerbaron), operetta by Johann Strauss II. Libretto by Ignaz Schnitzer. Première: Theater-an-der-Wien, October 24, 1885. This operetta, besides being a staple in the musical theater, is sometimes performed in the opera house. The story concerns Sandor Barinkay, taken from his ancestral home as a child. Returning to it as a man, he finds it overrun with gypsies. He falls in love with one of them, Saffi. The overture, the *Schatz* (Treasure) waltzes, and the tenor aria "Ja, das Alles auf Ehr" are celebrated.

H

Hába, Alois, composer. Born Vyzovice, Moravia, June 21, 1893. An experimental composer who has specialized in quarter-tone music, Hába has written several operas in this idiom. His music study took place at the Prague Conservatory and, after World War I, privately with Franz Schreker. Hába began writing quarter-tone music in 1921. He founded a class in quarter-tone music at the Prague Conservatory; devised his own musical notation; had instruments manufactured capable of performing his works. He wrote the first opera in quarter tones, *Die Mutter* (The Mother), in 1929. It was introduced in Munich two years later. Subsequent quarter-tone operas were: *The*

Unemployed (1932); *New Earth* (1935); *The Kingdom Come* (1940).

Habanera, (1) a slow dance in 2/4 time, said to have originated in Havana, Cuba, and long popular in Spain. An example in opera is Carmen's song "L'amour est un oiseau rebelle" in Bizet's *Carmen*.

(2) Opera by Raoul Laparra. Libretto by the composer. Première: Opéra-Comique, February 26, 1908. The setting is Spain. Pedro has married Pilar, but Pedro's brother, Ramon, is in love with her. Jealous of Pilar's happiness as she dances a habanera with her bridegroom, Ramon kills his brother and escapes before anyone can identify him as the murderer. Haunted

by Pedro's ghost, Ramon reveals his crime to Pilar at Pedro's grave. Pilar sinks dead on the grave. Impetuously, Ramon plays a snatch from the fateful Habanera, then destroys his guitar and flees into the darkness.

Habet Acht! Brangäne's warning to Tristan and Isolde in Act II of Wagner's *Tristan und Isolde.*

Hab' mir's gelobt, trio of the Marschallin, Octavian, and Sophie in Act III of Richard Strauss's *Der Rosenkavalier.*

Hadji, a Hindu slave (tenor) in Delibes's *Lakmé.*

Hadley, Henry, composer. Born Somerville, Massachusetts, December 20, 1871; died New York City, September 6, 1937. After attending the New England Conservatory he served as conductor with the Schirmer-Mapleson Opera Company. Additional study took place in Vienna with Eusebius Mandyczewski, following which he returned to America and devoted himself to conducting. From 1904 to the end of his life he directed major symphony orchestras in America and Europe, serving at different times as the permanent conductor of the Seattle Symphony, the San Francisco Symphony, and the Manhattan Symphony, and as associate conductor of the New York Philharmonic. He was also a founder of the Berkshire Symphonic Festival, where he conducted the first concert in 1933. He composed works in every major form. His opera *Azora* was introduced by the Chicago Opera on December 26, 1917. A one-act opera, *Bianca,* won the William Hinshaw Prize in 1918. Hadley's most successful opera, *Cleopatra's Night,* was introduced by the Metropolitan Opera in 1920. Other operas: *Safie* (1909); *A Night in Old Paris* (1925).

Hageman, Richard, composer and conductor. Born Leeuwarden, Holland, July 9, 1882. His parents were professional musicians, his father the director of the Amsterdam Conservatory, his mother a concert singer. After attending the Brussels Conservatory he was appointed assistant conductor of the Amsterdam Royal Opera when he was only sixteen; two years later he became its principal conductor. In 1907 he came to the United States, and five years later became a conductor at the Metropolitan Opera. He remained with the Metropolitan nine years, returning for the 1935–1936 season. As a composer he is best known for his songs (the most famous being "Do Not Go, My Love" and "At the Well") and his opera *Caponsacchi,* introduced in Freiburg, Germany, in 1932, and five years later given at the Metropolitan. After leaving the Metropolitan, Hageman went to Hollywood to write music for motion pictures.

Hagen, half brother (bass) of Gunther and Gutrune in Wagner's *Die Götterdämmerung.*

Hagenbach, Wally's beloved (tenor) in Catalani's *La Wally.*

Hagith, one-act opera by Karol Szymanowski. Libretto by Felix Dörmann. Première: Warsaw Opera, May 13, 1922. This opera is based on an Oriental legend in which an old and cruel king tries to regain health and youth by making love to a young girl, Hagith.

Hahn, Reynaldo, composer and conductor. Born Caracas, Venezuela, August 9, 1875; died Paris, January 28, 1947. As a student of the Paris Conservatory he had one of his works published when he was fourteen, and some of his orchestral music performed a year later. His first opera, *L'ile de Rêve,* with a Polynesian setting, was introduced by the Opéra-Comique on March 23, 1898. Later operas were *La Carmélite* (1902), *Nausicaa* (1919), *La colombe de Bouddha* (1921) and *Le marchand de Venise* (1935). Famous as a song writer, Hahn also conducted opera performances in Paris, Salzburg, and Cannes; in 1945 he was appointed director of the Paris Opéra.

Halász, Lászlo, conductor and opera manager. Born Debrecen, Hungary, June 6, 1905. He attended the Royal Academy of Music in Budapest, and in 1928 made his debut in Budapest both as pianist and conductor. From 1930 to 1932 he was a conductor of the Prague Opera and from 1933 to 1936 of the Vienna Volksoper. During this period he participated in important festivals in Vienna and Salzburg. He came to the United States in 1936, making his debut as opera conductor in St. Louis with *Tristan und Isolde.* In 1938 he was appointed conductor of the Philadelphia Civic Opera Company, a year later musical director of the St. Louis Grand Opera. When the New York City Opera was founded in 1943, he became its artistic and musical director. He initiated a vital and progressive artistic program and helped launch the careers of many American singers (*see* NEW YORK CITY OPERA). Dismissed in December, 1951, after a disagreement with the management, he won a court verdict which found that his conduct as director and manager had not constituted a threat to the "prosperity and advancement" of the New York City Opera, as had been charged. He subsequently became the musical director of Remington Records.

Halévy, Jacques-François (born LEVY), composer. Born Paris, May 27, 1799; died Nice, March 17, 1862. He entered the Paris Conservatory at age ten, where he won numerous prizes, including the Prix de Rome (1819). In Italy, he devoted himself to the study and writing of operas. Returning to Paris, he made repeated but futile efforts to get some of his operas performed. However, a new work in a comic vein, *L'artisan,* was given at the Théâtre Feydeau in 1827. His first success came a year later with *Clari,* performed at the Théâtre des Italiens. His greatest triumph was achieved in 1835 with the opera by which he is today remembered, *La Juive,* given by the Opéra on February 23. In 1836 an unusual comic opera, *L'eclair,* requiring only four singers and no chorus, was also successful. Though Halévy wrote some twenty more operas, he never duplicated either the inspiration or the popularity of his masterwork, *La Juive.* In 1827 he began a long and successful career as teacher when he was appointed professor of harmony and accompaniment at the Conservatory. Subsequently he became professor of counterpoint, fugue, and advanced composition; his many pupils included Gounod and Bizet. In 1830 he became chorus master at the Opéra, a post he retained for sixteen years. In 1836 he was elected a member of the Institut de France, and in 1854 appointed its permanent secretary. His daughter, Geneviève, married Bizet in 1869. His most successful operas were: *Clari* (1828); *La Juive* (1835); *L'eclair* (1835); *La reine de Chypre* (1841); *Les mousquetaires de la reine* (1846); *La fée aux roses* (1849); *La dame de pique* (1850); *Le Juif errant* (1852); *L'inconsolable* (1855); *La magicienne* (1858).

Halka, Polish folk opera by Stanislaw Moniuszko. Libretto by Vladimir Wolski. Première: Warsaw Opera, February 16, 1854. This is the most famous opera to come out of Poland, and that country's most significant national opera. The plot revolves around the love of Halka for Pan Janusz who, in turn, is in love with Sophie. Janusz, however, seduces Halka and then forsakes her. She goes to his castle to denounce him, forces her way inside as Janusz and Sophie are about to be married, and kills herself before their eyes. The Peasant Ballet in the third act is celebrated. Another Polish composer, Wallek-Valevski, wrote a sequel to *Halka* entitled *Jontek's Revenge.*

Hallelujah to Wine, a vulgar parody sung by the juggler Jean in Act I of Massenet's *Le Jongleur de Notre Dame.*

Hallström, Ivar, composer. Born Stockholm, June 5, 1826; died there April 11, 1901. While studying law he collaborated with Prince Gustav of Sweden in writing an opera, *The White Lady of Drottningholm,* produced by the Stockholm Opera in 1847. After completing law study, Hallström became librarian to the Crown Prince Oscar. At the same time he taught the piano and subsequently (1861) became director of Lindblad's Music School. In 1874 he achieved a major success with his opera, *The Mountain King,* which, after its introduction in Stockholm, was heard in Munich, Copenhagen, and Hamburg. His operas are so strongly national in character that he has been described as "Sweden's most national scenic composer." Other operas are: *Bride of the Gnome* (1875); *Vikings' Voyage* (1877); *Nyaga* (1885); *Per Svinaherde* (1887); *Granada's Daughter* (1892); *Lilen Kcrin* (1897).

Haltière, Madame de la, Cinderella's stepmother (contralto) in Massenet's *Cendrillon.*

Hamburg Opera, one of the oldest opera institutions in Europe. It was founded in 1678 (in Hamburg, Germany) by Johann Adam Reincken and opened with Johann Theile's *Adam und Eva* in a theater (no longer existent) on the Gänsemarkt. Its heyday was between 1695 and 1706, during the artistic direction of Reinhard Keiser, who was responsible for making it the leading opera house in Germany. Kaiser wrote over a hundred operas for the theater, and it was during his regime that Handel's first opera, *Almira,* was written and produced. German opera went into decline after 1738, and in 1740 Italian opera became ascendant. For the next century, the Hamburg Opera assumed a secondary position among German opera companies. With the building of the Stadttheater in 1874, a new era began. B. Pollini was the first of the new artistic directors. The company's general musical directors have included such outstanding figures as Mahler, Weingartner, Klemperer, Josef Stransky, and Egon Pollack. The Stadttheater was hit by a bomb during World War II and was partially destroyed. Rebuilt after the war, it has been under the artistic direction of Gunther Rennert and the musical direction of Leopold Ludwig.

Hamlet, (1) tragedy by Shakespeare, the source of several operas (*see* SHAKESPEARE), the most famous being Thomas's *Hamlet.*

(2) Opera by Ambroise Thomas. Libretto by Michel Carré and Jules Barbier. Première: Paris Opéra, March 9, 1868. The story is essentially the same as Shakespeare's. Learning that his father has been murdered, Hamlet determines to seek revenge. In order to accomplish his ends more surely he simulates insanity. He has a group of players perform before the king and queen a mock re-enactment of the crime. King Claudius, the murderer, gives himself away, and Hamlet ultimately kills him. Meanwhile, Ophelia, in love with Hamlet, is so upset by his apparent madness that she truly loses her mind and dies.

While this is the most celebrated operatic treatment of Shakespeare's play, and while it enjoyed success when first heard, *Hamlet* has never acquired popularity. There is a lack of sustained musical interest, and the libretto is at times confused. However, there are memorable passages in the score: "O vin, dissipe la tristesse," Hamlet's drinking song in the second act; Ophelia's mad scene, "Partagez-vous mes fleurs" in the fourth act; and Hamlet's beautiful arioso, "Comme une pâle fleur," in the last act.

Hammer Song, *see* HO-HO! SCHMIEDE, MEIN HAMMER.

Hammerstein, Oscar, impresario. Born Stettin, Germany, May 8, 1846; died New York City, August 1, 1919. Coming to America a penniless immigrant, he realized a fortune from the invention and sale of devices for the manufacture of cigars. He then turned to the theater, wrote several plays, and had a hand in building a number of theaters in New York. He engaged in the managing of opera for the first time in 1906 when he built the Manhattan Opera House and established there a company competitive with the Metropolitan Opera. For four years Hammerstein presented brilliant performances, introducing many world-famous artists to America and emphasizing operatic novelties (*see* MANHATTAN OPERA HOUSE). In 1908 he built an opera house in Philadelphia where, several nights a week, he brought his New York company; in 1909 he brought opera to Baltimore. In April, 1910, Hammerstein sold his operatic interests to the Metropolitan Opera, thus bringing to an end four years of intense rivalry between the two companies. The agreement specified that for the following ten years Hammerstein was not to produce opera in America. Hammerstein, consequently, transferred his activity to London where, in 1911, he built the London Opera House in Kingsway. When this venture collapsed, he returned to New York and, in 1913, tried to evade his contract by building the Lexington Opera House and attempting to produce operas there. The Metropolitan restrained him by recourse to law. He was still planning to produce opera in New York when he died.

Handel, George Frideric, composer. Born Halle, Saxony, February 23, 1685; died London, April 14, 1759. Handel belongs with the greatest masters in music by virtue of his oratorios. He was also a prolific opera composer. His operas have not survived in the living repertory, but their writing and their production were a major facet of his career. Handel began the study of music formally with Friederich Zachow, organist at Halle, who for three years gave the boy a comprehensive training. But academic study was not abandoned. Handel attended the University of Halle for the study of law, at the same time serving as organist at the Dom-Kirche. He left the University in 1703 and went to Hamburg, then the foremost opera center in Europe. There he became a violinist in the opera orchestra. There, too, he wrote his first opera, *Almira,* produced with outstanding success on January 8, 1705.

Handel left Hamburg in 1706 and went to Italy. An oratorio, *La Risurezzione,* introduced in Rome, and an opera, *Agrippina,* given in Venice, made Handel famous in that country. He attracted the interest of Agostino Steffani, a kapellmeister from Hanover, then visiting Italy. Steffani prevailed on Handel to take over his Hanover post. Handel was not long in Hanover before he asked for and received a leave of absence to visit London. There he arrived in 1710. Within two weeks he had written a new opera, *Rinaldo.* When performed on February 24, 1711, *Rinaldo* was a sensation. It was given fifteen performances to sold-out houses, and Handel became the idol of the English opera public. He now returned to Hanover to fulfill his duties as kapellmeister, but did not remain there long. In the fall of 1712 he received a second leave and returned to London. This time he remained to become one of the most celebrated musicians in England, and one of the most popular opera composers. Queen Anne made him court composer. When the Elector of Hanover succeeded her to the throne, he appointed his Hanoverian kapellmeister royal music master. In 1717, Handel went to Cannons, near London, to serve as music master for the Duke of Chandos. He worked here for three years, writing many works, including *Acis and Galatea.* He was back in London in 1720, filling the important post of artistic director of the newly founded Royal Academy of Music. He gathered some of the leading singers of Europe for his company, and wrote some of his most successful operas, the first being *Radamisto,* performed on April 27, 1720.

Handel was now at the peak of his popularity and success, a favorite with royalty and the common man alike. But he also had powerful enemies. There were those who resented him because he was a foreigner, or because he was a musical tyrant, because he was a pet of nobility, or because he was a man of boorish manners. The enemies gathered around the powerful figure of the Earl of Burlington, and to offset Handel's popularity, they imported the celebrated opera composer Giovanni Maria Bononcini. Bononcini's operas, given by the Academy, were enthusiastically received; but when Handel countered with his opera *Ottone,* a triumph, his rival went into permanent retreat. Handel did not long enjoy his victory. In 1728 *The Beggar's Opera* was such a hit that the London public suddenly lost interest in serious opera and threw its support to this new type of musical entertainment. The decline in box office receipts spelled doom for the Academy, and it went into bankruptcy.

Handel now went into partnership with John Jacob Heidegger to organize a new opera company. For this company Handel wrote one opera after another, but none was able to duplicate his earlier successes. When a rival company, sponsored by the Prince of Wales, drew away Handel's best singers and his audiences, Handel had to admit defeat. It was now that he concentrated his enormous energies and gifts in another field in which he was to be incomparable, that of the oratorio. The greatest of these works, the *Messiah,* introduced in Dublin in 1742, rehabili-

tated Handel's position as a composer. While working on his last oratorio, *Jephtha,* Handel suddenly went blind. An operation was unsuccessful. Handel continued to compose, to give organ concerts, and to direct performances of his oratorios. It was while conducting the *Messiah* at Covent Garden that he collapsed with his last illness. He died a few days later and was buried in Westminster Abbey.

It is not difficult to understand why Handel saw his operas lose favor with his audiences and why, since his time, they have seldom been performed. Handel was content to write within the Italian patterns and according to traditions which even then were going out of style. Handel's operas, consequently, were too stilted, too formal, too lacking in dramatic interest to have sustained appeal. On the rare occasion that a Handel opera is revived in our time it appears like a museum piece. Such interest as these operas have for us now are found in some of their wonderful arias: "Ombra mai fù" (better known as Handel's "Largo") from *Serse,* "Cara sposa" from *Rinaldo,* "Care selve" from *Atalanta,* or "Caro amore" from *Floridante.*

The most successful of Handel's forty operas were: *Almira* (1705); *Rodrigo* (1707); *Agrippina* (1709); *Rinaldo* (1711); *Radamisto* (1720); *Acis and Galatea* (1720); *Floridante* (1721); *Giulio Cesare* (1723); *Tamerlano* (1724); *Rodelinda* (1725); *Scipione* (1726); *Admeto* (1727); *Siroe* (1728); *Partenope* (1730); *Poro* (1731); *Ezio* (1732); *Arianna* (1734); *Ariodante* (1735); *Alcina* (1735); *Atalanta* (1736); *Berenice* (1737); *Faramondo* (1738); *Serse* (1738).

Handlung, German for "action," a term used by Richard Wagner to describe his *Tristan und Isolde,* in order to distinguish it from his earlier, more conventional operas.

Hans, (1) Marie's sweetheart (tenor) in Smetana's *The Bartered Bride*.

(2) *See* HANS HEILING.

Hansel and Gretel (Hänsel und Gretel), fairy opera in three acts by Engelbert Humperdinck. Libretto by Adelheid Wette (the composer's sister), based on the fairy tale of Ludwig Grimm. Première: Hoftheater, Weimar, Germany, December 23, 1893. American première: Daly's Theater, New York, October 8, 1895.

Characters: Peter, a broommaker (baritone); Gertrude, his wife (mezzo-soprano); Hansel, their son (mezzo-soprano); Gretel, their daughter (soprano); the Witch (mezzo-soprano); the Sandman (soprano); the Dewman (soprano); angels; gingerbread children.

The orchestral prelude contains melodies from the opera. It opens and closes with the prayer theme. Other fragments from the third act follow, including the happy melody of the children at the end of the opera.

Act I. Peter's house. Hansel and Gretel are hungry. To distract her little brother, Gretchen teaches him to sing and dance ("Brüderchen, komm tanz' mit mir"). Their mother scolds them for playing instead of working. Now the children must go into the woods to pick strawberries for the evening meal. While they are gone their father returns from work with a bundle of food under his arm. Learning that the children have gone into the wood, he grows apprehensive, and joins his wife in searching for them.

Act II. In the forest. While Hansel is looking for strawberries, Gretel sits under a tree singing a folk song ("Ein Männlein steht im Walde"). Hansel returns with the berries and since the children are hungry, they yield to the temptation of eating them. Night begins to fall. The darkness brings terror, for the children are now unable to find their way home. The Sandman comes to put the children to sleep with a lullaby ("Der kleine Sandmann bin ich"). The children say their prayers ("Abends will ich schlafen gehn"). After they fall asleep, angels descend to provide the children with a protective ring ("Dream Pantomime").

Act III. The Witch's house. The Dewman sprinkles dewdrops on the children ("Der kleine Taumann heiss' ich"). Waking, the children find themselves in front of a gingerbread house. Excitedly, the hungry children begin to eat morsels of the house. Suddenly, the Witch emerges. Her magic keeps the children rooted to the ground. She locks Hansel in a cage and sets Gretel at housework. The Witch sings a gleeful song about her weird activities (Hexenlied: "Hurr, hopp, hopp, hopp"). Gretel, meanwhile, steals her wand and with it frees her brother. When the Witch orders Gretel to look into the flaming oven, Gretel simulates stupidity and asks the Witch to show her how this is done. When the Witch stands before the open door, Hansel and Gretel push her inside and slam the door. They now express their joy at being free (Gingerbread Waltz: "Knusperwalzer"). The Witch's oven explodes. This is the signal for all the Witch's previous victims to change from gingerbread back into little children. The parents of Hansel and Gretel appear, find their children free, and a celebration takes place.

With *Hansel and Gretel,* Humperdinck helped create the operatic fairy tale. There now arose a school of German opera composers who drew thei texts from German folklore and fairy tales and based their music on folk tunes. To this day, *Hansel and Gretel* remains the finest product of this school. It is lovable in its simplicity, infectious in its youthful spirit. Humperdinck did not originally plan *Hansel and Gretel* for the opera house. His sister had written a play for children based on the Grimm fairy tale, and she asked her brother to provide an appropriate musical setting. Even after Humperdinck finished this chore, the story continued to appeal to him, and he decided to extend his score to operatic proportions. He sent his finished manuscript to Richard Strauss, who accepted it for performance in Weimar. "Truly it is a masterwork of the first rank," Strauss wrote the composer.

Hans Heiling, opera by Heinrich Marschner. Libretto by Eduard Devrient. Première: Berlin Opera, May 24, 1833. The setting is the Harz Mountains in the sixteenth century. Hans Heiling, son of the Queen of the Spirits, assumes human form. He falls in love with Anna, but when she learns of his origin she abandons him for Konrad. Hans calls on the spirits for revenge. They fail to help him. He sinks into the earth, vowing that never again will a living mortal see him. "An jenem Tag," Hans's aria in Act I, is famous.

Hanslick, Eduard, music critic. Born Prague, Bohemia, September 11, 1825; died Baden, Austria, August 6, 1904. He wrote provocative and influential criticisms for the *Wiener Zeitung,* the *Presse,* and the *Neue freie Presse.* An avowed advocate of absolute music, he was a devoted champion of Brahms and a bitter antagonist of Wagner and the esthetics of the Wagnerian music dramas. In Vienna, in the closing decades of the nineteenth century, he was the spearhead for all attacks on Wagner. It is believed that when Wagner created the role of Beckmesser he had Hanslick in mind, and that he originally even intended calling this character Hans Lick. Hanslick wrote an eight-volume history of opera, *Die moderne Oper* (1875), and a two-volume autobiography, *Aus meinem Leben* (1894), among other books. A collection of his essays was published in the United States under the title of *Vienna's Golden Years of Music* (1950).

Hanson, Howard, composer. Born Wahoo, Nebraska, October 28, 1896. His musical education took place at Luther College in Wahoo, the Institute of Musical Art in New York, and Northwestern University. He then became professor of theory and composition, and subsequently Dean of the Conservatory of Fine Arts, at the College of the Pacific, San José, California. In 1921 he became the first Music Fellow to receive the Prix de Rome through a competition of the American Academy in Rome. His *Nordic Symphony* was successfully introduced in Italy. In 1924 he returned to the United States and became director of the Eastman School of Music at the University of Rochester, a post he has held since that time with great distinction. His opera *Merry Mount* was introduced by the Metropolitan Opera on February 10, 1934, and his fourth symphony became the first work in that form to receive the Pulitzer Prize.

Hans Sachs, (1) opera by Gustav Albert Lortzing. Libretto by the composer and Philip Reger, based on a play by Johann Ludwig Deinhardstein. Première: Leipzig Opera, June 23, 1840. This opera, based on the same subject as Wagner's *Die Meistersinger,* preceded Wagner's opera by over two decades.

(2) Cobbler-philosopher (bass or baritone) in Wagner's *Die Meistersinger.*

Happy Shade, a character (soprano) in Gluck's *Orfeo ed Euridice.*

Hardy, Thomas, novelist and poet. Born Upper Brockhampton, England, June 2, 1840; died Dorchester, January 11, 1928. Rutland Boughton's opera *The Queen of Cornwall* was derived from Hardy's play of the same name. Hardy's famous novel *Tess of the D'Urbervilles* is the source of Frederic d'Erlanger's opera *Tess. The Three Strangers* served for operas by Hubert Bath and Julian Gardiner, and

The Mayor of Casterbridge was made into an opera by Peter Tranchell.

Hark, the passing bell! Boris Godunov's death in the final scene of Mussorgsky's *Boris Godunov*.

Harlequin's Serenade, Beppe's song (as Arlecchino) "O Colombina" in Act II of Leoncavallo's *Pagliacci*.

Harmonie der Welt, Die (The Harmony of the World), opera by Paul Hindemith. Libretto by the composer. The opera was completed in 1952, and has for its central character the astronomer Johann Kepler. A symphony, adapted from the score, was introduced in Basel, Switzerland, in January, 1952.

Harriet, Lady, maid of honor (soprano) to Queen Anne, in Flotow's *Martha*.

Harun-al-Rashid, Caliph of Bagdad (bass) in Weber's *Oberon*.

Háry János, folk opera by Zoltán Kodály. Libretto by Béla Paulini and Zsolt Harsányi, based on a poem by János Garay. Première: Budapest Opera, October 16, 1926. The central character is the prodigious liar popular in Hungarian folk tales, Háry János. The opera is built around his boastful tales: how Marie Louise, daughter of Emperor Francis and wife of Napoleon, falls in love with him and wants him to join her in Paris, and how she competes for his love with the peasant woman, Orzse; how, after Napoleon declares war on Austria as a result of this scandal, Háry defeats the enemy singlehanded; how he is welcomed back to Vienna in triumph, where he announces his rejection of Marie Louise and his acceptance of Orzse as his wife. Portions of Kodály's brilliant music are familiar as an orchestral suite, frequently performed at symphony concerts.

Haug, Hans, composer. Born Basel, Switzerland, July 27, 1900. After studying at the Basel Conservatory with Egon Petri and Ernst Levy, and in Munich with Walter Courvoisier, he began a career as conductor. Between 1928 and 1934 he was one of the conductors of the Basel Municipal Theater. After 1934 he conducted various symphony and radio orchestras in Switzerland, Italy, and France. He has also been a teacher of theory at the Lausanne Conservatory. He has written several operas, the most successful of these being in a comic vein. His operas: *Don Juan in der Fremde* (1930); *Madrisa* (1934); *Tartuffe* (1937); *Ariadne* (1943); *Le malade immortel* (1946).

Hauk, Minnie, soprano. Born New York City, November 16, 1851; died Triebschen, Switzerland, February 6, 1929. After only a few months of study with Achille Errani in New York, she mastered the principal soprano roles of half a dozen leading operas. She was only fourteen when she made her debut, in Brooklyn, in *La sonnambula*, on October 13, 1866. When she was fifteen she created for America the role of Juliet in Gounod's *Roméo et Juliette,* at the Academy of Music in New York, only seven months after the opera's world première. Following additional appearances in the United States, she sang in the leading opera houses of Europe, her popularity mounting steadily. In 1877 she made her first appearance as Carmen, at the Théâtre de la Monnaie. She brought to the role such dramatic vitality and personal animation that she was a sensation. In 1878 she created the role for England and the United States. From this time on, Carmen was to be her most celebrated impersonation (she sang it over six hundred times in English, French, German, and Italian); it was largely due to her performances that Bizet's opera emerged from its original obscurity to world-wide acclaim. From 1878 on she divided her operatic activity between America and Europe. For America she created the role of Manon in Massenet's opera, and in this country she scored a personal triumph as Selika in *L'Africaine*. She made her Metropolitan Opera debut in the role of Selika on February 10, 1891. After a season at the Metropolitan she organized her own opera company and toured the United States. Though now at the height of her fame and ability, she suddenly decided to retire and spend the rest of her life at her beautiful villa near Lucerne, Switzerland, with her husband, Baron Ernst von Hesse-Wartegg. The death of her husband and the depletion of her fortune during World War I reduced her to poverty. In the last years of her life she suffered the additional affliction of blindness. Her only source of income was a fund created by a group of American opera lovers headed by Geraldine Farrar. She wrote an autobiography, *Memories of a Singer* (1925).

Hauptmann, Gerhart, dramatist. Born Obersalzbrunn, Silesia, November 15, 1862; died Schreiberhau, Silesia, June 6, 1946. One of the leading dramatists of Germany, and the winner of the 1912 Nobel Prize for Literature, Hauptmann was the author of several dramas which were turned into operas. *Die versunkene Glocke* (*The Sunken Bell*) was the source of operas by Alexei Davidoff, Ottorino Respighi, and Heinrich Zöllner. Paul Graener made an opera of *Hanneles Himmelfahrt* (*The Assumption of Hannele*); Vit Nejedly, of *Die Weber* (*The Weavers*); and Erwin Lendvai, of *Elga*.

Ha! wie will ich triumphieren! Osmin's aria in Act III of Mozart's *The Abduction from the Seraglio*.

Hawthorne, Nathaniel, author. Born Salem, Massachusetts, July 4, 1804; died Plymouth, New Hampshire, May 19, 1864. His classic, *The Scarlet Letter*, is the source of operas by Walter Damrosch and Vittorio Giannini. Other Hawthorne works used for operas include *The Maypole of Merry Mount* (Howard Hanson's *Merry Mount*); *Rappaccini's Daughter* (Charles Wakefield Cadman's *The Garden of Mystery*); *The Snow Image* (Ned Rorem's *A Childhood Miracle*).

Haydn, Franz Joseph, composer. Born Rohrau, Lower Austria, March 31, 1732; died Vienna, May 31, 1809. A giant figure in the development of the symphony, sonata, and string quartet, Haydn was a comparatively negligible influence in opera, though he wrote about twenty works for the stage. As a young composer in Vienna, he achieved a first measure of success with music for a Viennese farce, *Der krumme Teufel*, introduced at the Burgtheater in 1752; that success, however, was aborted when a powerful Viennese nobleman interpreted the play as a satire on himself and had it removed. As kapellmeister of the Eszterházy family, an appointment received in 1761 and continued for thirty years, Haydn directed all the musical performances and wrote innumerable works, among which were operas. Since the theater at the Eszterházy palace was comparatively small and the musical forces limited, these operas have the character of chamber works. They yielded to the prevailing Italian taste and conformed rigidly to the established traditions. Haydn's finest and most ambitious opera, *Orfeo ed Euridice*, was written in 1791 during the composer's first sojourn in London. Because of Haydn's impresario's legal difficulties, a performance was not possible; the work was first performed in Florence a century and a half after it was composed.

Haydn's operas: *La canterina* (1767); *Lo speziale* (1768); *Le pescatrici* (1770); *L'incontro improviso* (1775); *La vera costanza* (1776); *Il mondo della luna* (1777); *L'isola disabitata* (1779); *La fedeltà premiata* (1780); *Orlando Paladino* (1782); *Armida* (1784); *Orfeo ed Euridice* (1791).

Heart of Midlothian, The, *see* SCOTT, SIR WALTER.

Heart, the Seat of Soft Delight, Galatea's aria in Act II of Handel's *Acis and Galatea.*

Hebbel, Friedrich, dramatist and poet. Born Wesselburen, Holstein, Germany, March 18, 1813; died Vienna, December 13, 1863. The father of German social and naturalistic drama, Hebbel was the author of several plays used in operas. Among the operas are: Eugène D'Albert's *Der Rubin;* Max Ettinger's *Judith;* Felix Mottl's *Agnes Bernauer;* Emil von Rezniček's *Holofernes;* Max von Schillings' *Moloch;* Robert Schumann's *Genoveva;* Joseph Messner's *Agnes Bernauer.*

Hedwig, William Tell's wife (soprano) in Rossini's *William Tell.*

Heger, Robert, conductor and composer. Born Strassburg, Alsace, August 19, 1886. After musical study at conservatories in Strassburg, Zurich, and Munich, he received his first assignment as conductor in Strassburg in 1907. Various engagements followed until 1913 when he was appointed principal conductor of the Nuremberg Opera. While holding this post he wrote his first opera, *Ein Fest auf Haderslev.* From 1925 to 1933 he was conductor of the Vienna State Opera; subsequently, he held conductorial posts with the Berlin State Opera, Covent Garden, and other major European opera houses. Later operas include: *Bettler Namelos* (1932); *Der verlorene Sohn* (1935).

Heil dir, Sonne! The love duet of Brünnhilde and Siegfried in the final scene of Wagner's *Siegfried.*

Heil Sachs! The closing chorus in Wagner's *Die Meistersinger.*

Heimchen am Herd, Das, *see* CRICKET ON THE HEARTH, THE.

Heine, Heinrich, poet and dramatist. Born Düsseldorf, Germany, December 13, 1797; died Paris, February 17, 1856. One of the foremost German lyric poets, Heine was the author of a tragedy, *William Ratcliff,* which was the source of many operas, notably by Volkmar Andrae, César Cui, Xavier Leroux, and Mascagni. Kurt Atterberg's opera *Fanal* was based on Heine's ballad *Der Schelm von Bergen.* Wagner's *Der fliegende Holländer* was based on Heine's version of an old German legend.

Heldentenor, German for "heroic tenor"; a tenor with a large voice suitable for dramatic rather than lyric roles.

Helen (or Helen of Troy), (1) a character (soprano) in Boïto's *Mefistofele.*

(2) (Soprano) beloved of Paris in Gluck's *Paride ed Elena.*

(3) Wife (soprano) of Menelaus in Richard Strauss's *Die aegyptische Helena.*

Hempel, Frieda, soprano. Born Leipzig, Germany, June 26, 1885. After studying voice for three years with Mme. Nicklass-Kempner in Berlin she made her opera debut at the Berlin Opera in 1905 in *The Merry Wives of Windsor.* For two years she appeared at the Schwerin Opera, after which she returned to the Berlin Opera, remaining there five years and establishing her reputation. She specialized in the German repertory, scoring major successes in the Mozart operas. In 1911 she was Richard Strauss's personal choice for the role of the Marschallin in the Berlin première of *Der Rosenkavalier.* She also appeared at the Bayreuth and Munich Festivals. On December 27, 1912, she made her Metropolitan Opera debut in *Les Huguenots.* She was a member of the Metropolitan Opera Company seven seasons. Meanwhile, she was acclaimed in the concert hall, particularly for her "Jenny Lind" concerts, which she performed in costume. After leaving the Metropolitan she appeared in recitals in the United States and Europe for a decade and then retired.

Henrietta, Queen, widow (soprano) of Charles I in Bellini's *I Puritani*.

Henry the Fowler, King of Germany (bass) in Wagner's *Lohengrin*.

Henry IV, *see* KING HENRY IV.

Henry VIII, opera by Camille Saint-Saëns. Libretto by Léonce Détroyat and Armand Silvestre, based on Shakespeare. Première: Paris Opéra, March 5, 1883. The story is built around the love of Henry VIII for Anne Boleyn, whom he makes queen despite her love for the Spanish ambassador Gomez and despite the disapproval of Rome. The ballet music in Act II is popular.

Henze, Hans Werner, composer. Born Gütersloh, Germany, July 1, 1926. He attended the Braunschweig Conservatory. After studying composition with Wolfgang Fortner and René Leibowitz —the latter a disciple of Arnold Schoenberg—he attracted attention with his compositions in the atonal idiom. His concerto for piano and orchestra won the Robert Schumann Prize in Düsseldorf in 1951. His first full-length opera, *Boulevard Solitude,* aroused considerable controversy when introduced in Hanover in 1952. He has also written a one-act opera, *Das Wundertheater* (1948), and a radio opera *Ein Landarzt* (1951). His other works include ballets. From 1950 to 1952 he was musical adviser for ballet at the Wiesbaden Opera.

Herbert, Victor, composer. Born Dublin, Ireland, February 1, 1859; died New York City, May 27, 1924. He received an extensive musical education in Germany and became a fine cellist. After marrying Theresa Förster, a prima donna, he came to the United States in 1886. His wife joined the company of the Metropolitan Opera, and he played in the opera orchestra. He started writing operettas in 1892, scoring his first major success in 1895 with *The Wizard of the Nile.* After 1903 he produced a series of outstanding operetta successes, including *Babes in Toyland, Naughty Marietta,* and *Mlle. Modiste;* but he did not neglect serious music. His first opera, *Natoma,* was written on a commission from the Manhattan Opera Company, but this organization ended its career before the opera could be produced. It was first heard in Philadelphia, February 25, 1911, sung by artists of the Philadelphia-Chicago Opera Company. Three days later, performed by the same singers, it was heard at the Metropolitan Opera in New York. In 1913 Herbert wrote a one-act opera, *Madeleine.* It was first performed by the Metropolitan Opera on January 24, 1914. Both these operas were favorably received in their day, but it is as a composer of operettas that Herbert is chiefly remembered.

Hercules, a Greek god (bass) in Gluck's *Alceste.*

Her (or His) Majesty's Theatre, a theater in the Haymarket (London) known by this name since the accession of Victoria in 1837. Previously called the King's Theatre, and earlier the Queen's Theatre, the original structure had a history going back to 1705. It was here that most of Handel's operas and oratorios were introduced. Destroyed by fires, the theater was rebuilt in 1790 and 1868. It was pulled down and rebuilt in the 1890's. Many significant opera performances took place on this site, in one or another of the theaters. Verdi's *I masnadieri* took place here on July 22, 1847. In 1877, Mapleson used the theater as an opera house and during the next decade gave the first performance in England of Wagner's entire *Ring* cycle, *Mefistofele,* and *Carmen* (with Minnie Hauk). In 1887 he gave his last season, memorable for performances of *Fidelio* with Lilli Lehmann and *La traviata* with Patti.

Hermann, (1) an army officer (tenor) in love with Lisa in Tchaikovsky's *Pique-Dame.*

(2) Landgrave of Thuringia (bass) in Wagner's *Tannhäuser*.

Hero and Leander, *see* ERO E LEANDRO.

Herod (Hérode), (1) King of Galilee (baritone) in Massenet's *Hérodiade*. (2) Tetrarch of Judea (tenor) in Richard Strauss's *Salomé*.

Hérodiade, opera in four acts by Jules Massenet. Libretto by Paul Milliet and Henri Grémont, based on Flaubert's story *Hérodias,* in turn derived from an episode in the Bible. Première: Théâtre de la Monnaie, December 19, 1881. American première: New Orleans, February 13, 1892.

Characters: Salomé, daughter of Hérodias (soprano); Herod (Hérode), King of Galilee (baritone); Hérodias, his wife (contralto); Phanuel, a young Jew (bass); John the Baptist (tenor); Vitellius, a Roman proconsul (baritone); the High Priest; merchants, soldiers, priests. The setting is Jerusalem, A.D. 30.

Act I. Courtyard of Herod's palace. Salomé relates to Phanuel how John the Baptist saved her as a child in the desert ("Il est doux, il est bon"). Herod then appears, aroused by his passionate desire for Salomé. His wife, Hérodias, comes to tell him how John the Baptist has denounced her and demanded Herod's death. Knowing of the prophet's hold on the people, Herod is upset. At this point, John the Baptist enters and denounces Herod and Hérodias, who flee in horror. Salomé recognizes John the Baptist as her savior. She falls at his feet adoringly and begs for his love. The prophet advises her to turn to God and find spiritual love.

Act II, Scene 1. Herod's chamber. Slave girls sing and dance for Herod as he lies on his couch, pining for Salomé. A potion is brought with the promise that it will bring him a vision of the one he loves most. Herod drinks the draught and sees a vision of Salomé, dazzling in her beauty. Rapturously, he sings of his passion

("Vision fugitive"). The vision disappears, leaving Herod more disturbed than ever. Phanuel now brings news that the people are restive against the Roman occupation and want Herod to lead them in rebellion.

Scene 2. A public square. In leading the people against the Romans, Herod gains the support of John the Baptist. But when the Roman Vitellius arrives with his men, Herod once again cringes before the rulers. John the Baptist remains defiant, and impresses Vitellius with his courage and dedication. Hérodias warns Vitellius of the malicious influence of the prophet. Fearlessly, John denounces the Romans and prophesies their doom.

Act III, Scene 1. Phanuel's house. Hérodias, aware of Herod's passion for Salomé, comes to Phanuel to have him read her future in the stars. Phanuel can find only blood. He also reveals to Hérodias that her long-lost daughter is none other than Salomé.

Scene 2. The temple. John the Baptist has been imprisoned. Salomé comes to the temple to seek him out. Herod tries to make love to her, but she rejects him, declaring she loves another. Herod vows to destroy his rival. When Vitellius demands that John be tried, Herod orders the prophet brought to him. Salomé throws herself upon the prophet with such passion that Herod realizes that it is the prophet with whom Salomé is in love. Disgusted, he orders the death of both Salomé and John the Baptist.

Act IV, Scene 1. A dungeon in the Temple. John awaits his sentence, praying for divine guidance ("Adieu donc, vains objets"). Salomé, pardoned by Herod, comes to the prophet, ready to die with him. They fall in one another's arms, conscious of their love ("Il est beau de mourir en s'aimant"). The priests, coming to take the prophet to his execution, separate them.

Scene 2. A hall in the palace. A brilliant festival is taking place before Herod and Hérodias. Salomé implores Herod to free John the Baptist or, failing this, to allow her to die with him. But it is too late. The executioner comes with the news that the prophet is dead. Horror-stricken, Salomé rushes toward the triumphant Hérodias with a dagger. To save herself, Hérodias reveals to Salomé that she is her mother, whereupon Salomé stabs herself.

Massenet's *Hérodiade* and Richard Strauss's *Salome*—based on the same Biblical episode—are in remarkable contrast. In Strauss's opera, both story and music are filled with savage eroticism. Salome's love is made to pass from sensuality to perversion. Massenet's opera is on a more elevated plane. The sensuality is there, but it is treated with refinement and restraint, finding expression in noble French lyricism, characteristic in its Massenet sweetness. It is never permitted to lapse into decadence.

Herodias, Herod's wife (contralto) in Massenet's *Hérodiade* and (mezzosoprano) Richard Strauss's *Salome*.

Hérold, Louis Joseph Ferdinand, composer. Born Paris, January 28, 1791; died Les Ternes, January 19, 1833. He entered the Paris Conservatory in 1806 and six years later won the Prix de Rome. He became interested in opera in Italy and completed his first opera, *La jeunesse de Henry V,* successfully performed in Naples in 1815. Returning to Paris, Herold collaborated with Boieldieu in writing *Charles de France,* and followed it with an opéra-comique of his own, *Les rosières.* In the next twelve years he wrote fourteen operas, the finest and the most successful being *Zampa,* in 1831, and *Le pré aux clercs,* a year later. While none of his opéracomiques has survived in the permanent repertory, Herold was a major figure in the evolution of that form, a link from its early masters, Auber and Boieldieu, to Offenbach. His most celebrated works were: *Les rosières* (1817); *L'amour platonique* (1819); *Le muletier* (1823); *Le Roi René* (1824); *Le lapin blanc* (1825); *Marie* (1826); *Zampa* (1831); *La médicine sans médicin* (1832); *Le pré aux clercs* (1832); *Ludovic* (completed by Halévy).

Herr Kavalier! Letter scene of Baron Ochs and Annina in Act II of Richard Strauss's *Der Rosenkavalier.*

Hertz, Alfred, conductor. Born Frankfort-on-the-Main, Germany, July 15, 1872; died San Francisco, California, April 17, 1942. After graduating with honors from the Raff Conservatory he received his first appointment as conductor, at the Halle State Theater in 1891. Various engagements followed, leading to his arrival in the United States, where he was engaged to conduct the German repertory at the Metropolitan Opera. His American debut took place on November 28, 1902, with *Lohengrin.* He remained with the Metropolitan for thirteen seasons, a period in which he directed 27 different operas. In 1903 he led the first stage performance of *Parsifal* outside Bayreuth. This act, regarded by many Wagnerites as a violation of the master's wish to confine *Parsifal* to Bayreuth, henceforth closed Bayreuth to him, and all leading German opera houses. In 1907 he conducted the American première of *Salome,* and in 1910 the world première of *Die Königskinder.* He also conducted the premières of all American operas performed at the Metropolitan in those years. After leaving the Metropolitan in 1915, he served for fifteen years as conductor of the San Francisco Symphony, and guest conductor of the San Francisco Opera.

Herzeleide Scene, duet of Parsifal and Kundry in Act II, Scene 2, of Wagner's *Parsifal.*

Hester Prynne, *see* PRYNNE, HESTER.

heure Espagnole, L' (The Spanish Hour), one-act comic opera by Maurice Ravel. Libretto by Franc-Nohain. Première: Opéra-Comique, May 19, 1911. This, one of two operas by Ravel, is a jewel of wit and satire in both text and music. The setting is Toledo, Spain, in the eighteenth century. While Torquemada, a clockmaker, attends to the clocks of the town, his wife Concepcion, entertains a succession of lovers in his shop. As one lover arrives after another, the earlier ones are concealed in grandfather clocks. Torquemada, returning from his work, discovers the various men but is willing to accept their explanation that they are only customers. The complicated situation ends with everybody in gay spirits.

Hexenlied (Witch's Song), *see* HURR, HOPP, HOPP, HOPP.

Hexenritt (Witch's Ride), the prelude to Act II of Humperdinck's *Hansel and Gretel.*

Hidroat, King of Damascus (bass) in Gluck's *Armide.*

Hier soll ich dich denn sehen, Belmonte's aria in Act I of Mozart's *The Abduction from the Seraglio.*

Hiller, Johann Adam, composer. Born Wendisch-Ossig, Prussia, December 25, 1728; died Leipzig, Germany, June 16, 1804. He established the operatic form of the singspiel (which see), forerunner of German comic opera. His music study took place at the Kreuzschule in Dresden. In 1754 he became musical tutor to the household of Count Brühl. Four years later he settled permanently in Leipzig, becoming a major figure of its musical life. He stimulated the revival of musical activity following the Seven Years' War by conducting concerts of orchestral and oratorio music, founding a singing school, establishing the Concerts Spirituels, and serving as cantor of the Thomasschule (the post once held by Johann Sebastian Bach). During this period he wrote numerous musical works, his major

contribution being in the field of opera. His most important operas and singspiels were: *Der Teufel ist los* (1766); *Lottchen am Hofe* (1767); *Der Dorfbarbier* (1770); *Die Jagd* (1770); *Der Krieg* (1772); *Die Jubelhochzeit* (1773); *Das Grab des Mufti* (1779).

Hindemith, Paul, composer. Born Hanau, Germany, November 16, 1895. After completing his music study at the Hoch Conservatory in Frankfort, he became concertmaster and eventually conductor of the Frankfort Opera orchestra. During this period (1915–1923) he wrote many chamber-music works which were successfully given at leading German and Austrian festivals. He also founded and played in the Amar String Quartet. In 1926 his opera *Cardillac,* introduced in Dresden, made him famous throughout Europe. Even more controversial, and more successful, was *Neues vom Tage* (*News of the Day*), first seen in Berlin in 1929. Hindemith was now one of the most highly esteemed of the younger German composers. He combined creative activity with the teaching of composition at the Berlin Hochschule. He was also made a member of the German Academy. Soon after the rise of Hitler, Hindemith's opera *Mathis der Maler* became the center of a political controversy (*see* MATHIS DER MALER). Hindemith's music was banned in Germany, and the composer was compelled to leave the country. He came to the United States in 1937 and subsequently became a citizen. He joined the faculty of Yale University, and has given master classes in composition at the Berkshire Music Center and Harvard University. His operas: *Cardillac* (1926, extensively revised 1952); *Hin und zurück* (1927); *Neues vom Tage* (1929); *Wir bauen eine Stadt,* children's opera (1931); *Mathis der Maler* (1934); *Die Harmonie der Welt* (1952). Hindemith has written many pieces for pianola, radio, brass band, theater, motion pictures, and education (the opera *Wir bauen eine Stadt* is an example of the last), for which the term Gebrauchsmusik (or "functional music") was coined. See GEBRAUCHSMUSIK.

Hinrichs, Gustav, conductor. Born Mecklenburg, Germany, December 10, 1850; died Mountain Lakes, New Jersey, March 26, 1942. He directed the American premières of *Cavalleria rusticana, Manon Lescaut,* and *Pagliacci.* After completing his studies in Hamburg, he played in the Stadttheater orchestra. In 1870 he came to the United States, and soon after that became Theodore Thomas' assistant with the American Opera Company. In 1886 he organized, and subsequently managed, his own opera company in Philadelphia, and for a decade toured the eastern United States. It was with this group that he gave the American premières of many notable Italian and French operas. He made his debut at the Metropolitan Opera on February 12, 1904, with *Cavalleria rusticana,* and stayed with the company only a single season.

Hin und zurück (There and Back), opera by Paul Hindemith. Libretto by Marcellus Schiffer. Première: Baden-Baden, July 17, 1927. This amusing little opera, described by its authors as a "film sketch," exploits a trick. After the murder of an adulteress by her husband, the plot is put in reverse: the adulteress comes back to life, the husband puts his revolver back in his pocket, the physician backs out of the front door, and so forth.

Hippodameia, a trilogy of operas, or melodramas, by Zdeněk Fibich. Written between 1890 and 1892, the three are: *Pelops' Wooing, The Atonement of Tantalus,* and *The Death of Hippodameia.* This was the most ambitious of Fibich's attempts to realize a new musico-dramatic art form which he called a melodrama, and which aspired to a more complete marriage of poetry

and music than had heretofore been realized in opera.

Hippolyte et Aricie (Hippolytus and Aricia), opera by Jean Philippe Rameau. Libretto by Simon Joseph de Pellegrin, based on Euripides. Première: Paris Opéra, October 1, 1733. Hippolytus is in love with Aricia. Learning that he is loved by his stepmother, Phèdre, Hippolytus plans to flee with Aricia. He is finally destroyed in a sea storm. Ravaged by her conscience, Phèdre commits suicide.

Hoël, goatherd (baritone) in Meyerbeer's *Dinorah.*

Hölle Rache kocht in meinem Herzen, Der, aria of the Queen of the Night in Act II, Scene 3, of Mozart's *The Magic Flute.*

Hoffmann, poet (tenor) in Offenbach's *The Tales of Hoffmann.*

Hoffmann, Ernst Theodor Amadeus, composer, poet, author, critic. Born Königsberg, Prussia, January 24, 1776; died Berlin, June 25, 1822. Immortalized in Offenbach's opera *The Tales of Hoffmann,* Hoffmann was a man of extraordinary versatility. He wrote many fantasies and fairy tales that inspired musical works, including Hindemith's *Cardillac.* He himself wrote ten operas, the most celebrated being *Die lustigen Musikanten* (1805) and *Undine* (1816).

Hofmannsthal, Hugo von, poet, dramatist, librettist. Born Vienna, February 1, 1874, died Vienna, July 15, 1929. He was for many years Richard Strauss's librettist. They collaborated upon the following works: *Arabella, Die aegyptische Helena, Ariadne auf Naxos, Elektra, Die Frau ohne Schatten,* and *Der Rosenkavalier.* Hofmannsthal also wrote the libretto for Egon Wellesz' *Alkestis,* while his play *Die Hochzeit der Sobeide* was used by Alexander Tcherepnin for his opera of the same name.

Hof-und-National-Theater (or Munich Opera), the leading opera house of

Munich, Germany. The building was erected in 1818 on the site of a Franciscan convent. It burned down in its initial year, was rebuilt with funds provided by King Ludwig I, and reopened on January 2, 1825, at which time it received its present name. Here some of Wagner's most important dramas were introduced: *Tristan und Isolde* in 1865; *Die Meistersinger* in 1868; *Das Rheingold* in 1869; and *Die Walküre* in 1870. Later premières of operas have included Wolf-Ferrari's *Le donne curiose* (1903) and *The Secret of Suzanne* (1909); Hans Pfitzner's *Palestrina* (1917); Albert Coates's *Samuel Pepys* (1929); Alois Hába's *The Mother* (1931); Carl Orff's *Der Mond* (1939); Richard Strauss's *Der Friedenstag* (1938); and *Capriccio* (1942). Principal conductors of the opera house have included (chronological order) Franz Lachner, Hans von Bülow, Hermann Levi, Richard Strauss, Felix Mottl, Bruno Walter, Hans Knappertsbusch, and Clemens Krauss.

After World War II the opera house was reopened with Georg Hartmann as intendant and Georg Solti as principal conductor. Other conductors in this recent period have included Hans Knappertsbusch, Eugen Jochum, and Robert Heger.

Ho! He! Ho! The sailors' chorus (Marinesca) in Act II of Ponchielli's *La Gioconda.*

Ho-Ho! Schmiede, mein Hammer, the Hammer Song—Siegfried's narrative in Act I of Wagner's *Siegfried.*

Ho-Jo-To-Ho! Brünnhilde's battle cry in Act II of Wagner's *Die Walküre.*

Holberg, Ludvig, dramatist and writer. Born Bergen, Norway, December 3, 1684; died Copenhagen, Denmark, January 28, 1754. The works of the founder of Danish literature were the source of several operas, including Carl Nielsen's *Maskarade,* Othmar Schoeck's *Ranudo,* and Julius Weismann's *Die pfiffige Magd.*

Holbrooke, Josef (or JOSEPH), composer. Born Croydon, England, July 5, 1878. Educated at the Royal Academy of Music in London, he received recognition as composer with *The Raven,* an orchestral work introduced in 1900. His magnum opus is an operatic trilogy based on Celtic legends and entitled *The Cauldron of Anwyn.* The first opera in this trilogy, *The Children of Don,* was introduced in 1911. The other two, *Dylan* and *Bronwen,* were heard in 1913 and 1929. Holbrooke's other operas: *Pierrot and Pierrette* (1909); *The Wizard* (1915); *The Stranger* (1924).

Holland Music Festival, a festival inaugurated in June, 1949, with performances of opera, symphony, choral, and chamber music in Amsterdam, The Hague, and Scheveningen. Opera has been a basic part of the programs since the first year, with performances by the Netherlands Opera and various visiting groups (including the Vienna State Opera, La Scala, the English Opera Group under Britten, and the Municipal Opera of Essen). World premières at this festival have included: Hendrik Andriessen's *Philomela* and Françaix's *L'apostrophe.* Novelties have included: Bartók's *Bluebeard's Castle;* Berg's *Lulu;* Falla's *El retablo de Maese Pedro* and *La vida breve;* Janáček's *Jenufa;* Rossini's *La cenerentola;* Stravinsky's *Le rossignol.*

Holst, Gustav, composer. Born Cheltenham, England, September 21, 1874; died London, May 25, 1934. After completing music study at the Royal College of Music he was appointed musical director of the St. Paul's Girls' School and professor at the Royal College of Music, retaining both posts for over two decades. During World War I he was engaged by the YMCA to organize the musical activities among British troops in Salonika, Constantinople, and points in Asia Minor. In 1923 he visited the United States, where he lectured at the University of Michigan. Later, his health deteriorated and he was forced to give up all musical activities except composing. One of the early influences in his music was Sanskrit literature, a phase that saw the composition of a chamber opera, *Savitri*, and a grand opera, *Sita*. Subsequently, English folk song affected his writing; in this vein he wrote some of his finest works, including an opera, *At the Boar's Head*. Toward the end of his life he began experimenting with nonharmonic counterpoint and free tonalities. His operas, with years of composition: *Sita* (1906); *Savitri* (1908); *The Perfect Fool* (1921); *At the Boar's Head* (1924); *The Tale of the Wandering Scholar* (1929).

Holy Grail, The, a legendary cup supposed to have been used at the Last Supper. It appears in Wagner's *Parsifal*.

Homer, Greek epic poet who lived about the 9th century B.C., but whose specific dates and places of birth and death are uncertain. His epics, *The Iliad* and *The Odyssey* were sources for the following operas: Robert Heger's *Der Bettler Namelos;* Vittorio Gnecchi's *Cassandra;* August Bungert's cycle of six operas *Homerische Welt;* Gabriel Fauré's *Pénélope;* Hermann Reutter's *Odysseus*.

Homer, Louise (born LOUISE DILWORTH BEATTY), contralto. Born Sewickley, Pennsylvania, April 28, 1871; died Winter Park, Florida, May 6, 1947. For almost two decades she was one of the great stars of the Metropolitan Opera, an imperial figure during the "golden age" of opera in America. The daughter of a Presbyterian minister, she showed marked musical talent as a child. While attending the New England Conservatory she met and fell in love with the composer Sidney Homer, her teacher in theory and harmony. They were married in 1895. Conscious of his wife's talent, he took her to Paris, where she studied with Mme. Fidèle

König and Paul Lhérie. Her opera debut took place in Vichy, in 1898, in *La favorita*. She next sang at Covent Garden (1899–1900). Following an engagement at the Théâtre de la Monnaie, she returned to the United States. Her American debut took place on November 14, 1900, in the role of Amneris, in San Francisco, during a national tour of the Metropolitan Opera. A month later she appeared in the same role in New York. She was a participant in many of the historic events at the Metropolitan during the early years of the present century. She was the Voice in the first performance of *Parsifal* outside Bayreuth, in 1903. She was Maddalena in the performance of *Rigoletto* in 1903 in which Caruso made his American debut. When Arturo Toscanini appeared at the Metropolitan Opera for the first time, she was Amneris. Some of the most important premières and revivals in which she was featured included *Armide* (Gluck), *La dame blanche, The Gypsy Baron, Hansel and Gretel, Die Königskinder, Madama Butterfly, Manru,* and *Orfeo ed Euridice* (Gluck).

Her resignation from the Metropolitan Opera in 1919 by no means ended her career. She appeared with other major American opera companies (notably the Chicago Opera, where she sang for three seasons) and on December 13, 1927, she was back on the stage of the Metropolitan for a guest appearance in *Aïda*. In 1929 she returned there for the last time, in a performance of *Il trovatore*. As she closed her career there were few to deny that she had been a worthy partner of the great—that she could indeed be considered among the greatest. She combined the highest artistic integrity with remarkable versatility, beauty of voice and diction with a majestic stage presence. She was a prima donna in the grand manner. After her retirement, she devoted herself to her family.

The story of the Homers' happy marriage was told by Sidney Homer in *My Wife and I* (1939).

Home, Sweet Home, a celebrated song with words by John Howard Payne and music by Sir Henry Rowley Bishop. It originated as an aria in Bishop's comic opera *Clari, the Maid of Milan* (1823). Another version of the same melody is found in Donizetti's *Anne Boleyn*. Some prima donnas have interpolated Bishop's song in the Letter Scene of *The Barber of Seville*.

Honegger, Arthur, composer. Born Le Havre, France, March 10, 1892. Educated in the Conservatories of Zurich and Paris, he did not become known until after World War I, when his name was linked with those of five young French composers (including Milhaud and Poulenc). This avant-garde group was for some years known as the "Six." Honegger achieved his first substantial success with an oratorio, *Le Roi David,* in 1921. One of his most important works of this period was the opera *Judith,* introduced in Monte Carlo in 1926. His later works reveal a deep religious feeling and mysticism. He visited the United States for the first time in 1929, appearing as guest conductor of his own works. He remained in Paris during World War II. In 1947 he returned to America to conduct a class in composition at the Berkshire Music Center. He is married to the concert pianist Andrée Vaurabourg. His operas: *Judith* (1925); *Antigone* (1927); *Amphion* (1928); *L'Aiglon* (1937) (a collaboration with Jacques Ibert); *Nicolas de Flue* (1939); *Charles le Tréméraire* (1944). His dramatic oratorio *Jeanne d'arc au bûcher* has sometimes been performed as an opera.

Hoo-Chee, Hoo-Tsin's son (contralto) in Leoni's *L'Oracolo.*

Hooker, Brian, dramatist and librettist. Born New York City, 1880; died there December 28, 1946. Best known for his translation of *Cyrano de Bergerac,*

Hooker was also the author of librettos for Horatio Parker's operas *Fairyland* and *Mona.* He also wrote the libretto for Louis Gruenberg's radio opera, *Green Mansions,* based on the novel of W. H. Hudson.

Hoo-Tsin, a wealthy merchant (bass) in Leoni's *L'Oracolo.*

Hopak (or Gopak), a spirited Russian dance with two beats to the measure. A celebrated example in opera occurs in Mussorgsky's *The Fair at Sorochinsk.*

Horch, die Lerche, Fenton's romance in Act II of Nicolai's *The Merry Wives of Windsor.*

Ho sete! Ho sete! Love duet of Avito and Fiora in Act II of Montemezzi's *L'amore dei tre re.*

Hotter, Hans (baritone). Born Offenbach-on-the-Main, Germany, January 19, 1909. As a boy he sang in church choirs; as a young man he served as church organist and choir master. He pursued the study of church music at the Munich Academy of Music, after which he decided to enter the operatic field. His initiation took place in Troppau and Breslau, after which he was engaged by the Berlin Opera. After a season at the Prague Opera, he was engaged by the Munich Opera as principal baritone, achieving such success there that he was given the honorary title of Kammersänger. He appeared in the premières of Richard Strauss's *Der Friedenstag* and *Capriccio.* Still a member of the Munich Opera, he has made guest appearances in Barcelona, Amsterdam, Antwerp, and at Covent Garden. He joined the Metropolitan Opera in 1950. He has also distinguished himself on the concert stage as an interpreter of lieder.

Hua-Quee, a nurse (contralto) in Leoni's *L'Oracolo.*

Hubay, Jenö, violinist and composer. Born Budapest, September 14, 1858; died Vienna, March 12, 1937. After studying with Joseph Joachim at the Berlin Hochschule he made appearances as violinist throughout Europe. He combined a career as virtuoso with that of teacher, serving as professor of the violin class at the Brussels Conservatory and the Budapest Conservatory; from 1919 to 1934 he was the director of the latter institution. His first opera, *Alienor*, was produced in Budapest in 1891. Two years later his most successful opera, *Der Geigenmacher von Cremona*, was produced in Budapest and acclaimed. Other operas: *Der Dorflump* (1896); *Moosröschen* (1903); *Lavottas Liebe* (1906); *Anna Karenina* (1915). Hubay was knighted in 1907.

Hüsch, Gerhard (baritone). Born Hanover, Germany, February 2, 1901. He started singing lessons in his nineteenth year with a local teacher, continuing his studies at the Opera School of the Hanover Conservatory. His opera debut took place in 1923 at the Osnabrück Theater. He then became a member of the Bremen State Theater and the Cologne Opera, distinguishing himself in both places in Mozart's operas. In 1930 he was engaged as leading baritone of the Charlottenburg Opera in Berlin. He subsequently made numerous appearances in the leading opera houses of Europe, including Bayreuth, where in 1931 he was selected by Toscanini for the role of Wolfram. Hüsch continued his career in Europe after World War II.

Hugh the Drover, ballad opera by Ralph Vaughan Williams. Libretto by Harold Child. Première: London, July 4, 1924. In the Cotswold, about 1812, Mary is in love with Hugh the Drover, but her father wants her to marry John the Butcher. In a fight to decide who will win Mary, Hugh is the victor, but before he gains his prize he is accused by Mary's father of being a Napoleonic spy. When a sergeant recognizes Hugh as a friend and a loyal subject, he absolves him and instead conscripts John the Butcher. Mary's father is now willing to have her marry Hugh.

Hugo, Victor, dramatist, novelist, poet. Born Besançon, France, February 26, 1802; died Paris, May 22, 1885. A leading figure in the French romantic movement, Hugo wrote many novels and dramas that have been effectively adapted as operas. These include: *Angelo, le tyran de Padoue* (Ponchielli's *La Gioconda*, César Cui's *Angelo*, Alfred Bruneau's *Angelo, le tyran de Padoue*); *Hernani* (Verdi's *Ernani*); *Le roi s'amuse* (Verdi's *Rigoletto*); *Lucrezia Borgia* (Donizetti's *Lucrezia Borgia*); *Marie Tudor* (Balfe's *The Armourer of Nantes*, and an opera by Vladimir Kashperov); *Marion Delorme* (operas by Giovanni Bottesini, Carlo Pedrotti, Ponchielli); *Mazeppa* (opera by Felipe Pedrell); *Notre Dame de Paris* (Dargomizhsky's *Esmeralda*, William Fry's *Notre Dame de Paris*, Felipe Pedrell's *Quasimodo*, Arthur Goring Thomas' *Esmeralda*); *Ruy Blas* (opera by Filippo Marchetti).

Huguenots, Les (The Huguenots), opera in five acts by Giacomo Meyerbeer. Libretto by Eugène Scribe and Emile Deschamps. Première: Paris Opéra, February 29, 1836. American première: Théâtre d'Orléans, New Orleans, April 29, 1839.

Characters: Count de St. Bris, a Catholic nobleman (bass); Count de Nevers, another Catholic nobleman (baritone); Raoul de Nangis, a Huguenot nobleman (tenor); Marcel, his servant (bass); Marguerite de Valois, betrothed to Henry IV of Navarre (soprano); Valentine, the daughter of St. Bris (soprano); Urbain, Marguerite's page (soprano or contralto); ladies, gentlemen, citizens, soldiers, students, monks. The action takes place in Touraine and Paris in 1572.

Act I. The House of Count de Nevers. The brief orchestral prelude consists largely of the Lutheran chorale, "Ein' feste Burg," which is quoted throughout the opera as symbol of militant Protestantism. In an effort to reconcile the Catholics and the Huguenots, the Count de Nevers has invited to his house the Huguenot nobleman, Raoul de Nangis, to meet his Catholic friends. The men begin to speak about their favorite ladies. Raoul describes one whom that very morning he saved from danger, and whose identity is unknown to him (Romanza: "Plus blanche que la blanche hermine"—"Bianca al par hermine"). Raoul's servant then tactlessly sings "Ein' Feste Burg," after this a battle ditty, continuing with a song about the Huguenots' ultimate triumph over monks and priests (Chanson Huguenote: "Pour les couvents c'est fini"—"Finita è per frati"). Suddenly, a veiled lady comes seeking the Count. Raoul recognizes her as the woman he saved earlier. He does not know that she has come to beseech the Count, her fiancé, to release her from her marriage vows since she is now in love with Raoul. After the woman disappears a page salutes the nobleman (Page's Song: "Nobles seigneurs salut!"—"Lieti signori, salute"). He has come for Raoul and makes the strange request that Raoul accompany him blindfolded to an unspecified destination.

Act II. The garden of Marguerite's castle at Chenonçeaux. Marguerite de Valois, betrothed to Henry IV, rhapsodizes over the beauty of the Touraine countryside ("O beau pays de la Touraine"—"O vago suol della Turrena"). Valentine arrives to inform her that Count de Nevers has consented to breaking their engagement, and that she (Valentine) is now free to marry Raoul. Marguerite is delighted, for through this marriage she hopes to cement the friendship of the Catholics and the Huguenots. Raoul enters blindfolded. When he discovers he is in the presence of Marguerite, he offers her his services and is told that he is to marry Valentine. Raoul consents to do so, even though he does not know who Valentine is. When Valentine is

brought to him, Raoul denounces her, for, recognizing her as the veiled lady, he is convinced she is the mistress of Count de Nevers. His behavior outrages the Catholic nobleman Count de St. Bris, who challenges Raoul to a duel. Catholics confront Huguenots menacingly.

Act III. A square in Paris. Valentine, spurned by Raoul, is marrying Count de Nevers. The bridal procession comes to the square to the strains of an "Ave Maria." Marcel brings the Count Raoul's challenge to a duel. Meanwhile, within the chapel, Valentine overhears a Catholic plot to assassinate Raoul. Frantically, she appeals to Marcel to warn his master. A fight between Huguenots and Catholics threatens and is only stopped by the timely arrival of Marguerite. Raoul learns from Marguerite that Valentine is innocent. Too late, he realizes how deeply he has wronged the woman he loves.

Act IV. A room in the castle of Count de Nevers. Valentine is heartbroken for her lost Raoul ("Parmi les pleurs"—"In preda al duol"). Raoul suddenly appears: he has risked his life to see Valentine. When De Nevers, St. Bris, and other Catholic leaders arrive, Raoul hides and overhears their plot to massacre the Huguenots. Only Count de Nevers refuses to be a party to the slaughter, and for this stand he is taken into custody. Three monks come to bless the Catholics (Benediction of the Swords: "Glaives pieux, saintes épées" —"Nobili acciar, nobili e santi"). When the Catholics depart, Raoul emerges from hiding and he and Valentine exchange ardent vows of love ("Oh ciel, où courez-vous?"—"O ciel, dove vai tu?"). He then leads Valentine to the window where, after a signal of tolling bells, the massacre begins. Valentine collapses. (The Italian version of the opera ends at this point, the lovers perishing in a burst of firing from the street.)

Act V. In the original French version the opera continues with scenes showing the course of the St. Bartholomew's Day Massacre, the hasty Catholic marriage of Valentine and Raoul, and finally their deaths at the hands of a mob led by St. Bris. Only at the end does St. Bris realize that he has been responsible for the death of his beloved daughter.

In *Les Huguenots* Meyerbeer is a showman par excellence. He has a blood and thunder story, and his music matches the text in drama, splendor, and ceremony. But it would be a mistake to consider the opera only a spectacle, for it combined theatricalism with profound depths of feeling. Wagner, whose artistic thinking was antithetical to that of Meyerbeer, had to concede that the fourth act (particularly the love duet) was among the finest moments in opera. Others, too, consider this act the high point of Meyerbeer's art. To Arthur Hervey, if Meyerbeer had written nothing else, "he would still be entitled to rank as one of the greatest dramatic composers of all time."

Les Huguenots was Meyerbeer's second French opera, coming five years after *Robert le Diable*. It had a brilliant première in Paris and was an outstanding success.

Humperdinck, Engelbert, composer. Born Siegburg, Germany, September 1, 1854; died Neusterlitz, September 27, 1921. He wrote *Hansel and Gretel,* one of the earliest operatic fairy tales and to this day one of the best. Originally a student of architecture, when he was twenty-five he heard music he had written for a Goethe play performed, and this convinced him to become a professional musician. Four years of study at the Cologne Conservatory followed. Winning the Mozart Award enabled him to go to Munich for additional study with Franz Lachner and

Josef Rheinberger at the Royal School. The Mendelssohn Prize, in 1879, provided him with funds for travel in Italy. During this trip he met Richard Wagner who, from this point on, exerted a powerful influence on the young man. He invited Humperdinck to Bayreuth to assist him in preparing the first performance of *Parsifal.* Humperdinck served as stage manager and helped write out the orchestral parts of *Parsifal.* He also conducted Wagner's youthful symphony, a performance arranged by Wagner to honor his wife Cosima during their visit to Venice.

Humperdinck's career as a teacher began in 1885 when he became a professor at the Barcelona Conservatory. From 1890 to 1896 he taught at the Hoch Conservatory in Frankfort-on-the-Main. It was during this period that he wrote the work that made him famous, *Hansel and Gretel,* introduced in Weimar on December 23, 1893, with phenomenal success. The opera was soon heard throughout Germany, as well as in London and New York. This success enabled Humperdinck to give up teaching and concentrate on composition. But in 1900 he returned to pedagogy as director of the Akademieschule in Berlin. In the same year he was elected to the Senate of the Royal Academy, becoming its president in 1913. During the next few years Humperdinck wrote several more operas, all of them failures. Discouraged, he abandoned opera for a while to write incidental music for various theatrical productions. One of these scores was later augmented into an opera, *Die Königskinder,* whose world première took place at the Metropolitan Opera in 1910, with the composer present.

His operas: *Hansel and Gretel* (1893); *Dornröschen* (1902); *Die Heirat wider Willen* (1905); *Die Königskinder* (1910); *Die Marketenderin* (1914); *Gaudeamus* (1919).

Hunchback of Notre Dame, The (Notre Dame de Paris), see HUGO, VICTOR.

Hunding, Sieglinde's husband (bass) the slayer of Siegmund in Wagner's *Die Walküre.*

Huntsmen's Chorus, see WAS GLEICHT WOHL AUF ERDEN.

Huon de Bordeaux, Sir, Duke of Guienne (tenor) in Weber's *Oberon.*

Hurr, hopp, hopp, hopp, the Witch's song (Hexenlied) in Act III of Humperdinck's *Hansel and Gretel.*

Hurricane, The (L'uragano), opera by Lodovico Rocca. Libretto by E. Possenti, based on a play by A. N. Ostrowski. Première: La Scala, February 8, 1952. The Ostrowski play from which this opera was derived was also the basis of an opera by Janáček, *Kata*

Kabanová. The gloomy text describes the disintegration of Catherine, wife of a mentally disturbed drunkard, and sister of a sex-ridden neurotic bent on bringing about the former's ruin. Rocca's chief musical device, as explained by Newell Jenkins, is his "use of insistent ostinato rhythms in which neither the melodic reiteration nor the harmonic scheme remain the same for two consecutive phrases, thereby avoiding tiresome repetition."

Hymne de joie, the Hebrews' song of victory in Act I of Saint-Saëns' *Samson et Dalila.*

Hymn to Champagne, see DIE MAJESTÄT WIRD ANNERKANT.

Hymn to the Sun (Salut à toi, soleil), aria of the Queen of Shemakha in Act II of Rimsky-Korsakov's *Le coq d'or.*

I

Iago, Otello's lieutenant (baritone) in Verdi's *Otello.*

Ibert, Jacques, composer. Born Paris, August 15, 1890. His study of music was intermittent until his twenty-first year, when he entered the Paris Conservatory. Here his teachers included Roger-Ducasse and Fauré, and he won the Prix de Rome in 1919. Success came to him soon after his return from Italy with *The Ballad of Reading Gaol,* a symphonic poem, introduced in Paris in 1922. His most famous orchestral work, *Escales,* followed two years later. In 1937 Ibert became the first musician appointed as director of the Académie de France in Rome. After World War II he combined this office with that of assistant director of the Paris Opéra. Ibert visited the United States in 1950 to conduct master classes in composi-

tion at the Berkshire Music Center at Tanglewood, Massachusetts. In conjunction with this visit, his opera *Le Roi d'Yvetot* received its American première. Ibert's operas: *Andromède et Persée* (1920); *Angélique* (1927); *Le Roi d'Yvetot* (1930); *Gonzague* (1935); *L'Aiglon* (1937); (written in collaboration with Arthur Honegger); *La famille cardinal* (1938); *Barbe-Bleue* (1943); *Le chevalier errant* (1949).

Ibsen, Henrik, playwright and poet. Born Skien, Norway, March 20, 1828; died Christiana (now Oslo), May 23, 1906. One of the earliest important social dramatists, Ibsen was the author of *Peer Gynt,* the source of operas by Werner Egk, Leslie Heward, and Victor Ullmann. Wilhelm Stenhammer's opera *The Feast at Solhaug* is derived from the Ibsen drama of the same name.

Ibbetson, Colonel, Peter Ibbetson's uncle (baritone) in Deems Taylor's *Peter Ibbetson.*

Ice-Heart, see QUEEN ICE-HEART.

Ich baue ganz auf deine Stärke, Belmonte's aria in Act III of Mozart's *The Abduction from the Seraglio.*

Ich gehe, doch rathe ich dir, duet of Osmin and Blonde in Act II of Mozart's *The Abduction from the Seraglio.*

Ich sah das Kind an seiner Mutter Brust, Kundry's monologue in Act II, Scene 2, of Wagner's *Parsifal.*

Idamante, son (mezzo-soprano) of the King of Crete in Mozart's *Idomeneo.*

Idomeneo, rè di Creta (Idomeneo, King of Crete), opera by Mozart. Libretto by Giambattista Varesco, based on a French libretto by Danchet for an opera by André Campra, *Idoménée.* Première: Munich, January 29, 1781. Returning from the Trojan wars, the fleet of the King of Crete is ravaged by a storm. Emerging from this crisis, the king vows to sacrifice the first person meeting him when he reaches home. The person turns out to be his son, Idamante. In an effort to save his child, the king sends him away; but just as Idamante embarks on his ship, a terrible storm erupts. Idomeneo, realizing that the gods are punishing him, confesses everything to the High Priest. At this point, Idamante is ready to sacrifice himself, but his beloved, Ilia, offers herself in his stead. The gods announce that they will forgive everything if Idomeneo abdicates, if Idamante becomes the new ruler, and if Ilia marries the new king. The opera is rarely performed, but the overture and ballet music are sometimes performed at concerts. Among the more significant vocal numbers are Idamente's aria "Non ho colpa" in Act I, Scene 1; Idomeneo's aria, "Vedrommi intorno," in Act I, Scene 2; and Ilia's aria "Zeffiretti lusinghieri," in Act III, Scene 1.

Ifigenia in Aulide, see IPHIGENIA IN AULIS.

Ifigenia in Tauride, see IPHIGENIA IN TAURIS.

Igor, Prince, father (baritone) of Vladimir and husband of Yaroslavna in Borodin's *Prince Igor.*

I got plenty o' nuttin', Porgy's song in Act II of Gershwin's *Porgy and Bess.*

I have attained the highest power, Boris Godunov's aria in Act II, Scene 2, of Mussorgsky's *Boris Godunov.*

Il balen del suo sorriso, Count di Luna's aria in Act II, Scene 2, of Verdi's *Il trovatore.*

Il cavallo scalpita, Alfio's aria in Mascagni's *Cavalleria rusticana.*

Il core vi dono, duet of Guglielmo and Dorabella in Act II, Scene 2, of Mozart's *Così fan tutte.*

Il est beau de mourir en s'aimant, duet of John the Baptist and Salomé in Act IV, Scene 1, of Massenet's *Hérodiade.*

Il est des Musulmans, Mârouf's aria in Act I of Rabaud's *Mârouf.*

Il était une fois à la cour d'Eisenach, the legend of Kleinzach, Hoffmann's aria in the prologue of Offenbach's *The Tales of Hoffmann.*

Il était un roi de Thulé, the ballad of the King of Thule, Marguerite's aria in Act III of Gounod's *Faust.*

Il faut partir, mes bons compagnons (Convien partir, o miei compagni d'arme), Marie's farewell to her regiment in Act I of Donizetti's *The Daughter of the Regiment.*

Ilia, beloved (soprano) of Idamante in Mozart's *Idomeneo.*

Il lacerato spirito, Fiesco's aria in Act I of Verdi's *Simon Boccanegra.*

Iliad, The, see HOMER.

Illica, Luigi, librettist. Born Castellarquato, Italy, 1857; died there December 16, 1919. He began his literary career as a journalist in Milan. After 1892 he wrote the librettos for Catalani's *La Wally,* Franchetti's *Germania* and *Cristoforo Colombo,* Giordano's *Andrea Chénier,* Mascagni's *Iris, Isa-*

beau, and *Le Maschere,* and Gnecchi's *Cassandra,* besides various librettos for the composers Alfano and Montemezzi. In collaboration with Giuseppe Giacosa he wrote the librettos for four of Puccini's operas: *La Bohème, Madama Butterfly, Manon Lescaut,* and *Tosca.*

Il mio tesoro, Don Ottavio's love song in Act II, Scene 2, of Mozart's *Don Giovanni.*

Il partit au printemps, Griselidis's aria in Act II of Massenet's *Griselidis.*

Il segreto per essere felice (Brindisi), Orsini's drinking song in Act II of Donizetti's *Lucrezia Borgia.*

Il tabarro, *see* TABARRO, IL.

Imbroglio, an Italian term meaning an intricate or complicated situation. It is used in music to designate a passage whose combination of themes suggests confusion but is actually carefully contrived. Examples of imbroglios in opera are found at the end of the first and second acts of *Die Meistersinger.*

Im Mohrenland gefangen war, Pedrillo's serenade in Act III of Mozart's *The Abduction from the Seraglio.*

Immortal Hour, The, opera by Rutland Boughton. The libretto is the play of the same name by Fiona Macleod (William Sharp). Première: Glastonbury, England, August 26, 1914. Etain, from fairyland, and the dreamer king of Ireland, Eochaidh, are to be married. Midir, a fairy prince, kisses Etain's hand and enchants her with his legends and songs. She comes under his spell and follows him to the Land of Heart's Desire. Eochaidh falls dead when he is touched by the Shadow God, Dalua.

Impresario, the manager, or director, of an opera company.

Impresario, The, *see* SCHAUSPIELDIREKTOR, DER.

Impressionism, a style of composition in which a sensation or an impression created by a subject is emphasized rather than the subject itself; which is more concerned with subtle nuances and effects than with substance and structure. The style and term originated in the paintings of the Frenchmen Manet, Monet, Pissarro, and Renoir and was applied particularly to the musical style developed by Claude Debussy, whose opera *Pelléas et Mélisande* is the ideal realization of impressionist writing. Dukas's *Ariane et Barbe-Bleue* and Delius' *A Village Romeo and Juliet* are two more operas written in an impressionist style.

Inaffia l'ugola! (Brindisi), Iago's drinking song in Act I of Verdi's *Otello.*

incoronazione di Poppea, L' (The Coronation of Poppea), opera by Claudio Monteverdi. Libretto by Francesco Busenello. Première: Teatro SS. Giovanni e Paolo, Venice, 1642. This is Monteverdi's last opera and one of his most expressive scores. The story concerns Nero's mistress, Poppea, who replaces the Empress and assumes her place on the throne.

Indes galantes, Les (The Indigo Suitors), ballet-opera by Rameau. Libretto by Louis Fuzelier. Première: Paris Opéra, August 23, 1735. This was Rameau's third opera, and a major success. The story concerns four different tales of love, each taking place in a different and remote part of the world.

In des Lebens Frühlingstagen, Florestan's aria recalling his happy days with his wife, Leonore, in Act II, Scene 1, of Beethoven's *Fidelio.*

In dieser feierlichen Stunde, Sophie's prayer in Act II of Richard Strauss's *Der Rosenkavalier.*

In diesen heil'gen Hallen, Sarastro's aria in Act II, Scene 3, of Mozart's *The Magic Flute.*

Indy, Vincent d', *see* D'INDY, VINCENT.

Inez, (1) Leonora's confidante (soprano) in Donizetti's *La favorita.*

(2) Don Diego's daughter (soprano) in Meyerbeer's *L'Africaine.*

(3) Leonora's confidante (soprano) in Verdi's *Il trovatore.*

In fernem Land, Lohengrin's narrative in Act III of Wagner's *Lohengrin.*

Inghelbrecht, Désiré, conductor and composer. Born Paris, September 17, 1880. After completing music study at the Paris Conservatory he began his conductorial career with the orchestra of the Société Nationale de Musique. In 1913 he became music director of the Théâtre des Champs Elysées, where ballets and operas were performed. In 1919 he was engaged as conductor of the newly organized Swedish Ballet. Five years later he became a conductor of the Opéra-Comique, becoming principal conductor in 1932. After World War II he became principal conductor of the Paris Opéra. He wrote one opera: *La nuit vénitienne* (1908).

In grembo a me, *see* SUR MES GENOUX, FILS DU SOLEIL.

In mia man alfin tu sei, Norma's duet with Pollione in Act II of Bellini's *Norma.*

Inigo, Don, a banker (bass) in Ravel's *L'heure espagnole.*

In preda al duol, *see* PARMI LES PLEURS.

In quelle trine morbide, Manon's aria in Act II of Puccini's *Manon Lescaut.*

Inquisitore, L', the Grand Inquisitor of Spain (bass) in Verdi's *Don Carlos.*

Inspector General, The, *see* GOGOL, NIKOLAI.

Intendant, German for "manager" or "director"—the director of a German opera house. The term was applied particularly to opera directors at German courts, these men formerly having been dilettantes of high birth or station.

Intermède, French for "intermezzo."

Intermezzo, (1) in the early Italian theater, a little play with music interpolated between the acts of a serious drama to allow the actors to rest and to permit for changes of scenery. The form became increasingly popular and developed into the opera buffa of the early eighteenth century. Nicola Logroscino was a famous writer in this form. Pergolesi was another: his *La serva padrona* is an intermezzo.

(2) In the nineteenth century the name intermezzo was given to an instrumental interlude played in an opera with the curtain raised, the music denoting a passage of time. Celebrated examples are those in *Cavalleria rusticana, Goyescas,* and *The Jewels of the Madonna.*

(3) Opera or "domestic comedy" by Richard Strauss. Libretto by the composer. Première: Dresden Opera, November 4, 1924. The composer based his text on an actual incident from his own life: a misunderstanding that arose between himself and his wife when an unknown female admirer wrote him an effusive love letter. Convinced that her husband was unfaithful, Frau Strauss announced her intention to seek a divorce. Only when it was discovered that the letter writer had sent her note to the wrong man—she had intended it for one of Strauss's colleagues—was the matter straightened out. At the opera's first performance the principal singers were made up to look like Herr and Frau Strauss.

In the Pasha's Garden, one-act opera by John Laurence Seymour. Libretto by H. C. Tracy, based on a story by H. G. Dwight. Première: Metropolitan Opera, January 24, 1935. The Pasha's wife, Helene, is carrying on a love affair with Etienne. When the Pasha comes unexpectedly, she hides her lover in a large chest. The Pasha and Helene partake of a meal, using the chest as a table. When the meal is over, the Pasha orders that the chest be buried in the garden.

In the Storm, opera by Tikhon Khrennikov. Libretto by the composer, based on a novel by Virta. *Solitude.* Première: Moscow, May 31, 1939. This successful Soviet opera has for its setting the Kulak rebellion in Tambov in 1919–1921; its basic theme is the love of two young collectivists. The opera is a revision of an earlier work entitled *The Brothers*.

In the Town of Kazan, Varlaam's aria In Act II, Scene 1, of Mussorgsky's *Boris Godunov*.

Introduction, in some operas, the opening number that follows the overture.

Introduzione, Italian for introduction.

In uomini, in soldati, Despina's aria in Act I, Scene 3, of Mozart's *Così fan tutte*.

Invano Alvaro, Don Carlo's duet with Alvaro in Act IV, Scene 1, of Verdi's *La forza del destino*.

Invisible City of Kitezh, The, opera by Rimsky-Korsakov. Libretto by I. Bielsky. Première: Maryinsky Theater, St. Petersburg, February 20, 1907. Prince Vsevolod, son of King Yury and joint ruler of Kitezh, meets Fevronia in a forest, falls in love with her, and asks for her hand in marriage. They marry in Kitezh. The Tartars descend on the city and capture the bride. Two Tartars fight for her, but the drunkard Grisha helps her escape. Fevronia flees through a forest haunted by dancing devils and goblins. Exhausted, she sinks to the ground. The spirit of the Prince arrives to lead her back to the holy city where bride and groom are welcomed back by the king and his people.

invocation à la nature, L', *see* NATURE IMMENSE.

Io sono l'umile ancella, Adriana's aria in Act I of Cilèa's *Adriana Lecouvreur*.

Io vidi la luce nel campo guerrier, *see* AU BRUIT DE LA GUERRE.

Io vidi miei signori, *see* J'AI VU, NOBLES SEIGNEURS.

Ipanov, Count Loris, Fedora's lover (tenor) in Giordano's *Fedora*.

Iphigenia in Aulis, tragedy by Euripides, the source of operas by many composers, including Gluck, Antonio Caldara, Karl Heinrich Graun, Niccolò Jommelli, and Nicola Zingarelli. For plot, *see* IPHIGENIE EN AULIDE.

Iphigenia in Tauris, tragedy by Euripides, the source of operas by Gluck, Baldassare Galuppi, Niccolò Jommelli, and Nicola Piccinni, among others. For plot, *see* IPHIGENIE EN TAURIDE.

Iphigénie en Aulide (Iphigenia in Aulis), lyric tragedy in three acts by Christoph Willibald Gluck. Libretto by Bailli du Roullet, based on a drama by Racine, in turn derived from Euripides. Première: Paris Opéra, May 19, 1774.

Characters: Agamemnon, King of the Greeks (baritone); Klytemnestra, his wife (mezzo-soprano); Iphigenia, their daughter (soprano); Achilles, her betrothed (tenor); Patrocolos (bass); Calchas, high priest (bass); Arkas, captain of the guards (bass); Artemis, or Diana, a goddess (soprano). The action takes place in Aulis after the Trojan War.

The famous overture opens with a majestic subject interpreted by Richard Wagner as an invocation for deliverance. Two other basic themes represent to Wagner, in turn, an assertion of the will, and the maidenly tenderness of Iphigenia.

Act I. The camp of Agamemnon. The goddess Artemis is angered and prevents the Greek fleet from leaving for Troy. To appease her, Calchas demands the sacrifice of Iphigenia. When Iphigenia and her mother arrive at the camp, they are welcomed with song and dance. It is now that Klytemnestra tells her daughter that Achilles, betrothed to Iphigenia, has been unfaithful; but when Achilles arrives he reassures his beloved.

Act II. Agamemnon's palace. The marriage of Achilles and Iphigenia is being celebrated. Achilles is about to lead his bride to the altar when Arkas informs him that she must be sacrificed. Klytemnestra entreats Achilles to save Iphigenia, but Agamemnon insists that the sacrifice must take place.

Act III, Scene 1. Agamemnon's tent. With the Greeks demanding that Iphigenia be given up to the gods, Achilles comes to Agamemnon's tent to save her. But Iphigenia insists on her own death.

Scene 2. Altar of Artemis. The goddess announces that Iphigenia's life is, after all, to be spared. Joyfully, Achilles embraces his wife.

Iphigénie en Aulide was the first opera Gluck wrote for Paris after his arrival from Vienna in the fall of 1773. In it he once again carried out the ideals and principles previously established in *Orfeo ed Euridice* and *Alceste:* noble simplicity, humanity, dramatic truth, and integration of music and play. In Paris, as in Vienna, Gluck found opponents of his new ideas, particularly among adherents of the Italian school. These enemies did everything they could to block the impending première of *Iphigénie*. Had it not been for the personal intervention of Marie Antoinette, the opera might not have been produced. Once given, the opera was a sensation. It became the subject of discussion in all Parisian salons. "We can find nothing else to talk about"— so wrote Marie Antoinette; "You can scarcely imagine what excitement reigns in all minds in regard to this event." Commenting on one of the airs, Abbé Arnaud said: "With that air one might found a religion."

Iphigénie en Tauride (Iphigenia in Tauris), opera by Christoph Willibald Gluck. Libretto by François Guillard, adapted from the drama of Euripides. Première: Paris Opéra, May 18, 1779. The opera continues the story begun in *Iphigenia in Aulis*. Iphigenia has become a priestess of the Scythians at Tauris, where the gods, angered, must be appeased with a sacrifice. For this sacrifice two strangers from Greece are chosen, after they are shipwrecked on the shores of Tauris; they are Orestes and Pliades. Since Iphigenia recalls her Greek origin, she is incapable of sanctioning their death. She finally consents to have only one of them sacrificed, and the choice falls on Orestes. Only now

does she discover that Orestes is her brother. When a band of Greeks, headed by Pliades, attacks and defeats the Scythians, Iphigenia and Orestes are able, with the approval of Diana, to return to their native land. One of the opera's most celebrated arias is that of Pliades in Act II, "Unis dès la plus tendre enfance."

Irene, (1) Rienzi's sister (soprano) in Wagner's *Rienzi*.

(2) Jean Gaussin's cousin (contralto) in Massenet's *Sapho*.

Iris, opera by Pietro Mascagni. Libretto by Luigi Illica. Première: Teatro Costanzi, Rome, November 22, 1898. In Japan, the libertine Osaka wants Iris, daughter of the blind man Cieco. He manages to get her absorbed in a puppet show, then has her abducted. Her father, told that she has entered a house of ill repute, believes she has gone there of her own free will and curses her. In the luxurious setting of Yoshiwara, Iris resists Osaka. Her father comes and, finding her, throws mud at her. Humiliated, Iris jumps out of the window into the sewer below. Dying, she bemoans her fate and wonders why tragedy should have befallen her. Then, caressed by the rays of the rising sun and embraced by opening flowers, she is lifted heavenward.

Irma, a seamstress (soprano) in Charpentier's *Louise*.

Irving, Washington, author and humorist. Born New York City, April 3, 1783; died Tarrytown, New York, November 28, 1859. His famous tale "Rip Van Winkle," from *The Sketch Book*, was the source of operas by George Frederick Bristow and Reginald De Koven and into an operetta by Robert Planquette. Spohr's opera *Der Alchemist* was derived from another Irving story.

Isabella, (1) A beautiful Italian lady (contralto) in Rossini's *L'Italiana in Algeri*.

(2) Princess of Sicily (soprano) in Meyerbeer's *Robert le Diable*.

Isepo, a public letter writer (tenor) in Ponchielli's *La Gioconda.*

Island God, The, opera by Gian-Carlo Menotti. Libretto by the composer (originally in Italian but translated into English by Fleming McLeish). Première: Metropolitan Opera, February 20, 1942. On a Mediterranean island, the gods command Ilo to rebuild a temple now in ruins. As he is busy at his task, his beloved Telea falls in love with Luca. Luca and Telea succeed in enmeshing Ilo in a fisherman's net, and before he can extricate himself they escape. Feeling the gods have abandoned him, Ilo destroys what he has built, and the gods destroy him. This is one of Menotti's few operas to originate in a formal opera house; it is also the only one of his operas that has failed to win a following.

Isolde, princess of Ireland (soprano) in love with Tristan in Wagner's *Tristan und Isolde.*

Ist ein Traum, the duet of Sophie and Octavian in Act III of Strauss's *Der Rosenkavalier.*

It ain't necessarily so, Sportin' Life's cynical approach to religion in Act II of Gershwin's *Porgy and Bess.*

Italia! Italia! e tutto il mio ricordo! Archibaldo's aria in Act I of Montemezzi's *L'amore dei tre re.*

Italiana in Algeri, L' (The Italian Lady in Algiers), opera by Rossini. Libretto by Angelo Anelli. Première: Teatro San Benedetto, Venice, May 22, 1813. The Italian woman of the title is Isabella, with whom the Mustafa of Algiers is in love. Complications arise when Isabella falls in love with Lindoro, favorite slave of the Mustafa. The Mustafa is distracted by being initiated into a secret society dedicated to sensuality. While going through the ridiculous rites, Lindoro and Isabella make their escape. The overture is famous—a staple of the concert hall. The outstanding vocal numbers are two arias of Isabella in the second act, "Per

lui che adoro," sung with quartet, and "Cruda sorte, amor tiranno."

Italian Opera House, the first theater in New York built especially for opera. It was erected in 1833 by Lorenzo da Ponte on Church and Leonard Streets, and opened on November 18 with *La gazza ladra.* After a season of opera, the theater was used for spoken drama. It was destroyed by fire in 1835.

Ivanhoe, (1) romantic novel by Sir Walter Scott, the source of several operas, notably Otto Nicolai's *Il Templario,* Heinrich Marschner's *Der Templer und die Jüdin,* and Sir Arthur Sullivan's *Ivanhoe.* The central character is Wilfrid, knight of Ivanhoe, who, during the reign of Richard I, loves, woos, and after varied complications wins, Rowena.

(2) Opera by Sir Arthur Sullivan. Libretto by Julian Sturges based on Scott's novel. Première: Royal English Opera House, London, January 31, 1891. This was Sullivan's solitary excursion into grand opera, and it was a failure. It was the opening performance of the newly founded and short-lived English Opera House, created as the home for native English opera.

Ivan Susanin, *see* LIFE FOR THE CZAR, A.

Ivan the Terrible (The Maid of Pskov), opera by Nikolai Rimsky-Korsakov. Libretto is the drama of the same name by Lev Mey. Première: Maryinsky Theater, St. Petersburg, January 13, 1873. In sixteenth-century Russia, the tyrannical Czar Ivan inflicts terror on the city of Novgorod. A similar fate awaits Pskov. Its subjects join to fight the Czar and in the surprise attack that follows Olga and her lover Tutcha— both natives of Pskov—are killed. Olga turns out to have been the secret daughter of the Czar. The overture and the storm music of Act III are occasionally performed at symphony concerts.

Ivogün, Maria (born INGE VON GÜN-

THER), coloratura soprano. Born Budapest, November 11, 1891. She studied with Irene Schlemmer in Vienna, and in 1913 became principal soprano of the Munich Opera, where she remained for twelve years specializing in coloratura roles. In 1916, Richard Strauss selected her to create the role of Zerbinetta in *Ariadne auf Naxos,* in which she also made her Covent Garden debut in 1924. From 1925 to 1927 she was principal soprano of the Berlin State Opera. After this she appeared extensively in leading European opera houses and with the Chicago Opera. She withdrew from opera in the 1930's because of failing eyesight that soon resulted in total blindness. She henceforth devoted herself to teaching, joining the faculty of the Vienna academy of music in 1948, and the Berlin High School of Music in 1950.

ivrogne corrigé, L' (The Reformed Drunkard), comic opera by Gluck. Libretto by Louis Anseaume, based on a fable of La Fontaine. Première: Burgtheater, Vienna, April 1760. A peasant with an extraordinary love for liquor is cured of this vice by going through an ordeal arranged by his wife and daughter in which he believes he is brought to judgment in Hell before Pluto and the Furies.

J

Jackrabbit, Billy, an Indian (bass) in Puccini's *The Girl of the Golden West.*

Ja, das Alles auf Ehr', Sandor's aria in Johann Strauss's *The Gypsy Baron.*

Jadlowker, Hermann, tenor. Born Riga, Latvia, July 5, 1879; died Tel Aviv, Israel, May 13, 1953. After attending the Vienna Conservatory he made his opera debut in Cologne, in 1899, in Kreutzer's *Das Nachtlager von Granada.* For five years he appeared in principal tenor roles at the Berlin Opera and for several seasons after that at the Vienna Opera. He made his American debut at the Metropolitan Opera on January 22, 1910, in *Faust.* He remained at the Metropolitan for three seasons, creating the principal tenor roles in *Le donne curiose, Die Königskinder,* and *Lobetanz.* In 1913 he returned to the Berlin Opera for another six seasons. After leaving the operatic stage, he served as cantor and professor of singing in Riga. In the last fifteen years of his life he taught singing in Israel.

Jagel, Frederick, tenor. Born Brooklyn, New York, June 10, 1897. After study with Portanova in New York and Cataldi in Italy he made his opera debut in Leghorn in 1924 in *La Bohème,* billed under the Italianized name of Federico Jeghelli. During the next four years he appeared 194 times in leading Italian opera houses. On November 8, 1927, he made his Metropolitan Opera debut in *Aïda.* For over two decades he was seen there in a large variety of roles, including the world première of *In the Pasha's Garden* and the American premières of *La notte di Zoraima* ar.d *The Fair at Sorotchinsk.* With another company he appeared in the American première of Rocca's *The Dybbuk* in 1935, and a year or so later he was the first American tenor to appear in a leading role at the Teatro

Colón on the opening night of the Buenos Aires opera season.

J'ai le bonheur dans l'âme, duet of Hoffmann and Antonia in Act III of Offenbach's *The Tales of Hoffmann.*

J'ai vu, nobles seigneurs (Io vidi, miei signori), Vasco da Gama's aria in Act I of Meyerbeer's *L'Africaine.*

Jake, a fisherman (baritone) in Gershwin's *Porgy and Bess.*

Ja! lasst uns zum Himmel die Blicke erheben, concluding chorus of Weber's *Der Freischütz.*

Janáček, Leoš, composer. Born Hukvaldy, Moravia, July 3, 1854; died Prague, August 12, 1928. Sometimes called the "Mussorgsky of Moravia," Janáček evolved in his operas a personal musical speech based on Moravian peasant music and the inflections of the Bohemian language which he termed "melodies of the language." After music study in Brünn and at the College of Organ Playing in Prague, Janáček settled in Brünn as a teacher of music. In 1878 he visited Vienna for additional study, after which he assumed a major position in the musical life of Moravia by establishing and conducting public concerts and founding and subsequently directing the Organ School and Conservatory in Brünn. In 1896 he paid his first visit to Russia. His interest in the Russian language and literature was now aroused; henceforth their impact on his musical thinking was profound. A second and even more significant influence was that of Moravian folk music, of which he made an intensive study. This study led him to create a musical system of his own, derived from folk elements, in which melody and rhythm were molded after the inflections and rhythms of speech. The crystallization of this new style is found in his greatest work, the opera *Jenufa,* which took him seven years to write. Janáček wrote orchestral, choral, and chamber works in addition to operas, and he achieved

considerable recognition. In 1925 he received an honorary degree from the University of Brünn, and on the occasion of his seventieth birthday a cycle of his operas was performed in Brünn. Two years after his death, an extensive cycle of his operas was performed throughout Czechoslovakia. His operas: *Sarka* (1887); *Jenufa* (1903); *Fate* (1905); *The Excursions of Mr. Brouček* (1914); *Kate Kabanová* (1921); *The Cunning Little Vixen* (1923); *The Makropoulos Affair* (1924); *From a House of the Dead* (*Aus einem Totenhaus*), completed by Břetislav Bakala (1930).

Janssen, Herbert, baritone. Born Cologne, Germany, September 22, 1895. He studied with J. Daniel in Berlin and in 1924 made his debut with the Berlin State Opera, where he appeared for several years in principal baritone roles of the Germany repertory. After successful guest appearances in Europe and South America he made his American debut on January 24, 1939, at the Metropolitan Opera as Wotan in *Siegfried.* For the next decade he gave distinguished performances at Covent Garden, the Paris Opéra, the Vienna State Opera, and other major opera houses.

Janusz, Halka's beloved (baritone) in Moniuszko's *Halka.*

Jaquino, the jailer's assistant (tenor) in Beethoven's *Fidelio.*

Jaroslavna, Prince Igor's wife (soprano) in Borodin's *Prince Igor.*

Jasager, Der (The Boy Who Said Yes), children's opera by Kurt Weill. Libretto by Bertholt Brecht. Première: Berlin, 1930. This is a school opera meant to be performed by children. After its première in Berlin it was played in five hundred schools throughout Germany. The subject, derived from the Japanese, emphasizes the necessity of sacrificing the weaker individuals of society for the good of humanity.

Jazz, a style of American popular music which grew out of the ragtime playing of Negro bands and reached its heyday in the 1920's. Significant elements of jazz, distinguishing it from other kinds of American popular music, include its marked syncopations against a steady four-beat rhythm, prominence of certain "blue" notes (combined major and minor thirds and the flatted seventh), characteristic "breaks" or comments by solo instruments, and the cultivation of strange tone qualities. During the 1920's there was a vogue in Germany for writing operas in the jazz idiom, and among the works produced there at that time were George Antheil's *Transatlantic,* Paul Hindemith's *Neues vom Tage,* Ernst Křenek's *Jonny spielt auf,* and Kurt Weill's *The Rise and Fall of Mahogany* and *The Threepenny Opera.*

Jean, a juggler (tenor or soprano) in Massenet's *Le Jongleur de Notre Dame.*

Jeanne d'Arc au bûcher (Joan of Arc at the Stake), dramatic oratorio by Arthur Honegger. Libretto by Paul Claudel. Première: Basel, Switzerland, May 10, 1938. This work is intended for concert performance; but it has often been performed with scenery and costumes and presented as an opera—or, in the composer's designation, as a "mimodrama." Given thus, it is usually performed on two separate stages: on one, Joan is seen fastened to the stake throughout the entire performance, while the action of the play proceeds on the other. In 1953 the San Carlo Opera of Naples presented this work in a stage production directed by Roberto Rossellini and starring Ingrid Bergman, in the central speaking role. In 1954 the San Francisco Opera gave a similar performance with Dorothy Maguire.

See also JOAN OF ARC.

Je connais un pauvre enfant (Styri-enne), Mignon's aria in Act II, Scene 1, of Thomas's *Mignon.*

J'écris à mon père (Duo de la lettre), duet of Des Grieux and Manon in Act II of Massenet's *Manon.*

Je crois entendre, Philine's aria in Act II, Scene 1, of Thomas's *Mignon.*

Je crois entendre encore, Nadir's aria in Act I of Bizet's *Les pêcheurs de perles.*

Je dis que rien ne m'épouvante, Micaëla's air in Act III of Bizet's *Carmen.*

Je marche sur tous les chemins, Manon's aria in Act III, Scene 1, of Massenet's *Manon.*

Je me souviens, sans voix, inanimée, Gerald's aria in Act III of Delibes's *Lakmé.*

Jemmy, William Tell's son (soprano) in Rossini's *William Tell.*

Jenufa, folk opera by Leoš Janáček. Libretto by Gabriela Preissová. Première: Brünn Opera, January 21, 1904. This is not only Janáček's greatest work but one of the most significant Bohemian folk operas. Jenufa is a peasant girl who bears a child to her stepbrother, Stewa. Stewa no longer loves her, but his brother, Laca, is willing to marry Jenufa and accept the child as his own. Jenufa's mother, refusing to let Laca make such a sacrifice, murders the child, then tells Jenufa that it died a natural death and was quickly buried. Jenufa and Laca are married. During the ceremony, the dead body of the child is discovered. The mother confesses her crime and is arrested. The opera's musical style features recitatives shaped from speech patterns, but a more formal lyricism is not abandoned as, for example, Jenufa's "Ave Maria" in Act I. For many years *Jenufa* enjoyed an immense success in European opera houses, following a successful revival at the Prague Opera in 1916 and a Viennese performance in 1918.

Jepson, Helen, soprano. Born Titusville, Pennsylvania, November 25, 1906. A scholarship student, she attended the Curtis Institute in Philadelphia for five years, and made her opera debut in 1928 with the Philadelphia Civic Opera as Marcellina in *The Marriage of Figaro*. After appearances with the Philadelphia Grand Opera and the Montreal Opera, she made her Metropolitan Opera debut on January 24, 1935, in the world première of *In the Pasha's Garden*. She was a featured soprano at the Metropolitan for the next seven years. During this period she also appeared with the Chicago Civic Opera (where she scored a major success as Thaïs, a role she studied with Mary Garden) and with the San Francisco Opera. She left the Metropolitan after the 1946–1947 season; since 1948 she has appeared as a lecturer about that institution. She has also been a faculty member of Fairleigh-Dickinson College and has taught voice privately.

Je ris de me voir, the Jewel Song, Marguerite's aria in Act III of Gounod's *Faust*.

Jeritza, Maria (born JEDLITZKA), soprano. Born Brünn, Moravia, October 6, 1887. She studied singing at the Brünn Musikschule and privately with Auspitzer. In 1910 she made her debut with the Olmütz Opera in *Lohengrin*. A year later she appeared with the Vienna Volksoper, where she scored a striking success as Elisabeth. On October 25, 1912, she created the title role in Richard Strauss's *Ariadne auf Naxos* in Stuttgart, and in the summer of the same year she appeared in a special performance of *Die Fledermaus* before the Emperor. In the fall of 1912 she appeared for the first time at the Vienna Royal Opera in an opera written expressly for her, Max von Oberleithner's *Aphrodite*. She was so successful that the Royal Opera bought out her contract with the Volksoper. She now became one of the stars of the Royal Opera (later State Opera) remaining until 1932, achieving personal triumphs in operas by Puccini, Korngold, and Richard Strauss, as well as in the Wagner repertory. Puccini regarded her as the ideal Tosca, and she was Richard Strauss's favorite interpreter of his leading female roles. On November 19, 1921, she made her American debut at the Metropolitan Opera in the American première of *Die tote Stadt*. A few days later she created a sensation in *Tosca*. She remained at the Metropolitan ten years, scoring striking success in *Thaïs, Turandot, Fedora,* and *Der Rosenkavalier*. She created for America the principal female roles in *Jenufa, Madonna Imperia, Turandot,* and *Violanta;* she was featured in revivals of *Boccaccio, Donna Juanita, The Girl of the Golden West, Der fliegende Holländer,* and *Thaïs*. A woman of striking beauty and great charm, possessed of a voice of singular beauty, she brought to the operatic stage an unforgettable presence. In 1932 she resigned from the Metropolitan and for the next few years made appearances in Europe and America. In 1935 she married the motion-picture executive Winfield Sheehan and withdrew from opera. Several years later, however, she made a few more appearances, singing for a single evening at the Metropolitan, in 1951, in *Die Fledermaus*. Her autobiography, *Sunlight and Song,* was published in 1924.

Jerum! Jerum! (Cobbler's Song), Hans Sachs' song in Act II of Wagner's *Die Meistersinger*.

Jerusalem Delivered, epic poem by Tasso, the source of numerous operas. *See* ARMIDA; RINALDO; TASSO.

Jessner, Irene (soprano). Born Vienna, about 1910. She attended the Vienna Conservatory, after which she made her opera debut in Teplitz as Elsa. For two years she was a member of the Prague Opera. On December 25, 1936, she made her American debut at the

Metropolitan Opera in *Hansel and Gretel.* She has remained with the Metropolitan Opera company for over a decade, appearing in leading soprano roles in the German, French, and Italian repertories.

Jessonda, opera by Ludwig Spohr. Libretto by Eduard Heinrich Gehe, based on a tragedy by Antoine Martin Lemièrre, *La veuve de Malabar.* Première: Kassel, July 28, 1823. Following a custom in India, Jessonda is fated to be burned alive with the dead body of her husband, the Rajah. She is saved by the Portuguese general, Tristan d'Acumba, whom she had loved years before and whom she suddenly meets again when the Portuguese attack her city.

Together with *Euryanthe, Jessonda* is one of the earliest German operas to employ accompanied recitatives instead of spoken dialogue.

Je suis encore tout étourdie, Manon's aria in Act I of Massenet's *Manon.*

Je suis l'oiseau, Alain's aria in Act III of Massenet's *Grisélidis.*

Je suis Titania, Philine's Polonaise in Act II, Scene 2, of Thomas's *Mignon.*

Je veux vivre dans ce rêve, Juliette's waltz song in Act I of Gounod's *Roméo et Juliette.*

Jewels of the Madonna, The (I gioielli della Madonna), opera in three acts by Ermanno Wolf-Ferrari. Libretto by the composer, with verses by Carlo Zangarini and Enrico Golisciani. Première: Kurfürstenoper, Berlin, December 23, 1911. American première: Chicago Auditorium, January 16, 1912; a performance by the Chicago Opera Company under the composer's direction.

Characters: Gennaro, a blacksmith (tenor); Carmela, his mother (mezzo-soprano); Maliella, Carmela's adopted daughter (soprano); Rafaele, leader of the Camorrists (baritone); Biaiso, a public letter writer (tenor); Cicillo, a Camorrist (tenor); Rocco, a Camorrist (bass); Stella, a Camorrist (soprano); Concetta, a Camorrist (soprano); Serena, a Camorrist (soprano); citizens of Naples. The action takes place in Naples at the beginning of the nineteenth century.

Act I. A public square. While the festival of the Madonna is being celebrated, Maliella is coquettishly flirting with Rafaele, who is madly in love with her. He stands ready to meet any test to prove his love—even to stealing the jewels of the Madonna image that has just been carried through the streets. Gennaro, also in love with Maliella, overhears this rash promise of Rafaele's and realizes that he has the information to help him win for himself the woman he loves.

Act II. A garden. When Gennaro tries to make love to Maliella, he is rudely rejected. He leaves in despair. Rafaele and some of his Camorrist friends now appear. Rafaele sings a serenade ("Aprila, o bella"). He wins over Maliella completely and she promises to run away with him the following day. When Rafaele and his friends depart, Gennaro returns. He has stolen the Madonna's jewels, hoping thereby to gain Maliella's interest. Maliella puts on the jewels and is so taken with them that, in gratitude, she gives herself to Gennaro.

Act III. A Camorrist hide-out. Reveling, the Camorrists dance a tarantella ("Dance of the Camorristi"). Maliella, sobbing wildly, comes to Rafaele and confesses what has happened. Rafaele is furious; he throws her savagely to the ground. As she falls, the jewels of the Madonna scatter and superstitious terror grips the Camorrists as they recognize them. When the remorseful Gennaro arrives, Maliella points him out as the thief. A sudden storm blows out the candles. Hysterical in the darkness, Maliella rushes away to drown herself in the sea. The Camorrists disperse. Gennaro gathers the scattered jewels and offers them to an image of the Madonna, then kills himself with a dagger.

While the name of Wolf-Ferrari is most often associated with opera buffa, a form in which he was an acknowledged master, he succeeded in producing a tragic opera that has survived. *The Jewels of the Madonna* belongs to the naturalistic school of Italian opera known as "verismo." Most of the impact of the score comes from the realistic manner in which the music portrays a storm, a carnival, or the suicides of Maliella and Gennaro. It is in its dramatic power, rather than its lyricism, that this work is of paramount interest.

Jewel Song, *see* JE RIS DE ME VOIR.

Joan of Arc, the fifteenth century French heroine and martyr whose military leadership of the French compelled the English to lift the siege of Orleans, and who finally was burned at the stake by the English. She appears as the principal character in many operas, notably Balfe's *Joan of Arc,* Norman Dello Joio's *The Triumph of Joan,* Honegger's *Jeanne d'Arc au bûcher,* Rezniček's *Die Jungfrau von Orleans,* Tchaikovsky's *The Maid of Orleans,* and Verdi's *Giovanna d'Arco.*

Jochum, Eugen, conductor. Born Babenhausen, Germany, November 1, 1902. He studied music at the Augsburg Conservatory and conducting with Siegmund von Hausegger in Munich. After holding various minor posts as conductor he became music director of the city of Duisburg in 1930. In 1934 he succeeded Karl Muck as conductor of the Hamburg Opera and Philharmonic. From 1939 to 1949 he was the general music director of the city of Hamburg. He has appeared as conductor in most of the major European opera houses and at the Bayreuth and Munich Festivals. Since 1949 he has conducted the Munich Radio Symphony, which he helped reorganize. In 1950 he was elected president of the German section of the Bruckner Society.

John, (1) Dot's husband (baritone) in Goldmark's *The Cricket on the Hearth.*

(2) A butcher (bass-baritone) in Vaughan Williams' *Hugh the Drover.*

John of Leyden, leader of the Anabaptists (tenor) in Meyerbeer's *Le Prophète.*

Johnson, Dick, an outlaw (tenor) alias Ramerrez, in Puccini's *The Girl of the Golden West.*

Johnson, Edward, tenor and opera manager. Born Guelph, Ontario, August 22, 1881. For over a decade he was a leading tenor of the Metropolitan Opera, and after that its general manager for fifteen years. He began to study music seriously in New York with Mme. von Feilitsch. For a while he earned his living singing in churches. In 1908 he appeared in New York in the leading role of the Oscar Straus operetta, *A Waltz Dream.* He then went to Europe, where for two years he studied with Caruso's coach, Vincenzo Lombardi. He made his opera debut in Padua in 1912, singing in *Andrea Chénier,* billed as Edoardo di Giovanni. After appearing for two seasons in various Italian theaters, he made his debut at La Scala in the first Italian performance of *Parsifal.* Johnson was the leading tenor of La Scala for five seasons, during which time he created the leading tenor roles in *Fedra, La Nave,* and *L'Ombra di Don Giovanni* and appeared in the Italian premières of *Gianni Schicchi* and *Il tabarro.* His American debut took place in 1920 in *Fedora,* with the Chicago Opera. After three years with this company he was engaged by the Metropolitan Opera, making his debut on November 16, 1922, as Avito. He sang at the Metropolitan Opera for the next thirteen years, featured in the world premières of *The King's Henchman, Merry Mount,* and *Peter Ibbetson,* and in such important American premières and

Metropolitan revivals as *Fra Gherardo,
Pelléas et Mélisande, Sadko,* and *La
Vestale.* The role of Pelléas was considered by many to be one of his most
successful interpretations. In 1935
Johnson was appointed director of a
special spring season then inaugurated
by the Metropolitan. The sudden death
of Herbert Witherspoon, chosen to succeed Gatti-Casazza as general manager,
brought the directorial post to Johnson
that fall. Johnson remained the general
manager of the Metropolitan Opera for
the next fifteen years, a regime marked
by prosperity at the box office and by
many substantial artistic successes (*see*
METROPOLITAN OPERA HOUSE). Johnson resigned his post in 1950 and was
succeeded by Rudolf Bing. On February 28, a gala evening took place at the
Metropolitan to honor the departing
manager. It was climaxed by an opera
pageant in which many of the company's outstanding singers appeared in
roles from twelve of the most successful operas produced under Johnson's
direction. Twelve days earlier, the last
of these important revivals had taken
place: *Khovantchina.*

John the Baptist (The Prophet) (tenor)
in Massenet's *Hérodiade.*

Jokanaan (John the Baptist), the
Prophet (baritone) in Richard Strauss's
Salome.

Jommelli, Niccolò, composer. Born
Aversa, Italy, September 10, 1714;
died Naples, August 25, 1774. His
musical studies took place at conservatories in Naples, his teachers including Francesco Durante. Before turning
to opera he wrote some church music
and ballets. His first opera, *L'errore
amoroso,* was completed in 1737, and
was acclaimed. His second opera,
Odoardo, was well received in 1738,
and brought him a commission from
Cardinal Albani to write operas for performance at his palace. Jommelli went
to Rome in 1740 and wrote two operas
for the Cardinal. A year later another

opera, *Merope,* was such an immense
success in Venice that the Council of
Ten appointed him the director of the
Conservatorio degli Incurabili. In 1745
Jommelli visited Vienna to attend premières of several of his new operas. He
was greatly honored by the Empress
and others of high station. After he returned to Rome, he became assistant
maestro di cappella at St. Peter's, a post
he held from 1749 to 1754. He then
went to Stuttgart to fill the post of
kapellmeister for the Duke of Württemberg. During this period (nearly sixteen
years) he wrote twenty operas. In 1769
Jommelli returned to the city of his
birth to find that he no longer occupied
his former exalted position. His last
operas were failures, a circumstance
that broke his health and spirit. After
writing an occasional cantata and a
Miserere, he died of apoplexy.

Jommelli wrote over fifty operas.
When he adhered to Neapolitan traditions, his works were filled with supple
and pleasing melodies and ingratiating
harmonies that had an immediate appeal. The works of his German period
were in a richer and deeper vein. His
harmonic language became so varied,
his instrumental coloring so subtle, his
dramatic feeling so pronounced that he
acquired the sobriquet of the "Italian
Gluck." Thus, he wrote his most popular operas while still in Italy, his best
operas in the closing years of his life.
His most famous operas were: *Merope*
(1741); *Ezio* (1741, revised 1748 and
1771); *Achille in Sciro* (1745); *Eumene* (1747); *Didone* (1748); *Artaserse* (1749); *Demetrio* (1753); *Alessandro nell' Indie* (1757); *La clemenza
di Tito* (1758); *Armida* (1770); *Ifigenia in Tauride* (1773).

Jonas, an Anabaptist preacher (tenor)
in Meyerbeer's *Le Prophète.*

Jones, Brutus, ruler of a West Indian
island (baritone) in Gruenberg's *The
Emperor Jones.*

Jongleur de Notre Dame, Le (The Jug-

gler of our Lady), opera in three acts by Jules Massenet. Libretto by Maurice Léna, based on *L'etui de nacre,* a short story by Anatole France, derived from a medieval miracle play. Première: Monte Carlo Opera, February 18, 1902. American première: Manhattan Opera House, New York, November 27, 1908.

Characters: Jean, a juggler (tenor or soprano); Boniface, the monastery cook (baritone); the prior (bass); the monk painter (baritone); the monk musician (baritone); the monk poet (tenor); the monk sculptor (bass); citizens of the town. The setting is the Abby of Cluny, near Paris, in the fourteenth century.

Act I. A square before the abbey. A crowd has gathered to celebrate market day. Jean, a juggler, begs for attention, but the crowd, indifferent to his tricks, persuades him to render a profane song, "Hallelujah to Wine." For this offense, Jean is threatened with excommunication, the prior advising Jean that there is only one way he can find absolution: by entering a monastery. Jean expresses his great love for freedom ("O liberté ma mie!"). The monastery cook arrives, leading a mule laden with food. Starving, Jean is abruptly tempted to follow the prior's advice.

Act II. A hall within the abbey. His fellow monks continually mock Jean for his inability to master Latin and the hymns. Jean is mortified at his stupidity. Boniface consoles him by telling him the story of the sage flower that sheltered the child Jesus when the rose feared to ruin its petals. (Légende de la sauge: "Fleurissait une sauge"). Boniface emphasizes that even one so humble as Jean can serve God and be accepted by Him.

Act III. The chapel. Wearing his juggler's clothes, Jean performs his tricks and songs before the altar. The monks discover him in this performance and are outraged at the sacrilege. Boniface restrains them from removing him. Exhausted, Jean collapses at the feet of the Virgin. The face of the image begins to shine with a beatific light and the hands stretch in blessing. Jean dies joyously, a choir of angels receives his soul, the kneeling monks witnessing the miracle.

Since Massenet intended *Le Jongleur* as a "mystery," he wrote no parts for women in it. When the opera was first produced the part of Jean was sung by a tenor. Thus the composer who had become famous for his characterizations of women (Manon, Hérodiade, Thaïs, Sapho, and Grisélidis) produced an effective opera without a woman in a principal role. Subsequently, wishing to appear in the work, Mary Garden prevailed on Massenet to adapt the leading role for soprano voice. It was a happy change; the role of Jean proved even more effective. When Mary Garden appeared in *Le Jongleur* for the first time (at its American première) she scored a sensation; from then on her name was associated with the opera, and since her retirement, performances of *Le Jongleur* have been infrequent.

Jonny spielt auf (Johnny Strikes Up), opera by Ernest Křenek. Libretto by the composer. Première: Leipzig Opera, February 10, 1927. Křenek explained that this jazz opera was an interpretation of the "rhythms and atmosphere of modern life in this age of technical science." It is one of several similar operas that enjoyed sensational success in Europe in the 1920's. The story revolves around a Negro jazz-band leader, Jonny. He steals Daniello's violin and makes such irresistible music with it that he wins everybody who hears him. His jazz conquers the world, and he bestrides the world as a conqueror. Playing from the North Pole, he inspires a mass performance of the Charleston. Because of its novel subject matter and its jazz idiom, *Jonny* *spielt auf* was turned down by many companies before it was accepted by the Leipzig Opera. Within the next few years it was heard in about a hundred European cities, translated into eighteen languages. New York found it less wonderful—produced by the Metropolitan in 1929, *Jonny* lasted three performances. The opera has now vanished like the frenetic era it celebrated.

Jonson, Ben, dramatist. Born Westminster, England, about 1573; died there April 6, 1637. One of the most significant Elizabethan dramatists after Shakespeare and Marlowe, Jonson wrote several plays that have been made into operas. *Epiocene* was the source of Antonio Salieri's *Angiolina* and Richard Strauss's *Die schweigsame Frau,* and *Volpone* of operas of the same name by George Antheil, Louis Gruenberg, and Norman Demuth. Sir Edward Elgar wrote an uncompleted opera, *The Spanish Lady,* based on Jonson's *The Devil is an Ass.*

Jontek's Revenge, see HALKA.

José, see DON JOSÉ.

Joseph, opera by Etienne Nicolas Méhul. Libretto by Alexandre Duval, based on the Bible story. Première: Opéra-Comique, February 17, 1807. The story of Joseph, and his sale by his brothers to the Egyptians, is only partially told in this opera. The work opens in Egypt, where Joseph has attained high station through his wisdom in saving that country from famine. He has assumed the name of Cleophas, and rules as governor in Memphis. A famine in Palestine brings Joseph's blind father and brothers to Memphis to plead for food. They do not recognize him, but he knows them instantly. Eventually he reveals his identity, forgives his brothers, and prevails on his father to do likewise. Joseph's beautiful aria in Act I, "Champs paternels," is famous. Carl Maria von Weber wrote a set of piano variations on a theme from *Joseph,* and melodies from the opera were used by Ethel Smyth in her one-act opera *L'Entente Cordiale.*

Joseph and His Brethren, see MANN, THOMAS.

Jour naissait dans le bocage, Le (Venus Descendeth; Sorgeva il dì del bosco in seno), Maria's singing lesson in Act II of Donizetti's *The Daughter of the Regiment.*

Journet, Marcel (bass). Born Grasse, Alpes Maritimes, France, July 25, 1867; died Vittel, France, September 5, 1933. His music study took place at the Paris Conservatory. In 1891 he made his opera debut at the Théâtre de la Monnaie. Six years later he became a member of the Covent Garden company. On December 22, 1900, he made his American debut at the Metropolitan Opera as Ramfis. He remained at the Metropolitan eight years. From 1908 to 1914 he made star appearances in leading opera houses of Europe, and after 1914 was associated with the Chicago Opera, the Paris Opéra, and La Scala. His repertory included sixty-five French, twenty-seven Italian, and eight Wagnerian roles. He appeared in the world premières of *Monna Vanna, La Navarraise, Samson et Dalila, Nerone* (Boïto), and *Thaïs.*

Jouy, Victor Joseph Etienne de, dramatist. Born Jouy, near Versailles, 1764; died St. Germain-en-Laye, September 4, 1846. One of the most successful French dramatists of the early nineteenth century, De Jouy wrote many opera librettos, the most famous being that for Rossini's *William Tell,* which he wrote collaboratively with Hippolyte Bis. De Jouy also wrote the librettos for Rossini's *Moïse,* Spontini's *Fernand Cortez* and *La Vestale,* and for various other operas by Boieldieu, Cherubini, Méhul, and Rossini.

Juan, see DON JUAN.

Juarez and Maximilian, see WERFEL, FRANZ.

Juch, Emma, soprano and opera manager. Born Vienna, July 4, 1863; died

New York City, March 6, 1939. She was born while her parents, American citizens, were visiting Vienna. When she was four she was brought to the United States. She studied singing with her father and Mme. Murio-Celli. In her eighteenth year she joined Her Majesty's Theatre in London, then directed by Mapleson, making her debut as Philine. Returning to the United States, she made her American debut with the Mapleson company at the Academy of Music on October 21, 1881, once again as Philine. She was engaged by the Theodore Thomas Opera Company in 1885, and sang many principal roles. When that company was dissolved, she organized the Emma Juch Grand Opera Company. From 1889 to 1891 this company toured the United States, Mexico, and Canada. After 1891 Juch withdrew from opera to devote herself to concert appearances. In 1894 she married Francis L. Wellman, a United States District Attorney, retiring to live in New York City.

Juchhei, nun ist die Hexe todt! Duet of Hansel and Gretel in Act III of Humperdinck's *Hansel and Gretel.*

Judith, (1) opera by Eugène Goossens. Libretto by Arnold Bennett. Première: Covent Garden, June 25, 1929. Holofernes, oppressor of the Israelites, is taken with Judith's beauty and seeks to win her love. During a bacchanale, Judith feigns acceptance and at the height of a love scene she kills him.

(2) Opera by Arthur Honegger. Libretto by René Morax. Première: Monte Carlo Opera, February 13, 1926. This three-act opera is an extension of incidental music for Morax's play, introduced in Switzerland in 1925. Judith goes forth to the conquering Assyrians to plead for her people. The leader, Holofernes, tries to make love to her, and, as part of a plan to win him over, she submits. When Holofernes falls into a drunken stupor, Judith cuts off

his head. She brings it to her people, who are so inspired that they defeat the enemy. Judith gives thanks to God for the victory and puts on a widow's veil as mourning for the dead.

Juif Polonais, Le (The Polish Jew), opera by Camille Erlanger. Libretto by Henri Cain and Pierre Barthélemy Gheusi, based on the novel of the same name by Erckmann-Chatrian. Première: Opéra-Comique, April 11, 1900. This is essentially the story in which Henry Irving achieved a triumph on the English stage—his play was called *The Bells.* At an inn, Schmitt tells the story of the murder of a Polish Jew fifteen years earlier. As soon as he finishes his tale, a Polish Jew enters the inn. One of those who has heard Schmitt's story, Mathis the burgomaster, faints—for it is he who had murdered the Jew years ago, and the present stranger appears to him like an apparition from the grave. Mathis has a nightmare in which he sees himself hung for the crime. He cries out, then dies of a heart attack. The murderer has been punished.

Juive, La (The Jewess), opera in five acts by Jacques Halévy. Libretto by Eugène Scribe. Première: Paris Opéra, February 23, 1835. American première: Théâtre d'Orléans, New Orleans, February 13, 1844.

Characters: Cardinal Brogny (bass); Prince Léopold (tenor); Princess Eudoxie, his betrothed (soprano); Eléazar, a Jewish goldsmith (tenor); Rachel, his daughter (soprano); Ruggiero, chief bailiff (baritone); Albert, officer of the imperial guard (bass); courtiers; priests; soldiers; people. The action takes place in the city of Constance, in Baden, in the year 1414.

Act I. The square before the cathedral. The victory of Prince Léopold over the Hussites is being celebrated. However, Eléazar appears oblivious and goes on working. The crowd is angered at his indifference and drags him and his daughter out into the open square. The pair are about to be taken to prison when Cardinal Brogny emerges from the cathedral. He recognizes the Jew: many years before, he had instigated a pogrom in which Eléazar's two sons were killed. Some time after, the Cardinal had also known tragedy when his house mysteriously burned and his wife and daughter died in the flames. The Cardinal now makes a gesture of friendship, but Eléazar rejects it. The Cardinal prays for all nonbelievers, and entreats the crowd to replace hatred with tolerance ("Si la rigueur et la vengeance"). Eléazar and his daughter are let go, and the square becomes empty. Prince Léopold appears, disguised in simple clothes and calling himself Samuel. In love with Rachel, he knows the only way he can hope to win her is by concealing his true identity. Rachel, who returns Léopold's love, invites him to the Passover ceremonies at her house. When the crowd returns and resumes its hostility toward Eléazar, Léopold quickly exerts his authority. Rachel for the first time begins to suspect that her lover is no humble Jewish painter.

Act II. A room in Eléazar's house. The Passover feast, at which Léopold is a guest, is celebrated. A prayer is intoned ("O Dieu, Dieu de nos pères"), after which Eléazar begs for divine guidance and for the ultimate destruction of Israel's enemies ("Dieu, que ma voix tremblante"). Unexpectedly, Princess Eudoxie arrives to order a jewel she wants to present to Léopold the following day. To avoid being recognized, Léopold hides until she leaves. Puzzled by his strange behavior, Rachel insists on an explanation. Léopold confesses the truth. At first Rachel is shocked, but so great is her love that she is soon ready to elope with him. Growing furious, Eléazar rushes at Léopold with drawn dagger. Rachel intervenes and saves her lover. So elo-

quently does she plead for her right to love that Eléazar's anger turns to compassion. Reluctantly, he gives his consent to their marriage. But Léopold has been contemplating another relationship than marriage. Crying that he cannot wed a Jewess, he rushes out of the house.

Act III. The great hall in the imperial palace. At a royal feast, a ballet pantomime is performed, after which the people hail Prince Léopold ("Sonnez, clairons, que vos chants de victoire"). Eléazar and Rachel arrive to deliver the jewel Eudoxie has ordered. As Eudoxie presents it to Léopold, her betrothed, Rachel furiously denounces Léopold as unworthy, revealing to the startled assemblage that he has made love to her and has promised to marry her. Cardinal Brogny excommunicates Léopold for this offense ("Vous qui du Dieu vivant outragez la puissance"). Condemned to death, Eléazar, Rachel and Léopold are imprisoned.

Act IV. A hall in the court of justice. Eudoxie persuades Rachel to change her testimony and assume full responsibility for what has passed between herself and Léopold. Rachel finally consents. The Cardinal arrives, offering to pardon Eléazar and his daughter if they embrace Christianity; defiantly and proudly they both refuse. At this point, Eléazar reveals to the Cardinal that the latter's daughter is alive, having been rescued from his flaming home by a Jew. The Cardinal is deeply moved, but is unable to learn from Eléazar his daughter's whereabouts. After the Cardinal departs, Eléazar quietly contemplates the twist of fate whereby he is compelled to sacrifice Rachel, one whom he has raised as his own and loves dearly ("Rachel, quand du Seigneur").

Act V. A place of execution. A restive crowd eagerly awaits the punishment of Eléazar and Rachel, who are to be boiled in oil ("Quel plaisir, quelle joie!"). Eléazar now learns that Léopold will escape death because Rachel has falsely sworn to his innocence. The old Jew now pleads with Rachel to save herself by embracing Christianity, but again she refuses. As she is thrown into the boiling caldron, Eléazar cries to the horrified Cardinal: *"There* is your daughter!" Having had his moment of triumph, Eléazar goes to his death proudly.

Though Halévy wrote over thirty operas, only *La Juive* keeps his name alive. Into this work—one of the finest in the French lyric theater—the composer poured his most expressive lyricism; it is through melody that the dramatic conflicts and crises find expression. Though the central character is a Jew, and one of the scenes reproduces a Jewish ceremony, Halévy made no attempt at writing Hebrew music or enlisting Hebraic idioms to give his music authenticity. His melody remains characteristically French throughout in its refinement, objectivity, and sensitivity. The French tenor who created the role of Eléazar, Adolphe Nourrit, was a constructive influence while the opera was being written. Halévy originally planned the role of Eléazar for a high bass voice, but Nourrit convinced him that a tenor would be more appropriate. It was also Nourrit who convinced Halévy to include in the fourth act the opera's most famous aria: "Rachel, quand du Seigneur." The role of Eléazar was one of Caruso's favorites. In his history of the Metropolitan Opera, Irving Kolodin goes so far as to say that Caruso's performance of this role was "without doubt the most striking artistic triumph of his career.... The impersonation he finally presented was the product of more care and study, especially dramatically, than any of the 35 other roles he sang during his career in New York."

Julien, opera by Gustave Charpentier. Libretto by the composer. Première:

Opéra-Comique, June 3, 1913. This is a sequel to *Louise,* but it never equaled the success of the earlier opera. In this work Louise is dead, but she appears to Julien in a vision, a representation of beauty to reawaken his faith in art and beauty. Julien goes vainly in search of his soul, only to meet frustration and death.

Juliette, daughter of Capulet (soprano) in Gounod's *Roméo et Juliette.*

Julius Caesar, drama by Shakespeare about the conspiracy of Cassius and Brutus that led to the assassination of Caesar and the conspirators' subsequent death in battle. The drama has served as the basis of several operas. *See* GIULIO CESARE.

Jumping Frog of Calaveras County, The, opera by Lukas Foss. Libretto by Jean Karsavina, based on the story of the same name by Mark Twain. Première: Bloomington, Indiana, May 19, 1950. A comedy in two scenes (one in a saloon, the other outside), the opera is built around Twain's celebrated jumping contest. The score is made up of cowboy and other western folk tunes as well as broadly satirical music. It was successfully produced at the Venice Music Festival in September, 1953.

Jungfrau von Orleans, Die, *see* SCHILLER, FRIEDRICH.

Jupiter, a god (baritone) in Gounod's *Philémon et Baucis.*

Jurinac, Sena (soprano). Born Travnik, Yugoslavia, October 24, 1921. After attending the Zagreb Conservatory she made her debut at the Zagreb Opera on October 15, 1942. Some three years later she was engaged by the Vienna State Opera, where she scored her first major successes in a varied repertory of Italian, French, and German operas. In 1947 she appeared for the first time at the Salzburg Festival, and in the next few years was acclaimed not only at Salzburg but also at the Glyndebourne and Edinburgh Festivals.

K

Kabale und Liebe, *see* SCHILLER, FRIEDRICH.

Kabalevsky, Dmitri, composer. Born St. Petersburg, Russia, December 30, 1904. His music study took place at the Moscow Conservatory and with Miaskovsky. In 1925 he began writing songs and piano pieces, and success came in 1934 with his second symphony. He extended his reputation with the opera *Colas Breugnon,* in 1937. Later operas: *Near Moscow* (1942); *The Family of Taras* (1947); *Nikita Vershinin* (1954). In 1939 Kabalevsky was elected a member of the Presidium of the Organizing Committee of the Union of Soviet Composers. A year later he received the Order of Merit, and in 1946 the Stalin Prize for a string quartet. He has served as professor of composition at the Moscow Conservatory.

Kafka, Franz, author. Born Prague, Bohemia, July 2, 1883; died Kierling, near Vienna, June 3, 1924. The author of some extraordinary novels and short stories, many of which would have been lost after his death, except for the efforts of his literary executor, Max Brod, Kafka is perhaps best known for his haunting tale of persecution, *The Trial* (*Der Prozess*), the source of an opera by Gottfried von Einem.

Kärntnerthortheater, a theater in Vienna, situated behind the State Opera building which arose on the site of the Stadttheater when the latter burned down in 1761. Originally the home of spoken drama, the Kärntnerthor was frequently used for opera and operetta performances after 1790. In 1820 Schubert's operetta *Die Zwillingsbrüder* was introduced there. The heyday of this theater as a home for opera was between 1821 and 1828 when Domenico Barbaja was its artistic director. It was here that Rossini presented a cycle of his operas during his visit to Vienna in 1821. Weber's *Der Freischütz* received its Austrian première in this house, and it was for the Kärntnerthor that Weber wrote *Euryanthe,* which received its world première here in 1823. *Linda di Chamounix* and *Martha* were other operas to be introduced in this theater, in 1842 and 1847, respectively.

Kahn, Otto H., opera patron and financier. Born Mannheim, Germany, February 21, 1867; died New York City, March 29, 1934. A noted banker and member of the New York banking house of Kuhn, Loeb and Co., Otto Kahn joined the board of directors of the Metropolitan Opera in 1903. When that organization suffered financial reverses five years later he was one of fourteen subscribers raising a fund of $150,000. In 1924 he became president of the board of the Metropolitan, holding this post seven years. During this time he was a vital force in trying to get a new opera house for the Metropolitan, but his efforts were in vain. After he resigned as president, he remained a member of the board until the end of his life. He was also honorary director of Covent Garden and the Boston Opera.

Kaiser, Georg, dramatist. Born Magdeburg, Germany, November 25, 1878; died Ascona, Switzerland, June 5, 1945. A representative of the German expressionist theater, Kaiser wrote librettos for three operas by Kurt Weill: *Der Protagonist, Silbersee,* and *Der Zar lässt sich photographieren.* He also wrote the libretto for Max Ettinger's *Juana.*

Kalidasa, Sanskrit epic poet and dramatist. He lived about the fifth century after Christ and was the author of a drama, *Sakuntala,* that has inspired many musical works including the following operas: Franco Alfano's *La leggenda di Sacùntala,* Ignace Jan Paderewski's *Sakuntala,* Felix Weingartner's *Sakuntala,* and Louis Coerne's *Sakuntala.*

Kammeroper, *see* CHAMBER OPERA.

Kammersänger (feminine: Kammersängerin), German for "chamber singer." This is the highest honorary title bestowed by the German and the Austrian governments on leading opera singers.

Kann ich mich auch an ein Mädel erinnern, the Marschallin's monologue lamenting the loss of her youth, in Act I of Richard Strauss's *Der Rosenkavalier.*

Kapellmeister, *see* MAESTRO DI CAPPELLA.

Kapp, Julius, musicologist. Born Steinbach, Germany, October 1, 1883. He received his academic education in Marburg, Berlin, and Munich, and his doctorate in Munich in 1906. For several years he was stage director of the Berlin State Opera, and in 1921 he founded the periodical *Blätter der Staatsoper.* He has written many books on operatic subjects, including: *Richard Wagner und Franz Liszt* (1908); *Richard Wagner* (1910); *Richard Wagner und die Frauen* (1911); *Hector Berlioz* (1914); *Das Dreigestirn: Berlioz-Liszt-Wagner* (1920); *Meyerbeer* (1920); *Franz Schreker* (1921); *Das Opernbuch* (1922); *Die Oper der Gegenwart* (1922); *Weber* (1922); *Die Staatsoper 1919 bis 1925* (1925); *R. Wagner und seine erste Elisabeth* (1925); *185 Jahre Staatsoper* (1928); *Wagner und die Berliner Staatsoper* (1933); *Wagner in Bildern* (1933); *200 Jahre Staatsoper in Bild* (1942).

Kappel, Gertrude, soprano. Born Halle, Germany, September 1, 1884. After graduating from the Leipzig Conservatory she joined the Hanover Opera, where she made her debut in 1903 in *Il trovatore.* After a few seasons in Hanover she appeared in Vienna in leading Wagner roles and was so successful that she was engaged as principal soprano of the Munich Opera. For three seasons she appeared with immense success both in Vienna and Munich. On January 16, 1928, she made her American debut at the Metropolitan Opera as Isolde. She remained at the Metropolitan through the 1934–1935 season, appearing in the Wagner repertory and operas by Richard Strauss. Meanwhile, in 1933, she became a member of the San Francisco Opera Company. She retired from opera in 1937, establishing her home in Berlin.

Karajan, Herbert von, conductor. Born Salzburg, April 5, 1908. He received his musical education in Vienna, and his apprenticeship as conductor in Ulm. For five years he was general music director of the city of Aachen. In 1938 he was appointed a conductor at the Berlin State Opera, where he remained through the war years. After the war, he became musical director of the Vienna State Opera and the Vienna Philharmonic. His performances in Vienna and in leading European festivals (including those of Bayreuth, Salzburg, Munich, Aix-en-Provence, and Lucerne) placed him with the foremost conductors of Europe, and as the most significant conductor to emerge to fame after World War II. He has also been acclaimed in guest appearances in London and Paris. When the conductor Wilhelm Furtwaengler died before he could appear in the United States with the Berlin Philharmonic Orchestra in 1955, Von Karajan was appointed in his place, and his American debut took place with the Berlin Philharmonic in Washington, D.C., February 28, 1955. He was also appointed musical director of the Berlin Philharmonic.

Katerina, the murderess (soprano) in Shostakovich's *Lady Macbeth of Mtsensk.*

Kathinka, Kruschina's wife (soprano) in Smetana's *The Bartered Bride.*

Keilberth, Joseph, conductor. Born Karlsruhe, April 19, 1908. His association with opera began in 1925 when he was appointed coach at the Karlsruhe State Theater. Ten years later he assumed the post of principal conductor there, and in 1945 was engaged as the musical director of the Dresden Opera. In 1951 he became the conductor of the Hamburg Philharmonic Orchestra. A year later he made the first of several appearances at the Bayreuth Festivals and conducted the Hamburg Opera in guest performances at the Edinburgh Festival. He has since been widely heard in guest performances in leading European opera houses.

Keiser, Reinhard, composer and opera director. Born Teuchern, Germany, January 9, 1674; died Hamburg, September 12, 1739. Both as composer and director he was for over forty years the dominant figure in Hamburg, then the opera capital of Germany. His musical training took place at the Thomasschule in Leipzig. In 1692 he was appointed court musician at Brunswick, and here his first opera, *Basilius,* was performed. Two years later he became chief composer of the Hamburg Opera. During the next four decades he wrote over a hundred operas, many of them outstandingly successful in their time; he was probably the most celebrated opera composer of his time in Germany. Within formal Italian patterns he brought a rich fund of light and pleasing melodies together with dramatic expressiveness. Perhaps his greatest appeal lay in the fact that he used popular subjects, the first German composer to do so. His most successful operas were: *Mahmuth II* (1696); *Ismene* (1699); *Ottavia* (1705); *Almira* (1706); *Die Leipziger Messe* (1710); *L'inganno fedele* (1714); *Der Hamburger Jahrmarkt* (1725). In 1703 Keiser was appointed director of the Hamburg Opera, an institution he succeeded in raising to a position of first importance. It was while he held this post that Handel played the violin in the orchestra and wrote his first opera, *Almira,* produced by the company. Keiser's extravagance in mounting his productions, and his complete indifference to business matters, brought the opera house to the brink of ruin; for a period Keiser had to go into hiding to evade his creditors. His marriage to a wealthy woman, in 1709, helped to rehabilitate his fortune. Between 1719 and 1721 he was court composer in Stuttgart, and from 1722 to 1728 he lived in Copenhagen, serving much of this time as royal kapellmeister. After returning to Hamburg, he was engaged to supervise opera productions in St. Petersburg. He was en route from Russia to Italy, intending to engage singers for St. Petersburg, when he again visited Hamburg. He suddenly decided not to return to Russia, despite his contract, and never left Hamburg again. A year after his death his operas disappeared permanently from the stage.

Keller, Gottfried, poet and novelist. Born Zürich, Switzerland, July 19, 1819; died there July 15, 1890. His collection of portraits of Swiss provincial life, *Die Leute von Seldwyla,* contained a story, *Romeo und Julia auf dem Dorfe,* which Delius used for his opera *A Village Romeo and Juliet.* Another Keller story was the source of Zemlinsky's opera *Kleider machen Leute.*

Kellogg, Clara Louise, dramatic soprano and opera manager. Born Sumterville, South Carolina, July 12, 1842; died New Haven, Connecticut, May 13, 1916. Beginning in 1857, she received singing lessons from various New York teachers. After going on a concert tour, she made her opera debut on February 27, 1861, at the Academy of Music in New York as Gilda. She established her reputation in the next few years in about fifteen different roles. On November 2, 1867, she made her London debut at Her Majesty's Theatre in *Faust*. She toured the United States between 1868 and 1872. In 1872 she became an opera manager when she joined Pauline Lucca in founding and directing the Lucca-Kellogg Opera Company, in which both artists starred; the venture was a success. From 1874 to 1876 she traveled throughout the United States with another opera company under her own management, appearing in over a hundred performances in a single season. After 1876 she was a guest artist with many leading European opera companies. The following year she married her manager, Carl Strakosch, and soon after went into retirement in New Haven. Her autobiography, *Memoirs of an American Prima Donna*, was published in 1913.

Kelly, Michael, tenor. Born Dublin, Ireland, December 25, 1762; died Margate, England, October 9, 1826. He studied singing in Naples, after which he appeared successfully on the leading Italian opera stages. Moving to Vienna, he came to know Mozart well when, for four years, he sang as a member of the Court Theater. He appeared in the world première of *The Marriage of Figaro* as Basilio and Don Curzio. In 1787 he went to London and became a leading tenor at Drury Lane, singing there until he retired in 1811. His autobiography, *Reminiscences* (1826), contains valuable information about opera and Mozart.

Kenilworth, *see* SCOTT, SIR WALTER.

Kezal, marriage broker (bass) in Smetana's *The Bartered Bride*.

Khan Kontchak, *see* KONTCHAK.

Khivria, Tcherevik's wife (mezzosoprano) in Mussorgsky's *The Fair at Sorochinsk.*

Khovantchina, musical drama in five acts by Modest Mussorgsky (completed and revised by Rimsky-Korsakov). Libretto by the composer and Vladimir Stassov. Première: St. Petersburg, February 21, 1886 (amateur performance); Kiev, November 7, 1892 (professional performance). American première: Philadelphia, April 18, 1928.

Characters: Prince Ivan Khovantsky, commander of the Streltsy (bass); Andrei, his son (tenor); Prince Vassily Galitzin, a reformer, member of Young Russia (tenor); Dositheus, leader of the Old Believers (bass); Marfa, a young widow (mezzo-soprano); Emma, a young German girl (contralto); Shaklovity, a boyar (baritone); peasants; slaves; townspeople. The setting is Moscow during the reign of Peter the Great.

Act I. A square within the Kremlin. The orchestral prelude is entitled "Dawn on the Moskva River," and is a picture of Moscow at dawn; it consists of a folk melody with five variations. The Streltsy, a band of radicals, is plotting against the Czar's regime. Prince Ivan Khovantsky has inflamed them with a speech, since he, too, is ambitious to overthrow the Czar. The boyar Shaklovity bribes a letter writer to inscribe an accusation of treason against the Prince. The Prince's son, Andrei, now appears, pursuing the girl Emma. Marfa, a discarded mistress of Andrei, protects the girl, bitterly denounces Andrei and prophesies a terrible fate for him. When Andrei's father sees Emma he, too, is taken by her beauty and orders his men to arrest her. She is saved by Dositheus, leader of the

Old Believers, who rebukes the men and pacifies them.

Act II. Prince Galitzin's house. Marfa, reading the horoscope of Prince Galitzin, can find only tragedy. Terrified, the Prince quarrels with her and secretly orders one of his servants to drown her. When Prince Khovantsky arrives, Galitzin quarrels with him, too, and only the arrival of Dositheus saves the situation.

Act III. A street outside Khovantsky's house. Having been spared, Marfa is sitting outside the Prince's house recalling her love affair with Andrei. Night brings peace. Shaklovity passes by and remarks on the sleeping city ("Yes, the Streltsy are sleeping"). But while the city is quiet, her enemies are awake to plot her destruction. The Streltsy enter and are confronted by their women, upset by their husbands' activities. Heated words are exchanged. The letter writer suddenly appears with word that the Czar, with foreign aid, has suppressed an insurrection. The cause of Old Russia is lost; the Streltsy pray for divine help.

Act IV, Scene 1. The country house of Khovantsky. The Prince is being entertained by a lavish spectacle of Persian dances ("Dance of the Persian Slaves"). Shaklovity comes to summon him to a council of state. As the Prince is dressing, assassins murder him.

Scene 2. A square before the Church Vassili Blazheny. Between the scenes, a mournful entr'acte is played. Prince Galitzin is being led to exile. Dositheus laments the sad fate of Russia. His misery is heightened when Marfa tells him that foreign mercenaries have been enlisted to destroy the Old Believers. Rather than await such a fate, the Streltsy bring axes for their own execution. Before they destroy themselves, they receive word that they have been pardoned by the Czar.

Act V. A wood near Moscow. The Old Believers, true to their principles, are determined to kill themselves. A funeral pyre is built. When Marfa applies the torch, they solemnly march to their deaths in the flames, singing as they go.

In *Khovantchina* Mussorgsky was more concerned in creating a great historical panorama than in emphasizing characters or in pointing up dramatic incidents. As Rosa Newmarch wrote: "It reminds us of those early icons belonging to the period when the transport of pictures through the forests, bogs, and wilderness of Russia so restricted their distribution that the religious painter resorted to the expedient of representing on one canvas as many saints as could be packed into it." This is the essential weakness of the opera and accounts for its lack of integration and a focal point of interest. But while not essentially a great opera, it is the work of a great composer, with many pages that remind us of the best moments in *Boris Godunov*. The choral passages, the atmospheric entr'acte and the poignant closing scene more than compensate for the dramatic shortcomings of the text.

Khovantsky, Prince Ivan, commander of the Streltsy (bass) in Mussorgsky's *Khovantchina.*

Khrennikov, Tikhon, composer. Born Elets, Russia, June 10, 1913. He attended the Moscow Conservatory and achieved recognition as a composer with a piano concerto written when he was nineteen. His opera *The Brothers* (1936) added greatly to his reputation, as did his second symphony (1943). Khrennikov writes in a folk idiom, with a strong emphasis on rhythm and broad Russian melody. *The Brothers,* revised and renamed *In the Storm,* was reintroduced with outstanding acclaim in Moscow on May 31, 1939.

Kienzl, Wilhelm, composer. Born Waizenkirchen, Austria, January 17, 1857; died Vienna, October 3, 1941. His music study was pursued at the

Prague Conservatory and with Josef Rheinberger in Munich. A friendship with Wagner was a decisive and permanent influence. For a number of years Kienzl lived with Wagner at Wahnfried in Bayreuth, and as a result of this personal association absorbed many of the master's ideas and principles. He completed his first opera, *Urvasi*, in 1886. His greatest success came with *Der Evangelimann*, introduced in Berlin in 1895 and afterward produced with outstanding success throughout Germany and Austria. Kienzl was the director of German opera in Amsterdam and Krefeld, principal conductor of the Hamburg Opera from 1889 to 1892, and first conductor at the Munich Opera from 1892 to 1893. In 1893 he settled in Graz and devoted himself principally to composition. His operas: *Urvasi* (1886, revised 1909); *Heilmar der Narr* (1892); *Der Evangelimann* (1895); *Don Quixote* (1898); *In Knecht Rupprechts Werkstatt* (1907); *Der Kuhreigen* (1911); *Das Testament* (1916); *Hassan der Schwärmer* (1925); *Sanctissimum* (1925); *Hans Kipfel* (1928). He also completed Adolph Jensen's *Turandot*, and edited Mozart's *La clemenza di Tito*.

Kiepura, Jan, tenor. Born Sosnowice, Poland, May 16, 1902. He graduated from the University of Warsaw in 1924. His debut took place a year later with the Warsaw Opera in *Faust*. In 1926 he was engaged by the Vienna State Opera, where he appeared for two years. After 1928 he made successful appearances with leading European opera companies, including La Scala, the Berlin State Opera, and the Opéra-Comique. He made his first American appearance with the Chicago Opera in 1931. On February 10, 1938, he made his debut at the Metropolitan Opera in *La Bohème*. He stayed at the Metropolitan through the 1941–1942 season. Since then he has appeared primarily on the concert stage and in motion pictures.

Kilian, a peasant (tenor) in Weber's *Der Freischütz*.

King Henry IV, a two-part drama by Shakespeare. The two plays are concerned with the unsuccessful insurrections against the rule of Henry IV, first by the Percys of Northumberland, and then by the Earl of Northumberland and the Archbishop of York. In the end, Prince Hal succeeds his father, becoming Henry V. *King Henry IV* was the source for several operas: Holst's *At the Boar's Head,* Pacini's *La gioventù di Enrico V,* Verdi's *Falstaff* (which also uses material from *The Merry Wives of Windsor*).

King Lear, tragedy by Shakespeare. Lear, king of ancient Britain, divides his realm between two daughters who ungratefully drive him away. His third daughter, Cordelia, to whom he has given nothing, remains true. Cordelia is wrongfully imprisoned by the wicked general, Edmund, and later hanged, whereupon the aged Lear dies of grief. The two ruthless sisters also come to an unhappy end, one of them poisoning the other and then committing suicide. The drama was made into an opera by Vito Frazzi: *Re Lear.* After he finished *Aïda,* Verdi worked on and off on an opera based on *King Lear,* but he completed only random sketches. At his request, they were destroyed after his death.

King Louis VI, French king (bass) in Weber's *Euryanthe*.

King Mark, husband of Isolde (bass) in Wagner's *Tristan und Isolde*.

King of Clubs, The, a character (bass) in Prokofiev's *The Love for Three Oranges*.

King of Egypt, Amneris' father (bass) in Verdi's *Aïda*.

King of the Seas, The, a character (bass) in Rimsky-Korsakov's *Sadko*.

King Roger, opera by Karol Szymanowski. Libretto by the composer. Première: Warsaw Opera, June 19, 1926. The theme is the struggle between Western Christianity and Eastern paganism. The setting is medieval Sicily, where the Queen Roxane falls in love with a shepherd-prophet come from India to proclaim a new gospel of beauty and joy. The shepherd is denounced as a heretic. King Roger is finally converted to the new doctrine. The opera closes with a bacchanale in the ruins of a Greek temple. The Song of Roxane, in which the Queen pleads with the shepherd for clemency, is famous in a transcription by Jascha Heifetz for violin and piano.

King's Henchman, The, opera in three acts by Deems Taylor. Libretto by Edna St. Vincent Millay. Première: Metropolitan Opera, February 17, 1927. The setting—England in the tenth century—led one of America's outstanding poets to write her poetic play in archaic language. The story is a variation of the Tristan-Isolde theme. Eadgar of Wessex, King of England, sends his trusted friend Aethelwold to Devon to win for him the princess Aelfrida. Aethelwold, lost in a forest near Devon, falls asleep. Aelfrida, singing an incantation that will divine her future husband, comes upon Aethelwold, and the magic spell makes them fall in love with one another. He sends word to the King that the Devon princess is ugly. After Aethelwold and Aelfrida are married, Eadgar comes to Devon and discovers how beautiful Aelfrida really is. Overwhelmed by his own treachery, Aethelwold kills himself with a dagger, in the presence of his wife and King. The King mourns the death of his dear friend ("Nay, Maccus, lay him down").

King Winter, the Snow Maiden's father (bass) in Rimsky-Korsakov's *The Snow Maiden.*

Kipnis, Alexander, bass. Born Zhitomir, Ukraine, February 13, 1891. He was graduated from the Warsaw Conservatory in 1912 as a conductor. He was receiving vocal lessons in Berlin with Grenzbach when World War I began. As an enemy alien he was at first interned, but he was soon permitted to continue his studies. In 1915 he made his debut with the Hamburg Opera, and from 1916 to 1918 he was a member of the Wiesbaden Opera. After the war he made many successful appearances throughout Europe. He came to the United States for the first time in 1923 as a member of the visiting Wagner Festival Company, making his debut in New York as Pogner on February 12. He was immediately engaged by the Chicago Civic Opera, where he remained nine seasons. After leaving Chicago, Kipnis became principal bass of the Berlin State Opera. He also appeared with outstanding success at the Bayreuth Festivals. After the rise of Hitler, Kipnis abandoned Germany and joined the Vienna State Opera; he was also seen at the Paris Opéra, at Covent Garden, Glyndebourne, and the Salzburg Festivals. On January 5, 1940, he made his debut at the Metropolitan Opera as Gurnemanz. He remained with the Metropolitan through the 1945–1946 season, acclaimed in the Wagnerian repertory and in such roles as Boris Godunov, Sarastro, Arkel, and Rocco. Kipnis is also famous as an interpreter of lieder.

Kirchhofer, Werner, the trumpeter (baritone) in Nessler's *Der Trompeter von Säkkingen.*

Kirsten, Dorothy, soprano. Born Montclair, New Jersey, July 6, 1917. She studied singing at the Juilliard School of Music while working for the telephone company. An appearance over the radio attracted the interest of Grace Moore, who become her sponsor. Kirsten's studies now continued in Italy with Astolfo Pescia. After returning to the United States in 1940, she made her opera debut with the Chicago

Opera, after which she was heard in leading soprano roles with the New York City Opera and the San Carlo. On December 1, 1945, she made a successful debut at the Metropolitan Opera as Mimi. She has also appeared extensively over radio and television, and in the movies (Mr. Music, The Great Caruso).

Kiss, The, folk opera by Bedřich Smetana. Libretto by Eliska Krashnohorská, based on a story by Karolina Svetlá. Première: National Theater, Prague, November 7, 1876. Hanno, a young widower, is eager to kiss his bride, Marinka before their marriage—since popular belief has it that such a kiss arouses the anger of a dead wife. He succeeds, but only after complications. Marinka's lullaby, "Sleep My Child," in Act I, and Brigitta's Lark Song in Act II are familiar.

Klänge der Heimat (Csárdás), Rosalinde's aria with chorus in Act II of Johann Strauss's Die Fledermaus.

Klafsky, Katharina, dramatic soprano. Born St. Johann, Hungary, September 19, 1855; died Hamburg, Germany, September 22, 1896. She studied singing with Mathilde Marchesi in Vienna. After a career in comic opera, and in minor opera roles in Salzburg, she went into temporary retirement. She returned to the stage in 1876 and appeared in minor parts in the Wagnerian repertory with the Angelo Neumann Company. By 1881 she was appearing in leading parts. From 1886 to 1895 she was the principal soprano of the Hamburg Opera, and a guest performer at other leading European opera houses. Soon after her marriage to Otto Lohse, the conductor, she came to the United States and sang the principal Wagnerian roles for the Damrosch Opera Company; her husband appeared with the company as conductor. In the fall of 1896 she again appeared with the Hamburg Opera. On September 11 she sang her final per-

formance—her role was Leonore in Fidelio—and a few days later her career was cut short by death.

Kleiber, Erich, conductor. Born Vienna, August 5, 1890. He received a comprehensive academic education in Vienna and Prague while studying music privately and at the Prague Conservatory. In 1911 he became a coach with the Prague Opera. A year later he was appointed conductor of the Darmstadt Opera, where he remained eight years. He next served as general music director at Mannheim, where a remarkable performance of Fidelio brought him an appointment as musical director of the Berlin Opera. He remained in Berlin for over a decade, distinguishing himself both in the standard repertory and in modern works; among the notable world premières he conducted were Křenek's Das Leben des Orest, Berg's Wozzeck, and Milhaud's Cristophe Colomb. On October 2, 1930, he made his American debut as guest conductor of the New York Philharmonic. During the next decade he appeared frequently in America as a symphony conductor. After Hitler came to power, Kleiber was permitted to retain his post with the Berlin Opera; but after the controversy over the première of Mathis der Maler he resigned his post. (See MATHIS DER MALER; FURTWAENGLER.) Between 1934 and 1937 he conducted symphony and opera in Moscow. He also led notable performances at La Scala and the Salzburg Festivals. In 1938 he became the principal conductor of the National Theater in Prague. From 1939 to 1949 he conducted extensively in South America. After 1949 he appeared as guest conductor in leading European opera houses and at major opera festivals in Amsterdam, Zurich, and Prague. In January, 1951, he returned to conduct the Berlin Opera in East Germany. Upon reconstruction of the opera house and the development of plans for a grand reopening in 1955, Kleiber was appointed musical director. However, a few months after the appointment, and before the scheduled reopening, Kleiber resigned his position and moved to West Germany, accusing Communist officials of injecting propaganda into art. He was succeeded by Franz Konwitschny.

Klein, Herman, writer on music. Born Norwich, England, July 23, 1856; died London, March 10, 1934. For twenty years, up to 1901, he was the music critic of the London *Sunday Times*. He came to New York in 1902, where he remained seven years, writing criticisms for the *Herald* and promoting concerts. From 1927 to 1934 he was the music critic of the *Saturday Review* in London. He wrote a number of books on operatic subjects: *The Reign of Patti* (1920); *The Art of Bel Canto* (1924); *Great Women Singers of Our Time* (1931); *The Golden Days of Opera* (1933).

Kleist, Heinrich Wilhelm von, author and dramatist. Born Frankfort-on-the-Oder, Germany, October 18, 1777; died Wannsee, November 21, 1811. His novelette *Michael Kohlhaas,* one of the finest stories in German literature, was made into an opera by Paul von Klenau. His tragedies *Penthesilea* and *Der Prinz von Homburg* were made into operas, the first by Othmar Schoeck, the second by Paul Graener.

Klemperer, Otto, conductor. Born Breslau, Germany, May 14, 1885. His music study took place at the Hoch Conservatory in Frankfort-on-the-Main, and in Berlin with Xaver Scharwenka, James Kwast, and Hans Pfitzner. He held several minor posts as conductor between 1905 and 1907, when he attracted the interest of Gustav Mahler, who recommended him to the National Theater in Prague. Klemperer stayed there three years. Once again with Mahler's recommendation,

he now became principal conductor of the Hamburg Opera. Several important engagements in various German opera houses preceded his appointment in 1924 as principal conductor of the Berlin Volksoper, where his performances were so outstanding that in 1926 he was made musical director of the Berlin State Opera on a ten-year contract. During this period Klemperer gave many extraordinary performances, particularly of modern operas by Křenek, Schoenberg, and Hindemith. He made his American debut on January 24, 1926, as guest conductor of the New York Symphony Society. He returned to America for several additional seasons besides giving guest performances in Italy, Spain, the Soviet Union, and South America. When the Nazis came to power, Klemperer left his native land for good. In 1933 he was appointed permanent conductor of the Los Angeles Philharmonic Orchestra. Six years later poor health compelled him to resign from this post. Since then his appearances have been intermittent. In Europe, where he has been since the end of World War II, he has appeared chiefly as a conductor of symphonic concerts, although he conducted at the Budapest Opera from 1947 to 1950.

Klenau, Paul von, composer and conductor. Born Copenhagen, February 11, 1883; died there August 31, 1946. He studied with Max Bruch and Karl Halir in Berlin from 1902 to 1904 and after that with Ludwig Thuille in Munich. He was only twenty-one when he wrote *Inferno,* a successful work for orchestra. In 1907 he was appointed conductor of the Freiburg Opera and from 1908 to 1914 he was Max von Schillings' assistant at the Stuttgart Opera. In 1913 his one-act opera *Sulamith* was given in Munich. Of his later operas, *Michael Kohlhaas* was the most successful, first given in Stuttgart in 1933. In 1920 he founded the Vienna

Singakademie and the Konzerthausgesellschaft. His operas: *Sulamith* (1913); *Kjartan und Gudrun* (1918, revised as *Gudrun auf Island* (1924); *Die Lästerschule* (1927); *Michael Kohlhaas* (1933); *Rembrandt von Rijn* (1937).

Klingsor, a magician (baritone) in Wagner's *Parsifal.*

Kluge, Die (The Wise One), opera by Carl Orff. Libretto by the composer, based on a fairy tale by the Grimm brothers. Première: Frankfort-on-the-Main, February 20, 1943. Subtitled *The Story of the King and the Wise Woman,* the opera deals with a peasant woman who is so wise she is able to win the heart of, and marry, the king. The opera is written in singspiel style and includes spoken recitatives.

Klytemnestra, (1) wife of Agamemnon (mezzo-soprano) in Gluck's *Iphigénie en Aulide.*

(2) Mother of Elektra (mezzo-soprano) in Richard Strauss's *Elektra.*

Knappertsbusch, Hans, conductor. Born Elberfeld, Germany, March 12, 1888. His academic studies took place at the Bonn University, where he specialized in philosophy; his musical training was completed at the Cologne Conservatory. In 1912 he directed a festival of music dramas in Holland. A year later he became director of opera at Elberfeld, holding this post five years. In 1922, given a lifetime contract, he succeeded Bruno Walter as musical director of the Munich Opera. He now achieved recognition throughout Europe as an interpreter of Wagner, Mozart, and Richard Strauss. When the Nazis came to power, he refused to join the Nazi party or to subscribe to its political or musical doctrines. On Hitler's personal decision, he was removed from his Munich post. In 1936 he went to Vienna where he conducted at the State Opera, becoming its musical director in 1938. After the Anschluss he left Austria and gave guest performances in several of Europe's non-Fascist countries. Since the end of World War II he has directed in Munich, Vienna, at the Salzburg Festivals, and in other European centers.

Knote, Heinrich, dramatic tenor. Born Munich, November 26, 1870; died Garmisch-Partenkirchen, January 15, 1953. After completing his studies with E. Kirschner in Munich he joined the Munich Opera in 1892, remaining with that company a dozen years. During this period he appeared in most of the major German opera houses. He made his American debut at the Metropolitan Opera on December 3, 1904, in *Die Meistersinger.* He was so successful in this and subsequent Wagnerian performances that during the three seasons he sang at the Metropolitan his popularity rivaled that of Caruso and Jean de Reszke. From 1917 on he was the principal tenor of the Charlottenburg Opera. He retired from opera in 1924.

Knusperhäuschen, Das, see WITCH'S HOUSE, THE.

Knusperwalzer, see GINGERBREAD WALTZ, THE.

Koanga, opera by Frederick Delius. Libretto by Charles Francis Keary, taken from material in *The Grandissimes,* a novel by George Washington Cable. Première: Elberfeld Opera, March 30, 1904 (in German). The opera was written between 1895 and 1897. Koanga is an African chieftain who is captured and brought to Spain. There, the octoroon Palmyra falls in love with him; they both meet their doom in the jungle after Koanga invokes voodoo and black magic.

Kobbé, Gustav, writer on music. Born New York City, March 4, 1857; died Babylon, Long Island, July 27, 1918. His music study was pursued in Wiesbaden with Adolf Tagen and in New York with Joseph Mosenthal. Beginning in 1880 he served successively as music critic for several New

York newspapers, including the *Sun, World, Mail,* and *Herald.* He was sent to Bayreuth in 1883 to cover the first festival there. He wrote several books on opera: *Wagner's Life and Works* (1890); *Opera-Singers* (1901); *Wagner and his Isolde* (1906); *The Complete Opera Book* (1919, revised 1954 by the Earl of Harewood). He also wrote a novel about the world of opera, *Signora, a Child of the Opera-House* (1902).

Kobus, Fritz, principal male character (tenor) in Mascagni's *L'amico Fritz.*

Kodály, Zoltán, composer. Born Kecskemét, Hungary, December 16, 1882. He entered the Budapest Conservatory when he was eighteen, a fellow student of Béla Bartók. He accompanied Bartók on some of his expeditions through Hungary to study and record the folk songs and dances of different regions and, with Bartók, he was a powerful influence in bringing the folk music of his native land to public notice. This preoccupation had a far-reaching influence on his own music: his works are unmistakably Hungarian. Kodály has written interesting and significant works in many forms. His opera *Háry János* (1926) is his best-known composition—not because it is often performed but because of the popular concert suite drawn from its music. His later operas: *The Spinning Room of the Szekelys* (1932) and *Czinka Panna* (1943).

Kolodin, Irving, music critic. Born New York City, February 22, 1908. His studies were completed at the Institute of Musical Art, following which, in 1932, he became assistant music critic to W. J. Henderson on the New York *Sun.* Upon Henderson's death, in 1937, he became principal music critic, remaining in this post until the dissolution of the newspaper a decade later. He then became the music editor of the *Saturday Review of Literature.* He has written two different histories of the

Metropolitan Opera: *The Metropolitan Opera: 1883–1939* and *The Story of the Metropolitan Opera: 1883–1950.*

Königin von Saba, Die, *see* QUEEN OF SHEBA, THE.

Königskinder, Die (The Royal Children), fairy opera by Engelbert Humperdinck. Libretto by Ernst Rosmer (pen name of Elsa Bernstein). Première: Metropolitan Opera, December 28, 1910. The Goose Girl, who lives in the woods with a cruel witch, meets and falls in love with the King's son, who comes to her hut disguised as a beggar. Before he leaves her, he promises she will see him again when a star falls into a certain lily. When this miracle occurs, Goose Girl and the King's son meet again, only to die of a poisoned pastry prepared by the witch.

Konrad, a huntsman (tenor) in Marschner's *Hans Heiling.*

Konstanze, *see* CONSTANZA.

Kontchak, Khan, the Polovtsian chief (bass) in Borodin's *Prince Igor.*

Kontchakovna, Kontchak's daughter (contralto) in Borodin's *Prince Igor.*

Körner, Karl Theodor, poet and dramatist. Born Dresden, Germany, September 23, 1791; died Gadebusch, August 26, 1813. Dvořák's first opera, *Alfred,* was based on a drama by Körner. Körner's comedy *Die vierjährige Posten* was made into comic operas of the same name by Paul Graener, Carl Reinecke, and Franz Schubert. Flotow's *Die Bergknappen* had a libretto by Körner.

Korngold, Erich Wolfgang, composer. Born Brünn, Austria, May 29, 1897. The son of a noted Viennese music critic, Julius Korngold, Erich Wolfgang began to study music at an early age. He was extraordinarily precocious. When he was eleven he wrote a pantomime, *Der Schneemann,* performed by the Vienna Opera. On March 28, 1916, two one-act operas were introduced in Munich: *Der Ring des Polykrates* and *Violanta;* both were subsequently heard in America. His greatest success came with *Die tote Stadt,* introduced simultaneously in Hamburg and Cologne on December 4, 1920, and soon after given at the Metropolitan Opera. Later operas: *The Miracle of Heliane* (1927); *Kathrin* (1937). Korngold came to the United States in 1935 and has since devoted himself principally to writing scores for motion pictures. In one picture, *Give Us This Night,* he introduced an original one-act opera.

Kotzebue, August Friedrich von, dramatist. Born Weimar, Germany, May 3, 1761; died Mannheim, Germany, March 23, 1819. He was the author of over two hundred fairy, historical, and social plays, some of which were made into operas, notably Gustav Albert Lortzing's *Der Wildschütz* and Ernst Wolf's *Der Eremit auf Formentara.* Kotzebue wrote the libretto for Franz Schubert's operetta *Des Teufels Lustschloss* and for Johann Reichardt's *Das Zauberschloss.* Ludwig Spohr's *Der Kreuzfahrer* and François Boieldieu's *Beniowski* were based on plays by Kotzebue.

Koupava, a village girl (mezzo-soprano) in Rimsky-Korsakov's *The Snow Maiden.*

Kraus, Ernst, dramatic tenor. Born Erlangen, Bavaria, June 8, 1863; died Wörthersee, Austria, September 6, 1941. After studying with Cesare Galliera in Milan and Anna Schimon-Regan in Munich he made his debut in Mannheim as Tamino on March 26, 1893. In 1894 and 1895 he came to the United States and appeared as principal tenor of the Damrosch Opera Company. In 1896 he was appointed leading tenor of the Berlin Opera, where he remained for twenty-seven years, distinguishing himself in the German repertory. During this period he appeared as Siegmund at Bayreuth in 1901 and made his debut at the Metropolitan Opera, also as Siegmund, on

November 25, 1903. He retired from opera in 1924 and devoted himself thenceforward to teaching.

Kraus, Felix von, dramatic bass. Born Vienna, October 3, 1870; died Munich, October 30, 1937. He received a doctorate in musicology from the University of Vienna in 1894, but in singing he was mostly self-taught. His debut took place at Bayreuth in 1899 when he appeared as Hagen. He was subsequently heard regularly at the Bayreuth Festivals and in leading European opera houses, a specialist in the Wagner repertory. In 1908 he became artistic director of the Munich Opera and a professor at the Munich Conservatory. In 1899 he married the American contralto Adrienne Osborne, who also appeared in the Wagner dramas. He retired from the stage in 1924.

Krauss, Clemens, conductor. Born Vienna, March 31, 1893; died Mexico City, May 16, 1954. As a child he was a member of the Imperial Choir in which both Haydn and Schubert had been choristers. His music study took place at the Vienna Conservatory, after which he became chorus master at the Brünn Opera. In 1913 he became second conductor of the Riga Opera. Various appointments followed and in 1922 he succeeded Furtwaengler as director of the Tonkünstlerverein in Vienna. From 1924 to 1929 he was the artistic director of the Frankfort Opera, where he distinguished himself in performances of operas by Mozart and Richard Strauss. Appearances at the Munich and Salzburg Festivals extended his reputation. In 1929 he became artistic director of the Vienna State Opera, and from 1929 to 1934 he was a principal conductor at the Salzburg Festivals. He made his American debut in 1929 as a guest conductor of the Philadelphia Orchestra. When Furtwaengler was deposed as musical director of the Berlin State Opera in 1934, following his dispute with the Nazi authorities over the première of *Mathis der Maler,* Krauss was selected as his successor. He was not liked in Berlin and before long he had to shift his activities to Munich. In 1944 he directed the world première of Richard Strauss's *Die Liebe der Danae* in Salzburg. In 1947 he returned to the Vienna State Opera, and after that he directed performances at the Bayreuth Festival. Subsequently, he appeared in South America and Mexico as a conductor of opera, and it was in Mexico that he unexpectedly died.

Krauss, Gabrielle, soprano. Born Vienna, March 24, 1842; died Paris, January 6, 1906. She attended the Vienna Conservatory, then studied singing with Mathilde Marchesi. In July 1859 she made her debut at the Vienna Opera in *William Tell.* For several years she continued to appear there. On April 6, 1867, she made a strikingly successful debut at the Théâtre des Italiens in Paris in *Il trovatore.* She continued to sing there three seasons. During the next few years she was acclaimed in Italy and Russia. On January 5, 1875, she made her debut at the Paris Opéra, when its new building was inaugurated. She remained a principal soprano of the Paris Opéra for the next dozen years, after which she retired from the stage and devoted herself to teaching. She was a remarkable actress and was often described by her French public as "la Rachel chantante." She appeared in many world premières, including Gounod's *Polyeucte* and *Le tribut de Zamora,* Halévy's *Guido et Ginevra,* and Saint-Saëns' *Henry VIII.*

Krehbiel, Henry Edward, writer on music. Born Ann Arbor, Michigan, March 10, 1854; died New York City, March 20, 1923. He served his apprenticeship as critic with the Cincinnati *Gazette* from 1874 to 1880. Going to New York, he then became the music critic of the *Tribune,* retaining this post until his death. Of his many books, the following are of operatic interest: *Studies in the Wagnerian Drama* (1891); *Chapters of Opera* (1908); *A Book of Operas* (1909); *A Second Book of Operas* (1917); *More Chapters of Opera* (1919); Krehbiel translated the libretto of *Parsifal* into English.

Křenek, Ernest, composer. Born Vienna, August 23, 1900. After studying with Franz Schreker in Vienna and Berlin, he wrote his first opera, *Zwingburg,* in 1922. A second opera, *Der Sprung über den Schatten,* aroused interest when introduced at Frankfort-on-the-Main in 1925. A year later he was appointed conductor of the Kassel State Theater. During this period he completed *Jonny spielt auf,* an opera in the jazz idiom that achieved a sensational success throughout Europe after its introduction at the Leipzig Opera on February 11, 1927. After 1928 Křenek abandoned the jazz idiom to write in the romantic style of Schreker; later he adopted the atonal manner of Schoenberg. Křenek first visited the United States in 1937 as conductor of the Salzburg Opera Guild. After the German seizure of Austria, he settled permanently in this country, holding teaching positions with various universities and conservatories. Other operas by Křenek: *Orpheus und Eurydike* (1923); *Der Diktator* (1926); *Das geheime Königreich* (1927); *Das Leben des Orest* (1929); *Cefalo e Procri* (1933); *Karl V* (1933); *Tarquin* (1941); *What Price Confidence* (1946); *Pallas Athena Weint* (1952); *Dark Waters* (1954).

Kreutzer, Conradin, composer and conductor. Born Messkirch, Baden, November 22, 1780; died Riga, Latvia, December 14, 1849. He studied law in Freiburg but turned to music in 1800 and completed an operetta which was performed that year. In 1805 he went to Vienna, where for two years he studied with Johann Albrechtsberger and where his opera *Jery and Bätely* was successful when produced in 1810. In 1812 he was appointed court kapellmeister in Stuttgart, where he wrote eight operas. Five years later he held a similar post at Donaueschingen. He returned to Vienna in 1822. He was, on and off, a principal conductor of the Kärntnerthortheater between 1825 and 1840. From 1833 to 1837 he conducted opera at the Josephstädter Theater. From 1840 to 1846 he was musical director of the city of Cologne. He wrote about thirty operas. The most successful were: *Jery und Bätely* (1810); *Libussa* (1822); *Aesop in Phrygien* (1822); *Das Nachtlager in Granada* (1834); *Der Verschwender* (1836).

Kreutzer, Rodolphe, composer and violinist. Born Versailles, November 16, 1766; died Geneva, January 6, 1831. Kreutzer, to whom Beethoven dedicated his *Kreutzer Sonata,* was the composer of many once popular operas. He studied the violin with his father and Anton Stamitz and was only thirteen when he appeared in Paris, playing one of his own concertos. Three years later he was appointed first violinist of the Chapelle du Roi, and in 1790 solo violinist at the Théâtre des Italiens. In 1790 his first opera, *Jeanne d'Arc à Orléans,* was introduced at the Théâtre des Italiens, followed a year later by *Paul et Virginie.* He wrote about forty more operas, many of them produced at the Opéra, the Opéra-Comique, and the Théâtre des Italiens; the most famous was *Lodoïska,* heard in 1791. In 1795 he was appointed a professor of the violin at the Paris Conservatory, where he remained until 1825; in 1801 he was appointed first violinst at the Opéra; and after 1816 he conducted at the Opéra. Several of his operas were performed here, notably *Astyanax* (1801), *Aristippe* (1808), and *La mort d'Abel* (1810). In 1806 he became solo violinist to Emperor Napoleon, and in 1815

maître de chapelle to Louis XVIII. He retired from all musical activity after 1826.

Kronold, Selma, soprano. Born Cracow, Poland, 1866; died New York City, October 9, 1920. After studying with Arthur Nikisch at the Leipzig Conservatory she made her debut at the Leipzig Opera in 1882 as Agathe. She then joined the Angelo Neumann Opera Company, which toured Europe in the Wagner dramas. After an additional period of study with Desirée Artôt in Paris she came to the United States in 1888 and appeared in concerts. From 1889 to 1891 she sang with the Berlin Opera, and in 1891 with the Gustav Hinrichs Opera Company in the United States; as a member of the latter company she created for America the leading soprano roles in *Cavalleria rusticana, Pagliacci,* and *Manon Lescaut.* She made her debut at the Metropolitan Opera on February 6, 1891, in *Die Walküre.* She retired from opera in 1904 and henceforth devoted herself to the Catholic Oratorio Society, which she founded and directed. Subsequently, she entered a convent.

Kruschina, a Bohemian peasant (baritone) in Smetana's *The Bartered Bride.*

Kullman, Charles, tenor. Born New Haven, Connecticut, January 13, 1903. He graduated from Yale University in 1924, after which he specialized in music. He received some coaching from local teachers, then won a scholarship at the Juilliard School of Music in New York, where he stayed three years. Another scholarship enabled him to continue his studies at the American Conservatory in Fontainebleau, France. Back in the United States in 1928, he joined the music faculty of Smith College, where he appeared in several opera productions. He also appeared with the American Opera Company in leading tenor roles. After a period of studying the opera repertory in Berlin he was engaged by

Otto Klemperer for the Kroll Theater and made his European opera debut as Pinkerton on February 24, 1931. A year later he joined the Berlin State Opera. He was so popular there that he had to appear in *Madama Butterfly* twenty-five times in a single season. In 1934 Kullman made his debuts at the Vienna State Opera and at Covent Garden, and from 1934 to 1936 he appeared at the Salzburg Festivals in operas conducted by Toscanini. Meanwhile, on December 20, 1935, he made his Metropolitan Opera debut in *Faust*. He has since appeared at the Metropolitan in leading Italian and French roles. In 1955 he received an award from the Metropolitan to celebrate his twentieth anniversary there. He has also been successful in the concert hall, in radio performances, and in motion pictures.

Kundry, an enchantress (mezzo-soprano) in Wagner's *Parsifal.*

Kuno, Prince Ottokar's head ranger (bass) in Weber's *Der Freischütz.*

Kunz, Erich, bass. Born Vienna, May 20, 1909. After studying with Theodore Lierhammer in Vienna he became an understudy at the Glyndebourne Opera in England. He made his official opera debut as Osmin with the Troppau Opera in Troppau (now Opava), Czechoslovakia. After singing in Plannen and Breslau, he achieved an outstanding reputation at the Salzburg and Bayreuth festivals between 1941 and 1943. He subsequently performed in the leading opera houses of Europe and was particularly acclaimed for his Mozart interpretations. His American debut took place at the Metropolitan Opera on November 26, 1952, as Leporello.

Kurt, Melanie, dramatic soprano. Born Vienna, January 8, 1880; died New York City, March 11, 1941. She began the study of singing with Fannie Mütter in Vienna and in 1902 made her debut in Lübeck as Elisabeth. After appearances in Leipzig she withdrew from the stage to study with Lilli Lehmann in Berlin. She returned to opera in 1905 with appearances in Brunswick. From 1908 to 1912 she was principal soprano of the Berlin Opera, famous for her Wagnerian interpretations. When the Charlottenburg Opera opened in Berlin in 1912, she was engaged as principal soprano. On February 1, 1915, she made her American debut at the Metropolitan Opera as Isolde; she remained with this company two seasons. Her appearances in opera now grew infrequent as she specialized in concert appearances and teaching. She went into retirement in 1932, and at the outbreak of World War II came to live in New York.

Kurwenal, Tristan's servant (baritone) in Wagner's *Tristan und Isolde.*

Kurz, Selma, coloratura soprano. Born Bielitz, Austria, November 15, 1875; died Vienna, May 10, 1933. After studying voice with Hans Pless she made her debut at the Frankfort Opera. She was so successful that Gustav Mahler engaged her for the Vienna Opera. She remained in Vienna for over a quarter of a century, until 1926, scoring triumphs in most of the famous coloratura roles. In 1930 she was made an honorary member of the Vienna Opera, an honor previously held by only six singers. Shortly before her retirement in 1926 she came to the United States for a few concert appearances. As a member of the Vienna Opera, she appeared in successful guest performances in other major European opera houses, including Covent Garden.

Kyoto, a procurer (baritone) in Mascagni's *Iris.*

L

Lablache, Luigi, bass. Born Naples, December 6, 1794; died there January 23, 1858. His music study took place at the Conservatorio della Pietà de' Turchini in Naples. In 1812 he made his debut at the San Carlo in Naples in Valentino Fiorávanti's *La molinara.* After an additional period of study he was engaged as principal bass of the Palermo Opera, where he remained for five years. He then made his debut at La Scala in *La cenerentola.* After many other successful appearances throughout Italy, he appeared in Vienna in 1824, receiving a gold medal. Three years later he sang in a performance of Mozart's *Requiem* performed at Beethoven's funeral services. Schubert came to know Lablache at this time, dedicating his *Three Italian Songs* to him. In 1830 Lablache made his London debut in *Il matrimonio segreto,* and the following year he appeared for the first time in Paris in the same opera. For several years he appeared alternately in London and Paris, a great favorite in both cities. For a period he was Queen Victoria's singing master. Poor health forced him to retire in 1852. He had a voice exceptional in range and volume. His most famous characterization was that of Leporello.

Laca, Jenufa's stepbrother and lover (tenor) in Janáček's *Jenufa.*

La calunnia è un venticello, Basilio's aria about slander in Act I, Scene 3, of Rossini's *The Barber of Seville.*

La Charmeuse, a dancer (soprano) in Massenet's *Thaïs.*

Là ci darem la mano, duet of Don Giovanni and Zerlina in Act I, Scene 3, of Mozart's *Don Giovanni.*

La Cieca, La Gioconda's blind mother (contralto) in Ponchielli's *La Gioconda.*

La Ciesca, Marco's wife (soprano) in Puccini's *Gianni Schicchi.*

Lackland, Sir Gower, fiancé (tenor) of Lady Marigold in Howard Hanson's *Merry Mount.*

La donna è mobile, one of the most celebrated tenor arias in Italian opera, the Duke of Mantua's aria in Act IV of Verdi's *Rigoletto.*

Lady Macbeth of Mtsensk, opera by Dmitri Shostakovich. Libretto by the composer and A. Preiss, based on a story by Nikolai Leskov. Première: Leningrad, January 22, 1934. Katerina murders both her husband and her father-in-law in order to marry Serge, a clerk. During their wedding dinner, one of the corpses is found in the bedroom. Katerina and Serge are sentenced to hard labor in Siberia. Serge now falls in love with a prostitute. Katerina kills her and commits suicide. In the original Leskov story Katerina is portrayed as the cruel and ruthless woman she really is. In the opera the composer depicts her as the helpless victim of a callous society. The opera was produced with great success in Moscow, running two years to packed houses. In 1936 it suddenly became the target for a violent attack by the Soviet press. *Pravda* now described it as "the coarsest kind of naturalism. . . . The music quacks, grunts, growls, and suffocates itself in order to express the amatory scenes as naturalistically as possible." Because of this official censure, which continued on and off for about a year, Shostakovich became *persona non grata,* avoided by his fellow composers, looked down upon by others.

He later rehabilitated his position in Soviet music with his Fifth Symphony.

Lady Macbeth was produced in the United States (Cleveland and New York) under the direction of Artur Rodzinski in January and February, 1935, and was enthusiastically received. It might be noted that its favorable reception here preceded its abrupt denunciation by *Pravda*.

Laërtes, (1) Polonius' son (tenor) in Thomas's *Hamlet*.

(2) An actor (tenor) in Thomas's *Mignon*.

Laetitia, a servant (soprano) in Menotti's *The Old Maid and the Thief*.

La fatal pietra, Radames' duet with Aïda in Act IV, Scene 2, in Verdi's *Aïda*.

la Fontaine, Jean de, poet and fabulist. Born Château Thierry, France, July 8, 1621; died Paris, April 13, 1695. Famous for his *Contes* and *Fables,* La Fontaine wrote works that were the source for operas by Grétry, Hérold, Monsigny, Philidor, and numerous other composers, including Gluck (*L'ivrogne corrigé*), Gounod (*La colombe*), and Pierné (*La coupe enchantée*).

Lagerlöf, Selma, novelist. Born Mårbacka, Sweden, November 20, 1858; died Mårbacka, March 16, 1940. One of Sweden's most celebrated authors, Lagerlöf has written stories that have inspired a number of operas. Her masterwork, *Gösta Berling's Saga,* was made into an opera by Zandonai, *I cavalieri di Ekebu.* Manfred Gurlitt's *Nordische Ballade* was derived from *Arne's Treasure,* Oskar Lindberg's *Fredlös* from *The Outlaw,* and Franco Vittadini's *Nazareth* from one of the author's short stories.

Lakmé, opera in three acts by Léo Delibes. Libretto by Edmond Gondinet and Philippe Gille, based on Gondinet's *Le mariage de Loti.* Première: Opéra-Comique, April 14, 1883. American première: Academy of Music, New York, March 1, 1886.

Characters: Gérald, a British officer (tenor); Frédéric, his friend (baritone); Nilakantha, a Brahman priest (bass); Lakmé, his daughter (soprano); Mallika, her slave (mezzo-soprano); Ellen, the Governor's daughter (soprano); Rose, her cousin (mezzo-soprano) Mrs. Benson, their governess (mezzo-soprano); Hadji, a Hindu slave (tenor); Hindus; British officers; ladies; sailors; musicians; Brahmans. The setting is India in the middle nineteenth century.

Act I. A garden. The Brahman priest Nilakantha tells his followers that the English invader will soon be driven from India. From within the temple comes Lakmé's voice in prayer ("Blanche Dourge"). After the worshipers scatter, some English sightseers invade the garden, heedless of its sanctity. One of them is Gérald. He espies Lakmé's jewels and is so taken by them that, when his friends leave, he decides to sketch them ("Prendre le dessin d'un bijou"). Lakmé catches him in the act. They are immediately drawn to one another. Lakmé begs Gérald to leave the garden. He leaves, but only after he has told Lakmé how much he loves her. Nilakantha reappears, perceives that the place has been desecrated by a foreigner, and vows that the criminal must die.

Act II. A public square. Lakmé and Nilakantha are both in disguise, for the priest is searching for the stranger who profaned the temple. He orders Lakmé to sing, certain that the offender will reveal himself at the sound of her voice. Lakmé renders a haunting, exotic melody (Bell Song: "Où va la jeune Hindoue?"). Overcome by his love, Gérald rushes toward Lakmé. Nilakantha stabs him and escapes, thinking Gérald dead. Lakmé is overjoyed to find that Gérald's wound is not mortal.

Act III. A forest. Lakmé is nursing

Gérald back to health. He greets his savior and lover ("Je me souviens, sans voix"). Lakmé goes to get some water from a near-by sacred fountain; those who drink it will remain true in their love. While she is away, Gérald's friend Frédéric arrives and urges him to rejoin his regiment. As Lakmé returns with the magic water, martial music is heard in the distance. Noting her lover's reaction to it, Lakmé realizes that he will return to his proper place and she will lose him forever. Unable to think of life without him, she plucks a lethal datura blossom and eats it. Gérald is horrified but Lakmé welcomes death ("Tu m'as donné le plus doux rêve"). Her father appears. Incensed at finding the Englishman with his daughter, he orders his men to kill Gérald. Lakmé proudly tells her father that she will placate the gods by dying in his place. Singing of her love for Gérald, she expires.

The only one of Delibes's operas to survive, *Lakmé* is one of the most popular items in the French repertory. Its appeal is not difficult to explain. Its orientalism gives it a delightful flavor, and it boasts one of the most famous coloratura arias in all opera, the "Bell Song."

Lalla Roukh, a poem by Thomas Moore consisting of four tales about an Indian princess, Lalla Roukh, who goes to the valley of Kashmir to meet her beloved, the Sultan of Bucharia. Operas inspired by this poem include Félicien David's *Lalla Roukh,* Anton Rubinstein's *Feramors,* and Charles Stanford's *The Veiled Prophet of Khorassan.*

Lalo, Edouard, composer. Born Lille, France, January 27, 1823; died Paris, April 22, 1892. After preliminary studies in Lille, Lalo entered the Paris Conservatory in his sixteenth year. In 1867 he entered his first opera, *Fiesque,* in a competition sponsored by the Théâtre Lyrique. While it won only third prize, the director of the Paris Opéra accepted it. Due to a series of misfortunes, including the burning of the opera house, it was not performed. Lalo's recognition as a composer came in the 1870's with two works for violin and orchestra written for the famous violinist Sarasate: a concerto and the *Symphonie Espagnole.* His most important and most successful opera, *Le Roi d'Ys,* was given at the Opéra-Comique in 1888 and entered the permanent repertory of that theater. Lalo began a third opera, *La Jacquerie;* the bulk of it was finished by Arthur Coquard after Lalo's death.

L'altra notte, Marguerite's aria in Act III of Boïto's *Mefistofele.*

La mamma morta, Madeleine's aria in Act III of Giordano's *Andrea Chénier.*

Lamartine, Alphonse de, poet. Born Macon, France, October 21, 1790; died Paris, March 1, 1869. Benjamin Godard's *Jocelyn* and Jules Mazellier's *Graziella* were inspired by poems of this author.

Lament of Arianna, *see* LASCIATEMI MORIRE.

Lamoureux, Charles, conductor. Born Bordeaux, France, September 28, 1834; died Paris, December 21, 1899. His studies were completed at the Paris Conservatory after which he played the violin in various orchestras and was assistant conductor of the Paris Conservatory Orchestra. In 1876 he conducted at the Opéra-Comique, resigning a year later due to differences with the management over matters of interpretation. He then became conductor of the Opéra for the next two years, once again leaving after disagreements with the management. He now devoted himself principally to symphonic music, founding the renowned Concerts Lamoureux in 1881, which he directed for the remainder of his life. He was one of the earliest champions of Wagner in France and led the first performance of a Wagner opera to be heard in Paris after the fiasco of *Tannhäuser* in 1861. The opera was *Lohengrin,* introduced on May 3, 1887.

Lamperti, Francesco, teacher of singing. Born Savona, Italy, March 11, 1813; died Como, Italy, May 1, 1892. He attended the Milan Conservatory. Beginning in 1850, he taught there for a quarter of a century. Afterward, he taught privately. The long list of noted singers who were his pupils includes Emma Albani, Desirée Artôt, Italo Campanini, and Marcella Sembrich. His methods adhered to the old traditions of Italian singing; he wrote several valuable treatises. He was made a Commander of the Crown of Italy.

L'anima ho stanca, Maurice's aria in Act II of Cilèa's *Adriana Lecouvreur.*

La notte il giorno, the beggars' chorus in Act I of Giordano's *Andrea Chénier.*

Laparra, Raoul, composer. Born Bordeaux, France, May 13, 1876; died Paris, April 4, 1943. He attended the Paris Conservatory, winning the Prix de Rome in 1903. He completed his first opera in 1899, *La peau d'âne.* Success came with his second opera, *La Habanera,* introduced by the Opéra-Comique in 1908. He was music critic for *Le Matin* until 1933, after which he concentrated on composition. Later operas: *La jota* (1911); *Le joueur de viole* (1925); *Las Torreras* (1929); *L'illustre Fregona* (1931). Laparra made extensive use of Spanish and Basque folk elements in his operas. He was killed during an air raid.

Lara, Isidore de, *see* DE LARA, ISIDORE.

Largo, Handel's, the designation by which the aria "Ombrai mai fu" from Handel's opera *Serse* is today most familiar. In the opera it is a tenor aria describing the cool shade of a palm tree. It has been retitled "Largo" in countless instrumental transcriptions.

Largo al factotum, Figaro's patter aria in Act I, Scene 1, of Rossini's *The Barber of Seville.*

Larina, Madame, a landowner (mezzosoprano) in Tchaikovsky's *Eugene Onegin.*

La rivedrà nell' estasi, Riccardo's aria in Act I of Verdi's *Un ballo in maschera.*

Larsen-Todsen, Nanny, dramatic soprano. Born Hagby, Sweden, August 2, 1884. After completing her studies at the Stockholm Conservatory she made her debut at the Stockholm Royal Opera in 1907. During her association with this company she made many guest appearances in major European opera houses. On January 31, 1925, she made her American debut at the Metropolitan Opera as Brünnhilde in *Die Götterdämmerung.* She remained at the Metropolitan Opera three seasons, specializing in Wagnerian roles. She appeared at Bayreuth for the first time in 1927, singing there in the next four festivals, one of the few Bayreuth sopranos to be heard as Brünnhilde, Isolde, and Kundry. She was a guest singer at leading European opera houses up to World War II, after which her operatic appearances were few.

La Scala, *see* TEATRO ALLA SCALA.

Lascia ch' io pianga, a famous soprano aria in Handel's opera *Rinaldo.*

Lasciatemi morire, the celebrated Lament of Arianna, the only excerpt to survive from Monteverdi's opera *Arianna.*

Lassalle, Jean, baritone. Born Lyons, France, December 14, 1847; died Paris, September 7, 1909. Though originally intent on a mercantile career, he attended the Paris Conservatory and studied singing privately with Novelli. He made his debut in Liège in 1869 as St. Bris. In 1872 he was engaged by the Paris Opéra, where he made his debut on June 7 in *William Tell.* For the next twenty years he was an idol of the Paris Opéra audiences. His repertoire included sixty operas in the French, Italian, and German repertories. He created the principal baritone roles in Gounod's *Polyeucte,*

Massenet's *Le Roi de Lahore,* Emile Paladilhe's *La Patrie,* Ernest Reyer's *Sigurd,* Saint-Saëns' *Ascanio* and Henry *VIII,* and Thomas's *Françoise de Rimini.* On June 15, 1892, he made his American debut at the Metropolitan Opera as Nelusko and scored a major success. For the next half dozen years he appeared at the Metropolitan. In 1901 he settled in Paris as a teacher of singing, and in 1903 he became a professor at the Paris Conservatory.

Lasst mich euch fragen, the drinking chorus opening Act III of Flotow's *Martha.*

Last Days of Pompeii, The, *see* BULWER-LYTTON.

Last Rose of Summer, The, (Qui sola, vergin rosa), Harriet's song in Act II of Flotow's *Martha.* The song was not original with Flotow. It is an old Irish air, "The Groves of Blarney," to which was set a poem by Thomas Moore.

Lattuada, Felice, composer. Born Morimondo, Italy, February 5, 1882. He graduated from the Milan Conservatory in 1911 and four years later completed his first opera, *Sandha,* performed in Genoa in 1924. Recognition came with his *Don Giovanni,* which won a national prize in 1922, and success with his finest opera, *Le preziose ridicole* (after Molière), first given at La Scala in 1929 and soon heard throughout Italy, in South America, and at the Metropolitan Opera. His other operas: *La tempesta* (1922); *La caverna di Sálamanca* (1937).

Laubenthal, Rudolf, dramatic tenor. Born Düsseldorf, Germany, about 1890. He turned to singing after being educated in the sciences. His apprenticeship as an opera singer took place with the Berlin Opera, after which he achieved success as a Wagnerian tenor in leading German opera houses and at Covent Garden. On November 9, 1923, he made his American debut at the Metropolitan Opera in *Die Meistersinger.* He remained at the Metropoli-

tan for a decade, appearing in all the major Wagnerian dramas as well as in such important revivals and premières as *Die aegyptische Helena, The Bartered Bride, Der Freischütz, Jenufa,* and *Schwanda.* After leaving the Metropolitan he appeared in the major opera houses of Austria, Germany, and England.

Laughing Song, (1) an aria from Auber's *Manon Lescaut* ("C'est l'histoire amoureuse") frequently interpolated into the Lesson Scene of Rossini's *The Barber of Seville.*

(2) Adele's aria ("Mein Herr Marquis") in Act II of Johann Strauss's *Die Fledermaus.*

Laura, Alvise's wife (mezzo-soprano), in love with Enzo, in Ponchielli's *La Gioconda.*

Lauretta, Gianni Schicchi's daughter (soprano) in Puccini's *Gianni Schicchi.*

Lauretta mia, duet of Rinuccio and Lauretta in Puccini's *Gianni Schicchi*

Lauri-Volpi, Giacomo, tenor. Born Rome, December 11, 1894. After studying at the Santa Cecilia in Rome and privately with Enrico Rosati, he made his debut in 1920 at the Teatro Costanzi in *Manon.* Two years later he was engaged by La Scala, where he scored a major success. On January 27, 1923, he made his American debut at the Metropolitan Opera in *Rigoletto* and became an immediate favorite. He remained at the Metropolitan for the next decade, during which time he created for America the role of Calaf in *Turandot.* He toured South America extensively in 1926. After leaving the Metropolitan (1932) he appeared at La Scala and other major opera houses of Europe. Just before and after World War II he confined his appearances to Italy. His autobiography appeared in two volumes: *L'equivoco* (1938) and *A viso aperto* (1953).

Lausanne International Competition, an annual competition for opera singers organized in Lausanne, Switzerland, in 1950. Winners receive contracts with leading European opera houses. The judges have included Claude Delvincourt, Frederick Jacobi, Toti dal Monte, and Ninon Vallin. Among the winners have been Victoria de los Angeles, Rita Goor, Grace Hoffmann, Anne McKnight, and Teresa Stich-Randall.

La Vergine degli angeli, Leonora's scene with chorus in Act II, Scene 2, of Verdi's *La forza del destino.*

Lawrence, Marjorie, dramatic soprano. Born Deans Marsh, Australia, February 17, 1909. She first studied singing with the local pastor and afterward in Melbourne, where she won first prize in an opera competition. For the next three years she studied with Cécile Gilly in Paris, after which she made her debut in 1932 in Monte Carlo in *Tannhäuser.* She was immediately engaged by the Paris Opéra for leading Wagnerian roles, appearing there for four years. On December 18, 1935, she made her American debut at the Metropolitan Opera in *Die Walküre.* During her next six seasons at the Metropolitan she was recognized as one of the outstanding Wagnerian sopranos.

During a performance of *Die Walküre* in Mexico City in 1941 she was stricken with poliomyelitis. It was feared she would be paralyzed for life. Showing a remarkable will to recover, and aided by her husband, Dr. Thomas King (whom she had married shortly before the tragedy), she fought her sickness with supreme patience. In a few months she could move her muscles again; a few months more and she found she had regained her voice. In the fall of 1942 she appeared in a song recital in Town Hall, New York, seated in a wheel chair. Still unable to walk, she returned to the stage of the Metropolitan Opera on January 22, 1943, as Venus, singing the role in a reclining

position. A year later, she attempted the role of Isolde at the Metropolitan, strapped throughout the performance to a carefully camouflaged wheel chair. In Chicago, on December 11, 1947, she stood throughout an entire performance of *Elektra.* She then proceeded to fill a complete schedule of concert and opera appearances in America and Europe. She has told the story of her illness and recovery in *Interrupted Melody,* a book published in 1949 and made into a motion picture in 1955.

Lazzari, Virgilio, bass. Born Assisi, Italy, April 20, 1887; died Castel Gandolfo, Italy, October 4, 1953. After studying with Antonio Cotogni in Rome he joined the Vitale Light Opera Company. He transferred from light to serious opera for a season at the Teatro Costanzi. For three seasons he appeared in leading bass roles at the Teatro Colón in Buenos Aires. In 1916 he came to the United States, making his debut with the St. Louis Opera as Ramfis. From 1918 to 1933 he was a member of the Chicago Opera. On December 28, 1933, he made his Metropolitan Opera debut as Don Pedro in *L'Africaine.* For the next seventeen years he appeared at the Metropolitan in twenty-two major bass roles, his last appearance there being on December 5, 1950, as Leporello. Besides his appearances at the Metropolitan, Lazzari was heard at the Salzburg Festivals, Covent Garden, La Scala, and the Opéra-Comique. One of his most celebrated roles was that of King Archibaldo, which he had the benefit of studying with the composer, Montemezzi.

Leading motif, *see* LEITMOTIV.

Leander, the Prime Minister (baritone) in Prokofiev's *The Love for Three Oranges.*

Leben des Orest, Das (The Life of Orestes), opera by Ernst Křenek. Libretto by the composer, based on the tragedy of Euripides. Première: Leipzig, January 19, 1930. While the text follows the plot of the Euripides tragedy, the opera is an interesting experiment at modernizing the Greek drama through the application of a jazz style. *See* ORESTES.

Leb' wohl, Wotan's farewell to Brünnhilde in the closing scene of Wagner's *Die Walküre.*

Lefebvre, a police sergeant (tenor), later Duke of Danzig, in Giordano's *Madame Sans-Gêne.*

Le Flem, Paul, composer. Born Lézardrieux, France, March 18, 1881. He divided his studies between philosophy at the Sorbonne and music at the Paris Conservatory. After deciding to concentrate on music he continued his studies in Moscow and at the Schola Cantorum in Paris. He wrote a good deal of music in various forms before turning to the theater. His opera *Le rossignol de St. Malo* was introduced by the Opéra-Comique in 1942 and has become a fixture in its repertory. Later operas: *La clairière des fées* (1943); *La magicienne de la mer* (1947). Le Flem's operas are based on Breton legends, and his melodic ideas are derived from Breton folk music. The composer has taught counterpoint at the Schola Cantorum, written music criticisms for the periodical *Comoedia,* and been choral director at the Opéra-Comique.

Legend, The, one-act opera by Joseph Breil. Libretto by Jacques Byrne. Première: Metropolitan Opera, March 12, 1919. In a mythical Balkan kingdom Stephen, in love with Carmelita, is out to capture the notorious bandit, Black Lorenzo. He is not dissuaded from this mission even after learning that the bandit is his beloved's father, a rich nobleman. Carmelita stabs Stephen when he tries to seize the bandit. Soldiers then kill both the bandit and Carmelita.

Légende de la sauge; *see* FLEURISSAIT UNE SAUGE.

Legend of Czar Saltan, The, opera by Nikolai Rimsky-Korsakov. Libretto by Vladimir Bielsky, based on a poem by Alexander Pushkin. Première: Moscow, November 3, 1900. The Czar Saltan marries Militrissa, youngest of three sisters. When he goes off to war, the two envious sisters write him that Militrissa has given birth to a monster. Saltan orders his wife and child put in a casket and thrown into the sea. The casket drifts to an island which becomes the new home for Militrissa and her son. The boy grows up to be a magician. From the bottom of the sea he evokes a kingdom and proclaims himself its Czar. When the Czar Saltan finally discovers that his son is human, he becomes reconciled to him and Militrissa. This opera is known in America chiefly for its orchestral interlude in the third act, "The Flight of the Bumblebee."

Legend of Kleinzach, The, *see* IL ETAIT UNE FOIS A LA COUR D'EISENACH.

Lehmann, Lilli, dramatic soprano. Born Würzburg, Germany, November 24, 1848; died Berlin, May 17, 1929. One of the greatest Wagnerian sopranos of all time, she was raised in a highly musical atmosphere, started piano lessons when she was six, and a few years later studied singing with her mother, the opera singer Marie Loew. On October 20, 1865, Lilli Lehmann made her debut in Prague as the First Page in *The Magic Flute.* At the next performance of the opera the leading singer became indisposed, and Lehmann stepped into her role. She gave such a good account of herself that henceforth she was assigned leading roles. After her appearances in Danzig and Leipzig she was engaged by the Berlin Opera, where she made her debut on August 19, 1870, as Vielka in Meyerbeer's *Das Feldlager in Schlesien.* She remained with that company for many years, distinguishing herself in coloratura roles. After a period of study with Richard Wagner she appeared at the first Bayreuth Festival, creating the roles of the Forest Bird and Woglinde. After 1875 she was made life member of the Berlin Opera, with the title of Kammersängerin, and allowed frequent leaves of absence to appear with other European companies. She made her American debut on November 25, 1885, at the Metropolitan Opera. Her role was Carmen and she made a favorable impression even though Carmen was never one of her outstanding interpretations. A more accurate measure of her art came five days later when she was heard as Brünnhilde in *Die Walküre*. Henry Krehbiel described her as "a most statuesquely beautiful Brünnhilde," with a voice "clear and ringing, never out of tune, and full of feeling." Lehmann stayed at the Metropolitan Opera until 1889, an idol. She created for America the roles of Isolde, and Brünnhilde in *Siegfried* and *Die Götterdämmerung*. She was also acclaimed in *Don Giovanni, Euryanthe, Fidelio,* and *The Queen of Sheba*. She invested every role with nobility and dramatic fire, just as she brought to her singing a consummate technique and a profound understanding of style. She was one of the most versatile singers of any age, mastering about 170 roles in 119 operas of the French, Italian, and German repertories, even including parts in comic operas.

Because she stayed in America beyond her leave of absence from the Berlin Opera, she became for a time *persona non grata* in German opera houses. On her return to Germany, she had to concentrate on song recitals, becoming a pre-eminent interpreter of lieder. In 1891 the Emperor lifted the ban against her. She returned to the Berlin Opera, and to the foremost opera stages of Germany and Austria, and renewed her triumphs. In 1896 she appeared in all three Brünnhilde roles at Bayreuth. After 1905 she was associated with the Salzburg Festivals both as a leading soprano and as director; her performances in the Mozart operas became a criterion. She returned to the United States in 1891 and during the next eight years appeared at the Metropolitan Opera and with the Damrosch Opera Company. Her last appearance at the Metropolitan took place on March 25, 1899, in *Les Huguenots*.

Besides her fruitful career as a prima donna, Lehmann distinguished herself as one of the outstanding singing teachers of her generation. Many notable singers studied with her, including Geraldine Farrar and Olive Fremstad, both of whom received practically their entire training from her. During her first visit to the United States, Lehmann met the Wagnerian tenor Paul Kalisch (1855–1946). They were married in 1888, but divorced a few years later.

Lehmann was the author of several books, including a treatise on singing published in English as *How to Sing* (1902); a study on *Fidelio* (1904); an autobiography, published in English as *My Path Through Life* (1914). She translated Victor Maurel's *Dix ans de carrière* into German and edited a volume of arias by Mozart.

Lehmann, Lotte, dramatic soprano. Born Perlberg, Germany, February 27, 1888. In no way related to Lilli Lehmann (see above), Lotte Lehmann has occupied an imperial position in concert hall and opera house Her music study took place at the Berlin Royal Academy of Music, and privately with Mathilde Mallinger. Between 1910 and 1913 she appeared at the Hamburg Opera, making her debut there as Freia in *Das Rheingold*. In 1914 she was invited to make a guest appearance at the Vienna Opera as Agathe. She was so successful that she was immediately engaged on a permanent basis. It was in Vienna that she displayed the operatic characterizations that were to spread her fame throughout Europe: Sieglinde, the Marschallin, Leonore in *Fidelio*.

She became a favorite of Richard Strauss, who selected her for the roles of the Young Composer in *Ariadne auf Naxos* and Barak's wife in *Die Frau ohne Schatten* in the Vienna premières of these operas; he wrote *Arabella* for her, and invited her to create the role of Christine in *Intermezzo*. From the Austrian government she received the honorary title of Kammersängerin. Engagements in leading opera houses of Europe, and at the Salzburg Festivals, followed. On October 28, 1930, she made her American debut with the Chicago Opera in the role of Sieglinde. On January 11, 1934, she made her first appearance at the Metropolitan Opera, once again as Sieglinde. She appeared at the Metropolitan in all her famous roles until 1945, repeating the triumphs she had earned abroad.

When the Nazis came to power, Lotte Lehmann renounced her native land and settled in Austria. After the Anschluss she came to the United States and became an American citizen. She made her last appearance at the Metropolitan on March 29, 1945, in one of her greatest roles, that of the Marschallin. On February 16, 1951, concluding a song recital at Town Hall, New York, she quietly announced to the audience that she was retiring from the concert stage.

She has written several books, including a novel *Eternal Flight* (1937), and three autobiographical volumes, *Wings of Song* (1937), *Midway in My Song* (1938), and *My Many Lives* (1948).

Leiden des jungen Werthers, Die, *see* GOETHE, JOHANN WOLFGANG VON.

Leider, Frida (soprano). Born Berlin, April 18, 1888. She studied singing in Berlin and Milan, following which she made her opera debut in Halle, Germany. Successful appearances in opera and song recitals led to her engagement by the Berlin Opera, where for many years she appeared principally in the Wagnerian music dramas. In 1924 she appeared for the first time at Covent Garden, as Isolde. In 1928 singing for the first time at Bayreuth, she appeared as Brünnhilde and Kundry. She came to the United States in 1928 and for four seasons appeared with the Chicago Opera. On January 16, 1933, she made her debut at the Metropolitan Opera as Isolde, remaining with that company for two seasons. She then appeared with the Vienna State Opera and other major opera houses in Europe.

Leila, a priestess (soprano) in Bizet's *Les pêcheurs de perles*.

Leinsdorf, Erich, conductor. Born Vienna, February 4, 1912. He attended the Vienna Academy. In the summer of 1934 he became Bruno Walter's assistant at the Salzburg Festival, and the following year he was Toscanini's assistant there besides helping prepare performances for the Florence May Music Festival. In the fall and winter of 1936 he conducted opera and orchestra concerts in Italy. On January 21, 1938, he made his American debut at the Metropolitan Opera with *Die Walküre* and was acclaimed. When Artur Bodanzky fell ill just before the opening of the Metropolitan's 1939–1940 season, Leinsdorf assumed leadership of the entire Wagner repertory; upon Bodanzky's death, Leinsdorf became his successor. In 1943 he left the Metropolitan to concentrate on symphonic music. He first became musical director of the Cleveland Orchestra and then, in 1947, of the Rochester Philharmonic.

Leise, leise, fromme Weise, Agathe's prayer in Act II of Weber's *Der Freischütz*.

Leitmotiv, German for "leading motive," a melodic, harmonic, or rhythmic pattern or figure recurring throughout an opera to identify some person, thing, situation, feeling, or idea. The music dramas of Richard Wagner make extensive use of a series of leitmotivs. The device of a recurring musical motive was not new with Wagner. It is found, for instance, in the operas of Carl Maria von Weber, and Hector Berlioz used the technique in his *Symphonie fantastique*. However, no one before Wagner used the device so extensively and so adroitly. Wagner frequently built up various melodic, rhythmic, and contrapuntal combinations of leitmotives into elaborate symphonic textures.

Lel, a shepherd (tenor) in Rimsky-Korsakov's *The Snow Maiden*.

Lelio, Eleanora's husband (baritone) in Wolf-Ferrari's *Le donne curiose*.

Lemoyne, Jean-Baptiste (born MOYNE), composer and conductor. Born Eymet, France, April 3, 1751; died Paris, December 30, 1796. He studied in Berlin with Johann Philipp Kirnberger. For a while he was conductor for Frederick the Great at the Berlin Court Theater. After returning to Paris he falsely represented himself as Gluck's pupil, and in 1782 produced an opera, *Elektra,* written according to Gluck's ideas and principles. After *Elektra* failed, Gluck repudiated Lemoyne, insisting that he had never been the Frenchman's teacher. Out of revenge, Lemoyne began writing operas in the style of Nicola Piccinni, Gluck's esthetic rival. The most successful were *Phèdre* (1786) and *Nepthé* (1789).

Lensky, Eugene Onegin's friend (tenor) in Tchaikovsky's *Eugene Onegin*.

Leo, Leonardo, composer and teacher. Born San Vito degli Schiavi, near Brindisi, Italy, August 5, 1694; died Naples, October 31, 1744. He attended the Conservatorio della Pietà dei Turchini in Naples, where one of his teachers was Alessandro Scarlatti. His first success as an opera composer came with his *Sofonisba* in 1718. In all he wrote some fifty operas, many of them enormously popular in their time, a number being fine examples of the opera buffa form. *Demofoönte* (1735);

Farnace (1736); *L'Olimpiade* (1737); and *La contesa dell' Amore colla Virtù* (1744) were among his best productions. In 1725 Leo became a teacher at the Conservatorio di Sant' Onofrio in Naples. Here his pupils included Niccolò Jommelli, Giovanni Battista Pergolesi, Nicola Piccinni, and Antonio Sacchini.

Leoncavallo, Ruggiero, composer. Born Naples, Italy, March 8, 1858; died Montecatini, August 9, 1919. He was the composer of *Pagliacci*. Soon after his graduation from the Bologna Conservatory he wrote his first opera, *Chatterton*. Hoping to get it produced, he turned his savings over to an irresponsible impresario who absconded with the money. Leoncavallo now traveled extensively, earning his way by singing and playing the piano. In Paris he wrote music hall songs and played the piano in cabarets. Victor Maurel, the baritone, became interested in him and introduced him to the Italian publisher, Ricordi, who commissioned him to write an operatic trilogy set in the Renaissance. The first opera of the set, *I Medici,* was turned down as too expensive to produce. Leoncavallo decided to write a more modest work in the realistic style of *Cavalleria rusticana,* an opera then achieving a sensational success. Leoncavallo's, written in a space of four months, was *Pagliacci*. When it was introduced in Milan on May 21, 1892, with Maurel as Tonio and Arturo Toscanini conducting, it was a major triumph. Now famous and prosperous, Leoncavallo had his two earlier works performed. *I Medici* was given by La Scala in 1893 and *Chatterton* was given in Rome in 1896. Both were fiascos. Even the composer's later operas were poorly received. Indeed, only a single work of the dozen or so operas he wrote after *Pagliacci* made a favorable impression: *Zaza,* introduced in Milan in 1900 and soon after given in the world's major opera houses.

Leoncavallo's repeated failures embittered him, and he died a man broken in health and spirit. His operas: *Chatterton* (1877); *I Medici* (1888); *Pagliacci* (1892); *La Bohème* (1897); *Zaza* (1900); *Der Roland* (1904); *La jeunesse de Figaro* (1906); *Maia* (1910); *Malbruk* (1910); *La Reginetta delle Rose* (1912); *I Zingari* (1912); *La candidata* (1915); *Ave Maria* (1916); *Gioffredo Mameli* (1916); *Prestami tua moglie* (1916); *Edipo Re* (1918).

Leoni, Franco, composer. Born Milan, October 24, 1864; died London, February 8, 1949. He attended the Milan Conservatory, where he was a pupil of Amilcare Ponchielli. His first opera, *Raggio di Luna*, was produced in Milan in 1888. Four years later he settled in England. Here he produced *Rip Van Winkle* in 1897, and in 1905 the opera that made him internationally famous, *L'oracolo*. His other operas: *Ib and Little Christina* (1901); *Tzigana* (1908); *Francesca da Rimini* (1914); *Massemarello* (1920); *Le baruffe chiozzotte* (1920); *Falene* (1920); *La terra del sogno* (1921).

Leonora, (1) Stradella's sweetheart (soprano) in Flotow's *Alessandro Stradella*.

(2) The Marquis of Calatrava's daughter (soprano) in Verdi's *La forza del destino*.

(3) Manrico's beloved (soprano) in Verdi's *Il trovatore*.

Leonora de Guzman, King Alfonso's mistress (soprano) in Donizetti's *La favorita*.

Leonore, Florestan's wife (soprano) in Beethoven's *Fidelio*.

Leonore 40/45, opera by Rolf Liebermann. Libretto by Heinrich Strobel. Première: Basel, Switzerland, March 25, 1952. The libretto is in two languages, French and German, sometimes used alternately, sometimes simultaneously. The story concerns the love of a German soldier for a French girl, symbolizing the life and civiliza-

tion of Europe in the years between 1940 and 1945.

Leonore Overtures Nos. 1, 2, and 3, *see* FIDELIO.

Leopold, a prince (tenor) in love with Rachel in Halévy's *La Juive*.

Leporello, Don Giovanni's servant (bass) in Mozart's *Don Giovanni*.

Leroux, Xavier, composer. Born Velletri, Italy, October 11, 1863; died Paris, February 2, 1919. He attended the Paris Conservatory, winning the Prix de Rome in 1885. He began writing for the theater in 1890 by composing incidental music for a play by Victorien Sardou and Emile Moreau, *Cléopâtre*. Five years later his first opera, *Evangéline*, was produced in Brussels. His next opera, *Astarté*, introduced by the Paris Opéra in 1901, was a major success. Two later operas were acclaimed: *La Reine Fiammette,* given by Opéra-Comique in 1903 with Mary Garden as Orlanda; and *Le chemineau,* introduced by the Opéra-Comique in 1907. From 1896 to the time of his death Leroux taught harmony at the Paris Conservatory. His other operas: *William Ratcliffe* (1906); *Théodora* (1907); *Le carillonneur* (1913); *La fille de Figaro* (1914); *Les cadeaux de Noël* (1916); *1814* (1918); *Nausithoé* (1920); *La plus Forte* (1924).

Lert, Ernst, conductor, stage director, opera manager. Born Vienna, May 12, 1883; died Baltimore, Maryland, January 30, 1955. He received his doctorate from the University of Vienna in 1908. He then became associated with several German and Austrian opera companies as assistant director and coach. From 1920 to 1923 he was general director of the Frankfort-on-the-Main Opera, and from 1923 to 1929 he was the stage director of La Scala during Arturo Toscanini's artistic direction. He came to the United States in 1929 and for two seasons was stage director of the Metropolitan Opera. Subsequently, he directed opera performances in various American cities, and in Paris, Salzburg, Florence, Spain, and South America. From 1936 to 1938 he was head of the opera department of the Curtis Institute of Music in Philadelphia. In 1950 he became director of opera at the Peabody Institute in Baltimore. He is the author of *Mozart auf dem Theater* (1918) and of a biography of the conductor Otto Lohse (1918).

le Sage, Alain René, novelist and dramatist. Born Sarzeau, France, May 8, 1668; died Boulogne, November 17, 1747. His most celebrated novel, *Gil Blas,* was the source of Théophile Semet's opera of the same name, while an episode from that story was used in Jean-François Lesueuer's *La caverne.* Le Sage's drama *Le Diable boiteux* was the source for Balfe's *Satanella* and Haydn's *Der neue krumme Teufel.* Gluck's comic opera *La rencontre imprévue* came from a Le Sage comedy.

Lescaut, (1) cousin of Manon Lescaut, a guardsman (baritone) in Massenet's *Manon.*

(2) Manon Lescaut's brother (baritone) in Puccini's *Manon Lescaut.*

Lesueur, Jean-François, composer. Born Drucat-Plessiel, France, February 15, 1760; died Paris, October 6, 1837. He received some musical training in two monasteries, but for the most part he was self-taught. Going to Paris in 1784, he presently became maître de chapelle at Notre Dame, where he inaugurated ambitious performances of church music. He was forced to leave this post in 1788 when he was accused of extravagance in spending church funds for these concerts, and of degrading the dignity of the church. Turning to dramatic music, Lesueur wrote his first opera, *La caverne,* in 1793. It was successful, and so was his second, *Paul et Virginie.* He wrote many operas after this, some of them enormously popular. His best operas, besides those mentioned, were: *Télémaque* (1796);

Ossian (1804); and *La mort d'Adam* (1809). He was made an official of the Paris Conservatory at its founding in 1795. In 1804 he became maître de chapelle to Napoleon. In 1818 he became a professor of composition at the Conservatory: his pupils included Berlioz, Gounod, and Thomas. He succeeded André Grétry as a member of the Institut de France.

Let's Make an Opera, children's opera by Benjamin Britten. Libretto by Eric Crozier. Première: Aldeburgh Festival, 1949. In this simple and straightforward musical play for children, composer and librettist demonstrate how an opera is written and produced. The work is in two acts. In the first are discussed the problems facing composer, librettist, and producer—a discussion in which six children and five adults take part. In the second act, the opera these actors develop is performed. It is called *The Little Sweep.* One of the novel features of *The Little Sweep* is the presence of several choruses to be sung by the audience. The audience is taught its part during the intermission between acts.

Letter Duet, (1) *see* CHE SOAVE ZEFFIRETTO.

(2) *See* VOICI CE QU'IL ECRIT.

Letter Scene, Tatiana's aria in Act I, Scene 2, of Tchaikovsky's *Eugene Onegin.*

Leuchtende Liebe! The love duet of Brünnhilde and Siegfried in the closing scene of Wagner's *Siegfried.*

Levasseur, Nicolas Prosper, dramatic bass. Born Bresles, France, March 9, 1791; died Paris, December 7, 1871. He attended the Paris Conservatory and made his debut in 1813 at the Paris Opéra. After successful appearances at the King's Theatre in London he rejoined the Opéra in 1816 and for half a dozen years appeared in minor roles. In 1822 he was acclaimed in Milan for his performance in Meyerbeer's *Marguerite d'Anjou.* This success brought him a five-year contract with the Théâtre des Italiens in Paris. From 1828 to 1853 he was the principal bass of the Paris Opéra, creating the leading bass roles in several important operas including Meyerbeer's *Robert le Diable* (1832), *Les Huguenots* (1836) and *Le Prophète* (1849). In 1841 he was appointed professor of lyric declamation at the Paris Conservatory. After retiring from the opera stage in 1853 he concentrated on teaching, remaining at the Conservatory until a year before his death.

Le veau d'or, Méphistophélès' scornful song to the village crowd in Act II of Gounod's *Faust.*

Levi, Hermann, conductor. Born Giessen, Germany, November 7, 1839; died Munich, May 13, 1900. A celebrated conductor of Wagner, he directed the world première of *Parsifal.* After studying in Mannheim with Vincenz Lachner and at the Leipzig Conservatory, he became music director at Saarbrücken, holding this post from 1859 to 1861. For the next three years he conducted German operas in Rotterdam. In 1872 he was engaged as principal conductor of the Munich Opera, where he remained almost a quarter of a century, particularly distinguishing himself in the Wagner dramas. In 1882 he was invited by Wagner to conduct the première of *Parsifal* in Bayreuth. A year later he led the musical performance at Wagner's funeral services in Bayreuth. In 1894 he was appointed music director of the Munich Court Theater, but poor health compelled him to resign two years later. He edited Mozart's *Così fan tutte, Don Giovanni,* and *The Marriage of Figaro,* and translated into German the librettos of Berlioz' *Les Troyens* and Chabrier's *Gwendoline.*

Libiamo, libiamo (Brindisi), the drinking song of Alfred Germont and Violetta in Act I of Verdi's *La traviata.*

Libretto, an Italian term, literally "little book," but specifically a term used for the text, or play, of an opera. The earliest librettos for Italian operas were highly standardized, the subjects usually dealing with mythological or historical subjects. Such librettos were written at the beginning of the seventeenth century by the poet Ottavio Rinuccini for the very first composers of opera: Peri, Caccini, and Monteverdi. A century later Pietro Metastasio, favoring historical subjects over mythological, wrote some thirty dramatically sound librettos that, being set by composer after composer, made him one of the dominant figures of eighteenth century Italian opera. The poet Raniero de Calzabigi, author of the librettos for Gluck's *Alceste* and *Orfeo ed Euridice,* was the first influential opera dramatist to break away from the traditional forms established by Metastasio. The revolt was carried further by Lorenzo da Ponte, author of the librettos for Mozart's *Così fan tutte, Don Giovanni,* and *The Marriage of Figaro.* A French libretto tradition was established by Pierre Perrin and Philippe Quinault in their texts for operas by Jean-Baptiste Lully. Important later writers of librettos in France include Eugène Scribe, who provided texts for Auber, Meyerbeer, and Halévy; Henri Meilhac (texts for Massenet and Bizet); and the collaborators Jules Barbier and Michel Carré (for Meyerbeer, Gounod, Thomas, and Offenbach). The important Italian librettists of the nineteenth century include Felice Romani (texts for Bellini and Donizetti); Francesco Maria Piave (for Verdi); Arrigo Boïto (Verdi and Ponchielli); and the collaborators Giuseppe Giacosa and Luigi Illica (Puccini). The notable figure among German librettists is Hugo von Hofmannsthal, the author linked with the operas of Richard Strauss. The composer who wrote his own texts was, pre-eminently, Richard Wagner. Others include Gustave Charpentier (*Louise*) and Ruggiero Leoncavallo (*Pagliacci*). An outstanding present-day example of a librettist-composer is Gian-Carlo Menotti. Another source for an opera libretto is a spoken drama, set to music with little or no alteration. An instance is Claude Debussy's *Pelléas et Mélisande,* using Maurice Maeterlinck's poetic play of the same name.

Libuše, opera by Bedřich Smetana. Libretto by Joseph Wenzig. Première: National Theater, Prague, June 11, 1881. Two brothers, Chrudos and Stahlav, vie for the love of Krasava. They are brought to trial before Libuše, Queen of Bohemia. Chrudos is contemptuous about appearing before a woman and insults her, an incident that has wide repercussions. The Queen descends from her throne in favor of a man who can rule the land with iron hand. She marries such a man, and the new king effects a reconciliation of the two brothers after Chrudos apologizes to Libuše.

Libuše's aria "Eternal Gods" in Act I, and her prophecy in Act III are the two best-known vocal excerpts.

Licino, a captain in the Roman army (tenor) in love with Giulia in Spontini's *La Vestale.*

Liebe der Danae, Die (Danae's Love), opera by Richard Strauss. Libretto by Joseph Gregor. Première: Salzburg Festival, August 14, 1952. Strauss completed this opera in 1940. Stylistically he reverted to the Wagnerian influences of his youth. The libretto combines two mythological stories, that of Danae, and that of Midas. Danae, daughter of King Pollux, is sought by Midas and by Jupiter; the latter assumes Midas' form in trying to win her. Jupiter threatens Midas with the loss of his golden touch and a return to human status if he does not give up Danae. The latter is ready to share Midas' humble fate and rejects the mighty Jupiter. She becomes a house-wife in a dreary hut, where Jupiter reappears to tempt her with wealth and power. But once again she rejects him. Touched by her devotion, Jupiter finally gives the pair his blessing.

Liebermann, Rolf, composer. Born Zurich, Switzerland, September 14, 1910. He studied with Hermann Scherchen and Vladimir Vogel, after which he conducted various Swiss orchestras. He first came to prominence as a composer with the opera *Leonore 40/45,* which excited considerable controversy at its première in Basel in 1952. A second opera, *Penelope,* was well received when introduced at the Salzburg Festival on August 17, 1954.

Liebesnacht (O sink hernieder, Nacht der Liebe), the night-of-love duet of Tristan and Isolde in Act II of Wagner's *Tristan und Isolde.*

Liebestod (Mild und leise wie er lächelt), Isolde's love-death song in Act III of Wagner's *Tristan und Isolde.*

Liebesverbot, Das (Love's Prohibition), opera by Richard Wagner. Libretto by the composer, based on Shakespeare's *Measure for Measure.* Première: Magdeburg Opera, March 29, 1836. This was Wagner's second complete opera; it was a fiasco. Claudio, an aristocrat, is arrested for love-making (a legal crime), and is to be executed. His sister Isabella arouses a mob against the Governor, who is then forced to abrogate the unreasonable law.

Liederspiel, German for "song play." This term was sometimes used interchangeably with the more familiar one of singspiel for a popular musical play combining dialogue and songs in the German and Austrian theatre of the eighteenth century.

Lieti Signori, salute! *See* NOBLES SEIGNEURS, SALUT!

Life for the Czar, A (or Ivan Susanin), opera in five acts by Michael Glinka. Libretto by Baron von Rosen. Première: Imperial Theater, St. Petersburg, December 9, 1836.

Characters: Ivan Susanin, a peasant (bass); Antonida, his daughter (soprano); Vanya, Susanin's adopted son (contralto); Bogdan Sabinin, Antonida's betrothed (tenor); Sigismund, the Polish King (bass); a Russian soldier; a Polish messenger; peasants; soldiers; Polish ladies and gentlemen. The action takes place in Russia and Poland in the winter of 1612.

Act I. The village of Domnino, Russia. News arrives of the defeat of the Poles by the Russians. Antonida thinks of her love for Sabinin (Antonida's Cavatina: "My gaze is fixed on the fields"). She is ready to marry him as soon as peace returns to Russia.

Act II. The camp of the Poles. The Polish soldiers vow to f.ght on until they achieve victory. When they hear that Romanov has been made Czar of Russia they decide to advance against him.

Act III. Susanin's hut. On Antonida's wedding day, the Poles compel Susanin to lead them to the new Czar. Susanin secretly plans to defend Romanov with his life.

Act IV, Scene 1. Before a monastery. Vanya urges his followers to save Romanov from the Poles. In rushing forth, he rides his horse to its death. Scene 2. A wood. Susanin is leading the Poles through the snow, not to the Czar but on a false trail. He soon realizes that the Poles have guessed what he is up to ("They guess the truth"). The price of Susanin's heroic deed is his life.

Act V. Before the Kremlin. In a festive ceremony, the new Czar comes to Moscow. He honors Antonida, Sabinin, and Vanya for their loyalty. He also laments the death of the hero Susanin. The people raise their voices in a hymn to their Czar.

A Life for the Czar, Glinka's first opera, has greater historic than esthetic interest. Except for two fine arias (those of Antonida in Act I and Susanin in Act IV), several stirring choruses (particularly the closing hymn), and some delightful folk dances, the opera means little to present-day audiences outside Russia. The libretto is diffuse and confusing, the combination of Italian lyric style with Russian national elements is often disturbing, and many of the numbers are banal. But the opera's place in history is secure. As the first Russian national opera, it showed the way for such later masters as Dargomizhsky, Mussorgsky, Rimsky-Korsakov, and Borodin.

The première of *A Life for the Czar* was a success. But apparently the audiences of 1836 liked the opera for the wrong things. What they admired was the Italian-influenced arias; the typically Russian portions of the work were largely ignored. Some critics, speaking of Glinka's use of folk songs and dances, sardonically referred to the score as "coachman's music."

Light opera, a term used interchangeably with operetta to designate a romantic play featuring songs (and sometimes dances) and spoken dialogue.

Lincoln's Inn Fields Theatre, a theater which figured in Handel's career. It was built in 1714 by Christopher Rich and opened after his death by his son John. Here, in 1727, *The Beggar's Opera* was first performed. In 1732 John Rich moved his activities to Covent Garden and for two seasons the Lincoln's Inn Field Theatre was the home for various entertainments. In 1734 Italian opera was produced here in direct opposition to Handel's ventures; Niccolò Porpora was the leading composer, Senesino the principal singer. Subsequently, Handel used this theater for performances of some of his operas and oratorios. The building was last used as a theater in 1756. Ninety years later it was demolished and the site used by the College of Surgeons.

Lind, Jenny, soprano. Born Stockholm, October 6, 1820; died Wynd's Point, Malvern, England, November 2, 1887. The "Swedish Nightingale" was the most celebrated singer to come from Sweden and one of the most glamorous prima donnas of the nineteenth century. She began the study of singing at the school of the Stockholm Court Theater when she was nine. On March 7, 1838, she made her debut with the Stockholm Royal Opera as Agathe. Her successes during the next few years were brilliant enough, but in 1841, dissatisfied with her technique, she went to Paris to study with Manuel García. Meyerbeer heard her at this time and used his influence to gain her an appointment with the Berlin Opera. After making her Berlin debut as Norma, she appeared in the world première of an opera written for her by Meyerbeer, *Das Feldlager in Schlesien.* She was a sensation. Triumphant performances followed in principal German cities, Stockholm, and Copenhagen. In Vienna, where she made her debut as Norma in 1846, she aroused such a pitch of excitement that she had to take some thirty curtain calls; at a benefit performance of *La sonnambula,* the Empress tossed a royal wreath on the stage, an act without precedent. In London she was an idol after her first appearance there in *Robert le Diable* in 1849. Though outstanding in dramatic roles, she was particularly famous in coloratura parts, where the purity of her voice and its remarkable range and flexibility aroused the greatest enthusiasm.

After 1849 she withdrew from opera and devoted herself exclusively to concert recitals and appearances in oratorio. She visited the United States in 1850 and for two years made extensive concert tours under the management of P. T. Barnum. In 1852, in Boston, she married Otto Goldschmidt, her concert conductor. After 1856 they lived mostly in England. Jenny Lind made one of her last public appearances in 1870 at the Rhenish Music Festival in Düsseldorf in a performance of her husband's oratorio *Ruth.* In 1883 she became a teacher of singing at the Royal College of Music in London. Two years after her death a memorial medallion was unveiled in Westminster Abbey.

Linda di Chamounix, opera by Donizetti. Libretto by Gaetano Rossi. Première: Kärntnerthortheater, Vienna, May 19, 1842. Linda is in love with the young painter, Charles, who is actually a nobleman in disguise. When she thinks herself deserted she goes mad. But the apparent desertion was only a misunderstanding. After Charles returns to Linda and reminds her of their love by singing an old love song, she recovers her sanity. Linda's aria in Act I, "O luce di quest' anima," and her Mad Scene in Act II, "Linda! A che pensato," are famous.

Lindorf, a councilor (bass) in Offenbach's *The Tales of Hoffmann.*

Lindoro, (1) the name assumed by Count Almaviva to woo Rosina in Rossini's *The Barber of Seville.*

(2) A favorite slave (tenor) of the Mustafa in Rossini's *L'Italiana in Algeri.*

Linette, a princess (contralto) hidden in an orange in Prokofiev's *The Love for Three Oranges.*

Lionel, young man (tenor) in love with Harriet in Flotow's *Martha.*

Lisa, (1) an innkeeper (soprano) in love with Elvino in Bellini's *La sonnambula.*

(2) The Countess' granddaughter (soprano) in Tchaikovsky's *Pique dame.*

List, Emanuel, bass. Born Vienna, March 22, 1891. After appearing as boy soprano and later as a member of a vocal quartet in Europe, he came to the United States and received his first systematic vocal instruction from

Josiah Zuro in New York City. In 1921 he returned to Europe and a year later made his debut at the Vienna Volksoper as Méphistophélès in *Faust*. Two years later he was engaged by the Berlin Opera, where he remained a decade, successful in leading German bass roles, particularly those of Hunding and Baron Ochs. He also appeared at the festivals in Bayreuth, Salzburg, and Munich. On December 27, 1933, he made his debut at the Metropolitan Opera as Hermann in *Tannhäuser*. He remained at the Metropolitan more than a decade, appearing in the principal Wagner music dramas. He also made numerous guest appearances with other leading opera houses in America, Europe, and South America.

Little Sweep, The, *see* LET'S MAKE AN OPERA.

Litvinne, Félia (born FRANÇOISE-JEANNE SCHÜTZ), dramatic soprano. Born St. Petersburg, Russia, 1860; died Paris, October 12, 1936. She studied singing with Victor Maurel and Giovanni Sbriglia. She made her debut in 1882 at the Théâtre des Italiens in Paris in *Hérodiade*. She then appeared for a season with the Mapleson company in New York. In 1886 she made her debut at the Paris Opéra, establishing herself as a favorite in the Wagnerian music dramas. On November 25, 1896, she made her debut at the Metropolitan Opera in *Les Huguenots*. She remained at the Metropolitan only one season. Her subsequent career in Europe included appearances at La Scala and Covent Garden. After leaving the stage in 1917 she devoted herself to teaching, becoming a professor of singing at the American Conservatory at Fontainebleau in 1927. She wrote an autobiography, *Ma vie et mon art* (1933).

Lìu, a young slave girl (soprano), in love with Calaf in Puccini's *Turandot*.

Ljungberg, Göta, dramatic soprano. Born Sundsvall, Sweden, October 4,

1893; died Lidingo, near Stockholm, June 28, 1955. She studied singing at the Stockholm Royal Academy and the Royal Opera School. Her debut took place in 1920 at the Stockholm Royal Opera as Elsa. In 1925 she sang at Covent Garden. Eugène Goossens wrote his opera *Judith* for her. She next appeared for three seasons at the Berlin Opera, where she was acclaimed in the Wagnerian repertory. On January 20, 1932, she made her American debut at the Metropolitan Opera in *Die Walküre*. She appeared at the Metropolitan through the 1934–1935 season, specializing in the Wagner operas, and scoring outstanding success in the world première of *Merry Mount* and in a revival of *Salome* in which she herself performed the Dance of the Seven Veils. After leaving the Metropolitan she appeared extensively in Europe. For several years after 1945 she taught singing in New York, after which she returned to Sweden.

Lobetanz, opera by Ludwig Thuille. Libretto by Otto Julius Bierbaum. Première: Karlsruhe Opera, February 6, 1898. The setting is Germany in the middle ages. The minstrel Lobetanz wins the heart of a princess with his songs. For this offense he must suffer death. The princess falls into a coma of grief and is revived only by Lobetanz' songs. He is now given royal blessing for his marriage with the princess.

L'oca del Cairo (The Goose of Cairo), an unfinished opera by Mozart. *See* DON PEDRO.

Lodoletta, opera by Pietro Mascagni. Libretto by Gioacchino Forzano, based on *Two Little Wooden Shoes,* a children's story by Ouida. Première: Teatro Costanzi, April 30, 1917. In a Dutch village in 1853, Antonio gives his adopted daughter, Lodoletta, a birthday gift of new red shoes. After he dies, Lodoletta falls in love with a French painter, Flammən. She follows him to

Paris and arrives at his house on New Year's Eve while a party is in progress. She does not have the courage to enter. Aimlessly she wanders in the snow, and dies of cold and starvation. Flammen finds her and takes her in his arms, lamenting that he has always been in love with her. Two arias, both from Act III, have become popular. One is Flammen's "Ah! ritrovarla nella sua capanna," the other Lodoletta's "Flammen, perdonami."

Lodovico, Ambassador to the Venetian Republic (bass) in Verdi's *Otello.*

Loge, god of fire (tenor) in Wagner's *Das Rheingold.*

Logroscino, Nicola, composer. Born Bitonto, Italy, October, 1698; died Palermo, Sicily, about 1765. He studied with Francesco Durante and Gaetano Veneziano. From 1728 to 1731 he was an organist. After settling in Naples (about 1738) he became a deservedly popular composer of comic operas, his works remaining in vogue until the ascendancy of Piccinni. His finest operas were: *L'inganno per inganno* (1738); *La violanta* (1741); *Il governatore* (1747); *Tanto bene che male; Il vecchio marito; La furba burlata* (1760), the last a collaboration with Nicola Piccinni.

Lohengrin, opera in three acts by Richard Wagner. Libretto by the composer, based on medieval legends. Première: Weimar, August 28, 1850. American première: Stadt Theater, New York, April 3, 1871.

Characters: Henry the Fowler, King of Germany (bass); Frederick of Telramund, Count of Brabant (baritone); Ortrud, his wife (mezzo-soprano); Elsa of Brabant (soprano); Lohengrin, Knight of the Holy Grail (tenor); a herald (bass); Gottfried, Elsa's brother (silent role); nobles of Saxony and Brabant; gentlemen and ladies of the court; pages; attendants. The action takes place in and near Antwerp early in the tenth century.

Act I. The banks of the Scheldt River. The orchestral prelude, symbolizing the Holy Grail, is wrought entirely of the Grail theme. King Henry, desiring to form an army, finds the people of Brabant torn by dissension. Telramund reveals that much of the trouble arises from the suspicion that Elsa has murdered her brother in order to make a bid for the throne of Brabant. Called before King Henry to defend herself against this charge, Elsa tells of a strange dream in which a knight appeared to tell her he will be her protector (Elsa's Dream: "Einsam in trüben Tagen"). Telramund insists that his accusations against Elsa are well founded; he is ready to fight anyone who questions his veracity. When the King asks Elsa who her champion is to be, she mentions the knight of her dreams. The King's herald then calls on her champion to appear, but there is no answer. Elsa falls on her knees and prays for her deliverer to come (Elsa's Prayer: "Du trugest zu ihm meine Klage"). There now appears a swan-drawn boat bearing the Knight Lohengrin. After Lohengrin bids his swan farewell ("Nun sei bedankt, mein lieber Schwan") he approaches King Henry and announces his intention of championing Elsa. Elsa promises her hand in marriage to Lohengrin if he is victorious. In return, Lohengrin extracts from her the promise that she will not attempt to discover who he is or from where he has come. After King Henry offers a prayer for the contestants ("Mein Herr und Gott, nun ruf' ich dich"), the duel between Telramund and Lohengrin takes place. Lohengrin, the victor, generously spares the life of the defeated and shamed Telramund.

Act II. In the fortress of Antwerp. Telramund and his wife Ortrud are in disgrace. Ortrud conceives a method of defeating Lohengrin and regaining their lost station: Elsa must make the mysterious knight reveal his identity, thus depriving him of his magic powers. At this point Elsa appears and speaks of her happiness ("Euch Lüften, die mein Klagen"). Ortrud begs Elsa for forgiveness, and Elsa promises to do what she can to gain clemency for her and her husband. Dawn breaks. The square is filled with courtiers and knights who hail the day of Elsa's wedding. As Elsa and her retinue make their way from the fortress toward the cathedral, the people acclaim her. Elsa is about to mount the cathedral steps when she is stopped by Ortrud who accuses Lohengrin of being a black magician whose defeat of Telramund was achieved by foul means. Elsa denounces her for this slander, but the seeds of doubt have now been planted in her mind. Telramund creates suspicion among the people by repeating his wife's accusation. Lohengrin insists that the charge against him is false even though he may not reveal his true identity. Elsa now assures Lohengrin that she does not doubt him. The bridal procession continues into the cathedral.

Act III, Scene 1. The bridal chamber. A vigorous orchestral prelude describes the joy surrounding the marriage of Lohengrin and Elsa. To the strains of the celebrated Wedding March ("Treulich geführt") the bridal procession enters the chamber. When Lohengrin and Elsa are left alone, they embrace and speak of their love ("Das süsse Lied verhallt"). But doubt has entered Elsa's heart, and she longs to know who her husband really is. Lohengrin entreats her to desist from asking, but Elsa is insistent. At this point Telramund and four of his men burst into the chamber to attack Lohengrin. Lohengrin kills Telramund with a single blow of his sword, whereupon the henchmen abandon their evil mission. Sadly, Lohengrin reveals that all his happiness has ended ("Weh! Nun ist all unser Glück dahin!").

Act IV. The banks of the Scheldt. Lohengrin announces to King Henry and his retinue that the time has come to reveal his identity. He is a Knight of the Holy Grail (Gralserzählung: "In fernem Land"), the son of Parsifal, King of the Grail, and his power is that of destroying evil influences. Having revealed his secret, Lohengrin must return to Montsalvat, home of the Grail. He takes his bride Elsa in his arms and laments the fact that they are now to part forever. Lohengrin's swan appears. Ortrud triumphantly reveals that the swan is none other than Elsa's brother Gottfried, transformed by her own evil magic. Had Elsa kept faith with Lohengrin, she declares, she would not only have had him and his love—she could have saved her brother. Lohengrin rights one more wrong: he restores the boy Gottfried to human form, then departs in his boat, now drawn by the white dove of the Holy Grail.

Lohengrin, written between 1846 and 1848, was Wagner's last "opera." After *Lohengrin* he cut the cord that had tied him to the past; the new esthetics which he was to articulate were to be realized in his very next work, *Tristan und Isolde,* his first music drama. Standing as it does between *Tannhäuser* and *Tristan, Lohengrin* looks two ways. We find the past in the emphasis on the voice, which is still a dominating element; in many of the arias, duets and choral numbers cast in traditional mold; in such formal scenes as the bridal procession. The future is found in the increasingly ingenious use of leitmotivs, in the virtuoso handling of the orchestra, in the oneness of the conception, the singleness of the mood. But for all its glimpses into the future, *Lohengrin* remains a romantic opera. Having proved to his satisfaction that he had written the finest romantic opera of his generation, Wagner could abandon old paths and strike for new directions.

Since Wagner was *persona non grata* in Dresden after 1849—a political refugee—*Lohengrin* was turned down by the Dresden Opera, which had previously introduced *Der fliegende Holländer* and *Tannhäuser.* When hoped-for premières failed to materialize in Paris and London, Wagner sent the score to Liszt, then the kapellmeister in Weimar. It took courage for Liszt to produce a new work by a revolutionary in exile, and Liszt possessed that courage. The opera at first proved more or less a failure. Partly, it was too long and invited fatigue; partly, it was too new for immediate acceptance. But the opera did not have to wait long for recognition. Within the next few years it was heard in different parts of Germany, achieving an ever-mounting success. By 1860 *Lohengrin* had been performed so often that Wagner, living in exile in Switzerland, once complained that he was the only German alive who had not heard it.

Lohse, Otto, conductor. Born Dresden, Germany, September 21, 1859; died Baden-Baden, May 5, 1925. He studied with Hans Richter and Felix Draeseke at the Dresden Conservatory. In 1882 he was appointed conductor of the Wagner Society and the Imperial Russian Music Society, both in Riga. Seven years later he was made first kapellmeister of the Riga Stadttheater. In 1893 he became director of the Hamburg Stadttheater. Here he married the celebrated singer Katharina Klafsky. In the spring of 1896 he and his wife came to the United States and joined the Damrosch Opera Company. They were back in Germany a year later, and from then on Lohse held important conductorial posts: in Strassburg from 1897 to 1904; in Cologne from 1904 to 1911; with the Théâtre de la Monnaie in 1911–1912; and with the Leipzig Stadttheater from 1912 to 1923.

From 1901 to 1904 he also directed performances of the Wagnerian music dramas at Covent Garden. In 1916 he received the honorary title of Royal Professor. He wrote one opera, *Der Prinz wider Willen,* performed in Riga in 1890.

Lola, Alfio's wife (mezzo-soprano) in Mascagni's *Cavalleria rusticana.*

Lombardi, I (The Lombards), opera by Verdi. Libretto by Themistocles Solera, based on a romance by Tommaso Grossi. Première: La Scala, February 11, 1843. During the time of the Crusades the rivalry of two brothers, Arvino and Pagano, for the love of Viclinda results in Pagano's exile. He lives as a hermit in a mountain cave. When the Crusaders, headed by Arvino, attack Jerusalem, the hermit is brought to them. Pagano reveals his identity, begs his brother's forgiveness for their bitter rivalry, and dies. The trio of Griselda, Oronto, and Arvino, "Qui posa il fianco," and the Crusaders' war chorus, "O Signore, dal tetto natio," are of enduring interest.

London, George (originally BURNSON), bass-baritone. Born Montreal, Canada, May 30, 1921. The son of American parents, he was taken to Los Angeles in his boyhood. Here, while attending public schools, he sang in churches and various amateur productions. In 1947 he toured the United States with the Bel Canto Trio (whose other members were Mario Lanza and Frances Yeend). He made his opera debut in Europe in 1949 with the Vienna State Opera. His role was that of Amonasro. Subsequently he appeared in Vienna as Escamillo, Boris Godunov, and in all the four baritone roles in *The Tales of Hoffmann.* On November 13, 1951 (the opening night of the season), he made his debut at the Metropolitan Opera as Amonasro. On January 9, 1952, he made his debut at La Scala as Pizarro. After this, besides making appearances at the Metropolitan, he sang in the leading festivals of Europe, including those of Bayreuth, Salzburg, Munich, Holland, Glyndebourne, and Edinburgh. In 1955 London was honored by the President of Austria, Theodor Koerner, with the title of Kammersänger.

Longfellow, Henry Wordsworth, poet. Born Portland, Maine, February 27, 1807; died Cambridge, Massachusetts, March 24, 1882. Otto Luening's opera *Evangeline* was adapted from Longfellow's poem. *The Blind Girl of Castel-Cuillé* was the source of operas of the same name by Cornelis Dopper and Earl Ross Drake.

Lorek, surgeon (baritone) in Giordano's *Fedora*.

Loreley, Die, opera by Alfredo Catalani. Libretto by Angelo Zanardini and Carlo d'Ormville. Première: Teatro Regio, Turin, February 16, 1890. The setting is the banks of the Rhine in medieval times. Walter, governor of Oberwesel, betrothed to Anna, meets and falls in love with the Loreley. He decides to remain true to Anna. At their marriage ceremony, the Loreley appears. Walter rushes to her, but the Loreley eludes him by sinking into the Rhine. Anna dies of grief. Walter continues to pursue the Loreley and finds her singing atop her rock. When she once again evades him, he commits suicide by jumping into the river.

Lorenz, Max, tenor. Born Düsseldorf, May 10, 1901. He studied singing in Berlin with E. Grenzebach. In 1928 he was engaged by Fritz Busch for the Dresden Opera, where he scored a major success in *Die aegyptische Helena*. In 1931–1932 and again in 1933–1934 he appeared at the Metropolitan Opera in leading Wagnerian tenor roles. Between 1933 and 1943 he appeared at the Bayreuth Festivals; in 1937 he was engaged as principal tenor of the Vienna State Opera. Just before World War II he appeared in other countries; since the war he has sung not only at

the Vienna State Opera but at the Teatro Colón in Buenos Aires, in Italy, and at festivals in Paris, Amsterdam, and Zurich.

Lortzing, Gustav Albert, composer. Born Berlin, October 23, 1801; died there January 21, 1851. The son of professional actors, Lortzing spent his boyhood and youth traveling with his parents. His education, consequently, was haphazard, though at one time he did attend the Singakademie in Berlin. He wrote his first operetta, *Ali Pascha von Janina,* in 1822; it was produced two years later. He subsequently achieved considerable success as a tenor with the Leipzig Stadttheater. It was in this theater, too, that he was recognized as a composer with *Die beiden Schützen* in 1837. His most famous opera, *Zar und Zimmermann,* was an even greater success when introduced in Leipzig a year later and soon performed extensively throughout Germany and Austria. In 1842 he was again successful with a new opera, *Der Wildschütz.* In 1846 Lortzing was invited to Vienna to conduct the première of his *Der Waffenschmied.* Engaged as first conductor of the Theater-an-der-Wien, he lost the post during the revolutionary period of 1848. From then on he was unable to obtain a satisfactory position. He supported himself first as an actor in small German theaters, then by conducting ballet and vaudeville performances in Berlin. His health and spirit were broken; he died a poor and unhappy man. His most famous operas were in a comic vein, but he also wrote a romantic opera, *Regina.* It was first performed at the Berlin Opera in 1899, nearly fifty years after Lortzing's death.

Lo sposo deluso (The Deluded Spouse), unfinished opera by Mozart. *See* DON PEDRO.

Lothario, an aged minstrel (bass) who turns out to be a nobleman in Thomas's *Mignon.*

Loti, Pierre (originally LOUIS MARIE

JULIEN VIAUD), novelist. Born Rochefort, France, January 14, 1850; died Hendaye, France, June 10, 1923. His novels that were made into operas include: *Madame Chrysanthème* (André Messager); *Ramuntcho* (Deems Taylor); and *Le roman de Spahi* (Lucien Lambert's *Le Spahi*).

Louise, opera in four acts by Gustave Charpentier. Libretto by the composer. Première: Opéra-Comique, February 2, 1900. American première: Manhattan Opera House, New York, January 3, 1908.

Characters: Louise, a seamstress (soprano); her mother (contralto); her father (baritone); Julien, a painter (tenor); Irma, a seamstress (soprano); an errand girl; King of the Fools; peddlers; housekeepers; working people; street boys; grisettes; Bohemians. The setting is Paris towards the end of the nineteenth century.

Act I. The attic flat of Louise's family. Julien has written to Louise's parents requesting their permission for his marriage to their daughter; from his balcony across the way he urges her to elope with him if permission is denied. Louise is torn between her love for Julien and her duty to her parents. Julient recalls the time when first he fell in love with her ("Depuis longtemps j'habitais cette chambre"). As they are repeating to each other their expressions of love, Louise's mother appears and takes her daughter severely to task for encouraging a worthless Bohemian. When Louise's father comes home he reads Julien's letter. The mother insists that no consideration be given to Julien, but the father prefers a more cautious approach. This so infuriates the mother that she begins to quarrel with her daughter. The father consoles Louise ("O mon enfant"), and the young woman reluctantly promises not to see Julien again.

Act II, Scene 1. A street in Montmartre. Dawn breaks on Paris. Street cries are heard. Louise and her mother come to the establishment where Louise is employed. When her mother leaves, Julien appears and begs Louise to elope with him ("Ah! Louise, si tu m'aimes"). Once again Louise is torn between love and duty. She rushes into the shop, leaving Julien forlorn.

Scene 2. A dressmaker's workroom. (This scene is often omitted.) The other seamstresses do not fail to notice how disturbed Louise is. Irma suggests that Louise may be in love, then pronounces a rhapsody over love and Paris ("Une voix mystérieuse"). From outside the window comes Julien's unexpected serenade ("Dans la cité lointaine"). The seamstresses mock Julien, but Louise, unable to resist his appeals, rushes out to join him.

Act III. A cottage. Louise and Julien, now living together, are more in love than ever. Louise recalls the day when first she yielded ("Depuis le jour"). They embrace, and the gathering night envelops them. The beauty of Paris causes Julien to sing the praises of both the city and his beloved ("De Paris tout en fête"). A few moments later a group of Bohemians appear; they call to Julien and Louise, then crown Louise the Muse of Montmartre. The gaiety is at a climax when Louise's mother appears with the news that Louise's father is dying. Julien is willing to let Louise go to her father but only after the mother promises that she will be free to return.

Act IV. The attic flat. Louise has nursed her father back to health. She is still with him, her mother having refused to let her go back to Julien. The father laments the lot of a workman with ungrateful children. But he soon draws Louise to him and sings her a lullaby as if she were still a child (Berceuse: "Reste, repose-toi"). Since Louise is eager to return to her lover, a harsh quarrel erupts. Sternly, the father opens the door and orders her out. After Louise has gone, he cries after her. Then, shaking his fist at the city outside his window, he bitterly exclaims that it is the evil of Paris that has destroyed his home.

Naturalism entered French opera with *Louise,* a story concerned with the lives of everyday people. The central theme—the love affair of the artist and the seamstress—raised the then revolutionary question of a woman's right to live her life without dictation by parents or society. *Louise* was the first opera, moreover, to contain elements of socialist thinking. There is another important element in the work —a symbolic one—Paris. The spell of the city is made tangible through the musical tributes of Julien in the third act and Louise in the fourth, through the cries of street vendors, through the evocative orchestral interludes. It is the inescapable magic of the setting quite as much as the turbulent love of Julien and Louise that has enchanted opera audiences.

Two months after *Louise* was first heard, Marthe Rioton, the creator of the title role, fell ill during a performance. Her understudy, who had never yet sung before an audience, stepped into the part and gained an ovation. Her name was Mary Garden and the measure of her triumph was that she sang the role of Louise over two hundred times at the Opéra-Comique during the next few years.

On February 28, 1950, the fiftieth anniversary of *Louise* was celebrated in Paris. The composer, now in his ninetieth year, took over the baton for the closing scene. Climaxing the occasion, the President of France conferred on Charpentier the grade of Grand Officer of the Legion of Honor.

Luÿys, Pierre (born **Louis**), poet and novelist. Born Ghent, Belgium, December 10, 1870; died Paris, June 4, 1925. His famous novel *Aphrodité,* was the source of Camille Erlanger's opera of the same name. Another of his novels, *La femme et le pantin,* was used for Zandonai's *Conchita.*

Love for Three Oranges, The (L'amour des trois oranges), opera by Serge Prokofiev, libretto by the composer, based on a story by Carlo Gozzi. Première: Chicago Opera, December 30, 1921. Prokofiev's opera is a play within a play. A highly demonstrative audience of Cynics, Emptyheads, Glooms, and Joys watches the performances of a burlesque opera about a legendary prince. The young man, dying of gloom, can be cured only by laughter. A wicked sorceress, Fata Morgana, thwarts every attempt to lighten his spirits, but when she takes a ridiculous fall during a scuffle with palace guards, the prince laughs and is cured. The sorceress now decrees that he must find and fall in love with three oranges. When the prince finds the oranges in a desert, he learns that each contains a beautiful princess. Two of the young women perish of thirst. The Cynics of Prokofiev's audience revive the third with a bucket of water. After more trials, the prince and princess are united and the sorceress and her evil cohorts meet suitable justice. In the course of this gay work Prokofiev pokes good-natured fun at various absurdities of conventional opera plots. The best-known selections are the orchestral "Scène infernale," "March," and "Scherzo." "The March" has been used as the signature for the radio program, "The F.B.I. in Peace and War."

Love in her eyes sits playing, Acis' aria in Handel's *Acis and Galatea.*

Love of Three Kings, The, *see* **amore dei tre re, L'.**

Lualdi, Adriano, composer. Born Larino, Campobasso, Italy, March 22, 1887. After studying with Stanislao Falchi in Rome and Ermanno Wolf-Ferrari in Venice he became an opera conductor in 1908. In 1918 he settled in Milan, and from 1923 to 1927 was music critic of the periodical *Secolo.* In 1928 he became head of the music department of the Italian government. For several years he was artistic director of the Florence and Venice music festivals. In 1936 he succeeded Francesco Cilèa as director of the Naples Conservatory, holding this post until 1943. Next, he became director of the Cherubini Conservatory in Florence. He did considerable research in old Italian music, which led him in his own compositions to revive old Italian operatic forms and styles. In this vein he wrote *Il cantico,* a lyric intermezzo (1915); *Le furie di Arlecchino,* an intermezzo giocoso for marionettes (1915); *La morte di Rinaldo,* a dramatic scene (1916); and *Guerin meschino,* a medieval legend for marionettes (1920). His more traditional operas are: *Le nozze di Haura* (1908); *La figlia del re* (1922); *Il Diavolo nel campanile* (1925); *La Grancèola* (1930); *The Moon of the Caribbees.*

Lubava, Sadko's wife (mezzo-soprano) in Rimsky-Korsakov's *Sadko.*

Lubin, Germaine, soprano. Born Paris, February 1, 1890. She attended the Collège Sévigné in Paris with the intention of becoming a doctor, but a passion for music made her change her mind. In 1908 she entered the Paris Conservatory, where she won three first prizes in singing. In 1912 she joined the Opéra-Comique, where she appeared in the world première of Guy Ropartz' *Le pays.* In 1914 she was engaged by the Paris Opéra, where she has ever since remained, becoming principal dramatic soprano in 1938. She has distinguished herself particularly in the Wagnerian repertory, but has also had notable success in contemporary French operas and operas by Berlioz, Gluck, and Richard Strauss. In 1938 she was seen at Bayreuth as Kundry. She has also sung in London, Berlin, Vienna, Prague, and at several Salzburg Festivals.

Luca, Giuseppe de, *see* DE LUCA, GIUSEPPE.

Lucca, Pauline, dramatic soprano. Born Vienna, April 25, 1841; died there February 28, 1908. After studying with Uschmann and Richard Lewy in Vienna she joined the chorus of the Vienna Opera. She made her debut in Olmütz on September 4, 1859, in *Ernani,* and was so successful that she was made a principal soprano of the company. After her sensational performances as Norma and Valentine in Prague in 1860, Meyerbeer called her to Berlin to create for Germany the role of Selika. She was immediately engaged as permanent court singer there. From 1863 to 1872 she sang every season in London (except for 1869) and was a great favorite. She terminated her ties with Berlin in 1872 and embarked on a two-year tour of America. After returning to Europe she was associated with the Vienna Opera until 1889, when she retired from the stage.

Lucia, Turiddu's mother (contralto) in Mascagni's *Cavalleria rusticana.*

Lucia di Lammermoor, opera in three acts by Gaetano Donizetti. Libretto by Salvatore Cammarano, based on Sir Walter Scott's novel *The Bride of Lammermoor.* Première: San Carlo, Naples, September 26, 1835. American première: Théâtre d'Orléans, New Orleans, December 28, 1841.

Characters: Lord Enrico Ashton of Lammermoor (baritone); Lucia, his sister (soprano); Raimondo, chaplain of Lammermoor (bass); Edgardo, master of Ravenswood (tenor); Lord Arturo Bucklaw (tenor); Alisa, Lucia's companion (soprano); Normanno, follower of Lord Ashton (tenor); followers of Ashton; inhabitants of Lammermoor. The setting is Scotland towards the end of the seventeenth century.

Act I, Scene 1. A wood. Normanno informs Lord Ashton that there is a prowler on the grounds of Lammermoor Castle and that he suspects the intruder to be Edgardo. Normanno further discloses that Lord Ashton's sister, Lucia, has been meeting the intruder. Ashton vows to destroy Edgardo, his mortal enemy.

Scene 2. A park near the castle. Awaiting her lover, Lucia sings of an apparition she believes she has seen of a young woman long ago murdered by one of the Ravenswoods ("Regnava nel silenzio"). Her gloomy mood lightens as she thinks of Edgardo ("Quando rapita in estasi"). When Edgardo appears, it is with the news that he has been ordered to France. He suggests that he visit Lord Ashton and confess that he loves Lucia. Lucia insists such a mission would be futile. The lovers bid each other a passionate farewell ("Verrano a te sull' aure").

Act II, Scene 1. An anteroom in Lammermoor Castle. Lord Ashton is determined to smash the love affair of his sister and Edgardo, since he plans to solve his own financial problems by marrying his sister to wealthy Lord Arturo Bucklaw. He shows Lucia a letter he has forged in Edgardo's hand. Reading it, Lucia mistakenly believes that Edgardo has deserted her. Distraught with grief, Lucia consents to marry Bucklaw.

Scene 2. The castle's great hall. Before an assemblage of knights and ladies, Lucia signs the marriage contract that makes her Lord Bucklaw's wife. An armed stranger boldly stalks into the hall. Consternation prevails when he is recognized as Edgardo. Now begins the opera's famous sextet ("Chi mi frena?"). Edgardo wonders what restrains him from an act of vengeance; Lucia voices her despair at her brother's treachery; Enrico voices compassion for his sister's plight; Raimondo, the kindly chaplain, invokes the aid of heaven; Alisa, Lucia's companion, and Arturo Bucklaw, her husband, express the hope that there will be no bloodshed. When Edgardo finally realizes

how ruthlessly he has been treated, he curses the entire Lammermoor family and rushes away.

Act III, Scene 1. The tower of Ravenswood. Lord Ashton, bent on avenging the honor of his household, comes to Ravenswood castle to challenge Edgardo to a duel. As a storm rages, both Ashton and Edgardo vow vengeance.

Scene 2. The hall of Lammermoor. The wedding festivities of Lucia and Lord Bucklaw are being celebrated ("D'immenso giubilo"). Raimondo abruptly appears with the ghastly news that Lucia has slain her husband and gone mad ("Dalle stanze, ove Lucia"). As if in confirmation, Lucia enters the hall, dressed in a long white gown. She raves, unconscious of her surroundings (Mad Scene: "Ardon gl' incensi"). She believes that Edgardo is with her and that they are being married. She even mistakes her brother for Edgardo. Pathetically, she begs that a flower be placed on her grave, that no tears be shed ("Spargi d'amaro pianto"). Swooning, she falls into the arms of the faithful Alisa.

Scene 3. The burial ground of the Ravenswoods. Not knowing the fate that has befallen Lucia, Edgardo laments the fickleness of his loved one and longs for death (Tomb Scene: "Fra poco a me ricovero"). Mourners from Lammermoor pass. From them, Edgardo learns of Lucia's madness. He is about to rush to her side when a tolling bell announces that she is dead. Aware now that Lucia has never faltered in her love for him, Edgardo promises her spirit that they will never be parted again ("Tu che a Dio spiegasti l'ali"). He then stabs himself and dies.

Donizetti composed this opera with Fanny Persiani in mind. It was Persiani, one of the most brilliant coloratura sopranos of her day, who created the role of Lucia not only in Italy but England. Since her time virtually every great coloratura has appeared as Lucia and counted that role as one of her most grateful. These artists have included Adelina Patti, Jenny Lind, Marcella Sembrich, Nellie Melba, Luisa Tetrazzini, Amelita Galli-Curci, and Lily Pons. Containing two of the most famous numbers in all opera, the sextet and Lucia's mad scene, *Lucia* has remained popular with audiences quite as much because of the sustained beauty of its lyricism.

Lucinda, principal female character (soprano) in Wolf-Ferrari's *L'amore medico.*

Lucrezia Borgia, opera by Donizetti. Libretto by Felice Romani, based on Victor Hugo's play *Lucrèce Borgia.* Première: La Scala, December 26, 1833. In sixteenth century Italy, Don Alfonso, Duke of Ferrara, suspects that his wife, Lucrezia Borgia, is carrying on a love affair with young Gennaro. Gennaro is actually her son by a former marriage, though nobody but the mother knows this fact. When Alfonso has Gennaro arrested, Lucrezia arranges his escape. Infuriated by the insults of several Italians, she has them poisoned. To her horror, she discovers that one of those whom she has destroyed is her son. The most familiar excerpts are Gennaro's aria in the prologue, "Di pescatore ignobile," and Orsini's aria in Act II, "Il segreto per essere felice."

Lucy, a young lady fond of telephone conversations in Menotti's *The Telephone.*

Ludikar, Pavel, bass-baritone and opera manager. Born Prague, Bohemia, March 3, 1882. He turned to music after completing studies in law and philosophy. He made his opera debut in 1904 at the National Theater in Prague as Sarastro. Important appearances followed in leading European opera houses including La Scala and the Dresden Royal Opera. In 1913 he made his first opera appearances in America with the Boston Civic Opera

Company. On November 16, 1926, he made his debut at the Metropolitan Opera in the American première of Puccini's *Turandot.* He remained at the Metropolitan through the 1931–1932 season, singing about eighty roles in a dozen languages. In 1935 he was appointed manager of the National Theater in Prague, and three years later he created the title role in Ernst Křenek's *Karl V* in Prague. During World War II he was actively associated with Germany's musical life. Since the war he has appeared in Germany and Czechoslovakia.

Ludmilla, Svietosar's daughter (soprano) in Glinka's *Russlan and Ludmilla.*

Luigi, a longshoreman (tenor), Giorgetta's lover, in Puccini's *Il tabarro.*

Luisa Miller, opera by Verdi. Libretto by Salvatore Cammarano, based on Schiller's play *Kabale und Liebe.* Première: San Carlo, Naples, December 8, 1849. In eighteenth century Württemberg, Rodolfo must marry Federica, though he is in love with Luisa Miller. Rodolfo is put in jail by his father, who then compels Luisa to write Rodolfo that she is in love with somebody else. When Rodolfo is released upon his promise to marry Federica, he goes to Luisa's house to kill her. She tells him the truth, and the lovers die in each other's arms after drinking poison. Rodolfo's aria in Act II, "Quando le sere al placido," is one of the most popular numbers in the opera.

Lully, Jean-Baptiste, composer. Born Florence, Italy, November 28, 1632; died Paris, March 22, 1687. Though Italian by birth, Lully was the founder of French opera. He had little formal instruction in music. He was, however, so precocious that in 1646 Chevalier de Guise became interested in him and engaged him as a page for Mademoiselle de Montpensier, cousin of the King of France. Unattractive and clumsy, Lully was instead assigned to the kitchen. His preoccupation with music soon brought him a post with the house orchestra, and subsequently he was made a leader of a small band of violins. After six years with Mademoiselle de Montpensier, he was engaged by Louis XIV as ballet dancer, composer of ballet music, and director of the court orchestra. He soon became the king's favorite. A new orchestra was founded for him, "Les petits violons du roi." Under his direction it became one of the finest ensembles in France. In 1664 Lully began collaborating with Molière on opera-ballets, beginning with *Le mariage forcé*, which not only foreshadowed his own later procedures in writing operas but was also a predecessor of the opéra comique. These opera-ballets—*L'amour médecin* (1665) and *Le bourgeois gentilhomme* (1670) were two more— were very popular at court and made Lully one of France's first composers. Among the honors conferred upon him were a patent of nobility, an appointment as "secretaire du roi" and the rank of "maître de musique." So highly regarded was Lully at court that when, in 1662, he married Madeleine Lambert, the marriage contract was signed by the King, Queen, and the Queen Mother. Lully now used his power to acquire an exclusive patent to direct opera performances in Paris. For these performances he wrote a pastoral, *Les fêtes de l'amour*, in 1672. A year later came his first opera, *Cadmus et Hermione*, which can be regarded as the cornerstone on which French opera rests. In collaboration with the librettist Philippe Quinault, he now wrote some fifteen operas which established a tradition for French opera, introducing the so-called French overture, French declamation, accompanied recitatives, fully developed arias, and enriched harmonic and rhythmic vocabularies. Lully's best operas held the stage for almost a century after his death. They include: *Cadmus et Hermione* (1673); *Alceste* (1674); *Thésée* (1675); *Atys* (1676); *Isis* (1677); *Psyché* (1678); *Belle-rophon* (1679); *Proserpine* (1680); *Persée* (1682); *Phaëton* (1683); *Amadis de Gaule* (1684); *Roland* (1685); *Armide et Renaud* (1686); *Acis et Galatée* (1686). There is a story to the effect that Lully caused his own death from gangrene of the foot, brought on by a misplaced blow of the long stick he used for audible time-beating. It is a story that probably tells us more about how performances were directed at the Paris Opéra than it does about Lully's passing.

Lulu, unfinished opera by Alban Berg. Libretto by the composer, based on two tragedies by Frank Wedekind, *Der Erdgeist* and *Die Büchse der Pandora*. Première: Zurich, June 2, 1937. The opera begins with a prologue in which the animals of a circus are introduced with their tamer. Each animal symbolizes a character in the play that follows— Lulu being represented by a snake. The opera proper is concerned with the numerous illicit love affairs of Lulu, and the shocking consequence of each —murder, suicide. Eventually Lulu is slain by Jack the Ripper. Willi Reich observes that within the actions of Lulu and the tragedies they provoke, "there are the fundamental ethical principles, principles of eternal greatness, which could have been kindled into life only by a poet as gifted as Frank Wedekind was." Berg used a different approach in *Lulu* than in his earlier *Wozzeck*. For one thing, Lulu is written entirely in the Schoenbergian twelve-tone technique. The mold is also different. In *Wozzeck* there are many contrasting forms, some of them frankly instrumental in conception. In *Lulu* the style is more operatic, with a preference for arias, duets, trios, and ensemble numbers.

Luna, Count di, *see* DI LUNA, COUNT.

Lustigen Weiber von Windsor, Die, *see* MERRY WIVES OF WINDSOR, THE.

Luther, a tavern keeper (bass) in Offenbach's *The Tales of Hoffmann.*

Lyceum Theatre, a theater built in London in 1765. It was intended for opera, but since the necessary license could not be obtained it was first used for other entertainments. In 1809 the license was finally given. When Drury Lane burned down that year, its opera company moved to the Lyceum and gave four months of opera performances in English. Now named the English Opera House, the Lyceum was torn down and rebuilt in 1815. In the new house *Der Freischütz* received its English première in 1824. Destroyed by fire in 1830, the theater was again rebuilt. Ten years later it was again called the Lyceum. Through the years it was used for spoken drama as well as opera.

While Covent Garden was being rebuilt in 1856, the Royal Italian Opera Company appeared at the Lyceum. In 1876 and 1877 the Lyceum was used by the Carl Rosa Opera Company. Verdi's *Otello* was given its English première here in 1889. After 1900 the theater was chiefly a music hall, though in 1919 the Carl Rosa Opera Company again performed here.

Lycidas, *see* MILTON, JOHN.

Lyric drama, a synonym for opera in general.

Lyric opera, a term sometimes applied to an opera in which the lyrical element is more prominent than the dramatic, or in which the singing is of primary interest.

Lysiart, Count of Forêt (baritone) in Weber's *Euryanthe.*

Lysistrata, *see* ARISTOPHANES.

Lytton, *see* BULWER-LYTTON.

M

Macbeth (1) tragedy by William Shakespeare. Driven by an overpowering ambition, and abetted by his wife, Macbeth seizes the throne of Scotland by murdering Duncan. To maintain his position he must commit one murder after another; this succession of crimes affects Lady Macbeth's sanity and ultimately brings about Macbeth's destruction.

(2) Opera by Ernest Bloch. Libretto by Edmond Fleg, based on the Shakespeare tragedy. Première: Opéra-Comique, November 30, 1910.

(3) Opera by Verdi. Libretto by Francesco Piave and Andrea Maffei based on the Shakespeare tragedy. Première: Teatro della Pergola, Florence, March 14, 1847. This is the most im-

portant of Verdi's early operas, the first in which he became concerned with dramatic values. As Francis Toye wrote: "He was working for an expressiveness, an acute delineation of the human soul, never before realized. . . . The opera marks an especial striving after, and at times an especial realization of, new and noble ideals." Lady Macbeth's sleep-walking scene, "Una macchia è qui tuttora" is one of the opera's most memorable numbers.

McCormack, John, tenor. Born Athlone, Ireland, June 14, 1884; died Dublin, September 16, 1945. Though a concert favorite for many years, McCormack first achieved recognition in opera. After attending local schools, he was sent to Dublin to prepare for civil service examinations. While there he sang for Vincent O'Brien, director of the Marlboro Cathedral Choir, who urged him to devote himself to music. He gave McCormack some instruction, then engaged him to sing with his choir. In 1904 McCormack won a gold medal at the National Irish Festival in Dublin. After a period of study with Sabbatini in Italy he made his opera debut in Savona in 1905 in *L'amico Fritz*. He then went to London, where he earned his living singing in hotels and cabarets. An appearance at a Boosey Ballad Concert was so successful that he received several important engagements and an audition for Covent Garden. He made his Covent Garden debut on October 5, 1907, in *Cavalleria rusticana*. Oscar Hammerstein engaged him for the Manhattan Opera House and McCormack made his American debut here on November 10, 1909, in *La traviata*. From 1910 to 1912 he was the principal tenor of the Boston Opera, and from 1912 to 1914 of the Philadelphia Opera. Meanwhile, on November 29, 1910, he had made his first appearance at the Metropolitan Opera in *La traviata*. McCormack was again heard at the Metropolitan between 1912 and 1914, and between 1917 and 1919. The exquisite artistry of his phrasing and the purity of his voice made him a favorite, particularly in the operas of Puccini. His feeling for the classic style also made him an outstanding interpreter of Mozart.

After World War I, McCormack abandoned opera for the concert stage, becoming one of the most highly acclaimed and best-loved concert singers of the world, particularly admired for Irish songs and ballads. He also made numerous radio appearances and was starred in a Hollywood film, *Song of My Heart*. He went into retirement in 1938, but emerged during World War II to sing for the British Red Cross; he was soon ordered by his physician to give up all singing. Though he had been an American citizen since 1919, McCormack lived the last years of his life in his native Ireland, where he died in 1945 of bronchial pneumonia. Among the many honors that came to him were the Order of St. Gregory the Great, the Order of the Holy Sepulchre, and the vice-presidency of the Irish Royal Academy. On February 23, 1928, he was raised to papal peerage with the title of Count. In collaboration with Pierre Key, he wrote an autobiography, *John McCormack: His Own Life Story* (1919). After his death, his widow wrote his biography, *I Hear You Calling Me* (1949).

Maccus, Aethelwold's Master of Horse (bass) in Deems Taylor's *The King's Henchman*.

Macfarren, Sir George Alexander, composer. Born London, March 2, 1813; died there October 31, 1887. He completed his music study at the Royal Academy of Music, where in 1834 he became a professor. In 1875 he was appointed professor of music at Cambridge and a year later principal of the Royal Academy. The most notable of his twelve operas were: *The Devil's Opera* (1838); *Don Quixote* (1846); *King Charles II* (1849); *Robin Hood* (1860); *Jessy Lea* (1863); *She Stoops to Conquer* (1864); *The Soldiers' Legacy* (1864); *Helvellyn* (1864). In 1840 he edited Henry Purcell's *Dido and Aeneas*. He was knighted in 1883.

Machiavelli, Niccolò, statesman and writer. Born Florence, Italy, May 3, 1469; died there June 22, 1527. His powerful comedy, *La Mandragola*, a satire on the corruption of Italian society, was the source of operas by Mario Castelnuovo-Tedesco and Ignaz Waghalter.

Mackenzie, Sir Alexander, composer. Born Edinburgh, Scotland, August 22, 1847; died London, April 28, 1935. He attended the Sonderhausen Conservatory (Germany) and the Royal Academy of Music in London. In 1865 he returned to Edinburgh, where during the next decade he became prominent as violinist, teacher, and conductor. In 1879 he settled in Florence to concentrate on composition. Back in England in 1888, he was appointed principal of the Royal Academy of Music (from which he retired only in 1924), and in 1892 principal conductor of the Philharmonic Society. He was knighted in 1895 and in 1922 made Knight Commander of the Victorian Order. His operas: *Colomba* (1883); *The Troubadour* (1886); *His Majesty* (1897); *The Knights of the Road* (1905); *The Cricket on the Hearth* (1914); *The Eve of St. John* (1925).

Macpherson, James, writer and poet. Born Ruthven, Iverness-shire, Scotland, October 27, 1736; died there February 17, 1796. He published several volumes of translations by a supposed third-century Gaelic poet, Ossian. Though Macpherson was never able to prove the authenticity of this poetry, it had a tremendous vogue, not only in England but on the Continent. Several operas were derived from these poems of "Ossian," notably Edgar L. Bainton's *Oithona,* François Bathelemon's *Oithona,* Jean Georg Kastner's *Oskars Tod,* Méhul's *Uthal,* Edward Sobolewski's *Komola,* Ian Whyte's *Comola,* and Peter von White's *Colmal.* Ossian is the central character of Jean François Lesueur's opera *Ossian,* produced in 1804.

Ma dall' arido, Amelia's aria in Act II of Verdi's *Un ballo in maschera.*

Madama Butterfly, opera in three acts by Giacomo Puccini. Libretto by Luigi Illica and Giuseppe Giacosa, based on a play by David Belasco, in turn derived from a short story by John Luther Long. Première: La Scala, February 17, 1904. American première: Washington, D.C., October 15, 1906.

Characters: Cio-Cio-San (Madama Butterfly), a geisha girl (soprano); Suzuki, her servant (mezzo-soprano); B. F. Pinkerton, U.S. Navy lieutenant (tenor); Kate Pinkerton, his wife (mezzo-soprano); Sharpless, U.S. Consul at Nagasaki (baritone); Goro, a marriage broker (tenor); Prince Yamadori, in love with Cio-Cio-San (baritone); the Bonze, Cio-Cio-San's uncle (bass); friends and relatives of Cio-Cio-San. The setting is Nagasaki, Japan, in the early 1900's.

Act I. The exterior of Pinkerton's house. Pinkerton tells Sharpless of his infatuation for a Japanese girl and of his intention to marry her for "nine hundred and ninety-nine years," with the privilege of annulment when convenient. Pinkerton's levity upsets Sharpless, who tries to convince the lieutenant of the gravity of a relationship with a Japanese girl. Pinkerton repeats how intensely he loves her ("Amore o grillo"). Laughing voices of Japanese girls are heard and Cio-Cio-San appears. She introduces her relatives and friends to Pinkerton ("Spira sul mare"). Presently, she informs her beloved that for his sake she has renounced her religion. The marriage ceremony is interrupted when Cio-Cio-San's uncle appears to condemn his niece for renouncing her people. Contemptuously, her relatives spurn the girl and depart. Butterfly bursts into tears but is soon soothed by Pinkerton's tenderness. As night descends, the lovers are happy in each other's arms as they confide their passionate feelings ("Viene la sera").

Act II. Inside Butterfly's house. As Suzuki prays before an image of Buddha, Butterfly chides her gently for appealing to a Japanese god. Butterfly is faithful to Pinkerton, who has been forced to leave with the American fleet, and she is true to his religion and country, certain that some fine day he will come back to her ("Un bel di"). Sharpless brings Butterfly a letter which she is about to read when the marriage broker arrives with a wealthy suitor. Butterfly is deaf to all propositions. When Sharpless inquires what Butterfly would do if Pinkerton were to desert her, she answers gravely that she would kill herself. She now calls in the child, Little Trouble, who is the fruit of her love ("Chi vide mai a bimbo"). Sharpless now knows that a terrible tragedy is imminent. Suddenly there comes from the harbor the sound of a cannon shot. Cio-Cio-San seizes a telescope and learns that Pinkerton's ship has returned. In anticipation of her beloved's return, Butterfly helps Suzuki decorate the house with cherry blossoms (Flower Duet: "Scuoti quella fronda di ciliegio"). She then dons her wedding dress, but day passes into night with no sign of Pinkerton.

Act III. The same scene. A tender orchestral prelude recalls the love music of the first act. Dawn has come. Weary of her vigil, Butterfly goes to an inner room. While she is absent, Pinkerton and Sharpless arrive. Suzuki is overwhelmed with joy at the sight of Pinkerton, but when she sees an American woman at Pinkerton's side she senses the worst. Sharpless persuades Pinkerton to leave without seeing Butterfly. After a tender farewell to the house and his memories ("Addio fiorito asil"), Pinkerton departs. When Butterfly rushes into the room she finds not Pinkerton, but Sharpless and a strange woman. When she sees her servant in tears she begins to understand what has happened. The American woman—Pinkerton's wife Kate—implores Cio-Cio-San to turn over to her Pinkerton's child. At last, Cio-Cio-San is ready to do this—but only on condition that Pinkerton himself makes the request. When Sharpless and Kate leave to call Pinkerton, Butterfly raises a dagger to her throat. Little Trouble appears. Butterfly bids her child farewell ("Tu, tu piccolo Iddio!"). She gives him a doll and an American flag

to play with. Then she goes behind a screen with her dagger. A moment later she staggers out; by the time Pinkerton appears, she is dead. Pinkerton is overwhelmed with grief. Sharpless gently leads the motherless child from the house.

Madama Butterfly, today one of Puccini's best-loved operas, was a fiasco when first performed. The antagonism of the audience was such that the shouting often drowned out the music. Puccini finally appeared on the stage in an effort to restore order. He was jeered into the wings. There is good reason to believe that enemies of Puccini, envious of his mounting success, had helped organize this opposition. If so, they were also aided by the opera itself. Coming from a composer who had already endeared himself to his audiences with *Manon Lescaut, La Bohème,* and *Tosca, Madama Butterfly* was something of a shock. The exotic setting, the love affair of an American and a Japanese girl were not to Italian tastes. Puccini's score did not help matters either. With his sure dramatic instinct he had included unorthodox procedures calculated to project the background and atmosphere of the play, to accentuate the dramatic tensions: altered harmonies; provocative suspensions; melodies in oriental pentatonic scales; piercing dissonances; unusual rhythmic and instrumental effects. All this did not make for pleasant listening on first contact. A final irritant was the overlong second act (the opera was originally complete in two acts).

Believing that he had written his finest opera, Puccini did not lose faith. However, he allowed himself to be convinced by a few close friends (including Arturo Toscanini) to revise the work. He deleted some of the more objectionable and exotic vocal passages, made more of the tenor role, and divided the long second act to make a third. Three months after its première the opera was heard again. This time it was a triumph. A month later Toscanini introduced it to South America. From then on the opera aroused unqualified enthusiasm. Today, of course, it is one of the best-loved operas in the Italian repertory. It has something of the refinement and delicacy of a Japanese print. Puccini's lyricism, mainly Italian, is in his tenderest and sweetest vein, and his characterization of Butterfly is one of the most affecting in his gallery of operatic women. His ability to evoke the proper atmosphere and mood with a few strokes of the pen shows the hand of a master, and the oriental harmonic and instrumental effects no longer startle but provide a welcome suggestion of authenticity.

Madame Sans-Gêne, opera by Umberto Giordano. Libretto by Renato Simoni, based on the play of the same name by Victorien Sardou and Emile Moreau. Première: Metropolitan Opera, January 25, 1915. The setting is France during and following the French Revolution. Catherine, a laundress known as Madame Sans-Gêne, active in the revolution, marries the police sergeant Lefebvre. After the rise of Napoleon, Lefebvre becomes Duke of Danzig. Catherine arouses the displeasure of Napoleon's court with her frank and earthy ways, but Napoleon stands by her when she reminds him how she had been faithful to his cause. When Napoleon suspects his Empress of carrying on an affair with an Austrian count, Catherine becomes the means by which Napoleon is reassured of his wife's loyalty.

Madamina, il catalogo è questo, the "Catalogue Song" in Act I, Scene 2, of Mozart's *Don Giovanni* in which Leporello enumerates Don Giovanni's numerous conquests.

Maddalena, Sparafucile's sister (mezzosoprano) in Verdi's *Rigoletto.*

Madeleine, (1) one-act opera by Victor Herbert. Libretto by Grant Stewart, adapted from a play by Adrien Decourcelles and L. Thibault, *Je dine chez ma mère.* Première: Metropolitan Opera, January 24, 1914. The setting is Paris on New Year's Day, 1760. Madeleine Fleury, an opera singer, is heartbroken because her dinner invitations to various lovers and friends are turned down, all for the same reason: each happens to be dining with his or her mother. Finally, Madeleine removes the portrait of her own mother from the wall and announces that she, too, will dine with her mother.

(2) The Countess de Coigny's daughter (soprano) in Giordano's *Andrea Chénier.*

(3) A countess (soprano) in Richard Strauss's *Capriccio.*

Mademoiselle Fifi, *see* MAUPASSANT, GUY DE.

Madre, pietosa Vergine, Leonore's aria in Act II, Scene 2, of Verdi's *La forza del destino.*

Madrigal comedy, a musical form popular in Italy in the late sixteenth century, consisting of a series of madrigals (unaccompanied songs for several voices) so designed as to tell a dramatic story. The most famous example is Orazio Vecchi's *L'amfiparnaso,* published in 1597. The madrigal comedy was a forerunner of opera.

Mad Scene (*I Puritani*), *see* QUI LA VOCE SUA SOAVE.

Mad Scene (*Linda di Chamounix*), *see* LINDA DI CHAMOUNIX.

Mad Scene (*Lucia di Lammermoor*), *see* ARDON GL' INCENSI.

Maestro di cappella, an Italian term that first meant "master of the chapel," referring to the band of musicians playing or singing in a chapel or church. A maestro di cappella might be a choirmaster, a conductor, composer, or all three. The term later narrowed to denote a conductor. The German equivalent is kapellmeister; the French, maître de chapelle.

Maeterlinck, Maurice, author. Born Ghent, Belgium, August 29, 1862; died Nice, France, May 6, 1949. He won the Nobel Prize for literature in 1911. Several of his sensitive and at times mystical plays have been made into operas. The most celebrated is *Pelléas et Mélisande*, the text of Claude Debussy's impressionist opera. Others are: *Ariane et Barbe-Bleue* (Paul Dukas); *Monna Vanna* (operas by Henri Février and Emil Abrányi); *L'oiseau bleu* (Albert Louis Wolff); *Soeur Béatrice* (operas by Alexander Gretchaninoff and François Rasse); *La mort de Tintagiles* (operas by Lawrence Collingwood and Jean Nouguès).

Magda, (1) Ruggero's beloved (soprano) in Puccini's *La Rondine*.

(2) Heinrich's wife (soprano) in Respighi's *La campana sommersa*.

Magdalena, Eva's nurse (mezzo-soprano) in Wagner's *Die Meistersinger*.

Maggio Musicale Fiorentino, *see* FLORENCE MAY MUSIC FESTIVAL.

Maggiorivoglio, Marchioness of, the name for the Countess of Berkenfeld in the Italian version of Donizetti's *The Daughter of the Regiment*.

Magic Fire Scene (Feuerzauber), the concluding scene, in which Brünnhilde is surrounded by a circle of flames, of Wagner's *Die Walküre*.

Magic Flute, The (Die Zauberflöte), opera in two acts by Wolfgang Amadeus Mozart. Libretto by Johann Emanuel Schikaneder. Première: Theater-auf-der-Wieden, Vienna, September 30, 1791. American première: Park Theater, New York, April 17, 1833.

Characters: Sarastro, High Priest of Isis (bass); Queen of the Night, (soprano); Pamina, her daughter (soprano); Tamino, a prince (tenor); Papageno, a birdcatcher (baritone); Papagena, his sweetheart (soprano); Monostatos, a Moor (tenor); attendants of the Queen of the Night; priests; priestesses; slaves; warriors. The action takes place in Memphis, Egypt, in the days of Ramses I.

Act I, Scene 1. A lonely landscape. Tamino, fleeing from a serpent, is rescued by three attendants of the Queen of the Night. He hides as Papageno appears, preceded by his piping. Papageno sings a ditty ("Der Vogelfänger bin ich ja") that explains his preference for girls to birds. Tamino believes that it is Papageno who has saved him from the serpent: when he thanks the birdcatcher, Papageno does not disillusion him. Because of this deception, Papageno is punished by the Queen's attendants by having his lips sealed with a padlock. The attendants then show Tamino a portrait of the Queen's daughter, Pamina, whose beauty inspires the prince to rhapsody ("Dies Bildnis ist bezaubernd schön"). Now the Queen of the Night comes to tell Tamino that Pamina is the prisoner of a tyrant, that if Tamino will save her he can have her as his wife ("Zum Leiden bin ich auserkoren"). Tamino consents to save her. To help him in his adventure, the Queen provides him with a magic flute which, when played, will safeguard him from harm. Papageno is instructed to accompany Tamino to the palace of the abductor, Sarastro; the lock is removed from his lips and he is given a set of chimes whose magic property is similar to that of Tamino's flute.

Scene 2. A room in Sarastro's palace. Pamina is being guarded by Monostatos. The Moor flees in terror when Papageno appears, for he thinks the birdcatcher is the devil. Papageno convinces Pamina that she should follow him toward liberation and true love. They both sing the praises of love ("Bei Männern, welche Liebe fühlen").

Scene 3. A grove before the Temple of Isis. Tamino is told by a temple priest that Sarastro is no tyrant but a man of high ideals; someday, the priest declares, the reason for imprisoning Pamina will be made clear. When he hears the ringing of Papageno's chimes, Tamino goes searching for him. Papageno and Pamina have been trapped by Monostatos and his slaves and have been rescued by the magic chimes. Suddenly Sarastro enters. Pamina falls on her knees confessing that she had tried to escape. A moment later Tamino is brought in—he, too, has been seized by the slaves. Seeing each other for the first time, Tamino and Pamina rush toward one another. Sarastro reveals to the pair that they must now perform secret rites. Their heads are covered with veils.

Act II, Scene 1. A palm grove. Sarastro and his priests file in (March of the Priests). Sarastro pleads that Tamino be initiated into the final mysteries of their order, for his marriage to Pamina has been preordained. The priests give their consent. Sarastro then invokes the gods to bring the lovers the courage to meet their trials ("O Isis und Osiris").

Scene 2. The courtyard of the Temple of Isis. Tamino and Papageno are about to undergo several severe tests. For one thing, Tamino is denied the privilege of speaking to Pamina; for another, Papageno must not say a word to the bride selected for him by Sarastro. When the attendants of the Queen of the Night warn him to flee from this place of evil, Tamino turns a deaf ear.

Scene 3. A garden. While Pamina sleeps, Monostatos tries to steal a kiss. He is sent scurrying by the Queen of the Night, who then pleads with Pamina to avenge her ("Der Hölle Rache"). When she gives Pamina a dagger with which to kill Sarastro, Pamina recoils in horror. Later, Pamina appeals to Sarastro to have mercy on her mother. Sarastro replies that in a holy place there is no room for hate or vengeance ("In diesen heil'gen Hallen").

Scene 4. A hall in the Temple of Probation. Unable to hold his tongue, Papageno begins to blabber to an old woman who brings him water. A peal of thunder sends the old woman away in terror. Pages now arrive with a feast. Tamino, however, can think only of his beloved. He summons her with his flute. She appears, full of love, but is bewildered when Tamino refuses to speak to her. Convinced that she is no longer loved, Pamina seeks death ("Ach, ich fühl's"). Tamino and Papageno are summoned to their next test.

Scene 5. A place near the Pyramids. The priests and Sarastro call to the gods to witness the climax of Tamino's trials. Sarastro reassures both Tamino and Pamina that all will turn out well in the end. He then leads Tamino away. Papageno staggers in, tired and thirsty. Magically, a huge goblet of wine appears. Drinking his fill, Papageno becomes inebriated and expresses his wish for a wife ("Ein Mädchen oder Weibchen"). The old woman reappears and insists that she is the wife who has been selected for him. He must accept her— the alternative is a life with only bread and water. When he does so, she sheds her ragged clothing and is magically transformed into a young and beautiful bird-girl, Papagena. Papageno is about to take her in his arms when a voice warns him that he must first undergo another trial.

Scene 6. A garden. Pamina, convinced that she has lost Tamino's love, is about to kill herself. Her hand is stayed by three pages who reassure her that Tamino is still in love with her.

Scene 7. A wild mountain spot. Just before meeting another test of courage, Tamino is allowed to meet and embrace Pamina. Pamina takes Tamino's hand and conducts him through a cavern of fire as Tamino plays his magic flute for protection. After this the lovers negotiate a water cavern. The priests hail their success.

Scene 8. A garden. Papageno thinks

he has lost his beautiful girl. He is so upset that he wants to hang himself. The pages appear and remind him to use his chimes. Eagerly, Papageno tinkles out his magic tune. It brings Papagena, and the two embrace.

Scene 9. A gloomy spot near the Temple of Isis. The Queen of the Night, her three attendants, and Monostatos, are making one last attempt to destroy Sarastro and abduct Pamina. A burst of lightning and crash of thunder herald their descent into the depths of the earth.

Scene 10. The Temple. Tamino and Pamina are conducted before Sarastro. He pronounces them ready to serve Isis. All voices join in praise of Isis and Osiris.

The Magic Flute, written in the last year of Mozart's life, was commissioned by Emanuel Schikaneder, the impresario of a theater presenting singspiels. It was with the requirements of the singspiel in mind that Mozart wrote his music. Schikaneder wrote the kind of play Viennese liked. It was filled with good and diabolical forces in conflict; it had burlesque characters; it glorified the triumph of love over all obstacles; it had an Oriental setting. All this was made to serve as symbolism for the Masonic order of which both Schikaneder and Mozart were members. Confusion set in with last-minute revisions of characters and drastic alterations of plot made necessary by the production in Vienna of another opera with a story similar to Schikaneder's.

The popular character of the play— but certainly not its obfuscation—was carried out in Mozart's music. Interspersed with the dialogue are many tunes, particularly those sung by Papageno, that have the wholesomeness of folk songs. And there are arias that are among the most beautiful written by Mozart. Mozart's feeling for a comic situation was never surer. Yet at other

moments he could rise to grandeur and nobility, as in Sarastro's arias. Perhaps in no other of his operas is the full range of his genius so strongly evident as here—in this work intended as popular entertainment but transformed into a wondrous work of art.

Magic Garden Scene, the scene in Klingsor's Garden, Act II, Scene 2, of Wagner's *Parsifal*.

Magic opera (Zauberoper), a kind of opera popular in Vienna at the close of the eighteenth and the beginning of the nineteenth centuries. The text was usually on some fairy-tale subject, and included both broad comedy and elaborate scenic effects. The work was made up of spoken dialogue interspersed with music numbers. Mozart's *The Magic Flute* and Weber's *Der Freischütz* are the most celebrated examples. Franz Schubert's *Die Zauberharfe* is another.

Magnifico, Don, Cinderella's stepfather (bass) in Rossini's *La cenerentola*.

Mahabharata, The, a Hindu epic, the source of operas by Gustav Holst (*Savitri*) and Massenet (*Le Roi de Lahore*). Its theme is the great struggle between the houses of Kauravas and Pandavas for the control of a kingdom.

Mahler, Gustav, conductor and composer. Born Kalischt, Bohemia, July 7, 1860; died Vienna, May 18, 1911. Though Mahler wrote two operas in his youth, they were not published, and as a composer he is known only for his symphonies and his songs. However, he was one of the pre-eminent opera conductors of his time. He entered the Vienna Conservatory in 1875, his teachers including Robert Fuchs (harmony), Julius Epstein (piano), and Franz Krenn (composition). After leaving the Conservatory he conducted orchestras of several small-town theaters. In 1882 he was appointed conductor of the Olmütz Opera. This was followed by several other appointments, including one at the Prague Opera, where Mahler gave notable performances of the *Ring* cycle and several Mozart operas. He was next heard at the Budapest Opera, where his true stature as conductor was first appreciated. From Budapest, Mahler went on to the Hamburg Opera. There he continued to command the respect of discriminating musicians; his work was so outstanding that in 1897 several of Vienna's leading musicians (including Brahms and Guido Adler) recommended him for one of the most important conductorial posts in Europe, that of the Vienna Opera.

Mahler served as the music director of the Vienna Opera for a decade, creating one of the great epochs in the history of that company. Relentless in his dedication to perfection, he made the Vienna Opera the first opera house in Europe, if not in the world. He revitalized the repertory; he presented the old operas in new, restudied versions; he was meticulous about every detail of scenery and staging. He beat down mediocrity and opposition. "Certainly no operatic theater was ever directed on a more grandiose plan," wrote the Viennese critic Max Graf. There were those who called Mahler a saint because of his holy dedication to art. But he also had enemies who hated him for one or more reasons: because he was dictatorial, a Jew, lavish in his expenditures, or because they had been the object of his savage attacks when they were guilty of sloth or indifference. These enemies were indefatigable in putting obstacles in his way and creating intrigues against him. By 1907 Mahler decided he had to leave Vienna. As he explained to his coworkers: "I must keep on the heights. I cannot let anything irritate me or drag me down." He gave his last performance in Vienna—*Fidelio*—on October 15, 1907. That winter he came to the United States and on January 1, 1908, made his American debut at the Metropolitan Opera with *Tristan und Isolde*

(a performance in which Olive Fremstad appeared for the first time as Isolde). He was acclaimed. The following season he combined his activity at the Metropolitan Opera with duties as conductor of the New York Philharmonic Orchestra. He continued dividing his energies between the Metropolitan and the Philharmonic through the 1909–1910 season. The following season he devoted himself entirely to the Philharmonic. The strain of continuous work broke his health. He collapsed in New York, returned to Europe, where he led a few concerts of his own works, then died at the age of fifty.

Maid of Orleans, The, opera by Tchaikovsky. Libretto by the composer, based on a translation of Schiller's drama, *Die Jungfrau von Orleans.* Première: St. Petersburg, February 25, 1881. The opera is based on the story of Joan of Arc.

Maid of Pskov, The, *see* IVAN THE TERRIBLE.

Main de gloire, La, opera by Jean Françaix. Libretto by the composer, based on a story by Gérard de Nerval. Première: Bordeaux Festival, May 7, 1950. The text is compounded of fantasy, comedy, and tragedy. A young man has cast a spell over his hand so that he may be victorious in a duel. No longer in control of his hand, he finds that it leads him from one crime to another until he is punished with death.

Maintenant que le père de Pelléas est sauvé, Arkël's expression of relief in Act IV, Scene 2, of Debussy's *Pelléas et Mélisande.*

Maison, René, tenor. Born Traumeries, Belgium, November 24, 1895. He studied singing in Antwerp and at the Paris Conservatory after which he made his debut in Geneva in 1920 in *La Bohème.* Appearances followed in Monte Carlo in 1922 and after that elsewhere in Europe and South America. From 1927 to 1932 he appeared with the Chicago Civic Opera and the San Francisco Opera. In June, 1936, he made a successful debut at Covent Garden as Julien in *Louise*. From 1934 to 1937 he appeared at the Teatro Colón in Buenos Aires. He made his Metropolitan Opera debut on February 3, 1936, in *Die Meistersinger*, remaining at the Metropolitan for the next seven seasons. He then returned to sing in Europe.

Maître de chapelle, *see* MAESTRO DI CAPELLA.

Malatesta, Dr., physician (baritone) in Donizetti's *Don Pasquale*.

Malazarte, Brazilian folk opera by Oscar Lorenzo Fernandez. Libretto by the composer, based on a play by García Aranha. Première: Rio de Janeiro, September 30, 1941. The setting is colonial Brazil, the central character the legendary figure of Malazarte, master of evil arts. Music and text are filled with folk elements. Voodoo rites, magic, native dances play a prominent part. The Afro-Brazilian dance "Batuque," with which the first act ends, has been performed at symphony concerts in Brazil and the United States.

malheurs d'Orphée, Les, (The Misfortunes of Orpheus), opera by Darius Milhaud. Libretto by Armand Lunel. Première: Théâtre de la Monnaie, May 7, 1926. Orpheus is here a druggist in an unspecified town where his clients are animals. He falls in love with a gypsy girl, Eurydice, and takes her off to the mountains. Eurydice dies there; not even Orpheus' tender solicitude, the love of his animals, or the power of his drugs can save her.

Malibran, Maria Felicita (born GARCÍA), dramatic contralto. Born Paris, March 24, 1808; died Manchester, England, September 23, 1836. She was the daughter of the tenor Manuel García. At the age of five she appeared in Paër's opera *Agnese,* in Naples, playing a child's part. She then studied solfeggio with Auguste Mathieu Panseron in Naples and singing with her father. Her official debut took place in London, on June 7, 1825, as Rosina. She then came to New York, where for two years she appeared with the Manuel García opera company in works by Rossini and Mozart, and two operas written for her by her father. She went into temporary retirement in New York after marrying the French merchant Malibran. They soon separated. Returning to Paris, the singer made her debut at the Théâtre des Italiens in 1828. Her success was tremendous. She now appeared in opera in most of the major cities of Italy and in London. Later, she toured Europe in joint recitals with the violinist Charles de Bériot, whom she married early in 1836. A month later a fall from a horse gave her head injuries that she never overcame, and after a few more appearances in concerts her brief life came to an end. The American composer Robert Russell Bennett has made her the central character of his opera *Maria Malibran* (1935).

Maliella, Carmela's adopted daughter (soprano) in Wolf-Ferrari's *The Jewels of the Madonna.*

Malipiero, Gian Francesco, composer. Born Venice, March 18, 1882. He came from an aristocratic Venetian family that had included several notable musicians. When Malipiero was seven, his family left Italy. For several years father and son played in German and Austrian theater orchestras. In 1896 an Austrian nobleman provided funds for Malipiero's formal study. He now attended the Vienna Conservatory for one year, and after that the Liceo Benedetto Marcello in Venice; his studies were completed with Enrico Bossi in Bologna. In 1902 Malipiero began research in old Italian music, the results of which were his editions of the works of such masters as Vivaldi, Cavalli, Marcello, and Monteverdi. In

1913 he went to Paris; that city's musical life stimulated and affected him. At this point he entered five of his works in an Italian competition: he won four prizes. One of the five works was his first opera, *Canossa*, introduced in Rome on January 24, 1914. Just before the outbreak of World War I, Malipiero settled in Asolo. He has since lived there whenever his teaching and editorial duties have allowed him to. He became director of the Liceo Benedetto Marcello in Venice in 1939. He has written numerous works for the stage. His style, classical in its serenity, combines the techniques and spirit of old Italian music with contemporary harmonic thought. His operas: *Canossa* (1913); *L'Orfeide*, a cycle of three operas including *La morte della maschere* (1922), *Sette canzoni* (1920), and *Orfeo* (1921); *Tre commedie goldoniani*, a cycle of three operas including *La bottega di caffe*, *Sior Todoro Brontolon*, and *Le baruffe chiozzotte* (1923); *Filomela e l'infatuato* (1925); *Merlino maestro d'organi* (1927); *Il mistero di Venezia*, a cycle of three operas including *Le Aquile d'Aquileia* (1929), *Il finto Arlecchino* (1927), and *I corvi di San Marco* (1929); *Torneo notturno* (1930); *La bella e il mostro* (1930); *La favola del figlio cambiato* (1933); *Giulio Cesare* (1936); *Antonio e Cleopatra* (1938); *Ecuba* (1939); *La vita e sogno* (1940); *I capricci di Callot* (1941); *L'allegra brigata* (1943); *Mondi Celesti e Infernali* (1949).

Mallika, Lakmé's slave (mezzo-soprano) in Delibes's *Lakmé.*

Mallinger, Mathilde, (born LICHTENEGGER), dramatic soprano. Born Agram, Croatia, February 17, 1847; died Berlin, April 19, 1920. Her studies took place at the Prague Conservatory, and with Richard Lewy in Vienna. On October 4, 1866, she made her debut in Munich as Norma. Two years later she created the role of Eva in the world première of *Die Meistersinger*. During this period she appeared successfully in Austria and Russia. From 1890 to 1895 she was professor of singing at the Prague Conservatory and after 1895 she taught at the Eichelberg Conservatory in Berlin.

Mal reggendo all'aspro assalto, Manrico's narrative in Act II, Scene 1, of Verdi's *Il trovatore.*

Malvolio, an assassin (baritone) in Flotow's *Alessandro Stradella.*

mamelles de Tirésias, Les (The Breasts of Tirésias), an "opera-burlesque" by François Poulenc. Libretto by Guillaume Apollinaire. Première: Opéra-Comique, June 10, 1947. The opera concerns the changing of sexes by wife and husband, with the husband bearing forty thousand children. In the end both characters revert happily to their original sexes and advise the audience to proliferate in order to rehabilitate France and avoid future wars.

Ma mère, je la vois, duet of Don José and Micaëla in Act I of Bizet's *Carmen.*

Mamma Lucia, *see* LUCIA.

Mamma, quel vino è generoso, Turiddu's farewell to his mother in Mascagni's *Cavalleria rusticana.*

Mancinelli, Luigi, conductor and composer. Born Orvieto, Italy, February 5, 1848; died Rome, February 2, 1921. He studied the cello in Florence. After playing in various orchestras he became a conductor. In 1874 he was appointed conductor of the Rome Opera. Seven years later he settled in Bologna as conductor of the Teatro Communale and director of the Liceo Musicale. In 1886 he went to London, where for two years he was principal conductor at Covent Garden during its spring seasons. There he conducted the first German-language performance in England of *Tristan und Isolde*. From 1887 to 1895 he directed opera performances at the Royal Theater in Madrid. He came to the United States in 1893, making his debut at the Metropolitan

Opera on November 27 with *Faust*. He remained a principal conductor of the Metropolitan through the 1902–1903 season, conducting German, Italian, and French operas. He was a distinguished Wagnerian conductor, one of the first in Italy to sponsor the Wagner music dramas. In 1906 he helped inaugurate the Teatro Colón in Buenos Aires, conducting there until 1912. He then retired to his villa at Lake Maggiore. He wrote the following operas: *Isora di Provenza* (1884); *Ero e Leandro* (1897); *Paolo e Francesca* (1907); *Sogno di una notte d'estate* (1916).

Manfred, see BYRON, GEORGE NOLL GORDON, LORD.

Manfredo, King Archibaldo's son (baritone), husband of Fiora, in Montemezzi's *L'amore dei tre re*.

Manhattan Opera Company, company established by Oscar Hammerstein at the Manhattan Opera House on Thirty-fourth Street in New York City in 1906. For four years it was one of the great opera institutions of the world. Hammerstein built the Manhattan Opera House to make it a home for opera in English. Before the house was completed he changed his plans and decided to present great operas in their original languages and with the world's foremost singers. The new company and the new opera house were first seen on December 3, 1906. The opera was *Norma*. Cleofonte Campanini was artistic director. A glittering parade of opera stars appeared there, including Nellie Melba, Lillian Nordica, Luisa Tetrazzini, Ernestine Schumann-Heink, Giovanni Zenatello, Lina Cavalieri, John McCormack, Mary Garden, Alessandro Bonci, Charles Dalmorès, Maurice Renaud—many of them appearing in America for the first time. The emphasis was on French operas, then being neglected by the competitive Metropolitan Opera, and on provocative novelties. Among the latter were *Pel-*

léas et Mélisande with Mary Garden, *Elektra, Louise,* and *Sapho*. The success of the Manhattan Opera provided serious, even damaging, competition to the Metropolitan Opera, with the result that Heinrich Conried had to resign as director of the Metropolitan, where a drastic reorganization took place. After giving 463 performances of 49 different operas, the Manhattan Opera suddenly closed in 1910. What caused Hammerstein to withdraw from opera was at the time a mystery. Since then, however, it has been revealed that a contractual agreement was made between Hammerstein and the directors of the Metropolitan by which, for a cash sum of over one million dollars, Hammerstein agreed not to stage opera performances in America for a period of ten years.

Mann, Thomas, author. Born Lübeck, Germany, June 6, 1875; died Zurich, Switzerland, August 12, 1955. He won the Nobel Prize for literature in 1929. His monumental series of four novels on the Biblical story of Joseph, *Joseph and His Brethren,* was made into a cycle of four operas by Hilding Rosenberg. Mann's story *The Transposed Heads,* was used in an opera of the same name by Peggy Glanville-Hicks.

Manners, Charles (born SOUTHCOTE MANSERGH), bass and impresario. Born London, December 27, 1857; died Dublin, Ireland, May 3, 1935. His music study took place at the Royal Irish Academy in Dublin, the Royal Academy in London, and in Florence. He made his stage debut in the première of *Iolanthe* by Gilbert and Sullivan. Subsequently he joined the Carl Rosa Opera, and in 1890 he made his Covent Garden debut in *Robert le Diable*. In 1893 he appeared in the United States as vocal soloist with the Anton Seidl orchestra. Three years later he sang in South Africa. In 1897, with his wife Fanny Moody, an opera soprano, he formed the Moody-Manners Opera Company. Its aim was the presentation of opera in English. The company gave successful seasons at Covent Garden, Drury Lane, and in the English provinces; it was the first company to present Wagner's music dramas in English. In 1904 and 1906 Manners directed opera festivals in Sheffield, the proceeds going toward the founding of a university there. He retired from all operatic activity in 1913.

Manon, opera in five acts by Jules Massenet. Libretto by Henri Meilhac and Philippe Gille, based on Abbé Prévost's story *L'histoire du chevalier des Grieux et de Manon Lescaut.* Première: Opéra-Comique, January 19, 1884. American première: Academy of Music, New York, December 23, 1885.

Characters: Chevalier des Grieux (tenor); Count des Grieux, his father (bass); Manon, an adventuress (soprano); Lescaut, her cousin (baritone); Guillot de Morfontaine, Minister of France (bass); De Brétigny, a nobleman (baritone); actresses, travelers, soldiers, townspeople, vendors, gamblers. The action takes place in Amiens, Paris, and Le Havre early in the eighteenth century.

Act I. Courtyard of an inn at Amiens. Lescaut awaits the arrival of his unknown cousin, Manon. She arrives by coach and tells Lescaut of her journey ("Je suis encore tout étourdie"). The roué Guillot de Morfontaine tries to impress Manon with his wealth. She rebuffs him. Seemingly, she is not interested in men, since she is about to enter a convent; however, the sight of three prettily dressed young women sets her reflecting on the sadness of thus rejecting life and love ("Voyons, Manon, plus de chimères"). Her revery is interrupted by the arrival of the young Chevalier des Grieux. He is struck by her beauty—she is attracted to him. When Guillot puts his coach at Manon's disposal, hoping she will receive his attentions, Manon impetuously suggests

to Des Grieux that they use it to go to Paris together ("Nous vivrons à Paris tous les deux").

Act II. The apartment of Chevalier des Grieux in Paris. With the help of Manon, Des Grieux is writing to his father asking his consent to marry Manon (Duo de la lettre: "J'écris à mon père"). Lescaut arrives, angered that his cousin has been abducted; however, when Des Grieux shows him the letter he is writing, evidence of his honorable intentions, the visitor is placated. De Brétigny, who has come with Lescaut, is instantly drawn to Manon, and he soon tries to induce her to go off with him for a life of pleasure and wealth; to convince Manon, he informs her that Des Grieux's father will soon come to take his son away. Manon wavers as she recalls her happiness with her beloved; but she finally gives in. While Des Grieux is out mailing his letter, she bids farewell to the table where they both have enjoyed so many happy meals ("Adieu, notre petite table"). When Des Grieux returns he finds her in tears. He tries to console her by revealing one of his dreams, in which he lives with Manon in their own home (Le Rêve: "En fermant les yeux"). When Des Grieux goes to the door to answer a knock, he is abducted by his father's men.

Act III, Scene 1. A street in Paris during a festival day. Before the rise of the curtain, a minuet is played as an entr'acte. The festive crowd includes De Brétigny and Guillot de Morfontaine. When Manon appears she is instantly surrounded by her admirers. Gaily she tells them of her devil-may-care life ("Je marche sur tous les chemins"); she also voices her philosophy that life is meant for song and dance (Gavotte: "Obéissons quand leur voix appelle"). When Des Grieux's father arrives, he discloses to De Brétigny that his son is at Saint-Sulpice, about to enter the priesthood. The news reawakens Manon's love for the Chevalier. She rushes off to Saint-Sulpice.

Scene 2. A parlor in the Seminary of Saint-Sulpice. Des Grieux's father has come to beg his son not to renounce life; he finds the young man deaf to his pleas. After his father leaves, the Chevalier bids the world farewell, but even while doing this he is unable to free his mind of memories of Manon ("Ah! fuyez, douce image"). Now Manon herself enters. At first Des Grieux rejects her, but Manon's beauty and tenderness are overpowering. At last he takes her in his arms, confessing that he loves her more than ever.

Act IV. A fashionable gambling room in Paris. The Chevalier and Manon arrive. Manon extols the joys of gold ("Ce bruit de l'or rire"). Des Grieux begins to gamble; since his luck is phenomenal, he is accused of cheating. In the ensuing disturbance, the police are summoned. Des Grieux is saved from an unpleasant situation by his father, but Manon is apprehended as a woman of ill repute.

Act V. A lonely spot on the road to Le Havre. Manon is being sent by coach to exile. Des Grieux bribes an officer for permission to speak to her. Passionately, he tries to convince Manon to run away with him, but Manon, ill and exhausted, has lost the will to live. She grows weaker; her mind begins to wander. She falls into Des Grieux's arms, emits a cry, and dies.

Manon is both one of the most popular French operas and one of the most characteristic. It contains delightful dances in antique style which help evoke the background of eighteenth century France. It is also filled with a lyricism of such grace and refinement that only a Frenchman would have written it. There are some critics who consider this lyric style so characteristic of the composer that they have come to describe it as "Massenetique." But the dramatic element is not sacrificed, and for this reason *Manon* is one of Massenet's most effective works. The climactic scenes, while never aspiring to grandeur and bigness of those in Italian opera, do not fail to affect audiences everywhere. It was to accentuate the drama that Massenet borrowed from Wagner the leitmotiv technique. And it was also to serve the drama that he resorted to the innovation of utilizing the spoken dialogue traditional to opéra comique against an effective orchestral background; the dialogue keeps the action fluid, while the accompaniment emphasizes and intensifies the mood of the play. A decade after finishing *Manon* the composer wrote a one-act sequel, to a libretto by Georges Boyer, entitled *Le portrait de Manon*.

Manon Lescaut, (1) opera by Daniel François Auber. Libretto by Eugène Scribe, based on the story of Abbé Prévost. Première: Opéra-Comique, February 23, 1856.

(2) Opera in four acts by Giacomo Puccini. Libretto by Luigi Illica, Giuseppe Giacosa, Giulio Ricordi, Marco Praga, and Domenico Oliva, based on the story of Abbé Prévost. Première: Teatro Regio, Turin, February 1, 1893. American première: Grand Opera House, Philadelphia, August 29, 1894.

Characters: Chevalier des Grieux, (tenor); Manon Lescaut (soprano); Lescaut, her brother, sergeant of the King's Guards (baritone); Géronte de Ravoir, a Parisian gallant (bass); Edmondo, a student (tenor); students; citizens; courtesans; sailors; dancers; police; ladies; gentlemen. The settings are Amiens, Paris, Le Havre, and Louisiana; the time, early in the eighteenth century.

Act I. Before an inn in Amiens. The Chevalier des Grieux is sad. His friends chide him for being in love and he replies with a mocking serenade to all women ("Tra voi, belle"). When a coach arrives, it brings Manon, her brother, and Géronte. Des Grieux is struck by Manon's beauty. When he addresses her he learns she is about to enter a convent. Manon is called away by her brother. Des Grieux grows rhapsodic over Manon's fascination ("Donna non vidi mai"). The libertine Géronte plans to abduct Manon. Des Grieux learns of this. Upon Manon's reappearance, Des Grieux implores her to run away with him to Paris. Now herself in love, Manon agrees. They escape in Géronte's coach, to the humiliation of the old man, who vows to use his wealth to win Manon away from her lover.

Act II. An apartment in Géronte's house. Manon has deserted Des Grieux to live with Géronte. But she has not forgotten her former lover; she complains to her brother that wealth is no substitute for love ("In quelle trine morbide"). A band of musicians now entertains her with a madrigal ("Sulla vetta tu del monte"). When Géronte and his gay friends arrive, Manon dances a minuet for them. Des Grieux appears unexpectedly. At first he and Manon exchange bitter words ("Ah! Manon, mi tradisce"), but resentment soon turns to passionate love. When Géronte returns he finds Manon in Des Grieux's arms. He simulates indifference but as he leaves he mutters a threat. Lescaut bursts into the room to warn Manon that Géronte has gone to the police with a complaint against her. Quickly, Manon gathers her jewels and secretes them under her cloak. When the police arrive to seize her, Manon is so terrified she allows her cloak to fall off her shoulders and the jewels to scatter. The police arrest her.

Act III. A square in Le Havre. A brief orchestral intermezzo recalls Manon's love for Des Grieux and her despair at the tragedy befalling her. Lescaut and Des Grieux have come to save Manon from being deported to Louisiana. They are unable to bribe the guard. Des Grieux pleads with the captain to allow him to accompany Manon ("No! pazzo son! guardate!"). The captain is won over, and Des Grieux is smuggled aboard ship.

Act IV. A desolate plain near New Orleans. The countryside is gloomy; night is gathering. Des Grieux and Manon are seeking shelter. Ill and exhausted, she repents having brought such disaster to the man she loves ("Tutta su me ti sposa"). When Manon falls, Des Grieux continues his search alone. In the darkness, Manon grows increasingly terrified ("Sola, perduta, abbandonata"). Returning, Des Grieux finds Manon dying. He takes her in his arms, and there Manon breathes her last.

When we speak of Manon Lescaut, the opera that usually comes to mind is Massenet's. While Massenet's is the more popular and the finer work of the two, there is much in Puccini's to recommend it; and it surely would have won a wider and more enthusiastic acceptance if it did not have to compete with one of the finest creations of the French lyric theater. While *Manon Lescaut* is a comparatively early work of Puccini (it was his third opera), it already presents the qualities that endear Puccini to us: a lyricism of incomparable sweetness and charm which, in the big arias, becomes dominating and commanding; a compassion for the leading characters; a variety of harmony that sustains musical interest throughout the work; a diversity of mood and feeling. No wonder that when George Bernard Shaw first heard *Manon Lescaut* he wrote: "Puccini looks to me more like the heir of Verdi than any of his rivals." It took a strong faith on the part of an obscure and impoverished composer who thus far had written only two minor works to compete with Massenet on his own ground. Massenet's *Manon* was ten years old and already an established favorite in

European opera houses. Yet the act was justified by the results. *Manon Lescaut* was such a triumph at its first performance that Puccini had to take fifty curtain calls. Soon after, it was given throughout Italy, then the rest of Europe. The opera lifted its composer from his obscurity to that pinnacle of success he was henceforth to occupy

Manrico, the troubadour (tenor) in Verdi's *Il trovatore.*

Manru, opera by Ignace Jan Paderewski. Libretto by Alfred Nossig, based on a novel by Kraszewski. Première: Dresden Opera, May 29, 1910. Against her mother's wishes, Ulana marries the gypsy Manru. When Ulana feels she is losing her husband's love, she revives it with a love potion prepared by the dwarf Urok. Asa, a gypsy girl, lures Manru back to his people. When Ulana commits suicide, the gypsy kills Manru.

Man Without a Country, The, opera by Walter Damrosch. Libretto by Arthur Guiterman, based on the story of the same name by Edward Everett Hale. Première: Metropolitan Opera, May 12, 1937. The story is familiar. Lieutenant Philip Nolan betrays his country by joining Aaron Burr's conspiracy to found a new empire in the United States. He repents and expiates his crime by dying for his country in a naval engagement with the Berber pirates off Tripoli. The original story was changed by the librettist to include a love interest in the person of Mary Rutledge, Philip's sweetheart.

Manzoni, Alessandro, novelist and poet. Born Milan, March 7, 1785; died there May 22, 1873. His masterwork, the novel *I promessi sposi,* was the source of operas by Franz Gläser, Enrico Petrella, and Amilcare Ponchielli. Verdi wrote his famous Requiem on the occasion of Manzoni's death.

Mapleson, James Henry, impresario. Born London, May 4, 1830; died there November 14, 1901. After studying at the Royal Academy of Music in Lon-

don he sang in opera performances in Verona under the name of Enrico Manriani. His managerial career began in 1861 when he took over the Lyceum Theatre in London and presented Italian operas. From 1862 to 1867 he directed operas at Her Majesty's Theatre. When this house burned down, he transferred his activities to Drury Lane. In 1877 he reopened Her Majesty's Theatre. During his career in England "Colonel" Mapleson (as he was called) introduced many notable operas, among them *The Abduction from the Seraglio, Carmen, The Damnation of Faust, Faust, La forza del destino, Un ballo in maschera, Mefistofele, The Merry Wives of Windsor, The Sicilian Vespers,* and the complete *Ring* cycle. Singers who made their English debuts under his direction included Italo Campanini, Etelka Gerster, Lilli Lehmann, Christine Nilsson, and Jean de Reszke.

In 1878 Mapleson became the manager of the Academy of Music in New York; for the next eight years he helped shape operatic history in America. The extent of his repertory and the brilliance of his productions were not matched by any other opera company in the United States at that time. One after another the great operatic stars of Europe were introduced to America on the stage of the Academy of Music: Emma Albani, Italo Campanini, Etelka Gerster, Minnie Hauk, Clara Louise Kellogg, Pauline Lucca, Emma Nevada, Lillian Nordica, Adelina Patti, Victor Maurel and many others. During this period Mapleson gave 167 performances of 19 operas. The following were his important American premières: *Aïda, Carmen, Rigoletto, La traviata, Il trovatore.*

The competition offered by the newly opened Metropolitan Opera (1883) spelled doom for the Academy of Music. Receipts fell off; many of Mapleson's stars joined the new company. In 1886 Mapleson left New York to carry on his operatic activities in England. He returned in 1896 for another try at the Academy of Music but was unable to gain a permanent foothold. His autobiography, *The Mapleson Memoirs*, appeared in 1888.

Mapleson's nephew, Lionel S. Mapleson, was librarian of the Metropolitan Opera for nearly fifty years. He left an important collection of letters, autographs, photographs, clippings, scores, and first recordings (made by himself) of Metropolitan Opera performances. This collection is preserved in the Metropolitan's library.

M'appari, Lionel's aria in which he speaks of his hopeless love for Martha, in Act III of Flotow's *Martha.*

Marcel, Raoul de Nangis's servant (bass) in Meyerbeer's *Les Huguenots.*

Marcellina, (1) Rocco's daughter (soprano) in Beethoven's *Fidelio.*

(2) Figaro's mother (contralto) in Mozart's *The Marriage of Figaro.*

Marcello, a painter (baritone) in Puccini's *La Bohème.*

March, music used to accompany a parade or march, usually in 4/4 time, though sometimes in 2/4 or 6/8. Since the days of Lully, opera composers have used marches for scenes of pageantry and as a convenient device to bring large groups on stage, or off. Among the most famous marches in opera are the "Triumphal March" in *Aïda,* the "Coronation March" in *Le Prophète,* and the "March of the Guests" in *Tannhäuser.* The best known wedding march in opera is the "Bridal Procession" in *Lohengrin,* while the most eloquent funeral march is Siegfried's Death Music in *Die Götterdämmerung.* Other interesting examples of marches in opera are found in *La Bohème* (at the end of the second act); *Le coq d'or* ("Bridal Procession"); *The Damnation of Faust* ("Rakóczy March"); *A Life for the Czar; The Love for Three Oranges; The Magic*

Flute ("March of the Priests"); Handel's *Scipio* (to this day used by the Grenadier Guards of Britain).

Marche du couronnement, *see* CORONATION MARCH.

Marchesi, Mathilde de Castrone (born GRAUMANN), teacher of singing. Born Frankfort - on - the - Main, Germany, March 24, 1821; died London, November 17, 1913. She studied singing with Otto Nicolai in Vienna and Manuel García in Paris. After appearing in concerts she married the Italian baritone Salvatore Marchesi (see below) in 1852. Together they toured Europe in concert and opera. From 1854 to 1861 and again from 1869 to 1878 she taught singing at the Vienna Conservatory. She was also a member of the faculty of Cologne Conservatory, and taught singing privately in Paris. Many of her pupils became opera stars, including Emma Calvé, Emma Eames, Mary Garden, Etelka Gerster, and Nellie Melba. Marchesi wrote a vocal method, twenty-four volumes of studies, and an autobiography *Marchesi and Music* (1897).

Marchesi, Salvatore, baritone and teacher of singing. Born Palermo, Sicily, January 15, 1822; died Paris, February 20, 1908. He combined the study of music (with Francesco Lamperti and Pietro Raimondi) with that of law. Involved in the 1848 revolution in Italy, he was exiled. He came to America and made his opera debut in New York in *Ernani.* He then returned to Europe for study with Manuel García, after which he established himself in London both as singer and teacher. After marrying Mathilde Graumann in 1852 (see above) he made opera and concert appearances with her throughout Europe. Their daughter, Blanche Marchesi (1863–1940), was also a celebrated singer and teacher. He taught at the Conservatories of Vienna and Cologne, translated the librettos of many German and

French operas into Italian, and published a vocal method.

Marco, Simone's son (baritone) in Puccini's *Gianni Schicchi.*

Maretzek, Max, impresario and conductor. Born Brünn, Moravia, June 28, 1821; died Staten Island, New York, May 14, 1897. He began his musical career after studying medicine. After playing the violin in, and conducting, orchestras in Germany and England he came to the United States to conduct at the Astor Place Opera. Later, he formed his own company and presented opera. He was forced to suspend operations temporarily when the success of Jenny Lind deflected audiences from his theater. He later continued his activities at the Astor Place Opera House, Niblo's Garden, the Crosby Opera House in Chicago, and in Mexico and Havana. A highly volatile person, intransigent and dictatorial, he was often in violent disagreement with members of his various companies and with the critics. He wrote two operas: *Hamlet* (1843) and *Sleepy Hollow* (1879). He also wrote two books of reminiscences: *Crotchets and Quavers* (1885) and *Sharps and Flats* (1890).

Marfa, a widow (mezzo-soprano) in Mussorgsky's *Khovantchina.*

Margared, Rozenn's sister (mezzo-soprano) in Lalo's *Le Roi d'Ys.*

Margaret, Marie's neighbor (contralto) in Berg's *Wozzeck.*

Margaretha, the Baron of Schönau's daughter (soprano) in Nessler's *Der Trompeter von Säkkingen.*

Margiana, the Caliph's daughter (soprano) in Cornelius' *The Barber of Bagdad.*

Marguerite, Faust's beloved in Berlioz' *The Damnation of Faust,* Boïto's *Mefistofele,* and Gounod's *Faust.* In all three operas she is a soprano.

Maria, (1) Simon Boccanegra's daughter (soprano) in Verdi's *Simon Boccanegra.*

(2) Werner Kirchhofer's betrothed

(soprano) in Nessler's *Der Trompeter von Säkkingen.*

Marianne, Faninal's housekeeper (soprano) in Richard Strauss's *Der Rosenkavalier.*

Marie, (1) Wozzeck's sweetheart (soprano) in Berg's *Wozzeck.*

(2) Vivandière (canteen manager) (soprano) of the French 21st Regiment in Donizetti's *The Daughter of the Regiment.*

(3) Hans's beloved (soprano) in Smetana's *The Bartered Bride.*

Marietta, a dancer (soprano) in Korngold's *Die tote Stadt.*

Marina, a Polish landowner's daughter (mezzo-soprano) in Mussorgsky's *Boris Godunov.*

Marinaresca, *see* HO! HE! HO!

Marino Faliero, *see* BYRON, GEORGE NOEL GORDON, LORD.

Marinuzzi, Gino, (or GIUSEPPE), conductor and composer. Born Palermo, Sicily, March 24, 1882; died Milan, Italy, August 17, 1945. He made his debut as conductor in Catania after graduating from the Palermo Conservatory. He then conducted in many opera houses in Italy and Spain before going to La Scala, where he remained three seasons. From 1915 to 1919 he was director of the Liceo Musicale in Bologna. In 1919 he was appointed principal conductor of the Teatro Costanzi. In the same year he came to the United States to succeed Cleofonte Campanini as artistic director of the Chicago Opera. After 1921 he confined his conducting to Europe and South America. He wrote two operas: *Barberina* (1903); *Jacquerie* (1918).

Mario (born GIOVANNI MATTEO), tenor. Born Cagliari, Sardinia, October 17, 1810; died Rome, December 11, 1883. Born to a noble family, he was trained for the army. After completing his studies at the Turin Military Academy he joined the Piedmontese Guard, in which his father was colonel. In 1836 he fled from Italy with a ballet dancer.

Reaching Paris, he began the study of singing at the Paris Conservatory. On November 30, 1838, he made his debut at the Opéra in *Robert le Diable*. Strikingly handsome and endowed with an exquisitely beautiful voice, he was an immediate success. In 1839 he made a sensational debut in London in *Lucrezia Borgia*, and in 1840 he joined the Italian Opera in Paris. For the next quarter of a century he appeared both in Paris and London, recognized as one of the supreme operatic tenors of his generation, particularly in romantic roles. He frequently appeared with the soprano Giulia Grisi, whom he married in 1844. His last appearance was in *La favorita*, at Covent Garden in 1871. He then retired to Rome, where he was soon reduced to such poverty that, in 1880, his friends arranged a concert for his benefit in London.

Mario, Queena (born TILLOTSON), soprano. Born Akron, Ohio, August 21, 1896; died New York City, May 28, 1951. She studied singing with Marcella Sembrich and Oscar Saenger in New York and made her debut on September 4, 1918, with the San Carlo Opera in New York as Olympia. Her debut at the Metropolitan Opera took place on November 30, 1922, as Micaëla. She remained at the Metropolitan eighteen years. In 1931 she took over Mme. Sembrich's classes at the Curtis Institute. She also taught singing at the Juilliard School of Music and conducted a summer school and opera workshop at her farm in Bethel, Connecticut. In 1925 she married Wilfred Pelletier, conductor of the Metropolitan; they were divorced three years later. She wrote three mystery novels, one of them with an opera setting, *Murder in the Opera House*.

Mariola, an orphan (soprano) in love with Fra Gherardo in Pizzetti's *Fra Gherardo*.

Marionette Opera. Opera performances with marionettes are believed to have originated in Florence in the seventeenth century, when Filippo Acciajuoli produced such performances; the first marionette opera is believed to have been *Il Girello*, music by Jacopo Melani and Alessandro Stradella. Acciajuoli gave his marionette performances throughout northern Italy.

London had a puppet theater in the arcade of Covent Garden in 1713 where complete operas were given. Opera was also seen in an open-air marionette theater in Vienna at about this same time. There was a fully equipped marionette theater at the palace of Prince Esterházy for which Joseph Haydn wrote *Der Götterrat*, a prologue to his opera *Philemon und Baucis*. Subsequently, he wrote three more marionette operas, all now lost.

Notable marionette theaters have flourished in Europe in recent times, and in these theaters operas have been given. Examples are the Théâtre Guignol in Paris, the Teatro dei Piccoli in Rome, the Salzburg Marionette Theater, Ivo Puhonney's Marionette Theater in Germany, the Swiss Marionette Theater.

Several contemporary composers have written marionette operas. They include Mario Castelnuovo-Tedesco (*Auccasin et Nicolette*); Manuel de Falla (*El retablo de Maese Pedro*); Adriano Lualdi (*Le furie di Arlecchino,* and *Guerin Meschino*); and Ottorino Respighi (*The Sleeping Beauty*).

Marlowe, Christopher, dramatist and poet. Born Canterbury, England, February 6, 1564; died Deptford, England, May 30, 1593. One of the most distinguished of the Elizabethan dramatists, he appears as the central character in Wilfred Mellers' opera *The Tragical History of Cristopher Marlowe*. Ferruccio Busoni's opera *Doktor Faust* was based on Marlowe's *Dr. Faustus*.

Marmontel, Jean François, librettist. Born Bort, Limousin, France, July 11,

1723; died Abloville, France, December 31, 1799. He wrote librettos for operas by Cherubini, Grétry, Méhul, Piccinni, Rameau, and Spohr, among others, including Cherubini's *Démophon*, Grétry's *Céphale et Procris* and *Zémire et Azor*, and Piccinni's *Atys* and *Didon*. Marmontel was prominent in the clash in Paris between the forces supporting Gluck's ideas about opera and those on the side of Italian tradition and Piccinni; he allied himself with the Italian group.

Mârouf, comic opera by Henri Rabaud. Libretto by Lucien Népoty, based on a story from the *Arabian Nights*. Première: Opéra-Comique, May 15, 1914. The cobbler Mârouf escapes from his humdrum existence with a termagent wife by taking to sea. He becomes the solitary survivor of a shipwreck. Reaching the shores of Khaitan, he is rescued by an old boyhood friend, Ali, now a wealthy merchant. Ali dresses Mârouf in silks and finery and introduces him as the world's richest merchant. The Sultan offers Mârouf the run of his palace, and his daughter as wife. Mârouf takes advantage of this situation by depleting the Sultan's treasury. Day after day Mârouf promises the Sultan that his mighty caravans will soon arrive. At last, Mârouf confesses the truth to the princess, who continues to love the cobbler. They elope to the desert. Here they encounter a peasant who, through the power of a magic ring, turns into a genie, while the lovers' humble abode is transformed into a palace. The Sultan and his men, pursuing Mârouf, find him surrounded by wealth. Mârouf is forgiven. Two of Mârouf's arias are of interest: "Il est des Musulmans" in Act I and "A travers le désert" in Act II. One of the most effective portions of the opera is the ballet in Act III.

Marquis, The, Grisélidis's husband (baritone) in Massenet's *Grisélidis*.

Marriage, The, (1) a comedy by Nikolai Gogol, the source of operas by Gretchaninoff, Martinu, and Mussorgsky (the Mussorgsky opera was not finished).

(2) Opera by Alexander Gretchaninoff. Libretto by the composer. Première: Paris, October 8, 1950.

(3) Opera by Bohuslav Martinu. Libretto by the composer. Première: NBC Television network, February 7, 1953.

Marriage of Figaro, The (Le nozze di Figaro), opera buffa in four acts by Wolfgang Amadeus Mozart. Libretto by Lorenzo da Ponte, based on Beaumarchais's *Le mariage de Figaro*. Première: Burgtheater, Vienna, May 1, 1786. American première: possibly as early as 1799 in New York, though a presentation at the Park Theater in New York on May 10, 1824, was then claimed to be the first in America.

Characters: Count Almaviva (baritone); Countess Almaviva (soprano); Cherubino, the Count's page (soprano); Figaro, the Count's valet (baritone); Dr. Bartolo (bass); Don Basilio, a music master (tenor); Susanna, head waiting woman to the Countess (soprano); Marcellina (contralto); Antonio, gardener (bass); Barbarina, his daughter (soprano); Don Curzio, a counselor-at-law (tenor); peasants; townspeople; servants. The action takes place at Count Almaviva's chateau near Seville, Spain, in the seventeenth century.

A sprightly overture made up of two themes—one brisk, the other lyrical—sets the mood for this vivacious opera.

Act I. The apartment assigned to Figaro and Susanna. Figaro is about to be married to Susanna, and the Count has assigned them quarters conveniently near his own apartment. When Susanna suggests the reason for this, Figaro is at first concerned; then he remarks lightly that he knows how to handle his master ("Se vuol ballare"). There are other complications to Figaro's marriage. He has borrowed money from Marcellina and signed a contract promising to marry her if he fails to repay her. Bartolo and Marcellina arrive discussing this contract and the best way of implementing it. Susanna, suspicious of Marcellina, makes her feelings known to her rival. They exchange bitter words ("Via resti servita"), after which Marcellina leaves in a huff. Cherubino enters lamenting the fact that the Count is about to send him away for embracing Barbarina. The page, however, is secretly in love with the Countess. He eagerly sings a love song he has just written ("Non so più cosa son") for one of the Countess' ribbons now in Susanna's possession. When the Count abruptly appears, Cherubino hides behind a chair. Basilio's arrival sends the Count behind a chair, too. Eventually both are discovered and the Count angrily orders Cherubino to enlist in his regiment. Figaro mockingly gives Cherubino advice on how to behave as a soldier ("Non più andrai").

Act II. The apartment of the Countess. The Countess laments that the Count no longer loves her and is unfaithful ("Porgi amor"). Susanna and Figaro contrive a plan to revive the Count's interest in his wife by arousing his jealousy. The Count will be made to discover a letter seemingly sent to the Countess; at the same time a rendezvous will be arranged in which Susanna will appear disguised as the Countess. They also plan to make the Count ridiculous by having him meet Susanna at a tryst, with Cherubino dressed as Susanna. Cherubino enters the apartment thinking about the meaning of love ("Voi che sapete"). He is dressed in Susanna's clothes. As the Count appears Cherubino is hastily locked in a closet. The Count goes searching for a crowbar with which to force the door. While he is gone,

Cherubino escapes out the window and Susanna takes his place. When the Count finds Susanna he is effusive in his apologies until the gardener comes to tell him that somebody has just jumped out the window and trampled the flower bed. The gardener has also found a piece of paper dropped by the culprit—Cherubino's commission in the Count's regiment. Figaro assuages the Count's suspicions by insisting that he had Cherubino's commission in his own pocket, and that he is the man who jumped out the window. Marcellina now arrives to demand that Figaro go through with his bargain to marry her.

Act III. A hall. The Count, seeking a rendezvous with Susanna, threatens that he will insist on Marcellina and Figaro marrying if she declines. Susanna makes a pretense of yielding and the two arrange a meeting ("Crudel! perchè finora"). But, notwithstanding his agreement with Susanna, the Count is bent on punishing Figaro by forcing him to honor his agreement with Marcellina. Negotiations begin between Marcellina and her lawyer, and Figaro and the Count, during which the astonishing discovery is made that Figaro is actually Marcellina's long-lost son. The obstacle to Figaro's marriage to Susanna has thus been removed. Meanwhile the Countess, alone, recalls the time when the Count was in love with her ("Dove sono"). When Susanna arrives, the Countess dictates a letter arranging a rendezvous between the Count and Susanna (Letter Duet: "Che soave zeffiretto"). The Countess decides to take Cherubino's place in the affair—in other words, to disguise herself as Susanna. The marriage formalities of Figaro and Susanna are now taken care of. Guests enter to the strains of a march. Afterward, they dance a fandango. The Count, receiving the letter from Susanna, happily invites everyone to attend a gala celebration to be held later in the evening.

Act IV. The garden. Susanna and the Countess appear, each in the other's clothing. Figaro, who now believes (mistakenly) that his bride is about to yield to the Count, lurks in hiding and listens to Susanna sing an invitation to her absent lover ("Deh vieni, non tardar"). Cherubino is also in the garden, since he has an appointment with Barbarina. Seeing the Countess and believing her to be Susanna, he attempts to steal a kiss. He is sent packing by the Count, who now makes love to his wife under the apprehension that she is Susanna. Trying to awaken the jealousy of the Countess, Figaro learns that the Countess is really his own wife in disguise. The confusion is finally resolved, the Count obtains his wife's forgiveness, and the entire group enters the chauteau to get on with the celebration of Figaro's marriage.

The French dramatist Beaumarchais wrote a trilogy of plays in which the central character is Figaro. The first of these, *The Barber of Seville*, was made into operas by Paisiello and Rossini; the second, *The Marriage of Figaro*, was Mozart's inspiration. Beaumarchais's comedies were a pointed attack against the decadent aristocracy of his day—so much so that Napoleon described them as "the revolution already in action." Consequently, when Da Ponte and Mozart decided to collaborate on *The Marriage of Figaro*, the Austrian Emperor was not in favor of the project. Only when Mozart and Da Ponte promised to purge the play of political and social implications did the Emperor give his consent. Thus, Da Ponte's libretto became a farce rather than a social satire. Da Ponte scrambled his characters and their amatory designs with a lightness of touch that made for highly effective comedy. Following his librettist's suggestions, Mozart composed one of his most vivacious scores, chameleon-like in its rapidly changing hues, penetrating in

its psychological understanding of the characters. The music is sometimes sentimental and poetic, sometimes noble, sometimes touched with mockery. No wonder, then, that *The Marriage of Figaro* is sometimes called the "perfect opera buffa." It is, as Eric Blom wrote, "Italian comic opera in its final stage of perfection . . . as great as a whole as it is captivating in detail."

Marschallin, The, the Princess von Werdenberg (soprano) in Richard Strauss's *Der Rosenkavalier*.

Marschner, Heinrich, composer. Born Zittau, Germany, August 16, 1795; died Hanover, Germany, December 14, 1861. He studied law at the Leipzig University, but Johann Friedrich Rochlitz convinced him that he should embrace music. After studying with Johann Gottfried Schicht, he wrote his first opera, *Der Kyffhäuserberg*, produced in Vienna in 1816. Settling in Pressburg as a music teacher, he wrote two more operas, one of them, *Heinrich IV und d'Aubigné,* presented by Carl Maria von Weber in Dresden. Largely due to the popularity of this opera, Marschner was, in 1823, appointed joint-kapellmeister (with Weber and Francesco Morlacchi) of the Dresden Opera. He rose to the post of music director one year later but resigned when Weber died. He went on to Leipzig, where he became kapellmeister of the Leipzig Opera. There he wrote and had produced the opera that spread his fame throughout Europe: *Der Templer und die Jüdin* based on Sir Walter Scott's *Ivanhoe*. In 1831 he was appointed court kapellmeister in Hanover. He held this post until his retirement twenty-eight years later, when he was given the honorary title of Generalmusikdirektor. Marschner wrote his most famous opera in Hanover, *Hans Heiling,* a triumph at its première in Berlin on May 24, 1833. Marschner's operas are significant historically in that they carry on the German Romantic movement launched by Weber. His operas: *Der Kyffhäuserberg* (1816); *Saidar* (1819); *Heinrich IV und d'Aubigné* (1820); *Der Holzdieb* (1825); *Lucretia* (1826); *Der Vampyr* (1828); *Der Templer und die Jüdin* (1829); *Des Falkners Braut* (1832); *Hans Heiling* (1833); *Das Schloss am Aetna* (1835); *Der Bäbu* (1837); *Adolf von Nassau* (1843); *Austin* (1851); *Hjarne der Sängerkönig.*

Martern aller Arten, Constanza's aria in Act II of Mozart's *The Abduction from the Seraglio.*

Martha, (1) Marguerite's mother (contralto) in Boïto's *Mefistofele.*

(2) A village girl (soprano) in D'Albert's *Tiefland.*

(3) Marguerite's friend (mezzosoprano) in Gounod's *Faust.*

(4) Opera in four acts by Friedrich von Flotow. Libretto by Friedrich Wilhelm Riese, extended from a ballet-pantomime, *Lady Henriette,* by Vernoy de Saint-Georges, to which Flotow had contributed a portion of the music. Première: Vienna, November 25, 1847. American première: Niblo's Garden, New York, November 1, 1852.

Characters: Lady Harriet, maid of honor to Queen Anne (soprano); Nancy, her maid (contralto); Sir Tristan Mickleford, Lady Harriet's cousin (bass); Plunkett, a wealthy farmer (bass); Lionel, his foster brother, later the Earl of Derby (tenor); Sheriff of Richmond (bass); Lady Harriet's servants, other servants, farmers, pages, hunters, ladies. The setting is Richmond, England, during the reign of Queen Anne.

Act I, Scene 1. Lady Harriet's boudoir. The sound of happy voices outside her window gives Lady Harriet the idea for an amusing escapade: she will join her servants in disguise and accompany them to Richmond Fair. She and her maid Nancy don peasant garb and assume fictitious names: Lady Harriet becomes Martha, Nancy becomes Julia. Lady Harriet's cousin, Sir Tristan goes along as a farmer named John.

Scene 2. The Fair. Harriet and Nancy meet two young men, Lionel and Plunkett, who offer to hire them as servants. In a spirit of fun, the girls accept, binding themselves to their masters for a year. They soon regret this bargain but are unable to break it.

Act II. Plunkett's farm. The two girls decide to make their employers' lives intolerable. Meanwhile, Lionel has fallen in love with Martha. When he begs her to sing for him, she complies ("The Last Rose of Summer"). Lionel falls on his knees and confesses his love and willingness to marry her even if she is only a servant. The situation is relieved by Plunkett's arrival. He, too, loves his servant. The four now engage in a game of coquetry as they bid each other good night (Good Night Quartet: "Schlafe wohl! und mag dich reuen"). After Lionel and Plunkett retire, Sir Tristan taps at the window. Learning that he has a carriage waiting, the ladies escape.

Act III. A hunting park in Richmond Forest. Lionel and Plunkett are attending the Queen's hunt. Plunkett and a group of farmers sing the praises of British ale ("Lasst mich euch fragen"). When Nancy arrives with a company of court ladies, Plunkett immediately recognizes her as "Julia" and insists that she return to his service; the ladies rudely send him away. Lionel appears. He is sad, for he is thinking of his lost love ("M'appari," "Ach so fromm"). His reflections are interrupted by the sudden appearances of Lady Harriet. Lionel is confused at finding her dressed as a lady; but he is so overjoyed that he reveals his inmost feelings. Lady Harriet rejects him scornfully, since she does not want to give way to her true emotions. When

Lionel insists that she return to work for him, Lady Harriet has her men arrest him.

Act IV, Scene 1. Plunkett's farm. Lionel has been freed. Meanwhile, his possession of a certain ring has disclosed that he is really the Earl of Derby. Lady Harriet is now willing to concede that she loves him, but Lionel rejects her. Even her attempt to awaken his love by singing "The Last Rose of Summer" is futile. Brushing her aside, he leaves. Plunkett and Nancy, reunited, plan a method whereby Lionel and Lady Harriet will be reconciled.

Scene 2. Richmond Park. Lady Harriet has set up a replica of Richmond Fair. She hopes that if Lionel revisits the scene of their first meeting, his love may be revived. Lady Harriet and Nancy reappear in peasant dress. The strategy works. Seeing Lady Harriet again as Martha, in the setting of the Fair, Lionel is moved to happiness. The two pairs of lovers express their joy in a final rendition of "The Last Rose of Summer."

Martin, Riccardo (born HUGH WHITFIELD MARTIN), tenor. Born Hopkinsville, Kentucky, November 18, 1874; died New York City, August 11, 1952. He was one of the first American-born singers to appear in leading tenor roles at the Metropolitan Opera. An endowment enabled him to go to Paris in 1901 to study with Giovanni Sbriglia and Jean de Reszke, and later to complete his study with Vincenzo Lombardi in Florence. His debut took place in Nantes as Faust in 1904. Two years later he made his American debut with the San Carlo Opera, then visiting New Orleans. On November 20, 1907, he made his debut at the Metropolitan Opera in *Mefistofele,* a performance in which Feodor Chaliapin made his American bow. Martin remained at the Metropolitan Opera through the 1914–1915 season, appearing in leading tenor roles; he returned for the season of 1917–1918. He created the leading tenor roles in three American operas: *Cyrano de Bergerac, Mona,* and *The Pipe of Desire.* After leaving the Metropolitan he appeared with various opera companies in America and Europe, including three seasons with the Chicago Civic Opera.

Martinelli, Giovanni, tenor. Born Montagnana, Italy, October 22, 1885. Though musical as a child, he did not begin formal music study until comparatively late. While he was serving in the army in his twentieth year, the bandmaster recognized that he had an unusual voice and arranged for him to go to Milan for an audition. As a result, a sponsor was found to finance his study with Mandolini. After a concert debut in Milan, Martinelli made his bow in opera at the Teatro Dal Verme in Milan in *Ernani* on December 29, 1910. Puccini was so impressed that he engaged the singer for the European première of *The Girl of the Golden West* (Rome, 1911). After successful appearances throughout Europe, Martinelli made his American debut on November 3, 1913, with the Chicago-Philadelphia Opera Company in *Tosca.* On November 20 he appeared for the first time with the Metropolitan Opera. The opera was *La Bohème.* By the time Caruso's career came to its untimely end in 1920, Martinelli was recognized as his successor. During the more than three decades he was associated with the Metropolitan, Martinelli was heard in over fifty leading tenor roles of the French and Italian repertories. He sang in such notable world and American premières as those of *La campana sommersa, Francesca da Rimini, Goyescas,* and *Madame Sans-Gêne.* On November 24, 1939, he made one of his rare appearances in German opera when he was none too happily cast by the Chicago Opera as Tristan. His own favorite roles were Eleazar, Otello, and Radames. He left

the Metropolitan Opera after the
1944–1945 season. He was then occasionally heard in recitals, as soloist with
orchestras, over the radio, and as
master of ceremonies for the television
program "Opera Cameos."

Martini, Nino, tenor. Born Verona,
Italy, August 8, 1905. He received
singing lessons from a local choirmaster and after that from Giovanni Zenatello. He made his debut in Milan in
Rigoletto. Shortly after, he was acclaimed for his interpretation of the
role of Pollione. He sang it in the original key, something not ordinarily
done. Martini came to the United States
in 1929. In 1931 he made several appearances with the Philadelphia Opera
Company, but his first substantial success in this country came from his radio
appearances. On December 28, 1933,
he made his debut at the Metropolitan
Opera in *Rigoletto.* He remained at the
Metropolitan for a decade. He was also
heard in recitals, and was starred in
several musical motion pictures. Martini now makes his home in Italy.

Martinu, Bohuslav, composer. Born
Polička, Czechoslovakia, December
8, 1890. He graduated from the Prague
Conservatory in 1913. For ten years
he earned his living playing the violin
in the Czech Philharmonic Orchestra.
During this period he wrote several orchestral works, including a ballet, *Istar,*
successfully produced in Prague in
1922. In 1923 he went to Paris where
he remained until 1940. It was in Paris
that he began writing operas. His first,
The Soldier and the Dancer, was introduced at the Brünn National Theater
in 1928. His most important opera of
this period was *The Miracle of Our
Lady,* performed in Brünn in 1934.
Soon after the German invasion of
France Martinu came to the United
States and made it his permanent home.
His orchestral and chamber works have
been widely performed in this country.
In his later works, Czech influences are

combined with French precision, refinement, and restraint, usually within
classical forms. Besides the two already
mentioned, Martinu has written the following operas: *The Day of Charity*
(1930); *The Voice of the Forest*
(1935); *Comedy on a Bridge* (1936);
The Suburban Theater (1936); *Alexander Bis* (1937); *Juliette* (1938); *The
Marriage* (1952); *What Men Live By*
(1953); *Locandiera* (1954).

Martín y Soler (or Solar), Vicente, composer. Born Valencia, Spain, January
18, 1754; died St. Petersburg, Russia,
January 30, 1806. He was at first a
choirboy at the Valencia Cathedral,
and then an organist in Alicante. His
first opera, *I due avari,* was produced in
Madrid in 1766. He then visited Italy,
where some of his operas were so
well received that he became a favorite
with Italian audiences, a serious rival
of Cimarosa and Paisiello. *Una cosa
rara,* produced in Vienna in 1785, was
so popular that it succeeded in obscur-
ing Mozart's *The Marriage of Figaro.*
Mozart later quoted a number from
Una cosa rara in his *Don Giovanni:* it
is one of the little pieces played by the
Don's band during the supper scene.
From 1788 to 1801 Martín y Soler
directed Italian operas in St. Petersburg. When the vogue for Italian opera
gave way to French, the composer devoted his efforts to teaching. Besides
the operas mentioned, Martín y Soler
wrote the following successful works:
Ifigenia in Aulide (1781); *La donna
festeggiata* (1783); *Ipermestra* (1784);
and *L'arbore di Diana* (1787).

Marullo, a courtier (baritone) in
Verdi's *Rigoletto.*

Mary, (1) *see* Duchess of Towers.

(2) Senta's nurse (contralto) in
Wagner's *Der fliegende Holländer.*

Masaniello, Neapolitan   fisherman
(tenor) in Auber's *La muette de Portici.*

Mascagni, Pietro, composer. Born Leghorn, Italy, December 7, 1863; died

Rome, August 2, 1945. As a student of the Cherubini Institute in Leghorn he wrote a symphony and a choral work that were performed. His talent attracted the interest of Count Florestano de Larderel, who financed his further study at the Milan Conservatory. Unhappy at the Conservatory, Mascagni left it and supported himself by conducting a traveling opera company. He then settled in the town of Cerignola as piano teacher. In 1889 he wrote his first opera—it was *Cavalleria rusticana*—for the competition sponsored by the publishing house of Sonzogno. The opera not only won the prize but was a sensation at its première at the Teatro Costanzi on May 17, 1890. The acclaim was repeated wherever the opera was heard; by 1892 it had been performed not only throughout Italy, but also in Paris, Berlin, London, and New York. Mascagni became a household name in Italy, a man of influence and wealth. He wrote fourteen operas after *Cavalleria*. Several were minor successes: *L'amico Fritz* (1891), *Iris* (1898), and *Lodoletta* (1917). But Mascagni could never duplicate the success of his first opera. As the composer himself remarked: "It is a pity I wrote *Cavalleria* first. I was crowned before I became king."

Mascagni combined his career as composer with that of conductor. In 1902 he toured the United States directing performances of his operas; this tour ended disastrously, due to mismanagement. In 1911 he conducted his operas in South America. In 1929 he succeeded Arturo Toscanini as musical director of La Scala. He wrote several works glorifying fascism and Mussolini, notably the opera *Nerone*. During World War II, Mascagni came upon unfortunate times. His property was confiscated, and he himself was held in contempt for his avowed Fascist sympathies. The last year of his life was spent in poverty and disgrace in a small room at the Hotel Plaza in Rome.

His operas: *Cavalleria rusticana* (1890); *L'amico Fritz* (1891); *I Rantzau* (1892); *Guglielmo Ratcliff* (1895); *Silvano* (1895); *Zanetto* (1896); *Iris* (1898); *Le maschere* (1901); *Amica* (1905); *Isabeau* (1911); *Parisina* (1913); *Lodoletta* (1917); *Il piccolo Marat* (1921); *Scampolo* (1921); *Nerone* (1935).

Ma se m'è forza perderti, Riccardo's aria in Act III of Verdi's *Un ballo in maschera.*

Masetto, a peasant (baritone) in Mozart's *Don Giovanni.*

Masked Ball, A, *see* BALLO IN MASCHERA, UN.

Masque, an elaborate theatrical entertainment combining song, dance, poetry, and pageantry, in vogue between 1600 and 1800, and most frequently serving to entertain the aristocracy. The subject matter was usually mythological or allegorical. The form first became popular in Italy, but its heyday was reached in England. Some of England's foremost writers provided the texts, including Beaumont, Fletcher, Dekker, Ford, Ben Jonson, and Milton. The music was created by such outstanding English composers as Campion, Gibbons, Lawes, Locke, Purcell and Handel. One of the most celebrated of English masques was John Milton's *Comus,* with music by Henry Lawes, written for performance at Ludlow Castle in 1634. Later distinguished examples include Matthew Locke's and Cristopher Gibbons' *Cupid and Death* (1653), John Blow's *Venus and Adonis* (1680), and Thomas Arne's *Alfred* (1740).

Massé, Victor (born FELIX-MARIE MASSÉ), composer. Born Lorient, France, March 7, 1822; died Paris, July 5, 1884. He attended the Paris Conservatory, where he won the Prix de Rome in 1884. In Rome, he wrote his first opera, *La favorita e la schiava.*

Soon after returning to Paris he completed his first French opera, *La chambre gothique,* introduced at the Opéra-Comique in 1849; it was acclaimed. His greatest success came with the opéra comique *Les noces de Jeannette,* given at the Opéra-Comique in 1853. In 1860 he was engaged as chorus master of the Opéra, and in 1866 he was appointed professor of composition at the Paris Conservatory. A serious illness compelled him to go into retirement in 1876. Meanwhile, in 1872 he had succeeded Daniel Auber as member of the Institut de France. Among his other successful operas were *Paul et Virginie* (1876) and *Une nuite de Cléopâtre,* performed posthumously at the Opéra-Comique in 1885.

Massenet, Jules Emile Frédéric, composer. Born Montaud, France, May 12, 1842; died Paris, August 13, 1912. The composer of *Manon* and *Thaïs,* Massenet was a dominating figure in the French lyric theater of the late nineteenth century. He entered the Paris Conservatory in his tenth year. Under the sympathetic guidance of such teachers as Napoléon-Henri Reber and Ambroise Thomas, he won several prizes, including the Prix de Rome in 1863. Soon after returning to Paris from Rome, he completed a one-act opera, *La grand'tante,* produced by the Opéra-Comique in 1867 After the Franco-Prussian War, in which he participated as a member of the National Guard, Massenet attracted the limelight with his incidental music for *Les Erinnyes* (it included a section later to become popular as the song "Elegie") and an oratorio, *Marie Magdeleine.* His position in French music became secure with the opera *Le Roi de Lahore,* successfully performed at the Opéra on April 27, 1877. In 1879 he was elected to the Académie des Beaux-Arts, the youngest man ever to receive this honor.

Massenet's finest operas were pro-duced between 1880 and 1900. In 1881 there was *Hérodiade,* first seen in Brussels. Then there followed the operas that made Massenet a leading representative of French romantic opera: *Manon* (1884), *Le Cid* (1885), *Werther* (1892), *Thaïs* (1894), and *Sapho* (1897). In 1894 the composer wrote a one-act sequel to *Manon, Le portrait de Manon,* to a libretto by Georges Boyer. The finest traits of the French lyric theater are found in these works: a deep poetic feeling; tenderness of melody; delicacy of style; irresistible charm. Though Massenet wrote many operas after 1900, he never equaled the quality of his earlier work. Eager to maintain his popularity, he stunted his artistic growth by repeating the mannerisms and imitating the style that had made him famous; he remained a champion of an old and dying romantic tradition while younger composers were finding new approaches and esthetics.

Between 1878 and 1896 Massenet was professor of advanced composition at the Paris Conservatory. His many pupils included Gustave Charpentier, Gabriel Pierné, Florent Schmitt, and Henri Rabaud. Twenty-two years after Massenet's death, his bust was placed in the Opéra-Comique; on that occasion his *Don Quichotte* was revived with Feodor Chaliapin in the title role. Massenet's operas, besides those already mentioned: *Don César de Bazan* (1872); *Esclarmonde* (1889); *Le Mage* (1891); *La Navarraise* (1894); *Cendrillon* (1899); *Grisélidis* (1901); *Le jongleur de Notre Dame* (1902); *Chérubin* (1905); *Ariane* (1906); *Thérèse* (1907); *Bacchus* (1909); *Don Quichotte* (1910); *Roma* (1912); *Panurge* (1913); *Cléopâtre* (1914); *Amadis* (1924).

Master Peter's Puppet Show, *see* RE-TABLO DE MAESE PEDRO, EL.

Mastersingers (Meistersinger), the name given to the German guilds of poet-musicians that flourished from the thirteenth century to the sixteenth, with Hans Sachs their most noted figure. The mastersingers carried on the traditions of the earlier minnesingers, whose troubadour-influenced movement had its beginnings in the twelfth century. The songs of the mastersingers, usually on Biblical subjects, followed strict musical rules. Wagner's opera *Die Meistersinger* is concerned with the lives and activities of these musicians.

Mastersingers of Nuremberg, The, *see* MEISTERSINGER VON NÜRNBERG, DIE.

Materna, Amalie, dramatic soprano. Born St. Georgen, Styria, July 10, 1844; died Vienna, January 18, 1918. She created the roles of Brünnhilde (in *Siegfried* and *Die Götterdämmerung*) and Kundry. Her singing career began with concert appearances, her opera debut taking place in Graz in 1864. After marrying the actor Karl Friedrich, she was engaged by the Karl Theater in Vienna for appearances in operettas. A period of study with Heinrich Proch preceded her debut at the Kärntnerthortheater in Vienna in 1869 as Selika. She was a major success, and was engaged by the Vienna Opera, where she remained up to the time of her retirement in 1897. Besides appearances in the Italian and French repertory she was pre-eminently successful in the Wagner dramas. In 1876 Wagner selected her to sing the three Brünnhilde roles at the first Bayreuth Festival. In 1877 she sang under Wagner's direction at a Wagner Festival in London, and in 1882 she returned to Bayreuth to appear in the world première of *Parsifal*. In 1882 she visited the United States for the first time, appearing as soloist with the Theodore Thomas Orchestra in Wagner programs. On January 5, 1885, she made her American opera debut at the Metropolitan Opera as Elisabeth. She returned to the United States in 1894 to appear with the Damrosch Opera Com-

pany. After her retirement she devoted herself to teaching singing in Vienna.

Mathilde, Gessler's daughter (soprano) in Rossini's *William Tell*.

Mathis der Maler (Mathis the Painter), opera by Paul Hindemith. Libretto by the composer, based on the life of the sixteenth century painter Mathias Grünewald. Première: Zürich, May 28, 1938. The background is the Peasants' War of 1524. Grünewald becomes the spearhead for the peasants' uprising against the Church. But once he becomes involved in the struggle he sees so much oppression and tyranny on his own side that he loses faith in the cause. He escapes with his beloved, Regina, to Odenwald. At first, he is haunted by ugly visions. But then beautiful apparitions come to him, and these succeed in bringing him back to his art. He gives up Regina and the outside world to dedicate himself completely to his artistic mission. This opera figures in the political history of Nazi Germany. Wilhelm Furtwaengler, music director of the Berlin Opera, scheduled the première for 1934. But the Nazi authorities objected to the portrayal of peasants rising against authority, and expressed contempt for a composer who was married to a non-Aryan and whose works were "degenerate." Furtwaengler wrote a vehement letter to Goering protesting the ban, insisting that, as music director, he was the sole authority for the repertory of the Berlin Opera. He also published a heated defense of Hindemith in the *Deutsche Allgemeine Zeitung*. Because of his stand, Furtwaengler was relieved of all his official musical duties and sent into temporary retirement. Hindemith was forced to leave the country.

The music of *Mathis der Maler* first became known through a "symphony" adapted from the score by the composer. It is made up of three movements: "The Concert of the Angels" (the overture); "The Entombment" (the sixth scene); and "Temptation of Saint Anthony" (intermezzo of the final scene).

Mathisen, an Anabaptist preacher (bass) in Meyerbeer's *Le Prophète.*

Matho, a Libyan mercenary (tenor) in Reyer's *Salammbô.*

matrimonio segreto, Il (The Secret Marriage), opera buffa by Domenico Cimarosa. Libretto by Giovanni Bertati, based upon *The Clandestine Marriage,* by George Colman the elder and David Garrick. Première: Burgtheater, Vienna, February 7, 1792. Carolina, daughter of the rich and greedy merchant Geronimo, is secretly married to the lawyer Paolino. In order to mollify his father-in-law when he uncovers this marriage, Paolino arranges a match between the merchant's older daughter, Elisetta, and a rich friend named Count Robinson. But matters become complicated when Elisetta falls in love with Paolino, while the rich friend falls in love with Carolina. Paolino and Carolina run away, but are intercepted by Geronimo. Though the merchant is at first horrified to learn that they are married, he finally gives his belated blessings. And Count Robinson and Elisetta decide they are really meant for each other.

Il matrimonio segreto is a classic opera buffa, one of the early successful examples of this form, and an important predecessor of Rossini's *The Barber of Seville.* Verdi considered it the model of what an opera buffa should be; and it is the only Italian opera buffa between those of Pergolesi and Rossini that has survived. The sprightly overture is often played. The delightful arias and ensemble numbers include Carolina's aria "Questa cosa" and Geronimo's air "Udite tutti" both in Act I, and Paolino's song "Ah no, che tu così morir mi fai" in Act II.

Ma tu, o Re, tu possente, Amonasro's plea in Act II, Scene 2, of Verdi's *Aïda.*

Matzenauer, Margarete, dramatic so-prano and contralto. Born Temesvár, Hungary, June 1, 1881. After studying with Antonia Mielke and Fritz Emerich in Berlin she made her debut in 1901 in Strassburg in the role of Puck in *Oberon.* After three years in Strassburg, she was engaged as principal contralto of the Munich Opera, where she remained until 1911. There she scored major successes in the Italian and Wagnerian repertories. During this period she appeared as guest artist in many European opera houses, including the Bayreuth Festival, where, in the summer of 1911, she appeared as Waltraute, Flosshilde, and the Second Norn. On November 13, 1911 (the opening night of the season), she made her American debut at the Metropolitan Opera as Amneris. Later the same season she was acclaimed when she appeared as Kundry for the first time in her career, substituting for Olive Fremstad without a single rehearsal. Matzenauer remained at the Metropolitan through the 1929–1930 season. During the 1930's she appeared in concerts and as soloist with symphony orchestras and in oratorios. She has been in retirement since World War II. Though she had appeared as a contralto in Munich, she was heard in both leading contralto and soprano roles at the Metropolitan. Her greatest successes came in the roles of Leonore in *Fidelio,* the three Brünnhildes, Kundry, Isolde, Donna Elvira, Selika, Orfeo, Carmen, Dalila, and Amneris. She appeared in many significant world, American, and Metropolitan premières, including Liszt's *Saint Elizabeth* (in a stage presentation), Janáček's *Jenufa,* Spontini's *La Vestale,* and Verdi's *Un ballo in maschera* and *Don Carlos.*

Maudite à jamais soit la race, air of the High Priest in Act I of Saint-Saëns' *Samson et Dalila.*

Maupassant, Guy de, author. Born Château de Miromesnil, France, August 5, 1850; died Paris, July 6, 1893.

His celebrated short story *Mlle. Fifi* was made into operas by César Cui (*Mam'zelle Fifi*) and Reinhold Glière (*Rachel*). Benjamin Britten's opera *Albert Herring* was derived from Maupassant's *Le rosier de Mme. Husson.*

Maurel, Victor, baritone. Born Marseilles, France, June 17, 1848; died New York City, October 22, 1923. He first attended the Ecole de Musique in Marseilles and afterward the Paris Conservatory, where he received first prize upon his graduation in 1867. In the same year he made his debut at the Paris Opéra in *Les Huguenots.* He did not make much of an impression. When he was assigned only minor roles, he left the Opéra and during the next few years appeared at La Scala (where he was heard in the world première of Gomez' *Il Guarany*) and Covent Garden (where he created for England the Wagnerian roles of Telramund, Wolfram, and the Dutchman). On November 26, 1873, he made his American debut in the American première of *Aïda* at the Academy of Music in New York. He stayed only a single season in New York, after which he went into temporary retirement. He returned to the Paris Opéra on November 28, 1879, in *Hamlet;* for the next fifteen years he sang regularly at the Opéra, one of its most brilliant stars. Maurel's greatest triumphs came in 1887 and 1893 when he created the roles of Iago and Falstaff in the world premières of Verdi's *Otello* and *Falstaff.* Maurel's interpretations of these two roles were regarded as definitive. He was called upon to create the role of Iago in France and England, and that of Falstaff in France, England, and the United States. In 1894 and again in 1899 Maurel appeared at the Metropolitan Opera. After 1909 he settled in New York as a teacher of singing. In 1919 he designed the scenery for the Metropolitan production of Gounod's *Mireille.* He wrote four books on singing, and an autobiography, *Dix ans de carrière* (1897). Though Maurel had a comparatively limited voice, he used it with exquisite artistry and combined it with dramatic power.

Mavra, one-act opera by Igor Stravinsky. Libretto by Boris Kochno, based on Alexander Pushkin's poem *The Little House at Kolomna.* Première: Paris Opéra, June 2, 1922. When Parasha's mother laments the loss of her cook, Parasha brings a replacement in the form of Vassily, her suitor, who has assumed woman's disguise and the name of Mavra. The ruse is uncovered when Parasha and her mother find their cook—shaving. The mother faints. Vassily jumps out a window and escapes. Stravinsky has provided the following explanation about this opera: "*Mavra* is in direct tradition of Glinka and Dargomizhsky. I wanted merely to try my hand at this living form of opera buffa." The scoring is for twelve woodwinds, twelve brasses, and a doublebass. The recitative is dispensed with completely, and the emphasis is on a broad, at times a tender, lyricism, and on occasional satirical and burlesque effects in the orchestral accompaniment.

Max, (1) a composer (tenor) in Křenek's *Jonny spielt auf.*

(2) A ranger (tenor), principal male character in Weber's *Der Freischütz.*

Maximilien, opera by Darius Milhaud. Libretto by Armand Lunel, based on Franz Werfel's drama *Juarez and Maximilian.* Première: Paris Opéra, January 5, 1932. The opera depicts the career of Maximilian, the Austrian archduke who became Emperor of Mexico in 1864 and three years later was overthrown by the republican army headed by Juarez.

May Night, *see* GOGOL, NIKOLAI.

Maypole Dances, dances in Act II of Hanson's *Merry Mount.*

Mayr, Richard, bass. Born Henndorf, Austria, November 18, 1877; died Vienna, December 1, 1935. While attending the University of Vienna he sang for Gustav Mahler, who persuaded him to embrace a musical career. Mahler contracted Mayr for the Vienna Opera, but Mayr's debut took place in Bayreuth in 1902 in the role of Hagen. His Vienna debut took place the same year in *Ernani*. For the rest of his career Mayr remained the principal bass of the Vienna Opera, successful in the German repertory, acclaimed particularly for his interpretation of the role of Baron Ochs, probably his most celebrated characterization. It was in this role that he made his Covent Garden debut in 1924. His American debut took place at the Metropolitan Opera on November 2, 1927, in the role of Pogner. He remained at the Metropolitan Opera through the 1929–1930 season. He was also a great favorite at the Salzburg Festivals. Shortly before his death he retired from the Vienna Opera at his own request and received a pension.

Mazeppa, *see* TASSO.

Mazurka, a Polish national dance in triple time, usually with a strong accent on the second or third beat. There is a delightful Mazurka in Glinka's *A Life for the Czar.*

Meader, George, tenor. Born Minneapolis, Minnesota, July 6, 1888. He graduated from the University of Minnesota with a degree in law. While vacationing in Germany he decided to become a singer. After studying with Anna Schoen-René he made his concert debut in London in 1908, his opera debut in Leipzig in 1910 as Lionel. In 1911 he became a member of the Stuttgart Opera, remaining here eight years. His American debut took place at the Metropolitan Opera on November 19, 1921, in the American première of *Die tote Stadt.* He remained at the Metropolitan Opera a decade, appearing in a great variety of German, Italian, and French operas, particularly acclaimed for his interpretations of David and

Mime. After leaving the Metropolitan he appeared on Broadway in several musical and legitimate productions, and was soloist at the St. Thomas Church in New York.

Measure for Measure, *see* SHAKE-SPEARE.

Meco all' altar di Venere, Pollione's aria in Act I of Bellini's *Norma.*

Medea, a sorceress and murderess in Greek legend. She helps her husband Jason gain the Golden Fleece and flees with him to Corinth, afterward to Athens. Her story appears in several operas, notably ones by Marc-Antoine Charpentier, Luigi Cherubini, and Darius Milhaud. *See* MEDEE.

médecin malgré lui, Le, *see* MOLIERE.

Médée, (1) opera by Marc-Antoine Charpentier. Libretto by Thomas Corneille, based on Greek legend. Première: Paris Opéra, December 4, 1693.

(2) Opera by Luigi Cherubini. Libretto by Benoît Hoffmann, based on the tragedy of Corneille. Première: Théâtre Feydeau, Paris, March 13, 1797. This is sometimes referred to as the "first modern opera" since it is the first in the tradition of romantic grand opera. In the nineteenth century, recitatives were added by Franz Lachner; it is in this version that the opera is occasionally revived.

(3) Opera by Darius Milhaud. Libretto by Madeleine Milhaud (the composer's wife). Première: Antwerp, October 7, 1939 (in a Flemish translation).

Meditation, the orchestral entr'acte with violin obbligato prefacing Act II, Scene 1, of Massenet's *Thaïs.* It expresses Thaïs' renunciation of a life of pleasure for that of the spirit.

Medium, The, opera by Gian-Carlo Menotti. Libretto by the composer. Première: New York City, May 8, 1946. The setting is Madame Flora's parlor; the time, the present day. Flora is a medium who defrauds her clients by pretending to contact the spirits.

Her associates are her daughter, Monica, and a mute, Toby. During one of her seances, Flora feels a clammy hand on her throat. Panic-stricken, she confesses to her clients that she is a fraud, but they do not believe her. Convinced that it is Toby who is trying to drive her mad, she beats him and sends him away. She then seeks escape from her terrors in drink. Toby, in love with Monica, returns to the house to find her, and hides in a closet. Flora, aware that someone is concealed, seizes her revolver and shoots through the closet door, killing the mute.

Mefistofele, opera by Arrigo Boïto. Libretto by the composer, based on Goethe's *Faust.* Première: La Scala, March 5, 1868. Boïto's treatment of Goethe's drama differs markedly from that of Gounod: it attempts to incorporate the entire drama instead of merely the first part. After Marguerite's death there are scenes symbolizing the union of Greek and German ideals through the bringing together of Helen of Troy and Faust, and Faust's ultimate redemption. The opera's notable arias include: "Dai campi, dai prati," sung by Faust in Act I; Mefistofele's "Ballata del fischio," in Act I; Marguerite's song "L'altra notte" in Act III. The Peasant Waltz in Act I is also familiar.

At the première, a violent demonstration took place between Boïto's friends and admirers, who recognized the deeper intellectual veins tapped by the composer for Italian opera, and his enemies, who accused him of cerebralism and succumbing to the spell of Wagner. The second performance also aroused a demonstration, with the result that Milan's chief of police ordered the work withdrawn from the repertory. Boïto then revised and shortened his opera, and seven years later it was a great success when given in Bologna.

Méhul, Etienne Nicolas (or ETIENNE HENRI), composer. Born Givet, Ardennes, France, June 22, 1763; died Paris, October 18, 1817. As a boy he studied the organ. A wealthy amateur provided him with funds to go to Paris. There, in 1777, he studied with Johann Friedrich Edelmann. He wrote several operas before one was performed: an opéra comique, *Euphrosine et Coradin,* seen at the Théâtre des Italiens in 1790. The work was such outstanding success that Méhul became famous. In the next four years he extended his reputation with two more operas: *Stratonice* (1792) and *Horatius Coclès* (1794). During the French Revolution Méhul allied himself with the republicans by writing many patriotic songs, one of which, *Le chant du départ,* rivaled the *Marseillaise* in popularity. He was now the recipient of many honors. In 1795 he became a member of the Institut de France and an inspector of the Paris Conservatory. In 1800 he was commissioned to write a special work commemorating the storming of the Bastille.

With Napoleon's rise to power, Méhul managed to maintain his eminent position in French music. He was, indeed, a favorite of Napoleon. His most celebrated work, the opera *Joseph,* was a considerable success when introduced at the Théâtre Feydeau on February 17, 1807. Poems were written in its honor; Napoleon gave Méhul a prize of five thousand francs. The decline of Méhul's fortunes coincided with the fall of Napoleon. Méhul was demoted at the Conservatory from inspector to professor, with a corresponding reduction in salary. His operas were no longer performed, and his former fame was obscured by the rising popularity of other composers, notably Spontini. The last years of his life were unhappy. He died of consumption.

Méhul wrote some thirty operas. His comic operas were significant in helping establish the traditions of opéra comique; they are characterized by gaiety, verve, sparkling melodies, and effective ensemble numbers. In his serious operas, of which *Joseph* is the best, Méhul reveals a strong dramatic sense and a dignified, expressive lyricism. Besides those already mentioned, Méhul wrote the following successful works: *Le jeune Henri* (1797); *Adrien* (1799); *Ariodant* (1799); *Bion* (1800); *Une folie* (1802); *Le trésor supposé* (1802); *Joanna* (1802); *Helena* (1803); *Les deux aveugles de Tolède* (1806); *Uthal* (1806); *Gabrielle d'Estrées* (1806).

Mein Herr Marquis, Adele's "Laughing Song" in Act II of Johann Strauss's *Die Fledermaus.*

Mein Herr und Gott, nun ruf' ich Dich, King Henry's prayer in Act I of Wagner's *Lohengrin.*

Meister, Wilhelm, a student (tenor) in Thomas's *Mignon.*

Meistersinger, *see* MASTERSINGERS.

Meistersinger von Nürnberg, Die (The Mastersingers of Nuremberg), opera in three acts by Richard Wagner. Libretto by the composer. Première: National Theater, Munich, June 21, 1868. American première: Metropolitan Opera, New York, January 4, 1886.

Characters: Hans Sachs, cobbler (bass or baritone); David, his apprentice (tenor); Pogner, a goldsmith (bass); Eva, his daughter (soprano); Magdalena, her nurse (soprano); Walther von Stolzing, a knight (tenor); Beckmesser, a town clerk (bass); mastersingers; journeymen; apprentices; guildspeople; girls. The setting is Nuremberg in the middle of the sixteenth century.

The stirring orchestral prelude contains five major themes from the opera, beginning with the majestic march of the mastersingers and including the "Prize Song."

Act I. The church of St. Katharine. The chorale (Kirchenchor: "Da zu Dir der Heiland kam") brings the services to a close. Eva, as she leaves the church, is stopped by Walther von Stolzing.

She tells him she will be the wife of the guild master who wins the song contest soon to be held. In love with Eva, Walther is determined to win. He has David teach him some of the rules, but David only manages to confuse him. The mastersingers now file into the church for a musical test. Pogner, father of Eva, announces that his daughter is to be the prize in the song contest (Pogner's Address: "Das schöne Fest"). The mastersingers ask Walther where he learned the art of song. He tells them that his knowledge came from Nature and the ancient minstrels ("Am stillen Herd"). Walther is now asked to demonstrate his ability, and Beckmesser prepares to mark his errors on a slate. Walther improvises a song ("Fanget an! So rief der Lenz in den Wald"); the frequent scratchings on Beckmesser's slate betray the abundance of errors. Only Hans Sachs senses how much talent there is in Walther's song; the other mastersingers reject him rudely.

Act II. A street. Outside his shop, Hans Sachs soliloquizes on the beauty of Walther's song ("Wie duftet doch der Flieder"). Eva appears; coquettishly, she suggests that since she cannot have Walther for a husband she would accept Hans Sachs ("Gut'n Abend, Meister"). Hans Sachs is in love with Eva, but he knows that he is too old for her. When Walther arrives, he and Eva retire a little to plot elopement. Their planning is interrupted when Beckmesser comes to serenade Eva under her window. As Beckmesser sings his song, Hans Sachs rudely interrupts him with a ditty of his own ("Jerum! Jerum!"), banging loudly with his hammer as he sings. The noise attracts Magdalena to the window, and since Beckmesser mistakes her for Eva his singing becomes more passionate. The din awakens the townspeople. David, in love with Magdalena, sees Beckmesser serenading her and gives him a sound thrashing. The townspeople pour into the street, and pandemonium prevails. Walther and Eva decide that this is the moment to make their escape, but they are gently restrained by Sachs. When the din subsides, a watchman passes through the now silent streets proclaiming that all is well.

Act III, Scene 1. Inside Sachs's shop. The orchestral prelude has for its core Sachs's monologue "Wahn, Wahn," and a quotation from the "Prize Song." It is early morning of the following day. When David enters the shop, Sachs asks him to sing a hymn to St. John, whose festival day is soon to be celebrated. After completing the hymn ("Am Jordan Sankt Johannes stand"), David leaves. Sachs soliloquizes philosophically over the sad state of the world which, to him, has gone mad ("Wahn! Wahn! Überall Wahn!"). Walther now arrives to tell Sachs of a dream in which a song of great beauty came to him. When he sings a portion of it—it is the "Prize Song"—Sachs is impressed and puts it down on paper. Later, when Beckmesser slips into the shop, he finds this paper; thinking it is one of Sachs's songs, he decides to steal it and use it in the contest. After his escape, Eva arrives to have her shoes repaired; while she is present, Walther, Magdalena, and David also appear. Walther sings Eva a part of his dream song. Eva, Magdalena, David, and Sachs join in to express their individual reactions (Quintet: "Selig wie die Sonne").

Scene 2. A field beside the Pegnitz River. Here the song contest is to be held. Apprentices dance with their girls ("The Dance of the Apprentices"). The various guilds march in with flying banners. After Hans Sachs is acclaimed by the people ("Wach' auf, es nahet gen den Tag"), he announces the opening of the contest. Beckmesser is called first. Performing from the stolen manuscript, he becomes so confused that he arouses derisive laughter. Sachs now summons Walther, who sings his song ("Morgenlich leuchtend im rosigen Schein"). The people acclaim the singer and his song. Walther is the winner, and Eva is his. The people now once again acclaim their beloved cobbler, Hans Sachs ("Heil Sachs! Hans Sachs!").

It is a far different Wagner we meet in *Die Meistersinger* than in the *Ring* and *Tristan und Isolde*. *Die Meistersinger* is Wagner's only comedy. With his supreme command of musical resources and his infallible instinct for the theater, Wagner created a work whose salient features are humor, gentleness, glowing warmth, joyfulness. In *Die Meistersinger* we are no longer in the world of gods and legendary heroes, but in that of human beings whose problems are those of the real world: the world of success and failure (symbolized by the song contest); a world in which there is both frustration (for Sachs) and fulfillment (for Walther and Eva). There is a human quality here that we find nowhere else in Wagner. It appears not only in the remarkable text but in the radiant score, which often reminds us of old German chorales, of street songs and lute songs. *Die Meistersinger,* completed in 1867, came eight years after *Tristan und Isolde* and more than a decade after the first two dramas of the *Ring* cycle. Thus, it was conceived and completed when Wagner's ideas about the music drama were fully crystallized. In some respects, *Die Meistersinger* represents a retreat from these ideas: it returns to older concepts of opera with its formal arias, ensemble numbers, choral numbers, processional march, dances, and so forth. But the integration of these operatic elements is achieved with such skill, and the various elements are so essential to the dramatic context, that Wagner's basic concept of opera as a synthesis of the arts is still realized.

In planning *Die Meistersinger,* Wagner wanted to give comic treatment to a song contest, just as he had given it dramatic treatment in *Tannhäuser.* As the idea germinated, he planned to use the contest as a symbol of his own artistic struggle against rules and formal procedures. Thus Beckmesser, the ridiculous advocate of the *status quo* in art, became the symbol of the critics who continually attacked Wagner and his esthetics—in particular, the Viennese critic Eduard Hanslick. Walther, achieving a new artistic truth by iconoclastically destroying stultifying laws and concepts, arrives at a new freedom of expression, just as Wagner himself did.

Melba, Nellie (born HELEN MITCHELL), coloratura soprano. Born Richmond, near Melbourne, Australia, May 19, 1859; died Sydney, Australia, February 23, 1931. One of the most brilliant coloratura sopranos of her generation, her musical education was comprehensive rather than specialized, including piano, organ, and theory, as well as voice. She sang and played the organ in local churches until her marriage to Captain Charles Nesbit Armstrong in 1882. After that she concentrated on singing, studying in Melbourne, and for a year with Mathilde Marchesi in Paris. On October 12, 1887, Melba made her opera debut at the Théâtre de la Monnaie as Gilda. She was sufficiently impressive for Covent Garden to engage her. She appeared there for the first time on May 24, 1888, as Lucia and received a tumultuous ovation. The following year she made her debut at the Paris Opéra where for the next two seasons she was an outstanding attraction. After further European triumphs she came to the United States and made a sensational debut at the Metropolitan Opera, on December 4, 1893, as Lucia. "Her voice is . . . ex-

quisitely beautiful," wrote H. E. Krehbiel. "Added to . . . this she has the most admirable musical instincts." She continued to gather accolades at the Metropolitan during the next three seasons. On December 30, 1896, she made an unfortunate attempt to extend her repertory by appearing as Brünnhilde in *Siegfried.* This effort taxed her so severely that she was compelled to go into temporary retirement and give her voice a complete rest. She emerged in 1897 appearing with the Damrosch Opera Company. She then continued her dazzling career. In America she sang with the Manhattan Opera from 1907 to 1909 and after that with the Chicago Opera. Her last American appearance took place in the spring of 1920; her farewell opera appearance took place at Covent Garden in 1926. After that she retired to Melbourne, where she became director of the Melbourne Conservatory.

Her voice—exceptional in flexibility, clarity, and precision—was heard to best advantage in such coloratura roles as Lakmé, Gilda, Lucia, Violetta, and Rosina, but she was also an outstanding artist in the more lyrical parts of Mimi, Desdemona, and Marguerite. In 1918 she was made a Dame of the British Empire. She wrote an autobiography, *Melodies and Memories* (1925), and was the subject for a screen biography starring Patrice Munsel, *Melba* (1952).

Melchior, Lauritz, dramatic tenor. Born Copenhagen, March 20, 1890. For a quarter of a century one of the world's foremost Wagnerian tenors, he entered the Royal Opera House School in 1912, and on April 2, 1913, made his debut at the Royal Opera in the baritone role of Silvio. After several appearances in Denmark, he toured Sweden with Mme. Charles Cahier in *Il trovatore.* Mme. Cahier convinced Melchior to retrain his voice as a tenor. After several years of study Melchior made his return debut in Copenhagen on October 8, 1918, this time as a tenor, in *Tannhäuser*. A year later he sang in London, where the novelist Hugh Walpole urged him to specialize in the Wagnerian repertory. Still another period of study and readjustment took place while he worked on the Wagnerian repertory with Anna Bahr-Mildenburg. On May 14, 1924, Melchior appeared at Covent Garden as Siegmund, from then on specializing in the principal tenor roles of the Wagner music dramas. On July 23, 1924, he appeared for the first time at Bayreuth, singing the role of Parsifal. On February 17, 1926, he made his American debut at the Metropolitan Opera as Tannhäuser. He remained the principal Wagnerian tenor of the Metropolitan until 1950.

As a Wagnerian tenor Melchior made opera history. He sang the role of Tristan over two hundred times (his foremost predecessor, Jean de Reszke, had appeared in that role less than fifty times); as Tristan he appeared in sixteen different opera houses, and under twenty-two different conductors. On February 22, 1934, his hundredth performance as Siegfried was commemorated on the stage of the Metropolitan. His twentieth year at the Metropolitan was celebrated on February 17, 1946, when he appeared in several scenes from different Wagner dramas. His last appearance at the Metropolitan was in *Lohengrin* on February 2, 1950. While still a member of the Metropolitan, Melchior frequently appeared in motion pictures and on the radio. After retiring from opera, he continued these appearances, also singing in concerts, on television, and in night clubs. Many honors have been bestowed on him, including the Knighthood of Denneborg, the Knighthood of Bulgaria, the Saxonian Order of Knights, the Silver Cross of Denmark and the "Ingenio et Arti," which has been given to only three men

in Denmark. He also received the Carl Eduard Medal, first class, from Saxe-Coburg-Gotha for his outstanding services at Bayreuth.

Melchthal, Arnold's father (bass) in Rossini's *William Tell*.

Mélesville (born ANNE HONORE DUVEYRIER), dramatist and librettist. Born Paris, November 13, 1787; died there November 1865. With the collaboration of various writers, including Eugène Scribe, Pierre Carmouche, E. C. de Boirie, and J. T. Merle, she wrote numerous librettos for operas by Adam, Auber, Cherubini, Donizetti, Lortzing, and others. The libretto for Hérold's *Zampa* was her work alone, and Brüll's *Das goldene Kreuz* was derived from one of her plays.

Mélisande, (1) Arkel's wife (soprano), in love with Pelléas, in Debussy's *Pelléas et Mélisande*.

(2) One of Bluebeard's wives (soprano) in Dukas's *Ariane et Barbe-Bleue*.

Melitone, Fra, a friar (baritone) in Verdi's *La forza del destino*.

Melodies of the Language, *see* JANACEK, LEOS.

Melodrama, (1) an operatic passage or scene in which the singer recites his part while a musical commentary on the situation appears in the orchestral accompaniment. Examples of such melodramas are the grave-digging scene in *Fidelio* and the bullet-casting scene in *Der Freischütz*.

(2) An operatic form, similar to the above, in which the opera's entire text is spoken, not sung, to an orchestral accompaniment. The form was developed by Georg Benda in the eighteenth century. Its most ambitious practitioner was Zdeněk Fibich (1850–1900), who wrote a trilogy of melodramas, *Hippodameia*, in which he tried to realize a closer unity between poetry and music than had heretofore been achieved.

Melot, a courtier (tenor) in Wagner's *Tristan und Isolde*.

Melton, James, tenor. Born Moultrie, Georgia, January 2, 1904. After studying singing at Vanderbilt University he went to New York and appeared successfully over the radio and at the Roxy Theater. He made his opera debut with the Cincinnati Zoo Opera on June 28, 1938, as Pinkerton. His first appearance at the Metropolitan Opera took place on December 7, 1942. He remained with the Metropolitan for several years. He has also made many successful appearances in concerts and on radio and television.

Melville, Herman, author. Born New York City, August 1, 1819; died there September 28, 1891. His story *Billy Budd* has been made into operas by Benjamin Britten and Giorgio Ghedini.

Mendelssohn, Felix, composer. Born Hamburg, Germany, February 3, 1809; died Leipzig, November 4, 1847. This celebrated composer wrote several operas, none of them significant. As a boy he wrote a comic opera, *The Two Nephews,* whose only performance took place at his home to celebrate his fifteenth birthday. Two years later he completed *Die Hochzeit des Camacho,* based on an episode in *Don Quixote,* a fiasco when introduced in Berlin on April 29, 1827. In 1829 he completed an operetta, *Die Heimkehr aus der Fremde,* performed privately. Years after the composer's death it was given its first public performance, in England, under the title *Son and Stranger.* In his last year Mendelssohn worked on an opera entitled *Lorelei,* but he completed only a few excerpts: the finale to the first act, an "Ave Maria," and a "Vintage Song." The libretto was subsequently used by Max Bruch.

Mendès, Catulle, author. Born Bordeaux, France, May 22, 1841; died Saint-Germain, France, February 8, 1909. One of the most distinguished French poet-dramatists of the late nineteenth century, Mendès wrote the librettos for Emanuel Chabrier's *Gwendoline,* Camille Erlanger's *Le fils de l'étoile,* Reynaldo Hahn's *La Carmélite,* and Massenet's *Ariane* and *Bacchus.* His "conte dramatique" *La Reine Fiammette* was made into an opera by Xavier Leroux. Mendès wrote a book about Wagner (1886); in his novel *Le roi vierge* (1880) Wagner appears as a character.

Menelaus, king of Sparta (tenor), husband of Helen, in Richard Strauss's *Die aegyptische Helena.*

Menotti, Gian-Carlo, composer. Born Cadigliano, Italy, July 7, 1911. Between 1923 and 1928 he attended the Milan Conservatory and was so precocious that after a year there he wrote a three-act opera, *The Death of Pierrot,* libretto as well as music. He came to the United States in 1928 and continued his study at the Curtis Institute of Music. The fruit of this study was a one-act opera, *Amelia Goes to the Ball,* introduced in Philadelphia and New York by members of the Curtis Institute under Fritz Reiner in 1937. It was so successful that the National Broadcasting Company commissioned him to write a radio opera, *The Old Maid and the Thief,* introduced in 1939. Meanwhile, in 1938, the Metropolitan Opera presented *Amelia Goes to the Ball.* In 1942 the Metropolitan introduced *The Island God,* Menotti's only failure. After winning a thousand dollar grant from the American Academy and National Institute of Arts and Letters in 1945 and a Guggenheim Fellowship a year later, Menotti completed *The Medium* on a commission from the Ditson Fund. At its première in 1946 it received such acclaim that it was decided to produce it on Broadway. Together with Menotti's new one-act opera, *The Telephone,* it opened on Broadway on May 1, 1947, and was an outstanding success. Subsequently, *The Medium* received over a thousand performances in this country by various groups; it was made into a motion picture, directed by the composer; and it was heard in leading European opera houses. Menotti's next opera, *The Consul,* was also a formidable success. Opening on Broadway on March 16, 1950, it was acclaimed by the critics and was a box-office hit. It gathered several honors, including the Pulitzer Prize and the New York Drama Critics' Award. Subsequently, it was produced in about a dozen countries and in eight different languages. In 1950 it became the first opera written and first produced in America to be performed at La Scala. In 1951 Menotti wrote the first opera expressly intended for television transmission, *Amahl and the Night Visitors.* It was presented on Christmas Eve, 1951, and since then it has often been heard as a holiday feature over television, besides receiving successful stage presentations, including one at the Florence May Music Festival in 1953. In 1954 Menotti's *The Saint of Bleecker Street* was introduced on Broadway, winning the composer his second Pulitzer Prize. Besides writing his own librettos, Menotti is also his own stage and casting director. His musical style is eclectic. It can be popular or esoteric, realistic or romantic, cacophonous or lyrical, poetic or sardonic—in any case it meets demands of good theater with remarkable effectiveness.

Me pellegrina ed orfana, Leonora's aria in Act I of Verdi's *La forza del destino.*

Mephisto's Serenade, *see* VOUS QUI FAITES L'ENDORMIE.

Mephistopheles, the devil in Berlioz' *The Damnation of Faust,* Boïto's *Mefistofele,* and Gounod's *Faust.* In all three the role is for a bass.

Mercadante, Saverio, composer. Born Altamura, near Bari, Italy, September 1795; died Naples, December 17, 1870. He attended the Collegio di San Sebastiano in Naples, where he was one of Niccolò Zingarelli's star students. His first work for the stage was *L'apoteosi*

d'Ercole given with outstanding success at the San Carlo in Naples on August 19, 1819. International fame came with an opera buffa, *Elisa e Claudio,* given at La Scala in 1821. During the next forty-five years he wrote another fifty or more operas, in the best of which he instituted some major reforms by emphasizing harmonic and orchestral writing. His greatest successes, besides those already mentioned, were: *I briganti,* produced in Paris in 1836; *Il giuramento,* given at La Scala in 1837, and revived there in a revised version a century later; and *Il bravo,* introduced in Milan in 1839. His last opera, *Virginia,* was given in Naples in 1866. In 1833 Mercadante was appointed maestro di cappella at the Novara Cathedral. While holding this post he lost one eye, an infirmity that in 1862 resulted in complete blindness. In 1839 he became maestro di cappella at Lanciano, and a year afterward succeeded Zingarelli as the director of the Naples Conservatory, a post held until the end of his life.

Mercédès, Carmen's gypsy friend (soprano) in Bizet's *Carmen.*

Mercè, dilette amiche, Elena's bolero in Act V of Verdi's *The Sicilian Vespers.*

Merchant of Venice, The, *see* SHAKE-SPEARE.

Mercutio, Romeo's friend (baritone) in Gounod's *Roméo et Juliette.*

Mérimée, Prosper, author. Born Paris, September 28, 1803; died Cannes, France, September 23, 1870. Mérimée's short stories and novels provided subjects for many operas. The most famous is *Carmen,* the source of Bizet's famous opera. Others are *La carrosse du Saint-Sacrement* (operas by Henri Büsser and Lord Berners, and Jacques Offenbach's opéra comique *La Perichole*); *Colomba* (operas by Giovanni Pacini, Henri Büsser, and Alexander Mackenzie); *La dame de pique* (opera by Jacques Halévy); *Mateo Falcone* (opera by César Cui); *L'occasion* (opera by Louis Durey); *La Vénus d'Ille* (Othmar Schoeck's *Venus* and Hermann Wetzler's *The Basque Venus*).

Merlin, opera by Karl Goldmark. Libretto by Siegfried Lipiner. Première: Vienna Opera, November 19, 1886. Merlin, the enchanter, aids King Arthur and the Knights of the Round Table to defeat the attacking Saxons. He finally comes to his doom by falling in love with a beautiful mortal, Vivien, who, in grief, kills herself.

Merola, Gaetano, conductor and opera director. Born Naples, Italy, January 4, 1881; died San Francisco, California, August 30, 1953. He attended the Royal Conservatory in Naples. In 1899 he came to the United States and became assistant conductor of the Metropolitan Opera. In 1903 he conducted the Henry Savage Opera and in 1906 at the Manhattan Opera House. After a period of conducting in London he returned to the United States and in 1923 helped found the San Francisco Opera, which he directed up to the time of his death, helping make it the second leading opera institution in this country (*see* SAN FRANCISCO OPERA COMPANY).

Merrill, Robert, baritone. Born Brooklyn, New York, June 4, 1919. After receiving instruction from teachers in New York he appeared in hotels and at the Radio City Music Hall. In 1945 he won the Metropolitan Auditions of the Air, and on December 15 of that year made his Metropolitan Opera debut in *La traviata*. Since then he has been the principal baritone of the Metropolitan Opera. He has also sung in other leading opera houses of America and Europe. In 1946 he was the singer selected to appear at the Roosevelt Memorial before both houses of Congress. Arturo Toscanini selected him for the leading baritone roles in his broadcasts of *La traviata* and *Un ballo in maschera*.

Merry Mount, opera by Howard Hanson. Libretto by Richard L. Stokes, based on a New England legend and Nathaniel Hawthorne's *The Maypole of Merry Mount*. Première: Metropolitan Opera, February 10, 1934. The setting is New England in 1625. Pastor Bradford, a Puritan clergyman, is tortured by sensual dreams and unfulfilled desires. Spurning a Puritan girl, he falls in love with a Cavalier woman, Lady Marigold Sandys. The marriage of Lady Marigold to Gower Lackland is interrupted by the Puritans. There is a battle—the Cavaliers are defeated and some nominally friendly Indians are enraged. Later, Bradford makes advances to Lady Marigold. Attempting to aid her, Gower Lackland is struck by a guard and slain. When Bradford falls asleep, he dreams that he is in hell, that Gower is Lucifer, and that Lady Marigold becomes his. He awakes to find that the Indians have set fire to the settlement. The Puritans propose to stone Bradford and Lady Marigold, the fancied source of their trouble. Bradford, aghast at the tragedies his lusts have precipitated, sweeps Lady Marigold into his arms and springs into the flaming church. The composer has made an orchestral suite of the overture, a Children's Dance, the love duet of Bradford and Lady Marigold, the prelude to Act II, and the Maypole Dances.

Merry Wives of Windsor, The (Die lustigen Weiber von Windsor), comic opera by Otto Nicolai. Libretto by Hermann Salomon Mosenthal, based on Shakespeare's comedy. Première: Berlin Opera, March 9, 1849. The opera follows the Shakespeare play with minor modifications: Falstaff's followers (Bardolph, Pistol, and Nym) are omitted; the love of Fenton and Anne is touched upon only in passing. The opera emphasizes Falstaff's efforts to make love simultaneously to Mistress Ford and Mistress Page and the comic episodes befalling him in this attempt; the climax is reached in Windsor Park, where Falstaff becomes the victim of playful revenge on the part of the ladies and their husbands. The overture is well-known. Among the better-known vocal numbers are Fenton's Romance, "Horch, die Lerche" and Mistress Ford's aria "Nun eilt nerbei" in Act II.

See also FALSTAFF; SIR JOHN IN LOVE; SHAKESPEARE.

Mes amis, écoutez l'histoire (Ronde du Postillon), Chapelou's aria in Act I of Adam's *Le Postillon de Longjumeau.*

Me sedur han creduto, see C'EN EST DONC FAIT ET MON COEUR VA CHANGER.

Mesrour, chief of the harem guards (speaking part) in Weber's *Oberon.*

Messa di voce, a term applied to the singing of a gradual crescendo, followed by a gradual decrescendo, on a long-sustained note.

Messager, André Charles Prosper, composer and conductor. Born Montluçon, France, December 30, 1853; died Paris, February 24, 1929. He attended the Ecole Niedermeyer and concluded his studies with Camille Saint-Saëns. He then held various posts in Paris as organist and choirmaster. He first attracted attention as a composer with a symphony that won a gold medal in 1875, and a ballet, *Fleur d'orange,* introduced at the Folies Bergères in 1878. His first success in opera came in 1885 when the Bouffes-Parisiens presented his opéra comique, *La Béarnaise;* it was soon presented in London, where it had a long run. Later opéras comique established his reputation in France. These included *La Basoche* (1890), *Les petites Michu* (1897), and *Véronique* (1898). Messager also distinguished himself as an opera conductor. From 1898 to 1903 he was the conductor of the Opéra-Comique where, in 1902, he led the

world premiére of *Pelléas et Mélisande* (which is dedicated to him). From 1901 to 1907 he was artistic director of Covent Garden, and from 1907 to 1919 director and principal conductor of the Paris Opéra. He returned to the Opéra-Comique for the single season of 1919–1920. He also led the Concerts du Conservatoire after 1908, touring the United States with this orchestra in 1918. In 1926 he was elected a member of the Académie des Beaux Arts.

Besides the operas and opéras comique already mentioned, he wrote: *La fauvette du temple* (1885); *Le bourgeois de Calais* (1887); *Isoline* (1888); *Le mari de la reine* (1889); *Madame Chrysanthème* (1893); *Miss Dollar* (1893); *Mirette* (1894); *Le Chevalier d'Hermental* (1896); *La fiancée en loterie* (1896); *Les dragons de l'impératrice* (1905); *Fortunio* (1907); *Béatrice* (1910).

Metastasio, Pietro (born TRAPASSI), dramatist, poet, librettist. Born Rome, January 3, 1698; died Vienna, April 12, 1782. His poetic dramas on classical and Biblical subjects were used by an entire generation of opera composers; more than any other single influence, they were responsible for maintaining stylistic traditions in Italian opera. As a boy, the poet was adopted and supported by a patron, Gian Vincenzo Gravina, at whose request he changed his name to Metastasio; Gravina ultimately left him his fortune. In 1730 Metastasio went to Vienna and became court poet and dramatist, holding this post until the end of his life more than half a century later. He wrote a great number of poetic dramas in the grand manner favored by the Viennese court. Of particular interest were his twenty-seven "drammi per musici." All of these were set to music, some of them an astonishing number of times. *Artaserse* was made into forty different operas between 1724 and 1823; *La clemenza di Tito* into six operas (one by Mozart). Johann Adolph Hasse set all of Metastasio's dramas to music, some of them several times. A few other composers who used his dramas were Gluck, Handel, Haydn, Jommelli, Piccinni, and Porpora. These dramas were filled with intricate plots, flowery speeches, and grandiose climaxes, all appealing strongly to eighteenth century taste. It was against the Metastasio tradition that Gluck rebelled when he produced such operas as *Orfeo ed Euridice* and *Alceste.*

Metastasio's most popular works included: *Achille in Sciro; Adriano in Siria; Alessandro nell' Indie; Antigono; Artaserse; La clemenza di Tito; Demetrio, Demofoönte; Didone abbandonata; Ezio; L'isola disabitata; Olimpiade; Partenope; Il rè pastore; Il Ruggiero; Semiramide riconosciuta; Siroe; Il trionfo di Clelia.*

Metropolitan Auditions of the Air, a weekly half-hour radio program of the ABC network, instituted in 1936 to audition likely singers for the Metropolitan Opera. A special committee from the Metropolitan Opera selects each season those worthy of appearing with the regular company. The first winner was Thomas L. Thomas, who made his Metropolitan Opera debut on May 16, 1937, in *Pagliacci.* Other singers who stepped into the ranks of the Metropolitan Opera from these auditions have been: Marilyn Cotlow, Frances Greer, Frank Guarrera, Margaret Harshaw, Clifford Harvuot, Thomas Hayward, Raoul Joubin, Robert Merrill, Patrice Munsel, Regina Resnik, Eleanor Steber, Risë Stevens, and Leonard Warren.

Metropolitan Opera Guild, an organization founded in 1935 to help sell subscriptions to the Metropolitan Opera and to increase attendance. Its first president was Mrs. August Belmont. The first year the membership

numbered two thousand. The Guild has since extended its membership to over forty thousand and expanded its activities to include assistance in the Metropolitan's various fund-raising campaigns, and the provision of money for new productions, scenery, and equipment. It also instituted, in 1936, a youth series, giving school children the chance to attend special performances at reduced prices, with the Guild making up the difference. Since 1936 over two hundred and fifty thousand school children have attended the Metropolitan. During the opera season the Guild publishes the *Opera News,* edited by Mrs. John De Witt Peltz, providing listeners to the Metropolitan's weekly radio broadcasts complete information about the current opera, as well as general information about operas, their composers, and their performers.

Metropolitan Opera House, the home of the Metropolitan Opera, more properly the Metropolitan Opera Association, on Broadway between Thirty-Ninth and Fortieth Streets in New York City. For over seventy years the Metropolitan Opera has been the foremost operatic institution of the United States, and one of the great opera organizations of the world. It was founded by several leading New York financiers who, unable to procure boxes for the operas at the Academy of Music, decided to sponsor a house of their own, further uptown. They subscribed eight hundred thousand dollars for the purpose. With Henry E. Abbey as the first artistic director, the new opera house opened on October 22, 1883, with *Faust,* starring Italo Campanini and Christine Nilsson. During the first season there were sixty-one representations of nineteen operas. Though the rich of New York filled the boxes, the opera suffered a deficit of six hundred thousand dollars. With the stockholders now taking over the management, Leopold Damrosch was made artistic director. The emphasis was on German opera. The few works of non-German origin were sung in appropriate translations. Damrosch's sudden death before the season's end brought his young son Walter as a hurried replacement. For the third season, Anton Seidl was engaged as principal conductor, with Walter Damrosch as his assistant. This regime continued through the 1890–1891 season, all performances being heard in German. Important singers of the period included Max Alvary, Emil Fischer, and Lilli Lehmann. The following American premières took place: *Die Meistersinger, Tristan und Isolde, Das Rheingold, Siegfried,* and *Die Götterdämmerung.*

From 1892 to 1898 the Metropolitan was guided by the directorial triumvirate of Henry E. Abbey, Maurice Grau, and Edward Schoeffel. The German policy was now abandoned. The company sang the French and Italian repertories in the original languages; German operas, given occasionally, were performed in Italian. Notable new singers included Emma Calvé, Emma Eames, Edouard and Jean de Reszke, Pol Plançon, and Nellie Melba. A fire ravaged the theater in 1892, necessitating extensive alterations. There was no opera season in the fall and winter of that year. The reconstructed house reopened on November 27, 1893, with *Faust.* There was no 1897–1898 season at the Metropolitan, the death of Henry Abbey in 1896 having precipitated a reorganization of the company. Maurice Grau now became manager. A new era was launched with *Tannhäuser* on November 29, 1898. The five seasons of Grau's direction have been described as "the golden age of opera." With some of the greatest singers of that generation in his company—Giuseppe Campanari, Eames, Johanna Gadski, Louise Homer, Lilli Lehmann, Victor Maurel, Lillian Nordica, Plan-

çon, the De Reszkes, Thomas Salignac, Marcella Sembrich, Ernestine Schumann-Heink, Milka Ternina, Ernst Van Dyck—Grau assembled incomparable all-star casts. The Grau management was an outstanding success both artistically and financially.

When poor health compelled Grau to withdraw, Heinrich Conried took over the reins from 1903 to 1908. A new reorganization of the company took place with Conried's arrival. Twelve directors, who assumed all financial responsibility, created the Conried Metropolitan Opera Company on a stockholding basis. The Conried regime saw many brilliant new members added to the company: Alessandro Bonci, Enrico Caruso, Feodor Chaliapin, Geraldine Farrar, Olive Fremstad, and the conductors Gustav Mahler and Felix Mottl. Placing less importance on individual singers than on integrated productions, Conried made notable advances in such matters as costuming and staging. Two outstanding events of his era were the American premières of *Parsifal* and *Salome,* each of which attracted considerable publicity and aroused passionate feelings. *Parsifal* came in Conried's first season. It stirred nation-wide controversy, since it was the first stage production of the opera outside Bayreuth, and it was known that Wagner had intended *Parsifal* for his festival theater alone. When Bayreuth representatives went to court to keep the performance from taking place, New York split into two camps, one siding with Cosima Wagner, the other feeling that a work of art belonged to the world. Conried won the legal fight, and the announced performance stirred the anticipation of opera lovers throughout the country. A special "Parsifal" train was run from Chicago. The première on December 24, 1903, was a tremendous artistic and financial success. For the remainder of the season, *Parsifal* remained the most exciting opera in the repertory, always playing to sold-out houses; its eleven performances brought in almost two hundred thousand dollars. The première of *Salome* (January 22, 1907) also brought on a tempest. A righteous-minded citizenry descended on Conried for permitting such a display of obscenity on the stage. The clergy and press joined in the battle. Before a second performance could take place, the Metropolitan directors decided that further performances of the opera were "detrimental to the best interests" of the company, and the work was removed from the repertory.

Other notable events during Conried's regime had happier results. Humperdinck and Puccini were invited to attend performances of their operas. Lavish productions were given *The Queen of Sheba, Die Fledermaus,* and *The Gypsy Baron.* Conried resigned in 1908 on the grounds of ill health, but it was no secret that, having come upon evil days at the box office, due to the competition of the newly founded Manhattan Opera Company, the management of the Metropolitan felt that a new deal was called for. The Metropolitan Opera Company was formed in 1908 with Giulio Gatti-Casazza of La Scala as general manager and Andreas Dippel as administrative manager. Gustav Mahler was engaged to conduct German works, and Arturo Toscanini was added to the staff of conductors.

Gatti-Casazza remained general manager of the Metropolitan for a quarter of a century. He sensitively gauged the wishes of his audiences and catered to them; his regime was prosperous. Yet the ideals of a great operatic institution were not discarded. He was responsible for over a hundred novelties, many of them world premières. It was due to him that an American opera was given at the Metropolitan for the first time, Frederick Converse's *The Pipe of Desire* (1910), and that

the Metropolitan offered a ten thousand dollar prize for another American opera, won in 1912 by Horatio Parker's *Mona.* Gatti-Casazza also gave seasonal cycles of the Wagnerian music dramas, frequently without cuts. He enriched every department of the opera house, he made possible the weekly broadcasts, and he maintained the Metropolitan as one of the world's great musical centers.

After the 1934–1935 season Gatti-Casazza retired and was succeeded by Herbert Witherspoon. Witherspoon's sudden death, even before his first season began, placed the direction of the Metropolitan in the hands of Edward Johnson, for many years a principal tenor of the company, and in 1935 the director of a special popular-priced spring season. For the next fifteen years Johnson directed the Metropolitan with great distinction. He helped develop American singers: such outstanding American artists as Mimi Benzell, Dorothy Kirsten, Grace Moore, Patrice Munsel, Jan Peerce, Eleanor Steber, Blanche Thebom, Helen Traubel, Richard Tucker, and Leonard Warren made their bow. He also inaugurated the Metropolitan Auditions of the Air and encouraged the founding of the Metropolitan Opera Guild. He was responsible for consummating a deal with Columbia Records for the recording of Metropolitan Opera performances. His regime emphasized performances of operas in English, in fresh new translations. He was responsible for many significant American premières and revivals. When Johnson resigned his post in 1950, he was succeeded by Rudolf Bing, artistic director of the Glyndebourne Opera and the Edinburgh Festival. Bing helped to revitalize performances of Metropolitan Opera staples through restudied productions, modernized stage methods, new scenery, and fresh approaches. In

1953, for example, he had *Faust* presented in nineteenth century costuming. He employed such experienced theatrical figures as Garson Kanin, Margaret Webster, and Alfred Lunt to stage various productions. He was the first Metropolitan director to engage Negro singers. Many more operas were given in new English translations. An important American première was that of Igor Stravinsky's *The Rake's Progress.* Bing helped to open new vistas for his company. In 1951 he created a special touring company to take *Die Fledermaus* throughout the country. In 1952 he organized a special unit, under Herbert Graf, to study television techniques and to prepare productions for television broadcast; the first operas televised by the Metropolitan Opera were *Die Fledermaus* and *La Bohème,* both in English, during the 1952–1953 season. He also supported the innovation of televising opera from the stage of the Metropolitan through a closed circuit to theaters throughout the country; the opening night performance of the 1954–1955 season, which for the first time in the Metropolitan's history was made up of acts from different operas, was thus heard and seen by audiences throughout the country.

By the end of the 1954–1955 season the Metropolitan had presented a total of 210 different operas in a total of 9,125 performances. The following operas had their world premières at the Metropolitan: *The Blue Bird; The Canterbury Pilgrims; Cleopatra's Night; Cyrano de Bergerac; The Emperor Jones; The Girl of the Golden West; Gianni Schicchi; Goyescas; In the Pasha's Garden; The Island God; The King's Henchman; Die Königskinder; The Legend; Madeleine; The Man Without a Country; Merry Mount; Mona; Peter Ibbetson; The Pipe of Desire; Shanewis; Suor Angelica; Il tabarro; The Temple Dancer; The Warrior.*

Me voici dans son boudoir, Frederick's gavotte in Act II, Scene 1, of Thomas's *Mignon.*

Meyerbeer, Giacomo (born JAKOB LIEBMANN BEER), composer. Born Berlin, September 5, 1791; died Paris, May 2, 1864. Though of German birth and Italian training, Meyerbeer was one of the creators of the French grand opera tradition. He changed his last name when a rich relative named Meyer left him a legacy; his first name was Italianized when he started writing Italian operas. His wealthy parents encouraged him in his musical interests, and he made remarkable progress. He appeared in concerts as a prodigy pianist. His music study took place with Muzio Clementi, Carl Friedrich Zelter, and Anselm Weber. In 1810 he went to live with Abbé Vogler as his pupil and household guest. Vogler's intensive training led Meyerbeer to write his first ambitious works, including his first opera, *Jephtha's Vow,* a dismal failure when performed in Munich. In 1812 Meyerbeer left Vogler and went to Vienna. His second opera, *Wirth und Gast*—introduced in Stuttgart—was now performed in Vienna and was such a fiasco that for a while Meyerbeer thought seriously of giving up composing for good. The celebrated composer and court musician Antonio Salieri convinced him he needed more study. For several years Meyerbeer traveled in Italy, absorbing the Italian tradition, and writing operas in the Italian manner. One of these, *Romilda e Costanza,* was a triumph when introduced in Padua in 1817. He now received commissions from several major Italian opera houses for new works. These operas—including *Semiramide riconosciuta* (1819), *Eduardo e Cristina* (1820), and *Margherita d'Anjou* (1820) —made Meyerbeer one of the most popular opera composers in Italy. After a prolonged visit to Berlin, where he wrote *Il crociato in Egitto,* performed with outstanding success in Venice in 1824, Meyerbeer went to Paris in 1926. Acquaintance with such important composers of French operas as Halévy, Auber, and Cherubini, and the assimilation of the ideals and techniques of French opera, made Meyerbeer dissatisfied with the kind of music he had written up to now. For a while he stopped writing altogether. When he returned to composition, he completely discarded his Italian identity and became French. His first opera in the French style was *Robert le Diable,* produced at the Opéra on November 21, 1831. It created a sensation. *Les Huguenots* followed in 1836, and *Le Prophète* in 1849—operas that made Meyerbeer not only the most famous opera composer in Europe at that time but the outstanding exponent of French opera. Meyerbeer glorified, as had no French opera composer before him, heroic drama, stage action, ballet, pomp, stunning visual effects, and overpowering climactic scenes. But his passion for stage effects was combined with immense dramatic power, an inspired lyricism, and a remarkable orchestral virtuosity. His ability to dramatize musical writing made Hugo Riemann refer to him as "one of the most important steps to Wagner's art." And Wagner himself expressed his indebtedness to the best pages in Meyerbeer.

In 1842 the King of Prussia appointed Meyerbeer kapellmeister in Berlin. There he completed and produced a new opera, *Das Feldlager in Schlesien* (1840), written with Jenny Lind in mind (she appeared in it a few weeks after the opera's première). Despite his many activities as conductor and composer in Berlin, Meyerbeer was able to visit different parts of Europe. He was in Paris in 1849 for the première of *Le Prophète,* and once again there in 1854 when his *L'Etoile du Nord* was thunderingly acclaimed at its première at the Opéra-Comique. In 1859 he completed another successful work produced by the Opéra-Comique, *Le pardon de Ploërmel,"* subsequently famous under its later title of *Dinorah.* Meyerbeer's last opera, *L'Africaine,* occupied him for a quarter of a century. Feeling it was his greatest work, he gave it a devotion bestowed on no other opera. Even when the opera was in rehearsal he kept on making painstaking revisions. He did not live to see it performed. He died in May, 1864, and *L'Africaine* was introduced at the Opéra almost a year later, on April 28.

Besides the operas already mentioned, Meyerbeer wrote the following works: *Emma di Resburgo* (1820); *L'esule di Granata* (1822); *Das Brandenburger Thor* (1823).

Mezza aria, literally a half aria—that is, an aria having in part the character of a recitative.

Mezza voce, singing with half voice—the volume and force reduced for quiet utterance.

Mezzo-soprano, the female voice intermediate between contralto and soprano, partaking of the qualities of each.

Micaëla, a peasant girl (soprano) in love with Don José in Bizet's *Carmen.*

Micaëla's Air, see JE DIS QUE RIEN NE M'EPOUVANTE.

Micha, a wealthy landowner (bass), father of Hans and Wenzel, in Smetana's *The Bartered Bride.*

Michael Kohlhaas, see KLEIST, HEINRICH WILHELM VON.

Michele, a skipper (baritone) in Puccini's IL TABARRO.

Mi chiamano Mimi, Mimi's aria in Act I of Puccini's *La Bohème.*

Mickleford, Sir Tristram, Lady Harriet's cousin (bass) in Flotow's *Martha.*

Midsummer Night's Dream, A, see SHAKESPEARE.

Mignon, opéra comique in three acts by Ambroise Thomas. Libretto by Michel Carré and Jules Barbier, based on Goethe's novel *Wilhelm Meisters Lehrjahre.* Première: Opéra-Comique, November 17, 1866. American première: Academy of Music, New York, November 22, 1871.

Characters: Mignon, a girl kidnaped by gypsies (mezzo-soprano); Philine, an actress (soprano); Wilhelm Meister, a student (tenor); Frederick, another tudent (tenor); Laërtes, an actor tenor); Lothario, a wandering minstrel (bass); Giarno, leader of a gypsy band (bass); Antonio, a servant (bass); gypsies; townspeople; peasants; actors; actresses; ladies; gentlemen; servants. The action takes place in Germany and Italy in the late eighteenth century.

The familiar overture is made up of two of the opera's best-known arias: that of Mignon, "Connais tu le pays?" and Philine's polonaise, "Je suis Titania."

Act I. The courtyard of a German inn. Lothario, long bereft of his memory through grief at the loss of his daughter to gypsies, is following the life of a wandering minstrel. He tells a group of merrymakers of his continuing search for his child ("Fugitif et tremblant"). Gypsies arrive and entertain the crowd. Their leader, Giarno, asks Mignon to dance. When she refuses he threatens to strike her. Lothario and Wilhelm Meister leap to her defense. After Mignon thanks them, Wilhelm, a happy-go-lucky student, speaks of his personal philosophy: he wants only to wander in freedom and enjoy pleasures ("Oui, je veux par le monde"). When he asks Mignon about herself, all she can tell him is that she comes from a distant land from which she was taken by gypsies when a child ("Connais-tu le pays?"). Touched by her story, Wilhelm buys her freedom from the gypsies and engages her as his servant. When Lothario approaches Mignon to bid her farewell, she recalls the land of sunshine and swallows ("Légères hirondelles"). Wilhelm Meister, meanwhile, has become acquainted with the actress Philine, to whom he is immediately at-

tracted. She invites him to a party at the castle of Baron Rosenberg.

Act II, Scene 1. A boudoir in the castle. Before the rise of the curtain, there is a delicate gavotte as an entr'acte. On entering Philine's boudoir, Laërtes sings a madrigal about the way her charms affect his lover's heart ("Belle, ayez pitié de nous"). Soon Wilhelm appears, followed by his gypsy servant. Laërtes informs him that the actors will present *A Midsummer Night's Dream,* with Philine as Titania. With anguish Mignon notices how adoringly Wilhelm regards Philine as the actress puts on her make-up and sings a ditty on how all men are attracted to her ("Je crois entendre"). After Wilhelm and Philine depart, Mignon muses about a gypsy lad she once knew (Styrienne: "Je connais un pauvre enfant"). She then tries on one of Philine's gowns, and applies Philine's cosmetics. While she is momentarily out of the room, Frederick, who is in love with Philine, enters the boudoir through a window, singing of his joy at being near his sweetheart ("Me voici dans son boudoir"). When Wilhelm returns and finds Frederick there is a quarrel. Mignon intervenes to stop it. Seeing how attractive she is in Philine's dress, Wilhelm sadly tells Mignon that she can be his servant no longer ("Adieu, Mignon, courage"). Upset that she must lose her master, Mignon tears off Philine's dress and dons her gypsy garb.

Scene 2. The castle gardens. The distraught Mignon is contemplating suicide. The demented Lothario appears and listens to the girl's tale of sorrow and her hope that lightning will strike the castle and spoil the triumph of her rival, Philine. After Lothario wanders off toward the castle, the performers and guests stream into the garden. When Philine is praised by her admirers she sings them a stirring polonaise ("Je suis Titania"). Lothario returns and tells Mignon that her vengeance has been realized: he has set the castle afire. When Philine orders Mignon to enter the castle and fetch a bouquet, Mignon meekly obeys. Suddenly the cry is heard that the castle is burning. Wilhelm rushes into the flames and saves Mignon.

Act III. A castle in Italy. Wilhelm and Lothario are nursing the ailing Mignon. Lothario sings her a lullaby (Berceuse: "De son coeur, j'ai calmé la fièvre"). Now aware that he loves Mignon, Wilhelm gives voice to his inmost feelings (Romance: "Elle ne croyait pas"). When the girl awakes, Wilhelm convinces her that he loves her alone. Meanwhile, Lothario, finding himself in familiar scenes, has recovered his memory. He is overjoyed to find that this castle to which Wilhelm has brought him is actually his own and that he is Count Lothario ("Mignon! Wilhelm! Salut à vous!"). Then his ancient sorrow returns—the memory of his lost daughter Sperata. The name stirs Mignon's memory, she recognizes her surroundings, and father and daughter are joyously reunited.

Mignon is an opera of which it can be said that its parts are greater than the whole. Several arias and numbers are deservedly famous, for they are representative of the French lyric theater at its best. It is these high points that keep *Mignon* alive and make its performance a rewarding evening. But not even the most enthusiastic admirer of this opera would call it a masterpiece. The libretto is pedestrian, and the music too often descends to the level of the play. Thomas himself thought that it would be a failure. However, the première, with Galli-Marié as Mignon, was a triumph. And the composer lived to attend the thousandth performance of *Mignon* at the Opéra-Comique. At this institution the original version of *Mignon,* with spoken dialogue, is still given. Other opera houses prefer to use the recitatives which Thomas himself prepared for the English première at Drury Lane in 1870.

Milanov, Zinka (born KUNC), soprano. Born Zagreb, Jugoslavia, May 17, 1906. She attended the Zagreb Conservatory for five years after which she made her debut at the Zagreb Opera in *Il trovatore* in 1927. After nine years with that company she was engaged by Bruno Walter for the Vienna State Opera, where she appeared in the Italian repertory. Guest appearances in Germany, Czechoslovakia, and at the Salzburg Festival spread her reputation throughout Europe. On December 17, 1937, she made her American debut at the Metropolitan Opera in *Il trovatore.* She has been a principal soprano of the Metropolitan since that time, one of the few principal singers to inaugurate two successive Metropolitan Opera seasons, those of 1951 and 1952; in 1940 she had also appeared in the opening-night performance at the Metropolitan.

Mildenburg, Anna von, dramatic soprano. Born Vienna, November 29, 1872; died there January 27, 1947. After completing her studies at the Vienna Conservatory she made her opera debut in Hamburg in 1895. She was quickly recognized as a leading interpreter of the Wagnerian dramas, and in 1897 she was invited to appear at Bayreuth. In 1898 she was appointed a member of the Vienna Opera, where she remained two decades, achieving great success in the Wagnerian repertory. After her retirement in 1917 she settled in Munich, where for two years she taught singing and dramatics at the State Academy. In 1921 she was appointed stage director of the Munich National Theater, and in 1926 she founded her own singing school. In 1938 she transferred her activities to Berlin, where she taught at the German Institute of Music. Together with her husband, the dramatist Hermann Bahr, whom she married in 1909, she wrote *Bayreuth und das Wagner Theater* (1910).

Milder-Hauptmann, Pauline Anna, dramatic soprano. Born Constantinople, Turkey, December 13, 1785; died Berlin, May 29, 1838. Beethoven wrote for her the role of Leonore in *Fidelio*. When she was a girl of fourteen Joseph Haydn found that she had a voice "as big as a house" and gave her some instruction. Then, encouraged by the Viennese impresario Emmanuel Schikaneder, she studied with Tomascelli, a singing master, and Antonio Salieri, the court kapellmeister. She made her debut in 1803 in Franz Xaver Süssmayer's opera *Der Spiegel von Arkadien* and was engaged as principal soprano of the Vienna Opera. On November 20, 1805, she appeared in the first performance of *Fidelio*. Between 1816 and 1829 she was the principal soprano of the Berlin Opera, where she was particularly successful in Gluck's operas. She left Berlin after differences with the director, Gasparo Spontini, and went on a tour of Russia, Sweden, and Denmark. She made her farewell appearance in 1836 in Vienna.

Mild und leise wie er lächelt, *see* LIEBESTOD.

Milhaud, Darius, composer. Born Aix-en-Provence, France, September 4, 1892. His music study took place at the College of Aix and the Paris Conservatory. While still a Conservatory student Milhaud wrote an opera, *La brebis egarée*, produced a decade later by the Opéra-Comique. In 1917 Milhaud accompanied Paul Claudel, French Ambassador to Brazil, to South America as an attaché of the Legation. There he became interested in Brazilian folk and popular music, and began incorporating elements from it in his music. He also had the collaboration of Claudel on several ambitious works which included *Oreste,* a trilogy of operas based

on tragedies of Aeschylus: *Agamemnon, Les Choéphores,* and *Les Euménides.* Milhaud returned to Paris in 1919 and soon became famous, his name being associated with five other young French composers in a group known as "Les Six." His reputation grew in the era between the two World Wars; after Ravel's death, it was generally conceded that Milhaud was France's leading composer. His works, in every branch of composition, embraced a wide variety of styles ranging from the popular to the esoteric. Among the most significant of these works is a second trilogy of operas, this time on American subjects, described by Virgil Thomson as "a monument of incomparable grandeur": *Christophe Colomb* (1928); *Maximilien* (1930); and *Bolivar* (1943). Still another opera, *Médée* (1938), was performed at the Paris Opéra just before the Nazi occupation. At that time, Milhaud left France and settled in America, teaching at Mills College in California, and writing many symphonic and chamber works. Milhaud returned to Paris after the war, and has since divided his time between the United States and France. Besides those already mentioned, Milhaud has written the following operas: *Les malheurs d'Orphée* (1924); *Esther de Carpentras* (1925); *Le pauvre matelot* (1927); *David* (1954). The following are children's operas: *A propos de botte* (1932); *Un petit peu de musique* (1932); *Un petit peu d'exercise* (1934).

Milton, John, poet. Born London, December 9, 1608; died there November 8, 1674. His epic, *Paradise Lost,* was the source of an opera of the same name by Anton Rubinstein. His masque *Comus,* was set to music first by Henry Lawes, later by Thomas Arne. He is the central character in an opera by Gasparo Spontini, *Milton.* Spontini planned a sequel, *Milton's Death,* but it was never completed.

Mime, a Nibelung (tenor) in Wagner's *Das Rheingold* and *Siegfried.*

Mimi, a maker of artificial flowers (soprano), in love with Rodolfo, in Puccini's *La Bohème.*

Mimi's Farewell, see ADDIO, SENZA RANCOR.

Mimodrama, a musical drama utilizing pantomime. Honegger's *Jeanne d'Arc au bûcher* is an example.

Minnesingers, poet-musicians of the twelfth and thirteenth centuries, the German equivalent of troubadours, they specialized in writing love songs (minnelieder). The minnesingers were succeeded in the fourteenth century by the mastersingers. Tannhäuser, the central figure in Wagner's opera of the same name, was a minnesinger.

Minnie, owner of the Polka saloon (soprano) in Puccini's *The Girl of the Golden West.*

Minnie della mia casa, Jack Rance's aria in Act I of Puccini's *The Girl of the Golden West.*

Minuet, a dance in triple time, usually of moderate speed, first popular at the court of Louis XIV of France, after which it spread throughout Europe to become easily the predominant dance form of the eighteenth century. Such vast numbers of minuets were turned out that Joseph Haydn is reported to have said that a truly individual piece in this restricted form would be a fair measure of a composer's worth. Jean-Baptiste Lully (1632–1687) was the first composer to introduce minuets in his operas. Operas by the later masters Rameau and Handel also contain minuets. One of the most celebrated of operatic minuets is that occurring in Mozart's *Don Giovanni* (Act I, Scene 5). Later notable examples in operas are Berlioz' "Minuet of the Will-o'-the-Wisps" (see below), the opening of the festival scene in Act III of Massenet's *Manon,* and the dance of the courtiers in Act I of *Rigoletto.*

Minuet of the Will-o'-the-Wisps, the music with which Mephistopheles summons evil spirits to surround Marguerite's abode in Part III of Berlioz' *The Damnation of Faust.*

Miracle, Dr., the magician (baritone) in Offenbach's *The Tales of Hoffmann;* his other identities in the opera are Dapertutto and Coppélius.

Mira, o Norma, duet of Norma and Adalgisa in Act III of Bellini's *Norma.*

Mireille, opera by Charles Gounod. Libretto by Michel Carré, based on *Mireio,* a poem by Frédéric Mistral. Première: Paris Opéra, March 19, 1864. In the province of Millaine, in legendary times, Mireille is in love with Vincent. Complications—the opposition of Mireille's father, and the presence of a dangerous rival for Mireille, Ourrias—are finally overcome, and the lovers are joyfully united. Mireille's waltz in Act I, "O légère hirondelle," the choral Farandole in Act II, and Vincent's cavatina in Act III, "Anges du paradis," are well known.

Mir ist so wunderbar! The quartet, written in the form of a canon, of Marcellina, Jacquino, Leonore, and Rocco in Act I of Beethoven's *Fidelio.*

Miserere, see AH! CHE LA MORTE OGNORA.

Missail, a monk (tenor) in Mussorgsky's *Boris Godunov.*

Mistral, Frédéric, poet. Born Maillane, France, Sptember 8, 1830; died there March 25, 1914. He won the Nobel Prize for Literature in 1904. His epic poem *Mireio* was the basis for Gounod's opera *Mireille,* while his poem of bygone Avignon, *Nerto,* was the source of an opera by Charles Widor.

Mit Gewitter und Sturm (Steuermannslied), the Steersman's song in Act I of Wagner's *Der fliegende Holländer.*

Mit mir, mit mir, Baron Ochs's waltz in Act II of Richard Strauss's *Der Rosenkavalier.*

Mitropoulos, Dimitri, conductor. Born Athens, Greece, March 1, 1896. He attended the Athens Conservatory, where he wrote an opera, *Beatrice*, in 1919, which won him a scholarship to study in Brussels with Paul Gilson and in Berlin with Ferruccio Busoni. His studies over, Mitropoulos became an assistant conductor of the Berlin Opera. He returned to Athens in 1924, where he directed several opera performances and became principal conductor of the Athens Symphony. Later he made significant appearances in guest performances throughout Europe and the Soviet Union. In 1936 he made his American debut as guest conductor of the Boston Symphony. In 1937 he was appointed principal conductor of the Minneapolis Symphony, and in 1949 he came to the head of the New York Philharmonic-Symphony. At his symphony concerts, he has given notable performances (in concert versions) of such significant operas as Busoni's *Arlecchino*, Milhaud's *Les Choëphores* and *Cristophe Colomb*, Schoenberg's *Erwartung*, Ravel's *L'heure espagnole*, Monteverdi's *Orfeo*, and Berg's *Wozzeck*. He has conducted opera performances at La Scala, the Florence Music Festival, and (during the 1954–1955 season) the Metropolitan Opera.

Mizguir, a Tartar merchant (baritone) in love with the Snow Maiden in Rimsky-Korsakov's *The Snow Maiden.*

Moïse, *see* MOSE IN EGITTO.

Molière (born JEAN BAPTISTE POQUELIN), playwright. Born Paris, January 15, 1622; died there February 17, 1673. France's master of comedy and satire was a veritable reservoir of opera texts. Among his plays made into operas are *Les amants magnifiques* (opera by Lully); *L'amour médecin* (operas by Lully, Ferdinand Poise, and Wolf-Ferrari); *Le bourgeois gentilhomme* (opera by Lully; also Richard Strauss's *Ariadne auf Naxos*); *Georges Dandin* (Lully); *Le médecin malgré lui* (operas by Marc-Antoine Charpentier, Charles Gounod); *Amphitrion* (Grétry); *L'amour peintre* (Bizet); *Le mariage*

forcé (operas by Lully, Marc-Antoine Charpentier); *Le malade imaginaire* (opera by Marc-Antoine Charpentier, and Hans Haug's *Le malade immortel*); *Les précieuses ridicules* (operas by Felice Lattuada and Otakar Zich); *Monsieur de Pourceaugnac* (Franchetti); *La Princesse d'Elide* (Lully); *Le Sicilien* (Lully); *Tartuffe* (operas by Hans Haug and Yuri Shaporin).

Mona, opera by Horatio W. Parker. Libretto by Brian Hooker. Première: Metropolitan Opera, March 14, 1912. This opera won the ten thousand dollar prize offered in 1911 by the Metropolitan Opera for an American work. The setting is Britain toward the end of the first century. Mona is a British princess in love with Quintus, son of the Roman Emperor. During the confusion attending the Roman conquest, Mona slays Quintus in the mistaken belief that she is aiding the cause of peace between her people and the conquerors. Critics and public applauded the opera but it was dropped after only four performances.

Mona Lisa, opera by Max von Schillings. Libretto by Beatrice Dovsky. Première: Stuttgart Opera, September 26, 1915. In the prologue, a tourist and his wife visit a Carthusian monastery in Florence. They hear from a lay brother the story of Mona Lisa, wife of Giocondo, and her love for Giovanni, this story becoming the core of the opera. In the epilogue, the tourists are revealed to be modern reincarnations of Mona Lisa and Giocondo, while the lay brother is Giovanni.

Mon coeur s'ouvre à ta voix, Dalila's song of love in Act II of Saint-Saëns' *Samson et Dalila.*

Monforte, governor of Sicily (baritone) in Verdi's *The Sicilian Vespers.*

Monica, the medium's daughter (soprano) in Menotti's *The Medium.*

Moniuszko, Stanislaus, composer. Born Ubiel, Poland, May 5, 1819; died Warsaw, June 4, 1872. He studied in

Warsaw with August Freyer and in Berlin with Carl Rungenhagen. In 1840 he left Berlin for Vilna, where he became a church organist and taught piano. In 1846 his opera buffa, *The Lottery*, was successfully performed in Warsaw. Two years later there took place in Vilna a concert performance of his masterwork, the folk opera *Halka;* ten years later *Halka* received its first stage presentation at the Warsaw Opera. Between 1868 and 1892 it was acclaimed in leading European opera houses, and in 1905 it was given in the United States. In 1858 Moniuszko settled in Warsaw, where for many years he was director of the Opera and a teacher at the Conservatory. His last two operas were failures, contributing to bring about his untimely death. In all he wrote fifteen operas. Besides those already mentioned, his greatest successes were: *The Haunted Castle* (1865); *The Countess; The Raftsman; The Pariah* (1869); *Beata* (1872).

Monna Vanna, opera by Henri Février. Libretto is the drama of the same name by Maurice Maeterlinck. Première: Paris Opéra, January 13, 1909. The setting is Pisa in the middle ages, under siege by the Florentine army. The commander of the invading army, Prinzivalle, offers to lift the siege if Monna Vanna, wife of the Pisan commander Guido, will come to his tent at night. Guido refuses, but Monna Vanna is ready to sacrifice herself for her people. The Florentine commander turns out to be a childhood friend of Monna; he respects her love for her husband, treats her with courtesy, and orders the siege lifted. For this, his men regard him as a traitor. He escapes with Monna Vanna to Pisa. Guido refuses to believe his wife innocent and orders Prinzivalle thrown into a dungeon. This unjust act turns Monna Vanna against her husband. Now in love with Prinzivalle, she gets the key to his cell and together they effect an escape. The composer Emil Abrányi also wrote an opera derived from this play, using his own libretto.

Monodrama, a musical drama for a single character. Monodramas originated in the eighteenth century and were seen chiefly in France and Germany. The single character was usually a woman, the heroine of the dramatic situation in question. Such monodramas were written by Jean Jacques Rousseau in France and Johann Freidrich Reichardt and Georg Benda in Germany, among others. Berlioz' *Lélio* —for actor, supplemented by solo singers, chorus, piano, and orchestras—was described by the composer as a "lyric monodrama." In the twentieth century the form was revived by Arnold Schoenberg in his *Erwartung*.

Monostatos, a Moor (tenor) in Mozart's *The Magic Flute*.

Monsalvat, the castle of the Holy Grail in Wagner's *Parsifal*.

Monsigny, Pierre-Alexandre, composer. Born Fauquembergue, France, October 17, 1729; died Paris, January 14, 1817. He was one of the earliest masters of the opéra comique. After receiving his early musical training at the Jesuit College in St. Omer, he went to Paris in his eighteenth year and became a clerk. Hearing Pergolesi's *La serva padrona* sent him back to music study. After five months of training with Pietro Gianotti, he wrote his first opéra comique, *Les aveux indiscrets,* introduced at the Théâtre de la Foire in 1759. His first major success was *Le Cadi dupé* in 1761. During the next fifteen years Monsigny was elevated to a position of first importance among the composers of opéra comique of that period, his succession of triumphs including *Rose et Colas* (1764), *Le déserteur* (1769), *Le faucon* (1772), *Le rendezvous bien employé* (1774), and *Félix* (1777). Though at the height of his popularity and creative power when he completed *Félix,* he never wrote another opera. In 1768 he was given the sinecure of maître d'hotel in the household of the Duke of Orleans, a post enabling him to live in luxury. When the Duke died in 1785, his son appointed Monsigny administrator of his affairs. The French Revolution wiped out Monsigny's fortune. His poverty was relieved in 1798 when the Opéra-Comique gave him an annuity, increased by Napoleon a few years later. In 1800 Monsigny was appointed an inspector of the Paris Conservatory. Proving insufficiently experienced for this post, he relinquished it two years later. In 1813 he succeeded Grétry as a member of the Institut de France.

Montano, Otello's predecessor as governor of Cyprus (bass) in Verdi's *Otello.*

Montemezzi, Italo, composer. Born Vigasio, Italy, August 4, 1875; died Verona, Italy, May 15, 1952. The composer of *L'amore dei tre re,* he completed his high-school education in Verona and went to Milan intending to enter the University. En route he decided to specialize in music. After three trials he was admitted to the Milan Conservatory, from which he later graduated with honors. His first opera, *Giovanni Gallurese,* was produced in Turin in 1905 with such success that it was repeated seventeen times in one month. A second opera, *Hellera,* was given in Turin in 1909. His third opera, *L'amore dei tre re,* presented at La Scala on April 10, 1913, made him world famous; it has since been accepted as one of the finest works of the twentieth century Italian lyric theater. Montemezzi's *La nave,* introduced at La Scala in 1918, was also a major success; it was selected by the Chicago Opera to inaugurate its 1919–1920 season, a performance the composer conducted. His later operas: *La notte di Zoraima* (1930); *L'incan-*

tesimo (1943). In 1939 Montemezzi settled in California, where he resided for a decade, appearing as guest conductor at the Metropolitan Opera in 1949 for performances of *L'amore dei tre re.* He returned to Italy in 1949.

Monterone, Count, father (baritone) of one of the victims of the Duke of Mantua in Verdi's *Rigoletto.*

Monteux, Pierre, conductor. Born Paris, April 4, 1875. After studying at the Paris Conservatory he played the violin in various orchestras, including that of the Opéra-Comique. In 1911 he became the principal conductor of Sergei Diaghilev's Ballet Russe. As such he led the first performances of Igor Stravinsky's *The Rites of Spring* and Maurice Ravel's *Daphnis and Chloé.* In 1913 and 1914 he was a principal conductor of the Paris Opéra; during this period he was also a guest conductor at Covent Garden and Drury Lane. He saw action for two years in the infantry during World War I, then was recalled from the front to propagandize the Allied cause in the United States. He visited America for the first time in 1916 as conductor of the Swedish Ballet. In 1917 he was appointed principal conductor of French operas at the Metropolitan Opera, making his debut there on November 17 with *Faust.* He remained at the Metropolitan two seasons, leading the first performance of Henry F. Gilbert's ballet *The Dance in Place Congo,* and the American premières of *Le coq d'or, Mârouf,* and *La Reine Fiammette.* Monteux returned to the Metropolitan Opera thirty-five years later, on November 17, 1953, opening the new season with a new production of *Faust.* Best known as a symphonic conductor, Monteux founded, and for many years directed, the Paris Symphony. Between 1919 and 1924 he was the conductor of the Boston Symphony, and from 1935 to 1951 music director of the San Francisco Symphony.

Monteverdi, Claudio, composer. Born Cremona, Italy, May 1567; died Venice, November 29, 1643. The first great figure in the history of opera, Monteverdi studied music with Marc Antonio Ingegneri in Cremona. In 1583 he began publishing the first volumes of his vocal compositions. In 1589 he became a violinist and singer at the court of the Duke of Mantua. The Duke took the composer with him on various trips through Europe, which enabled Monteverdi to broaden his musical horizons. A hearing of the Florentine composer Peri's *Euridice* stimulated Monteverdi's interest in opera. In 1606, now the Duke's maestro di cappella, Monteverdi was commissioned to write a work honoring the marriage of the Duke's son to the Infanta of Savoy. For this occasion Monteverdi produced his first opera, *Orfeo,* a work that was to enjoy wide popularity. With his very first opera the composer carried the new art form to artistic significance, producing a wealth of feeling and dramatic power it had not known with Peri. In 1608 Monteverdi wrote a second opera, *Arianna.* The "Lament" from this work (the only fragment that has not been lost) was one of the most celebrated vocal numbers of that day. Monteverdi later arranged it as a madrigal for five voices.

Monteverdi left Mantua in 1612, and in 1613 was appointed maestro di cappella of St. Mark's Cathedral in Venice, where he remained for the rest of his life. He kept on writing operas, helping Venice to become the operatic center of Italy. In 1637 the first public opera house was opened in Venice. Between 1641 and 1649 twenty different operas were performed there, many of them by Monteverdi. Of his operas written in Venice the most celebrated is *L'incoronazione di Poppea* (1642). In 1643 Monteverdi decided to revisit the city of his birth. He never reached his destination. Taken ill en route, he was brought back to Venice, where he soon died. He was given a funeral of the sort usually reserved for princes.

Monteverdi was one of music's most significant pioneers. He endowed the then prevailing style of opera, the recitative, with such expressiveness that it is sometimes said that the aria was born with him. He was the first composer to use ensemble numbers and purely instrumental passages in operas. He extended the resources of the orchestra and made many experiments in matters of orchestration: he claimed as his own discovery the string tremolo and pizzicato. The effectiveness of Monteverdi's operas may be emphasized by pointing out that they are the first in history which can still afford pleasure to present-day audiences. Besides operas already mentioned, Monteverdi wrote: *Il ballo delle ingrate* (1608); *La favola di Peleo e di Theti* (1617); *Il matrimonio d'Alceste con Admeto* (1617); *La vittoria d'amore* (1619); *Andromeda* (1619); *Il commento d'Apollo* (1620); *Il combattimento di Tancredi e di Clorinda* (1624); *La finta pazza Licori* (1627); *Adone* (1639); *Le nozze d'Enea con Lavinia* (1641); *Il ritorno d'Ulisse* (1641).

Montfleury, an actor (tenor) in Walter Damrosch's *Cyrano de Bergerac.*

Moore, Douglas, composer. Born Cutchogue, New York, August 10, 1893. His academic education took place at Yale University, where he took music courses with David Stanley Smith and Horatio Parker. He continued his music study in Europe with Vincent d'Indy and Nadia Boulanger. After winning a Pulitzer Fellowship in music in 1925 he became a member of the music department of Columbia University, where he has since remained, becoming head of the department in 1940. His first opera, *White Wings,* was completed in 1935 but it was not heard until 1949. His second,

The Devil and Daniel Webster, written in 1938, received a highly successful performance in New York in 1939. Moore completed a third opera in 1950, *Giants in the Earth,* introduced in New York on March 28, 1951.

Moore, Grace, soprano. Born Slabtown, Tennessee, December 1, 1901; died Copenhagen, Denmark, January 26, 1947. She studied voice at Ward-Belmont College in Tennessee and the Wilson-Green Music School in Maryland. She made her debut in 1918 in a concert with Giovanni Martinelli in Washington, D.C. She then went to New York, where she appeared in night clubs and musical comedies, becoming a star. She abandoned Broadway in 1926 and for eighteen months studied with Richard Barthelmy and Mary Garden in Europe. She made her opera debut at the Metropolitan Opera on February 7, 1928, as Mimi. She was received enthusiastically by the audience, but the critics felt she still required more training and experience. On September 29, 1928, she made her debut at the Opéra-Comique. Appearances there and in other European opera houses added to her reputation, but it was only after her resounding success in the motion picture *One Night of Love* (1934) that she became world famous as a prima donna. In 1935 she made her debut at Covent Garden in a command performance before Queen Mary. Soon after, she gave twelve command performances, six for kings and six for presidents; four nations decorated her; and she became the only American singer to have her name inscribed on a golden plaque outside the Opéra-Comique.

Although critics were never completely won over by her artistry, her vitality and glamor made her a favorite in opera houses everywhere. One of her greatest artistic successes came on January 28, 1939, at the Metropolitan Opera, when she sang the role of Louise, having previously studied it with the composer. She died in an airplane crash at a time when her career was at its peak. She wrote an autobiography, *You're Only Human Once* (1944), and her career was made the subject of a motion picture, *So This Is Love,* starring Kathryn Grayson.

Moore, Thomas, poet. Born Dublin, Ireland, May 28, 1779; died Bromham, England, February 25, 1852. His metrical romance *Lallah Roukh* was the source of operas by Félicien David, Daniel Kashkin (*The One-Day Reign of Nourmahal*), Anton Rubinstein (*Feramors*), Gasparo Spontini (*Nurmahal*), and Charles Stanford (*The Veiled Prophet of Kohrassan*). Arthur Goring Thomas' *The Light of the Harem* was based on Moore's poem of the same name. After 1802 Moore frequently wrote melodies as well as poems, and between 1807 and 1834 he published a collection of Irish tunes to his own words. In 1811 he wrote the text for an opera, *M.P.,* the music by Charles Edward Horn. Produced in London, it was a failure. Moore's poem "The Last Rose of Summer"—set to the melody "The Groves of Blarney" —was introduced by Flotow in his opera *Martha.*

Morales, a sergeant (bass) in Bizet's *Carmen.*

Morfontaine, Guillot de, Minister of France (bass) in Massenet's *Manon.*

Morgenlich leuchtend im rosigen Schein, Walther's Prize Song in Act III, Scene 2 of Wagner's *Die Meistersinger.*

Morrò, ma prima in grazia, Amelia's aria in Act III of Verdi's *Un ballo in maschera.*

Mort de Don Quichotte, *see* ECOUTE, MON AMI.

Moscona, Nicola, bass. Born Athens, Greece, September 23, 1907. After completing his studies at the Athens Conservatory he made his opera debut at the National Theater in Athens in

1930 as Sparafucile. Between 1931 and 1937 he made many operatic appearances in Greece and Egypt. A scholarship from the Greek government enabled him to go to Milan for additional study. He was there only two months when he was heard by Edward Johnson, who engaged him for the Metropolitan Opera. Moscona made his American debut at the Metropolitan on December 13, 1937, in Aïda. Since then he has appeared in principal bass roles in over four hundred performances in New York. His repertory includes over a hundred operas in seven languages. In addition to his Metropolitan Opera appearances he has been heard with major European companies and at major European festivals.

Mosè in Egitto (Moses in Egypt), opera by Rossini. Libretto by Andrea Tottola. Premiére: San Carlo, Naples, March 5, 1818. While originally an oratorio, this work has often been given stage presentations. Rossini rewrote it for Paris, with a French text by Jouy and Balocchi, and as Moïse it was introduced at the Paris Opéra on March 26, 1827. The story of both versions follows the Biblical story of the struggle between Pharaoh and the Jews, the infliction of the plagues on the Egyptians, and the exodus from Egypt, culminating in the destruction of the Egyptian army in the Red Sea. Love interest is introduced between Pharaoh's son and a Hebrew girl, Anaïs, an episode invented to provide a vehicle for the brilliant prima donna Isabella Colbran, Rossini's mistress, later his wife.

Mosenthal, Salomon Hermann von, novelist, dramatist, librettist. Born Kassel, Germany, January 14, 1821; died Vienna, February 17, 1877. This distinguished German writer contributed librettos to several operas, including Ignaz Brüll's Das goldene Kreuz and Der Landfriede; Friedrich von Flotow's Albin; Karl Goldmark's The Queen of Sheba; Edmund Kretsch-

mer's Die Folkunger; Theodor Leschetizky's Die erste Falte; and Otto Nicolai's The Merry Wives of Windsor. Sir George Macfarren's Helvellyn was based on Mosenthal's novel Der Sonnenwendhof.

Moses and Aron, oratorio-opera by Arnold Schoenberg. Libretto by the composer. Première: Hamburg, Germany, March 12, 1954. Schoenberg wrote this opera in the early 1930's. The text is based on the story of Moses as told in Exodus. Schoenberg completed only two acts and some fragments for other acts, and the opera was introduced in this incomplete version.

Mother, The (Die Mutter), opera by Alois Hába. Libretto by the composer. Première: Munich, May 17, 1931. This is the first opera written in quartertones. The composer ordered a special quarter-tone piano, two quarter-tone clarinets, and two quarter-tone trumpets built to perform his music, explaining: "Harmonically I have used combinations ranging from two to twenty-four different sounds. Melodically I use multiples of quarter-tones: ¾ tones, 5/4 tones, neutral thirds, sixths, fourths, and sevenths." The vocal parts require quarter-tone singing.

Mother of Us All, The, opera by Virgil Thomson. Libretto by Gertrude Stein. Première: New York City, May 7, 1947. This was the second collaboration of Thomson and Stein. The first, Four Saints in Three Acts, had subject matter but no plot. The Mother of Us All has for its central character Susan B. Anthony, pioneer in the woman suffrage movement, and the opera traces her career from her initial struggles to her final victory. Other historical characters appear, including Ulysses S. Grant, Daniel Webster, and Andrew Jackson; also two characters identified as Virgil T. and Gertrude S. Thomson wrote this opera on a commission from the Alice M. Ditson Fund. After its première it was given a special citation by the New York Music Critics Circle; a regular award was prohibited because Thomson was a member of the Circle.

Mottl, Felix, conductor. Born Unter St. Veit, Austria, August 24, 1856; died Munich, Germany, July 2, 1911. He graduated from the Vienna Conservatory with highest honors. His first important appointment was as conductor of the Akademische Richard Wagner Verein in Vienna. In 1875 he was Wagner's assistant at Bayreuth, preparing the first festival. Five years later he became court kapellmeister in Karlsruhe, rising to the position of general music director in 1893. He distinguished himself there not only in the works of Wagner but in his presentation of a cycle of Berlioz' operas, including the complete *Les Troyens*. In 1886 he was principal conductor of the Bayreuth Festival. In 1903 he left Karlsruhe and became general music director in Munich; in 1907 he combined this post with that of director of the Court Opera. Mottl was engaged by the Metropolitan Opera in 1903 to direct the American première of *Parsifal*. When Bayreuth instituted legal action to prevent this performance Mottl withdrew and Alfred Hertz substituted for him. However, Mottl conducted other performances at the Metropolitan Opera that season, making his debut there on November 25 with *Die Walküre*. Besides the Wagner dramas, he directed operas by Mozart, Bizet, Boieldieu, and Gounod. Mottl wrote three operas: *Agnes Bernauer* (1880); *Fürst und Sänger* (1893); and *Ramin*. He edited the vocal scores of all the Wagner music dramas.

Moussorsky, see MUSSORGSKY.

Mozart, Wolfgang Amadeus, composer. Born Salzburg, Austria, January 27, 1756; died Vienna, December 5, 1791. Like every other branch of musical composition, opera was greatly enriched by the genius of Mozart. His

Don Giovanni, The Marriage of Figaro, and *The Magic Flute* are milestones in the evolution of opera. Mozart began studying the harpsichord when he was only four, being taught by his father, Leopold, an eminent musician in his own right. The boy immediately revealed sure instincts and a phenomenal capacity to assimilate everything taught him. His gifts for sight-reading and improvisation aroused the awe of all who heard him. Quite as wonderful were his creative powers. He wrote minuets when he was five, a sonata when seven, a symphony when eight. In his sixth year he and his sister (a skilled harpsichord player) were taken by their father to the Electoral court in Munich, where the young performers won the hearts of royalty and were showered with gifts. The success of this trip encouraged the father to undertake others. For the next several years he exhibited his children throughout Europe. In Paris, four of Mozart's violin sonatas were published; in London, his first symphonies were performed, and his harpsichord playing amazed and pleased the Queen's music master, Johann Christian Bach.

In Vienna, in 1768, the Emperor commissioned Mozart to write an opera buffa, *La finta semplice*. The artists at the opera house (led by Gluck) refused to participate in a performance of an opera by a child, and they prevented its performance. Another little Mozart opera, *Bastien and Bastienne,* was given privately in the garden theater of Dr. Anton Mesmer, the hypnotist. Mozart and his father embarked on an extended tour of Italy in 1770. In Rome the fourteen-year-old boy gave a remarkable demonstration of his genius by writing down from memory, after a single hearing, the complete score of Gregorio Allegri's celebrated *Miserere*. In Italy, as elsewhere, Mozart received numerous honors and tributes. In Milan he was commissioned to write an opera. The work, *Mitridate, Rè di Ponto,* was introduced on Christmas Day, 1770; it was so successful that it was given twenty times. The following year Mozart's *Ascanio in Alba,* a serenata, was also a major success at its première in Milan. The venerable opera composer Johann Adolph Hasse, remarked prophetically: "This boy will throw all of us into the shade." In 1774 Mozart again returned to Milan to supervise the première of his comic opera, *La finta giardiniera (The Pretended Gardener).*

The period between 1772 and 1777 was spent principally in Salzburg, under unhappy conditions. A new Archbishop, Hieronymus von Colloredo, had come to Salzburg. He had little appreciation of Mozart's genius, and the young man was treated as servants were, with imperious authority and personal abuse. The masterworks that Mozart was creating were ignored. A welcome avenue of escape finally came in 1777 when Mozart and his mother set off for Paris hoping to find there some advantageous post; the father, denied a leave of absence by the Archbishop, had to remain behind. The visit to Paris did not prove rewarding. There seemed to be no further interest in Mozart now that he had outgrown the appeal of childhood. A tragic circumstance of the journey was the sudden death of Mozart's mother. Mozart had to return to Salzburg, to his drab existence in a post that paid poorly and in which he suffered so many indignities. The insufferable situation was accentuated on January 29, 1781, when, on a trip to Munich, he was acclaimed for his new opera, *Idomeneo,* his first work that gave an indication of his coming powers as a stage composer. Mozart now knew he would have to make a permanent break with the Archbishop and make his way elsewhere. That break came in 1782 when Mozart visited Vienna with the Archbishop's en-

tourage. Denied permission to appear at some benefit concerts, Mozart flew into a rage, denounced his employer, and was summarily dismissed. From now on, to the end of his life, Mozart lived in Vienna. He did not have to wait long for recognition. The Emperor commissioned him to write a new opera for the court theater, *The Abduction from the Seraglio (Die Entführung aus dem Serail).* It was given on July 16, 1782, and despite the intrigues organized against it by envious composers, headed by Antonio Salieri, it was a triumph. One of the best comic operas before those of Rossini, *The Abduction* is today the oldest opera in the German language that is still performed.

Confident of his future, Mozart married Constance Weber on August 4, 1782. He expected a profitable post at court, but was kept waiting. The Emperor was lavish with praise and commissions, but niggardly about opening his purse strings. To earn a living, Mozart gave lessons, which brought him a pittance. Frequently he was subjected to the humiliation of begging friends for loans. But his frustrations and disappointments did not arrest the flow of his compositions: masterworks bringing new dimensions to the symphony, concerto, and string quartet, as well as opera. A few in Vienna recognized the grandeur of this music. One was Joseph Haydn, then the most celebrated composer in Europe, who described Mozart as "the greatest composer I know, either personally or by name."

A meeting with Lorenzo da Ponte in 1785—he had recently been appointed poet of the Viennese court theaters—resulted in three of Mozart's greatest operas. Da Ponte wrote admirable librettos for all three. The first was *The Marriage of Figaro (Le nozze di Figaro),* given at the Burgtheater on May 1, 1786. Once again the anti-Mozart forces in Vienna rallied to sabotage the performance; only the personal intervention of the Emperor thwarted this maneuver. The opera was a success of the first magnitude; so many numbers were encored that the length of the first performance was almost doubled. But Mozart's enemies were not defeated. They presented at the Burgtheater a catchy little opera—Martín y Soler's *Una cosa rara*—to deflect the interest and enthusiasm of the Viennese from Mozart's new work. They succeeded: *The Marriage of Figaro* closed after only nine performances. The following year Mozart again collaborated with Da Ponte. Their new work was *Don Giovanni,* whose première in Prague on October 29, 1787, was another triumph. The city went mad over Mozart's melodies. "Connoisseurs and artists say that nothing like this has been given in Prague," reported a contemporary journal. The last of these three collaborations was *Così fan tutte,* first given in Vienna on January 26, 1790. Considered a failure, the opera was given only ten performances.

The year of 1791, the last of Mozart's life, brought no end to the composer's personal misfortunes. While he had finally been given a permanent post as court composer and chamber musician (in succession to Gluck) he received such a small salary that it neither relieved him of his debts nor provided for the necessities of life. Impoverished, sick in body and spirit, Mozart gave way to despair. Yet this last year was a period of wonderful creation, yielding two operas and the *Requiem. The Magic Flute (Die Zauberflöte)* and the *Requiem* were the results of commissions. The first was ordered by Johann Emanuel Schikaneder, who wanted a popular German opera for his Theater-auf-der-Wieden. He supplied his own libretto. The opera was introduced on September 30 and in time became so popular that it was given a hundred performances. The *Requiem* had a dramatic history. A

stranger appeared at Mozart's house and engaged him to write the work on condition that Mozart make no attempt to learn the identity of his patron. Actually, the stranger was a representative of Count Franz von Walsegg, an amateur musician who used to order music and present it as his own. Unaware of this, tortured by poverty and failing health, Mozart realized that he was penning his own requiem. And so it proved. The afternoon before he died Mozart summoned three friends and with them sang portions of the composition. At the beginning of the "Lacrimosa" he was so overcome with sorrow he had to put the music by. After Mozart's death, incidentally, Count Walsegg copied the *Requiem* in his own hand and claimed the authorship when he had it performed.

Mozart completed twenty-five works for the stage, counting serenatas, intermezzi, operettas, comedies and plays with music, and comic and serious operas. Two—*The Abduction from the Seraglio* and *The Magic Flute*—were the first important operas written to German texts, consequently the foundation on which all later German operas were built. Mozart's other operas, settings of Italian texts, for the most part conformed to prevailing Italian styles and tastes. But even here Mozart wrote a special chapter in opera history. His greatest Italian operas—*The Marriage of Figaro, Don Giovanni, Così fan tutte*—may be within formal patterns of Italian opera, but something decidedly new was added. No composer before Mozart had his gift for musical characterization: a sudden accent, the injection of a rhythmic figure, a change of orchestral color, the introduction of a new melodic idea—and we suddenly get a new insight into the idiosyncrasies and hidden motivations of the characters. No one before Mozart had his amazing gamut of musical expression: from levity to grandeur and nobility, from malice to the most eloquent outbursts of feeling and passion. Mozart's operas: *Bastien and Bastienne* (1768); *La finta semplice* (1768); *Mitridate, Rè di Ponto* (1770); *Ascanio in Alba* (1771); *Il sogno di Scipione* (1772); *Lucio Silla* (1772); *La finta giardiniera* (1774); *Idomeneo, Rè di Creta* (1781); *Die Entführung aus dem Serail* (*The Abduction from the Seraglio*) (1782); *Der Schauspieldirektor* (1786); *Le nozze di Figaro*, (*The Marriage of Figaro*) (1786); *Don Giovanni* (1787); *Così fan tutte* (1790); *Die Zauberflöte*, (*The Magic Flute*) (1791); *La clemenza di Tito* (1791).

Mozart was the central character of an opera by Rimsky-Korsakov, *Mozart and Salieri*, based on Alexander Pushkin's dramatic poem. The theme is the historic rivalry between these two composers, and the unfounded rumor, circulated after Mozart's death, that Salieri had poisoned him.

Much Ado About Nothing, a comedy by William Shakespeare, revolving around the confusions and misunderstandings complicating the love affair of Claudio and Hero, natives of Messina. See BEATRICE ET BENEDICT; SHAKESPEARE.

Muck, Karl, conductor. Born Darmstadt, Germany, October 22, 1859; died Stuttgart, March 3, 1940. One of the foremost conductors of his generation, Muck was particularly eminent as an interpreter of Wagner. He received an extensive academic education in classical philology at the Universities of Heidelberg and Leipzig before he specialized in music. His musical training took place at the Leipzig Conservatory. After serving as chorus master at the Zurich Opera, he conducted in smaller Austrian opera houses. In 1886 he became principal conductor of the Angelo Newman Opera Company in Prague. Three years later he attracted attention with his performances of the Wagnerian music dramas in Russia. In 1892 he was appointed first conductor of the Berlin Opera, and it was here that he first became famous throughout Europe for his searching and fastidiously prepared performances. Further triumphs were gathered in Covent Garden in 1899, at Bayreuth beginning in 1901 (he appeared regularly at Bayreuth until 1931), and in Vienna. In 1906 Muck visited the United States for the first time, making his American debut as conductor of the Boston Symphony Orchestra on October 12, 1906. After returning to Berlin, he was made musical director of the Berlin Opera. The twenty years of his regime was one of the most resplendent periods in the history of that institution. In that time Muck directed 1,071 performances of 103 different operas, including 35 novelties. He set an artistic standard for Berlin that was rarely equaled elsewhere.

In 1912 Muck returned to Boston to become permanent conductor of the Boston Symphony Orchestra. When America entered World War I, his position in this country became increasingly embarrassing. It was known he was a friend of the Kaiser, from whom he had received many honors, and his sympathies had been publicized as resting with his countrymen. On March 25, 1918, Muck was arrested. He remained a political prisoner several months, after which he was deported. Back in Europe, Muck continued to conduct in Berlin, Bayreuth, and other major German opera houses. From 1922 to 1933 he was the musical director of the Hamburg Philharmonic Orchestra. After 1933 he lived in retirement in Stuttgart.

Müller, Maria, soprano. Born Leitmoritz, Bohemia, January 29, 1898. She studied singing at the Prague Conservatory and with Max Altglass in New York. Her debut took place in 1919 in Linz, Austria, as Elsa. From

1921 to 1923 she appeared with the Prague Opera and in 1923–1924 with the Munich Opera. On January 21, 1925, she made her American debut at the Metropolitan Opera as Sieglinde. She stayed at the Metropolitan eleven years, scoring her greatest successes in the German repertory, and starring in such important premières and revivals as Alfano's *Madonna Imperia*, Montemezzi's *Giovanni Gallurese*, Pizzetti's *Fra Gherardo*, and Weinberger's *Schwanda*. While a member of the Metropolitan she appeared extensively throughout Europe, including Covent Garden, the Berlin Opera, the Vienna State Opera, and the festivals in Bayreuth and Salzburg.

muette di Portici, La (also known as MASANIELLO), opera by Daniel Auber. Libretto by Eugène Scribe and Germaine Delavigne. Première: Paris Opera, February 29, 1828. In seventeenth century Naples, Fanella, the mute sister of Masaniello, is imprisoned by Alfonso. Masaniello heads an army in a successful revolt against Alfonso, and is given the crown of Naples. Poisoned by a former friend, Masaniello goes mad, a development enabling Alfonso to quell the revolt. Masaniello is killed, and his sister commits suicide. This opera had far-reaching political repercussions when it was introduced in Brussels in 1830. Its theme—political revolution—made such a profound impression on the Belgians that it sparked their revolt against Dutch rule, resulting in the constitution of Belgium as an independent state. The overture and the barcarolle in Act II are favorites of salon and "pop" orchestras.

Muff, a comedian (tenor) in Smetana's *The Bartered Bride*.

Mugnone, Leopoldo, conductor. Born Naples, September 29, 1858; died there December 22, 1941. His music study took place at the Naples Conservatory. In 1874 he was appointed conductor of the Teatro la Fenice in Venice, but he first achieved major recognition at the Teatro Costanzi in Rome, where he led the world première of *Cavalleria rusticana* in 1890. He now became principal conductor of La Scala where, in 1893, he conducted the première of *Falstaff*. He was equally distinguished in the Italian and Wagnerian repertories and was one of the earliest advocates of Wagner in Italy. In 1905 he led successful performances at Covent Garden, an opera house he frequently returned to, leading there the first English performances of *Fedora* (in 1906) and *Iris* (in 1919). Mugnone also conducted at the San Carlo in Naples, the Augusteo in Rome, and in the United States. He wrote several operas, the best known being *Il Biricchino* (1892) and *Vita brettona* (1905).

Munich Opera, *see* HOF-UND-NATIONAL-THEATER.

Munsel, Patrice, soprano. Born Spokane, Washington, May 14, 1925. She began studying singing when she was twelve. After an additional period of study with William P. Herman in New York, she won the Metropolitan Opera Auditions of the Air in 1943. Her debut at the Metropolitan Opera took place on December 4, 1943, in *Mignon*. She has since appeared in leading soprano roles at the the Metropolitan. In 1948 she made her European debut at the Copenhagen Opera. In 1952 she appeared in a motion picture dramatization of the life of Nellie Melba. She has also frequently been heard in recitals, and over radio and television.

Muratore, Lucien, tenor. Born Marseilles, France, August 29, 1876; died Paris, July 16, 1954. He was first a dramatic actor in Paris and Monte Carlo, appearing in leading roles with Rejane at the Odéon. Albert Carré, director of the Opéra-Comique, induced him to leave the dramatic stage for opera. After a period of preparation at the Paris Conservatory he made his debut at the Opéra-Comique on December 16, 1902, in the world première of Reynaldo Hahn's *La Carmélite*. He was so successful at the Opéra-Comique that in 1905 he was engaged by the Paris Opéra as principal tenor. For the next half dozen years he was one of the most highly acclaimed singers in France. Massenet selected him to appear in the world première of *Ariane* in 1906, and in the premières of his later operas, *Bacchus* and *Roma*. Muratore also created the leading tenor roles in Février's *Monna Vanna*, Georges Huë's *Le Miracle*, and Saint-Saëns' *Déjanire*. He made his American debut in 1913 with the Boston Opera. He then joined the Chicago Opera where, except for a single season when he saw action in the French Army during World War I, he remained a principal tenor through the 1921–1922 season. After his retirement from opera he devoted himself to teaching singing in Europe, becoming a member of the faculty of the American Conservatory in Fontainebleau in 1938. Muratore was married to the prima donna, Lina Cavalieri, in 1913; it was his second marriage.

Muse, The, a character (soprano) in Offenbach's *The Tales of Hoffmann.*

Musetta, Marcel's sweetheart (soprano) in Puccini's *La Bohème.*

Musetta's Waltz, *see* QUANDO ME'N VO' SOLETTA.

Musette, a simple pastoral melody in moderate tempo, usually associated with a drone effect, especially popular in France in the seventeenth and eighteenth centuries. The music derives its name from the instrument for which it was originally intended, the musette being a small, sweet-toned member of the bagpipe family. Compositions entitled musettes are found in many seventeenth and eighteenth century operas. Noteworthy examples include the mussette in Act IV of Gluck's *Armide,* and musettes in Rameau's *Acanthe et*

Céphise and *Les fêtes d'Hébé* and Handel's *Il pastor fido.*

Musica parlante, literally "speaking music"—a term used by the first Florentine composers of opera to describe the recitative style of their vocal writing.

Music drama, a term used by Richard Wagner and some of his successors to designate an opera of serious nature in which the close integration of text and music departed from the older, more formal concepts of opera.

Musset, Alfred de, poet and playwright. Born Paris, November 11, 1810; died there May 1, 1857. Many of de Musset's plays have been adapted for operas, including: *Carmosine* (Février); *La coupe et les lèvres* (Puccini's *Edgar*); *Fantasio* (Ethel Smyth); *Fortunio* (Messager); *Namouna* (Bizet's *Djamileh*); *On ne badine pas avec l'amour* (Pierné); *La Rosiera* (Gnecchi); *Les caprices de Marianne* (Chausson).

Mussorgsky, Modest, composer. Born Karevo, Russia, March 21, 1839; died St. Petersburg, March 28, 1881. A member of the Russian nationalist school, Mussorgsky created the most distinguished folk opera to come out of Russia, *Boris Godunov*. The son of a wealthy landowner, Mussorgsky was originally directed by his parents to the army. While in uniform he met the composers Dargomizhsky and Balakirev, who stimulated his latent musical interests. After a brief period of study with Balakirev, Mussorgsky decided, in 1858, to give up the army for music. His debut as composer took place in 1860 with an orchestral work performed in St. Petersburg. He now began planning and writing more ambitious works, including a symphony and an orchestral fantasy. From 1860 on he was a passionate advocate of musical nationalism, in which ideal he associated himself with his colleagues Balakirev, Cui, Rimsky-Korsakov, and

Borodin in a group since known as the "Five." The abolition of serfdom in 1861 put an end to Mussorgsky's financial security. The necessity of earning a living made him accept a clerkship in the Ministry of Transport in 1863. For the next seventeen years (virtually to the end of his life) he worked as a clerk, relegating musical composition to the status of an avocation. The death of his mother, in 1865, was a shattering blow. Previously having revealing unmistakable symptoms of nervous disorders and melancholia, Mussorgsky now became a victim of alcoholism. He was to suffer physically and mentally for the remainder of his life, but this suffering only seemed to strengthen his creative resources. He planned and outlined an opera, *The Marriage,* in which he tried to introduce a new concept of musical realism with melodies patterned after the inflections of human speech. He abandoned this project after finishing a single act, but only because a new and greater venture began to absorb him completely: the opera *Boris Godunov,* completed in 1870. Unorthodox in approach and style, *Boris* was at first turned down by the Imperial Theater of St. Petersburg. But after the work was revised along somewhat more formal lines, three scenes were presented at the Imperial Theater on February 17, 1873. A complete performance of the opera followed on February 8, 1874, and was more or less a failure.

After 1873, Mussorgsky's physical and moral disintegration became complete. He sometimes existed as a beggar might, and there were periods when he was in a state of intoxicated stupefaction. Yet in his periods of mental clarity he produced several remarkable works, including the opera *Khovantchina.* In 1879 he undertook a concert tour of southern Russia with the singer Daria Leonova. Back in St. Petersburg he returned to his old ways. Just be-fore his death, he gave indication of losing his mind.

His colleagues sometimes looked upon his music with condescension because, having been poorly trained, he was guilty of harmonic and instrumental crudities. But he possessed one of the greatest talents of his day, and his best music is marked with a vitality that was hardly matched by that of his contemporaries. Mussorgsky's aim was not exclusively the creation of a national art. He also sought truth, the identification of music with life. His most significant innovations are found in his operas, in his efforts to bring opera closer to life through melodic recitatives in which the lyricism approximated speech. "What I want to do," he once said, "is to make my characters speak on the stage as they would in real life and yet write music that is thoroughly artistic." In this quest, his music often avoided the more traditional concepts of beauty, giving way to ugly discords, primitive rhythms, abrupt transitions, strange chord sequences. Many of these effects, we realize today, were not due to ignorance but to a calculated attempt by Mussorgsky to broaden the expressiveness of his music. This fact was not clear to Mussorgsky's colleagues when, after his death, they edited his works and tried to remove the imperfections. Mussorgsky's operas: *Salammbô* (unfinished); *The Marriage* (unfinished); *Boris Godunov* (1870); *Khovantchina* (1873), completed by Rimsky-Korsakov; *The Fair at Sorochinsk* (1874), different versions completed by César Cui, Nikolai Tcherepnin, and Vissarion Shebalin.

Mustafa, Bey of Algiers (bass) in Rossini's *L'Italiana in Algeri.*

Mutter, Die, *see* MOTHER, THE.

Muzio, Claudia, dramatic soprano. Born Pavia, Italy, 1892; died Rome, May 24, 1936. Her father was, at various times, assistant stage manager at Covent Garden and at the Metropolitan Opera. The young singer received her vocal training with Mme. Casaloni, after which she made her opera debut in Arezzo, on February 7, 1912, in the title role of *Manon Lescaut.* Numerous appearances followed in leading Italian opera houses. She made her American debut at the Metropolitan Opera on December 4, 1916, as Tosca. She remained with the Metropolitan six years, during which period she created the role of Giorgetta in *Il tabarro* and appeared in many notable American premières and revivals, including *L'amore dei tre re, Andrea*

Chénier, Die Loreley, and *Eugene Onegin.* After leaving the Metropolitan Opera, she became a member of the Chicago Opera company; later, she returned to the Metropolitan for the single season of 1933–1934.

Mylio, Rozenn's childhood sweetheart (tenor) in Lalo's *Le Roi d'Ys.*

My Man's Gone Now, Bess's lament in Act III of Gershwin's *Porgy and Bess.*

Myrtale, a slave girl (mezzo-soprano) in Massenet's *Thaïs.*

Mystery, a form of sacred drama with music, popular in Europe in the fifteenth and sixteenth centuries.

Mytyl, a child (soprano) in Wolff's *The Blue Bird.*

N

Nabucco (or Nabucodonosor), opera by Verdi. Libretto by Themistocles Solera. Première: La Scala, March 9, 1842. As Francis Toye wrote: *"Nabucco* is probably the most satisfactory of all the early Verdi operas . . . not until *Rigoletto* did the composer produce again an opera so satisfactory as an artistic whole."* Nabucco (Nebuchadnezzar) is the Babylonian king who conquers Jerusalem and takes the Hebrews into captivity. Returning to Babylon, he goes mad and is imprisoned. His daughter Abigaille, supported by the priests, sets herself up as regent. In prison, Nabucco sees his other daughter, Fenena, being taken to her execution. He prays to Jehovah for help. Suddenly, his sanity returns. With the aid of his guards he saves Fenena. Abigaille commits suicide, and Nabucco resumes his rule, now faithful to Jehovah. The overture, the chorus of the Levites in Act II, and the chorus "Va,

pensiero sull' ali dorate" in Act III are among the finest pages in the opera.

Nachbaur, Franz, tenor. Born Weiler Giessen, Württemberg, Germany, March 25, 1835; died Munich, March 21, 1902. After studying singing with Johann Baptist Pischek in Stuttgart he made opera appearances in Mannheim, Hanover, Prague, and Vienna. In 1866 he was appointed principal tenor of the Munich Opera, remaining here until his retirement a quarter of a century later. He created several Wagnerian roles, including that of Walther in 1868, and Froh in 1869.

Nachtlager von Granada, Das (The Night Camp at Granada), opera by Conradin Kreutzer. Libretto by Karl Johann Braun, based on a play by Friedrich Kind. Première: Josefstadt Theater, Vienna, January 13, 1834. The setting is Spain in the sixteenth century. The Crown Prince, disguised as a hunter, makes love to the shepherdess

Gabriele. His rivals plot to kill him, but Gabriele saves his life. In gratitude, the Prince unites Gabriele with Gomez, the man of her choice.

Nadir, a pearl fisher (tenor) in Bizet's *Les pêcheurs de perles.*

Naiad, a character (soprano) in Richard Strauss's *Ariadne auf Naxos.*

Naina, an evil fairy (mezzo-soprano) in Glinka's *Russlan and Ludmilla.*

Nancy, Lady Harriet's maid (contralto) in Flotow's *Martha.*

Nannetta, Ford's daughter (soprano) in Verdi's *Falstaff.*

Napoleon Bonaparte, Emperor of France (baritone) in Giordano's *Madame Sans-Gêne.*

Nápravník, Eduard, conductor and composer. Born Beischt, Bohemia, August 24, 1839; died St. Petersburg, Russia, November 23, 1916. He studied privately with local teachers and with Jan Bedřich Kittel, and at the Modern School and the Organ School, both in Prague. He went to Russia in 1861, and in 1869 was appointed principal conductor of the St. Petersburg Opera, where he had been assistant conductor since 1863. For forty years or so he was one of the most significant opera conductors in Russia, responsible for making the St. Petersburg Opera one of the finest in that country. He directed more than three thousand performances, many of them of Russian operas. Napravnik also wrote several operas: *The Citizens of Nizhni* (1868); *Harold* (1886); *Dubrovsky* (1895); *Francesca da Rimini* (1903).

Narraboth, captain of the guards (tenor) in Richard Strauss's *Salome.*

National Opera Company, an opera company organized in New York in 1886 to succeed the recently defunct American Opera Company. With Theodore Thomas, Gustav Hinrichs, and Arthur Mees as conductors, the company toured the country during its initial season. The following season it presented opera performances at the Metropolitan Opera, remaining the resident company until 1889. The company then dissolved, to be reorganized as the Emma Juch Grand Opera Company, which toured the United States, Canada, and Mexico until 1891.

Natoma, opera by Victor Herbert. Libretto by Joseph Redding. Première: Philadelphia, February 25, 1911. The setting is early California. Natoma, an Indian princess, is in love with Lieutenant Merrill; her rival is a young American girl, Barbara. When the princess overhears the plot of Alvarado to kidnap Barbara, she is willing to sacrifice her own happiness for that of the man she loves. She kills Alvarado, then atones for this crime by entering a convent.

Naturalism, a movement in twentieth century opera attempting to portray life as authentically as possible. It was preceded by the verismo school in Italian opera. The outstanding apostles of naturalism were Gustave Charpentier and Alfred Bruneau.

Nature immense (L'invocation à la nature), Faust's invocation in Part Four of Berlioz's *The Damnation of Faust.*

Navarraise, La (The Girl from Navarre), opera by Massenet. Libretto by Jules Claretie and Henri Cain, based on Claretie's story, *La Cigarette.* Première: Covent Garden, June 20, 1894. In Spain, during the Carlist war, Anita is in love with Araquil, but his father opposes the marriage since the girl has no dowry. To gain the money, Anita helps General Garrido of the Royalist troops capture and kill his Carlist enemy, Zuccaraga. Araquil is horrified to learn of this, and is killed when he follows Anita to Zuccaraga's camp. Anita loses her mind out of grief.

nave, La (The Ship), opera by Italo Montemezzi. Libretto by Tito Ricordi, based on Gabriele d'Annunzio's tragedy of the same name. Première: La Scala, November 3, 1918. The text is a plea for a unified Italy, with Italy represented by a ship about to be launched; the play is filled with other symbols of church and state.

Nay, Maccus, Lay Him Down, Eadgar's aria in the concluding scene of Deems Taylor's *The King's Henchman.*

Nedda, Canio's wife (soprano) in Leoncavallo's *Pagliacci.*

Neipperg, Count, an Austrian nobleman (tenor) in Giordano's *Madame Sans-Gêne.*

Nella, Gherardo's wife (soprano) in Puccini's *Gianni Schicchi.*

Nelusko, a slave (baritone) in Meyerbeer's *L'Africaine.*

Nemico della patria, Gérard's monologue in Act III of Giordano's *Andrea Chénier.*

Nemorino, a young peasant (tenor) in Donizetti's *L'elisir d'amore.*

Nero, the tyrannical emperor of Rome, persecutor of the Christians, is the central character in operas by Boïto, Mascagni, and Anton Rubinstein, all three entitled either *Nero* or *Nerone.*

Nerone, opera by Arrigo Boïto. Libretto by the composer. Première: La Scala, May 1, 1924. Nero has murdered his mother and comes to bury her ashes. When this is done, the snake-charmer, Asteria, appears and confesses to Simon that she loves the Emperor. Together they try to win Nero's confidence, but Nero believes Asteria is a fury risen to avenge the murder, and sends her away; he also imprisons Simon. To effect his escape, Simon has the city set afire. As it burns, the Christians are doomed, and Simon perishes in the flames.

Nerto, *see* MISTRAL, FREDERIC.

Nerval, Gérard de, (born GERARD LABRUNIE) poet, dramatist, translator. Born Paris, May 22, 1808; died there January 25, 1855. His French translation of Goethe's *Faust* was used for Berlioz' *The Damnation of Faust.* Claude Arrieu's *Les deux rendezvous,*

Jean Françaix's *La main de gloire,* and Gounod's *La Reine de Saba* were operas derived from Nerval's poetic dramas.

Nessler, Victor, composer. Born Baldenheim, Alsace, January 28, 1841; died Strassburg, May 28, 1890. He planned to enter the church, but the success of his first operetta, *Fleurette,* in 1864 convinced him to become a composer for the stage. He held various musical posts, including that of choral director of the Leipzig Stadttheater, beginning in 1870, and conductor of the Caroltheater, also in Leipzig, after 1879. His most successful operas were: *Dornröschens Brautfahrt,* a fairy opera (1868); *Der Rattenfänger von Hameln* (1879); *Der wilde Jäger* (1881); *Der Trompeter von Säkkingen* (1884); *Otto der Schütz* (1886); *Die Rose von Strassburg* (1890).

Nessun dorma, *see* QUESTA NOTTE.

Neuendorff, Adolf, conductor. Born Hamburg, Germany, June 13, 1843; died New York City, December 4, 1897. He came to the United States in his twelfth year. After studying music with J. Weinlich and G. Matzka he became in 1867 the conductor of the New York Stadttheater. There, on April 15, 1871, he led a visiting European troupe in the American première of *Lohengrin.* In 1872 he helped direct a season of Italian opera at the Academy of Music; he returned there in 1875 for a season of German operas. In 1876 he attended the first Bayreuth Festival as a correspondent for the New York *Staats-Zeitung.* A year later he led a Wagner festival in New York which featured the American première of *Die Walküre* on April 3, 1887. He wrote four comic operas: *The Rat Charmer of Hamelin* (1880); *Don Quixote* (1882); *Prince Woodruff* (1887); *The Minstrel* (1892).

Neues vom Tage (News of the Day), opera by Paul Hindemith. Libretto by Marcellus Schiffer. Première: Kroll

Theater, Berlin, June 8, 1929. The racy, comic, often farcical text concerns a marital dispute between Laura and Edward. They decide on divorce and to prove adultery they engage the handsome Mr. Hermann to be corespondent. Mr. Hermann falls in love with Laura, arousing Edward's jealousy. Hermann and Laura become so notorious that they are booked as an attraction in a variety theater. In the end Laura and Edward discover they really love each other, and decide to live together again. But newspaper subscribers complain that, since Laura and Edward are news of the day, they thus belong to the news-reading public and must go ahead with their divorce.

Neumann, Angelo, tenor and impresario. Born Vienna, August 18, 1838; died Prague, December 20, 1910. He abandoned a mercantile career to become an opera singer. After studying with Stilke-Sessi he made his opera debut in 1859, after which he sang in various European theaters. From 1862 to 1876 he was principal tenor of the Vienna Opera. In 1876 he became the director of the Leipzig Opera, a post held for six years. He then organized a traveling company which presented the Wagner music dramas throughout Europe, a significant force in popularizing Wagner. From 1882 to 1885 he was director of the Bremen Opera, and from 1885 until his death director of the Landestheater in Prague. He wrote a volume of reminiscences of Wagner, *Erinnerungen an R. Wagner* (1907).

Nevada, Emma (born WIXOM), soprano. Born Alpha, California, February 7, 1859; died Liverpool, England, June 20, 1940. She visited Europe in her fifteenth year and decided to stay there for concentrated vocal training. She studied with Mathilde Marchesi in Vienna from 1877 on, and on May 17, 1880, made her debut at Her Majesty's Theatre in London as Amina. It was on this occasion that she assumed her stage name of "Nevada." She next went to Italy, where she sang successfully in several cities. Verdi heard her, and arranged for her to appear at La Scala. On May 27, 1887, she appeared for the first time at the Opéra-Comique. In 1884 she returned to America and appeared with Mapleson's company at the Academy of Music, singing on alternate evenings with Patti. A year later she married a London physician, Raymond Palmer. In 1885 and 1889 she appeared in opera festivals in Chicago. After retiring from the stage she devoted herself to teaching. Alexander Mackenzie wrote his opera *Rose of Sharon* for her, and she created the title role at Covent Garden in 1884. Her daughter, Mignon Nevada (1886–) made successful appearances as a soprano at Covent Garden, La Scala, and other major European opera houses.

Nevers, Count de, (1) *see* ADOLAR.

(2) Catholic nobleman (baritone) in Meyerbeer's *Les Huguenots.*

Newman, Ernest, musicologist. Born Liverpool, England, November 30, 1868. Music was for a long time an avocation while Newman prepared to enter Indian Civil Service, and then decided to engage in business activities. In 1903 he entered music professionally when he became professor of music at the Midland Institute in Birmingham. Two years later he became the music critic of the *Manchester Guardian,* and from 1906 to 1919 of the *Daily Post.* He made his home in London in 1919, and from 1920 was the principal music critic of the *Sunday Times.* In 1924–1925 he was a guest critic for the *New York Post.* He has written many books on opera, the most important being his four-volume *The Life of Richard Wagner* (1933–1946), the fruit of a lifetime of research, and now the definitive, most completely documented Wagner biography. His other books on operatic subjects are: *Gluck and the*

Opera (1895); *A Study of Wagner* (1899); *Wagner* (1904); *Richard Strauss* (1908); *Stories of the Great Operas* (1929); *Fact and Fiction of Wagner* (1931); *Opera Nights,* published in the United States as *More Stories of Famous Operas* (1943); *The Wagner Operas,* published in England as *Wagner Nights* (1949); *Seventeen Famous Operas,* published in England as *More Opera Nights* (1955). Newman has translated into English the librettos of the Wagner music dramas, and has edited Berlioz' *Memoirs.*

New York City Opera Company, a company organized in 1943 as a unit of the City Center of Music and Drama to provide good opera at popular prices. The original company was made up of fifteen singers, two conductors, and one stage director; scenery was borrowed from the St. Louis Municipal Opera. The opening performance was *Tosca* on February 21, 1944; the first music director was Lászlo Halász. Three operas were given the first season: *Carmen, Martha,* and *Tosca.* The first production which was entirely designed and mounted by the new company was *Ariadne auf Naxos,* in the fall of 1946, a presentation that received wide critical acclaim. The company now gives two six-week seasons, one in the spring, the other in the fall; each season embraces about forty performances of twelve or fifteen operas. Besides performing in New York, the company has given annual seasons in Chicago and Detroit, and has toured the Midwest and the Atlantic seaboard. It has given the following world premières: Aaron Copland's *The Tender Land;* William Grant Still's *The Troubled Island;* David Tamkin's *The Dybbuk.* American premières: Béla Bartók's *Bluebeard's Castle;* Gottfried von Einem's *The Trial.* Revivals and novelties of unusual interest have included: Alban Berg's *Wozzeck;* Jerome Kern's *Show Boat* (presented for the first time by an opera company in its regular repertory); Massenet's *Werther;* Menotti's *Amahl and the Night Visitors* (first stage presentation); Prokofiev's *The Love for Three Oranges* (first American performance since its première); Rossini's *La cenerentola* (first performances in New York since 1831); Richard Strauss's *Ariadne auf Naxos* (first professional performance in America); Nicolai's *The Merry Wives of Windsor;* Tchaikovsky's *Eugene Onegin.*

Many outstanding singers have achieved their first recognition at the New York City Opera and gone on to sing in other opera houses. Among those who first sang here and then appeared at the Metropolitan Opera: Mario Berini, Regina Resnik, Gertrude Ribla, Norman Scott, Giuseppe Valdengo, Ramón Vinay. In May, 1946, Camilla Williams became the first Negro to appear with a major opera company in America—she was heard as Cio-Cio-San.

Lászlo Halász was summarily dismissed from his post on December 21, 1951, on the grounds that his conduct was "a threat to the prosperity and advancement of the City Center." This action of the Center's administration led to a law suit in which Mr. Halasz won a substantial settlement. The post of artistic and musical director was filled by Josef Rosenstock, whose initial season was made memorable by the first repertory production in America of *Wozzeck* in the spring of 1952. In July, 1953, the organization was the recipient of a grant from the Rockefeller Foundation to be used for new productions within a period ending July 1, 1956.

Nibelhelm, the abode of the Nibelungs in Wagner's *Das Rheingold.*

Nibelung Ring, The, *see* RING DES NIBELUNGEN, DER.

Nibelung Saga, The (Das Nibelungenlied), an epic of ancient Teutonic times, the source of Wagner's *Der Ring des Nibelungen*. Several other operas were derived from this saga, among them Heinrich Dorn's *Die Nibelungen* (1854); Felix Draeske's *Gudrun* (1884); and Ernest Reyer's *Sigurd* (1884).

Nicias, a voluptuary (tenor) in Massenet's *Thaïs*.

Nicklausse, Hoffmann's friend (mezzo-soprano) in Offenbach's *The Tales of Hoffmann*.

Nicolai, Carl Otto, composer and conductor. Born Königsberg, Germany, June 9, 1810; died Berlin, May 11, 1849. The composer of the delightful comic opera, *The Merry Wives of Windsor*, studied music with Carl Friedrich Zelter and Bernhard Klein. Settling in Berlin, between 1830 and 1833 he completed several works for orchestra and for chorus. In 1834 he went to Italy, where he was employed as organist in the Prussian Embassy in Rome, and where he became vitally interested in opera. His first opera was a failure. His second, *Il Templario,* was acclaimed at its première in Turin in 1840; after that it was heard in leading European cities and in New York. Meanwhile, in 1837, he was appointed principal conductor of the Kärntnerthortheater in Vienna. In 1841 he became principal conductor of the Vienna Royal Opera. A year later he helped found the Vienna Philharmonic. After six years of distinguished services as conductor in Vienna he was appointed kapellmeister of the Berlin Opera; two years after this, on March 9, 1849, his masterwork, *The Merry Wives of Windsor,* was introduced there. Though seriously ill, he conducted the first four performances. He died of apoplexy the day he was elected a member of the Berlin Academy. His operas: *Rosmonda d'Inghilterra* (1839); *Il Templario* (1840); *Odoardo e Gildippe* (1841); *Die Heimkehr des Verbannten* (1844); *Die lustigen*

Weiber von Windsor (*The Merry Wives of Windsor*) (1849).

Nicolette, a princess hidden in an orange (mezzo-soprano) in Prokofiev's *The Love for Three Oranges.*

Nicolini (born NICOLA GRIMALDI), male soprano. Born Naples, April 1673; died there January 1, 1732. A castrato, his first appearance in opera took place in Naples when he was only twelve. By 1690 he was acknowledged the best soprano in all Naples, and he was engaged for the royal chapel. While holding his post, he appeared with outstanding success in opera houses in Naples, Rome, and Venice; in the last-named city he was decorated with the Order of St. Mark. In 1708 he went to England, making his debut in London on December 14 in Alessandro Scarlatti's *Pirro e Demetrio*. He sang in England for some years, scoring particularly in Handel's *Rinaldo* in 1711 and Handel's *Amadigi* in 1715. In 1717 he returned to Italy, where he again appeared extensively, particularly in Venice. In 1731 he returned to his native city to appear in Pergolesi's first opera, *Salustia,* but he fell ill during rehearsals and died before the première.

Nielsen, Alice, soprano. Born Nashville, Tennessee, June 7, 1876; died New York City, March 8, 1943. Her debut on the musical stage took place with the Boston Opera Company in *The Mikado.* Success came in operetta, particularly in Victor Herbert's *The Fortune Teller* in 1898. Determined to make her way in opera, she went to Italy for further study and then made her opera debut at the Bellini Theater in Naples on December 6, 1903, as Marguerite. In the spring of 1904 she appeared for the first time at Covent Garden. Her role was Zerlina. Back in the United States, she was heard in New York City in 1905 in *Don Pasquale.* In 1908 she appeared with the Henry Russell San Carlo Opera Company in New Orleans, and from 1909 to 1913 with the Boston Opera. Her first appearance at the Metropolitan Opera took place on November 19, 1909, in *La Bohème*. After abandoning opera, she returned to operetta and was starred in Friml's *Kitty Darlin'* in 1917.

Niemann, Albert, dramatic tenor. Born Erxleben, Germany, January 15, 1831; died Berlin, January 13, 1917. In 1849 he sang in the chorus and appeared in minor parts at the Dessau Opera. After a protracted period of study with Friedrich Schneider, Nusch, and Gilbert Duprez, he appeared in smaller German opera theaters. In 1860 he was engaged as principal tenor of the Hanover Opera. Six years later he joined the Berlin Opera, where he appeared with outstanding success in tenor roles up to the time of his retirement in 1889. Wagner selected him to create for Paris the role of Tannhäuser in 1861, and that of Siegmund for Bayreuth in 1876. On November 10, 1886, Niemann made his American debut at the Metropolitan Opera as Siegmund. While appearing at the Metropolitan he created for America the roles of Tristan (1887) and Siegfried in *Die Götterdämmerung* (1888). His correspondence with Wagner was published in 1924.

Nie werd' ich deine Huld verkennen, concluding chorus of Mozart's *The Abduction from the Seraglio*.

Nikisch, Arthur, conductor. Born Lébényi Szent Miklos, Hungary, October 12, 1855; died Leipzig, Germany, January 23, 1922. A child prodigy, he entered the Vienna Conservatory when he was eleven. Two years after leaving the Conservatory he became a conductor of the Leipzig Opera, making his bow there on February 11, 1878. He rose to post of principal conductor in 1882, and for seven years his distinguished performances attracted attention throughout Germany. From 1889 to 1893 Nikisch was principal con-

ductor of the Boston Symphony. In 1893 he assumed the post of principal conductor of the Budapest Opera, where he remained two years. After 1895, though he made intermittent guest appearances in various European opera houses, he specialized in conducting symphonic music, particularly in his posts as principal conductor of the Leipzig Gewandhaus and the Berlin Philharmonic orchestra. From 1902 to 1907 he was also director of the Leipzig Conservatory. In 1912 he toured the United States with the London Symphony Orchestra.

Nikolaidi, Elena, contralto. Born Smyrna, Turkey, June 13, 1909. She entered the Athens Conservatory when she was fifteen and during her last year there made her debut as soloist with the Athens State Orchestra under Dimitri Mitropoulos. She became a member of the Athens Lyric Theater, where she was heard in *Carmen, Samson et Dalila,* and other operas. In Vienna, where she went for further study, she won an international vocal competition, entitling her to a recital at the Konzerthaus. Engaged by Bruno Walter for the Vienna State Opera, she made her debut there on the opening night of the 1936 season, singing Princess Eboli in Don Carlos. She became so successful in Vienna that in 1947 she received the honorary title of Kammersängerin. Subsequently appearing throughout Europe, she then made her American debut in a New York recital on January 20, 1949. On the opening night of the San Francisco Opera season in 1950 she appeared as Aïda. It was in the same role that she made her Metropolitan Opera debut on the opening night of the season in 1951. She has since appeared in leading contralto roles there. She has also sung mezzo-soprano roles with outstanding success. A special act of Congress conferred on her the right to remain a permanent resident of the United States.

Nilakantha, Brahman priest (bass), father of Lakmé, in Delibes's *Lakmé.*

Nilsson, Christine, soprano. Born Wexiö, Sweden, August 20, 1843; died Stockholm, November 22, 1921. She studied singing with Franz Berwald and Baroness Leuhusen in Sweden, and with Pierre François Wartel, and Enrico delle Sedie in Paris. Her debut took place at the Théâtre Lyrique in Paris on October 27, 1864, her role being Violetta. She remained a principal soprano there for three years, afterward appearing with the Paris Opéra. Here, her performance of Marguerite in a revival of *Faust,* ten years after its première, was a determining factor in establishing the success of that opera. Between 1870 and 1872 she toured the United States. Between 1872 and 1877 she appeared annually at the Drury Lane Theatre, creating for England the principal soprano roles in *Lohengrin, Mefistofele* and Balfe's *Talismano.* On October 27, 1883, she appeared at the Metropolitan Opera in the performance of *Faust* which opened that opera house. She went into retirement in 1891.

Ninette, a princess hidden in an orange (soprano) in Prokofiev's *The Love for Three Oranges.*

Niun mi tema, Otello's death scene in Act IV of Verdi's *Otello.*

Nobili acciar, nobili e santi, *see* GLAIVES PIEUX, SAINTES ÉPÉES.

Nobles Seigneurs, salut! (Lieti Signori, salute!), the Page's song in Act I of Meyerbeer's *Les Huguenots.*

Noëmi, Cinderella's stepsister (soprano) in Massenet's *Cendrillon.*

Nolan, Philip, principal character (tenor) in Damrosch's *The Man Without a Country.*

Non ho colpa, Idamante's aria in Act I, Scene 1, of Mozart's *Idomeneo.*

Non imprecare, umiliati, trio of Padre Guardiano, Leonora, and Don Alvaro,

the closing scene of Verdi's *La forza del destino.*

Non la sospiri la nostra casetta, duet of Tosca and Cavaradossi in Act I of Puccini's *Tosca.*

Non mi dir, Donna Anna's aria in Act II, Scene 4, of Mozart's *Don Giovanni.*

Non mi resta che il pianto, Suzel's aria in Act III of Mascagni's *L'amico Fritz.*

Non più andrai, Figaro's taunt to Cherubino in Act I of Mozart's *The Marriage of Figaro.* Mozart has Don Giovanni's private band play a portion of this aria as Don Giovanni eats supper in the climactic scene of *Don Giovanni.*

Non siate ritrosi, Guglielmo's aria in Act I, Scene 3, of Mozart's *Così fan tutte.*

Non so più cosa son, Cherubino's aria in Act I of Mozart's *The Marriage of Figaro.*

No! pazzo son! guardate, Des Grieux's plea to the ship captain in Act III of Puccini's *Manon Lescaut.*

Nordica, Lillian (born NORTON), dramatic soprano. Born Farmington, Maine, May 12, 1857; died Batavia, Java, May 10, 1914. One of the most celebrated Wagnerian sopranos of her generation, she began her music study with John O'Neal at the New England Conservatory in Boston, and continued in Milan with Antonio Sangiovanni. On March 8, 1879, she made her debut at the Manzoni Theater in Milan in *Don Giovanni,* using the stage name of Nordica. Numerous performances followed in Italy, Russia, and Germany, leading to her successful debut at the Paris Opéra on April 22, 1882, as Marguerite. Her American debut took place at the Academy of Music in New York on November 23, 1883, once again as Marguerite. For the next four seasons she appeared in New York and other American cities under Colonel Mapleson's direction. After making her Covent Garden debut on March 27, 1887, she appeared in London each season until 1893, either at Covent Garden or Drury Lane. Back in America, she made her bow at the Metropolitan Opera on March 27, 1890, in *Il trovatore.* In 1894 she appeared in Bayreuth as Elsa. Her success was so striking that then and there she decided to specialize in Wagnerian roles. After a period of study with Julius Kriese in Bayreuth, she appeared for the first time as Isolde in 1895, at the Metropolitan Opera. Henceforth she appeared in all the principal soprano roles of the Wagnerian music dramas, acclaimed for her stirring interpretations of the three Brünnhilde roles, Isolde, and Kundry. She was an actress of outstanding magnetism who brought nobility to each of her characterizations; and her voice combined beauty with dramatic force. Her last appearance at the Metropolitan took place on December 8, 1909, in the role of Isolde. For a while she appeared with other opera companies in America. A nervous breakdown compelled her to leave the stage, and she concentrated for a while on recitals. In the fall of 1913 she embarked on a tour of the world, planning it as her farewell to a professional career. Her ship was wrecked in the Malay archipelago. Taken to Java, Nordica died there.

At one period in her life Nordica was fired with the ambition to create an American Bayreuth at Harmon-on-the-Hudson, New York. The project never grew beyond the planning stage. In 1927 the Nordica Memorial Association was organized to buy and renovate her birthplace in Maine and make it a museum for mementos of her career.

Noréna, Eidé (born KAJA HANSEN EIDE), soprano. Born Oslo, Norway, April 26, 1884. She began her career as concert singer after a period of music study in Oslo. Her opera debut took place with the Oslo Opera, in Gluck's *Orfeo ed Euridice.* She then returned to more study of singing in Weimar, London, Italy, and Paris. A successful audition with Arturo Toscanini brought her a contract for La Scala, where she made her debut in *Rigoletto;* it was for this occasion that she permanently assumed the stage name of Eidé Noréna. After appearances at La Scala, Covent Garden, and the Paris Opéra, she came to the United States and for six years was a member of the Chicago Opera. On February 9, 1933, she made her debut at the Metropolitan Opera in *La Bohème.* She remained at the Metropolitan through the 1937–1938 season, after which she returned to Europe to sing at the Paris Opéra and other leading opera houses.

Norina, a young widow (soprano) in Donizetti's *Don Pasquale.*

Norma, opera in four (originally two) acts by Vincenzo Bellini. Libretto by Felice Romani, based on a tragedy by L. A. Soumet. Première: La Scala, December 26, 1831. American première: Park Theater, New York, February 25, 1841.

Characters: Norma, high priestess of the Druid Temple of Esus (soprano); Oroveso, the Archdruid, her father (bass); Clotilda, her confidante (soprano); Adalgisa, virgin of the Temple of Esus (mezzo-soprano); Pollione, Roman proconsul (tenor); Flavio, his centurion (tenor); priests; warriors; virgins of the temple; the two children of Norma and Pollione. The setting is Gaul during the Roman occupation, about 50 B.C.

Act I. Night in the Druid's sacred forest. Norma performs a sacred rite as the Gallic soldiers and Druid priests implore the gods for aid in destroying the Roman oppressors ("Dell' aura tua profetica"). When the Druids depart, Pollione reveals to Flavio that Norma has violated her holy vow by bearing him two sons. Pollione further confesses that he is in love with Adalgisa ("Meco all' altar di Venere"). At the sudden return of the Druids, Pollione and Flavio conceal themselves. Norma appeals to her people not to revolt, for the time is not yet ripe. She then prays for peace ("Casta diva") and laments that the hatred of her people for the Romans must inevitably mean hatred for her beloved Pollione ("Ah! bello a me ritorna"). When she departs, followed by the Druids, Adalgisa appears. Tormented by her love for Pollione, she begs the gods to rescue her ("Deh! proteggimi, o Dio!"). Pollione approaches her. At first she resists him, but he is persuasive ("Va, crudele") and she cannot resist him. She rushes into his arms, and they decide to escape to Rome.

Act II. Norma's dwelling. Norma reveals to Clotilda that Pollione is about to return to Rome, a development that causes her no little anxiety. Adalgisa, unaware of Norma's love for Pollione, comes to confess that she has fallen in love and must therefore desert the Temple. Norma is sympathetic ("Ah! si, fa core e abbraccia"). When Pollione appears, Norma and Adalgisa realize that they love the same man. Norma curses the proconsul for his treachery, while Adalgisa begs for some explanation ("Oh! di qual sei tu"). Pollione entreats Norma to direct her anger solely against him. Norma is summoned to the temple for the performance of her rites, and Pollione rushes away.

Act III. Norma's dwelling. Tortured by Pollione's infidelity, Norma decides to kill him and their children and then to destroy herself on a funeral pyre. She is, however, incapable of summoning the strength to kill the children. She begs Adalgisa to take care of them after her own death ("Deh! con te li prendi"). Adalgisa entreats Norma not to commit suicide, but to be influenced by her maternal instincts and love ("Mira o Norma"). Further, she insists that she will renounce Pollione

and urge him to return to the mother of his children.

Act IV. In the woods, near the Temple. The Gallic warriors and Druids have gathered to plan military action against the Romans. Clotilde informs Norma that Pollione has refused Adalgisa's request that he return to Norma. Aroused, Norma calls on her people to wage relentless war on the Romans, and the people respond with cries of vengeance ("Guerra! le Galliche selve!"). Pollione is brought to Norma. She snatches a dagger and advances toward him, claiming that now he is in her power ("In mia man alfin tu sei"). But she is unable to kill the man she loves. She promises him his freedom if he renounces Adalgisa. He refuses to do so. Norma now turns to her people and confesses that she had desecrated her vows and must be sacrificed. To Pollione she repeats her love vows, and her confidence that they will be reunited in death ("Qual cor tradisti"). Profoundly moved, Pollione now asks to die with Norma. Norma takes his hand and leads him into the funeral pyre.

Of all of Bellini's operas, *Norma* is the one that has remained a favorite. It was also the composer's favorite. He once said: "If I were shipwrecked at sea, I would leave all the rest of my operas and try to save *Norma*." While the work has scenes of pageantry and moments of ringing grandeur, it rises to its greatest heights in its lyricism, as in the justly celebrated "Casta diva" aria of Norma, and the poignant opening scene of the third act. The simplicity and directness of Bellini's art led Wagner to express enthusiasm for *Norma*, which reminded him "of the dignity of the Greek tragedy. . . . The music is noble and great, simple and grandiose in style." *Norma* was a failure at its première, but La Scala kept on performing the work until it became a triumphant success.

Normanno, Lord Ashton's follower (tenor) in Donizetti's *Lucia di Lammermoor.*

Nothung! Nothung! Neidliches Schwert! Siegfried's narrative while forging his sword in Act I of Wagner's *Siegfried.*

Notre Dame de Paris, novel by Victor Hugo, the source of operas by Alexander Dargomizhsky, William Henry Fry, Felipe Pedrell, Franz Schmidt, and Arthur Goring Thomas, among others. *See* ESMERALDA; HUGO, VICTOR.

Nouguès, Jean, composer. Born Bordeaux, France, April 25, 1875; died Paris, August 28, 1932. He wrote his first opera, *Le Roi du Papagey,* when he was only sixteen. After a period of music study in Paris he completed his first mature opera, *Yannha,* performed in Bordeaux in 1897. In 1904 he produced *Thamyris.* The greatest success of his career was *Quo Vadis,* his finest work, introduced in Nice on February 9, 1909, and soon after acclaimed in Paris, London, Milan, Philadelphia, and New York. His other operas: *La mort de Tintagiles* (1905); *Chiquito* (1909); *L'auberge rouge* (1910); *La Vendetta* (1911); *L'Aigle* (1912); *L'Eclaircie* (1914).

Nourrit, Adolphe, tenor. Born Paris, March 3, 1802; died Naples, March 8, 1839. The creator of principal tenor roles in some of the outstanding French operas of the early nineteenth century, he was the son of Louis Nourrit, a leading tenor of the Paris Opéra. After some study with Manuel García, Adolphe made his opera debut at the Paris Opéra on September 10, 1821, in Gluck's *Iphigénie en Tauride.* In 1826 he succeeded his father as principal tenor of the Opéra. Within the next decade he appeared in numerous world premières, many of these operas being written for him. Among the roles he created were the leading tenor parts in Auber's *La muette de Portici,* Halévy's *La Juive,* Meyerbeer's *Robert le Diable* and *Les Huguenots,* and Rossini's *Moïse* (revised French version) and *William Tell.* It was on Nourrit's advice that Halévy interpolated into *La Juive* Eleazar's celebrated aria, with Nourrit himself writing the text; Nourrit also suggested to Meyerbeer the abrupt ending to the popular duet in *Les Huguenots.* When, in 1836, the Paris Opéra engaged Gilbert Duprez as a principal tenor, Nourrit regarded this as a personal affront and tendered his resignation. He appeared at the Opéra for the last time on April 1, 1837. He then went to Italy to study the Italian style and traditions with Donizetti, and to make appearances at La Scala and in Naples. The immense popularity of Duprez in Paris embittered him, a feeling that was heightened by his feeling that he did not receive in Italy the acclaim he deserved. His depression led to his suicide in Naples. In the years before he left France Nourrit taught at the Paris Conservatory. He also wrote scenarios for ballets performed by Fanny Elssler and Maria Taglioni.

Nous vivrons à Paris, duet of Manon and Des Grieux concluding Act I of Massenet's *Manon.*

Novák, Vítěszlav, composer. Born Kamenitz, Bohemia, December 5, 1870; died Skuteč, Slovakia, July 18, 1949. He combined the study of law with that of music, the latter at the Prague Conservatory. His teacher in composition, Antonin Dvořák, induced him to specialize in music. After 1900 Novák began writing music with a strongly Bohemian character. In this style he completed his first opera, *The Sprite of the Castle* (1914). Later operas: *Karlstein* (1916); *The Lantern* (1922); *John the Fiddler* (1925). From 1909 to 1925 he was professor of composition at the Prague Conservatory. In 1918 he became professor at the Master School there, and in 1919, director. In 1946 he received the honorary title of National Artist from the Republic of Czechoslovakia.

Novotna, Jarmila, soprano. Born Prague, Bohemia, September 23, 1908. She was fifteen when she sang for Emmy Destinn, who encouraged her to study music seriously. After a year of training, Novotna made her opera debut at the National Theater in Prague on June 27, 1926, in *La traviata*. She then went to Milan for additional study and for appearances in various Italian opera houses. In 1928 she was engaged by the Berlin Opera. As a member of this company, she made many guest appearances in Vienna and Paris and at the Salzburg Festival. In 1939 she made her American debut with the San Francisco Opera in *Madama Butterfly*. Her Metropolitan Opera debut took place on January 5, 1940, in *La Bohème*. She remained at the Metropolitan for seven seasons, and for several seasons after that made intermittent appearances there. She has also made occasional appearances in non-singing dramatic roles on the stage, in motion pictures, and in television.

Nozze di Figaro, Le, *see* MARRIAGE OF FIGARO, THE.

nozze di Teti e Peleo (The Wedding of Thetis and Peleus), opera by Cavalli. Libretto by Orazio Persiani. Première: Teatro San Cassiano, Venice, 1639. This is the first work in operatic history specifically called an opera. The composer's designation was "opera scenica." All preceding musico-dramatic

works had been called "dramma per musica."

Nuit d'hyménée, love duet of Roméo and Juliette in Act IV of Gounod's *Roméo et Juliette*.

Nuitter, Charles Louis (born TRUINET), writer on music, librettist, and translator. Born Paris, April 24, 1828; died there February 24, 1899. A prolific writer and a devotee of the theater, he wrote librettos for nine of Offenbach's operas comique. For performances at the Théâtre-Lyrique and the Paris Opéra he translated into French many operas, including *Aïda, Lohengrin, The Magic Flute, Oberon, Rienzi,* and *Tannhäuser.* He also revised the libretto of Verdi's *Macbeth* for French presentation. In 1865 he was appointed archivist of the Opéra. He wrote several books of operatic interest, including *Le nouvel opéra* (1875), *L'histoire et description au nouvel opéra* (1884), and, with Ernest Thoinon, *Les origines de l'opéra français* (1886).

Nume, custode e vindice, Ramfis' scene with chorus in Act I, Scene 2, of Verdi's *Aïda.*

Nun eilt herbei, Mrs. Ford's aria in Act II of Nicolai's *The Merry Wives of Windsor.*

Nun sei bedankt, mein lieber Schwan, Lohengrin's farewell to his swan in Act I of Wagner's *Lohengrin.*

Nureddin, young man (tenor) in love with Margiana in Cornelius' *The Barber of Bagdad.*

Nuri, a village girl (soprano) in D'Albert's *Tiefland.*

O

O amore, o bella luce, Fritz's aria in Act III of Mascagni's *L'amico Fritz.*

O beau pays de la Touraine (O vago suol della Turrena), Marguerite de Valois's aria in Act II of Meyerbeer's *Les Huguenots.*

Obéissons, quand leur voix appelle, Manon's Gavotte in Act III of Massenet's *Manon.*

O belle enfant! Je t'aime, Faust's love song in Act II of Gounod's *Faust.*

Oberon, opera in three acts by Carl Maria von Weber. Libretto by James Robinson Planché, based on an old French romance, *Huon de Bordeaux.* Première: Covent Garden, April 12, 1826. American première: Park Theater, New York, October 9, 1828.

Characters: Oberon, king of the fairies (tenor); Titania, his queen (speaking part); Puck, his attendant (contralto); Harun-al-Rashid, Caliph of Bagdad (bass); Rezia, his daughter (soprano); Fatima, her attendant (mezzo-soprano); Sir Huon de Bordeaux (tenor); Sherasmin, his squire (baritone); Babekan, a Persian prince (baritone); Mesrour, chief of the harem guards (acting part); Almanzor, Emir of Tunis (baritone); Charlemagne (bass); Droll (contralto); Abdallah, a corsair (baritone); Roschana, Almanzor's wife (contralto); elves, nymphs, sylphs, mermaids, spirits; ladies, gentlemen, servants; Moors and pirates. The action takes place in France, Bagdad, and Tunis at the beginning of the ninth century. The famous overture opens with a horn call and has for its core Rezia's aria, "Ocean, thou mighty monster."

Act I, Scene 1. Oberon's palace. After a quarrel, Queen Titania vows never to return to her husband until he finds two lovers who will remain true in spite of every obstacle. Puck suggests a candidate to Oberon: Sir Huon de Bordeaux, knight of Charlemagne's court, who is in disgrace for having killed Charlemagne's son. For his penance, Huon is sent to Bagdad to kill the man who sits at the Caliph's right hand, and then claim the Caliph's daughter, Rezia, as bride. Oberon conjures a vision of Rezia for Huon; at the same time he gives Rezia a vision of

Huon. The two fall in love. To aid Huon in his quest, Oberon provides him with a magic horn. Huon is then transported to Bagdad.

Scene 2. The Caliph's harem. Rezia refuses to consider Prince Babekan as a husband, since she is in love with Huon. Fatima announces the arrival of Huon, who is ready to save Rezia from the Prince.

Act II, Scene 1. The Caliph's palace. Prince Babekan is sitting at the Caliph's side. Huon forces his way into the palace, finds Rezia, and takes her in his arms. Prince Babekan opposes him and is killed. Sounding his magic horn, Huon is able to escape.

Scene 2. The palace garden. After Sherasmin and Fatima vow to love each other, Oberon comes to help Huon and Rezia escape from Bagdad. He warns them against being unfaithful, and brings them to the harbor from which the lovers can proceed to Greece ("Over the dark blue waters").

Scene 3. A cave on a desolate island. Oberon must now test the fidelity of his lovers. Puck orders their ship wrecked. Huon carries Rezia ashore. While Huon is searching for help, Rezia voices her awe at the might of the ocean ("Ocean, thou mighty monster"). Pirates capture Rezia and sell her to the Emir of Tunisia. Since Huon has lost his magic horn, he is unable to save her.

Act III, Scene 1. Garden of the Emir's palace. Fatima, now a slave of the Emir, is joined by her beloved, Sherasmin. They inform Huon that Rezia is also a slave and advise him to assume the disguise of a gardener in order to save her.

Scene 2. A hall in the palace. Rezia is grief-stricken at her fate ("Grieve my heart"). The Emir forces his love on Rezia, but she is saved by the sudden appearance of Huon.

Scene 3. The Emir's palace. Huon and Rezia are to be burned alive. But

Sherasmin has found Huon's magic horn. When Huon blows on it, his enemies become spellbound and motionless. Oberon and Titania, now reconciled, appear and save the lovers, blessing them for their fidelity. Huon and Rezia are then transported back to the court of Charlemagne, where the ruler forgives Huon.

While *Oberon* is a cornerstone on which German romantic opera rests, it is a work more often discussed than performed. The sad truth is that it is incapable of sustaining interest, however well it is performed. It has magnificent pages: the remarkable overture, which invoked for romanticism a new fairy world; the vocal scena, "Ocean, thou mighty monster," surely one of the finest pages of dramatic writing before Wagner; such delightful examples of romantic song as Huon's aria, "From boyhood trained." But for all these intermittent flights toward greatness, the score as a whole does little to lift the absurd and cumbersome play above a prevailing level of mediocrity. Yet the première of *Oberon* was the crowning triumph of Weber's career. He wrote it on a commission from Covent Garden (which explains why it has an English text). Weber himself conducted the première and received an unprecedented acclaim, "an honor which no composer had ever before obtained in England," he wrote. It was his last taste of success; less than a month later he was dead.

Oberthal, Count, ruler of Dordrecht (baritone) in Meyerbeer's *Le Prophète*.

Obigny, *see* D'OBIGNY.

O Carlo, ascolta, Rodrigo's farewell in Act IV, Scene 2, of Verdi's *Don Carlos*.

Ocean, thou mighty monster, Rezia's aria in Act II, Scene 3, of Weber's *Oberon*.

Ochs, Baron, the Marschallin's cousin (bass) in Richard Strauss's *Der Rosenkavalier*.

O ciel! dove vai tu? *See* OH CIEL, OU COUREZ-VOUS?

O cieli azzurri, the opening words of Aïda's aria "O patria mia" in Act III of Verdi's *Aïda*.

O Colombina, Harlequin's serenade in Act II of Leoncavallo's *Pagliacci*.

Octavian, a young gentleman (mezzosoprano), in Richard Strauss's *Der Rosenkavalier*.

O del mio dolce ardor, Paris' aria in Act I of Gluck's *Paride ed Elena*.

Ode to the Evening Star, *see* O DU MEIN HOLDER ABENDSTERN.

O Dieu, Dieu de nos pères, the Passover scene of Eléazar, Rachel, and chorus in Act II of Halévy's *La Juive*.

O! di qual sei tu, trio of Norma, Pollione, and Adalgisa in Act II of Bellini's *Norma*.

O du mein holder Abendstern, Wolfram's ode to the evening star in Act III of Wagner's *Tannhäuser*.

Odyssey, The, *see* HOMER.

Oedipus, the subject of two tragedies by Sophocles. Oedipus, king of Thebes, discovers that Laius, whom he has murdered, was actually his own father; also that he, Oedipus, is married to his own mother, Jocasta. In his horror, Oedipus plucks out his eyes, while Jocasta hangs herself. The Oedipus story has been made into operas by Georges Enesco and Ruggiero Leoncavallo, among others. Igor Stravinsky's oratorio *Oedipus Rex* has sometimes been performed as an opera.

See SOPHOCLES.

Offenbach, Jacques (born EBERST), composer. Born Cologne, Germany, June 20, 1819; died Paris, October 4, 1880. A master of French comic opera (opéra bouffe), he also produced an outstanding serious opera, *The Tales of Hoffmann*. The son of a synagogue cantor, his musical talent led to his being enrolled in the Paris Conservatory—the rule forbidding admission to foreigners was waived for his benefit. Offenbach was unhappy with the discıpline at the Conservatory and remained there only a year. He continued to study privately: cello with Pierre Norblin, composition with Jacques Halévy. His studies ended, he began playing in the orchestra of the Opéra-Comique. In 1850 he was appointed musical director of the Comédie Française, continuing in this capacity for five years. During this period he began writing comic operas. Unable to get them performed, he decided to open a theater of his own. This was the Bouffes Parisiens, which opened on July 5, 1855 with his musical satire, *Les deux aveugles.* Offenbach wrote for his theater twenty-five musical satires, farces, and comic operas within a three-year period. He became an idol of Parisian theatergoers. In 1858 his greatest opéra bouffe, *Orpheus in the Underworld (Orphée aux enfers),* was introduced. The public did not at first respond favorably to this satire on the Olympian gods. But when the critic Jules Janin attacked it for blasphemy and profanation, curiosity was piqued and the opera began playing to crowded houses. After 1861 Offenbach gave up his own theater but continued writing outstandingly successful works for other managers, including *La belle Hélène* (1865), *La vie parisienne* (1866), and *La Périchole* (1868). In 1872 Offenbach opened a new theater, the Gaîté, which failed three years later, overwhelming Offenbach with debts. To satisfy his creditors he allocated to them all his income for the next few years; when this did not prove sufficient to pay his bills, he undertook an American tour in 1876, which was helpful. Back in Paris, he continued writing comic operas; but the old touch was no longer there, and none of them found favor. Offenbach now avoided the society of his one-time friends and admirers and lived in comparative seclusion. He began working on a serious opera for the first time in his life: *The Tales of Hoffmann.* He

lived to complete it, but not to witness its première at the Opéra-Comique on February 10, 1881.

Essentially a composer of music in the lighter vein, Offenbach had the gifts of verve, spontaneity, wit, satire, and a ready flow of lovable melodies. But his one serious opera gave proof that he might also have been one of France's foremost writers of grand opera.

O grandi occhi, Fedora's aria in Act I of Giordano's *Fedora.*

O grido di quest' anima, Enzo's duet with Barnaba in Act I of Ponchielli's *La Gioconda.*

Oh ciel, où courez-vous (O ciel! dove vai tu?), the duet of Raoul and Valentine in Act IV of Meyerbeer's *Les Huguenots.*

O holdes Bild, the love duet or "dove duo" of Nureddin and Margiana in Act II of Cornelius' *The Barber of Bagdad.*

Oh! qu'est-ce que c'est? . . . tes cheveux, Pelléas' apostrophe to Mélisande's hair in Act III, Scene 1, of Debussy's *Pelléas et Mélisande.*

O inferno, Amelia qui! Adorno's aria in Act II of Verdi's *Simon Boccanegra.*

oiseau bleu, L', *see* BLUE BIRD, THE.

oiseaux dans la charmille, Les, Olympia's Doll Song in Act I of Offenbach's *The Tales of Hoffmann.*

O Isis und Osiris, Sarastro's invocation in Act II, Scene 1, of Mozart's *The Magic Flute.*

Old Maid and the Thief, The, one-act comic opera by Gian-Carlo Menotti. Libretto by the composer. Première: NBC network, April 22, 1939 (radio version); Philadelphia Opera Company, February 11, 1949 (stage version). Menotti wrote this opera expressly for radio presentation (it had been commissioned by the National Broadcasting Company). Miss Todd, an old maid, welcomes a tramp into her home as a permanent lodger. She treats him royally, but in her efforts to satisfy his ever-increasing needs, she resorts to stealing. When neighbors suspect the tramp of being the thief, Miss Todd urges him to escape. But the tramp insists that since he is innocent he will remain, and that Miss Todd must pay for her crimes. Disillusioned, Miss Todd goes to the police with her sad story. While she is gone, the tramp escapes with Miss Todd's maid, carrying away everything portable. Thus (as the subtitle of the opera remarks), "a virtuous woman makes a thief of an honest man."

O légère hirondelle, Mireille's waltz in Act I of Gounod's *Mireille*.

Olga, Tatiana's sister (contralto) in Tchaikovsky's *Eugene Onegin*.

O Liberté ma mie, Jean's aria in Act I of Massenet's *Le jongleur de Notre Dame*.

Olivier, a poet (baritone), in love with the Countess, in Richard Strauss's *Capriccio*.

O Lola bianca, Turiddu's Siciliana in Mascagni's *Cavalleria rusticana*.

O luce di quest' anima, Linda's aria in Act I of Donizetti's *Linda di Chamounix*.

Olympia, a mechanical doll (soprano), one of Hoffmann's loves, in Offenbach's *The Tales of Hoffmann*.

O ma lyre immortelle, Sapho's aria in Act III of Gounod's *Sapho*.

Ombra mai fu, the aria in Handel's *Serse*, today known as Handel's "Largo."

Ombre légère (Ombra leggiera), Dinorah's Shadow Dance aria in Act II of Meyerbeer's *Dinorah*.

O Mimi, tu più non torni, Rodolfo's duet with Marcello in Act IV of Puccini's *La Bohème*.

O mio babbino caro, Lauretta's aria in Puccini's *Gianni Schicchi*.

O mio Fernando, Leonora's aria in Act III of Donizetti's *La favorita*.

O mon enfant, aria of Louise's father in Act I of Charpentier's *Louise*.

O monumento, Barnaba's soliloquy in Act I of Ponchielli's *La Gioconda*.

O namenlose Freude, duet of Florestan and Leonore in Act II, Scene 1, of Beethoven's *Fidelio*.

Onegin, *see* EUGENE ONEGIN.

Onégin, Sigrid (born HOFFMANN), contralto. Born Stockholm, Sweden, June 1, 1891; died Magliasco, Switzerland, June 16, 1943. After studying singing with Resz, E. R. Weiss, and Di Ranieri, she made her opera debut in Stuttgart in 1912 as Carmen. She was so successful that she was given twelve starring roles that season. In 1919 she was engaged by the Munich Opera, where she was acclaimed in the Wagner repertory. Her American opera debut took place at the Metropolitan Opera on November 22, 1922, in *Aïda*. She remained at the Metropolitan through the 1923–1924 season. From 1926 to 1933 she was principal contralto of the Berlin Opera. She returned to the United States in 1934 for several opera and concert appearances.

O'Neill, Eugene, dramatist. Born New York City, October 16, 1888; died Boston, Massachusetts, November 27, 1953. He won the Nobel Prize for literature in 1936, and the Pulitzer Prize for drama three different times. Three of his plays were made into operas. The most notable was Louis Gruenberg's *The Emperor Jones*. This play was also set by the Czech composer Miroslav Ponc. *Before Breakfast* was adapted as a monodrama by Erik Chisholm, and renamed *Dark Sonnet*. *The Moon of the Caribbees* was made into a one-act opera by Adriano Lualdi. *The Rope* was used for an opera by Louis Mennini.

On ne badine pas avec l'amour, *see* MUSSET, ALFRED DE.

O nuit, étends sur eux ton ombre, Méphistophélès' aria in Act III of Gounod's *Faust*.

O padre mio, Manfredo's aria in Act I of Montemezzi's *L'amore dei tre re*.

O Paradis (O Paradiso), Vasco da Gama's rapturous tribute to the island of Madagascar in Act IV of Meyerbeer's *L'Africaine*.

O patria mia, Aïda's aria recalling her homeland in Act III of Verdi *Aïda*.

Opera, a dramatic performance, with costumes, scenery, and action, wholly or mostly sung to an orchestral accompaniment. In some operas the music is interrupted by passages of spoken dialogue; in others it is continuous, consisting of recitatives, arias, duets, trios and other ensemble numbers, choruses and ballets; in still others the music continues without definite demarcation into numbers. Opera represents a collaboration of text and music in which storytelling is combined with music's power to arouse emotions and create mood and atmosphere. Together they can achieve an expression which neither can achieve alone. At times in the history of opera the play has served merely as an excuse for the music—a convenient hook on which the composer hung his vocal and instrumental numbers—but in general the evolution of opera shows a continuous attempt to realize the musico-dramatic ideal of an integrated, artistic unity of good theater and music.

ORIGINS OF OPERA. Though opera was formally born in 1597 with the performance of Jacopo Peri's *Dafne* in Florence, the new art form was actually the creation of several men, a group including the noblemen Giovanni Bardi and Jacopo Corsi, the poet Ottavio Rinuccini, and the composers Vincenzo Galilei, Jacopo Peri, and Emilio del Cavalieri. This group, known as the Camerata ("those who meet in a chamber"), aspired to restore the forms of Greek drama, including the dramatic use of music. A major problem was that the music of the Camerata's day was chiefly polyphonic. Giving equality to a group of voices, it was not suited for singing by a solo voice. Also, the complex texture of the music made clear articulation of the words difficult. The Camerata finally realized that (according to Giovanni Battista Doni, a seventeenth century Florentine) "means must be found in the attempt to bring music closer to that of classical times, and to bring out the chief melody prominently so that the poetry could be clearly understandable." Obviously, a new music had to be created. Going for guidance to the Greeks, the Camerata came upon a treatise by Aristoxenus that said song should be patterned after speech. In Plato they read: "Let music be first of all language and rhythm, and secondly tone, and not vice versa." Gradually, the members of the group evolved a single-voiced melody, the "stilo rappresentativo," or recitative. In 1590 Emilio del Cavalieri used the new style in a series of musical scenes or pastorals. Soon after, Galilei created settings of the *Lamentations of Jeremiah*. To Jacopo Peri went the assignment of using the new style in a dramatic presentation restoring the Greek drama. In 1597 he completed *Dafne* which he described as a "dramma per musica." As the first stage work to be set to music from beginning to end, *Dafne* is the first opera ever written. But since the music of *Dafne* has been lost, musical historians usually point to Peri's second "dramma per musica," *Euridice*, as the first opera. *Euridice* was introduced in Florence on February 9, 1600, to acclaim. The Camerata felt that its restoration of classic Greek drama was successful; it did not realize that it had evolved a new art form. Giulio Caccini followed Peri's lead in writing works in the new form. Their operas consisted of a continuous flow of recitatives: that is, the poetic lines of the text were sung with exaggerated inflections and accompanied by a small orchestra of lutes and harpsichord. Occasionally, brief choruses and ballets were introduced to provide variety. Though the music was often impressive in its declamation, the use of the recitative style throughout an opera inevitably resulted in monotony.

EARLY DEVELOPMENTS IN ITALY. After the Florentine originators, Claudio Monteverdi was the first significant composer of operas. His *Orfeo,* first heard in Mantua in 1607, is the earliest opera still occasionally performed. Monteverdi brought an expressiveness and dramatic impact to opera that it had not known before. His innovations are discussed in this volume in his biographical entry. Monteverdi's presence in Venice, where he settled in 1612, made that city the operatic center of Italy. The next notable figure of the Venetian school was Cavalli (Monteverdi's pupil), the first composer to use the term "opera" for one of his productions. About 1685 a new school of opera composers emerged in Italy, the Neapolitan school, whose influence predominated until 1750. Its founder and guiding spirit was Alessandro Scarlatti, and its most significant members were Francesco Durante, Leonardo Leo, Niccolò Jommelli, Baldassare Galuppi, and Nicola Piccinni. In the typical opera of this school the play was based on episodes from history and legend. The songs, or arias, became the major musical element, and the Neapolitans established the importance of the da capo aria. The virtuoso singer was glorified as composers outdid each other in writing decorative melodies for displays of vocal dexterity. Ensemble music was also emphasized, and the overture acquired increasing importance. In Naples, too, a new type of opera was cultivated: the opera buffa, or comic opera, opposed to the opera seria, or serious opera. Giovanni Battista Pergolesi was the first important composer of this new type.

THE GOLDEN AGE OF ITALIAN OPERA. The popularity of Neapolitan opera spread throughout Europe, and its traditions became everywhere firmly rooted. The Italian poet and dramatist Pietro Metastasio was one of the most influential forces in the dominance of the Neapolitan style, chiefly through the overwhelming popularity of his numerous librettos. While certain specific abuses in Italian opera—the outlandish absurdity of many of the texts, the exaggerated importance of singer and song, the excessive ornamentation of melodies—were modified by the reforms of Cristoph Willibald Gluck, Italian composers generally continued to conform to established patterns. Within their limits, they produced works of unquestioned genius, dressing their stories with musical inspiration of a high order. Rossini and Donizetti brought to both their comic and serious operas—and Bellini exclusively to serious operas—a wealth of melodic beauty and a freshness of musical thought that often lifted the humdrum librettos to works of stature. Melodic beauty and freshness of musical thought were combined with the demands of effective theater in the operas of Giuseppe Verdi, the greatest operatic figure produced by Italy, a genius who dominated the world of opera for half a century. Verdi brought the century-old traditions of Italian opera to their highest stages of technical and artistic development, at the same time introducing a dramatic vigor previously unknown. In his last two operas, *Otello* and *Falstaff,* he abandoned many of the more formal Italian patterns to create a synthesis of music and drama never before achieved by an Italian. With Pietro Mascagni's *Cavalleria rusticana* (1890) a new movement entered Italian opera —verismo, or realism—emphasizing a more realistic sort of drama and greater naturalism in the music. Followers of this movement included Ruggiero Leoncavallo and the most successful Italian opera composer after Verdi, Giacomo Puccini.

BEGINNINGS OF FRENCH OPERA. Jean-

Baptiste Lully, Italian by birth, was the first major figure in French opera. He made several significant departures from the Italian style, placing greater emphasis on ballet and the activity on the stage, giving a new importance to the recitative while simplifying the aria, enriching the harmonic and instrumental writing. These tendencies were further developed by the next outstanding composer of French opera, Jean Philippe Rameau. In his search for dramatic truth and his indefatigable efforts to extend the horizons of harmony and orchestration, Rameau precipitated an acrimonious debate between the partisans of French opera and Italian. Rameau's ultimate victory was a major step in the advance of French opera. Eleven years after Rameau's death, however, the basic issues of the argument were still contested, this time centering around the personalities of Gluck and Piccinni. The decisive triumph of Gluck's operas was a vindication of Rameau's esthetics.

GLUCK'S REFORMS. The most significant break with the formulas of Italian opera was made by Gluck, a development discussed in more detail in his biographical entry in this volume. Arriving at a new humanity, simplicity, dramatic truth, and realizing a closer bond between text and music than had been previously achieved, Cristoph Willibald Gluck brought a new age for opera with works like *Orfeo ed Euridice, Alceste,* and *Iphigénie en Aulide.* His was a major revolution, setting the stage for Weber and Wagner.

ADVANCES IN FRENCH OPERA. Despite the recognition in France of the validity of the kind of operas written by Rameau and Gluck, the Italians continued to prosper there for many years. The new operas in vogue in Paris were those written by such highly favored Italians as Luigi Cherubini and Gasparo Spontini. These works catered to the French appetite for grandiose scenes, ceremonials, ballets, melodrama. Spectacle and melodrama were emphasized in the works of Meyerbeer, a German composer who achieved such popularity in France that his works became the models for subsequent French operas in the grand manner. But it should be noted that the operas with which Meyerbeer impressed France— *Robert le Diable, Les Huguenots, L'Africaine*—combined Italian styles and techniques with French temperament and tastes. French grand opera became an institution with the advent of Meyerbeer. There now appeared a number of composers writing in a similar style, though with the economy and restraint imposed on them by their French temperament. A new kind of lyricism—more refined and delicate than the Italian—a growing interest in characterization, and a mounting concern for human emotions now prevailed in French operas. The climax of this trend was reached in the works of Ludovic Halévy, Charles Gounod, Jules Massenet, Georges Bizet, and Camille Saint-Saëns, to mention only the leading figures. Naturalism came to French opera in 1900 with Gustave Charpentier's *Louise* and the operas of Alfred Bruneau—works which stressed a close identification with the problems of everyday life. At the same time there appeared a musical style called impressionism. The first impressionist opera, Claude Debussy's *Pelléas et Mélisande,* influenced not only French opera but that of the rest of the world. Paralleling the growth of serious French opera were the developments of opéra comique and opéra bouffe, discussed elsewhere under these headings.

DEVELOPMENT OF GERMAN OPERA. Heinrich Schütz's *Dafne,* composed in 1627, was the first opera written in the German language. It was a long time later, however, that a distinctly German tradition came into being. Italian opera meanwhile remained supreme in the courts of Germany and Austria. The Italian style was slavishly imitated by such leading German composers as Reinhard Keiser, Karl Heinrich Graun, and Johann Adolph Hasse. When, in Austria, Gluck parted company with the Italians, he was laying the groundwork for German opera, which was to be distinguished by its pronounced dramatic interest, enriched musical resources, poetic expressiveness, and intimate connection of music and libretto. But even Gluck's *Orfeo ed Euridice* and *Alceste* were settings of Italian texts, as were nearly all the major operas of Mozart, Gluck's most significant follower in the German school. *Don Giovanni* and *The Marriage of Figaro* are offsprings of Italian opera seria and opera buffa. Both operas, however, show such a marked step forward in the musico-dramatic concept, and such an advance in musical characterization, that they are key works in the shaping of the Germanic style. The right of Mozart to belong with the German school is enhanced by the fact that he was the first eighteenth century composer to write operas to German texts. His *The Abduction from the Seraglio* (1782) is the oldest German comic opera still performed. It was followed, in 1791, by *The Magic Flute*. Both these works derive their character not from the Italian opera buffa but from a German variety of popular musical theater known as the singspiel. They must, then, be considered the beginnings of German comic opera, a genre that subsequently yielded such works as *The Barber of Bagdad, Die Meistersinger,* and *Der Rosenkavalier.* Beethoven's *Fidelio* was another offspring of the singspiel, containing, as it does, spoken dialogue. But *Fidelio* is a powerful music drama, the first such in the German language. The accent is on intense emotional and psychological conflicts, and the music acquires a poetic expressiveness not found even in Mozart. Beethoven's single opera was the transition from Mozart to Carl Maria von Weber, with whom German folk opera came into existence. Both the texts and the music of *Der Freischütz* and *Euryanthe* drew copiously from German folk sources; in both, German romanticism came to flower. Richard Wagner was the climactic figure in German opera. In his works the ideal of the German music drama is finally realized. Wagner's esthetics and achievements are discussed in greater detail in his biographical entry. Here it is only essential to recall that Wagner conceived an operatic form and style incomparable for spaciousness and sublimity. Wagner's effect on operatic music was cataclysmic: once and for all the center of the opera world shifted from Italy to Germany. Inevitably, many German composers began using his language and esthetics. Of his immediate successors the most important was Richard Strauss: in his masterworks, *Salome* and *Elektra,* the shadow of the mighty Wagner is in evidence. The "Wagnerzauber" also prevailed in the operas of Engelbert Humperdinck and Hans Pfitzner.

NATIONALISM IN OPERA. The reaction to Wagner's influence was reflected in several ways, including the impressionism of Claude Debussy and the expressionism of Arnold Schoenberg. It was also to be seen in the national movement that arose in Russia among a group of composers sometimes referred to as the "Five." Inspired by examples set by Michael Glinka, the composers Mili Balakirev, Alexander Borodin, Modest Mussorgsky, César Cui, and Nikolai Rimsky-Korsakov opposed making their music a reflection of Western music, especially that of Germany. In their quest for individuality they arrived at a style that had its roots deep in Russian idioms and folk sources. Thus, they created a kind of music that can be mistaken for that of no other country. Outstanding in the work of these composers are three folk operas: Borodin's *Prince Igor,* Mussorgsky's *Boris Godunov,* and Rimsky-Korsakov's *Sadko.* The success of the "Five" inspired composers in other countries to create a national art. Notable examples of these aspirations are to be seen in such diverse works as Smetana's *The Bartered Bride* (Bohemia), Janáček's *Jenufa* (Moravia), Moniuszko's *Halka* (Poland), Kodály's *Háry János* (Hungary), and Vaughan Williams's *Hugh the Drover* (England).

OPERA IN ENGLAND. *The Siege of Rhodes* (1656) was the first play in England to be set throughout to music. No less than five composers were responsible for the score: Henry Lawes, Matthew Locke, Thomas Cooke, Edward Coleman, and George Hudson. The first great English opera composer was not long in appearing; Henry Purcell, who, about 1690, wrote the first significant English opera, *Dido and Aeneas,* a work so remarkable in dramatic and musical content that it is still capable of moving audiences today. The next great figure in English opera was not a native son but the Saxon-born George Frideric Handel. He lived in England from 1712 to the end of his life, and his many operas written during this period, all representative of the Italian school, were the most important ones of their era. Reaction against the dominant Italian style was manifested in the tremendous popularity of *The Beggar's Opera* (1728), one of the first ballad operas, a form that was to be in vogue for the rest of the century. In the middle nineteenth century the ballad opera became the inspiration of such opera composers as Michael William Balfe, Julius Benedict, and William Vincent Wallace. More in the Italian grand opera style, but with the infiltration of English personality, were operas by the later composers Alexander

Mackenzie and Charles Stanford. Gustav Holst and Ralph Vaughan Williams are twentieth century composers who have written operas that are unmistakably English in origin. Finally, in the years since World War II, Benjamin Britten has written a number of highly effective operas based on a wide range of moods and subjects.

OPERA IN AMERICA. The first work by an American composer that can reasonably be called an opera was Francis Hopkinson's *The Temple of Minerva,* produced in 1781. Hopkinson, a signer of the Declaration of Independence, wrote the words as well as the music for his "oratorial entertainment," as it was called at the time; the music has since been lost. The next American opera, a ballad opera, set forth the career of the Cherokee chieftain whose name was the production's title: *Tammany,* with libretto by Anne Julia Hatton, music by the popular and prolific James Hewitt. It was produced in New York in 1794 under the auspices of the Tammany Society. Two more American operas were introduced in 1796: Victor Pelissier's *Edwin and Angelina* and Benjamin Carr's *The Archers, or Mountaineers of Switzerland* (a work derived from Schiller's drama *Wilhelm Tell*). The first American work of grand-opera proportions was *Leonora,* adapted from Bulwer-Lytton's *The Lady of Lyons,* with music by William Henry Fry. In the style of operas by Donizetti and Bellini, *Leonora* has considerable value. It was introduced in Philadelphia in 1845, was well received, and was revived thirteen years later at the Academy of Music in New York. Fry later wrote another successful work, *Notre Dame de Paris,* based on Victor Hugo's novel. Well received in 1855 was George Frederick Bristow's opera *Rip Van Winkle,* a work using an American subject. An attempt to encourage the writing of operas on American themes was made by Ole Bull in 1855. He had recently become manager of the Academy of Music in New York, and in this capacity he offered a prize of a thousand dollars for "the best original grand opera by an American composer on an American subject." Unfortunately, Bull did not stay long enough at the Academy to conclude this competition successfully. Another effort in this direction was made thirty years later by Theodore Thomas, then the director of the American Opera Company. One of the announced aims of the new company was the presentation of authentic American operas, but in the year or so of the company's existence no such works were found that merited performance. Early in 1900 there was talk of producing an American opera produced at the Metropolitan Opera; the work selected was John Knowles Paine's *Azara*. Nothing came of this project. *Azara* was never performed on the stage, though it received a concert performance in 1903. The first American opera performed at the Metropolitan was Frederick Shepherd Converse's *The Pipe of Desire* in 1910. A year later Gatti-Casazza announced a competition for American operas. The winner was Horatio W. Parker's *Mona*, produced by the Metropolitan in 1912 and awarded a ten thousand dollar prize. A number of American composers have written operas that exploited peculiarly American idioms. Victor Herbert and Charles Wakefield Cadman produced operas (*Natoma* and *Shanewis*) that featured their conceptions of American Indian music; Louis Gruenberg (in *The Emperor Jones*) and George Gershwin (in *Porgy and Bess*) used Negro subjects and musical materials; Marc Blitzstein has used jazz styles and techniques (*The Cradle Will Rock*). However, most American opera composers have used an eclectic style owing little to American folk or popular sources, and in this vein some notable

operas have been written: Deems Taylor's *The King's Henchman*, Douglas Moore's *The Devil and Daniel Webster*, Gian-Carlo Menotti's *The Medium* and *The Consul*.

The following American operas are among those discussed in their proper alphabetical order in this encyclopedia: *Amahl and the Night Visitors; The Canterbury Pilgrims; Caponsacchi; The Consul; The Cradle Will Rock; Cyrano de Bergerac; The Devil and Daniel Webster; Down in the Valley; A Drumlin Legend; The Emperor Jones; Evangeline; Four Saints in Three Acts; The Island God; The Jumping Frog of Calaveras County; The King's Henchman; Madeleine; The Man Without a Country; The Medium; Merry Mount; Mona; The Mother of Us All; Natoma; The Old Maid and the Thief; The Pipe of Desire; Porgy and Bess; Regina; The Saint of Bleecker Street; The Scarlet Letter; Shanewis; The Taming of the Shrew; The Telephone; The Temple Dancer; The Tender Land; The Tree on the Plains; Trouble in Tahiti; Volpone; White Wings; A Witch of Salem.*

Opéra, L' (Académie de Musique), the oldest opera house in France, and the most celebrated. It is subsidized by the French government. In 1669 a grant was given by Louis XIV to Pierre Perrin, Robert Cambert, and the Marquis de Sourdéac to organize a theater in Paris for opera and ballet performances. On March 19, 1671, the Académie de Musique was launched at the Jeu de Paume with a pastoral by Cambert, *Pomone,* now accepted as the first French opera. When Perrin was put in debtors' prison, Jean Baptiste Lully bought out the license and took over the direction and supervision of the Académie in 1672. A year later, when Molière's death left the theater at the Palais Royal vacant, the opera company transferred there, remaining ninety years until the theater was destroyed by fire in 1763. Lully headed the opera company until his death in 1687 and created the first significant period in the history of the Paris Opéra; during his regime some twenty of his operas and opera-ballets were introduced, exerting a far-reaching influence on the early evolution of French opera. The second great period in the history of the Opéra came between 1737 and 1760, when many of the outstanding operas of Rameau were given. From 1773 to 1779 the Opéra was dominated by the personality and genius of Gluck, then visiting from Vienna, a visit climaxed by the victory of his operas over those of the Italian opposition. In 1794 the Académie de Musique moved to the Rue de Richelieu and temporarily became known as the Théâtre des Arts. After the French Revolution, the new regime removed from the repertory many works considered too aristocratic, substituting new operas by Méhul, Gossec, Monsigny, and Philidor. The Empire helped restore many of the prohibited operas, but the opera-ballets of the preceding century had now lost their audience and were replaced by French historical operas. In 1821 the Académie occupied the Salle Favart, and a year later it moved to a theater in Rue Le Peletier, where operas by Rossini, Weber, Donizetti, and Mozart were added to the repertory. When this theater burned, a movement was launched to build a permanent home for the Académie. In 1861 the foundation for the new building, designed by Charles Garnier, was finally laid; the building itself, erected at the head of the Avenue de l'Opéra, was not completed for fourteen years. It opened on January 5, 1875, with an orchestral concert. The first opera performed there was *La Juive,* three days later. The first opera new to the repertory, *Jeanne d'Arc,* by Auguste Mermet, was given on April 5, 1876.

Since its beginnings as the Académie de Musique, the Paris Opéra has presented almost a thousand different works. Many of the première performances have been of the foremost operas written in France: Cherubini's *Anacréon;* Lully's *Armide* and *Atys;* Berlioz' *Benvenuto Cellini;* Rameau's *Castor et Pollux, Dardanus, Hippolyte et Aricie,* and *Les Indes galantes;* Massenet's *Le Cid, Le Roi de Lahore,* and *Thaïs;* Saint-Saëns' *Henri VIII;* Halévy's *La Juive;* Auber's *La muette di Portici;* Février's *Monna Vanna;* Gounod's *Sapho.* Among the most important premières by non-French composers have been Meyerbeer's *L'Africaine, Les Huguenots, Le Prophète,* and *Robert le Diable;* Grétry's *Andromaque;* Gluck's *Armide, Iphigénie en Aulide,* and *Iphigénie en Tauride;* Verdi's *Don Carlos;* Donizetti's *La favorita;* Rossini's *William Tell.*

Opéra bouffe, a French comic opera—light, trivial, and frequently farcical in character. It was an outgrowth of the opéra-comique of the middle nineteenth century (see below) which then acquired a more serious and artistically ambitious character. The term "opéra bouffe" came from the theaters in which these pieces were played, called "bouffes." Thus, when Offenbach opened his theater for musical satires and farces he called it the Bouffes Parisiens. Offenbach was the foremost exponent of the opéra-bouffe, and his *Orpheus in the Underworld* is a classic in this genre.

Opera buffa, an Italian comic opera, with dialogue in recitativo secco (*see* RECITATIVE). It is distinguished from the more serious variety of Italian opera (opera seria) in its use of a comic rather than serious subjects. But there are other basic differences. The opera seria generally used historical or legendary characters; the opera buffa preferred human beings in everyday situations, human-interest stories filled with farcical situations. The opera buffa is usually concerned with love intrigues involving cuckolds, deceiving wives, and scheming servants. The opera seria utilized complex, highly ornamented arias to exploit the virtuosity of individual singers; the opera buffa preferred simple melodies and tunes. The opera seria emphasized massive scenes of pageantry; the opera buffa required only a few characters moving against a simple setting. Musically, the opera buffa popularized certain stylistic devices by which it was to be identified: the swift alternation of light and shade for contrast; the use of rhythmic, staccato passages to emphasize coquettish moods; the exploitation of patter songs; the extended finales concluding each act.

The opera buffa developed from the intermezzo, a brief comic scene set to music, popular in Italy in the late sixteenth and early seventeenth centuries (*see* INTERMEZZO). A special theater was opened in 1709 in Naples to perform these intermezzi, whose texts were in the Neapolitan dialect. Such figures of the Neapolitan school as Alessandro Scarlatti and Nicola Logroscino were among the most popular writers of these comic pieces. Another Neapolitan, Giovanni Battista Pergolesi, wrote intermezzi; and in 1733 he completed a more extended work in a similar style which can be regarded as the first opera buffa in history: *La serva padrona,* introduced in Naples on August 28, 1733. Pergolesi's delightful one-act comic opera not only established the form, style, and esthetic approach which subsequent works in this genre were to assume; its tremendous popularity in Italy inspired composers there to produce works in a similar vein, thus bringing universal acceptance to the new medium. In Italy, Pergolesi's immediate successors were Domenico Cimarosa (*Il matrimonio segreto*), Baldassare Galuppi (*Il filosofo di campagna*), Giovanni Paisiello

(*The Barber of Seville*), and Nicola Piccinni (*La Cecchina*). These composers represent the natural transition from Pergolesi to one of the greatest of all opera buffa composers, Rossini, whose masterwork, *The Barber of Seville,* was the model followed by all later composers of opera buffa. Of Rossini's successors, the most notable was Donizetti (*L'elisir d'amore* and *Don Pasquale*), while Ermanno Wolf-Ferrari carried the style into the twentieth century with *The Secret of Suzanne.* The first opera of the American composer Gian-Carlo Menotti is also in the opera buffa tradition: *Amelia Goes to the Ball.*

Opéra comique, a type of French opera utilizing spoken dialogue. Its present-day meaning is not—as a literal translation of the name might imply—a comic opera. Many celebrated works in this genre are tragic—for example, *Carmen.* The form was evolved early in the eighteenth century. At that time the Académie de Musique (or Paris Opéra, as it is now known) had a monopoly on all opera performances. To circumvent this monopoly, a new type of opera was evolved: a light form of theatrical entertainment utilizing singing. These musical plays, with spoken dialogue, were first seen at Paris fairs. It was at the Foire St. Germain, in 1715, that the term "opéra comique" was used for the first time. Originally, the opéra comique was true to its name by being a work of humorous character, simple in appeal, designed exclusively as entertainment. The first of these works were often parodies of the serious operas given at the Académie de Musique.

A powerful stimulus to French composers was the performance in Paris of *La serva padrona* in 1752. Frenchmen set out to write works in a similar style. In the same year of 1752 Jean Jacques Rousseau—the philosopher and musician—completed and had performed a delightful little opéra comique, *Le devin du village*, which he confessed was in frank imitation of the Pergolesi opera. A twenty-three-year-old composer, Pierre-Alexandre Monsigny, was so moved by *La serva padrona* that he decided henceforth to write only comic operas. André Grétry was another composer who was similarly influenced. In time, these two were succeeded by the triumvirate who made the opéra comique an institution: François Boieldieu, Adolphe Adam, and Daniel Auber.

In the middle of the nineteenth century the character of the opéra comique changed radically. The term was henceforth applied to operas—frequently tragic in theme—in which there were passages of spoken dialogue. The works formerly designated as opéras comiques now became known as opéras bouffes. The Opéra-Comique was established as the home for this new type of opera.

Opéra-Comique, L', one of the two national lyric theaters of France (the other being the Académie de Musique or, as it is now known, L'Opéra), situated on the Rue Favart in Paris. It was established in 1715 after a special agreement with the Académie de Musique that it would present only operatic works with spoken dialogue. Originally, the Opéra-Comique gave its performances at the Foire St. Germain, where it was so successful that rival theater managers combined to have it closed in 1745. When the Opéra-Comique resumed operations in 1752, it combined with the Comédie Italien in presenting performances at the Mauconseil. In 1783 the Opéra-Comique moved to the Rue Favart, but it did not stay there long. A competitive company was organized on the Rue Feydeau in 1791. With audiences divided between the theaters, both suffered financial reverses and had to close in 1801. They then combined forces

and gave opera performances on the Rue Feydeau until 1829, from 1829 to 1832 at the Théâtre Ventadour, and from 1832 to 1840 at the Théâtre des Nouveautés. In 1840 the Opéra-Comique moved to its present site on the Rue Favart, and the present theater opened on December 7, 1898.

The list of world premières at the Opéra-Comique includes some of the most celebrated works of the French lyric theater: *Ariane et Barbe-Bleue; Carmen; Cendrillon; Fra Diavolo; Grisélidis; La Habanera; L'heure Espagnole; Joseph; Le Juif Polonais; Lakmé; Lalla Roukh; Louise; Manon; Mârouf; Mignon; Les noces de Jeannette; Pelléas et Mélisande; Le postillon de Longjumeau; Le Roi d'Yvetot; Le Roi d'Ys; Sapho* (Jules Massenet); *The Tales of Hoffmann.* Operas by foreign composers introduced at the Opéra-Comique have included: *The Daughter of the Regiment; Dinorah;* Ernest Bloch's *Macbeth.*

It should be pointed out that many works now performed at the Opéra-Comique are traditional in form, the company having abandoned its original policy of confining its activities exclusively to operas with spoken dialogue.

Opera performance. (1) EUROPE. The first operas by members of the Florentine "Camerata" and their immediate successors were performed privately in the palaces of Italy's nobility. The first public opera house came into existence in 1637 when one of Venice's noblest families, the Trons, opened the Teatro San Cassiano with a performance of an opera by a now forgotten composer: Manelli's *Andromède.* The boxes were rented annually by Venetian nobility and foreign princes; the general public gained admission to the spacious parterre by paying approximately twenty cents. This new theater was a tremendous success, and many operas were given there. Between 1641 and 1649, for example, thirty different works were performed, including several by Claudio Monteverdi.

By the end of the seventeenth century opera was such a favored form of entertainment that there were sixteen different opera houses in Venice alone, each run by a different Venetian family. An opera theater opened in Naples in 1684, the Teatro San Bartolomeo, where works by Alessandro Scarlatti were performed and where *La serva padrona* was introduced. The most celebrated of all Italian opera houses, La Scala, in Milan, was opened in 1776. In Rome, where opera was long forbidden by papal edict, performances were restricted to the homes of royalty.

An opera theater was inaugurated in London as early as 1656. Not until the Restoration, four years later, when the Puritan ban on theatrical entertainment was removed, did opera begin to flourish. The first major opera house in London, Drury Lane Theatre, opened in 1696, followed by His Majesty's Theatre in 1705 and Covent Garden in 1732. However, when Handel's first English opera was performed in 1711 (*Rinaldo* —a triumph giving impetus to opera in England), it was given in a smaller house, the Queen's Theatre, in Haymarket. Handel helped found and was artistic director of his own opera company in 1721, the Royal Academy of Music; when that failed, he organized still another company in 1729 at the King's Theatre.

An opera house came into existence in Hamburg, Germany, in 1678. From 1695 to 1705 it was directed by Reinhard Keiser, who was responsible for making Hamburg the first major operatic center of Germany. It was here that Handel made his bow as an opera composer. The Paris Opéra, or Académie de Musique, was organized in 1669. The first opera house in Prague opened in 1725, the first in Berlin in 1742. In Vienna, the Burgtheater came into existence in 1741 with a performance of Carcano's *Amleto,* and soon became the home for operas by Gluck and Mozart.

The history of opera performances is to be found in the history of the world's leading opera houses. For additional material in the present volume, consult entries on individual opera houses and on major opera festivals.

(2) UNITED STATES. The first opera performance in the Colonies took place in a courtroom in Charleston, South Carolina, on February 8, 1735. The production was an English ballad opera, *Flora.* Ballad operas continued to be the only operas performed in the New World up to 1791. In that year an opera troupe, under the direction of Louis Tabery, visited New Orleans and presented French operas in private homes. A year later a theater was built for this group, the Théâtre le Spectacle de Rue St. Pierre. This was the first of several opera theaters in New Orleans. In the French Opera House, opened in 1859, most of the famous French operas of the eighteenth and nineteenth centuries were introduced to the United States.

The first Italian opera heard in America was Paisiello's *The Barber of Seville,* sung in Baltimore in 1794 (in an English translation). Rossini's *The Barber of Seville,* also sung in English, was given in New York City in 1819. In 1825 the same work was sung in Italian, the first Italian language performance in America. In the same year, Manuel García visited New York with an Italian opera company and performed several Rossini operas new to this country. García's collaborator in this venture was Lorenzo da Ponte who, with a French tenor named Montrésor, organized another season of Italian operas in New York a year later. Lorenzo da Ponte was a primary force in the erection of the first permanent opera house in New York, the Italian Opera House. The Astor Place Opera, opening in 1847, and its successor, the Academy of Music, opening in 1854, carried on the major operatic activity in New York. German operas in the original tongue were heard for the first time in New York in 1855, when a visiting troupe gave *Der Freischütz, Martha,* and *Zar und Zimmermann.* The first Wagner opera heard in this country was *Tannhäuser,* at the Stadttheater in New York in 1859.

Opera first came to Chicago with a performance of *La sonnambula* in 1850. Nine years later Chicago had its first opera season, under the direction of Maurice Strakosch, and in 1865 the first opera house in Chicago was erected, the Crosby Opera. Opera was heard for the first time in San Francisco in 1852 when the Pellegrini Opera Troupe gave *La sonnambula.*

The history of opera performances in America is told further in the stories of individual opera houses.

(3) RADIO. Even while radio was in its earliest experimental stages it was interested in opera. In the early 1900s, Lee De Forest broadcast from the stage of the Manhattan Opera an aria from *Carmen* sung by Mariette Mazarin. On January 13, 1910, there took place an experimental broadcast from the stage of the Metropolitan Opera: parts of *Cavalleria rusticana* with Emmy Destinn and Riccardo Martin, and *Pagliacci,* with Enrico Caruso and Pasquale Amato. This broadcast was picked up by some fifty radio amateurs in or near New York City, and by the *S.S. Avon* at sea.

In 1925, WEAF (New York) began weekly broadcasts of operas, sung in their original languages by professional performers directed by Cesare Sodero. This was the first time a series of operas came to radio, and the program was continued successfully for several years. On September 7, 1925, the first stage performance of a regular opera performance was broadcast by a commercial radio station when WJZ transmitted *Aïda* as given by the Boston Civic Opera at the Manhattan Opera House in New York.

When the first radio network came into existence on November 15, 1926— the National Broadcasting Company network—the event was celebrated with a performance by two opera stars, Mary Garden singing in Chicago, Titta Ruffo, in New York.

On January 21, 1928, excerpts from opera performances at the Chicago Auditorium became a regular radio feature. Soon after the first broadcast (the Garden Scene from *Faust*), a manufacturer of radio supplies sponsored these broadcasts; complete acts were henceforth transmitted through nineteen associated stations.

A new day dawned for opera broadcasting on March 16, 1930, when a performance of *Fidelio* was transmitted to America from the stage of the Dresden Opera, the first transatlantic broadcast of an opera. In 1931 a performance was relayed to America from Covent Garden.

The Metropolitan Opera became affiliated with radio in 1931. Gatti-Casazza, the manager, consented to a trial broadcast, the quality of which would influence his decision as to whether Metropolitan performances should be broadcast regularly. A performance of *Madama Butterfly* was privately relayed to Gatti-Casazza and his musical staff in an NBC studio. Gatti-Casazza was so impressed that he consented to the transmission of *Hansel and Gretel* from the stage of the Metropolitan on Christmas Day, 1931. This led to the weekly Saturday-afternoon broadcasts from the Metropolitan, first over the NBC network, subsequently over the ABC. For six years these broadcasts were a sustaining feature. During the next three years various sponsors were associated with the broadcasts. In 1940 the Texas Company became the permanent sponsor. In the first twenty years of broadcasting, over four hundred performances of over seventy different operas were heard. The three heard most often were: *Tristan und Isolde, Aïda,* and *Carmen.*

In 1933, radio recognized its responsibility to opera in general, and the American composer in particular, by commissioning the first opera ever written directly for radio, *The Willow Tree,* by Charles Wakefield Cadman, broadcast over the NBC network on October 3. Previously, on April 17, 1930 NBC had given the world première of Charles Skilton's one-act Indian opera, *Sun Bride.* In 1937 CBS began commissioning operas. Among these were: *Flora, Beauty and the Beast,* and *Blennerhasset,* all three by Vittorio Giannini; and Louis Gruenberg's *Green Mansions.* In 1939 NBC commissioned Gian-Carlo Menotti to write *The Old Maid and the Thief.*

This résumé of opera over the radio would be incomplete without mention of the extraordinary concert performances directed by Arturo Toscanini. Among the operas presented by Toscanini with the NBC Symphony Orchestra were: *Aïda, Un ballo in maschera, La Bohème, Falstaff, Fidelio, Orfeo ed Euridice, Otello,* and *La traviata.*

(4) RECORDINGS. Opera singers and excerpts from operas played a prominent role in the early history of recorded music. The phonograph was born on December 15, 1877, when Thomas A. Edison filed a patent application for a reproducing machine. The records for this primitive instrument were cylindrical. In the 1890's several prominent opera stars recorded arias, among them Calvé, Maurel, Nordica, and Tamagno. Meanwhile, in 1887, the disk record was invented by Emile Berliner. The Berliner Gramophone Company issued recordings by some lesser opera singers in 1895. By 1900 the disk had replaced the cylinder, and the recording industry was ready to go into high gear.

In 1898 the Gramophone Company was organized in Europe. Soon after the turn of the century it embarked on an ambitious program to record great music in outstanding performances. In 1901 it despatched engineers to Russia to record eminent singers of the Imperial Opera, including Feodor Chaliapin. In March, 1902, the same company contracted Enrico Caruso to record ten numbers; and soon after this, Emma Calvé, Mattia Battistini, Maurice Renaud, and Pol Plançon joined the company.

In 1903 the Columbia Phonograph Company of New York introduced the Grand Opera Series with stars of the Metropolitan Opera, including Suzanne Adams, Edouard de Reszke, Ernestine Schumann-Heink, Marcella Sembrich, and Antonio Scotti. The venture was a failure and was temporarily abandoned, but it spurred the Victor Talking Machine Company (incorporated in 1901) to competition. In 1903 Victor set up a recording studio in Carnegie Hall. The first record made with the now-famous Red Seal Label was "Caro mio ben," sung by Ada Crossley. Giuseppe Campanari, Louise Homer, Plançon, and Scotti were soon making records for Victor; at the same time Victor was issuing in America the best releases of the European Gramophone Company. Lacking an outstanding tenor on its list, Victor signed Caruso to an exclusive contract in 1903. Caruso recorded ten numbers early in 1904 at a fee of four hundred dollars a record. These records proved so popular that after a second recording session, in 1905, Caruso's fee was raised to a thousand dollars a record. The records made at the third session, in 1906, were recently re-released by RCA Victor in its Treasury Series.

Geraldine Farrar, Alma Gluck, and John McCormack were other leading singers added to the Victor list after 1907. Influenced by Victor's success, the Columbia company revived its program by engaging Mary Garden, Lillian Nordica, Olive Fremstad, and Alessandro Bonci, among others.

Despite the high price for these records—a typical single twelve-inch disk, recorded on only one side, cost as much as seven dollars—sales soared. Within a decade the assets of Victor rose to eight million dollars; in two decades this figure doubled. Caruso's income from his recordings eventually amounted to more than two million dollars.

The early recordings were made acoustically. The artist would sing into a horn which would converge the sound and give it sufficient power to cut a track on a wax disk. This method did not permit much clarity of reproduction. In the 1920's a revolution took place. The radio introduced new techniques and implements, among them the electric microphone. The artist's voice was now amplified and registered electrically. From the field of radio, too, the phonograph acquired a new type of loud speaker in which amplification was achieved through radio tubes. Electricity brought an altogether new fidelity to recordings. A boom now took place in the sale of phonographs and records. Recordings themselves became more ambitious: for the first time an entire opera was recorded, when in 1928 Victor issued *Rigoletto*. In 1929 Columbia followed suit with its own first complete opera, *Carmen*. For some years European companies were the dominant makers of operatic records, but a change came in 1947 when Columbia announced a contract with the Metropolitan Opera to record parts of operas or complete operas from the stage of that house. The first of these recordings was the "Liebes-nachtmusik" from *Tristan und Isolde,* sung by Helen Traubel, Herta Glaz, and Torsten Ralf, with Fritz Busch conducting. The first complete opera was *Hansel and Gretel.*

The next major development in the recording of operas came in 1948 when Columbia Records announced its long-playing record. Thirty to forty-five minutes of music could now be put on one complete record. This innovation, soon adopted by all other major companies, made it possible to record entire operas on from two to five records. The increase in record quality and the concomitant reduction in price led to a tremendous increase in record buying. In the first full year of long-playing record sales (1949) the total amounted to over two million dollars. Numerous new companies came into existence, and the new releases each year reached prodigious numbers. Not only were all the major operas recorded in performances by several different companies, the less familiar works of the repertory —and sometimes complete strangers— were made available. An indication of the sale of complete opera recordings is to be found in RCA Victor's announcement in 1954 that, since 1949, it had sold four hundred and fifty thousand opera albums at a total price of eight and a half million dollars.

(5) TELEVISION. While experiments in televising opera took place in London as early as 1936, and opera was televised by the BBC in 1937–1938, it was not until 1941 that opera came to television in the United States. On March 10, 1941, there took place the first American opera telecast: a tabloid version of *Pagliacci* transmitted from Radio City with Metropolitan Opera performers. The next major event in televised opera took place on November 29, 1948, when the Metropolitan Opera transmitted its opening night performance (*Otello*) over NBC. Subsequent opening-night performances

that the Metropolitan televised were: *Der Rosenkavalier* (1949) and *Don Carlos* (1950).

In 1950 a weekly program devoted to abridged operas (on films) entitled "Opera Cameos" was begun on WPIX, produced by Carlo Vinti. This series has since then been televised over a network.

An experiment of far-reaching implications was undertaken on December 11, 1952, when an actual subscription performance of *Carmen* at the Metropolitan Opera was televised over a closed circuit to thirty-one theaters in twenty-seven cities. It was estimated that some seventy thousand persons paid over a hundred thousand dollars in admissions to hear this performance. This was the first time that an actual stage production of any kind was sent from a point to different parts of the country. The experiment's success pointed out at least one direction which televised opera could take, and in the fall of 1954 the Metropolitan signed a three-year agreement with Theater Network Television to televise on a closed circuit its opening-night performances. The first transmission under this agreement took place on the opening night of the 1954–1955 season when, for the first time in its history, the Metropolitan Opera presented as its first attraction acts from four different operas, instead of one complete opera.

A significant advance in televising opera was made in 1949–1950 with the formation of the NBC Television Opera Theater, with Samuel Chotzinoff as producer, and Peter Herman Adler as musical and artistic director. Each season, over the NBC network, the Theater has presented eight operas, all of them sung in English. Besides standard works, the Theater has presented such novelties as *Salome, Pique Dame,* Puccini's *Il Trittico,* Vittorio Giannini's *The Taming of the Shrew,* the world

première of Bohuslav Martinu's *The Marriage,* the first American performance of Benjamin Britten's *Billy Budd,* and the first professional performance of Leonard Bernstein's *Trouble in Tahiti.* In 1954, the Theater commissioned two operas, Stanley Hollingsworth's *La Grande Breteche,* and an opera by Lukas Foss. Martinu had written *The Marriage* directly for the television medium; but this was not the first opera so written. The distinction belongs to Gian-Carlo Menotti's *Amahl and the Night Visitors,* introduced by NBC on Christmas Eve, 1951. Since the program was a sponsored one, *Amahl* became the first opera whose world première was sponsored by a commercial organization.

On February 1, 1953, the Metropolitan Opera was presented by the Omnibus program of the Ford Foundation in *Die Fledermaus;* the Metropolitan later returned to Omnibus with a performance of *La Bohème,* in a new English translation. Subsequently Omnibus presented a revival of George Gershwin's one-act opera *135th Street,* the American première of Ottorino Respighi's *The Sleeping Beauty,* and the world première of Alec Wilder's *Chicken Little.*

Opera was televised in color for the first time on October 31, 1953, with an NBC presentation of *Carmen.* Soon after this, *Amahl and the Night Visitors* and *The Taming of the Shrew* were also given in color.

Opera seria, a serious or tragic opera, as opposed to an opera buffa, or comic opera. The term was usually applied to Italian operas of the seventeenth and early eighteenth centuries, or operas resembling them, including those of Handel and such works as Mozart's *Idomeneo.*

Operetta (or Light opera), a romantic play containing songs, musical numbers, and dances. An operetta differs from a musical comedy in that the musical score is of a more ambitious character.

Ophelia, Polonius' daughter (soprano), in love with Hamlet, in Thomas's *Hamlet.*

Ophelia's Mad Scene, *see* PARTAGEZ-VOUS MES FLEURS!

O prêtres de Baal, Fidès' prison scene in Act V, Scene 1, of Meyerbeer's *Le Prophète.*

O pur bonheur, quartet of Juliette, Romeo, Friar Laurence, and Gertrude in Act III, Scene 1, of Gounod's *Roméo et Juliette.*

O qual soave brivido, love duet of Riccardo and Amelia in Act III of Verdi's *Un ballo in maschera.*

Oracolo, L', (The Oracle), one-act opera by Franco Leoni. Libretto by Camilio Zanoni, based on *The Cat and the Cherub,* a story by Chester Bailey Fernald. Première: Covent Garden, June 28, 1905. In the Chinese section of San Francisco, about 1900, the New Year is being celebrated. Win-Shee, a learned doctor, consults his books and finds that tragedy awaits Hoo-Chee, son of a wealthy merchant. Chim-Fen, owner of an opium den, overhears this prophecy and decides to help destiny by kidnaping the boy. Win-Shee's son follows the boy to the opium den, where Chim-Fen kills him. Win-Shee avenges the death of his son by strangling the murderer, after the kidnaped boy is returned to his father.

Ora e per sempre addio sante memorie, Otello's farewell to his peace of mind in Act II of Verdi's *Otello.*

Ora soave, sublime ora d'amore, love duet of Andrea Chénier and Madeleine in Act II of Giordano's *Andrea Chénier.*

Or co' dadi, *see* SQUILLI, ECHEGGI LA TROMBA GUERRIERA.

Ordgar, Thane of Devon (bass) in Deems Taylor's *The King's Henchman.*

Oreste, a trilogy of operas by Darius Milhaud. Libretto by Paul Claudel, based on the tragedies of Aeschylus.

The three operas are *Agamemnon, Les Choephores,* and *Les Euménides.* Edward Lockspeiser explains that Milhaud's approach to the Greek tragedies is "not in the spirit of raging conflict . . . but in a mood of tolerance and charity. He does not disclose the fears and anxieties of his characters . . . however terrifyingly these characters are portrayed. *See below:* ORESTES.

Orestes, (1) in Greek legend, the son of Klytemnestra and Agamemnon, and brother of Elektra. He avenges the murder of his father by killing his mother and her lover, Aegisthus. The fates, or Erinnyes, torment him until a trial absolves him. The dramatist Aeschylus wrote a trilogy with Orestes as the central character, and these have been made into numerous operas (*see* AESCHYLUS). Richard Strauss's *Elektra* was based on Sophocles' tragedy on the same subject; Gluck's *Iphigénie en Tauride* and Ernest Křenek's *Das Leben des Orest* were derived from Euripides' treatment of the Orestes story.

(2) Son of Klytemnestra and Agamemnon (baritone) in Richard Strauss's *Elektra.*

(3) Iphigenia's brother (baritone) in Gluck's *Iphigénie en Tauride.*

Orest! Es rührt sich niemand! Elektra's Recognition Scene in Richard Strauss's *Elektra.*

Orfeo, opera, or "favola in musica," by Claudio Monteverdi. Libretto by Alessandro Striggio. First performance: Mantua, February 24, 1607. The libretto follows the familiar legend. When Eurydice dies, her husband Orpheus is so stricken with grief that he is permitted to go to Hades to rescue her. But he is commanded not to look at her face until he has made his return to earth. Eurydice, thinking Orpheus loves her no longer, pleads so eloquently for Orpheus to look at her that, helplessly, he complies. Eurydice dies a second time. The god of love pities Orpheus, revives Eurydice, and allows them to return to earth. Monteverdi's opera is a vibrant work of art. The tentative vocal writing of the earlier composers Peri and Caccini has been superseded. Melody emerges for the first time in such moving arias as "Ecco pur ch'a voi ritorno," which opens the second act; recitatives are supplemented by duets and trios. Monteverdi employed a larger, more varied orchestra than had been used previously, and the instrumental portions of the score took part in projecting atmosphere and mood and in heightening the dramatic effect through effects of harmony and tone color. As Henri Prunières wrote: "Monteverdi turned the aristocratic spectacle of Florence into modern musical drama, overflowing with life and bearing in its mighty waves of sounds the passions which make up the human soul."

Orfeo ed Euridice (Orpheus and Eurydice), (1) opera in four acts by Cristoph Willibald Gluck. Libretto by Raniere de Calzabigi, based on the Greek legend. Première: Burgtheater, Vienna, October 5, 1762. American première: New York, the Winter Garden, May 25, 1863.

Characters: Orfeo, or Orpheus, Greek musician (contralto); Euridice, or Eurydice, his wife (soprano); Amor, god of love (soprano); Happy Shade (soprano); Blessed Spirits, Furies, shepherds, shepherdesses, heroes, and heroines. The setting is legendary Greece.

Act I. The tomb of Euridice. Shepherds, shepherdesses, and nymphs are mourning ("Ah! se intorno a quest' urna funesta") and Orfeo mourns with them. Amor, touched by his grief, instructs him to descend to the lower world and lead Euridice back to earth. But he is to do this only on the condition that he does not look at her.

Act II. Tartarus. The Furies confront Orfeo, demanding to know who dares enter their realm. They perform a demoniac dance to frighten him. Orfeo pleads for mercy ("Deh! placatevi con me") and wins them over. The gates of the lower world open. Orfeo passes through and the Furies continue their infernal dance ("Dance of the Furies").

Act III. The Elysian Fields. The Blessed Spirits perform a serene and radiant dance ("Dance of the Blessed Spirits"). Euridice describes the peace and beauty of Elysium ("E quest' asilo ameno e grato"). When Orfeo arrives, he too is spellbound by the heavenly beauty ("Che puro ciel"). Finding his beloved Euridice, he entreats her to follow him; as she does so, he refuses to look back at her.

Act IV. A forest. Euridice is heartbroken, for she interprets Orfeo's failure to look at her as an indication that he loves her no more. She renews her entreaties until Orfeo can resist no longer. He takes his wife passionately into his arms; as he does so, she dies. Orfeo is grief-stricken ("Che farò senza Euridice"). Amor once again comes to his aid, reviving Euridice and allowing the pair to return safely to earth. In the Temple of Love, Orfeo sings the praises of Amor. Dances of rejoicing take place. Then Orfeo, Euridice, and Amor sing a hymn to true love ("Gaudio son al cuore queste pene dell' amor").

Orfeo ed Euridice is Gluck's first opera in which he purposefully departed from the stilted formulas, the meretricious texts, the sawdust historical characters, and the ornate music of the Italians; for these he substituted simplicity, economy, and deep human feeling. To this day, *Orfeo* remains a vital and poignant work of art. There is probably no opera in the repertory that accomplishes so much with so few means. There are only two main characters, and only four solo voices in all; very little happens throughout the four acts; there are no elaborate scenes or overpowering climaxes. Yet the opera never fails to have dramatic appeal, never fails to touch the heart. As remarkable as its simplicity is its dramatic truth. Text and music are wonderfully integrated as Gluck portrays the terrors of Hell with acrid dissonances, and recreates the rapture of Elysium with some of the most beatific music ever written.

The role of Orfeo was written (following the custom of Gluck's time) for performance by a castrato; today the part is taken by a woman.

(2) Opera by Joseph Haydn. Libretto by Bedini. Première: Florence May Music Festival, May 1951. This is Haydn's finest opera. He completed it in 1791 for performance in London during his visit to that city, but due to managerial difficulties it was never produced. The score was subsequently dismembered, and the work was known only through excerpts. Haydn scholars recently reassembled the various parts, and the work was performed for the first time in Vienna in 1950 for a phonograph recording. Its first stage performance took place a half year later.

Orff, Carl, composer. Born Munich, Germany, July 10, 1895. He graduated from the Akademie der Tonkunst in Munich in 1914. For four years he worked as coach and conductor in several German theaters. In 1921 he returned to study as a pupil of Heinrich Kaminski. In 1925 he received recognition through his adaptation of Monteverdi's *Orfeo*. After 1925 he taught at the Gunther School of Music and conducted the Bavarian Theater orchestra.

In 1935 Orff rejected all the works he had written up to then and began writing operas exclusively—operas in which the paraphernalia of costuming, scenery, and staging were discarded for a return to basic essentials. His first opera in this new style was *Car-*

mina Burana, based on anonymous medieval poems. It was presented in Frankfurt in 1937. *Carmina Burana* was the first work in a trilogy that eventually included *Catulli Carmina* and *Trionfo di Aphrodite.* The entire trilogy, *Trionfi,* was introduced at La Scala early in 1953. These and later Orff operas are based on texts that reach into the remote past, not only to medieval poetry, but also to old Bavarian legends and literature of ancient Greece. Orff's *Antigone* was introduced at the 1949 Salzburg Festival, his *Die Bernauerin* at the 1950 Munich Festival. These operas dispense with traditional lyricism, substituting for it a kind of rhythmic declamation, sometimes unaccompanied by instruments. Besides the works already mentioned, Orff has written *Der Mond* (1936, revised 1945); *Die Kluge* (1942); *Astutuli* (1946).

Orford, Ellen, schoolmistress (soprano) in Britten's *Peter Grimes.*

O riante nature, Baucis' aria in Act II of Gounod's *Philémon et Baucis.*

O Richard, O mon Roi! Blondel's aria in Grétry's *Richard Coeur-De-Lion.*

Orlovsky, Prince, a nobleman (generally sung by a mezzo-soprano) in Johann Strauss's *Die Fledermaus.*

Oroveso, chief of the Druids (bass) in Bellini's *Norma.*

Orpheus (or Orfeo), (1) a Thracian musician celebrated in Greek legend. The most famous story concerns his mission to Hades to retrieve his dead wife, Eurydice. This theme has been a favorite with composers since the beginnings of opera. Peri's *Euridice* (1600) and Caccini's *Euridice* (1602) were early settings of the story. Monteverdi's *Orfeo* (1607) was another. Joseph Haydn wrote an opera called *Orfeo e Euridice* thirty years after Gluck's work of the same title. Jacques Offenbach wrote a satire on the Orpheus story, *Orpheus in the Underworld.* The twentieth century composer

Francesco Malipiero has written a trilogy of operas entitled *L'Orfeide*, and Alfredo Casella, his contemporary, wrote *La favola d'Orfeo*. Ernst Křenek composed an *Orpheus und Eurydike* (1923).

(2) Euridice's husband in Gluck's *Orfeo ed Euridice*, Joseph Haydn's opera of the same title, and Claudio Monteverdi's *Orfeo*. In the first work the role is for a contralto, in the second a tenor, in the third a baritone.

Orpheus in the Underworld (Orphée aux Enfers), opéra bouffe by Jacques Offenbach. Book by Hector Crémieux. Première: Bouffes Parisiens, Paris, October 21, 1858. This is a burlesque on the Olympian gods and, incidentally, on the legend of Orpheus and Eurydice. Orpheus, a teacher of music in Thebes, is the husband of Eurydice. Both find their love elsewhere. Orpheus is attracted to Chloe, a shepherdess, while Eurydice loves the shepherd Aristeus, actually Pluto in disguise. When Eurydice elopes to Hades with Aristeus, Orpheus is delighted, but convention decrees that he try to reclaim her. He calls upon Jupiter for help. Jupiter commands Pluto to surrender Eurydice to her husband. Orpheus is ordered not to look at his wife until he has passed the Styx. Now it is Jupiter who falls in love with Eurydice. He hurls a bolt of lightning at Orpheus which so frightens the musician that he momentarily looks at his wife, thus losing her. Jupiter takes Eurydice as a Bacchante and Orpheus happily returns to Chloe.

The overture is a staple in the repertory of salon orchestras and "pop" concerts.

When introduced, *Orpheus in the Underworld* was not well received and seemed doomed to failure. But when the critic Jules Janin attacked it as a profanation of "holy and glorious antiquity" he aroused so much curiosity that the work suddenly attracted capacity audiences. It stayed on for 228 performances, and closed only because the cast needed a rest.

Orsini, Maffio, young nobleman (contralto) in Donizetti's *Lucrezia Borgia*.

Orsini, Paolo, head of the house of Orsini (bass) in Wagner's *Rienzi*.

Ortensio, the Countess of Berkenfeld's servant (bass) in Donizetti's *The Daughter of the Regiment*.

Ortrud, Telramund's wife (mezzo-soprano) in Wagner's *Lohengrin*.

O ruddier than the cherry, Polyphemus' aria in Part II of Handel's *Acis and Galatea*.

Orzse, childhood sweetheart (soprano) of Háry János in Kodály's *Háry János*.

Osaka, rich libertine (tenor) in Mascagni's *Iris*.

O Signore, dal tetto natio, war chorus in Act III of Verdi's *I Lombardi*.

O sink hernieder, Nacht der Liebe, *see* LIEBESNACHT.

Osmin, overseer of the Pasha's house (bass) in Mozart's *The Abduction from the Seraglio*.

O soave fanciulla, love duet of Mimi and Rodolfo in Act I of Puccini's *La Bohème*.

O sommo Carlo, trio of Elvira, Ernani, and Don Carlo, with chorus, in Act III of Verdi's *Ernani*.

O Souverain! O Juge! O Père, Rodrigo's aria in Act III of Massenet's *Le Cid*.

Ossian, *see* MACPHERSON, JAMES.

Otello (Othello), opera in four acts by Giuseppe Verdi. Libretto by Arrigo Boïto, based on the Shakespeare tragedy. Première: La Scala, February 5, 1887. American première: Academy of Music, New York, April 16, 1888.

Characters: Otello, a Moor, general in the Venetian army (tenor); Iago, his aide (baritone); Cassio, Otello's lieutenant (tenor); Roderigo, a Venetian gentleman (tenor); Lodovico, ambassador of the Venetian Republic (bass); Montano, Otello's predecessor as governor of Cyprus (bass); Desdemona, Otello's wife (soprano); Emilia, Iago's wife (mezzo-soprano); a herald (bass); soldiers, sailors, Venetians, Cypriots. The setting is a seaport of Cyprus toward the end of the fifteenth century.

Act I. Outside Otello's castle. A celebration honors the arrival of Otello, the people hailing his victory over the Turkish fleet ("Esultate!"). Among those present is Iago, who is resentful that Otello had chosen Cassio as his lieutenant. Cassio is here, too, in the company of Roderigo, who is in love with Otello's wife. Iago advises Roderigo to be patient, for surely Desdemona will tire of her husband. A festive bonfire is now lit, hailed by the people ("Fuoco di gioia!"). As the fire burns, Iago sings a drinking song (Brindisi: "Inaffia l'ugola!"). He induces Cassio to drink wine. When Cassio becomes inebriated, Iago provokes him to fight a duel with Montano, in which the latter is wounded. A riot develops, quelled only by the arrival of Otello. Otello punishes Cassio by removing him from his command. After the crowd disperses Otello is joined by his wife. Tenderly they recall how they came to fall in love ("Già nella notte densa").

Act II. A hall in the castle. Bent on destroying Otello, Iago philosophizes that God is a cruel being who meant man to be cruel ("Credo in un Dio crudel"). He encourages Cassio to address Dedemona, then when Otello apears, he fans the latter's jealousy by ointing out how Cassio and Desdemona are in consultation in the near-by garden. To appease his suspicions, Otello questions his wife about Cassio; his suspicion is aroused when Desdemona insists that Cassio is innocent of all wrongdoing. When Desdemona wipes her brow with a handkerchief which Otello had previously given her as a gift, he throws it passionately on the ground. This handkerchief is passed on by Emilia to Iago, who contrives to use it in his plot against Otello. Mean-

while, Otello bids farewell to his tranquility of mind and his hopes for the future ("Ora e per sempre addio sante memorie"). Iago comes, ostensibly to console Otello, but actually to arouse him further. He confides that he has found Desdemona's handkerchief in Cassio's room. Frantic with anger, Otello falls on his knees and swears to seek revenge ("Sì, pel ciel marmoreo giuro!")

Act III. The great hall of the castle. On further questioning, Desdemona insists to Otello that both she and Cassio are innocent ("Dio ti giocondi"). Otello asks Desdemona for her handkerchief, but the one she offers him is not the one he wants, and he sends her off to her room to fetch it. While she is gone, he bitterly laments that all his illusions have been shattered ("Dio! mi potevi scagliar"). Iago appears and urges Otello to secrete himself behind a column. When Otello does so, Iago accosts Cassio and induces him to boast about a love affair, which Otello assumes is with Desdemona. Cassio reveals to Iago that he has found Desdemona's handkerchief in his room, and wonders how it came there. The sight of the handkerchief is the final proof to Otello that his wife is guilty; he determines to kill her. Lodovico, the Venetian ambassador, arrives to tell Otello that he must return to Venice and turn over his place in Cyprus to Cassio. As Otello imparts this news to Desdemona, he becomes blind with rage and hurls her angrily to the ground. Desdemona bewails the fact that she has lost her husband's love forever ("A terra! Sì nel livido fango"). Otello himself becomes so overwrought that he falls fainting to the ground.

Act IV. Desdemona's bedroom. As Emilia prepares Desdemona for bed, Desdemona tells her of a song she learned in childhood (Canzone del Salce—Willow Song: "Salce! salce!"). After Emilia departs, Desdemona falls on her knees to pray to the Madonna for protection ("Ave Maria"). Otello arrives, once again to subject her to relentless questioning about Cassio. Her insistence that she is innocent arouses his fury. Mad with rage he chokes and kills his wife. Emilia and Lodovico burst into the room to find Desdemona dead. It is only now that Otello learns the truth: that Iago has created the fiction of Desdemona's infidelity in order to destroy him. Otello mourns the sad fate of his wife, then kills himself with his dagger (Otello's Death: "Niun mi tema").

With *Otello,* Verdi broke a self-imposed silence as composer for the stage that had lasted fifteen years: since 1871, when he had completed *Aïda.* The stimulus sending him back to opera was the powerful and moving libretto that Arrigo Boïto had fashioned from Shakespeare's tragedy. It was a far different Verdi who wrote *Otello* from the one who had gained fame with *Rigoletto, La traviata, Il trovatore,* and *Aïda.* Since his libretto called for a different musical approach—and since Verdi had assimilated Wagner's music dramas—he emerged in *Otello* as a supreme musical dramatist, whose music was continually the handmaid to the play, serving it completely. There was no attempt to imitate Wagner, but there was a conscious effort to bring to Italian lyricism a greater expressiveness, a deeper psychological insight into characters, and a greater dramatic force and truth. Verdi's score was an integrated, indivisible texture—no longer a collation of attractive sections.

The première of Otello attracted wide attention. Verdi himself was uncertain of his opera's worth, and he reserved the right to withdraw it if he found it unsatisfactory in rehearsal. A thunder of applause greeted the composer when the opera ended, and he had to take over twenty curtain calls. Some in the audience wept. Verdi's

carriage was dragged by his admirers to his hotel. Until five in the morning, his public continued to shout: "Viva, Verdi!"

That première profited by having the roles of Otello and Iago performed by Francesco Tamagno and Victor Maurel, each to become world-famous for these particular interpretations. Later distinguished Otellos were Leo Slezak, Giovanni Zenatello, and Giovanni Martinelli.

Otello, il Moro di Venezia (Othello, the Moor of Venice), opera by Rossini. Libretto by Francesco di Salsa, based on the Shakespeare tragedy. Première: Teatro del Fondo, Naples, December 4, 1816. The plot is essentially that of the opera discussed above.

Otello's Death, *see* NIUN MI TEMA.

O temple magnifique (O tempio sontuoso), Selika's aria in Act V of Meyerbeer's *L'Africaine.*

O terra, addio, the duet of Aïda and Radames, their farewell to the world, in Act IV, Scene 2, of Verdi's *Aïda.*

Othello, a Moor, general in the Venetian army, in Rossini's *Otello, il Moro di Venezia* and Verdi's *Otello.* In both operas the role is for a tenor.

O transport, O douce extase (Dove Son? O qual gioia), love duet of Vasco da Gama and Selika in Act IV of Meyerbeer's *L'Africaine.*

Ottavio, (1) Donna Anna's beloved (tenor) in Mozart's *Don Giovanni.*

(2) A wealthy Venetian (bass) in Wolf-Ferrari's *Le donne curiose.*

Ottokar, Prince of Bohemia (baritone) in Weber's *Der Freischütz.*

O tu che in seno agli angeli, Alvaro's aria in Act III, Scene 1, of Verdi's *La forza del destino.*

Oui, je veux par le monde, Wilhelm Meister's philosophy of life in Act I of Thomas's *Mignon.*

Ourrias, Mireille's suitor (bass) in Gounod's *Mireille.*

Où va la jeune hindoue? Lakmé's

celebrated Bell Song in Act II of Delibes's *Lakmé*.

O vago suol della Turrena, *see* O BEAU PAYS DE LA TOURAINE.

Over the dark blue waters, quartet of Rezia, Huon, Sherasmin, and Fatima in Act II, Scene 2, of Weber's *Oberon*.

Overture, an instrumental preface or prelude to an opera. The earliest operas did not have instrumental overtures, but began with extended vocal prologues. The use of brief instrumental preludes was soon begun; the early Italian composers calling these pieces "sinfonias" or "toccatas." It was early in the eighteenth century that Alessandro Scarlatti introduced what has since been called the "Italian overture," henceforth employed extensively by the Neapolitan school and its imitators. This type of overture comprised three sections: the first fast, frequently in fugal style, the second slow, the third fast. In France, Lully popularized a different kind of overture, now known as the "French overture." It began with a majestic slow section, was followed by a fast part, generally in fugal style, and concluded with a popular dance, such as a minuet.

These types were utilized by composers up to the time of Gluck's *Orfeo ed Euridice*. The music of the overtures bore no direct relationship to the operas they preceded. Since they were independent pieces, indeed, it was sometimes the practice of composers to use the overture of one opera for another. Cristoph Willibald Gluck was the first composer to create a relationship of mood between his overture and his opera. In his *Alceste* and *Iphigénie en Aulide* the overture establishes the atmosphere of the opera, and its concluding bars lead directly into the opening scene. Mozart was one of the first composers to use materials from his opera in the overture. This effort by Gluck and Mozart to create an intimate bond between overture and opera was furthered by Beethoven, Weber, and Wagner. The overtures of these composers were frequently miniature tone poems expressing the dramatic and emotional substance of the opera. Some of Wagner's overtures are called "preludes" to emphasize their role in creating the emotional setting for the opening scene.

O vin dissipe la tristesse (Chanson bachique), Hamlet's drinking song in Act II of Thomas's *Hamlet*.

O welche Lust! The prisoners' chorus in Act I of Beethoven's *Fidelio*.

Ozean, du Ungeheuer! *See* OCEAN, THOU MIGHTY MONSTER.

P

Pace e gioia, Almaviva's sardonic greeting to Bartolo in Act III of Rossini's *The Barber of Seville*.

Pace, pace, mio Dio! Leonora's aria in Act IV, Scene 2, of Verdi's *La forza del destino*.

Pace t' imploro, Amneris' plea for peace in the closing scene of Verdi's *Aïda*.

Paco, a gypsy (tenor) in Falla's *La vida breve*.

Paër, Ferdinando, composer. Born Parma, Italy, June 1, 1771; died Paris, May 3, 1839. After studying with Gasparo Ghiretti, he wrote his first opera, *La locanda dei vagabondi*, produced in Parma in 1789. He now wrote other operas, both comic and serious, all in the traditional Italian style, but after settling in Vienna in 1797, and becoming acquainted with Mozart's operas, his style became more refined, richer in content, with increasing dramatic interest. In 1802 he was appointed court kapellmeister in Dresden. During this period he wrote *Eleonora,* based on the same story as Beethoven's *Fidelio,* and introduced in Dresden in 1805. In 1807 he went to Paris where he became maître de chapelle to the court of Napoleon and conductor at the Opéra-Comique. In 1812 he succeeded Spontini as director of the Théâtre des Italiens, holding this post for fifteen years. In 1832 he was appointed conductor of royal chamber music. He wrote forty-three operas, all of them now forgotten. The most popular were: *I Molinari* (1793); *La Griselda* (1796); *La Sofonisba* (1796); *Achille* (1801); *Eleonora* (1805); *Didone abbandonata* (1810); *Agnes* (1819); *Le maître de chapelle* (1821).

Pagano, Arvino's brother (baritone) in Verdi's *I Lombardi.*

Page, husband (baritone) of one of Sir John Falstaff's prospective loves in Nicolai's *The Merry Wives of Windsor.* When the opera is sung in its original language, German, Page is known as Herr Reich.

Page, Mistress, (1) one of Sir John Falstaff's prospective loves (mezzo-soprano) in Nicolai's *The Merry Wives of Windsor.*

(2) The same (soprano) in Verdi's *Falstaff.*

Page's Song, *see* NOBLES SEIGNEURS, SALUT.

Pagliacci, opera in two acts by Ruggiero Leoncavallo. Libretto by the composer. Première: Teatro dal Verme, Milan, May 21, 1892. American première: Grand Opera House, New York, June 15, 1893.

Characters: Canio, head of a theatrical troupe (tenor); Nedda, his wife (soprano); Tonio, a clown (baritone);

Beppe, another clown (tenor); Silvio, a villager (baritone); peasants; villagers. The setting is Montalto, a village in Calabria, during the Feast of Assumption.

Prologue. A vivacious orchestral prelude describes the gaiety attending a village festival. Tonio steps before the curtain to explain that the play about to be witnessed is a real story with real people ("Si può?").

Act I. Entrance to the village of Montalto. The villagers hail the arrival of an itinerant theatrical company ("Viva Pagliaccio"). Canio announces a performance for that evening. Just before entering the inn, he invites Tonio to accompany him. When the clown refuses, a villager suggests slyly that perhaps Tonio wants to stay behind so that he can be alone with Canio's wife, Nedda. For a moment Canio is disturbed, but he brushes away his anxiety. As the church bells ring, the villagers assemble for vespers (Coro delle Campane—Chorus of the Bells: "Din, don, suona vespero"). The crowd departs, and Nedda is alone, at first troubled by Canio's momentary display of jealousy, but then dispelling her troubles with thoughts about the carefree flight of birds (Ballatella: "Oh! che volo d'augelli"). Tonio appears and tries to make love to her, but she is repelled by the ugly, deformed clown and drives him away with a whip. A villager, Silvio, is more pleasing to Nedda, and she is receptive to his lovemaking. They are overheard by the clown, who quickly summons Canio to witness the scene; but before Canio can identify his rival, Silvio escapes. When Nedda refuses to divulge the identity of her lover, Canio attacks her with a dagger, but she is saved by Beppe. Overwhelmed by the realization that his wife is unfaithful, Canio sobs bitterly and remarks on his tragic plight, appearing in a comic play while his heart is breaking ("Vesti la giubba").

Act II. The same scene. Villagers gather before the little traveling theater to witness the evening's performance. The curtain rises on a play which, by unhappy coincidence, concerns a situation similar to that involving the principal performers. Harlequin (played by Beppe) serenades Columbine (Nedda) outside her window, while her husband is away ("O Columbina"). Taddeo (Tonio) enters and tries to make love to Columbine ("E casta al par di neve!"), but he is soon driven away by Harlequin who enters through the window. Harlequin and Columbine are interrupted in their tryst by the sudden appearance of Columbine's husband, Pagliaccio (Canio). Harlequin makes his escape through the window. All at once, Canio forgets he is playing a part in a play. He demands the name of Nedda's lover. When she refuses to give it, he kills her with his dagger. From the audience, Silvio rushes to the stage to help Nedda, but it is too late. He, too, is slain by Canio, who then remarks to the horrified audience: "La commedia è finita" —"The comedy is ended."

Pagliacci belongs to the "verismo" school of Italian opera, which presented everyday characters in everyday situations. This new school of realism had been introduced by Mascagni's *Cavalleria rusticana* only two years earlier; and Leoncavallo's opera contributed handsomely to popularize its aims.

It is customary to speak of *Pagliacci* and *Cavalleria rusticana* in the same breath: the two operas are usually performed on the same bill. They have striking points of similarity in musical style, and in the emotional turmoil of their stories. But it is important to notice their points of difference. To his opera, Leoncavallo brought a refinement of writing and a poetic feeling, as well as occasional comic relief, not found in Mascagni's *Cavalleria.*

Victor Maurel, who created the role of Tonio, brought more than his vocal artistry to bear in making *Pagliacci* a sensational success at its première, for it was he who suggested to the composer the idea of the Prologue—an afterthought that Leoncavallo turned into one of the opera's finest arias.

Paillasse, the title by which *Pagliacci* is known in France.

Paintings and operas. The following operas were inspired by paintings and artists: Gian-Carlo Menotti's *Amahl and the Night Visitors* by Hieronymous Bosch's *The Adoration of the Magi;* Richard Strauss's *Friedenstag* by Velásquez' *The Surrender of Breda;* Enrique Granados' *Goyescas* by paintings of Goya; Igor Stravinsky's *The Rake's Progress* by William Hogarth's series of paintings of the same title; Josef Rheinberger's *Die sieben Raben* by a set of paintings by Moritz von Schwind.

The painter Thomas Gainsborough is the central character in Albert Coates' opera *Gainsborough;* Mathias Gruenewald in Paul Hindemith's *Mathis der Maler;* Rembrandt in Paul von Klenau's *Rembrandt von Rijn* and Henk Badings' opera of the same name.

Paisiello, Giovanni, composer. Born Taranto, Italy, May 9, 1740; died Naples, June 5, 1816. For nine years he attended the Conservatorio San Onofrio in Naples, beginning in 1754, his teachers including Francesco Durante and Geronimo Abos. For a while he specialized in writing choral music, but in 1763 he wrote a comic intermezzo that was so successful when introduced at the Conservatory that he received a commission for an opera, *La Pupilla,* a pre-eminent success when introduced in Bologna in 1764. During the next dozen years Paisiello wrote over fifty operas, many of them extensively performed; he became one of the most celebrated opera composers in Italy. In 1776 he was invited to Russia by Catherine II. He remained in St.

Petersburg eight years, acting as the Empress' music master. It was during this period that he wrote and had produced his best and most popular work, the opera buffa *The Barber of Seville*. Paisiello returned to Naples in 1784. For the next fifteen years he was maestro di cappella for Ferdinand IV, writing such successful operas as *L'Olimpiade* (1786), *La molinara* (1788), and *Nina* (1789).

When a republican government was temporarily established in Naples, in 1799, Paisiello became associated with the new regime as Composer to the Nation. After the Restoration, the court refused to reinstate him because of his republican associations. For two years Paisiello remained in Naples without employment. Then he was called to Paris by Napoleon to serve as maître de chapelle. Because of his wife's poor health, Paisiello returned to Naples two years later where, once again, he was the object of honor and acclaim, and was reinstated as maestro di cappella. His good fortune terminated when Ferdinand IV returned to Naples. From then on, Paisiello suffered neglect and poverty.

In all, Paisiello wrote over a hundred operas. The best abound in pleasing, graceful melodies and ingratiating comedy, and at times contain unusual instrumental and dramatic effects. His most successful operas were: *La Pupilla* (1764); *Demetrio* (1765); *Don Chisciotte* (1769); *Achille in Sciro* (1778); *La finta amante* (1780); *Il barbiere di Siviglia* (1782); *Andromeda* (1784); *L'Olimpiade* (1786); *La molinara* (1788); *Nina* (1789); *Proserpina* (1803).

Palemon, an old monk (bass) in Massenet's *Thaïs*.

Palestrina, opera by Hans Pfitzner. Libretto by the composer. Première: Munich Opera, June 12, 1917. This is the composer's most famous opera, and one of the last products of German romanticism. The theme is the legendary saving of the art of contrapuntal music from banishment by the Church in the sixteenth century through the success of Palestrina's most celebrated composition, the *Missa Papae Marcelli*. In the opera, Palestrina is told by Cardinal Borromeo that all sacred music except the Gregorian plain song is to be prohibited by the Council of Trent. Refusing to fight the edict, Palestrina vows he will never compose again. But he is visited by the spirits of nine composers who prevail on him to return to composition; and an angel sings him a theme which becomes part of the *Missa Papae Marcelli*. Palestrina falls asleep as he writes his music; his son and his pupil gather the manuscript pages. The mass is shown to the Pope, who is profoundly impressed. Thus, contrapuntal music is saved.

Though originally highly successful, and given frequent performances, *Palestrina* has virtually passed out of the permanent repertory, except in Germany. The reasons for this are its lack of love interest (there are no female characters), archaic flavor, and its long stretches of dullness. Occasionally, the three orchestral preludes, each prefacing an act, are heard at symphony concerts.

Pamela, *see* CECCHINA, LA.

Pamina, the Queen of the Night's daughter (soprano) in Mozart's *The Magic Flute*.

Pandolphe, Cinderella's father (bass) in Massenet's *Cendrillon*.

Pang, the General Purveyor (tenor) in Puccini's *Turandot*.

Panizza, Ettore, conductor. Born Buenos Aires, Argentina, August 12, 1875. Of Italian parents, Panizza was sent to Italy to study at the Milan Conservatory, from which he was graduated in 1900. After a rigorous apprenticeship as conductor of symphony orchestras and opera companies throughout Italy he was appointed principal conductor of Italian operas at Covent Garden in 1907: he held this post six years. In 1916 he made his debut at La Scala, and in 1921 he assisted Arturo Toscanini there as principal conductor. His first appearance in America took place with the Chicago Civic Opera, where he served as principal conductor for several years. When Tullio Serafin resigned as conductor of Italian operas at the Metropolitan Opera, Panizza was chosen as his successor. He made his debut at the Metropolitan Opera on December 22, 1934, with *Aïda*, remaining there through the 1941–1942 season, during which period he led the world premières of *In the Pasha's Garden* and *The Island God*. He wrote the following operas: *Il fidanzato del mare* (1897); *Medio Evo Latino* (1900); *Bisanzio* (1939).

Pantalone, a club member (baritone) in Wolf-Ferrari's *Le donne curiose.*

Panza, Sancho, Don Quixote's squire (baritone) in Massenet's *Don Quichotte.*

Paolino, a lawyer (tenor) in Cimarosa's *Il matrimonio segreto.*

Paolo e Francesca, opera by Luigi Mancinelli. Libretto by Arturo Colautti, based on Dante. Première: Teatro Communale, Bologna, November 11, 1907. The theme is the tragic romance of Paolo and Francesca as described by Dante in the fifth canto of the "Inferno" in *The Divine Comedy.* See FRANCESCA DA RIMINI.

Papagena, Papageno's sweetheart (soprano) in Mozart's *The Magic Flute.*

Papageno, bird catcher (baritone), Tamino's attendant, in Mozart's *The Magic Flute.*

Papi, Genarro, conductor. Born Naples, Italy, December 21, 1886; died New York City, November 29, 1941. After completing studies at the Naples Conservatory he held various posts as chorus master and conductor. He came to the United States in 1913 and joined the conducting staff of the Metropoli-

tan Opera as Arturo Toscanini's assistant. After Toscanini left the Metropolitan, Papi was elevated to principal conductor, making his debut in that capacity on November 16, 1916, with *Manon Lescaut.* From 1916 to 1927 he conducted at the Metropolitan, resigning to become first conductor of the Chicago Civic Opera. In 1935 he returned to his old post at the Metropolitan. He died just before he was to conduct a performance of *La traviata.* Another conductor was found, but the news of Papi's death was withheld from the cast until the performance ended. (This was a performance, broadcast by radio, in which Jan Peerce was making his debut.)

Paquiro, a toreador (baritone) in Granados' *Goyescas.*

Paradise Lost, see MILTON, JOHN.

Parassia, Tcherevik's daughter (soprano) in Mussorgsky's *The Fair at Sorochinsk.*

pardon de Ploërmel, Le, see DINORAH.

Paride ed Elena (Paris and Helen), opera by Cristoph Willibald Gluck. Libretto by Raniero de Calzabigi. Première: Burgtheater, Vienna, November 3, 1770. The libretto is based on the Greek mythological tale in which Paris, son of the King of Troy, sets sail for Greece. There he meets Helen, wife of Menelaus; he takes her off to Troy, an act that precipitates the Trojan War. In Gluck's opera, Helen is not married to Menelaus, and the central interest is in the passionate love of Helen and Paris. Paris' aria in Act I, "O del mio dolce ardor," is one of the most beautiful Gluck wrote.

Paride ed Elena was the third of Gluck's operas in his later, fully developed style (*Orfeo ed Euridice* and *Alceste* were the preceding ones). Its failure in Vienna led Gluck to leave that city and go to Paris.

Parigi, o cara, duet of Alfredo and Violetta in Act III of Verdi's *La traviata.*

Paris, a nobleman (baritone) in Gounod's *Roméo et Juliette.*

Paris Opéra, *see* OPERA, L'.

Parker, Horatio William, composer. Born Auburndale, Massachusetts, September 15, 1863; died Cedarhurst, Long Island, December 18, 1919. His music study took place in Boston with Stephen Emery, John Orth, and George W. Chadwick and in Munich, Germany, with Josef Rheinberger and Ludwig Abel. After returning to the United States in 1885 he became director of musical instruction in the schools of St. Paul and St. Mary in Garden City, Long Island; after this he served as organist in various New York churches, and as a teacher of counterpoint at the National Conservatory of Music, then under the direction of Antonin Dvořák. In 1894 he became a professor of music at Yale University, a position he retained until his death. Considerable recognition came to him a year earlier when his oratorio *Hora Novissima* was first heard. It was soon performed in a number of cities in the United States and in England. In 1911 Parker won a ten thousand dollar prize offered by the Metropolitan Opera for American opera. His opera was *Mona,* introduced by the Metropolitan on March 14, 1912, played four times to favorable response, and then dropped from the repertory. Two years later Parker won another ten thousand dollar prize, offered by the National Federation of Women's Clubs, with his opera *Fairyland,* produced in Los Angeles on July 1, 1915.

Park Theater, a theater in Park Row, New York City, where the first season of Italian opera in the United States was given by Manuel García in 1820. The original theater bearing this name had been built in 1798 as a home for spoken drama, but burned down in 1820. It was in a new building on the same site that García's company offered its productions. This theater was also destroyed by fire, in 1848.

Parlando (or Parlante), Italian for "speaking." In vocal music the term is an indication for the singer to imitate the sound of speech.

Parmi les pleurs (in preda al duol), Valentine's aria in Act IV of Meyerbeer's *Les Huguenots.*

Parmi veder le lagrime, the Duke's aria in Act II (or, in some versions, Act III) of Verdi's *Rigoletto.*

Parpignol, a toy vendor (tenor) in Puccini's *La Bohème.*

Parsifal, a "stage-consecrating festival drama" in three acts by Richard Wagner. Libretto by the composer, based on a medieval legend and a poem by Wolfram von Eschenbach. Première: Bayreuth, July 26, 1882. American première: Metropolitan Opera, New York, December 24, 1903.

Characters: Titurel, former King of the Knights of the Grail (bass); Amfortas, his son and successor (baritone); Gurnemanz, another Knight of the Grail (bass); Parsifal, a "guileless fool" (tenor); Kundry, half-woman, half-sorceress (soprano); Klingsor, a magician (bass); Knights of the Grail, flower maidens, squires, boys. The setting is in and about the Castle of Monsalvat in the Spanish Pyrenees in the Middle Ages.

Act I, Scene 1. A forest near Monsalvat. An orchestral prelude establishes the spiritual atmosphere of the opera; it is built from several of the drama's motives, beginning with the "Last Supper," and continuing with the "Grail," "Faith," and "Lance" motives.

Gurnemanz and his young squires kneel in prayer. Gurnemanz then tells the squires that the ailing Amfortas must be helped when he comes to bathe his spear wound in the lake. Kundry, a servant of the Grail Knights, gives Gurnemanz a vial of oil for the King's wound. When Amfortas is brought in on a litter, he expresses his despair of ever being cured. After he has gone to bathe, Gurnemanz tells his squires the history of the King's wound. Amfortas had been enticed into the garden of Klingsor, the magician who has determined to secure both the Holy Grail and the Spear which had pierced the body of Jesus. Wresting the Spear from Amfortas, Klingsor had wounded the King with it. Amfortas' wound, it is believed, can be healed only by recovery of the sacred relic, and the one destined to make this recovery is some "guileless fool." As Gurnemanz ends his narrative a wild swan falls to the ground, slain by an arrow. Parsifal appears, bow in hand. To Gurnemanz' queries, he reveals all that he knows about himself: the forest is his home, and his mother is named Heart's Sorrow. Gurnemanz, recognizing the boy as the "guileless fool," conducts him to Monsalvat. The scene changes to the stately music of the "Transformation Scene."

Scene 2. The hall of the Holy Grail. When Gurnemanz and Parsifal arrive, the hall is empty, but soon the Knights of the Holy Grail appear, singing as they file in ("Zum letzten Liebesmahle"). Amfortas is then brought in and, as the Knights partake of Communion, he uncovers the Holy Grail. Parsifal watches the spectacle but is unmoved, uncomprehending. Disgusted at his stupidity, Gurnemanz rudely drives him away.

Act II, Scene 1. A tower atop Klingsor's castle. The world of the magician is evoked in a brief prelude that includes the "Enchantment" and "Kundry" motives. Klingsor summons Kundry and orders her to seduce Parsifal so that he may be eliminated as the magician's opponent. Since Kundry is under Klingsor's spell, she yields helplessly to his command.

Scene 2. Klingsor's magic garden. Klingsor's flower maidens dance enticingly about Parsifal ("Flower

Maidens' Scene"). Kundry appears, no longer a hag but a beautiful woman. She reveals to Parsifal that it was she who gave him his name, and she tells him of his parents and his birth ("Ich sah das Kind an seiner Mutter Brust"). Moved, Parsifal at first yields to Kundry's kisses and embraces. But he remembers Amfortas, and he senses that it was in just such a garden as this that, tempted by a woman's beauty, the King received his wound. He pushes Kundry away. When Kundry calls on Klingsor for help, the magician hurls the sacred Spear at Parsifal. Protected by magic powers, Parsifal is unharmed: the Spear remains suspended in midair. Parsifal grasps the weapon, makes the sign of the cross with it, and declares Klingsor's power ended. Kundry falls to the ground with a cry of anguish— and Klingsor's castle collapses in ruin.

Act III, Scene 1. A hermit's hut near Monsalvat. A brief prelude depicts the gloom surrounding the Knights of the Grail; the main motives are those of "Kundry," "Spear," "Grail," "Promise," and "Enchantment." Years have passed. Now an aged hermit, Gurnemanz encounters the repentant Kundry. Parsifal, now an armored knight, appears carrying the sacred Spear. Gurnemanz recognizes him and tells him of the present sad state of the Knighthood. He sprinkles water on Parsifal's head; Kundry bathes Parsifal's feet and dries them with her hair. The countryside becomes radiant: it is Good Friday (Charfreitagszauber—Good Friday Spell). Tolling bells summon the Knights to prayer. Parsifal proceeds with Kundry and Gurnemanz to Monsalvat.

Scene 2. The great hall at Monsalvat. The Knights enter with the bier of Titurel, who has died in despair of having the Grail ceremony restored as of old. The ailing Amfortas is helped to his throne. In the depths of his misery he vows never again to uncover the

Grail; he implores the Knights to kill him and end his suffering. Parsifal appears, touches Amfortas' wound with the Spear, and heals him ("Nur eine Waffe taugt"). Parsifal then sinks in prayer before the Grail. The Grail glows with holy light. A beam of light descends upon Parsifal, and a dove flutters above his head. Kundry, absolved of her sins, dies at Parsifal's feet. Gurnemanz and Amfortas bow before Parsifal as he raises the Holy Grail in a renewal of consecration.

Parsifal, Wagner's last drama, had a special significance for him. He insisted, because of its religious content, that it never be presented in an ordinary opera house but confined to Bayreuth, where it should be presented as a kind of religious service. The theme of this religious drama was "enlightenment coming through conscious pity by salvation." Salvation had been the theme of some of Wagner's earlier works, but in them salvation had come through love, sacred and profane. In *Parsifal,* salvation comes through compassion, renunciation, and suffering.

There is no question that *Parsifal* is a moving, and at times inspiring, spectacle; but it cannot be said that it is a great music drama. The play is too static, the characters are too often lifeless, the monologues too numerous and attenuated. The work as a whole lacks artistic validity. It is most effective in its spiritual sections, less so in its more human and earthy scenes. It remains, as Wagner intended it to be, a great religious spectacle, but it is not great musical theater.

For several years after its première *Parsifal* remained Bayreuth's exclusive property. Concert versions were permitted, and these took place throughout the music world after Wagner's death, but Wagner's widow was scrupulous about not permitting *Parsifal* to be presented on any stage outside Bayreuth. The first such presentation away from Bayreuth took place at the Metropolitan Opera in 1903. For the dramatic circumstances surrounding this performance, *see* METROPOLITAN OPERA.

Partagez-vous mes fleurs! Ophelia's Mad Scene in Act IV of Thomas's *Hamlet.*

Pasha, The, *see* SELIM.

Pasquale, Don, an old bachelor (bass) in Donizetti's *Don Pasquale.*

Passacaglia, a dance of Spanish origin dating from the early seventeenth century. In stately triple time, it is characterized by a theme played in the bass and repeated throughout the composition, while the upper parts provide variations on the theme. Passacaglias were ordinarily so similar to chaconnes in form and style that the names were used interchangeably in the seventeenth and eighteenth centuries. For occurrences in opera *see* CHACONNE.

Passepied, a lively dance of French origin, similar to the minuet, though played considerably faster. It was used in early French ballets, and is occasionally found in operas of the late seventeenth and early eighteenth centuries. There are notable examples of this dance in Gluck's *Iphigénie en Aulide,* Mozart's *Idomeneo,* and Paisiello's *Proserpina.*

Passo a sei (Dance in Six), a dance in Act I of Rossini's *William Tell.*

Passover Scene, *see* O DIEU, DIEU DE NOS PERES.

Pasta, Giuditta (born NEGRI), dramatic soprano. Born Saronno, Italy, April 9, 1798; died Como, Italy, April 1, 1865. One of the most celebrated sopranos of the early nineteenth century, for whom Bellini wrote the roles of Norma and Amina, she entered the Milan Conservatory in her fifteenth year, where she studied with Bonifacio Asioli. Two years later she started appearing in smaller Italian opera houses. In 1817 she sang in London where she met and married the tenor, Pasta. After an additional period of study with Scappa in Italy, she returned to the opera stage in Venice in 1819. Her brilliant career, however, did not begin to unfold until 1822 in Paris, where she created a sensation. Her voice was faulty in quality and production, but she had such a remarkable range, expressiveness, and dramatic power that her shortcomings were disregarded. She repeated her Parisian triumph in London in 1824 in a series of Rossini operas. After her return to Italy she appeared in Giovanni Pacini's *Niobe,* which the composer wrote for her. In the early 1830's she created the principal soprano roles in Bellini's *Norma* and *La sonnambula* and Donizetti's *Anna Bolena,* all written expressly for her. By 1837 her voice suffered complete deterioration; nevertheless, she continued to appear in London and St. Petersburg. She went into retirement in 1850, devoting herself to teaching a few select pupils at her Como estate.

Pasticcio, Italian for "pie." The word was applied to a form of operatic entertainment popular throughout Europe in the eighteenth century. A pasticcio was made up of parts from several different operas, sometimes the operas of a single composer, but frequently of different composers. The intention was to give audiences the maximum number of familiar or exceptional songs in one performance. New words might be fitted to the songs to give cohesion to the whole. A case in point is Gluck's *Orfeo ed Euridice.* When this work was introduced in London in 1770, it was given the "benefit" of additional choruses, recitatives, and arias composed by John Christian Bach to words by Pietro Guglielmi, the program book calling attention to this pair's "enrichment" of the opera. In this case, the basic structure of Gluck's opera was preserved; in other instances, wholly new operas were concocted

from the writings of half a dozen or more composers. The term pasticcio was also extended to collaborations of several composers. *Muzio Scevola* (1721) was such an opera, its first act by Filippo Amadei, its second by Giovanni Bononcini, its third by Handel— and each act preceded by its own overture.

Pastorale, a term in use between the fifteenth and eighteenth centuries for a dramatic work on a pastoral or mythological subject, frequently allegorical in treatment. At various times pastorales were made up of arias, recitatives, choruses, and ballets. As such, they were one of the forerunners of opera. Especially popular in France, they continued to be written by such composers as Robert Cambert and Jean-Baptiste Lully well after the establishment of opera as a distinctive form.

Patti, Adelina (born ADELA PATTI), coloratura soprano. Born Madrid, Spain, February 10, 1843; died Brecknock, Wales, September 27, 1919. One of the most celebrated coloratura sopranos of all time, she was the daughter of the Italian singer Salvatore Patti and Caterina Barili. While still a child, she was brought to the United States, where she made her first public appearance when she was only seven. From 1851 to 1855 she concertized extensively under the management of Maurice Strakosch (who became her brother-in-law). At the age of twelve she ceased singing in public and began intensive study of music; her voice teachers were Ettore Barili and Strakosch. On November 24, 1859, she made her opera debut in New York as Lucia; on this occasion she was billed as "the little Florinda." She was acclaimed. Her London debut took place on May 14, 1861, in *La sonnambula.* Her popularity in England was such that it rivaled that of the sensational Giulia Grisi. Patti appeared at Covent Garden for twenty-three years, and another two years at Her Majesty's

Theatre, throughout this quarter of a century the idol of the English opera public. She duplicated her London successes in Paris in 1862 and at La Scala in 1877. She appeared in about forty roles and scored her greatest successes in operas by Bellini, Donizetti, Meyerbeer, and Rossini. She was not exceptional either as a musician or an actress, but no one could rival the beauty, purity, and sweetness of her singing.

She withdrew from opera in 1895 and for another decade appeared extensively in concerts. Her last concert took place in London in 1906, her last appearance in public (at a London Red Cross benefit) in 1914. Her older sister, Carlotta Patti (1835–1889) was also a successful singer, though not of her sister's stature. Due to a chronic infirmity of lameness, Carlotta's appearances were intermittent, but after her concert debut in 1861 she made successful appearances in Europe and the United States.

Patzak, Julius, tenor. Born Vienna, Austria, April 9, 1898. He attended the Vienna University and School of Music. His debut took place at the Reichenberg State Opera on April 3, 1926. After appearances at the Brünn Opera he became principal tenor of the Munich State Opera in 1928. During the next two decades (except for an interval during World War II) he distinguished himself, particularly in Mozart's operas, in Munich, Berlin, Vienna, Covent Garden, La Scala, and the Salzburg Festivals. In 1946 he was the first Austrian singer to be engaged by the British Broadcasting Commission following the war. In 1947 he appeared in the world première of Gottfried von Einem's *Dantons Tod* in Salzburg. He made his American debut at the Cincinnati Music Festival in May 1954.

Paul, the principal character (tenor) in Korngold's *Die tote Stadt*.

Paul et Virginie, a novel by Bernardin de St. Pierre, the source of operas by Pietro Carlo Guglielmi, Jean François Lesueur, and Victor Massé. The setting is Africa in the eighteenth century, where the lovers, Paul and Virginia, are separated when Virginia is sent home to France. Her ship is wrecked, and her body is washed ashore, to be found by the disconsolate Paul.

Pauly, Rosa, dramatic soprano. Born Eperjes, Hungary, March 15, 1895. After studying with Rosa Papier in Vienna, she made her debut in Hamburg in *Aïda*. From Hamburg she went to sing in Cologne. From 1927 to 1931 she appeared at the Kroll Opera in Berlin in many major roles, and in virtually every significant première. During this period she was also heard in guest appearances in Budapest and Paris. After 1931 she appeared with outstanding acclaim at the Vienna State Opera (hailed for her performances in dramas by Wagner and Richard Strauss and receiving the honorary title of Kämmersängerin), the Salzburg Festivals, and all the major opera houses of Italy. In Italy she was regarded so highly as an actress that she became known as the "German Duse," and her portrait was hung in the Hall of Fame at the Verdi Opera House, between those of Duse and Moissi. Her American debut took place with the New York Philharmonic-Symphony in 1937 in a concert version of *Elektra*. On January 7, 1938, she made her American opera debut at the Metropolitan Opera, once again as Elektra. Pauly remained at the Metropolitan until 1940. After World War II she appeared at Covent Garden, and in Germany and Austria.

Paur, Emil, conductor. Born Czernowitz, Austria, August 29, 1855; died Mistek, Moravia, June 7, 1932. After attending the Vienna Conservatory, he began his career in Königsberg in his twenty-first year. In 1880 he was appointed conductor of the Mannheim

Opera, and from 1891 to 1893 he conducted opera performances at the Leipzig Stadttheater. He came to America in 1893 to succeed Arthur Nikisch as conductor of the Boston Symphony orchestra. In 1899–1900 he directed performances of the Wagner music dramas at the Metropolitan Opera, making his debut there on December 23, 1899, with *Lohengrin*. In the spring of 1900 he conducted German operas at Covent Garden. In 1904 he became principal conductor of the Pittsburgh Symphony, holding this post six years. In 1912 he succeeded Karl Muck as musical director of the Berlin Opera, but due to a dispute with the Intendant, he held this post only two months. He remained in Berlin, directing symphony concerts for the rest of his career.

pauvre matelot, Le (The Poor Sailor), one-act opera by Darius Milhaud. Libretto by Jean Cocteau. Première: Opéra-Comique, December 12, 1927. The libretto was inspired by a newspaper account of an actual event. A sailor returns home after a year's absence and is not recognized by his wife. He decides to test her fidelity by telling her he is her husband's rich friend; then he tries to win her love. Late at night, while he is asleep, his wife murders him so that she may have his money and thus bring her husband home.

Pearl Fishers, The, see PECHEURS DE PERLES, LES.

Peasants' Ballet, the ballet in Act III of Moniuszko's *Halka*.

Peasants' Waltz, a waltz in Act I of Boïto's *Mefistofele*.

pêcheurs de perles, Les, (The Pearl Fishers), opera by Georges Bizet. Libretto by Michel Carré and Eugène Cormon. Première: Théâtre Lyrique, Paris, September 30, 1863. In Ceylon, in early times, a new tribal chieftain, Zurga, is chosen by the fishermen. A veiled priestess comes to pray for the people; Zurga promises the priestess a priceless pearl if she remains chaste, but death if she violates her purity. This priestess is Leila, whom Zurga and his friend Nadir had loved as youths. Nadir and Leila recognize each other, and their one-time love is revived. When Zurga discovers Nadir and Leila embracing he orders both to die. But Zurga is still in love with Leila. To save her, he sets the homes of his people aflame. When his act is discovered, the people force him to die in the flames, which spread and bring destruction to all, including Nadir and Leila.

One of Bizet's most famous tenor arias appears in this opera: Nadir's romance, "Je crois entendre encore" in Act I. Other familiar excerpts include the duet of Nadir and Zurga in Act I, "Au fond du temple," and Leila's cavatina in Act II, "Comme autrefois dans la nuit sombre."

Pedrillo, Belmonte's servant (tenor) in Mozart's *The Abduction from the Seraglio*.

Pedro, (1) a shepherd (tenor) in D'Albert's *Tiefland*.

(2) Pilar's bridegroom (tenor) in Laparra's *La Habanera*.

(3) A burlesquer (soprano) in Massenet's *Don Quichotte*.

See also DON PEDRO.

Peerce, Jan (born JACOB PERELMUTH), tenor. Born New York City, June 3, 1904. After studying the violin and playing in jazz bands, he was engaged in 1932 as a singer for the Radio City Music Hall. For five years his singing was a major attraction; he appeared nearly twenty-five hundred times on the stage and many times in radio programs originating at the theater. Presently, he became a star of his own radio program, "Great Moments in Music," and in 1936 was selected by a national poll as the leading male radio singer. Eager to embark on more serious musical endeavors, he began a period of study with Giuseppe Boghetti. In 1938 he sang in Beethoven's Ninth Symphony, conducted by Arturo Toscanini. In 1939 he made his first concert tour, and his bow in opera with the Columbia Opera Company, appearing as the Duke in *Rigoletto*.

His debut at the Metropolitan Opera took place on November 29, 1941, as Alfredo in *La traviata*. Peerce has since been a principal tenor of the Metropolitan Opera, distinguishing himself in the French and Italian repertories. He has appeared also with the San Francisco Opera as well as other major companies in America and Europe. He has sung under Toscanini in radio performances of several operas, including *La traviata, La Bohème,* and *Un ballo in maschera.*

Peer Gynt, (1) a poetic parable by Henrik Ibsen, the source of operas by Egk (see below), Leslie Heward, and Viktor Ullmann. Peer Gynt, a lustful, impetuous youth, is the symbol of moral degeneration. He abducts and then abandons the beautiful Solveig. Roaming to foreign lands, he makes love to the daughter of the Troll King. He returns home to be at the bedside of his dying mother, Ase, but then is off for further adventures and escapades. He returns home, old and wasted, to die in Solveig's arms.

(2) Opera by Werner Egk. Libretto by the composer, based on Ibsen's poetic drama. Première: Berlin Opera, November 24, 1938. When this opera was introduced, the Nazi press condemned it as unfit "for the National Socialist outlook on the world." When Hitler saw it, however, he expressed his enthusiasm; the official reaction to the opera changed overnight. It received a government prize of ten thousand marks and was performed extensively throughout Germany.

Pelléas et Mélisande, opera in five acts by Claude Debussy. Libretto is Maurice Maeterlinck's poetical drama of the same name. Première: Opéra-Comique, April 30, 1902. American pre-

mière: Manhattan Opera House, New York, February 19, 1908.

Characters: Arkel, king of Allemonde (bass); Geneviève, his daughter-in-law (soprano); Golaud, her older son (baritone); Pelléas, his brother (tenor); Mélisande, a princess (soprano); Yniold, young son of Golaud (soprano); blind beggars; servants; a physician. The setting is a legendary land in legendary times.

Act I, Scene 1. A forest. Golaud comes upon Mélisande, who is weeping. She answers his questions vaguely, refuses to reveal her identity, and will not allow him to recover the crown she has lost in a spring. Golaud persuades her to follow him to a place of shelter.

Scene 2. A hall in Arkel's castle. Six months have passed. Geneviève reads the blind Arkel a letter from Golaud telling of his marriage to Mélisande (Duo de la lettre: "Voici ce qu'il écrit"). Pelléas comes to ask permission to visit a sick friend. But the king reminds Pelléas that his own father is also sick and needs his attention. Arkel now orders Pelléas to show a signal lamp for the returning Golaud.

Scene 3. A terrace before the castle. When Geneviève goes to look after little Yniold, Pelléas and Mélisande are left alone. Pelléas reveals that he must leave the following day, news that brings Mélisande a stab of pain.

Act II, Scene 1. A fountain in the park. Pelléas and Mélisande come to a deserted fountain which is believed to have the power of opening the eyes of the blind. The fountain exerts its magic by opening the eyes of Pelléas and Mélisande to their love. As Mélisande playfully tosses her wedding ring into the air, it falls into the depths of the fountain. She is distraught. Pelléas urges her to tell Golaud the truth of its loss.

Scene 2. Golaud's chamber. At the moment Mélisande lost her ring in the fountain, Golaud hunting, was thrown from his horse. He is in bed, recovering. Mélisande, tending to him, bursts into tears. When he takes her into his arms he notices she is not wearing her wedding ring. As Golaud becomes more insistent in his questioning, Mélisande finally tells him she lost it in a grotto near the sea. Golaud orders her to go with Pelléas to look for it.

Scene 3. A grotto by the sea. It is night. Pelléas and Mélisande have come on their mock search for the ring. A sudden flood of moonlight reveals three blind beggars huddled in a corner of the cave. Mélisande is terrified, and Pelléas leads her back to the castle.

Act III, Scene 1. A tower of the castle. As she combs her hair at a window, Mélisande sings an ancient song ("Saint Daniel et Saint Michel"). Upon Pelléas' arrival on the path below, she gets him to promise not to leave on the morrow. As a reward for this promise, she extends her hand for Pelléas to kiss. Her long hair falls and covers Pelléas' face. The touch of her hair makes him ecstatic (Scène des cheveux: "Oh! qu'est-ce que c'est? . . . tes cheveux"). Golaud discovers them and scolds them for behaving like children.

Scene 2. The castle vaults. Golaud conducts Pelléas to a stagnant pool where the stench is one of death. Pelléas grows apprehensive at Golaud's strange behavior. The two leave the vaults in tense silence.

Scene 3. A terrace. Pelléas emerges from the caverns, sighing with relief ("Ah! je respire enfin"). Golaud warns him not to participate in any more childish games with Mélisande, nor to disturb her in any way, since she is about to become a mother.

Scene 4. Before the castle. Suspicious of the conduct between Pelléas and Mélisande, Golaud cautiously questions Yniold about their behavior. When the child is vague, Golaud's sus-

picions increase. A light appears in Mélisande's window. Golaud lifts his son to the window to see if Pelléas is inside. When Yniold reveals that this is so, Golaud is sure that his worst suspicions are well founded.

Act IV, Scene 1. A room in the castle. Pelléas tells Mélisande he is going away; they arrange a last rendezvous near the fountain. After they separate, Arkel appears. He is overjoyed that Pelléas' father has recovered, for he feels that the gloom of the castle will now be dispelled ("Maintenant que le père de Pelléas est sauvé"). Golaud appears. He is looking for his sword; there is blood on his brow. The innocence in Mélisande's eyes arouses his anger to a fever pitch. He seizes her by the hair and drags her across the floor until stopped by the entreaties of Arkel.

Scene 3. The Park. Waiting at the fountain, Pelléas muses on the strange destiny that has made him fall in love with his brother's wife. When Mélissande appears, all his doubts vanish. They embrace ecstatically (Duo de la fontaine: "Viens ici, ne rest pas au bord du clair de lune"). Mélisande hears a sound in the shadows. Golaud, who has been concealed there, rushes at Pelléas and kills him. He then pursues the fleeing Mélisande with drawn sword.

Act V. Mélisande's chamber. Mélisande, who has given birth to a daughter, lies in bed, grievously ill. Golaud is with her, penitent and forgiving ("Mélisande, as-tu pitié de moi?"). Mélisande forgives him. She cannot deny she loved Pelléas, but she insists that their love had been innocent. Arkel brings Mélisande her child. The servants file into the room. Suddenly, they fall on their knees: they sense that Mélisande is dead. The grief-stricken Arkel and Golaud leave the chamber with the motherless infant, who must now live in Mélisande's place.

Romain Rolland wrote that *Pelléas*

et Mélisande is "one of the three or four red-letter days in the calendar of our lyric theater." For this was a new kind of opera. The action was static; there were few emotional climaxes and no big scenes; the entire effect arose from subtle impressions. The music was as seemingly amorphous as the play, which was set in a dream world filled with symbolic suggestions and peopled with characters who were like shadows. Debussy's music caught the essence of Maeterlinck's drama. He wrote no arias or ensemble numbers, but a continual flow of declamation that resembled speech. This was set against a rich but restrained orchestral background that played as important a role in creating the over-all effect as the singing. The result was a fusion of play and music so complete that there are few equals in operatic literature.

Debussy took ten years to write his only opera. If the composition did not come easily, the production was also destined to bring him anguish. Maeterlinck's greatest interest in the opera lay in the decision of both poet and composer to feature Georgette Leblanc, at that time Maeterlinck's common-law wife, as Mélisande. But Albert Carré, director of the Opéra-Comique, had other plans. He sensed that Mary Garden, who had come to stardom in the role of Louise two years earlier, would be the ideal Mélisande. Without consulting Maeterlinck, he announced her for the part. Suspecting that Debussy was responsible for the decision, Maeterlinck threatened to beat him, and even thought of challenging him to a duel. Now an avowed enemy of Debussy, Carré, the Opéra-Comique, and the new opera, he did everything he could to discredit the work. He wrote a letter to *Le Figaro* before the première denouncing the management of the Opéra-Comique and expressing the wish for the "immediate and emphatic failure" of the opera. He was

probably responsible for a malicious parody of the play which was distributed outside the theater before the dress rehearsal, calculated to reduce the opera to ridicule. Thus, scandal preceded the première. To make matters still worse, the rehearsals went badly. The men in the orchestra had trouble deciphering their parts, prepared by a careless copyist. There were difficulties with scene designers, who insisted that the many changes were not feasible. A small portion of the opera was censored by a government official.

The public's reaction to this strange and revolutionary work was mixed. Hisses and guffaws mingled with applause and cheers. The critical opinion was also divided. Some thought the work was "without life . . . a continual dolorous melopoeia . . . deliberately shunning all semblance of precision." Others, such as Gustave Bret and André Corneau, did not hesitate to call the opera a masterwork. Of one thing there was no question: Mary Garden was the perfect Mélisande. From the very beginning she made the role her own, and as long as she sang it, both in Europe and America, she had no rivals. Notable Mélisandes after Garden were Maggie Teyte and Lucrezia Bori.

Penelope, (1) lyric drama by Gabriel Fauré. Libretto by René Fauchois, based on an episode from *The Odyssey*. Première: Monte Carlo, March 4, 1913. The story of Ulysses' return to Penelope, to find she has been faithful to him through the years, received from Fauré a restrained musical treatment in which there is little action or characterization, but only a sustained flow of lyrical beauty.

(2) Opera by Rolf Liebermann. Libretto by Heinrich Strobel. Première: Salzburg Festival, August 17, 1954. In this work the story of Ulysses and Penelope takes place in two times: in classical antiquity, and in the present day when Penelope is remarried to a

wealthy man after she is convinced that her first husband is dead. At the end of the opera Ulysses appears and makes a plea for pacifism.

Penthesilea, *see* KLEIST, HEINRICH WILHELM VON.

Pepa, Paquiro's sweetheart (contralto) in Granados' *Goyescas.*

Pepusch, Johann Christoph (or JOHN CRISTOPHER) composer. Born Berlin, Germany, 1667; died London, England, July 20, 1752. After holding a Prussian court post from 1681 to 1697, he moved to London in 1700. From 1712 to 1718 he was Handel's predecessor as organist and composer to the Duke of Chandos, and after 1713 he was for many years music director of the Lincoln's Inn Theatre. He wrote the music for several masques performed at this establishment: *Venus and Adonis* (1715); *Apollo and Daphne* (1716); *The Death of Dido* (1716); *The Union of Three Sisters* (1723). In 1718 Pepusch married the wealthy singer Marguerite de l'Epine. Ten years later he arranged the music for John Gay's sensationally successful ballad opera, *The Beggar's Opera* (*see* BEGGAR'S OPERA). The following year he arranged music for two more ballad operas, *The Wedding* and *Polly,* the latter a sequel to *The Beggar's Opera.* From 1737 until his death he was the organist at the Charter House.

Perchè ciò volle, il mio voler possente, Gerard's aria in Act III of Giordano's *Andrea Chénier.*

Perchè v'amo, *see* DEPUIS L'INSTANT OU DANS MES BRAS.

Perfect Fool, The, comic opera by Gustav Holst. Libretto by the composer. Première: British National Opera Company, London, May 14, 1923. This opera, an allegory in Elizabethan style, parodies opera conventions, particularly those of Wagner and Verdi. The Wizard, the Troubadours, and the Traveler all try to win the heart and hand of the Princess, but she rejects them. But the Fool, with the aid of a magic potion which he expropriates from the Wizard, wins the Princess and annihilates the Wizard when he comes to seek revenge.

Pergola, *see* TEATRO DELLA PERGOLA.

Pergolesi, Giovanni Battista, composer. Born Jesi, Italy, January 4, 1710; died Pozzuoli, Italy, March 16, 1736. The composer of *La serva padrona* studied music with Francesco Santini and Francesco Mondini, after which he entered the Conservatorio dei Poveri di Gesù Cristo in Naples in his sixteenth year. There, his teachers included Gaetano Greco, Francesco Durante, and Francesco Feo. His first major work, a sacred drama, *La conversione di San Guglielmo d'Aquitania,* was so successful when introduced in Naples in 1731 that he received a commission from the court for a new opera, *La Salustia,* performed the same year. Several of Pergolesi's next operas were comic interludes, or intermezzi, designed to be played between the acts of serious operas. In 1733 Pergolesi wrote an intermezzo that was soon performed independently and within a few months became an Italian favorite. This was *La serva padrona.* The first important opera buffa, *La serva padrona* established the traditions for later works in this form. (*See* OPERA BUFFA; SERVA PADRONA, LA.) Among Pergolesi's other operas, comic and serious, were: *Lo frate inamorato* (1732); *Il prigionero* (1733); *Adriano in Siria* (1734); *L'Olympiade* (1735); *Flaminio* (1735). All these works and others were finished within a brief period of creative activity before Pergolesi died of consumption at the age of twenty-six. In his serious works as well as his comic ones, Pergolesi's style was characterized by elegance and grace as well as lyric beauty. His best melodies often remind one of Mozart's in their aristocratic beauty, and on occasion (in *La serva padrona,* for example) we find something of Mozart's gift for creating characterizations and pointing up incidents by musical means.

Peri, Jacopo, composer. Born Rome (or Florence), August 20, 1561; died Florence, August 12, 1633. As a member of the Camerata which created opera, he wrote *Dafne,* the first opera in musical history. Born to a noble Florentine family, he received his musical training from Cristoforo Malvezzi. At different periods of his life he served as music master at the courts of Ferdinando I and Cosimo II de' Medici. When the Florentine Camerata evolved a musico-dramatic form calculated to revive ancient Greek drama (*see* CAMERATA; OPERA), Peri wrote *Dafne* in 1597 to a text by Ottavio Rinuccini. Though described by its composer as a "dramma per musica," *Dafne* is the first opera ever written, since it is the first play which is set throughout to music. *Dafne* was received so enthusiastically when introduced in Florence that, in 1600, Peri wrote a second opera, *Euridice,* performed in conjunction with the marriage ceremonies of Henry IV of France and Maria de' Medici. Later Peri operas: *Tetide* (1608); *Guerra d'amore,* a collaboration (1615); *Adone* (1620); *La precedenza delle dame* (1625). In 1608 Peri wrote the recitatives for *Arianna,* for which Claudio Monteverdi wrote the arias; and in 1628 he provided music for the part of Clori in Marco da Gagliano's *Flora.* Though Peri wrote a considerable amount of music, most of it has been lost, including that of the history-making *Dafne.*

Perlea, Jonel, conductor. Born Ograda, Rumania, December 13, 1900. After completing his music study in Munich with Paul Graener and Anton Beer-Walbrunn, and at the Leipzig Conservatory with Otto Lohse, he was appointed to a conducting post in Bucharest. In 1926 he was engaged by the Bucharest Opera, and after 1934 he was its musical director. He also became director of the Bucharest Conservatory. After making successful appearances as guest conductor of the San Francisco Symphony, Perlea made his debut at the Metropolitan Opera on December 1, 1949, with *Tristan und Isolde.* In 1955 he became conductor of the Connecticut Symphony Orchestra.

Per me ora fatale, Count di Luna's aria in Act II, Scene 2, of Verdi's *Il trovatore.*

Per pietà, ben mio perdona, Fiordiligi's aria in Act II, Scene 2, of Mozart's *Così fan tutte.*

Perrault, Charles, poet and writer of fairy tales. Born Paris, January 12, 1628; died there May 16, 1703. His collection of fairy tales, *Les contes de ma mère l'oye* (*Mother Goose*) (1697) is world famous. A number of operas have come from these stories: Nicolo Isouard's *Cendrillon,* Jules Massenet's *Cendrillon,* Rossini's *La Cenerentola,* Wolf-Ferrari's *La Cenerentola*—all these adaptations of *Cinderella;* Louis Aubert's *La forêt bleue;* Georges Huë's *Riquet à la Houppe;* Wolf-Ferrari's *Das Himmelskleid;* Ottorino Respighi's *La bella addormentata;* André Danican Philidor's *Le bûcheron;* Francesco Malipiero's *La bella e il mostro;* Engelbert Humperdinck's *Dornröschen;* Wheeler Beckett's *The Magic Mirror.*

Perrin, Emile Césare, impresario. Born Rouens, France, January 19, 1814; died Paris, October 8, 1885. One of the most significant opera impresarios of his time, he became manager of the Opéra-Comique in 1848, holding this post until 1857. During this period he was also manager of the Théâtre Lyrique for the single season of 1854–1855. Under his management of the Opéra-Comique such notable singers as Léon Carvalho, Jean-Baptiste Fauré, and Galli-Marié were introduced. From 1862 to 1870 Perrin was the manager of the Paris Opéra, where he introduced such outstanding operas as *L'Africaine, Don Carlos,* and *Hamlet,* and the distinguished singer Christine Nilsson. From 1870 until his death he was the manager of the Théâtre Français.

Perrin, Pierre, poet and librettist. Born Lyons, France, about 1616; died Paris, April 25, 1675. In 1668, with Robert Cambert, he obtained from Louis XIV a patent to organize the Académie de Musique, forerunner of the Paris Opéra, for the production of operas and ballets. He retained this patent until 1672 when, being accused of mismanagement, he was forced to relinquish it to Lully. He wrote librettos for the earliest French operas, all by Cambert: *Ariane, La Pastorale,* and *Pomone.*

Persephone, melodrama by Igor Stravinsky. Text by André Gide, based on Homer's *Hymn to Demeter.* Première: Paris Opéra, April 30, 1934. Described as a "melodrama," *Persephone* combines many different elements of the theater: music, singing, dancing, miming, and recitation. Gide's poetic text tells of Persephone breathing the aroma of narcissus, which gives her a vision of the lower regions where she is to be a queen. She descends to that world, grows bored with it, and returns to her former life. When introduced in Paris, *Persephone* received a full stage production, with Ida Rubinstein in the title role. It is more usually given in a concert version.

Persiani, Fanny (born **Tacchinardi**), coloratura soprano. Born Rome, October 4, 1812; died Neuilly, France, May 3, 1867. Her father, Niccolò Tacchinardi, taught her singing, and in her eleventh year she sang in opera performances in a little theater built by her father for his students. When she was eighteen she married an opera composer, Giuseppe Persiani, and two years afterward made her formal opera debut in Leghorn in Emile Fournier's *Francesca da Rimini.* Engagements in leading Italian opera houses followed. Donizetti was so impressed by the crystalline perfection of her voice that he wrote *Lucia di Lammermoor* with her in mind; she created the role of Lucia in Naples on September 26, 1835. In 1837 she made her first appearance in Paris, as Lucia, and in 1838 she made her London debut, as Amina. For the next decade she was a favorite in both cities. In 1850 she toured Holland and Russia. She made her farewell appearance at the Drury Lane Theatre in 1858, then went into retirement.

Persians, The, *see* AESCHYLUS.

Pertile, Aureliano, tenor. Born Montagnana, near Padua, Italy, November 9, 1885; died Milan, January 11, 1952. For four years he studied singing with Vittorio Oretice, after which he made his opera debut in Vicenza in *Martha,* in 1911. An additional period of study in Milan with Manlio Bavagnoli preceded his first major success, in the Italian première of Nougues' *Quo Vadis,* at the Teatro dal Verme in Milan, in 1912. For one season, during 1921–1922, he was a member of the Metropolitan Opera Company, making his debut there on December 1, 1921, in *Tosca.* Two years later, Toscanini engaged him for La Scala where, during the next decade and a half, he scored some of his greatest triumphs, notably in *Il trovatore, Manon,* and in Boïto's *Mefistofele* and *Nerone.* He made his farewell appearance on the operatic stage in 1940 in *Otello,* and five years after that became professor of singing at the Milan Conservatory, retaining this post until the time of his death.

Pescator, affonda l'esca, Barnaba's barcarolle in Act II of Ponchielli's *La Gioconda.*

Peter, a broommaker (baritone), father of Hansel and Gretel in Humperdinck's *Hansel and Gretel.*

Peter Grimes, opera by Benjamin Britten. Libretto by Montagu Slater, based on *The Borough,* a poem by George Crabbe. Première: Sadler's Wells, London, June 7, 1945. Peter Grimes, an English fisherman, is unjustly accused of having killed his apprentice. He is brought to trial. Though found innocent, the feeling persists among his neighbors that he is really guilty. This suspicion raises a wall of antagonism between Grimes and his neighbors. He grows hostile to everybody, even to Ellen Orford who loves him. When Grimes engages a new apprentice, the townspeople begin to speculate whether Grimes will kill him, too. They march in a body to Grimes's hut on the beach. When Grimes sees them approaching, he runs to his boat, and the apprentice, following him, slips and dies. The people are now convinced that Grimes is a murderer, and insist upon taking justice in their own hands. Tormented by his fears and rage, Grimes loses his mind, sails out on a calm day, and sinks his boat to find escape in death.

Peter Grimes made Britten internationally famous, and placed him in the front rank of contemporary opera composers. Its première was a gala event. The war in Europe had just ended, Sadler's Wells was being reopened for the first time since 1940, and *Peter Grimes* was the first new English opera in several years. All these circumstances combined to provide excitement. A line appeared in front of the theater twenty-four hours before curtain time. By the time the curtain rose, tickets for all scheduled performances had been sold. Notables from the world of music, and correspondents from the world's foremost newspapers, attended. The tension mounted during the performance itself, and erupted into pandemonium at the final curtain. Showers of bouquets descended on the stage. *Peter Grimes* was soon heard throughout the world of opera, translated into eight different languages, and wherever

it was given it was acclaimed as one of the major new operas of our time.

Though Britten used many different elements—sea chanteys, polytonal duets, simple jigs, broad arias, stark recitatives, realistic tone painting in the orchestra—integration was not sacrificed. With the skill of a master, Britten fused his diverse material into a gripping psychological drama.

Peter Ibbetson, opera by Deems Taylor. Libretto by the composer in collaboration with Constance Collier, based on the novel of the same name by George Du Maurier. Première: Metropolitan Opera, February 7, 1931. This was Taylor's second opera, completed four years after *The King's Henchman.* Peter Ibbetson is victimized by a tyrant uncle and finds relief in dreams. He kills his uncle and is imprisoned for life. In prison, he again succumbs to dreams. These bring up his past, and his childhood sweetheart, Mary. After thirty years in prison, Ibbetson learns that Mary is dead. Dreams no longer serve him; he has lost the will to live. He dies; the prison walls disintegrate. Peter is young again, and Mary is waiting for him.

Petrarch (born FRANCESCO DI PETRACCO), poet. Born Arezzo, Italy, July 20, 1304; died Arqua, near Padua, July 19, 1374. This celebrated figure of the Middle Ages is the central character in two operas: Enrique Granados' *Petrarca* and Johann Cristoph Kienlen's *Petrarca und Laura.*

Petroff, Ossip, dramatic basso. Born Elisabethgrad, Russia, November 15, 1807; died St. Petersburg, March 14, 1878. He was singing at a market fair in Kursk, in 1830, when the director of the St. Petersburg Opera heard him and engaged him. His debut took place that year as Sarastro. Recognition was immediate. To the end of his life, Petroff was without a rival, either histrionic or vocal, on the Russian stage. He created the leading bass roles in many Russian operas including *Boris Godunov, A Life for the Czar, The Maid of Pskov, Rusalka, Russlan and Ludmilla,* and *The Stone Guest.* His last appearance took place four days before his death.

Petrovich, a captain (baritone) in Tchaikovsky's *Eugene Onegin.*

Petruchio, nobleman of Verona (baritone) in Goetz's *The Taming of the Shrew.*

Pfitzner, Hans, composer. Born Moscow, Russia, May 5, 1869; died Salzburg, Austria, May 22, 1949. His parents, who were German, took him to their native land when he was a child. There he studied with his father (director of the Frankfurt Municipal Theater), and at the Hoch Conservatory. For a while he taught piano, conducted, and composed; in 1893 there took place in Berlin a successful concert devoted to his works. Two years later his first opera, *Der arme Heinrich,* was outstandingly successful when introduced in Mainz; it was heard in many major German opera houses during the next decade. In 1897 he settled in Berlin and was engaged as professor of the Stern Conservatory. From 1903 to 1906 he was the kapellmeister at the Theater des Westens; in 1907–1908 he conducted the Kaim Orchestra in Munich; in 1908 he was appointed director of the Strasbourg Conservatory, and in 1910 director of the Strasbourg Opera. Meanwhile, his second opera, *Die Rose vom Liebesgarten,* was introduced in Elberfeld in 1901, and his third opera, *Christelflein,* was produced in Munich, December 11, 1906.

Pfitzner settled temporarily in Munich in 1916, and it was here that he completed his masterwork, the opera *Palestrina,* first given on June 12, 1917, to great acclaim. It was sent on tour throughout Germany (despite travel restrictions imposed by World War I),

and it had many revivals after the war, particularly at the Munich Festivals.

In 1920 Pfitzner returned to Berlin, where he served as director of the master class in composition at the Academy of Fine Arts. He spent the last two decades of his life in Munich, where he was one of the city's major musical figures, distinguished as teacher, conductor, and composer. Pfitzner societies were formed in different parts of Europe to promote his works. When the Nazis came to power, Pfitzner enthusiastically allied himself with the new order and became one of its musical spokesmen. After the German defeat, Pfitzner became poverty stricken. He was found in a Munich home for the aged by the President of the Vienna Philharmonic, who had him brought to Austria, where he was supported by the orchestra.

Pfitzner wrote only one opera after *Palestrina* (though he revised his *Christelflein* in 1917). This was *Das Herz,* completed in 1931, and introduced simultaneously by the Berlin Opera and the Munich Opera on November 12 of that year.

Phanuel, a young Jew (tenor) in Massenet's *Hérodiade.*

Philémon et Baucis, opera by Charles Gounod. Libretto by Jules Barbier and Michel Carré, based on Ovid. Première: Théâtre Lyrique, Paris, February 18, 1860. The setting is Phrygia in mythical times. An old couple, Philémon and Baucis, provide hospitality to two strangers, who are really Vulcan and Jupiter come to punish Phrygia. The Phrygians are visited by a storm, but Jupiter saves Philémon and Baucis. Complications set in when Jupiter restores the youth of his benefactors, only to fall in love with the now beautiful Baucis. Touched by her devotion to Philémon, he finally permits the pair to enjoy their restored youth. Two arias from this work are popular: Vulcan's song in Act I, "Au bruit des lourds marteaux," and Baucis's aria in Act II, "O riante nature."

Philidor, François André Danican, composer. Born Dreux, France, September 7, 1726; died London, England, August 24, 1795. One of the most prolific and successful composers of opéras comiques of his time, he came from a family which for generations had been professional musicians. After studying music with André Campra, Philidor turned to chess and became a master of the game, leaving a name that is well known to chess players of the present day. After 1754 Philidor combined his activity in chess with composition, and in 1759 his first opéra comique, *Blaise le Savetier,* was a brilliant success. His later works, all extremely popular, included: *Le triomphe du temps* (1761); *Tom Jones* (1764); *L'amant déguisé* (1769); *Le bon fils* (1773); *Zémire et Mélide* (1773); *Les femmes vengées* (1775); *Persée* (1780); *Thémistocle* (1786); *La belle esclave* (1787). Philidor's reputation in London was only second to his fame in Paris. In 1767 he wrote a serious opera, *Ernelinde.* A success when introduced, it was revised two years later, renamed *Sandomir,* and performed again.

Philine, an actress (soprano) in Thomas' *Mignon.*

Philip II, king of Spain (bass) in Verdi's *Don Carlos.*

Phillips, Adelaide, dramatic contralto. Born Stratford-on-Avon, England, 1833; died Carlsbad, Bohemia, October 3, 1882. She was brought to the United States as a child of seven, where she appeared in several stage productions. Her voice teacher, Sophie Arnould, directed her to opera. Jenny Lind heard the young singer and suggested study with her own teacher, Manuel García. After two years with García in London, Phillips made her debut in Milan on December 17, 1854, as Rosina. After successful appearances in Italy she returned to the United States and made her American debut at the Academy of Music in New York, on March 17, 1856, as Azucena. She remained at the Academy of Music five years. In 1861 she toured Europe, making her Paris debut as Azucena at the Théâtre des Italiens. Subsequently, she appeared in the United States with a company directed by Carl Rosa and with the Boston Ideal Opera Company. She then formed the Adelaide Phillips Opera Company. It toured the United States and was a financial failure. After a farewell appearance in Cincinnati in 1881 she went to Europe to recover her health, dying a few months later.

Philomela, opera by Henrik Andriessen. Libretto by Jan Engelman, based on an episode in Ovid's *Metamorphoses.* Première: Holland Music Festival, Amsterdam, June 23, 1950. Sweet-singing Philomela is seduced by her brother, Tereus, who then cuts out her tongue to prevent her from betraying him. Philomela revenges herself by killing Tereus' child and serving it to his father at a feast. Through the intervention of the gods, all three principals in this gruesome tale are transformed into birds; now a nightingale, Philomela can sing again.

Phonograph recordings, see OPERA PERFORMANCE—RECORDINGS.

Piave, Francesco Maria, librettist. Born Mureno, Italy, May 18, 1810; died Milan, March 5, 1876. An intimate friend of Giuseppe Verdi, Piave wrote the librettos for nine of his operas, including *Ernani, La forza del destino, Macbeth,* and *Rigoletto.*

Piccinni, Nicola (sometimes NICCOLO), composer. Born Bari, Italy, January 16, 1728; died Passy, France, May 7, 1800. His operas were matched against those of Gluck during the esthetic war between Paris' "Gluckists" and "Piccinnists." Piccinni entered the Conservatorio San Onofrio in Naples in 1742, remaining there twelve years. His teachers included Francesco Durante and Leonardo Leo. After leaving the Conservatory he was sponsored by the Prince of Ventimiglia, who arranged to have his first opera, *Le donne dispettose,* produced in Naples in 1754. The opera was such a triumph that envious composers instigated cabals to discredit Piccinni. In spite of their efforts, his next operas continued to enjoy success; one of them, *Il curioso del proprio danno,* first heard in 1755, had several highly successful revivals. In 1758 Piccinni went to Rome to produce there his new opera, *Alessandro nell' Indie.* He now became one of the most celebrated composers in that city. In 1760 he produced a comic opera, *La Cecchina,* which was not only an instantaneous success but was one of the finest comic operas before those of Rossini. In 1762 Piccinni wrote six operas which were performed simultaneously by six leading theaters. His esteemed position in Rome was suddenly lost in 1773 with the rise in popularity of his pupil, Pasquale Anfossi. This eclipse so embittered Piccinni that his health broke down, and for a while he refused to compose. When he recovered he vowed never again to write an opera for the Roman public. He returned to Naples, where his popularity was still at its height, and where his next opera, *I Viaggiatori,* was acclaimed. In 1776 he was invited to Paris to write operas for the French stage. He enjoyed high favor at Versailles and was engaged to give singing lessons to the Queen. The first opera he produced in Paris was *Roland* (using a libretto which had also been set by Gluck). This marked the beginning of the bitter feud between the followers of Piccinni and those of Gluck. Recognizing the publicity value of this rivalry, the director of the Paris Opéra commissioned both composers to write music for the same libretto, *Iphigénie en Tauride.* Gluck's opera was performed first, in 1779, and was received so enthusiasti-cally that Piccinni tried to withdraw his own work. Piccinni's opera, received less well than Gluck's, nevertheless had seventeen consecutive performances. Still, there could no longer be a question in Paris as to the victory of Gluck's principles and esthetics over those of his Italian rival. However, when Gluck left Paris in 1780, Piccinni once again occupied the center of public interest. But three years later a new Italian composer came into fashion—Antonio Sacchini—and this led to a sharp decline in the popularity of Piccinni's operas. After the outbreak of the French Revolution, Piccinni returned to Naples where he received a commission from the king, and where several of his older operas were successfully revived. He lost favor when he was suspected of republican leanings. For four years he was a virtual prisoner in his own house. In 1798 he was allowed to return to Paris, where he received a government pension and shortly before his death was made an inspector at the Conservatory.

Piccinni wrote well over a hundred operas. His best works had a rich fund of pleasing melodies combined with skillfully written ensemble numbers and dramatic finales. Besides operas already mentioned, his most successful works were: *Le gelosie* (1755); *Il re pastore* (1760); *Berenice* (1764); *L'Olimpiade* (1768, revised 1774); *Didone abbandonata* (1770); *Antigono* (1771); *Atys* (1780); *Didon* (1783); *Le faux lord* (1783).

Pickwick, opera by Albert Coates. Libretto by the composer, based on Dickens' *Pickwick Papers.* Première: Covent Garden, November 20, 1936. The opera, like the Dickens novel, concerns the picaresque and frequently droll adventures of Mr. Samuel Pickwick, founder of the Pickwick Club, which culminate in his imprisonment. *See also* DICKENS, CHARLES.

Picture of Dorian Gray, The, *see* WILDE, OSCAR.

piège de Méduse, Le, *see* SATIE, ERIK.

Pierné, Gabriel, composer. Born Metz, France, August 16, 1863; died Ploujean, France, July 17, 1937. He attended the Paris Conservatory, where his teachers included Jules Massenet and César Franck and where he received the Prix de Rome in 1882. After his return to Paris from Rome he succeeded Franck as organist of Sainte-Clotilde, holding this post eight years. After 1903, he devoted himself to conducting, serving as musical director of the Colonne Orchestra (Paris) for a quarter of a century. As a composer he first achieved recognition with an oratorio, *The Children's Crusade,* in 1902, which won the City of Paris prize. He also wrote several operas: *La coupe enchantée* (1895, revised 1901); *Vendée* (1897); *La fille de Tabarin* (1901); *On ne badine pas avec l'amour* (1910); *Sophie Arnould* (1927).

Pierotto, a villager (contralto) in Donizetti's *Linda di Chamounix.*

Pilar, Pedro's bride (soprano) in Laparra's *La Habanera.*

Pilgrims' Chorus, the chorus of pilgrims on their journey to and from Rome in Acts I and III of Wagner's *Tannhäuser.*

Pilgrim's Progress, The, opera by Ralph Vaughan Williams. Libretto by the composer, adapted from Bunyan's *The Pilgrim's Progress.* Première: Covent Garden, April 26, 1951. This is not an opera in the accepted meaning of the term. The composer himself described it as a "morality," and the work consists of a series of picturesque tableaux which frequently suggest a religious ritual. The prologue shows Bunyan writing his book in Bedford Gaol, where the Pilgrim appears to him. The opera then describes the history of the Pilgrim as he goes through the City of Destruction, the Valley of Humiliation,

Vanity Fair, the Delectable Mountain to the Golden Gate of the Celestial City. In the epilogue, Bunyan reappears holding his book in hand. In the fourth act, Vaughan Williams incorporated a one-act opera, *The Shepherds of the Delectable Mountain,* which he had written in 1922.

Pimen, a monk (bass) in Mussorgsky's *Boris Godunov.*

Pinellino, a shoemaker (bass) in Puccini's *Gianni Schicchi.*

Ping, the Grand Chancellor (baritone) in Puccini's *Turandot.*

Pini-Corsi, Antonio, bass. Born Zara, Jugoslavia, June 1859; died Milan, Italy, April 21, 1918. He made his debut in Cremona in 1878. Subsequent appearances in Italian opera houses were so successful that Verdi selected him to appear in the première of *Falstaff* at La Scala in 1893. He made his American debut at the Metropolitan Opera on November 20, 1909, in *La Bohème,* remaining at the Metropolitan through the 1913–1914 season, appearing in the world premières of Damrosch's *Cyrano de Bergerac, The Girl of the Golden West, Die Königskinder,* and *Madeleine;* he was also seen in the American premières of *L'amore medico, Le donne curiose,* and *Germania.*

Pinkerton, U. S. Navy lieutenant (tenor), beloved of Cio-Cio-San, in Puccini's *Madama Butterfly.*

Pinkerton, Kate, Pinkerton's wife (mezzo-soprano) in Puccini's *Madama Butterfly.*

Pinkerton, Miss, a gossip (soprano) in Menotti's *The Old Maid and the Thief.*

Pinza, Ezio, bass. Born Rome, Italy, May 18, 1892. He originally planned to be a civil engineer, but abandoned engineering in his seventeenth year to become a professional bicycle racer. When he was eighteen he started studying singing with Ruzza, and continued his studies with Vizzani at the Bologna Conservatory. After service in the artillery during World War I, he made his debut at the Teatro Reale in Rome, in 1921, in the role of King Mark. After two seasons in Rome and another in Turin he was engaged by La Scala, where he appeared three seasons; during this period he appeared in the world première of Boïto's *Nerone*, in 1924.

Pinza made his American debut at the Metropolitan Opera on November 1, 1926, in Spontini's *La Vestale*, when Olin Downes described him as "a majestic figure on the stage" and a bass "of superb sonority and impressiveness." For the next quarter of a century Pinza was the principal bass of the Metropolitan, appearing in an extensive repertory, and achieving triumphs in such varied roles as Boris Godunov, Figaro (in *The Marriage of Figaro*), Don Giovanni, Frère Laurent, King Dodon, Oroveso, and Lothario. A dominating figure, he combined powerful characterizations with a voice remarkable for texture, range, volume, and flexibility. While a member of the Metropolitan, Pinza made numerous appearances throughout the world of opera; his interpretations of Don Giovanni and Figaro in *The Marriage of Figaro* were major attractions at the Salzburg Festivals before World War II.

In 1949 Pinza withdrew from opera to appear in motion pictures and in Broadway musical plays. He scored such a sensation in *South Pacific* that he became a matinee idol. He has since appeared in *Fanny*. He has also made numerous radio and television appearances. His daughter by his first marriage, Claudia Pinza, is also an opera singer. She made her debut in a San Francisco Opera production of *Faust*, appearing with her father (September 21, 1947). She subsequently appeared at the Metropolitan Opera.

Pipe of Desire, The, one-act opera by Frederick Shepherd Converse. Libretto by George Edward Burton. Première:

Boston, January 31, 1906. This was the first opera by an American to be performed at the Metropolitan Opera (March 18, 1910). The Pipe of Desire, owned by the Elf King, has magic powers. Because the peasant youth Iolan makes selfish use of the pipe he brings doom to himself and his beloved Naoia. The Elf King seizes the pipe from Iolan and plays on it. As he plays, Naoia falls dead at Iolan's feet. The Elf King plays again. This time Iolan dies, and his soul goes forth to meet that of his beloved.

Pique Dame (The Queen of Spades), opera by Peter Ilyitch Tchaikovsky. Libretto by Modest Tchaikovsky (the composer's brother), based on a story by Alexander Pushkin. Première: St. Petersburg, December 19, 1890. The Countess, a sinister old gambler nicknamed the Queen of Spades, knows the secret of winning at cards. Hermann, in love with her granddaughter, Lisa, is determined to learn the secret so that he may acquire the funds to marry. He goes to the Countess' bedroom to extort the information, but his sudden appearance causes the old woman to die of fright. The ghost of the Countess now advises Hermann to wager on three particular cards. As Hermann becomes obsessed with the idea of winning, Lisa becomes despondent and drowns herself. Hermann wins on the first and second cards, but loses everything on his third wager when the Queen of Spades turns up instead of an ace. The ghost of the Countess appears and taunts him. He commits suicide. Among the more familiar arias are Prince Yeletsky's proposal, "When you choose me for your husband," in Act II, and Lisa's aria in Act III, "It is near to midnight."

Pistol, Falstaff's henchman (bass) in Verdi's *Falstaff*.

Pitt, Percy, conductor. Born London, England, January 4, 1870; died there November 23, 1932. His studies took place at the Leipzig Conservatory and with Josef Rheinberger in Munich. He returned to England in 1893 and after holding various posts as organist and chorus master was appointed chorus master at Covent Garden in 1906. A year later he became an assistant conductor there, and in 1907 principal conductor. He remained at Covent Garden until 1915. From 1915 to 1920 he was conductor of the Beecham Opera Company, and from 1920 to 1924 director of the British National Opera. He returned to Covent Garden as principal conductor in 1924. Two years later he became general musical director of the British Broadcasting Corporation, holding this post until succeeded by Sir Adrian Boult in 1930.

Pittichinaccio, Giulietta's admirer (tenor) in Offenbach's *The Tales of Hoffmann.*

Pizarro, governor of the prison of Seville (baritone) in Beethoven's *Fidelio.*

Pizzetti, Ildebrando, composer. Born Parma, Italy, September 20, 1880. He entered the Parma Conservatory in his sixteenth year and while still a student wrote two operas. In 1901 he was engaged as assistant conductor of the Parma Opera. At this time he submitted a one-act opera, *Le Cid,* in a contest sponsored by Edoardo Sonzogno. His failure to win a prize was such a disappointment that for several years Pizzetti refused to write further for the stage. In 1908 he joined the faculty of the Parma Conservatory. A year later he became professor of theory and composition at the Florence Conservatory, where he remained for over a dozen years.

In 1905 Pizzetti entered the first significant phase of his creative career, a period when he was affected by the writings of Gabriele d'Annunzio. Many of his works were settings of D'Annunzio's writings. His major creation, the opera *Fedra,* was introduced at La Scala on March 20, 1915. While not initially successful, it subsequently proved one of the most popular of Pizzetti's works.

In his second phase, Pizzetti was influenced by the Bible. The opera of this period, *Debora e Jaele,* was given at La Scala on December 16, 1922. His last creative period was inspired by Italian history and backgrounds, and from these influences came *Fra Gherardo,* introduced at La Scala on May 16, 1928.

Pizzetti is essentially a dramatic rather than a lyric composer. His operas are characterized by expressive declamations molded after the inflections of the text, and are particularly notable for their choral pages.

In 1930 Pizzetti visited the United States and appeared as pianist and conductor in his own works. In 1936 he became professor of composition at the Santa Cecilia Academy in Rome (in succession to Ottorino Respighi), where he soon rose to the position of director. He resigned his directorial post in 1952. In 1950 he won the international Italia prize for a one-act opera, *Ifigenia.*

His operas: *Fedra* (1912); *Debora e Jaele* (1921); *Lo Straniero* (1925); *Fra Gherardo* (1927); *Orseolo* (1935); *L'Oro* (1942); *Vanna Lupa* (1947); *Ifigenia* (1950); *Cagliostro* (1952); *La figlia di Jorio* (1954).

Plançon, Pol-Henri, bass. Born Fumay, France, June 12, 1854; died Paris, August 11, 1914. Trained for a career in business, he turned instead to music. After studying with Gilbert Louis Duprez and Giovanni Sbriglia, Plançon made his debut in Lyons, in 1877, as St. Bris. Two seasons later, on February 11, 1880, he made his debut in Paris at the Théâtre de la Gaieté. Three years afterward he became a member of the Paris Opéra. He was a sensation as Méphistophélès in *Faust,* a role he sang over a hundred times in a decade. On June 3, 1891, he made his debut in London, as Méphistophélès, and became an instant favorite; he returned to London each season until 1904. On November 29, 1893, he made his American debut at the Metropolitan Opera as Jupiter in *Philémon et Baucis*. He remained at the Metropolitan through the 1907–1908 season, one of the stars of the golden age of great casts. He retired from the stage in 1908, having appeared in many world premières, including Mancinelli's *Ero e Leandro*, Massenet's *Ascanio, Le Cid*, and *La Navarraise*, and Stanford's *Much Ado About Nothing*.

Planquette, Jean-Robert, composer. Born Paris, July 31, 1848; died there January 28, 1903. A successful composer of opéras comiques, the most famous being *Les cloches de Corneville* (*The Chimes of Normandy*), he attended the Paris Conservatory, after which he earned his living writing songs for cafés. His first stage work was a one-act opera, *Paille d'avoine*, in 1874. Success followed in 1877 with *Les cloches de Corneville,* introduced at the Folies-Dramatiques, where it had a run of four hundred performances. His other works: *Rip van Winkle* (1882); *Surcouf* (1887); *Paul Jones* (1889); *Le talisman* (1892); *Panurge* (1895); *Mam'zelle Quat' Sous* (1897).

Pleurez, pleurez mes yeux, Chimène's aria in Act III of Massenet's *Le Cid*.

Plunkett, a wealthy farmer (bass) in Flotow's *Martha*.

Plus blanche que la blanche hermine (Bianca al par hermine), Raoul's romanza in Act I of Meyerbeer's *Les Huguenots*.

Poe, Edgar Allan, poet and story writer. Born Boston, Massachusetts, January 19, 1809; died Baltimore, Maryland, October 7, 1849. Adriano Lualdi's *Il Diavolo nel campanile* was derived from Poe's tale *The Devil in the Belfry*, and Vieri Tosatti's *Il sistema della dolcezza* was based on another of his stories. After 1908 Claude Debussy worked intermittently on an opera to his own libretto, based on *The Fall of the House of Usher,* but the work was never completed and only fragments of it exist. Franz Schreker prepared a libretto, *Der rote Tod*, after Poe's *The Masque of the Red Death*, but abandoned the work before he wrote the music.

Pogner, a goldsmith (bass), father of Eva, in Wagner's *Die Meistersinger*.

Poisoned Kiss, The, opera by Vaughan Williams. Libretto by Evelyn Sharp. Première: Cambridge, England, May 12, 1936. The text is a fable about a princess, endowed with magical powers, who falls in love with a handsome young man. The score contains pages that satirize various musical styles.

Polacco, Giorgio, conductor. Born Venice, Italy, April 12, 1875. His musical schooling took place at the Liceo Benedetto Marcello in Venice and at the Milan Conservatory. He began his career as conductor at the Shaftsbury Theatre in London, then presenting a season of opera. He subsequently directed opera performances in Italy, Portugal, Belgium, Poland, and Russia, and was acclaimed for his Wagner performances. After four seasons as principal conductor at the Teatro Colón in Buenos Aires, and seven more in Rio de Janeiro, he returned to Italy to become principal conductor at La Scala. There he directed the Italian première of *Pelléas et Mélisande*. In 1906 he directed several performances for the San Francisco Opera Company. On November 11, 1912, he made his Metropolitan Opera debut with *Manon Lescaut*. Three years later, when Arturo Toscanini left the Metropolitan, he became one of the leading conductors. He resigned from the Metropolitan in 1917, and for the next several years conducted opera performances in South America, Havana, and Paris. When the Chicago Civic Opera was organized in 1922, Polacco was engaged as its music director and principal conductor, retaining this post until 1930. Since then he has not conducted.

Polish Jew, The, (1) *see* JUIF POLONAIS, LE.

(2) Opera by Karel Weis. Libretto by Victor Leon and Richard Batka, based on the play of the same name by Erckmann-Chatrian. Première: Prague Opera, 1910. The theme of the opera is the same as in *Le Juif polonais*.

Polka, a fast Bohemian dance in duple time, usually with an accent on the second beat. There is a celebrated polka in the first act of Smetana's *The Bartered Bride*. Other operas containing polkas are Smetana's *The Kiss* and *Two Brides* and Weinberger's *Schwanda*.

Polkan, a general (baritone) in Rimsky-Korsakov's *Le coq d'or*.

Pollack, Egon, conductor. Born Prague, Bohemia, May 3, 1879; died there June 14, 1933. He attended the Prague Conservatory and began his musical career as chorus master of the Prague Landestheater. In 1905 he was appointed first conductor of the Bremen Opera, where he remained five years. Between 1910 and 1912 he was first conductor of the Leipzig Opera, and from 1912 to 1917 of the Frankfurt Opera. In 1914 he was invited to Covent Garden to direct the Wagner music dramas. Beginning in 1915 he was the principal conductor of the Wagner dramas with the Chicago Opera. From 1917 to 1933 he was music director of the Hamburg Opera. He was acclaimed not only for his Wagner performances but also for his interpretations of the Richard Strauss operas. He died of a heart attack while directing a performance of *Fidelio*.

Pollione, Roman proconsul (tenor) in love with Adalgisa, but loved by Norma, in Bellini's *Norma*.

Pollux, Castor's twin brother (baritone) in Rameau's *Castor et Pollux*.

Polly, ballad opera, dialogue and verses by John Gay, the music (ballads and popular songs) arranged by John Cristopher Pepusch. Première: Haymarket Theatre, London, June 19, 1777. Written in 1729 as a sequel to the extraordinarily popular *Beggar's Opera* (which see), *Polly* was kept from the stage, apparently as a reprisal for the satires of the earlier work. Gay had *Polly* printed, and in this form the ballad opera was widely admired, though not staged till nearly half a century later. Like *The Beggar's Opera*, Polly has been successfully performed in the twentieth century.

Polonaise, a Polish dance in triple time, characterized by accents on the half beat. Philine's aria "Je suis Titania" in Act II of *Mignon* is a polonaise; polonaises also appear in Act III of *Boris Godunov* and Act III of *Eugene Onegin*.

Polonius, chancellor of Denmark (baritone) in Thomas's *Hamlet*.

Polovtsian Dances, dances of the Polovtsian slaves before Prince Igor in Act II of Borodin's *Prince Igor*.

Polyeucte, a drama by Pierre Corneille, the source of operas by Donizetti (*Poliuto*) and Gounod (*Polyeucte*). Polyeucte, though deeply in love with his wife Pauline, refuses to recant his Christian vows and instead accepts martyrdom.

Polyphemus, the Cyclops (bass) in Handel's *Acis and Galatea*.

Pompeo, a character (baritone) in Berlioz' *Benvenuto Cellini*.

Ponchielli, Amilcare, composer. Born Paderno, Italy, August 31, 1834; died Milan, January 16, 1886. The composer of *La Gioconda*, he entered the Milan Conservatory in his ninth year and stayed there nine years. While still a student he wrote an operetta, *Il Sindaco Babbeo,* in collaboration with three other students. His studies ended, he became an organist in Cremona, and after that bandmaster in Piacenza. During this period he wrote his first opera, *I promessi sposi*, introduced in Cremona in 1856. For the opening of the Teatro dal Verme in Milan in 1872, Ponchielli was commissioned to write an opera. For this occasion he revised *I promessi sposi* and it was now acclaimed. His next opera, *Le due gemelle*, produced by La Scala in 1873, was also well received. Ponchielli became world-famous with *La Gioconda*, introduced at La Scala on April 8, 1876. The opera was a triumph at its première, and it was outstandingly successful when heard throughout Europe. None of the operas Ponchielli wrote after *La Gioconda* was able to repeat either the popular success or the consistently high level of dramatic and musical interest of that work.

In 1881 Ponchielli was appointed maestro di cappella of the Bergamo Cathedral, and from 1883 on he was professor of composition at the Milan Conservatory. His operas: *I promessi sposi* (1856); *La Savoiarda* (1861, revised as *Lina*, 1877); *Roderico* (1863); *La stella del monte* (1867); *Le due gemelle* (1873); *I Lituani* (1874, revised as *Alduna*, 1884); *La Gioconda* (1876); *Il figliuol prodigo* (1880); *Marion Delorme* (1885); *Bertrando de Bornio; I mori di Valenza* (completed by A. Cadora).

Pong, the Chief Cook (tenor) in Puccini's *Turandot.*

Pons, Lily, coloratura soprano. Born Cannes, France, April 12, 1904. She entered the Paris Conservatory in her thirteenth year as a student of piano. Plans for a virtuoso career were abandoned after a serious illness, following which she turned to singing. After World War I she appeared in minor singing roles at the Théâtre des Varietés, in Paris. Her first husband, whom she married in 1923, became convinced of her talent and engaged Alberti di Gorostiaga to teach her. In 1928 Pons made her debut at the Mulhouse Opera in Alsace as Lakmé. After appearances

in smaller French opera houses, she made her debut at the Metropolitan Opera on January 3, 1931, in *Lucia di Lammermoor.* Ever since, she has been a leading soprano there, acclaimed in the coloratura roles of the French and Italian repertories. Several operas were revived by the Metropolitan for her, including *The Daughter of the Regiment, Linda di Chamounix, Mignon, La sonnambula,* and *The Tales of Hoffmann.* In singing the role of Lakmé, she became the first singer in half a century to render the high F in the "Bell Song" that Delibes originally indicated.

In 1938 Pons married the conductor André Kostelanetz. She has often appeared as soloist in concerts directed by her husband. Her personal charm, capacity to invest each of her roles with engaging warmth, and brilliant coloratura range (two and a half octaves) have made her an outstanding drawing card in all the major opera houses of the world. Miss Pons has sung frequently on radio and television, and was starred in a motion picture, *I Dream Too Much.* She became an American citizen in 1940.

Ponselle, Carmela (born PONZILLO), mezzo-soprano. Born Schenectady, New York, June 7, 1892. the older sister of the celebrated soprano Rosa Ponselle (see below), she began studying singing after her sister had become successful. Her opera debut took place in New York City in 1923 in *Aïda.* On December 5, 1925, she made her debut at the Metropolitan Opera in the same opera. She remained at the Metropolitan until 1928, then returned in 1930 for an additional five seasons. On April 23, 1932, she appeared for the first time in an opera performance with her sister. The opera was *La Gioconda.* They sang it in Cleveland; in December of the same year the sisters again appeared in the same opera, this time at the Metropolitan in New York. Carmela left the Metropolitan in 1935 when Gatti-Casazza denied her permission to accept a radio contract. She specialized in radio appearances and teaching for several years.

Ponselle, Rosa (born ROSE PONZILLO), dramatic soprano. Born Meriden, Connecticut, January 22, 1894. She began singing early, and in her youth appeared in church choirs and as soloist in motion-picture theaters and vaudeville houses. She began to study intensively with William Thorner. Caruso became interested in her and gained her an audition at the Metropolitan. She made her debut there on November 15, 1918, in *La forza del destino,* when her voice was described by James Gibbons Huneker as "vocal gold, with its luscious lower and middle tones, dark, rich, and ductile; brilliant and flexible in the upper register." The following month she was heard in a revival of *Oberon* and later the same season in the world première of Joseph Breil's *The Legend.* In 1927 she achieved one of the outstanding triumphs of her career, singing the title role in a revival of Bellini's *Norma.* In 1930 she achieved another major success as Donna Anna, and in 1933 she was acclaimed at the Florence May Music Festival in *La Vestale.* In 1935 she gratified her life's ambition by appearing for the first time as Carmen, singing the role at the Metropolitan; but Carmen was never one of her best parts.

After marrying Carle A. Jackson in 1936, she decided to retire from opera, even though she was then at the height of her fame. Her last appearance at the Metropolitan took place on February 15, 1937, in *Carmen;* her voice—as remarkable for its richness and expressiveness as for its flexibility and range—had lost none of its magnetism or beauty.

In 1954, after a public silence of nearly seventeen years, Miss Ponselle made a recording of sixteen selected songs.

Ponte, Lorenzo da, *see* DA PONTE, LORENZO.

Porgi amor, Countess Almaviva's aria in Act II of Mozart's *The Marriage of Figaro.*

Porgy and Bess, opera in three acts by George Gershwin. Libretto by Du Bose Heyward and Ira Gershwin, based on the play *Porgy* by Du Bose and Dorothy Heyward. Première: Boston, September 30, 1935.

Characters: Porgy, a cripple (baritone); Bess, his girl (soprano); Crown (bass); Sportin' Life (tenor); Robbins (baritone); Serena, his wife (soprano); Clara (soprano); Jake, her husband (baritone); Frazier, a lawyer (baritone); natives; hucksters; policemen. The setting is the Negro section of Charleston, South Carolina, in the present time.

Act I, Scene 1. Catfish Row, a Negro tenement. Clara is singing a lullaby to her child ("Summertime") while a crap game is in progress. In the course of the game, Crown quarrels with Robbins and kills him. Crown escapes. Sportin' Life, who has always had an eye for Crown's girl, Bess, now feels free to invite her to New York. She turns him down, having found refuge in the home of the cripple, Porgy, who loves her unselfishly and devotedly.

Scene. Serena's Room. There is a wake for Robbins. Neighbors gather to drop coins in a saucer for his burial. Serena is distraught at her husband's death ("My man's gone now").

Act II, Scene 1. Catfish Row. Bess, completely reformed, is living with Porgy, who speaks of the happiness that it is ("I got plenty o' nuttin' "). They express their love for each other ("Bess, you is my woman now").

Scene 2. Kittiwah Island. During a lodge picnic, Sportin' Life entertains his friends with his cynical attitude toward religion ("It ain't necessarily so"). Crown, who has been hiding on the island, confronts Bess and persuades her to stay with him.

Scene 3. Catfish Row. A few days later, Bess returns to Porgy. She is sick and delirious. Porgy tenderly nurses her back to health. A hurricane blows up, and the women of Catfish Row are troubled over their men who have gone out fishing. Clara is particularly upset since she senses that her husband, Jake, is in danger. No one in Catfish Row can save Jake until Crown arrives unexpectedly and offers to do so, ridiculing the crippled Porgy for his inability to be of help.

Act 3. Catfish Row. Crown has returned after saving Jake; he is looking for Bess. Afraid of losing the woman he loves, Porgy reaches out of his window and fatally stabs his rival. The police are unable to discover the murderer, but since they suspect Porgy they take him off to jail. While Porgy is away, Sportin' Life finally persuades Bess to go off with him to New York, tempting her with a packet of dope. Back from jail a few days later, Porgy finds Bess gone. Heartbroken but undaunted, he sets out after her.

Porgy and Bess was Gershwin's last serious work, and his only full-length opera. It possesses that richness, vitality, and variety of melody, that vigor of rhythm, that spontaneity and freshness we associate with Gershwin's best music. It has much more, too. Of all Gershwin's serious works, it is the only one to reveal compassion, humanity, and a profound dramatic instinct. Rich in materials derived (but never quoted) from spirituals, shouts, and street cries, *Porgy and Bess* is truly a folk opera. Its roots are in the soil of the Negro people, whom it interprets with humor, tragedy, penetrating characterizations, dramatic power, and sympathy. Gershwin wrote his opera for the Theatre Guild, which organization produced it in New York in 1935 after a Boston

tryout. It was not at first successful. Some critics thought it was neither an opera nor a musical comedy, but a hybrid. Later revivals impressed audiences and critics alike with the beauty and originality of this work, and it was at length accepted as one of the finest operas written by an American. A return to Broadway in 1942 resulted in the longest run known up to then by a revival, and it was at this time singled out for special praise by the New York Music Critics Circle. The European première took place during World War II, at the Danish Royal Opera in Copenhagen (in a Danish translation) on March 27, 1943. Because of Gestapo antagonism to a successful American opera—Denmark was then occupied by the Nazis—the run had to end abruptly. In 1952, an American company began a several year tour of *Porgy and Bess* in Europe and the Near East. The success of the work has been greater than that of any other American opera played abroad. In February, 1955, the company appeared at La Scala, and thus *Porgy and Bess* became the first opera by a native American to be heard in that theater. The same company began a South American tour on July 8, 1955.

Porpora, Niccolò, singing teacher and composer. Born Naples, Italy, August 19, 1686; died there February 1766. While attending the Conservatorio di San Loreto in Naples he wrote his first opera, *Agrippina,* produced in Naples in 1708. Success came ten years later in Rome with *Berenice.* He held various court posts as music director before establishing in Naples, in 1712, a school of singing that became famous throughout Europe for the development of such singers as Caffarelli, Senesino, and Farinelli. In 1715 Porpora became singing master at the Conservatorio di San Onofrio in Naples. He held similar posts with two conservatories in Venice, after which he visited London in 1729, where for the next few years his operas were produced in competition with those by Handel. Handel's popularity finally overwhelmed him and he returned to Venice to become director of the Ospedaletto Conservatory (for girls). In 1747, he settled in Dresden as conductor and teacher, and, beginning with 1752, he spent two years in Vienna, where Haydn became his pupil and lackey. In 1760 he became choirmaster of the Naples Cathedral and director of the Conservatorio di San Onofrio. In the closing years of his life his fame suffered such a decline that he died a pauper. He wrote about fifty operas. The most successful were: *Flavio Anicio Olibrio* (1711); *Berenice* (1718); *Faramondo* (1719); *Eumene* (1721); *Semiramide riconosciuta* (1724); *Siface* (1725); *Tamerlano* (1730); *Ferdinando* (1734); *Temistocle* (1742); *Partenope* (1742).

Poulenc, Francis, composer. Born Paris, January 7, 1899. He studied the piano with Ricardo Viñes, composition with Charles Koechlin. He first became known about the end of World War I, when the French critic Collet linked his name with those of Darius Milhaud, Louis Durey, Arthur Honegger, George Auric, and Germaine Tailleferre in proclaiming these composers a new school of French music, henceforth known as "Les Six." Poulenc has written music in every form. Though some of his works have great intensity and expressiveness, he is probably best known for compositions filled with ironic statements and witty ideas. He has written the following operas: *Le gendarme incompris* (1920); *Les mamelles de Tirésias* (1947); *Les dialogues des Carmélites* (1953).

Poupelinière, Alexandre Jean-Joseph le Riche de la, music patron. Born Paris, 1692; died there December 5, 1762. Le Poupelinière was Rameau's patron, and through his influence Rameau was

able to produce some of his most important operas, beginning with *Hippolyte et Aricie* in 1733. For several years, from about 1727, Rameau lived in his palace, directed his orchestra, and taught him music. Some of the ariettas Le Poupelinière wrote were incorporated by Rameau in his operas. Rameau used his influence to have François Joseph Gossec appointed as music director of the palace concerts in 1751, after which these concerts became celebrated throughout France.

Pour Bertha moi je soupire, John's aria in Act II of Meyerbeer's *Le Prophète.*

Pour les couvents c'est fini (Finita è per frati), the "Chanson Huguenote," Marcel's aria in Act I of Meyerbeer's *Les Huguenots.*

Pour la Vierge, Boniface's aria in Act I of Massenet's *Le jongleur de Notre Dame.*

Pourquoi me réveiller? Werther's aria in Act III of Massenet's *Werther.*

Poveri fiori, Adriana's aria in Act IV of Cilèa's *Adriana Lecouvreur.*

Prague Opera, *see* STANDESTHEATER.

Pratella, Francesco, composer. Born Lugo di Romagna, Italy, February 1, 1880; died Ravenna, May 18, 1955. He attended the Liceo Rossini in Pesaro, then studied with Pietro Mascagni, under whose guidance he wrote a one-act opera, *Lilia,* which won honorable mention in the Sonzogno competition in 1903. After completing a second opera, *La Sina d'Vargöun,* produced in Bologna in 1909, he allied himself with the futurist movement in Italy, headed by F. T. Marinetti and Luigi Russolo. Pratella now began writing futurist operas, the first being on the subject of aviation, *L'aviatore Dro,* produced in Lugo on September 4, 1920. His later operas: *Il dono primaverile* (1923); *La ninna nanna della Bambola,* a children's opera (1923); *Fabiano* (1939). From 1926 he was director of the Istituto Musicale di

Lugo, and from 1927 to 1945, director of the Liceo Musicale of Ravenna.

précieuses ridicules, Les, see PREZIOSE RIDICOLE, LE; MOLIERE.

Prelude (German: VORSPIEL), a term frequently used in place of overture. Wagner, and many later composers, conceived of the prelude as suggesting the mood and content of the scene to follow, the prelude leading into the scene without a formal ending.

Prenderò quel brunettino, duet of Dorabella and Fiordiligi in Act II, Scene 1, of Mozart's *Così fan tutte.*

Prendre le dessin d'un bijou, Gérald's aria in Act I of Delibes' *Lakmé.*

Presago il core della tua condanna, duet of Aïda and Radames in Act IV, Scene 2, of Verdi's *Aïda.*

Près des remparts de Séville (Seguidilla), Carmen's invitation to Don José in Act I of Bizet's *Carmen.*

Preussisches Märchen (Prussian Fairy Tales), ballet-opera by Boris Blacher. Libretto by Heinz von Cramer. Première: Städtische Oper, Berlin, September 22, 1952. This satire on Prussian military bureaucracy was based on an actual event in Prussian history in 1900, when a servant of Wilhelm II, with the aid of a uniform and by forging the payroll, became a captain in the army.

Prévost, Abbé (ANTOINE FRANCOIS PREVOST D'EXILES), novelist. Born Hesdin, France, April 1, 1697; died Chantilly, France, November 23, 1763. He was the author of the celebrated romance *L'histoire du chevalier des Grieux et de Manon Lescaut,* the source for the following operas: Daniel Auber's *Manon Lescaut;* Hans Henze's *Boulevard Solitude;* Massenet's *Manon* and *Le portrait de Manon;* Puccini's *Manon Lescaut;* and Richard Kleinmichel's *Das Schloss de l'Orme.*

preziose ridicole, Le, (The Ridiculous Smart Women), opera by Felice Lattuada. Libretto by Arturo Rossata, based on Molière's *Les précieuses ridicules.*

Première: La Scala, February 9, 1929. The text is a satire on the affectations and precious language of the so-called intelligentsia of French society. Otakar Zich's opera *Les précieuses ridicules* was derived from the same Molière play.

Preziosilla, a gypsy (mezzo-soprano) in Verdi's *La forza del destino.*

prigioniero, Il (The Prisoner), one-act opera by Luigi Dallapiccola. Libretto by the composer, based on a short story by Villiers de l'Isle Adam, *La torture par l'espérance.* Première: Turin Radio, 1949; first stage presentation, Teatro Communale, Florence, May 20, 1950. During the reign of Philip II, a Flemish Protestant political prisoner is jailed and tortured. After making an unsuccessful effort to escape, he loses his mind and is executed. Dallapiccola has employed the twelve-tone technique for this opera.

Prima donna, Italian for "first lady," a term used since the end of the seventeenth century for the performer of the leading female role in opera. The term "prima donna assoluta" (absolute prima donna) was very much in vogue in the 19th century, with many female singers fighting over this designation.

Primo uomo, Italian for "first man." In early Italian operas this term was used for the performer of the leading male character, or for the castrato taking the leading female role.

Prince Igor, opera by Alexander Borodin (completed by Nikolai Rimsky-Korsakov and Alexander Glazunov). Libretto by Vladimir Stassov, based on an old Russian chronicle. Première: Imperial Opera, St. Petersburg, November 4, 1890. In the twelfth century, the Polovitzi—a Tartar race of Central Asia—capture Prince Igor and his son Vladimir. Khan Kontchak, the Polovtsian leader, assures Prince Igor that he is a guest and not a captive, and entertains him with a lavish feast and oriental dances. He offers him his freedom if he is ready to promise not to fight the Polovitzi. Igor cannot do this. He effects his escape and returns to his own people, where he is received with ceremony. His son, in love with the Khan's daughter, refuses to flee, and is accepted by the Khan as a son-in-law. The Polovtsian dances in the second act, with which the Khan entertains Prince Igor, are often performed at concerts. Two arias are popular: Vladimir's cavatina in Act II, "Daylight is fading away," and Igor's aria in the same act, "No sleep, no rest, for my afflicted soul."

Prince of Persia, The, a character (baritone) in Puccini's *Turandot.*

Princess Eboli, *see* EBOLI, PRINCESS.

Princesse d'Auberge, La, (The Princess of the Inn), opera by Jan Blockx. Libretto by Gustave Lagye. Première: Antwerp, October 30, 1896. In Brussels, in 1750, Rabo and Merlyn are rivals for the coquettish inn proprietress, Rita. Rita is partial to Merlyn. Rabo attacks and kills Merlyn and gives himself up to the authorities.

Princess on the Pea, The, *see* ANDERSEN, HANS CHRISTIAN.

Princess von Werdenberg, the Marschallin (soprano) in Richard Strauss's *Der Rosenkavalier.*

Printemps qui commence, Dalila's spring song in Act I of Saint-Saëns' *Samson et Dalila.*

Prinzivalle, commander of the Florentine army (tenor) in Février's *Monna Vanna.*

Prinzregenten Theater, the leading opera house in Munich, Germany. It was built in 1901 for the presentation of Wagner's music dramas, and it followed the general architecture of the Festspielhaus in Munich.

prise de Troie, La, (The Capture of Troy), opera by Hector Berlioz. Libretto by the composer, based on Virgil's *Aeneid.* Première: Théâtre Lyrique, Paris, November 4, 1863. This is the first of two operas which, to-

gether, are entitled *Les Troyens;* the second is *Les Troyens à Carthage.* In *La prise de Troie,* the legend of the wooden horse and the conquest of Troy is retold. Aeneas is advised by the spirit of Hector to go to Italy and found there a new kingdom.

Prisoner, The, *see* PRIGIONIERO, IL.

Prisoners' Chorus, *see* O WELCHE LUST!

Prison Scene, *see* O PRETES DE BAAL.

Pritzko, a young peasant (tenor) in Mussorgsky's *The Fair at Sorochinsk.*

Prize Song, *see* MORGENLICH LEUCHTEND.

Procession of the Knights, procession in Act III, Scene 2, of Wagner's *Parsifal.*

Procession of the Mastersingers, procession in Act III, Scene 2, of Wagner's *Die Meistersinger.*

Prodaná Nevěsta, original title of Smetana's *The Bartered Bride.*

Prokofiev, Serge, composer. Born Sontsovka, Russia, April 23, 1891; died Moscow, March 4, 1953. He was exceptionally precocious, completing the writing of three operas by the time he was twelve years old. He entered the St. Petersburg Conservatory in his thirteenth year, remaining there a decade; his teachers included Rimsky-Korsakov, Nikolai Tcherepnin, and Anatol Liadov. He graduated in 1914 with the Rubinstein Prize for his Second Piano Concerto. During World War I, Prokofiev was exempt from military service, being the only son of a widow. He now completed his first significant works, including the *Classical Symphony* and the ballet *Chout;* he also wrote an opera, *The Gambler.*

Soon after the outbreak of the Russian Revolution, Prokofiev came to the United States, where he appeared as pianist and conductor in performances of his own works. He was commissioned by the Chicago Opera Company to write a new opera. That work, *The Love of Three Oranges,* was produced in Chicago on December 30, 1921, and was a failure; when next presented—by the New York City Opera in 1949—it was a resounding artistic and box office success.

Prokofiev now settled in Paris, where he remained for the next half dozen years, completing many major orchestral works and ballets. In 1933 he returned to his native land, remaining there (except for an occasional tour) the rest of his life, identifying himself completely with the Soviet ideology, and writing many works that glorified it. Though acknowledged by Soviet leaders and press to be one of the great creative figures in the Soviet Union (he won the Stalin Prize for his seventh piano sonata), Prokofiev became in 1948 the object for violent attack by the General Committee of the Communist Party. The charge against Prokofiev (and other leading Soviet composers) was that Soviet music had become too esoteric and cerebral, that the musical thinking had grown decadent, representing "a negation of the basic principles of classical music." Prokofiev apologized publicly and promised to mend his ways. His first peace offering was an opera, *The Tale of the Real Man,* but this, too, was condemned for its modernistic and antimelodic writing. However, Prokofiev was eventually able to re-establish his honored position in Soviet music. In 1951 he won the Stalin Prize again, and a concert of his works was given to honor his sixtieth birthday. He died of a stroke two years later.

His principal operas: *The Gambler* (1917, revised 1928); *The Love for Three Oranges* (1919); *The Flaming Angel* (1923); *Simeon Kotko* (1939); *The Duenna* (1940); *War and Peace* (1948); *A Tale of the Real Man* (1948).

Promesse de mon avenir, Scandia's aria in Act III of Massenet's *Le Roi de Lahore.*

promessi sposi, I (The Betrothed), a romantic novel by Alessandro Manzoni. Its background is life in seventeenth century Italy, during the Spanish domination. The composers Enrico Petrella, Amilcare Ponchielli, and Franz Gläser have made it the basis of their operas.

Prometheus Bound, see AESCHYLUS.

Prophète, (The Prophet), opera in five acts by Giacomo Meyerbeer. Libretto by Eugène Scribe, based on the historical episode of the Anabaptist uprising in Holland in the sixteenth century. Première: Paris Opéra April 19, 1849. American première: New Orleans, April 2, 1850.

Characters: John of Leyden, prophet and leader of the Anabaptists (tenor); Fidès, his mother (mezzo-soprano); Bertha, his sweetheart (soprano); Count Oberthal, ruler of the Dordrecht region (baritone); Zacharias, an Anabaptist preacher (bass); Jonas, another Anabaptist preacher (tenor); Mathisen, a third preacher (basso); nobles; peasants; citizens; soldiers; prisoners. The action takes place in Holland and Germany in 1534.

Act I. A suburb of Dordrecht. Bertha comes to Count Oberthal to ask his permission to marry the innkeeper, John of Leyden. Before she gains admission, the castle is visited by three Anabaptists who are bent on rousing the people to rebellion. They are driven away soon after the Count makes his appearance. The Count finds Bertha so attractive that he wants her for himself. He denies her permission to marry John, and imprisons John's mother, Fidès, who had accompanied her.

Act II. An inn in the suburbs of Leyden. The three Anabaptists are struck by the resemblance of the innkeeper, John, to a likeness of King David. They beg him to be the leader of their movement. John tells them of a dream in which he was venerated by a crowd before a cathedral. This is further proof to the Anabaptists that John is their destined leader. But John turns them down, since he is interested only in marrying Bertha ("Pour Bertha, moi je soupire"). Bertha arrives, having escaped from the Count. The Count soon appears and warns John that unless he surrenders Bertha, his mother will be put to death. John reluctantly gives up his claim to Bertha. Released from prison, Fidès expresses her gratitude and love for her son ("Ah, mon fils!"). John is now ready to join the Anabaptist movement. To assure the success of their mission, the Anabaptists leave behind the impression that the innkeeper has been killed.

Act III. The camp of the Anabaptists. The people rally under John as their Prophet. Several prisoners are brought to the camp; one of them is the Count, who discloses to John that Bertha has escaped from him and is now in Münster. When the three Anabaptists are about to kill the Count, John intervenes and saves him; he wants Bertha to pass judgment and sentence on him. An element of John's following deserts him to attack Münster on its own. It is defeated. John now rallies the rest of his people and leads them to a victorious assault on Münster.

Act IV, Scene 1. A square in Münster. Fidès, reduced to beggary, meets Bertha and tells her the sad news that John is dead. Not knowing that John lives as the leader of the Anabaptists, Bertha is convinced that the death of her beloved was caused by the Prophet. She vows revenge.

Scene 2. The Münster Cathedral. Victorious, John enters in Münster triumph and is crowned king. The royal procession advances into the Cathedral with magnificent pomp (Marche du couronnement—Coronation March). Hiding behind a pillar, Fidès recognizes her son and rushes towards him tenderly. Since his followers believe him to be of divine origin, John must

denounce Fidès as a fraud. He accuses her of being insane. To protect her son, Fidès confesses she was mistaken.

Act V, Scene 1. A crypt. John's soldiers have imprisoned Fidès. She bewails the fact that her son has repudiated her and begs the gods to destroy him (Prison Scene: "O Prêtres de Baal"). John appears. He throws himself at his mother's feet and begs her forgiveness. At first Fidès is bitter. But she softens when John promises to give up his role and return to Leyden. Bertha suddenly makes her appearance. Seeing John, she is overjoyed to find him alive. But when she discovers that he is the Prophet, she curses him and kills herself with a dagger.

Scene 2. The great hall of the palace. John has been betrayed: the Anabaptists have begun to turn against him, and the Emperor's troops, led by Count Oberthal, have invaded the palace. John orders the palace gates secured. Then a holocaust is set. As the flames mount, Fidès arrives to join her son in death.

Le Prophète is one of the most colorful, stirring, and spectacular of Meyerbeer's operas. He completed it in 1849, eighteen years after Robert le Diable and thirteen after Les Huguenots. It is of particular interest because the composer gave the mother's role such prominence. It was necessity rather than artistic compulsion that led him to do this. At the time the opera was introduced there were no tenors at the Paris Opéra capable of doing justice to the role of John. Meyerbeer decided to balance this deficiency by casting one of the outstanding contraltos of the day, Pauline Viardot-García, as Fidès; and he built up the role to starring proportions. The two principal arias in the opera, among Meyerbeer's finest, are sung not by John or Bertha, but by Fidès: "Ah, mon fils" and "O Prêtres de Baal."

Protagonist, Der (The Protagonist),

one-act opera by Kurt Weill. Libretto by Georg Kaiser. Première: Dresden Opera, March 27, 1926. This was Weill's first opera, written in his twenty-fourth year. The central figure of Kaiser's expressionistic drama is an actor who confuses the stage with reality and commits suicide while enacting a play.

Prozess, Der, see TRIAL, THE.

Prunières, Henri, musicologist. Born Paris, May 24, 1886; died Nanterre, April 11, 1942. He was a student of Romain Rolland in music history and in 1913 received his doctorate with a thesis of major musicological importance: *L'opéra italien en France avant Lully.* He specialized in the music of the seventeenth and eighteenth centuries, producing two biographies of Lully (1910 and 1931), one of Monteverdi (1921), and a study of early Italian opera, *Francesco Cavalli et l'opéra venitien* (1931). In 1919 he founded and edited *La revue musicale,* one of the world's leading music journals. For ten years, up to 1934, he was the Paris correspondent on music for the *New York Times.* He was also the founder of the French section of the International Society for Contemporary Music.

Prynne, Hester, principal female character in Walter Damrosch's *The Scarlet Letter* and Vittorio Giannini's *The Scarlet Letter.* In both operas she is a soprano.

Prynne, Roger, Hester's husband (baritone) in Damrosch's *The Scarlet Letter.*

Puccini, Giacomo, composer. Born Lucca, Italy, December 22, 1858; died Brussels, Belgium, November 29, 1924.

The Puccini family had for several generations held musical posts in Lucca. As a boy, Giacomo was enrolled at the Istituto Musicale as a pupil of Carlo Angeloni. His musical progress was rapid, and he was soon able to fill the post of organist in near-by churches, and to write two ambitious choral

works enthusiastically received by the townspeople. Now something of a local celebrity, Puccini procured a small subsidy from Queen Margherita which, supplemented by additional funds from a great uncle, enabled him to go to Milan in 1880 for further study. He attended the Milan Conservatory, where his teachers included Antonio Bazzini and Amilcare Ponchielli. The latter aroused Puccini's love for the stage and directed him toward writing operas. He had a friend provide Puccini with a suitable libretto, then urged him to set it to music and submit it in the Sonzogno contest for one-act operas. Thus, Puccini completed his first opera, *Le Villi.* It did not win the prize, but when introduced at the Teatro dal Verme in Milan on May 31, 1884, it was so successful that La Scala accepted it for the following season, and Ricordi published the score. Ricordi also commissioned a second opera, *Edgar,* a failure when given at La Scala on April 21, 1889. But his third opera was a triumph—*Manon Lescaut,* introduced in Turin on February 1, 1893. Puccini found himself famous. His *La Bohème,* in 1896, and *Tosca,* in 1900, brought him universal recognition as Verdi's successor.

In 1903 Puccini was seriously hurt in an automobile accident, but confinement to an invalid's chair for eight months did not keep him from writing a new opera, *Madama Butterfly,* produced at La Scala on February 17, 1904. *Madama Butterfly* was, at its première, the greatest failure Puccini experienced; but in less than a year it turned out to be one of his best-loved and most widely performed works.

In 1907 Puccini came to America to assist at a Metropolitan Opera presentation of *Madama Butterfly.* Commissioned to write an opera for the Metropolitan, he produced a work with an American setting: *The Girl of the Golden West.* It was introduced by the

Metropolitan Opera on December 10, 1910, in one of the most exciting premières ever held by that house, the composer being present to receive an ovation.

During the next decade Puccini wrote several more operas: a light opera, *La Rondine;* a trilogy of one-act operas collectively entitled *Il Trittico;* and an opera with an oriental setting, *Turandot.* He did not live to complete the last work. Suffering from a cancer of the throat, Puccini underwent an operation in Brussels. A heart attack, following the operation, was fatal.

Puccini once said of himself that "the only music I can make is of small things." What he meant is that he was never intended by his talent or temperament to produce works in a large design, or ambitious in artistic purpose, in the manner of Wagner and Verdi. But if his world was, comparatively, a small one, he was its lord and master. In the writing of dramas appealing to the heart, filled with tenderness and beauty, he had few equals. He had a dramatic instinct that never failed him, a consummate knowledge of the demands of the stage, and a pronounced feeling for theatrical effect. His writing both for voice and orchestra was the last word in elegance, and his highly personal lyricism had incomparable sweetness, gentleness, and poignancy. His harmonic writing and instrumental colorations were often daring. His best operas brought to a culmination the naturalistic movement launched by *Cavalleria rusticana,* known as "verismo," in which the problems and conflicts of everyday life are treated realistically.

His operas: *Le Villi* (1884); *Edgar* (1889); *Manon Lescaut* (1893); *La Bohème* (1896); *Tosca* (1900); *Madama Butterfly* (1904); *La Fanciulla del West* (*The Girl of the Golden West*) (1910); *La Rondine* (1917); *Il*

Trittico, consisting of *Il tabarro, Suor Angelica,* and *Gianni Schicchi* (1918) *Turandot* (completed by Franco A' fano).

Puck, Titania's attendant (contralto) in Weber's *Oberon.*

Puente, Giuseppe del, *see* DEL PUENTE, GIUSEPPE.

Purcell, Henry, composer. Born London (?) about 1659; died Dean's Yard, Westminster, November 21, 1695. One of England's greatest composers, his *Dido and Aeneas* is a landmark in the early history of opera. His father was a Gentleman of the Chapel Royal. In 1669 Purcell became a chorister of the Chapel Royal, studying music with John Blow. In 1673 he left the choir and became "keeper of the instruments." Four years later he was appointed composer of the King's band and, in 1679, organist at Westminster Abbey. In 1682 he became one of three organists of the Chapel Royal. In the course of his brief life he composed a considerable amount of important instrumental and sacred music, secular songs, and incidental music and songs to plays of the day. His finest stage work was his opera *Dido and Aeneas* (which see). While some of the plays for which he wrote music have been classed as operas, it is only *Dido and Aeneas* that, strictly speaking, deserves the term.

Puritani, I, (The Puritans), opera by Vincenzo Bellini. Libretto by Carlo Pepoli, based on *Les Têtes Rondes et les Cavaliers* by Ancelot Saintine. Première: Théâtre des Italiens, Paris, January 25, 1835. The setting is England during the wars between Cromwell's followers and the Stuarts. Elvira's mother, Queen Henrietta Maria, imprisoned in a Plymouth fort, consents to the marriage of Elvira and Cavalier Lord Arthur Talbot. She has the lovers smuggled into her cell, where the marriage ceremony takes place. Lord Arthur effects the Queen's escape, and for this act he is imprisoned. Elvira, who does not know of these developments, suspects that Arthur has abandoned her. She goes mad with grief. Eventually, the victorious Cromwell releases Lord Arthur who, by explaining the true state of affairs to Elvira, helps bring about her complete recovery. The most important pages include Talbot's aria in Act I, "A te, o cara, amor talora," and Elvira's Mad Scene in Act II, "Qui la voce sua soave."

Pushkin, Alexander, poet. Born Moscow, June 6, 1799; died St. Petersburg, February 10, 1837. Russia's foremost literary figure was the author of many narratives that were made into operas. Among the composers indebted to Pushkin are: Boris Assafiev (*The Bronze Horseman* and *A Feast in Time of Plague*); César Cui (*The Captive in the Caucasus* and *A Feast in Time of Plague*); Alexander Dargomizhsky (*Russalka* and *The Stone Guest*); Glinka (*Russlan and Ludmilla*); Arthur Lourié (*A Feast in Time of Plague*); Modest Mussorgsky (*Boris Godunov*); Eduard Napravnik (*Dubrovsky*); Sergei Rachmaninoff (*Aleko* and *The Miserly Knight*); Nikolai Rimsky-Korsakov (*The Legend of Czar Saltan, Le coq d'or,* and *Mozart and Salieri*); Igor Stravinsky (*Mavra*); Tchaikovsky (*Eugene Onegin, Mazeppa,* and *Pique Dame*).

Pu-Tin-Pao, executioner (baritone) in Puccini's *Turandot.*

Pylades, Orestes' companion (tenor) in Gluck's *Iphigénie en Tauride.*

Q

Quadrille (or QUADRILLE DE CONTRE-DANSES), a French dance that made its first operatic appearance in Rameau's *Les fêtes de Polymnie,* in 1745. In the nineteenth century there was a tremendous vogue for quadrilles as social dances, and they appeared in numerous opéras bouffes and operettas. Usually, the music of these later quadrilles was not original, but consisted of opera melodies fitted to the five distinct sections of a typical quadrille. Emmanuel Chabrier carried the practice to its farthest development when he wrote a satirical quadrille on melodies from *Tristan und Isolde.*

Qual cor tradisti, final duet in Act IV of Bellini's *Norma.*

Quarter-tone operas, *see* HABA, ALOIS.

Quand apparaissent les étoiles, Don Quixote's serenade in Act I of Massenet's *Don Quichotte.*

Quand du Seigneur le jour luira, the church chorus in Act IV, Scene 2, of Gounod's *Faust.*

Quand le destin au milieu (Quando fanciulla ancor l'avverso), Marie's aria at the conclusion of Act II of Donizetti's *The Daughter of the Regiment.*

Quand'ero paggio del Duca di Norfolck, Falstaff's aria in Act II, Scene 2, of Verdi's *Falstaff.*

Quando le sere al placido, Rodolfo's aria in Act II of Verdi's *Luisa Miller.*

Quando me'n vo' soletta, Musetta's Waltz in Act II of Puccini's *La Bohème.*

Quando rapita in estasi, Lucia's aria concerning her love for Edgardo in Act I, Scene 2, of Donizetti's *Lucia di Lammermoor.*

Quanto è bella! Nemorino's aria in Act I of Donizetti's *L'elisir d'amore.*

quattro rusteghi, I (The Four Ruffians),

opera buffa by Ermanno Wolf-Ferrari. Libretto by Giuseppe Pizzolato, based on Carlo Goldoni's comedy. Première: Munich, March 19, 1906. The slender plot concerns Lunardo's refusal to permit his daughter and her lover to see each other before their wedding. Lunardo, however, is outwitted when the lover, Filipeto, dons a woman's garb.

Queen Ice-Heart, a character (mezzosoprano) in Weinberger's *Schwanda.*

Queen of Sheba, The (Die Königin von Saba), opera in four acts by Karl Goldmark. Libretto by Solomon Hermann Mosenthal, based on the Old Testament story. Première: Vienna Opera, March 10, 1875.

Characters: Queen of Sheba (soprano); King Solomon (baritone); High Priest (bass); Sulamith, his daughter (soprano); Assad, betrothed to Sulamith (tenor); Baal Hanan, overseer of the palace (baritone); Astaroth, the Queen's slave (soprano). The setting is Jerusalem and the Syrian desert during the reign of King Solomon.

Act I. Solomon's palace. Assad is to marry Sulamith, but he confides to King Solomon that he has fallen in love with an unknown woman he encountered while she was bathing in a stream. When the Queen of Sheba appears before Solomon, Assad recognizes her as that woman, but she ignores him.

Act II, Scene 1. The palace garden. Assad has a secret meeting with the Queen of Sheba. She is now receptive to his love-making. When she learns that he is to marry Sulamith, she urges him to run away with her.

Scene 2. The Temple. After the wedding of Assad and Sulamith is celebrated, Assad impulsively throws his wedding ring at the Queen's feet. But the Queen insists that Assad is a stranger to her. Assad, furious, curses Jehovah. For this he is condemned to death. The Queen intervenes and persuades King Solomon at least to spare his life.

Act III. The palace. A sumptuous ballet is seen during a reception honoring the Queen. The Queen begs Solomon to release Assad. When he refuses to do so, she angrily leaves him. Sulamith now comes to plead for Assad, promising to dedicate her life to God if he is freed.

Act IV. The desert. King Solomon has freed Assad, and he has become a wandering pilgrim. He is found by the Queen of Sheba. She pleads for his love, but he now feels only hate for her. During a sandstorm, Sulamith appears. Assad pleads for her forgiveness, obtains it, then, spent and exhausted, dies in her arms.

See also REINE DE SABA, LA.

Queen of Shemakha, a queen (soprano) in Rimsky-Korsakov's *Le coq d'or.*

Queen of Spades, The, *see* PIQUE DAME.

Queen of the Night, Pamina's mother (soprano) in Mozart's *The Magic Flute.*

Queen of the Spirits of the Earth, a character (soprano) in Marschner's *Hans Heiling.*

Que fais-tu, blanche tourterelle, Stephano's mocking serenade in Act III, Scene 2, of Gounod's *Roméo et Juliette.*

Quel plaisir! quelle joie! chorus in Act V of Halévy's *La Juive.*

Quentin Durward, *see* SCOTT, SIR WALTER.

Questa notte (Nessun dorma), Calaf's aria in Act III, Scene 1, of Puccini's *Turandot.*

Questa o quella, the Duke's aria in Act I of Verdi's *Rigoletto.*

Questo e quell pezzo, Despina's incantation as a mock doctor in Act I, Scene 4, of Mozart's *Così fan tutte.*

Questo sol è il soggiorno, *see* C'EST ICI LE SEJOUR.

Quickly, Dame, a character (mezzosoprano) in Verdi's *Falstaff.*

quiet Don, The (or Quiet Flows the Don), opera by Ivan Dzerzhinsky. Libretto by Leonid Dzerzhinsky, based on the novel of the same name by Mikhail Sholokhov. Première: Leningrad, October 22, 1935. The opera covers the period in Cossack history from 1914 to 1917. It centers around a Cossack, Gregor, who, upon returning home from war in 1914, finds his wife has become the mistress of a nobleman. He kills his rival, abandons his family and former masters, and leads the peasants in revolt.

The tremendous success of this opera came at a time when the Soviet press and political leaders were denouncing Shostakovich's *Lady Macbeth of Mtsensk* as "formalistic" and "degenerate." (By a curious paradox, Dzerzhinsky's opera was dedicated to Shostakovich.) The authorities used *The Quiet Don* as an example of what true Soviet opera should be; and it is largely as a result of this official blessing that the opera was acclaimed and extensively performed throughout the Soviet Union.

Qui la voce sua soave, Elvira's Mad Scene in Act II of Bellini's *I Puritani*.

Quinault, Philippe, poet and librettist. Born Paris, June 3, 1635; died there November 26, 1688. Regarded as the creator of French lyric tragedy, he was Jean-Baptiste Lully's librettist. Among the operas by Lully for which he wrote the texts were: *Alceste, Amadis, Armide, Atys, Cadmus et Hermione, Isis, Persée, Phaëton, Proserpine, Roland,* and *Thésée.* A number of his librettos were used by opera composers of a century later, notably Johann Christian Bach, Cristoph Willibald Gluck, and Nicola Piccinni.

Qui posa il fianco, trio of Giselda, Oronte, and Arvino in Act III of Verdi's *I Lombardi.*

Qui sola, vergin rosa, see LAST ROSE OF SUMMER, THE.

Quo Vadis, opera by Jean Nouguès. Libretto by Henri Cain, based on the novel of the same name by Henryk Sienkiewicz. Première: Nice, February 9, 1909. The setting is Rome in the age of Nero. Vincius is in love with the Roman hostage Lygia. At Nero's palace, when Vincius speaks to Lygia of his love, she rejects him because of their different religions. Nero orders a giant barbarian to fight a savage animal with Lygia strapped to his back. The barbarian wins the fight, and he and Lygia are freed. But Nero soon repents this act of mercy and sends Lygia to her doom.

R

Rabaud, Henri, composer and conductor. Born Paris, November 10, 1873; died there September 11, 1949. He attended the Paris Conservatory, where his teachers included Jules Massenet and André Gédalge, and where he won the Prix de Rome in 1894. As a conductor, he distinguished himself with performances at the Opéra between 1908 and 1914, and the Opéra-Comique in 1920; he also was conductor of the Concerts du Conservatoire, and, in 1918, of the Boston Symphony Orchestra. He achieved his first success as a composer with a tone poem, *La procession nocturne,* in 1899.

His most famous work was a comic opera, *Mârouf*, introduced at the Paris Opéra on May 15, 1914. In 1920 Rabaud became director of the Paris Conservatory. His other operas: *La fille de Roland* (1904); *Le premier glaive* (1908); *Antoine et Cléopâtre* (1917); *L'appel de la mer* (1922); *Rolande et le mauvais garçon* (1933).

Rabelais, François, physician and satirist. Born, probably in Chinon, France, about 1494; died, probably in Paris, about 1553. His masterwork was *Gargantua and Pantagruel,* a picture of the life, thought, and customs of the Renaissance. It was the source of several operas, including: André Grétry's *Panurge;* Antoine Mariotte's *Gargantua;* Massenet's *Panurge;* Pierre Alexandre Monsigny's *L'île sonnante;* and Claude Terrasse's *Pantagruel.*

Rachel, Eléazar's daughter (soprano) in Halévy's *La Juive.*

Rachel, quand du Seigneur, Eléazar's aria in Act IV of Halévy's *La Juive.*

Rachmaninoff, Sergei, composer, pianist, conductor. Born Onega, Russia, April 1, 1873; died Beverly Hills, California, March 28, 1943. Early in his career, after studying at the St. Petersburg and Moscow conservatories, he wrote three operas, none of them adding appreciably to his stature as a composer, and none of them performed today. His first, *Aleko,* written when he was only nineteen, was favorably received when introduced at the Moscow Opera on May 9, 1893. The other two were: *The Miserly Knight* (1905) and *Francesca da Rimini* (1905). He worked on a fourth opera, *Monna Vanna,* between 1906 and 1908, but abandoned it.

For a period, Rachmaninoff was a conductor of opera. He made his debut in this field soon after 1900 when he directed *A Life for the Czar* for the Mamontov Opera Company, with which organization he remained a year. In 1905 he was appointed conductor of the Bolshoi Theater in Moscow, making his debut there with Dargomizhsky's *Russalka;* during his year at the Bolshoi he conducted his *The Miserly Knight* and *Francesca da Rimini.*

Racine, Jean Baptiste, poet and dramatist. Born La Ferté-Milon, France, December, 1639; died Paris, April 26, 1699. Several of his dramas, which are the basic works of the classical French theater, were made into operas; notably: *Andromaque* (opera by André Grétry, and Rossini's *Ermione*); *Iphigénie en Aulide* (operas by Karl Heinrich Graun and Cristoph Willibald Gluck); *Mitridate* (Mozart); *Phèdre* (opera by Jean Baptiste Lemoyne); *Bérénice* (Albéric Magnard). Hervé's comic opera *Le nouvel Aladin* is a parody of Racine's *Bajazet.*

Radames, captain of the guards (tenor), in love with Aïda, in Verdi's *Aïda.*

Radames, è deciso il tuo fato, the priests' chorus in Act IV, Scene 1, of Verdi's *Aïda.*

Radio, *see* OPERA PERFORMANCE, section RADIO.

Räuber, Die, *see* SCHILLER, FRIEDRICH.

Rafaele, leader of the Camorrists (baritone) in Wolf-Ferrari's *The Jewels of the Madonna.*

Raimondo, chaplain (bass) in Donizetti's *Lucia di Lammermoor.*

Raisa, Rosa (born ROSE BURSTEIN), dramatic soprano. Born Bialystok, Poland, May 30, 1893. She studied singing with Eva Tetrazzini, and with Barbara Marchisio at the Naples Conservatory. Her opera debut took place in Parma on September 6, 1913, in Verdi's *Oberto.* After appearances at the Teatro Constanzi and Covent Garden, she made her American debut at the Chicago Opera on November 28, 1914, in *Aïda.* Soon afterward, she became one of the most adulated sopranos of that city. Besides appearing in the Italian, French, and (occasionally) the German repertories, she was seen in the world première of Franke Harling's *A Light of St. Agnes,* in 1925, and in the American premières of Mascagni's *Isabeau* and Montemezzi's *La nave.* At La Scala she created the roles of Asteria in Boïto's *Nerone* in 1924 and Turandot in 1926. In 1938 she created for America the role of Leah in Rocca's *The Dybbuk.* She was the wife of the late opera baritone Giacomo Rimini.

Rake's Progress, The, opera by Igor Stravinsky. Libretto by W. H. Auden and Chester Kallman. Première: Venice, September 11, 1951. The inspiration for this opera came from a set of eight paintings by William Hogarth. The setting is eighteenth century England, where Tom Rakewell deserts Anne Trulove for a life of carousing, lust, and the squandering of a fortune. He consigns his soul to Nick Shadow, otherwise the Devil. When the time comes for Tom to keep his bargain he engages Nick in a gamble, the prize being his soul. Tom wins, but then loses his mind and is confined in Bedlam. Just before he dies, Anne Trulove visits him. Purposely, the opera has an eighteenth century personality, with its traditional sequence of arias, recitatives, ensemble numbers, and choruses. Much of the music has a Mozartian economy, simplicity, and lyricism.

Rákóczy March, a march representing the advance of the Hungarian army at the close of Part I of Berlioz' *The Damnation of Faust.* It is made appropriate in the dramatic context by the fact that Faust is wandering through Hungary.

Rambaldo, a minstrel (tenor) in Meyerbeer's *Robert le Diable.*

Rameau, Jean-Philippe, composer and theorist. Born Dijon, France, September 25, 1683; died Paris, September 12, 1764. The first French-born master in opera, he was the son of a church organist. As a child he was taught the harpsichord, organ and violin. In 1701 he traveled through northern Italy, earning his living as an organist and violinist. When he returned to France he served as organist in Avignon and Clermont-Ferrand. During this period he began writing music for the harpsichord. About 1705 he settled in Paris, where he studied the organ with Louis Marchand and devoured every book on theory he could find. He published his first volume of harpsichord pieces a year later. Rameau first achieved prominence as a musical theorist, publishing in 1722 his monumental *Traité de l'harmonie,* even today a valuable work. Four years later he published a second volume, *Nouveau système de musique théorique.* He now came to the attention of the powerful patron, Riche de la Poupelinière, becoming the conductor of his private orchestra, and his organist. Meanwhile, in 1723, Rameau began writing music for the stage. Recognition as an opera composer came in 1733 when his *Hippolyte et Aricie* was performed at the Opéra. Rameau was soon accused of sacrificing melody for the sake of harmony and orchestration, and of placing too much emphasis on mere drama. With his subsequent operas—notably *Les Indes galantes* (1735), *Castor et Pollux* (1737), and *Dardanus* (1739)—the opposition grew increasingly hostile. Jean Jacques Rousseau, a believer in the Italian methods, wrote of Rameau that "the French airs are not airs at all, and the French recitative is not recitative." The musicians of the Opéra jeered at Rameau's complex orchestration. Friedrich Grimm derided Rameau for his excessive use of the ballet. Yet there were also those who acknowledged the composer's genius. "Rameau has made of music a new art," wrote Voltaire, and André Campra said, "He will eclipse us all."

The struggle between Rameau's followers and his enemies was climaxed in 1752 with a squabble known as the "guerre des bouffons." On August 1 a visiting Italian troupe performed Pergolesi's *La serva padrona*. Rameau's opponents proclaimed Pergolesi's opera the ideal opera, while violently condemning Rameau's operas for their intricacy and cerebralism. In the other camp were those who felt that Rameau was laying the foundations for a French operatic art. By the end of his life, Rameau was partially vindicated and was recognized as a master. Full vindication came after his death with the victory in Paris of Gluck's operas over the Italian operas of Nicola Piccinni.

The following are the most significant of Rameau's operas and opera-ballets: *Hippolyte et Aricie* (1733); *Les Indes galantes* (1735); *Castor et Pollux* (1737); *Dardanus* (1739); *Les fêtes de Polymnie* (1745); *Les fêtes de l'Hymen et de l'Amour* (1747); *Zoroastre* (1749); *Platée* (1749); *Acanthe et Céphise* (1751); *Zephire* (1754); *Anacréon* (1754); *Les Paladins* (1760).

Ramerrez, *see* JOHNSON, DICK.

Ramfis, High Priest of Egypt (bass) in Verdi's *Aïda*.

Ramiro, a muleteer (baritone) in Ravel's *L'heure Espagnole*.

Ramon, (1) a wealthy farmer (bass) in Gounod's *Mireille*.

(2) Pedro's brother and slayer (bass) in Laparra's *La Habanera*.

Ramuntcho, opera by Deems Taylor. Libretto by the composer, based on the novel of the same name by Pierre Loti. Première: Philadelphia, February 10, 1942. In the Basque village of Etchezar, in the early part of the present century, Gracieuse is ready to marry Ramuntcho, the smuggler, if he promises to give up his profession. Ramuntcho must first serve a three-year term in the army. His letters to Gracieuse are intercepted and destroyed by her mother. Convinced that her beloved has forgotten her, Gracieuse enters a convent. When Ramuntcho returns from the army, he discovers what has happened. Despite his pleadings, Gracieuse refuses to leave the convent, insisting that she is now God's bride. Taylor's score makes extensive use of Basque folk songs.

Rance, Jack, sheriff (baritone) in Puccini's *The Girl of the Golden West*.

Rangström, Ture, conductor and composer. Born Stockholm, Sweden, November 30, 1884; died there May 11, 1947. His studies took place in Berlin with Hans Pfitzner and in Stockholm. From 1907 he wrote music criticisms for various Swedish newspapers. After 1919 he taught at the Royal Academy of Music, from 1922 to 1925 was conductor of the Gothenburg Orchestra Society, and subsequently was stage director of the Stockholm Royal Opera. He wrote three operas: *Middle Ages* (1918); *Die Kronbraut* (1919); *Gilgamesh*.

Ranz des Vaches, a call played on the Alpine horn by Swiss herdsmen to summon their cattle. The call has numerous variations. A number of them appear in Rossini's *William Tell*, the Ranz des Vaches (for English horn) in the overture being the best known. Wilhelm Kienzl composed an opera entitled *Ranz des Vaches*.

Rape of Lucretia, The, opera by Benjamin Britten. Libretto by Ronald Duncan, based on a play by André Obey, *Le viol du Lucrèce*. Première: Glyndebourne, England, July 12, 1946. This is a chamber opera of modest proportions. There are two choruses, each consisting of a single person: one a man, one a woman. The cast is made up of six principals. The orchestra has only twelve instruments. The music is lean and concise, deriving much of its power from its concentration. The subject of the opera is the seduction of Lucretia, wife of the Roman general Collatinus, by the Etruscan prince,

Tarquinius, during the Etruscan domination of Rome.

Raskolnikoff, opera by Heinrich Sutermeister. Libretto by P. Sutermeister, based on Dostoyevsky's *Crime and Punishment.* Première: Stockholm Opera, October 14, 1948. This opera, like the Dostoyevsky's novel, emphasizes the psychology of a poor student who turns to murder and finds expiation in the love of a streetwalker. He finally gives himself up to the police to be exiled to Siberia.

Rataplan, (1) chorus of the French soldiers in Act I of Donizetti's *The Daughter of the Regiment.*

(2) Scene for Pueziosilla and chorus in Act III of Verdi's *La forza del destino.*

Ratmir, Ludmilla's suitor (contralto) in Glinka's *Russlan and Ludmilla.*

Ravel, Maurice, composer. Born Ciboure, France, March 7, 1875; died Paris, December 28, 1937. He received his musical training at the Paris Conservatory, and first attracted attention in 1902 with his piano pieces, *Pavane pour une infante défunte* and *Jeux d'eau.* During the next few years he twice became the object of critical attention in Paris: first, in 1905, for having failed for the fourth time to win the Prix de Rome; in 1907, when he was unjustly accused of being an imitator of Claude Debussy. But he was soon acknowledged a master of contemporary French music. His one-act opera, *L'heure Espagnole,* was introduced at the Opéra-Comique on May 19, 1911. Ravel later wrote another work, *L'enfant et les sortilèges* (1925), calling for pantomime, ballet, and singing.

Ravenswood, see EDGARDO.

Rebikov, Vladimir, composer. Born Krasnoyarsk, Siberia, May 31, 1866; died Yalta, Crimea, December 1, 1920. His music study took place at the Moscow Conservatory, in Berlin, and in Vienna. In 1894 he attracted attention with his opera *In the Storm,* produced

in Odessa. He evolved a personal theory in which music was "the language of emotion." He explained that, "Our feelings have no prepared and conventional forms and terminations," and that music should "give them corresponding expression." In line with this thinking, he wrote in an unorthodox style abounding in strange harmonies and tonalities. Sometimes called the "father of Russian modernism," he was one of the first composers to make extensive use of the whole-tone scale. He described some of his operas as musico-psychological dramas; notably, *Arachne; The Abyss; Alpha and Omega;* and *The Woman with the Dagger.* His other operas were: *In the Storm* (1894); *The Christmas Tree,* a children's opera (1903); *Prince Charming,* a fairy opera.

Recitative (Italian: RECITATIVO), the term denoting the declamatory portions of an opera (or kindred work), used principally to advance the narrative. A "dry" recitative (Italian: recitativo secco) has little or no melodic interest and simply follows the normal accentuation of the words; it is accompanied by a few fundamental chords played by harpsichord or piano. An accompanied recitative (recitativo accompagnato, or stromentato), is more developed, musically speaking, and has an orchestral accompaniment that may be simple or elaborate.

Recondita armonia, Cavaradossi's aria rhapsodizing Tosca's beauty in Act I of Puccini's *Tosca.*

Recordings, see OPERA PERFORMANCE, section RECORDINGS.

Re dell' abisso affrettati, Ulrica's incantation in Act II of Verdi's *Un ballo in maschera.*

Regina, musical play by Marc Blitzstein. Libretto by the composer, based on Lillian Hellman's play *The Little Foxes.* Première: Boston, October 11, 1949. The play (a great success on Broadway with Tallulah Bankhead in the leading role) deals with the rapacious Hubbard family, whose members are driven only by lust, hate, greed, deceit, and theft until they destroy each other and themselves. *Regina* was first seen in an opera house on April 2, 1953, performed by the New York City Opera Company. Previously, it had been staged on Broadway as a musical play.

Regnava nel silenzio, Lucia's aria in Act I, Scene 2, of Donizetti's *Lucia di Lammermoor.*

Reich, name for Page in the German-language version of Nicolai's *The Merry Wives of Windsor.* See PAGE.

Reichardt, Johann Friedrich, composer and conductor. Born Königsberg, Germany, November 25, 1752; died Giebichenstein, Germany, June 27, 1814. A pioneer writer of singspiels, he studied philosophy at the universities of Königsberg and Leipzig, music with C. G. Richter and Franz Adam Veichtner. Between 1775 and 1794 he was court conductor and composer for Frederick the Great and his successor, Friedrich Wilhelm II. In 1783 he founded the Concerts Spirituels for the performance of new music. He visited Paris in 1785, and was commissioned by the Opéra to write two works, *Tamerlan* and *Panthée.* When he was suddenly recalled to the court of Frederick the Great, the production of both operas was abandoned. In 1791 he was accused by Friedrich Wilhelm II of sympathy with the French Revolution and was suspended from his court post for three years, a period in which he traveled extensively. In 1794 the Emperor dismissed him permanently. He settled in Giebichenstein, where he became inspector of the Halle salt works. In 1808 he was appointed director of the Kassel Opera. In the same year he visited Vienna, where some of his singspiels were performed; because he overstayed his leave, he was dismissed from his post in Kassel. Besides being a composer and conductor, Reichardt was a

distinguished writer on musical subjects, and an editor of several music publications. His most important operas and singspiels: *Tamerlan* (1786); *Andromeda* (1787); *Protesilao* (1788); *Brenno* (1789); *Jery und Bätely* (1789); *L'Olimpiade* (1791).

Reichmann, Theodor, dramatic baritone. Born Rostock, Germany, March 15, 1848; died Marbach, Lake Constance, Switzerland, May 22, 1903. His music study took place in Berlin with Eduard Mantius, in Prague with Luise Ress, and in Milan with Francesco Lamperti. He made his opera debut in 1869 in Magdeburg. After appearing in various opera theaters in Germany and Holland he was engaged by the Munich Opera in 1874, where he distinguished himself in the Wagnerian repertory. In 1882, in Bayreuth, he created the role of Amfortas. For the next ten years he appeared regularly at the Bayreuth Festivals, but because of differences with the Wagner family, he did not appear there during the decade 1892–1902. From 1882 to 1889 he was a member of the Vienna Opera. On November 27, 1889, he made his American debut at the Metropolitan Opera in *Der fliegende Holländer.* He remained at the Metropolitan through the 1890–1891 season. In 1893 he returned to the Vienna Opera.

Reine de Saba, La, (The Queen of Sheba), opera by Charles Gounod. Libretto by Jules Barbier and Michel Carré. Première: Paris Opéra, February 28, 1862. The Hebrew sculptor Adoniram is in love with Balkis, the Queen of Sheba. Balkis leaves King Solomon to elope with Adoniram, but by the time she reaches him he has been murdered by one of his own men.

Reine Fiammette, La (Queen Fiammette), opera by Xavier Leroux. Libretto by Catulle Mendès, based on his own play. Première: Opéra-Comique, December 23, 1903. The setting is a mythical kingdom during the Renaissance. Cardinal Sforza, conspiring to get rid of Queen Fiammette, persuades Danielo to wield the assassin's knife, but Danielo recognizes the Queen as a young woman he had formerly loved. For his failure to kill the Queen, he is condemned to death. When he makes an attempt on the life of the wicked Sforza, both he and the Queen go to their doom.

Reiner, Fritz, conductor. Born Budapest, Hungary, December 10, 1888. He attended the National Academy in Budapest, after which, in 1909, he became chorus master of the Budapest Opera. In 1910 he was engaged as conductor of the Laibach National Opera and in 1911 of the People's Opera in Budapest. He was appointed first conductor of the Dresden Opera in 1914. When the war broke out, Reiner directed opera performances in Rome and Barcelona. In 1922 he was appointed music director of the Cincinnati Symphony, where he remained eight seasons. In 1930 he became head of the orchestra department of the Curtis Institute of Music. During this period he appeared as guest conductor of leading American orchestras and with several major opera companies. During the 1936 Coronation festivities in London, he scored a major success in performances of the Wagner music dramas. In 1938 he began a tenyear period as musical director of the Pittsburgh Symphony. After leaving Pittsburgh, Reiner became a principal conductor of the Metropolitan Opera, making his debut there on February 4, 1949, with *Salome*. He resigned from the Metropolitan in 1953 to become the music director of the Chicago Orchestra. In 1955 he was invited to conduct at the reopening of the Vienna State Opera.

Reiss, Albert, dramatic tenor. Born Berlin, Germany, February 22, 1870; died Nice, France, June 20, 1940. He appeared on the dramatic stage until 1897, when the opera impresario Bernhard Pollini urged him to turn to opera. Reiss's debut took place in Königsberg on September 28, 1897, in *Zar und Zimmermann*. He made his American debut at the Metropolitan Opera on December 23, 1901, in *Tristan und Isolde*. During the same season he was acclaimed for his interpretations of David in *Die Meistersinger* and Mime in *Siegfried;* for many years, these interpretations were considered a standard. He remained with the Metropolitan Opera through the 1919–1920 season, creating there the roles of Nick in *The Girl of the Golden West,* the Broommaker in *Die Königskinder,* Nial in *Mona,* and Richard II in *The Canterbury Pilgrims.*

He became an opera producer in 1916 when he presented, and appeared in, *Bastien and Bastienne* and *The Impresario,* both by Mozart. His success led him to direct a two-week season at the Lyceum Theater in the spring of 1916 and to organize the Society of American Singers in 1917. After leaving the Metropolitan, he appeared for a decade with the Charlottenburg Opera in Berlin. He was a frequent guest at Wagner festivals and at Covent Garden. He retired from opera in 1930.

Re Lear (King Lear), opera by Vito Frazzi. Libretto by the composer, based on the Shakespeare tragedy. Première: Florence May Music Festival, 1939. *See* KING LEAR.

Rembrandt van Rijn, painter. Born Leyden, Netherlands, July 15, 1607; died Amsterdam, October 4, 1669. The famous artist is the central character in two operas, both entitled *Rembrandt van Rijn,* one by Henk Badings, the other by Paul von Klenau.

Remendado, Le, a smuggler (tenor) in Bizet's *Carmen.*

Remigio, a farmer (bass) in Massenet's *La Navarraise.*

Renard, chamber opera, or opera-ballet, by Igor Stravinsky. Libretto by the composer, based on Russian folk tales. Première: Paris Opéra, June 3, 1922. The composer explains: *"Renard is to be played by buffoons, dancers, or acrobats, preferably on a trestle stage, with the orchestra placed behind. The players do not leave the stage. They enter together to the accompaniment of the little march that serves as an introduction and their exit is managed in the same way. The roles are dumb. The voices (two tenors and two basses) are placed in the orchestra."* The fable concerns two attempts on the part of Renard the fox to abduct the cock; the attempts are frustrated by the cat and the ram. The fox is finally conquered by the barnyard animals. A feature of Stravinsky's score is the inclusion of a cimbalom, an instrument found chiefly in Hungarian gypsy orchestras.

Renato, the king's secretary (baritone) in Verdi's *Un ballo in maschera.*

Renaud, Maurice, baritone. Born Bordeaux, France, July 24, 1861; died Paris, October 16, 1933. His musical training took place in the conservatories of Paris and Brussels. He made his opera debut at the Théâtre de la Monnaie in 1883. Until 1890 he was a leading baritone of that company, creating the title role in Reyer's *Sigurd* and the role of Hamilcar in the same composer's *Salammbô.* On October 12, 1890, he made his Paris debut at the Opéra-Comique in *Le Roi d'Ys,* and on July 17, 1891, his debut at the Opéra in *L'Africaine.* He was a principal baritone of the Opéra for over a decade. In 1897 he made his first appearance at Covent Garden, in a performance honoring Queen Victoria's Golden Jubilee. His American debut took place on January 4, 1893, at the French Opera House in New Orleans in *Samson et Dalila.* A decade later he was contracted by Maurice Grau to appear at the Metropolitan Opera, but when Heinrich Conried took over the direction,

Renaud abrogated his contract. He was heard again in America with the Manhattan Opera Company between 1906 and 1910, scoring triumphs in such roles as Athanael (which he created for America), Don Giovanni, and Scarpia. After the Manhattan Opera disbanded, he sang for a season with the Chicago Opera. On November 25, 1910, he made his bow at the Metropolitan Opera in *Rigoletto,* remaining with the company two years. Thereafter, he sang only in Europe.

Répétition, a French term meaning "rehearsal." A "répétition générale" denotes a dress rehearsal attended by critics and invited guests.

Rescue opera, a genre of French opera enjoying a brief span of popularity during the French Revolution. The typical libretto shows the hero or heroine being saved after many vicissitudes. Cherubini's *Les deux journées* is such an opera.

Residenztheater, an intimate opera house in Munich, built in rococo style, attached to the palace. It was opened in 1753. In 1781 it was the scene for the première of Mozart's *Idomeneo.* In recent times the theater has often been used for performances of Mozart's operas at the annual Munich festivals.

Respighi, Ottorino, composer. Born Bologna, Italy, July 9, 1879; died Rome, April 18, 1936. He graduated from the Bologna Liceo in 1899, after which he studied with Rimsky-Korsakov in St. Petersburg and Max Bruch in Berlin. He first attracted attention as a composer with an opera, *Re Enzo,* introduced in Bologna in 1905. While Respighi became famous for his orchestral music, and was a leading figure in the Italian movement to turn from opera to symphonic and chamber music, he did not abandon the stage. His first major operatic success came with *Belfagor,* given at La Scala on April 26, 1923. His most important opera, *La campana sommersa,* was introduced in Hamburg in 1927 and was given at the Metropolitan Opera a year later.

Respighi became professor of composition at the Santa Cecilia Academy in Rome in 1913. A decade later he was appointed director, holding this post two years. In 1932 he was appointed to the Royal Academy of Italy. He visited the United States in 1925, 1928, and 1932.

His operas: *Re Enzo* (1905); *Marie Victoire* (1909); *La bella addormentata* (1909); *Semirama* (1910); *Belfagor* (1923); *La campana sommersa* (1927); *Maria Egiziaca* (1932); *La Fiamma* (1933); *Lucrezia* (completed by his wife).

Reste, repose-toi, the berceuse of Louise's father in Act IV of Charpentier's *Louise.*

Resurrection, see RISURREZIONE, LA.

Reszke, see DE RESZKE.

retablo de Maese Pedro, El (Master Peter's Puppet Show), chamber opera for puppets by Manuel de Falla. Libretto by the composer, based on a scene from Part II of Cervantes' *Don Quixote.* Première: Seville, March 23, 1923 (concert version); Paris, June 25, 1923 (staged performance). This chamber opera is a play within a play. Don Quixote, Sancho Panza, and Master Peter are members of an audience watching a play in which the story of Melisandra is being enacted. The composer gave instructions that two sets of puppets be used: large puppets for the audience, small ones for the players. An alternative method was also suggested: actors wearing masks might replace the large puppets. The singers are placed with the orchestra in the pit.

Rethberg, Elizabeth (born SATTLER), lyric and dramatic soprano. Born Schwarzenburg, Germany, September 22, 1894. Her musical education took place at the Dresden Conservatory. Her debut took place in 1915 with the Dresden Opera in *The Gypsy Baron.* She remained a principal soprano of that company for seven seasons, at the same time making guest appearances at La Scala, the Vienna State Opera, and the Berlin Opera. Her American debut took place at the Metropolitan Opera on November 22, 1922, in *Aïda.* For two decades she was a principal soprano of the Metropolitan, appearing in the French, Italian, and German repertories. She was heard in the American première of *La campana sommersa* and in such major revivals as *The Barber of Bagdad, Der Freischütz, Iris,* and *The Magic Flute.* During her association with the Metropolitan, she was heard with the Chicago Civic Opera, the San Francisco Opera, Covent Garden, the Dresden Opera, and the Salzburg Festivals. In 1928 she created the title role in Richard Strauss's *Die aegyptische Helena;* two years later a Rethberg week was celebrated by the Dresden Opera, an occasion on which she was given honorary membership in all the state theaters of Saxony.

She left the Metropolitan Opera in 1942 after differences with the management over her contract. Her last appearance there took place on March 6, 1942, in *Aïda.* She has also been prominent as a concert singer and soloist in oratorios. In 1936 and 1938 she toured in joint recitals with Ezio Pinza. She received her American citizenship in 1939.

rêve, Le, *see* EN FERMANT LES YEUX.

Reyer, Ernest (born LOUIS ERNEST ETIENNE REY), composer. Born Marseilles, France, December 1, 1823; died Levandou, France, January 15, 1909. When he was sixteen he went to Algiers to live with an uncle, there he studied music (mostly by himself) and started composition. His extended stay in Algiers was responsible for his lifelong interest in Oriental subjects. In 1848 he returned to Paris, where his first major work, *Le Sélam,* for orchestra, was introduced two years later. In 1854 a one-act opera, *Maître Wolfram,* was given at the Théâtre Lyrique. It won praise from Halévy and Berlioz. Even more successful were a ballet-pantomime, *Sacountala,* given by the Opéra in 1858, and a comic opera, *La statue,* produced at the Théâtre Lyrique in 1861. Bizet regarded *La statue* the most important new French opera in a quarter of a century. Between 1865 and 1875 Reyer wrote music criticisms for French journals, and was a forceful voice for Wagner and the younger French composers. His criticisms were collected in two books: *Notes de musique* (1875) and *Quarante ans de musique* (1909). In 1876 he was appointed to the Institut de France in succession to Ferdinand David. Eight years later his most important work, the opera *Sigurd* (on which he had been working for a decade), was successfully given at the Théâtre de la Monnaie, and soon after performed at Covent Garden and the Paris Opéra. Reyer's last major work was the opera *Salammbô,* introduced at the Théâtre de la Monnaie in 1890. From 1866 on Reyer was the librarian of the Paris Opéra.

Rezia, the Caliph of Bagdad's daughter (soprano) in Weber's *Oberon.*

Rezniček, Emil von, composer. Born Vienna, May 4, 1860; died Berlin, August 2, 1945. He combined the study of music with that of law. His friendship with Ferruccio Busoni and Felix Weingartner led him to choose music. He entered the Leipzig Conservatory. After completing his studies he conducted theater orchestras in Austria and Germany for several years. All this while he was composing, but success did not come until 1894 when his opera *Donna Diana* was triumphantly introduced in Prague. Within a short period this opera was performed throughout Germany. From 1896 to 1899 Rez-

niček was the kapellmeister of the Court Theater in Mannheim. He settled in Berlin in 1901, where he founded a chamber orchestra and where, in 1906, he became a professor at the Scharwenka Conservatory. In 1907–1908 he was principal conductor of the Warsaw Opera and from 1909 to 1911 of the Komische Opera in Berlin. From 1920 to 1926 he was professor of composition at the Staatliche Hochschule in Berlin. He wrote eleven operas: *Die Jungfrau von Orleans* (1886); *Satanella* (1887); *Emmerich Fortunat* (1888); *Donna Diana* (1894, revised 1908 and 1933); *Till Eulenspiegel* (1902); *Eros und Psyche* (1917); *Ritter Blaubart* (1920); *Holofernes* (1923); *Satuala* (1927); *Spiel oder Ernst* (1930); *Der Gondolier des Dogen* (1931).

Rheingold, Das, *see* RING DES NIBELUNGEN.

Rhine maidens, three characters who appear in Wagner's *Das Rheingold* and *Die Götterdämmerung.* They are Flosshilde (contralto), Wellgunde (soprano), and Woglinde (soprano).

Riccardo, king of Sweden (tenor) in Verdi's *Un ballo in maschera.*

Richard Coeur de Lion (Richard the Lion-Hearted), opera by André Grétry. Libretto by Michel Jean Sedaine. Première: Comédie Italienne, Paris, October 21, 1784. For years, Blondel, Richard's minstrel, wanders about disguised as a blind singer. He discovers that his king is imprisoned and with the aid of an exiled English knight and Marguerite of Flanders he effects Richard's escape. A familiar aria is that of Blondel, "O Richard, o mon roi." Beethoven wrote a set of piano variations (Opus 184) on another air from this opera, "Une fièvre brûlante."

Richardson, Samuel, novelist. Born Derbyshire, England, 1689; died London, July 4, 1761. His most famous novel, *Pamela*—dealing with the dangers confronting a virtuous servant—

was made into several operas, including: Nicola Piccinni's *La cecchina* and *La buona figliuola maritata;* Pietro Generali's *Pamela nubile;* Tommaso Traetta's *La buona figliuola maritata.* Another Richardson novel, *Clarissa Harlowe,* was made into an opera by Georges Bizet.

Richepin, Jean, poet, novelist, playwright. Born Médéa, Algiers, February 4, 1849; died Paris, September 12, 1926. He adapted his own novel *Miarka* as a libretto for an opera by Alexandre Georges, and *Le mage* for an opera by Massenet. Other Richepin works used for operas include *Le chemineau* (Xavier Leroux), *Le filibustier* (César Cui) and *La glu* (Gabriel Dupont).

Richter, Hans, conductor. Born Raab, Hungary, April 4, 1843; died Bayreuth, Germany, December 5, 1916. One of the most celebrated Wagnerian conductors of all time, he began his career as a choirboy at the Vienna Court. From 1860 to 1865 he attended the Vienna Conservatory. In 1866 he went to live with Wagner at Lucerne. Here he copied the score of *Die Meistersinger* and played the organ and piano for Wagner. On Wagner's recommendation, he became chorus director of the Munich Opera, and on March 22, 1870, he directed the first Brussels performance of *Lohengrin.* From 1871 to 1875 he was a conductor of the Budapest Opera. In 1875 he was appointed principal conductor of the Vienna Opera, becoming music director in 1893. In 1876 he was invited by Wagner to direct the first complete performance of the *Ring* at Bayreuth. When the festival ended he was decorated with the Order of Maximilian by the King of Bavaria and the Falkenorder by the Grand Duke of Weimar. He remained a principal conductor at Bayreuth up to the time of his retirement.

In 1877 he visited London and alternated with Wagner in directing a festival of Wagner's music. His second visit to London in 1879 was such a success that the Richter Concerts were established, and continued for almost twenty years. In 1882 he led at Drury Lane the first performances in England of *Die Meistersinger* and *Tristan und Isolde.* After 1897 he was the principal conductor of the Hallé Orchestra. He led his last symphony concert with that orchestra on April 11, 1911, and his last opera performance a year later with the Vienna Opera (*Die Meistersinger*). He then went into retirement in Bayreuth.

Ricki, leading female character (soprano) in Franchetti's *Germania.*

Ricordi & Company, the leading music publishers of Italy. The house was founded in 1808 by Giovanni Ricordi who, originally, made his own engravings. As a friend of Rossini, Ricordi acquired the publishing rights to his operas. He recognized Verdi's genius when that composer was still largely unknown. Over the years the house of Ricordi had a far-reaching influence in establishing the reputations of other Italian composers. The Ricordi archives contain the manuscripts of over five hundred operas which the company has published. An American branch of the company was established in 1897.

Ride of the Valkyries, The, music accompanying the flight of the Valkyries on their steeds in the opening of Act III of Wagner's *Die Walküre.*

Riders to the Sea, opera by Ralph Vaughan Williams. Libretto is John Millington Synge's drama of the same name. Première: London, December 1, 1937 (amateur); Cambridge, February 22, 1938 (professional). The setting is a seacoast town of Ireland, where the sea destroys the husband and all the sons of Mauyra.

Rienzi, The Last of the Tribunes, opera by Richard Wagner. Libretto by the composer, based on the novel of the same name by Bulwer-Lytton. Première: Dresden Opera, October 20, 1842. In Rome, in the fourteenth century, Irene, sister of Cola di Rienzi, is abducted by the Orsinis and is rescued by Adriano Colonna, of a rival faction. After a struggle between the Orsinis and the Colonnas, Rienzi appears, decrying the degradation of Rome and the despotism of the nobles. He contrives the overthrow of the nobles at the hands of the people. Peace now prevails, with Rienzi the ruler of Rome. But Orsini and Colonna plot Rienzi's death. Their plot is frustrated and Rienzi nobly forgives them. Now acquiring the support of the Church, the followers of Orsini and Colonna stir the people to revolt. The Capitol is set afire and the heroic Rienzi and his sister perish, joined in death by Adriano, who loves but cannot save Irene.

Rienzi is the earliest of Wagner's operas which is still occasionally performed. He completed it in 1840, in his twenty-seventh year. When introduced in Dresden, the opera was an outstanding success. It became the most popular opera in the repertory that year and made Wagner's name known throughout Germany for the first time. The stirring overture is frequently performed. It contains materials from the opera itself, including Rienzi's prayer in Act V, "Allmächt'ger Vater, blick' herab."

Rigaudon, a lively dance, probably of Provençal origin, in either 2/4 or 4/4 time. It appears in eighteenth century operas; for example, in Handel's *Almira* and Rameau's *Dardanus* and *Platée.*

Rigoletto, opera in four acts by Giuseppe Verdi. Libretto by Francesco Piave, based on Victor Hugo's play, *Le roi s'amuse.* Première: Teatro la Fenice, Venice, March 11, 1851. American première: New York, Academy of Music, February 19, 1855.

Characters: Rigoletto, hunchback jester to the Duke of Mantua (baritone); Gilda, his daughter (soprano); Giovanna, her nurse (mezzo-soprano); Duke of Mantua (tenor); Sparafucile, a hired assassin (bass); Maddalena, his sister (contralto); Count Ceprano, a courtier (bass); Countess Ceprano, his wife (mezzo-soprano); Monterone, a nobleman (bass); Borsa, a courtier (tenor); Marullo, another courtier (baritone); courtiers, ladies, gentlemen, servants. The action takes place in Mantua in the sixteenth century.

Act I. The Duke's palace. During a party, the Duke confides to one of his courtiers that he is attracted by a girl who frequents a near-by church but whose identity is unknown to him. He also finds the beautiful Countess Ceprano alluring, and to the courtier, Borsa, he speaks flippantly of love (Ballata: "Questa o quella"). As the Duke's orchestra strikes up a minuet, the Duke dances with the Countess Ceprano, arousing the jealousy of her husband. The Duke tells his jester he would like to be rid of the Count. Rigoletto's sarcastic suggestions of imprisonment and murder anger the Duke. Count Monterone makes a sudden appearance, denouncing the Duke for ruining his daughter. The Duke orders his arrest. As the Count is led away, Rigoletto taunts him. Monterone turns to the hunchback and curses him so violently that Rigoletto recoils in horror.

Act II. A deserted street. Rigoletto is stopped by Sparafucile, who tells him his services are for hire. But Rigoletto is not interested. Sparafucile departs, leaving Rigoletto to muse on his objectionable employment with the Duke, and his unpleasant duty of having to find mistresses for him ("Pari siamo"). He now enters his own courtyard and his daughter Gilda comes to greet him. She begs him to tell her of her dead mother, and Rigoletto complies ("Deh! non parlare al misero").

He then warns her that they are surrounded by enemies and that she must be ever on her guard. Hearing a sound in the street, Rigoletto runs out to investigate. The Duke, disguised as a student, slips into the courtyard to woo Gilda—for it is she whom he has admired at her church. Rigoletto now takes leave of his daughter, unaware of the Duke's presence. Gilda is apprehensive at the appearance of the stranger, but he soothes her with a love song ("E il sol dell' anima"). Believing the Duke's word that he is a humble student, Gilda falls in love, and after the Duke leaves, she dreams of him ("Caro nome"). Some of the Duke's courtiers, headed by Ceprano, now come to avenge themselves on Rigoletto for his cruel taunts. Believing Gilda to be Rigoletto's mistress, they plan to abduct her. When Rigoletto returns, he is told by the masked men that they wish to abduct Ceprano's wife for the Duke's pleasure. Slyly, they enlist Rigoletto's help, tie his eyes, and place a ladder against the jester's house—the credulous Rigoletto believing that the house is Count Ceprano's. The courtiers abduct Gilda ("Zitti, zitti, moviamo a vendetta") and vanish. The jester removes the bandage from his eyes, discovers how he has been tricked, and realizes with horror that Monterone's curse of a father upon another father has begun to work its evil.

Act III. The Duke's palace. The Duke is upset, for, having visited Gilda again, he has found her gone ("Parmi veder le lagrime"). His courtiers try to amuse him with the story of their abduction of Rigoletto's "mistress," whom they have brought to the palace. The Duke goes to meet the "mistress." Rigoletto arrives, disheveled and distraught. He tries to force his way past the courtiers to the Duke's private chambers, crying that Gilda is his daughter. Gilda enters in tears, having yielded to the Duke. She rushes into her father's arms. But to her father's amazement she is not distressed, but rapturous, since—after her harrowing abduction—she has been united with her admirer of the courtyard. As Rigoletto realizes the hideous treachery that has been worked upon him, he swears vengeance upon the Duke.

Act IV. Sparafucile's inn. The Duke appears, disguised as a soldier. He calls for wine, cynically commenting on the fickleness of all women ("La donna è mobile"). He then attempts to make a conquest of Maddalena, Sparafucile's sister, who has been the means of luring him to this out-of-the-way spot. Rigoletto and Gilda watch the scene from outside, the jester desiring his daughter to see for herself the sort of man she has cherished. While Rigoletto speaks of his imminent revenge, and Gilda bitterly remarks on her lover's infidelity, Maddalena, within, skillfully leads the Duke on (Quartet: "Bella figlia dell' amore"). Rigoletto now sends his daughter off to shelter in Verona, after which he meets Sparafucile and gives him half his fee to deliver the Duke's body in a sack at midnight. But Maddalena has now become fond of the Duke, and she entreats her brother to spare him. Sparafucile, a man of his word, insists that the hunchback is entitled to a corpse, but he agrees to kill another man in the Duke's place—should another turn up before midnight. Gilda, having crept back to the inn to see her perfidious lover, overhears this evil agreement and sees in it the means of saving the Duke's life as well as ending her own disgrace and sorrow. Dressed as a cavalier for her intended journey to Verona, she enters Sparafucile's inn at the height of a thunderstorm, and in the flickering light is mistaken for a man. Soon, Rigoletto returns, pays the assassin the rest of his fee, and starts dragging the heavy sack toward the river. At this moment, within the inn, the Duke resumes his lighthearted song about women ("La donna è mobile"). Rigoletto tears open the sack and finds within it his daughter, dying. Grief-stricken, he takes her in his arms; father and daughter bid one another a last farewell ("Lassù in cielo"). The curse of Monterone has been realized.

Verdi wrote fifteen operas before *Rigoletto,* the first of his works destined to occupy a permanent place in the repertory of every leading opera house. *Rigoletto* was also the first of his operas to indicate the range of his lyric genius; it overflows with wonderful arias, duets, and ensemble numbers, one following the other in a seemingly endless procession of melodic beauty. Riches found in this score include one of the most celebrated tenor arias in all opera ("La donna è mobile"). one of the most brilliant of all coloratura arias ("Caro nome"), and one of the greatest vocal quartets ever written ("Bella figlia"). But *Rigoletto* has interest apart from its wonderful melodies. This is the first of Verdi's operas in which the composer forcefully revealed his dramatic gifts and his ability at trenchant musical characterization.

Rimsky-Korsakov, Nicolai, composer. Born Tikhvin, Novgorod, Russia, March 18, 1844; died St. Petersburg, June 21, 1908. A leading figure of the Russian national school, and the composer of several distinguished operas, Rimsky-Korsakov was trained for a naval career. A meeting with Mili Balakirev fired him with musical enthusiasm and he began writing a symphony, despite his inadequate knowledge. In the fall of 1862 he set out on a two-and-a-half year cruise as naval officer that brought him to the United States in 1864, and returned him to Russia in 1865. While stationed in St. Petersburg, he became intimate with Balakirev and his circle, and plunged into musical activity. He now com-

pleted the symphony he had begun a few years earlier. Introduced under Balakirev's direction on December 19, 1865, it was well received. The composer now planned several ambitious works, national in style and idiom, including the *Antar* symphony and his first opera, *The Maid of Pskov.* The opera was completed in 1872 and was acclaimed when introduced at the Maryinsky Theater on January 13, 1873. The work's success led the government to relieve Rimsky-Korsakov of most of his naval duties so that he could concentrate on music. The special post of Inspector of Naval Bands was created for him. During the next few years he not only led the band concerts but distinguished himself as a conductor of the Free Music Society and the Russian Symphony Concerts. He also became a professor at the St. Petersburg Conservatory. Between 1878 and 1881, Rimsky-Korsakov completed two operas: *May Night* and *The Snow Maiden.* For several years after 1881 he applied himself to the task of editing works by Glinka and Mussorgsky. In 1887 he was again productive as a composer, completing some of his most famous orchestral works (including *Scheherezade*) and the opera *Mlada.* Fatigue and inertia attacked him in 1891. For more than two years he was incapable of writing music. When this mental torpor passed, he completed an opera, *Christmas Eve,* in 1894. Of the numerous stage works that followed, the two best were *Sadko* (1896) and *Le coq d'or* (1907).

In 1905 Rimsky-Korsakov was dismissed from his Conservatory post for siding with the students in some protests against administrative and regulative practices. He was later reinstated, after the Conservatory had effected some needed reforms. Three years later the composer died of a heart attack.

While he is now best known for his orchestral music, it is in his operas that

Rimsky-Korsakov made his most imaginative and original contributions. His operas differed radically from those of his colleagues. Mussorgsky was the realist; Borodin, the Oriental. Rimsky-Korsakov created a make-believe world in which, as Gerald Abraham described, reality was "inextricably confused with the fantastic, naivete with sophistication, the romantic with the humorous, and beauty with absurdity."

His operas: *The Maid of Pskov* (or *Ivan the Terrible*) (1873, revised 1878 and 1893); *May Night* (1880); *The Snow Maiden* (1882); *Mlada* (1892); *Christmas Eve* (1895); *Sadko* (1898); *Mozart and Salieri* (1898); *Boyarina Vera Sheloga*, originally a prologue to *The Maid of Pskov*, but subsequently made into an independent opera (1898); *The Czar's Bride* (1899); *The Legend of the Czar Saltan* (1900); *Servilia* (1902); *Kastchei the Immortal* (1902); *Pan Voyevoda* (1904); *The Invisible City of Kitezh* (1907); *Le coq d'or* (1907).

Rinaldo, (1) opera by Handel. Libretto by Giacomo Rossi, after an episode in Tasso's *Jerusalem Delivered*. Première: Haymarket Theatre, London, February 24, 1711. This was the first opera Handel wrote in England. It was a major success, and its melodies and dances were heard throughout London. The aria "Cara sposa" became a favorite harpsichord piece, and the march was adopted as a regimental number by the London Life Guards. This march was also used for the highwaymen's chorus in *The Beggar's Opera*. The contralto aria "Lascia ch' io pianga" is still frequently heard. Gluck's opera *Armide* is based on the same Tasso episode (for plot, *see* ARMIDE).

(2) A knight (tenor), leader of the Crusades, in Gluck's *Armide*.

Ring des Nibelungen, Der (The Ring of the Nibelung), a cycle of four music dramas (called by the composer a te-

tralogy, the first drama being considered a prelude to the remaining three), comprising *Das Rheingold, Die Walküre, Siegfried,* and *Die Götterdämmerung*. Première (of entire cycle): August 13, 14, 16, and 17, 1876. American première (of entire cycle): Metropolitan Opera House, New York, March 4, 5, 8, 11, 1889.

(1) Das Rheingold (The Rhinegold), prelude (Vorabend) in one act (four scenes). Première: Hof-und-National-Theater, Munich, September 22, 1869. American première: Metropolitan Opera House, New York, January 4, 1889.

Characters: Wotan, ruler of the gods (bass-baritone); Donner, thunder god (bass); Froh, a god (tenor); Loge, fire god (tenor); Fricka, Wotan's wife (soprano); Freia, goddess of beauty and youth, sister of Fricka (soprano); Erda, earth goddess (contralto); Fasolt, giant (bass); Fafner, giant (bass); Alberich, king of the Nibelungs (baritone); Mime, Alberich's brother (tenor); Woglinde, a Rhine maiden (soprano); Wellgunde, Rhine maiden (soprano); Flosshilde, Rhine maiden (contralto). The settings are the bottom of the Rhine, mountain summits near the Rhine, and the caverns of Nibelheim, in legendary times.

Scene 1. The bottom of the Rhine. Three Rhine maidens guard a treasure of magic gold. He who gains the gold and fashions it into a ring may rule the world, but only if before making the ring he renounces love. Alberich, a misshapen dwarf, shouts his renunciation of love and makes off with the gold. The Rhine maidens bewail their loss.

Scene 2. A mountaintop. Wotan and Fricka learn that the new palace built for them by the giants Fasolt and Fafner is finished. Fricka reminds Wotan that the giants must be paid, and the payment promised is her sister, Freia. Wotan insists he was jesting when he had suggested Freia as the price. When the giants come for Freia, Loge suggests a substitute payment: the golden ring that Alberich has molded. The giants are willing to accept the ring.

Scene 3. Alberich's cave. Mime, Alberich's brother, has fashioned a helmet, the Tarnhelm, which will give its wearer any form he desires. Wotan and Loge enter the cave and through guile are able to get Alberich to put on the Tarnhelm and transform himself into a toad. He is then easily captured and taken to Valhalla.

Scene 4. A mountain slope near Valhalla. Alberich is compelled to bring up from his caverns all the wealth of the Nibelungs. When the gods insist that he also turn over the golden ring, he curses them. The ring, he cries, will bring disaster to its owner. When Fasolt and Fafner come for their payment, Erda rises from the earth to warn Wotan not to surrender the ring (Erda's Warning: "Weiche, Wotan, weiche"). But Wotan must stick to his bargain, and he hurls the ring at the giants. They fight over it, and Fasolt is killed; already the ring is fulfilling Alberich's curse. The gods now enter their new abode ("Entrance of the gods into Valhalla").

(2) Die Walküre (The Valkyrie), music drama in three acts. Première: Munich, Hof-und-National-Theater, June 26, 1870. American première: Academy of Music, New York, April 2, 1877.

Characters: Wotan, ruler of the gods (bass-baritone); Fricka, his wife (mezzo-soprano); Brünnhilde, his daughter (soprano); Siegmund, a Wälsung, son of Wotan by a mortal woman (tenor); Sieglinde, a Wälsung, Siegmund's twin sister (soprano); Hunding, her husband (bass); valkyries.

Act I. Interior of Hunding's house. A brief prelude describes a raging storm. Siegmund bursts into the house, seeking refuge. Exhausted, he stretches wearily before the fire. Sieglinde finds him there, brings him water, and urges him to be a guest. Upon Hunding's arrival, Siegmund is invited to partake of their meal, during which Siegmund tells his hosts all he knows about himself: During a hunt with his father, their house burned down and his twin sister disappeared; subsequently, his father died in combat, and he himself was fated to be a lonely wanderer. Hunding recognizes Siegmund as his enemy, but the laws of hospitality dictate that Siegmund be unharmed as long as he is under Hunding's roof. When Siegmund is left alone he laments the fact that he is in his enemy's house unarmed, and that a promise made by his father that he would find a powerful sword has not been kept ("Ein Schwert verhiess mir der Vater"). When Hunding is asleep, Sieglinde comes to Siegmund. She reveals how she was forced to marry Hunding and how, at the wedding feast, a one-eyed stranger plunged a sword, Nothung, into a tree, prophesying that a hero-warrior would someday remove it. Siegmund exclaims that he will withdraw the sword and avenge Sieglinde. Before he can do so, the door swings open and moonlight floods the room. Siegmund and Sieglinde embrace in a sudden recognition of their love for each other. He tells her of his love ("Winterstürme wichen dem Wonnenmond") and Sieglinde responds with equal ardor ("Du bist der Lenz"). They now know that they are Wälsungs, twin brother and sister, and that it was their father Wotan who left the sword in the tree. Siegmund withdraws the sword, and escapes with Sieglinde into the night.

Act II. A mountain pass. Brünnhilde, standing on a rocky peak, gives her battle cry ("Ho-jo-to-ho"), summoning the valkyries to aid Siegmund. Fricka is angered that Siegmund is to be helped, insisting that the unholy lovers must be punished. Sadly and reluctantly, Wotan gives in to Fricka. Now ordered to deny protection to Siegmund and Sieglinde, Brünnhilde is filled with despair. When Siegmund and Sieglinde appear, Brünnhilde tells them that Siegmund must die; that his magic sword has lost its power; that Hunding will destroy him. Siegmund maintains he will kill both himself and Sieglinde rather than permit their separation. This so moves Brünnhilde that she decides, in spite of her father's orders, to help him. When Hunding's horn is heard in the distance, Siegmund goes forth to meet him in battle. Brünnhilde tries to protect him. But Wotan intervenes and brings about Siegmund's death. Lifting Sieglinde and the pieces of Siegmund's broken sword, Brünnhilde hastens away. Wotan now destroys Hunding and wrathfully vows to punish his rebellious daughter.

Act III. The summit of a mountain. The prelude is the "Ride of the Valkyries" (Walkürenritt), depicting the flight of the valkyries on their magic steeds. Brünnhilde appears with Sieglinde, and tells her sisters how she has incurred her father's anger. The valkyries refuse to help her. Giving Sieglinde the pieces of her brother's sword, Brünnhilde sends her away to bear his child, destined one day to become a hero. Wotan appears. Brünnhilde pleads with her father: Has her sin, after all, been so grievous? (Brünnhildes Bitte: "War es so schmälich?"). Wotan's anger now turns to pity and love for his favorite daughter. But she must be punished. Deprived of her godhood, she will be put to sleep, protected by a circle of flame; the first man penetrating the fire and awakening her will become her husband. Embracing Brünnhilde, Wotan bids her a tender farewell (Wotan's Farewell: "Leb' wohl, du kühnes, herrliches Kind"). He places her on a rock, covers her with her shield, and orders Loge, god of fire to surround her with flames (Feuerzauber—Magic Fire Scene). Then, sadly, he departs.

(3) **Siegfried,** music drama in three acts. Première: Bayreuth, August 16, 1876. American première: Metropolitan Opera House, New York, November 9, 1887.

Characters: Siegfried, son of Siegmund and Sieglinde (tenor); Mime, a Nibelung (tenor); Alberich, his brother (baritone or bass); Wotan (bass-baritone); Brünnhilde, his daughter (soprano); Erda, earth goddess (contralto); Forest Bird (soprano); Fafner, the giant transformed into a dragon (bass).

Act I. Mime's cave. Mime is working at an anvil, trying to forge the broken sword, Nothung, which Sieglinde had left for her son, Siegfried. He becomes impatient as he fails to mend the parts, for he knows that if Nothung can be made whole, Siegfried will have a weapon that could slay the dragon Fafner, permitting recovery of the gold which the dragon guards. As the dwarf continues with his futile labors, Siegfried enters, leading a bear by a rope. Siegfried, who detests Mime, frightens him with the bear. Since Siegfried knows that Mime is not his father, he questions him about his origin. Terrified, Mime answers Siegfried by telling him that he is the son of Siegmund and Sieglinde, both now dead; that his mother left him the broken pieces of Nothung which, when mended, will be an invincible weapon; that he, Siegfried, has been raised by Mime in the forest. Ordering Mime to mend the sword, Siegfried leaves the cave. Wotan appears, disguised as a mortal, and Mime learns that only a man without fear can forge Nothung; also, that he who forges the sword will demand Mime's head as a prize. After Wotan leaves, Mime discovers to his horror that Siegfried is a man without fear.

And his terror mounts when Siegfried goes to work at the anvil to forge the sword himself (Forge Song: "Nothung! Nothung!"). Mime stealthily prepares a poison to destroy Siegfried after Fafner is overcome. The sword is finally forged; Siegfried triumphantly leaves with it.

Act II. Fafner's cave. Alberich, awaiting Siegfried at Fafner's cave, is told by Wotan that Siegfried will capture the magic Ring; Wotan urges Alberich to convince Fafner to give up the Ring before it is too late. Upset by these developments, Alberich departs, vowing to avenge himself against Wotan. Siegfried appears, followed by Mime, urging him on to destroy Fafner. While Mime waits at a near-by spring, Siegfried stretches out under a tree, enjoying the beauty of the forest and the songs of the birds (Waldweben —Forest Murmurs). Then, sounding his horn, he rouses Fafner. After he has killed the dragon, some of its blood burns his hand. Instinctively, Siegfried raises his hand to his lips. The blood has magic powers, and the taste of it enables Siegfried to understand the language of the birds. Listening to their songs, he learns that the cave contains the treasures of the Ring and the Tarnhelm. While Siegfried is in the cave, seeking them, Alberich and Mime appear, quarreling as to which shall now have Fafner's treasures. When Siegfried appears with the Ring and the Tarnhelm, Alberich hurries away. Mime tries to cajole Siegfried, but Siegfried is now able to understand Mime's true intentions. When Mime hands him a drinking horn containing poison, Siegfried slays him. He then lies to rest under a tree. The song of the bird reveals to him that Brünnhilde lies asleep on a rock, waiting to be awakened by a hero. Siegfried entreats the bird to lead him to her.

Act III, Scene 1. A wild glen. Wotan summons Erda to tell her he no longer fears doom since destiny lies in the hands of the hero, Siegfried. He then orders Erda back into the bowels of the earth. When Siegfried approaches, Wotan (wearing mortal guise) queries his grandson about Mime, the dragon, and the sword. Impatient with this colloquy, and angered at the way the mysterious wanderer blocks his way, Siegfried shatters Wotan's spear with his sword. Then, blowing his horn, he advances toward the sleeping Brünnhilde.

Scene 2. Brünnhilde's rock. Siegfried passes through the flames that ring the sleeping maiden. He bends down to kiss her on her lips. Brünnhilde awakens and greets the hero ecstatically. They embrace, and are transfigured by love ("Leuchtende Liebe! lachender Tod!").

(4) Die Götterdämmerung (The Twilight of the Gods), music drama in three acts and a prologue. Première: Bayreuth, August 17, 1876. American première: Metropolitan Opera House, New York, January 25, 1888.

Characters: Brünnhilde (soprano); Siegfried (tenor); Alberich (baritone); Gunther, chief of the Gibichungs (bass); Gutrune, his sister (soprano); Hagen, Gunther's half brother (bass); Waltraute, a valkyrie (mezzosoprano); three Norns; Rhine maidens; vassals; warriors; women.

Prologue. Brünnhilde's rock. The three Norns are spinning the fate of the world. When the thread breaks, they realize that doom is at hand. At dawn, Brünnhilde and Siegfried appear. Brünnhilde is leading her horse, Grane, while Siegfried is dressed in full armor. She is sending the hero off to seek adventure. Siegfried bids her farewell, vowing to love her forever; as a token of his love, he leaves her the Ring. Taking Grane, and carrying the Tarnhelm and Nothung, Siegfried sounds his horn and sets forth (Siegfried's Rhine Journey).

Act I, Scene 1. The hall of the Gibichungs. Concerned over the future of the Gibichungs, Hagen tells his half brother, Gunther, that he must marry Brünnhilde, and Siegfried must marry Gunther's sister, Gutrune. Hagen has a scheme to bring this about: when Siegfried comes to the hall, Hagen will make him drink a potion bringing on forgetfulness; Siegfried will then fall in love with Gutrune and help them gain Brünnhilde for Gunther. The sound of Siegfried's horn announces the arrival of the hero. After Siegfried is welcomed, he drinks the potion, instantly loses his memory, and falls in love with Gutrune. Gunther promises Siegfried he can have Gutrune but only if he will help him get Brünnhilde. After an oath of brotherhood, Gunther and Siegfried set forth.

Scene 2. Brünnhilde's rock. Waltraute comes to tell Brünnhilde that Wotan and the gods face a doom that can be prevented only if Brünnhilde gives up Siegfried's Ring (Waltraute's Narrative: "Seit er von dir geschieden"). Brünnhilde refuses to give up this symbol of Siegfried's love. After Waltraute goes, Siegfried arrives. Brünnhilde is shocked to see a stranger: for through the magic of the Tarnhelm, Siegfried has transformed himself into Gunther. He tears the Ring from her finger and seizes her.

Act II. Before the hall of the Gibichungs. Siegfried has come back to claim Gutrune as bride. Hagen sounds the call for his vassals and invites them to a marriage feast. The vassals acclaim Gunther when he arrives with his bride-to-be, Brünnhilde. When Brünnhilde sees Siegfried (now in his own form) she is overwhelmed with gloomy thoughts. The sight of the Ring on his finger convinces her that he has abandoned her for Gutrune. She is appalled when Siegfried acts as if he did not know who she was. When she learns that Hagen and Alberich are planning to kill Siegfried, she becomes their ally.

Act III, Scene 1. The bank of the Rhine. The Rhine maidens beg Siegfried, who has become separated from his hunting party, for the Ring. When he refuses to give it up, the maidens prophesy his doom. Gunther, Hagen, and their vassals catch up with Siegfried. The hero is given a potion that restores his memory of Brünnhilde. As Siegfried rapturously recalls his awakening of her, Hagen plunges his spear into Siegfried's back. With his dying breath the hero bids the absent Brünnhilde farewell ("Brünnhilde! heilige Braut"). His body is lifted and borne in a solemn procession (Siegfrieds Tod—Siegfried's Funeral Music).

Scene 2. The hall of the Gibichungs. Gutrune watches with horror as the vassals bring in Siegfried's body. At first, Gunther says that Siegfried was killed by a wild boar, but he finally names Hagen as the murderer. Hagen kills Gunther in a dispute over the Ring. When Hagen reaches for Siegfried's hand to tear off the Ring, the dead hand rises threateningly. Brünnhilde now orders a funeral pyre built. She sets it aflame, hails Siegfried, mounts her horse, and rides to her death in the flames (Immolation Scene: "Starke Scheite schichtet mir dort"). The Rhine rises, and out of its crest swim the Rhine maidens to retrieve the Ring. Hagen, attempting to save the Ring, is seized and carried beneath the flood. The river subsides. In the distance, Valhalla crumbles in flames, destroying the gods.

Wagner finished the texts of his four *Ring* dramas in reverse order. In 1848 he began sketching a single drama which he called *Siegfrieds Tod*. He found that it needed a second play to provide prefatory material; consequently, he began writing *Der junge Siegfried*. He then felt the need for a third, then a fourth, text for background explanations. Thus, *Die Götter-dämmerung* (the original *Siegfrieds Tod*) was written first, and *Das Rheingold* last. By 1853 the four texts were published. A year later, Wagner began writing the music for *Das Rheingold;* that for *Die Götterdämmerung* was at last completed in 1874. The *Ring* cycle, then, occupied Wagner for a quarter of a century—a period that also saw the completion and performance of *Tristan und Isolde* and *Die Meistersinger.* Wagner did not think he would live to see the *Ring* performed. He knew that a work requiring four full evenings for performance, and demanding tremendous musical and stage forces, would not find a sympathetic response from impresarios. Yet, even with the belief that his manuscript might lie untouched, he kept on composing—for the *Ring* represented the summit toward which he had all the time been climbing, his ultimate goal as a creative artist. The *Ring* was not only Wagner's most ambitious work, it was the one in which he most completely realized his theories about the music drama (see WAGNER). Nowhere was he more adventurous in projecting his ideas about stage direction and scene design. Nowhere was his music a more continuous flow of expressive melody. Nowhere was he more prodigal in his use of musical resources. The leitmotif technique appears in his other dramas, but in the *Ring* the leading motives are the spine and sinew of the musical texture.

The *Ring* was also Wagner's chief musical embodiment of his ethical and social thinking. In the pursuit of golden treasure, gods as well as men were destroyed. Siegfried was Wagner's conception of the Nietzschean Superman, come to redeem the world from avarice and fear.

Rinuccini, Ottavio, poet and librettist. Born Florence, Italy, January 20, 1562; died there March 28, 1621. As a member of the Florentine Camerata (which see) he wrote the librettos for the earliest operas: Caccini's *Euridice;* Gagliano's *Dafne;* Monteverdi's *Arianna* and *Il ballo delle ingrate;* Peri's *Dafne* and *Euridice.*

Rinuccio, Zita's nephew (tenor) in Puccini's *Gianni Schicchi.*

Rip Van Winkle, (1) a story by Washington Irving, originally published in *The Sketch Book.* The tale concerns the twenty-year enchanted sleep of the hero in the Catskills; he awakens an old man, the world around him completely changed. George Frederick Bristow's *Rip Van Winkle* was one of the first operas by an American on a native subject. The French composer of opéras bouffes, Robert Planquette, also wrote a work on this theme.

(2) Opera by Reginald De Koven. Libretto by Percy MacKaye, based on the Washington Irving story. Première: Chicago Opera Company, January 2, 1920. In this adaptation, Rip Van Winkle is the victim of a sleeping potion; and, instead of a nagging wife, he has a sweetheart.

Rise and Fall of Mahagonny, The, opera by Kurt Weill. Libretto by Bertolt Brecht. Première: Frankfort Opera, October 20, 1930. The story tells of three escaped criminals who found the city of Mahagonny in an American desert, a city in which the basest instincts of man are catered to. The immoral theme provoked a riot at the opera's first performance.

Risurrezione, La (Resurrection), opera by Franco Alfano. Libretto by Cesare Hanau, based on novel by Leo Tolstoy. Première: Teatro Vittorio Emanuele, Turin, November 30, 1904. Prince Dimitri, about to leave with his regiment, meets and falls in love with Katusha. She becomes pregnant, but the Prince is unaware of this. Her life ruined, Katusha, after her child's death, commits a murder and is sentenced to exile in Siberia. The Prince meets her in St. Petersburg just before her exile, is overwhelmed to learn that he has been the cause of her downfall. In Siberia, Dimitri is finally able to gain her pardon, but she no longer loves him, having found a devoted lover in the convict Simonson.

Ritorna vincitor! Aïda's aria in Act I, Scene 1, of Verdi's *Aïda*.

Ritornello, Italian for "little repetition." In early Italian operas, a ritornello was a brief instrumental passage played between scenes, or between the vocal portions of a song.

Robbins, Serena's husband (baritone) in Gershwin's *Porgy and Bess*.

Robert le Diable (Robert the Devil), opera by Giacomo Meyerbeer. Libretto by Eugène Scribe and Germain Delavigne. Première: Paris Opéra, November 21, 1831. The action takes place in thirteenth century Palermo. Robert, the Duke of Normandy, is the son of a mortal woman and a devil. That devil disguises himself as a man and assumes the name of Bertram. He follows his son with the hope of gaining his soul. In Sicily, Robert falls in love with the princess Isabella, and hopes to win her hand in a tournament. But the stratagems of the devil keep him from winning. Thinking that his cause is lost, Robert is willing to use diabolical means to win the woman he loves. At a midnight revel with ghosts, Robert acquires a magic branch with which he gains access to Isabella, and with which he hopes to win her against her will. Isabella pleads with him to break the branch, which he finally does, destroying its magic. He is about to sign a contract with his father for the consignment of his soul to hell when he is dissuaded by his foster sister. Robert denounces his father, who returns to the lower regions. The redeemed Robert and Isabella are now united in marriage. Some of the more popular excerpts are Alice's romance, "Vanne, disse," and the chorus of the Sicilian

knights, "Sorte amica," both in Act I; Bertram's invocation, "Suore, che riposate," and the chorus of the demons, "Demoni fatale," in Act III.

Robin Hood, (1) opera in three acts by George Alexander Macfarren. Libretto by John Oxenford. Première: London, Her Majesty's Theatre, October 11, 1860.

(2) Operetta by Reginald De Koven. Première: Chicago, June 9, 1890. This work contains the song now associated with weddings, "O promise me!"

Robinson, Count, Elisetta's rich suitor (bass) in Cimarosa's *Il matrimonio segreto*.

Rocca, Lodovico, composer. Born Turin, Italy, November 29, 1895. He attended the Milan Conservatory and the Turin University. His first opera, *La morte di Frine*, was produced in 1921. Success came in 1934 with *Il dibuk* (*The Dybbuk*), which won first prize in a contest sponsored by La Scala. In 1940 he became director of the Turin Conservatory. His other operas: *La corona di Re Gaulo* (1923); *Monte Ivnor* (1939); *In terra di leggenda; L'Uragano* (1952).

Rocco, (1) a jailer (bass) in Beethoven's *Fidelio*.

(2) A Camorrist (bass) in Wolf-Ferrari's *The Jewels of the Madonna*.

Rodelina, opera by Handel. Libretto by Nicolo Haym. Première: King's Theatre, Haymarket, February 13, 1725. Rodelina, queen of the Lombards, is threatened by Grimoaldo, who has usurped power: if she is not receptive to his love, he will kill her child. The exiled king, Bertarido, returns to the palace secretly, overthrows Grimoaldo's rule, and reassumes the throne. The baritone aria "Dove sei amato bene" is popular. The dungeon scene with Bertarido is sometimes mentioned in connection with the somewhat similar scene in Beethoven's *Fidelio*.

Roderigo, a Venetian gentleman (tenor) in Verdi's *Otello*.

Rodolfo, (1) a count (baritone) in Bellini's *La sonnambula.*

(2) A poet (tenor) in love with Mimi in Puccini's *La Bohème.*

(3) Son of a count (tenor) in love with Luisa Miller in Verdi's *Luisa Miller.*

Rodrigo, (1) Marquis of Posa (baritone) in Verdi's *Don Carlos.*

(2) The Cid, principal male character (tenor) in Massenet's *Le Cid.*

Rodrigue, the Cid—*see above,* RODRIGO.

Rogers, Bernard, composer. Born New York City, February 4, 1893. He studied with Arthur Farwell and Ernest Bloch; at the Institute of Musical Art; and in Europe with Nadia Boulanger and Frank Bridge. In 1929 he joined the faculty of the Eastman School of Music in Rochester, and it was in that city that his first works were performed. His first opera, *The Marriage of Aude,* was completed in 1932. His second, *The Warrior,* received the Alice Ditson Fund Prize in 1946 and was introduced at the Metropolitan Opera on January 11, 1947. Two more operas were completed in 1954: *The Veil* and *The Nightingale.*

Roi de Lahore, Le (The King of Lahore), opera by Jules Massenet. Libretto by Louis Gallet, based on the *Mahabharata.* Première: Paris Opéra, April 27, 1877. Alim, King of Lahore, and his minister, Scindia, are rivals for Sita's love. Scindia kills Alim, ascends the throne, and is about to marry Sita. The God Indra allows Alim to return to earth disguised as a beggar. He visits his palace and is recognized by Sita. She kills herself so that she may join Alim in paradise. The ballet music in Act III, and Scindia's arioso, "Promesse de mon avenir," are familiar excerpts.

Roi d'Ys, Le (The King of Ys), lyric drama by Edouard Lalo. Libretto by Edouard Blau, based on a Breton legend. Première: Opéra-Comique, May 7, 1888. The legend from which this opera was derived was also the inspiration of Debussy's piano prelude, *La cathédrale engloutie.* Mylio is in love with Rozenn, daughter of the King of Ys, but he is also loved by Rozenn's sister, Margared. On the wedding night, Margared is led by her jealousy to open the sea gates and flood the town of Ys. The panic-stricken townspeople rush to the hills for safety. Conscience-stricken, Margared commits suicide. The patron saint of Ys saves the city. The overture—which quotes principal melodic material from the opera and summarizes the main action—is a familiar concert number. The more familiar vocal excerpts include Rozenn's and Margared's duet in Act I, "En silence pourquoi souffrir," and Mylio's aubade in Act III, "Vainement, ma bien aimée."

Roi d'Yvetot, Le (The King of Yvetot), comic opera by Jacques Ibert. Libretto by Jean Limzon and André de la Tourrasse, based on a ballad by Pierre Jean de Beranger. Première: Opéra-Comique, January 15, 1930. The King of Yvetot is deprived of his rule by his freedom-loving subjects. The men of his realm must now assume command, and they no longer have time for love or work. The women band together to bring back their king. The king falls in love with a servant girl, and when he returns to the throne he makes her queen.

roi l'a dit, Le (The King Said So), opéra-comique by Léo Delibes. Libretto by Edmond Gondinet. Première: Opéra-Comique, May 24, 1873. This was Delibes's first opera. The Marquis de Montecontour, in a moment of confusion, tells Louis XIV that he has a son when actually he is the father of four daughters. Commanded to bring his son to court, the Marquis is forced to adopt a peasant boy and pass him off as a nobleman. The boy makes the most of his situation, to the dismay of the

Marquis, who now contrives a method by which he can be rid of him. All turns out well when the Marquis is made a duke to console him for the loss of his "son," and the adopted boy marries the maid with whom he is in love. The opera's duet for two sopranos, "Déjà les hirondelles," has become famous as a concert number.

Rolland, Romain, novelist, critic. Born Clamecy, France, January 29, 1866; died Vézelay, France, December 30, 1944. Winner of the Nobel Prize in 1915 for his novel about a musician, *Jean Christophe,* Rolland was educated at the École Normale Supérieure in Paris and the École Française in Rome. He received his doctorate with a thesis on the early history of opera, *Les origines du théâtre lyrique moderne,* which received the Prix Kastner-Bourgault in 1896. Later books by Rolland also contain significant material on the early history of opera and its composers. These include *Musiciens d'autrefois* (1908), published in the United States as *Some Musicians of Former Days,* and *Voyage musicale au pays du passé* (1909) (*A Musical Tour Through the Land of the Past*). In 1900 Rolland organized the first international congress for the history of music. Three years later he became president of the musical section of the Ecole des Hautes Etudes Sociales. He resigned in 1909 to devote himself to writing. From 1913 to 1938 he resided in Switzerland. Besides books already mentioned, he wrote a biography of Handel (1910), several volumes about Beethoven, an autobiography, *Journey Within* (1947), and an essay on seventeenth century Italian opera in Lavignac's Encyclopédie. His novel *Colas Breugnon* was adapted as an opera by Dmitri Kabalevsky.

Roller, Alfred, scene designer. Born Vienna, February 10, 1864; died there June 21, 1935. In 1903 he was engaged by Gustav Mahler to design new sets for many productions of the Vienna Opera, including the *Ring, Tristan und Isolde, Fidelio,* and *Don Giovanni.* He subsequently designed the scenery for the major productions of the annual Salzburg Festival. He was also director of the Vienna School of Commercial and Technical Arts.

Romance (Italian: ROMANZA), in operatic use, a designation for an aria of nondramatic nature, generally expressive of personal sentiments, or devoted to setting forth a narrative.

Romance d'Antonia, see ELLE A FUI, LA TOURTERELLE.

Romani, Felice, librettist. Born Genoa, Italy, January 31, 1788; died Moneglia, Italy, January 28, 1865. Though trained to be a lawyer, he turned to literature, writing librettos for over a hundred operas, and becoming the most significcan Italian librettist of his day. Among the operas for which he wrote the texts: Bellini's *Norma, Il pirata,* and *La sonnambula;* Bizet's *Parisina;* Donizetti's *L'elisir d'amore* and *Lucrezia Borgia;* Mercadante's *Normanni a Parigi;* Meyerbeer's *Margherita d'Anjou* and *L'esule di Granata;* Rossini's *Il Turco in Italia;* Verdi's *Un giorno di regno.* Late in life, becoming blind, Romani received a government pension.

Romeo and Juliet, (1) tragedy by Shakespeare, the source of many operas, the most famous being Gounod's (see below). The first opera written on this drama was *Romeo und Julia,* by Georg Benda, produced at Gotha on November 25, 1776. For others, *see* SHAKESPEARE.

(2) *Roméo et Juliette,* opera in five acts by Charles Gounod. Libretto by Jules Barbier and Michel Carré, based on Shakespeare. Première: Théâtre Lyrique, Paris, April 27, 1867.

Characters: Roméo, a Montague (tenor); Juliette, daughter of Capulet (soprano); Capulet, a nobleman (bass); Tybalt, Capulet's nephew (tenor); Paris, kinsman of Capulet (baritone);

Gregorio, another kinsman (baritone); Stephano, Roméo's page (soprano); Gertrude, Juliette's nurse (mezzo-soprano); Benvolio, Roméo's friend (tenor); Mercutio, another friend (baritone); Friar Laurence (bass); Duke of Verona (bass); Capulets, Montagues, retainers, guests. The setting is Verona in the fourteenth century.

Act I. Capulet's ballroom. Capulet is giving a ball in honor of Juliette. The masked guests include two uninvited ones from the rival house of Montague: Roméo and Mercutio. The latter tells of a dream he has had (Ballad of Queen Mab). Roméo forgets the danger of being in the house of his enemy when he catches sight of Juliette and falls in love as Juliette voices her pleasure in the happiness of the evening ("Je veux vivre dans ce rêve"). Approaching the yet unknown girl, Roméo begins to pay court, and finds Juliette not unreceptive ("Ange adorable"). Their colloquy is interrupted by Tybalt, who recognizes Roméo. Capulet intervenes and prevents a fight. Roméo and Mercutio leave, Roméo having learned that he has lost his heart to Capulet's daughter. Capulet and his guests resume their merrymaking.

Act II. Capulet's garden. Roméo serenades Juliette below her balcony ("Ah! lève-toi, soleil"). When Juliette appears, the lovers exchange tender sentiments. Finally, they must say good night ("De cet adieu si douce est la tristesse"). Juliette retires; Roméo lingers a few moments, musing on his passion.

Act III, Scene 1. Friar Laurence's cell. Roméo and Juliette are secretly married by Friar Laurence. The three, joined by Juliette's nurse Gertrude, give voice to their happiness ("O pur bonheur!").

Scene 2. A street. Stephano, seeking Roméo, sings a mocking serenade ("Que iais-tu blanche tourterelle?").

Rudely awakened, Gregorio, a Capulet, rushes out of his house to attack the Montague page. Mercutio and Tybalt, passing by, are involved in the fight. Mercutio is wounded. Believing that Mercutio is dying, Roméo kills Tybalt. The Duke of Verona, apprised of the situation, banishes Roméo.

Act IV. Juliette's room. Roméo comes secretly to bid Juliette farewell. Once again they speak of their love ("Nuit d'hyménée"). When Roméo has gone, Friar Laurence brings the news that the girl's marriage has been arranged. He counsels her to drink a potion which will induce a deathlike sleep; then, believing her dead, her family will place her in the family tomb, from which she can escape to join her lawful husband. Upon the approach of Paris and Capulet, Juliette drains the Friar's potion and falls into a trance.

Act V. Juliette's tomb. Outside the tomb, Friar Laurence learns to his horror that his message about Juliette's simulated death has not reached Roméo. Roméo arrives to mourn the death of his wife. Spellbound by her beauty ("Salut! tombeau! sombre et silencieux"), he joins her in death, as he thinks, by drinking poison. Before the poison takes effect, Roméo sees his wife stir. He realizes that she is not dead after all, but it is too late: Roméo is dying. Now it is Juliette's turn to die. She pierces her breast with a dagger, and the lovers die in one another's arms.

Roméo et Juliette, coming six years after Gounod's most famous opera, *Faust,* was such an extraordinary success that it was heard a hundred times in its first year. But the new opera never became as popular as its predecessor. It is not the sustained masterwork that *Faust* is, yet it has a refinement and delicacy that are less evident in *Faust.* When Arthur Hervey wrote that Gounod "created a musical language of his own, one of extraordinary sweetness," he must have had *Roméo et Juliette* in mind. Juliette's waltz in the first act, Roméo's serenade in the second act, and his radiant aria "Salut! tombeau!" in the last are among the finest examples of French lyricism. The opera may lack dramatic interest and variety of mood; but its finest moments make it a rewarding experience.

Romeo und Julia auf dem Dorfe, *see* VILLAGE ROMEO AND JULIET, A.

Romerzählung, Tannhäuser's Rome Narrative in Act III of Wagner's *Tannhäuser.*

Ronald, Sir Landon (born RUSSELL), conductor. Born London, June 7, 1873; died there August 14, 1938. He was the son of Henry Russell, the song composer, one of whose other sons, also named Henry Russell, was an opera impresario. Landon Ronald received his training at the Royal College of Music in London. In 1891 be became accompanist and coach at Covent Garden under Luigi Mancinelli. For two seasons he was conductor of the Italian Opera Company, directed by Augustus Harris. He then directed a season of English opera at Drury Lane. In 1894 he toured the United States as Nellie Melba's conductor and accompanist, and in 1895 he directed opera performances at Covent Garden. Afterward, he specialized in symphonic music, appearing as conductor with most of the great orchestras of the world. For a quarter of a century he was the principal of the Guildhall School of Music. Also a noted music critic, he was knighted in 1922.

Ronde du Postillon: *see* MES AMIS, ECOUTEZ L'HISTOIRE.

Rondine, La (The Swallow), opera by Giacomo Puccini. Libretto by Giuseppe Adami based on a German libretto by A. M. Willner and H. Reichert. Première: Monte Carlo, March 27, 1917. During the second French Empire, Magda, mistress of Rambaldo, a wealthy banker, is entertaining her guests. One of them is the poet Prunière, who suggests that sentimental love is returning to fashion. This encourages Magda to tell her friends about a romance she once had with a student she met in a dance hall, Le Bal Bullier. The poet then tells Magda's fortune: Like a swallow she has left her home, and like a swallow she will return. A young man from the country, the son of an old friend of Rambaldo, expresses a desire to visit Le Bal Bullier. Magda has a sudden impulse to follow him. The young man, Ruggero, is as attracted to Magda as she is to him. They take a villa on the Riviera and all goes well until Ruggero seeks his mother's permission to marry Magda. At this point, Magda's discretion prompts her to abandon her young lover and return to Rambaldo.

Rooy, Anton, *see* VAN ROOY, ANTON.

Rosa, Carl (born ROSE), impresario. Born Hamburg, Germany, March 21, 1842; died Paris, April 30, 1889. He attended the Conservatories of Leipzig and Paris, after which he made tours as a violinist and played in various orchestras. In 1866 he toured the United States, where he married the singer Euphrosyne Parepa. They formed an opera company which toured the United States. After his wife died in England in 1874, Rosa founded and directed the Carl Rosa Opera Company, which gave performances in English and became one of England's major operatic organizations. *See* CARL ROSA OPERA COMPANY.

Rosalinde, Baron von Eisenstein's wife (soprano) in Johann Strauss's *Die Fledermaus.*

Rosario, a lady of rank (soprano) in Granados' *Goyescas.*

Rosaura, Ottavio's daughter (soprano) in Wolf-Ferrari's *Le donne curiose.*

Roschana, Almanzor's wife (contralto) in Weber's *Oberon.*

Rose, Ellen's friend (mezzo-soprano) in Delibes's *Lakmé*.

Rosenberg, Hilding, composer. Born Bosjökloster, Sweden, June 21, 1892. He did not receive systematic musical training until, in his twenty-fourth year, he entered the Stockholm Conservatory. Travels to Dresden and Paris introduced him to modern idioms, and he now turned from romanticism to modernism. In this vein he wrote some chamber and orchestral music. In 1932 he completed his first opera, *Journey to America*. He subsequently completed several more operas, the most ambitious being a cycle of four opera-oratorios entitled *Joseph and His Brethren* (1949), whose text was derived from Thomas Mann's novel of the same name. His other operas: *The Marionettes* (1939); *The Isle of Felicity* (1945).

Rosenkavalier, Der (The Cavalier of the Rose), a "comedy for music" in three acts, by Richard Strauss. Libretto by Hugo von Hofmannsthal. Première: Dresden Opera, January 26, 1911. American première: Metropolitan Opera House, New York, December 9, 1913.

Characters: The Feldmarschallin, Princess von Werdenberg (soprano); Baron Ochs von Lerchenau, her cousin (bass); Octavian, a young nobleman (mezzo-soprano); Herr von Faninal, a wealthy merchant (baritone); Sophie, his daughter (soprano); Marianne, Faninal's housekeeper (soprano); Valzacchi, an intriguing Italian (tenor); Annina, his accomplice (contralto); a singer, flute player, notary, milliner, widow, innkeeper; orphans, waiters, musicians, guests; a major domo, and a Negro servant boy to the Princess. The setting is Vienna during the reign of Maria Theresa.

Act I. The Princess' bedroom. Young Octavian tells the Princess von Werdenberg how much he loves her. He is interrupted, and sent into hiding, by the sudden arrival of the Princess' cousin, Baron Ochs. The lecherous Baron has come with the news that he is about to marry Sophie Faninal. Octavian, disguised as a maid, emerges from concealment. It is not long before the Baron attempts to arrange a rendezvous with "her." The Baron then explains to his cousin that custom dictates that a silver rose, the pledge of love, be sent to his future bride; he begs the Princess to provide him with a suitable emissary. The major domo now announces that the Princess must attend to her various interviews of the day. As she does so, a tenor entertains her with an Italian serenade ("Di rigori armato"). When the room is emptied of all servants and visitors, and the Baron and Octavian have departed, the Princess reflects on her love affair with Octavian. In a poignant monologue ("Kann ich mich auch an ein Mädel erinnern") she laments the passing of her youth, and how futile it is for her to try to hold on to the love of one so young as Octavian. She then calls to her servant and instructs him to have Octavian bear Baron Ochs' silver rose to Sophie.

Act II. Faninal's house. Sophie prays that she may prove worthy as the wife of so exalted a man as the Baron Ochs ("In dieser feierlichen Stunde"). Octavian enters. With great dignity he presents the silver rose ("Presentation of the Rose"). The young people exchange meaningful glances; at that moment they realize how attracted they are to each other. The Baron arrives. Loud and vulgar, he repels Sophie. He tries to embrace her and repeatedly tells her how fortunate she is to gain him for a husband ("Mit mir, mit mir"). When Octavian and Sophie are left alone, they confide to each other their inmost feelings. Returning suddenly, the Baron catches them in an embrace. Enraged, he challenges Octavian to a duel. Scratched in the course of it, Ochs growls and bellows as if he had been mortally wounded. He forgets his pains when a perfumed note arrives proposing a rendezvous with the Princess' "maid"; the note has been sent, of course, by Octavian.

Act III. A private chamber in a disreputable hotel. Disguised as the maid, Octavian comes to keep his rendezous. The Baron follows, full of delicious anticipation. But a series of pranks, arranged by Octavian, harasses him. Strange faces appear in the windows. A distraught woman enters, followed by a brood of children who, she claims, belong to Ochs. Such a hubbub is raised that the police appear. They are about to arrest the Baron when the Princess makes her entrance. Octavian puts aside his disguise, and the Baron discovers how thoroughly he has been duped. With dignity, the Princess gives Octavian her blessing to love Sophie ("Hab' mir's gelobt, ihn lieb zu haben"). Octavian and Sophie embrace and give voice to their love ("Ist ein Traum, kann nicht wirklich sein").

Der Rosenkavalier represented a remarkable change of style for its composer, who had previously become famous with such lurid tragedies as Elektra and Salome. In Der Rosenkavalier the touch is light, and the mood consistently gay. Having always had a profound admiration for Mozart's comic operas, and a love for Johann Strauss's operettas, Richard Strauss wrote an opera which combined the best qualities of both composers: the infectious gaiety of the operetta composer, the penetrating wit and the contrasting humanity of Mozart. The wedding of Strauss's score, which traverses the gamut from broad burlesque to moving compassion, and Hugo von Hofmannsthal's libretto, one of the finest in all opera, resulted in a work which well deserves a place with the greatest comic operas of all time.

Rosenstock, Joseph, conductor. Born Cracow, Poland, January 27, 1895. He attended the Vienna Conservatory, then studied privately with Franz Schreker. In 1920 he became conductor of the Darmstadt Opera, rising to the position of general music director five years later. From 1925 to 1927 he was the principal conductor of the Wiesbaden Opera. On October 30, 1929, he made his American debut at the Metropolitan Opera, conducting Die Meistersinger, but he remained only a single season. From 1930 to 1933 he was music director of the Mannheim opera, and after this he conducted opera performances in Berlin and concerts in Tokyo. In the fall of 1951 he succeeded Lászlo Halász as artistic director and principal conductor of the New York City Opera Company.

Rosina, Bartolo's ward (soprano), in love with Almaviva, in Rossini's The Barber of Seville.

Rosing, Vladimir, tenor and impresario. Born St. Petersburg, Russia, January 23, 1890. After studying music with various teachers, including Giovanni Sbriglia and Jean de Reszke, he made his opera debut in St. Petersburg in 1912 in Eugene Onegin. A year later he went to London, studied with George Powell, and made his London debut in Pique Dame. In 1915 he directed a season of opera at the London Opera House. He became successful in London both in opera and as a recitalist. In 1923 he appeared in the United States for the first time, singing recitals. The same year he organized the American Opera Company, which for six years toured the United States, presenting operas in English. During this period he was also head of the opera department at the Eastman School of Music. The American Opera Company disbanded in 1929. Ten years later Rosing became director of another opera company in Los Angeles. Since then, he has joined the staff of the New

York City Opera Company as consultant and stage director.

Rossi, Gaetano, librettist. Born Verona, 1780; died there January 27, 1855. As the official playwright of the Teatro La Fenice in Venice, he wrote over a hundred opera librettos. They were set to music by many composers, including Donizetti, Meyerbeer, Nicolai, and Rossini. The most famous of these operas are Rossini's *Semiramide* and *Tancredi.*

rossignol, Le (The Nightingale), opera by Igor Stravinsky. Libretto by S. Mitousoff and the composer, based on Hans Christian Andersen's fairy tale. Première: Paris Opéra, May 26, 1914. In legendary times, a Chinese emperor acquires a nightingale whose beautiful singing moves him to tears. When three ambassadors from Japan bring the Emperor a gift of a mechanical nightingale, the live one disappears. The Emperor, on his deathbed, is magically cured when the real nightingale returns to sing for him. The courtiers, expecting to find him dead, discover him in the best of health. Stravinsky made three different versions of this story. This opera was the first. He then converted the opera into a ballet which was produced by Serge Diaghilev's Ballet Russe on February 2, 1920. Meanwhile, Stravinsky developed the musical material of the last two acts into a symphonic poem.

Rossi-Lemeni, Nicola, bass. Born Constantinople, 1922. He studied law, and after receiving his degree planned to enter diplomatic service. While in the Italian army, during World War II, he sang for the troops, and it was then that he decided to become a professional musician. After a period of intensive study he made his concert debut in Verona, and shortly after that made a highly successful debut at La Scala. Appearances in other major Italian and Latin American opera houses followed. His American debut took place at the San Francisco Opera on October 2, 1951, in the role of Boris Godunov, and his debut at the Metropolitan Opera followed on November 16, 1953 when he was heard as Mephistopheles in *Faust.*

Rossini, Gioacchino Antonio, composer. Born Pesaro, Italy, February 29, 1792; died Passy, France, November 13, 1868. A musical child, he entered the Liceo Musicale in Bologna at age twelve, and was an exceptional student; however, he was soon obliged to leave because of his family's financial difficulties. In 1810 he wrote his first opera, *La cambiale di matrimonio,* produced that year in Venice. He wrote a second opera in 1811, and three more in 1812, before achieving his first substantial success. This came with *La pietra del paragone,* introduced by La Scala in 1812, and given fifty times in its first season. *Tancredi* and *L'Italiana in Algeri,* both introduced in Venice in 1813, were even greater triumphs. Though only twenty-one, Rossini was already the idol of the Italian opera public.

In 1815 he was engaged by Domenico Barbaja to direct two opera companies in Naples and write new works for them. His first opera under this arrangement was *Elisabetta,* written expressly for the popular prima donna Isabella Colbran. Rossini was later to write several more operas for her. Since his contract permitted him to accept outside commissions, Rossini wrote his masterwork, *The Barber of Seville,* not for his companies in Naples but for production in Rome. It was introduced at the Teatro Argentina on February 20, 1816. A combination of unhappy circumstances spelled disaster for the première (see BARBER OF SEVILLE), but on the second evening the opera was acclaimed, and with each successive performance it gained new admirers.

In 1822, after marrying Isabella Colbran, Rossini left Italy for the first time, going to Vienna, where he and his operas became the rage. Two years later he went to Paris, to direct the Théâtre des Italiens. Rossini's popularity in Paris was so great that Charles X gave him a ten-year contract to write five new operas a year; at the expiration of the contract he was to receive a generous pension for life. Under the terms of this agreement, Rossini wrote *William Tell*, introduced at the Paris Opéra on August 3, 1829. Though discriminating musicians and some critics acclaimed it, the general public did not favor it, and the opera was a failure.

Though Rossini was only thirty-seven years old when he completed *William Tell*, and lived for another thirty-nine years, he never again wrote an opera. He was at the height of his creative powers, and a world-renowned figure, yet in the next four decades he produced only some sacred music, a few songs, some instrumental and piano pieces. This sudden withdrawal from the world of opera inspired many conjectures. Some said that Rossini's indolence had got the better of him now that he was a wealthy man. Others said that the failure of *William Tell* had embittered him. Still others found Rossini's neurasthenia, which became serious after 1830, the major cause. Whatever the reason, the most famous opera composer of his generation preferred to remain silent after his thirty-eighth opera.

During the next two decades or so, Rossini's life was complicated by prolonged legal battles over his contract with Charles X; by his neurasthenia; by his emotional attachment to Olympe Pélissier, whom he had loved for years, but could not marry until his first wife had died. After 1855, his life became easier. Now married to Olympe, he lived in a luxurious Paris apartment, and in a summer villa in Passy, entertaining his friends in the grand manner

and being a major figure in the social and cultural life of the city. His death was brought about by complications following a heart attack. He was buried in Père Lachaise cemetery in Paris, but at the request of the Italian government his body was removed to Florence and buried in the Santa Croce Church.

Rossini was a remarkably productive composer. He completed an average of two operas a year for nineteen years, in some years writing as many as four. This rate was made possible by an amazing creative facility, but what helped increase his output was his capacity for making compromises. He did not hesitate to use poor material when fresher and more original ideas required painstaking effort. He often borrowed ideas from his older operas. He even permitted other composers to interpolate numbers of their own in his works. But though he had the temperament of a hack, he was also a genius who could bring the highest flights of inspiration to his writing. A bold experimenter, some of his innovations changed opera procedures. He perfected what is today called the Rossini crescendo: a brief phrase in rapid tempo repeated over and over with no variation save that of volume. He was one of the first composers to write out cadenzas instead of allowing the singer to improvise them. He was a pioneer in accompanying recitatives with strings instead of piano. And he was one of the first Italian composers after Monteverdi to use orchestral effects and colors with such expressiveness and variety.

Rossini's most important operas were: *La scala di seta* (1812); *La pietra del paragone* (1812); *Il Signor Bruschino* (1813); *Tancredi* (1813); *L'Italiana in Algeri* (1813); *Elisabetta* (1815); *Il barbiere di Siviglia* (*The Barber of Seville*) (1815); *Otello* (1816); *La cenerentola* (1817); *La gazza ladra* (1817); *Armida* (1817);

Mosè in Egitto (1818); *La donna del lago* (1819); *Zelmira* (1822); *Semiramide* (1823); *Le siège de Corinthe* (1826); *Le Comte d'Ory* (1828); *Guillaume Tell* (*William Tell*) (1829).

Rostand, Edmond, poet and dramatist. Born Marseilles, France, April 1, 1864; died Paris, December 2, 1918. Rostand's most celebrated drama, *Cyrano de Bergerac,* was made into operas by Franco Alfano and Walter Damrosch. His *L'Aiglon* was the source for an opera written collaboratively by Arthur Honegger and Jacques Ibert; *La princesse lointaine,* for operas by Italo Montemezzi and Georges Witkowski; *La Samaritaine,* for an opera by Max d'Ollone; *Les Romanesques,* for Fritz Hart's *The Fantastics.*

Rothier, Léon, bass. Born Rheims, France, December 26, 1874; died New York City, December 6, 1951. He was trained as a violinist, but in his seventeenth year the director of the Paris Conservatory urged him to turn to singing. Rothier attended the Conservatory from 1894 to 1899. On October 1, 1899, he made his opera debut at the Opéra-Comique in *Philémon et Baucis.* Half a year later he appeared in the world première of *Louise.* Rothier remained with the Opéra-Comique until 1903, afterward appearing with opera companies in Marseilles, Nice, and Lyons. He made his American debut at the Metropolitan Opera on December 10, 1910, in *Faust.* For over a quarter of a century he was the principal bass of that company, heard in over a hundred different roles. He appeared in the world première of *Peter Ibbetson,* and the American premières of *Ariane et Barbe-Bleue, Boris Godunov,* and *L'oiseau bleu.* His last appearance at the Metropolitan was in *Manon* on February 25, 1939.

Rothier also distinguished himself as a teacher and opera coach. In 1944 he appeared in a Broadway play, *A Bell*

for Adano. In 1949 he celebrated his fiftieth anniversary as a singer with a New York recital. He received numerous honors, among them those of Chevalier of the Legion of Honor (France), Officier de l'Instruction Publique (France), and Chevalier of the Belgian Order of Leopold.

Rothmüller, Marko, baritone. Born Trnjani, Yugoslavia, December 31, 1908. He attended the Zagreb Music Academy, then went to Vienna and studied composition with Alban Berg and singing with Fritz Steiner. His opera debut took place with the Hamburg Opera in 1932. After two seasons with the Zagreb Opera, he became a member of the Zurich Opera, where he remained for thirteen years. He first became famous as a member of the New London Opera Company in England, which he joined in 1947, distinguishing himself in such roles as Rigoletto, Scarpia, and Jokanaan (in Richard Strauss's *Salome*). In 1946 he was engaged by the Vienna State Opera, where he has since been a leading baritone. In 1950 he scored major successes at Covent Garden and at the Glyndebourne Festival, and soon after that made his American debut with the New York City Opera.

Rothwell, Walter Henry, conductor. Born London, England, September 22, 1872; died Los Angeles, California, March 12, 1927. His music study took place at the Vienna Conservatory, and in Munich with Ludwig Thuille and Max von Schillings. He began his professional career as concert pianist, but abandoned this for conducting in 1895 when he was engaged as Gustav Mahler's assistant at the Hamburg Opera. He then conducted operas in various German theaters, and directed a season of German operas in Amsterdam in 1903–1904. In 1904 he was engaged by the Savage Opera Company to direct *Parsifal* (in English) in the United States. He remained with the company until 1908, during which period he led the American première of *Madama Butterfly*. Elisabeth Wolff, who appeared as Cio-Cio-San, became his wife. After 1908 he devoted himself to symphonic music, serving as conductor of the St. Paul Symphony and the Los Angeles Philharmonic.

rôtisserie de la Reine Pédauque, La, *see* FRANCE, ANATOLE.

Rousseau, Jean Jacques, philosopher, musical theoretician, and composer. Born Geneva, Switzerland, June 28, 1712; died Ermenonville, France, July 2, 1778. The celebrated philosopher was also an influential musician and an active participant in operatic activities. For some years he earned his living by copying music. He contributed articles on music to Diderot's *Encyclopédie,* and proposed some refinements in the system of musical notation. In 1747 he wrote his first opera, *Les muses galantes.* When Pergolesi's *La serva padrona* was introduced in Paris in 1752, Rousseau was so enchanted with it that he wrote a comic opera in a similar vein, *Le devin du village,* an outstanding success for many years. The controversy that arose over the differing merits of Pergolesi's comic opera and the more dramatic operas of Jean Philippe Rameau—the "Guerre des Bouffons"—found Rousseau in the Italian camp. His "Lettre sur la musique française," published in 1753, proved such a bitter attack that Rousseau was hanged in effigy by the artists of the Opéra. Rousseau's *Dictionnaire de musique,* which became internationally popular, was published in 1768. The composer worked on a third opera, *Daphnis et Chloé,* but did not finish it. He collected his vocal duets and romances in a volume entitled *Les consolations des misères de ma vie* (1781). His drama *Pygmalion* was the source for an opera by Giambattista Cimador, for which he himself wrote part of the music, while Mozart's *Bastien and Bastienne* was based on a parody of Rousseau's *Le devin du village.*

Roussel, Albert Charles, composer. Born Tourcoing, France, April 5, 1869; died Royan, France, August 23, 1937. He attended the Schola Cantorum, where his teachers included Vincent d'Indy. He first became famous with a ballet, *Le festin de l'araignée.* His opera-ballet *Padmâvatî,* inspired by a visit to India, was successfully given at the Paris Opéra on June 1, 1923. Roussel, also known for his instrumental music and his songs, wrote two operas: *La naissance de la lyre* (1924); *Le testament de la tante Caroline* (1933).

Rouvel, Baron, Fedora's friend (tenor) in Giordano's *Fedora.*

Roxane, Cyrano's beloved in *Cyrano de Bergerac,* an opera by Franco Alfano, and an opera by Walter Damrosch. In both she is a soprano.

Royal Palace, The, one-act opera by Kurt Weill. Libretto by Ivan Goll. Première: Berlin Opera, March 2, 1927. This opera is a mixture of play, pantomime, and motion pictures. At the Royal Palace, on the shores of an Italian lake, a fashionable woman is surrounded by three men: her rich husband who bores her, her shallow lover of a former day, her romantic lover of the future. The husband sends her on a trip through Europe, but this fails to cure her depression, and she drowns herself.

Roze, Marie (born PONSIN), soprano. Born Paris, March 2, 1846; died there June 21, 1926. She attended the Paris Conservatory, where she won two first prizes. Her debut took place at the Opéra-Comique on August 16, 1865, in the title role of Herold's *Marie.* She remained at the Opéra-Comique three years. After additional study with Pierre François Wartel she appeared at the Paris Opéra on January 2, 1870, as Marguerite and scored a major success. After 1872 she appeared in London and was a great favorite. From 1883 to 1889 she was a member of the Carl Rosa Opera Company. In 1890 she settled in Paris as a teacher of singing. Though she made her farewell tour in 1894, she appeared intermittently on the concert stage until 1903.

Rozenn, daughter (soprano) of the King of Ys in Lalo's *Le Roi d'Ys.*

Rubini, Giovanni Battista, tenor. Born Romano, Italy, April 7, 1794; died there March 2, 1854. One of the most celebrated opera singers of the early nineteenth century, he studied singing with Rosio in Bergamo and Andrea Nozzari in Naples, then began his career by appearing in minor roles with traveling companies. His increasing success brought him to the attention of Domenico Barbaja, who engaged him for his Naples company in 1816. He remained here many years, particularly distinguishing himself in Rossini's operas. During this period he made his first triumphant appearances in Vienna and Paris. Between 1831 and 1843 he appeared alternately in Paris and London, outstandingly successful in operas of Bellini, Donizetti, and Rossini. In 1843 he made a tour of Holland and Germany with Franz Liszt, then went on alone to St. Petersburg, where he was idolized. He retired in 1845, after having amassed a fortune. Rubini is remembered not only for the beauty of his singing but for his excessive use of such devices as the vocal sob (often called, in disparagement, the Rubini sob), contrasts between soft and loud, and the vibrato.

Rubinstein, Anton, composer and pianist. Born Vykhvatinets, Russia, November 28, 1829; died Peterhof, Russia, November 20, 1894. Taught the piano in childhood, he entered upon a concert career in his tenth year. In 1848 he began an eight-year period of intensive study and composition. Four years later his opera *Dmitri Donskoi* was produced in St. Petersburg and was such a suc-

cess that the Grand Duchess Helen became his patron. In 1857 he again toured as a pianist, now establishing himself as one of the world's leading concert artists. After a series of sensational piano recitals in St. Petersburg, he was appointed imperial concert director there in 1862. He helped found, and for five years he directed, the St. Petersburg Conservatory. His activity on the concert stage continued with little interruption; in 1872–1873 he made a triumphant tour of the United States. After 1890 he lived principally in Germany. He was the recipient of many honors, including the Order of Vladimir from the Czar, the Knighthood of the Prussian Order of Merit, and the title of Imperial Russian State Councilor. He wrote an autobiography in 1889 to commemorate his fiftieth anniversary as concert pianist; and in 1892 he issued a volume of essays on music, *Die Musik und ihre Meister,* translated into English as *A Conversation on Music.*

As a composer, Rubinstein belonged to the German romantic school. Once popular, few of his works are now played. His most celebrated opera was *The Demon.* For a long time it was in the repertory of several leading Russian opera companies. His operas: *Dmitri Donskoi* (1852); *Feramors* (1863); *The Demon* (1875); *The Maccabees* (1875); *Nero* (1879); *Sulamith* (1883); *Christus* (1888); *Moses* (1894).

Rudolf, Max, conductor. Born Frankfort-on-the-Main, Germany, June 15, 1902. After completing his music study at the Hoch Conservatory in his native city, he became an opera conductor in Freiburg in 1922. For thirteen years he conducted opera performances in Germany and Czechoslovakia, and for five years orchestral concerts in Sweden. He came to the United States in 1940 and five years later became a conductor at the Metropolitan Opera, making his debut on January 13, 1946, in a benefit concert; his first opera performance there was, on March 2, *Der Rosenkavalier.* He has been with the Metropolitan ever since, distinguishing himself particularly in Mozart's operas. He became an American citizen in 1946. In 1950 he published a treatise entitled *The Grammar of Conducting.*

Rühlmann, François, conductor. Born Brussels, January 11, 1868; died Paris, June 8, 1948. He completed his music study at the Brussels Conservatory. After holding various minor conductorial posts, he was engaged by the Théâtre de la Monnaie in 1908, where he remained until 1914. During this period, on September 6, 1905, he made his debut at the Opéra-Comique in Paris with *Carmen;* a year later he became a principal conductor of that company, holding the post until 1914. In 1914 he was engaged by the Paris Opéra, where he achieved recognition as one of France's leading opera conductors. Among the world and French premières he conducted were those of *Ariane et Barbe-Bleue, L'heure espagnole, Mârouf, Maximilien,* and *La vida breve.*

Ruffo, Titta (born RUFFO CAFIERO TITTA), baritone. Born Pisa, Italy, June 9, 1877; died Florence, July 6, 1953. After attending the Santa Cecilia Academy in Rome he studied privately with Cassini in Milan. He made his debut at the Teatro Costanzi in 1898 as the Herald in *Lohengrin.* His American debut took place with the Chicago-Philadelphia Opera in *Rigoletto* on November 4, 1912. He remained with this company two seasons. During World War I he served as mechanic in the Italian air force. After the war he returned to the Chicago-Philadelphia company, where he was now acclaimed one of the most significant opera singers of the time. On January 19, 1922, he made his debut at the Metropolitan Opera in one of his most brilliant roles,

that of Figaro in *The Barber of Seville.* He remained with the Metropolitan until 1929, acclaimed in such roles as Amonasro, Gérard, and Barnaba. Ruffo retired from opera in 1929 and entered the field of motion pictures. His last public appearance took place in 1932 at the opening of the Radio City Music Hall. After 1929 he lived mostly in France. Since he was an anti-Fascist (the Socialist deputy Giacomo Matteotti, murdered by the Fascists, was his brother-in-law), he lived in disfavor until the overthrow of Mussolini.

Ruggero, Magda's lover (tenor) in Puccini's *La rondine.*

Ruggiero, chief bailiff (baritone) in Halévy's *La Juive.*

Ruiz, a soldier (tenor) in Manrico's service in Verdi's *Il trovatore.*

Rusalka, opera by Antonin Dvořák. Libretto by Jaroslav Kvapil. Première: Prague National Opera, May 31, 1901. Rusalka, a water sprite, falls in love with a prince and enlists the help of a witch. The witch transforms the sprite into a beautiful woman. After the prince and the sprite are married, the prince commits an act of infidelity, which results in his wife being changed back into a sprite. When the prince finds her in a lake, he dies in her arms.

Russalka, opera by Alexander Dargomizhsky. Libretto by the composer, based on a play by Alexander Pushkin. Première: St. Petersburg, May 16, 1856. This is Dargomizhsky's best-known opera, and one in which the nationalistic ideals are fulfilled. The central character is Natasha, daughter of a miller, who, betrayed by a prince, commits suicide in a stream. She becomes transformed into a water sprite who lures men to their death. When the prince marries a woman of his own class, he hears the wail of the water sprite each time he tries to embrace his beloved. One day, wandering near the sprite's stream, he meets a child who tells him she is his daughter. Natasha's father, crazed by his misfortunes, throws the prince into the stream, where he is reunited with his one-time sweetheart and their child.

Russell, Henry, impresario. Born London, November 14, 1871; died there October 11, 1937. The son of Henry Russell, an eminent writer of songs, he was the brother of a noted conductor (*see* RONALD, SIR LANDON).

Russell planned to be a doctor, but turned to music after a permanent injury to his eye. After studying singing at the Royal College of Music, he joined its faculty and achieved considerable renown as a teacher of singing. Through his knowledge of anatomy, he was able to originate a new method of voice production, praised by Nellie Melba, Eleanora Duse, and others. His extensive association with singers enabled him to procure an appointment as manager of the Covent Garden opera season in 1903 and 1904. In 1905 he toured the United States with his own opera company. Four years later he became director of the newly founded Boston Opera Company, holding this post until the dissolution of the company in 1914. In the spring of 1914 he took his own company to Paris for a two-month season at the Théâtre des Champs Elysées. He then settled in London, but returned to the United States in 1921 to manage a lecture tour for Maurice Maeterlinck.

Russlan and Ludmilla, opera in five acts by Mikhail Glinka. Libretto by the composer, Shirkov, Bakhturin and others, based on a poem by Alexander Pushkin. Première: Imperial Theater, St. Petersburg, December 9, 1842.

Characters: Svietosar, Grand Duke of Kiev (bass); Ludmilla, his daughter (soprano); Russlan, a knight in love with Ludmilla (baritone); Ratmir, a second suitor of Ludmilla (contralto); Farlaf, a third suitor (bass); Gorislava, young lady in love with Ratmir (soprano); Finn, a sorcerer (tenor); Tchernomor, a dwarf; Naina, a witch (mezzo-soprano). The setting is legendary Russia.

Act I. The court of Svietosar. The Grand Duke is entertaining Ludmilla's three suitors. When festivities are at their height, Ludmilla mysteriously disappears. Svietosar promises his daughter to the suitor who finds her.

Act II. Finn's cave. Russlan learns from the sorcerer that Ludmilla had been abducted by the dwarf Tchernomor; he is also warned about the witch Naina, who is Farlaf's ally. Farlaf, meanwhile, visits the witch for help. Her advice is to allow Russlan to find Ludmilla, and then to kidnap her. The scene changes to a battlefield, where Russlan kills a breathing head, under which lies a magic sword.

Act III. Naina's domain. To help Farlaf, Naina uses her wiles to divert his rivals. Ratmir is imprisoned after he succumbs to the song of sirens. Russlan is about to experience the same fate, when he is saved by Finn.

Act IV. Tchernomor's house. The dwarf has imprisoned Ludmilla. He has arranged a ballet to amuse her, but she is bored. When Russlan comes to save his beloved, the dwarf puts Ludmilla to sleep. Russlan defeats the dwarf with his magic sword by cutting off his beard, in which his strength resides. Unable to rouse Ludmilla, Russlan carries her away.

Act V. The court of Svietosar. Finn again comes to Russlan's help, and Ludmilla is roused from her profound sleep. Svietosar happily gives his daughter to Russlan to wed.

Russlan and Ludmilla was Glinka's second and last opera. In his first, *A Life for the Czar,* he made his first experiments in creating a truly Russian opera. The opera had been a success chiefly because it was simply a Russianized Italian opera, and audiences could respond to its Bellinian melodies. But in *Russlan,* Glinka grew bolder in his attempts to achieve nationalism. He departed completely from the Italian style to produce an opera authentically Russian in spirit and music. *Russlan and Ludmilla* was revolutionary for its day; and, when first heard, was a failure. Not until its revival in 1859 did it receive the recognition it deserved. From then on, it was accepted in its native land as one of the most significant of all Russian operas. There was hardly a national composer in Russia after 1860 who was not influenced by it. The sprightly overture is a concert favorite. Several other excerpts are noteworthy: the "Bard's Song" in Act I; Farlaf's Rondo (or Patter Song) in Act II; and the "Persian Song" in Act III.

Rustic Chivalry, see CAVALLERIA RUSTICANA.

Rutledge, Mary, Philip Nolan's sweetheart (soprano) in Walter Damrosch's *The Man Without a Country.*

S

Saamschedine, Princess, the Sultan's daughter (soprano) in Rabaud's *Mârouf.*

Sabata, Victor de, conductor and composer. Born Trieste, Italy, April 10, 1892. He graduated from the Milan Conservatory in 1911, and soon afterward became known as an orchestral composer through performances in Italy, France, Belgium, and Russia. He scored an even greater success with an opera, *Il Macigno,* introduced at La Scala on March 30, 1917. He subsequently wrote two more operas: *Lysistrata* and *Mille e una notte.* Sabata also distinguished himself as one of the foremost conductors in Europe. After World War I, he was appointed first conductor of the Monte Carlo Opera, where he remained a dozen years. For the next twenty years he was first conductor of La Scala. He also conducted opera performances with other leading European companies, and at Bayreuth, Salzburg, and the Florence May Music Festival. He first visited the United States in 1927 as guest conductor of several orchestras, and has since returned a number of times.

Sacchini, Antonio, composer. Born Florence, Italy, June 14, 1730; died Paris, France, October 6, 1786. He attended the Conservatorio Santa Maria di Loreto in Naples, where he wrote an intermezzo, *Fra Donato,* which was acclaimed in 1756. In 1762 he achieved a major triumph in Rome with his opera *Semiramide;* he remained in Rome several years, rivaling the popularity of Nicola Piccinni. After the successful performance of his *Alessandro nell' Indie* in Venice, he was appointed director of the Conservatorio dell' Ospedaletto. Between 1770 and 1772 he lived in Germany and from 1772 and 1782 he was in London, where his operas were in vogue. Financial difficulties compelled him to flee to Paris in 1782. There he received royal patronage and was favored by the general public. In Paris he wrote two new operas in which he assimilated some of the progressive ideas and style of Gluck; one of these operas, *Oedipe* *à Colone,* his masterwork, received over six hundred performances between 1784 and 1844. He wrote over sixty operas, the most important being: *Semiramide* (1762); *L'Olimpiade* (1767); *Alessandro nell' Indie* (1768); *Ezio* (1770); *Tamerlano* (1771); *Armida e Rinaldo* (1772); *La Colonie* (1775); *Dardanus* (1784); *Oedipe à Colone* (1786).

Sachs, Hans, poet, composer, playwright. Born Nuremberg, Germany, November 5, 1494; died there January 19, 1576. The most celebrated of the mastersingers (which see), Sachs became the central character of Richard Wagner's *Die Meistersinger,* and of Gustav Albert Lortzing's *Hans Sachs.* Some of the poet's dramatic pieces were later the source of operas, including Joseph Forster's *Der dot Mon,* Bernhard Paumgartner's *Das heisse Eisen,* and Werner Wehrli's *Das heisse Eisen.*

Sachse, Leopold, stage director. Born Berlin, Germany, January 5, 1880. He attended the Cologne Conservatory, then studied singing with Benno Stolzenberg and Blanche Selva. In 1907 he became director of the Münster Stadttheater, and in 1915 intendant of the Hamburg Opera. After a period of staging musical productions in Paris, Sachse came to the United States. In 1935 he was appointed stage director of the Metropolitan Opera, and a year later teacher of stage techniques at the Juilliard School of Music. Subsequently he became the stage director of the New York City Opera.

Sacrifice, The, opera by Frederick Shepherd Converse. Libretto by the composer and John Macy. Première: Boston Opera House, March 3, 1911. In California—the year is 1846, and Mexico and the United States are at war—Bernal, a Mexican officer, is Captain Burton's rival for the love of Chonita. When the American officer realizes the extent of Chonita's love for the Mexican, he sacrifices his life to save theirs.

Sadko, opera by Nikolai Rimsky-Korsakov. Libretto by the composer and V. I. Bielsky. Première: Moscow, January 7, 1898. The story was derived from legends. Sadko is a wandering minstrel of the eleventh century, whose fantastic travels bring him to the Princess of the Sea. She eventually transforms herself into the Volkhova River, becoming the water route to the sea for the district of Novgorod. In Act II is heard the opera's best-known aria, the "Song of India." In the same act appears another familiar aria, the "Song of the Viking Guest."

Sadler's Wells Opera, a London opera company, the only permanent opera repertory theater in England dedicated to performances of opera in English. The original Sadler's Wells Theater was built in the eighteenth century in north London, and was used for plays, pantomimes, and musical productions. After about a century of use, the building was abandoned. In 1926 it was acquired by Lilian Baylis and other lovers of drama and opera as a branch for the Old Vic Theatre, which had been producing dramas and operas since 1914. After 1934, all the Old Vic's opera performances were given at Sadler's Wells. A famous ballet company developed at Sadler's Wells, originally under the direction of Ninette de Valois, which gave performances of its own, besides being seen in the opera productions. In addition to the usual French, Italian, and German operas, Sadler's Wells has produced such novelties as *Boris Godunov* (in the composer's original version), *The Bartered Bride, Eugene Onegin,* and *I Quatro Rusteghi;* also, such English operas as Lawrence Arthur Collingwood's *Macbeth,* Ralph Vaughan Williams' *Hugh the Drover* and *Sir John in Love,* and Benjamin Britten's *Peter Grimes.* The opera house closed during World War

II, reopening in 1945 with the world première of *Peter Grimes*. In 1955 Camilla Williams became the first Negro singer to appear at Sadler's Wells; she was seen as Cio-Cio-San.

Sailors' Chorus, *see* STEUERMANN! LASS DIE WACHT!

St. Bris, Count de, Catholic nobleman (bass) in Meyerbeer's *Les Huguenots.*

Saint Daniel et Saint Michel, Mélisande's ancient song at the opening of Act III of Debussy's *Pelléas et Mélisande.*

Saint-Georges, Jules Henri Vernoy de, novelist and librettist. Born Paris, November 7, 1801; died there December 23, 1875. One of the most significant and prolific French librettists after Eugène Scribe, he produced over a hundred librettos, which were set to music by Adam, Auber, Bizet, Flotow, Halévy, Hérold, and many others. Operas with his librettos include Auber's *Les diamants de la couronne;* Bizet's *La jolie fille de Perth* (libretto a collaboration with Jules Adenis); Donizetti's *The Daughter of the Regiment* (a collaboration with Alfred Bayard); Halévy's *L'Eclair* (a collaboration with F. A. E. Planard). Balfe's *The Bohemian Girl* was founded on Saint-Georges's ballet-pantomime *The Gypsy.*

Saint Julien l'Hospitalier, *see* FLAUBERT, GUSTAVE.

Saint of Bleecker Street, The, a music drama in three acts by Gian-Carlo Menotti. Libretto by the composer. Première: New York, Broadway Theater, December 27, 1954. The setting is New York's Italian quarter. Annina, a passionate religious mystic, receives the stigmata upon her palms, provoking the frenzied devotion of the Catholic neighborhood. Michele, her brother, a troubled agnostic, uses argument and force in his futile attempts to draw Annina away from what he regards as superstition and ignorance. At the same time he manifests a devotion to the sickly girl that prevents any decisive move. Desideria, Michele's worldly minded sweetheart, taunts him with being in love with his sister. In the furious quarrel that follows, Michele kills Desideria with a knife and becomes a hunted man. Later, he creeps back to Bleecker Street to watch his sister being accepted by the Church as the Bride of Christ. The emotional tension of the elaborate ceremony proves too much for the frail Annina, and she dies in her moment of greatest joy. The opera had a run of some ninety performances and was given the Pulitzer Prize, the Drama Critics Award, and the Music Critics Circle Award.

Saint-Pierre, Jacques Henri Bernardin de, writer. Born Le Havre, France, January 19, 1737; died Eragny-sur-Oise, France, January 21, 1814. His romantic novel *Paul et Virginie,* a French classic, was made into operas by Pietro Guglielmi, Rodolphe Kreutzer, Jean-François Lesueur, and Victor Massé. *See* PAUL ET VIRGINIE.

Saint-Saëns, Camille, composer. Born Paris, October 9, 1835; died Algiers, December 16, 1921. He was exceptionally precocious in music, began lessons when he was only three, and made his first public appearance as pianist when he was four and a half. Further study took place with Camille Stamaty and Pierre Maleden and later at the Paris Conservatory. In 1857 Saint-Saëns became organist of the Madeleine Church, remaining twenty years and achieving recognition as one of the foremost organists of his day. He also attracted attention as composer with his second symphony, which won first prize in a contest sponsored by the Société Sainte-Cécile. His first opera, *La princesse jaune,* was given by the Opéra-Comique on June 12, 1872. Five years later, his second, *Le timbre d'argent,* was produced at the Théâtre Lyrique. His *Samson et Dalila* was not accepted for performance in Paris because opera directors considered it too Wagnerian and too severe. It found an advocate in Franz Liszt, who used his influence to get it performed in Weimar, 1877. Not until thirteen years later, after it had been acclaimed in most of the rest of Europe, was it heard in Paris.

Saint-Saëns continued composing prolifically in every branch of musical composition for the remainder of his life. He had other activities, too: as concert pianist, conductor, organist, professor of piano at the Ecole Niedermeyer, editor, writer, and as founder of the Société Nationale de Musique. He also indulged his passion for travel. He visited the United States twice, in 1906 and 1916. His last appearance as pianist took place in Dieppe on August 6, 1921. Two weeks later he led his last orchestral concert. He died suddenly while on a visit to Algiers.

His operas: *La princesse jaune* (1872); *Samson et Dalila* (1877); *Le timbre d'argent* (1877); *Etienne Marcel* (1879); *Henry VIII* (1883); *Gabriella di Vergy* (1885); *Proserpine* (1887); *Ascanio* (1890); *Phryné* (1893); *Les Barbares* (1901); *Hélène* (1904); *L'Ancêtre* (1906); *Déjanire* (1910).

Sakuntala, a Sanskrit drama by Kalidasa, written in the fifth century. Several operas were derived from this drama, including Franco Alfano's *La leggenda di Sacuntala*, Ignace Jan Paderewski's *Sakuntala*, and Felix Weingartner's *Sakuntala*.

Salammbô, opera by Ernest Reyer. Libretto by Camille du Locle, based on Gustave Flaubert's novel of the same name. Première: Théâtre de la Monnaie, February 10, 1890. In an army camp outside ancient Carthage, Matho steals a magic veil (zaimph) that covers a holy statue at the shrine of the goddess Tamit. When Salammbô tries to recover the veil, Matho forces his love upon her. The Carthigians seize

Matho, imprison him for the theft, and condemn him to execution. To Salammbô, who has saved the veil, goes the honor of executing him. Salammbô uses the sword intended for Matho to commit suicide. Matho takes the dying Salammbô in his arms and kills himself with her sword.

Modest Mussorgsky began an opera on this story about 1860, but abandoned the project. The contemporary composer Josef Matthias Hauer has made Flaubert's novel into an opera that employs twelve-tone music.

Salce! Salce! Desdemona's Willow Song in Act IV of Verdi's *Otello*.

Salieri, Antonio, composer and conductor. Born Legnano, Italy, August 18, 1750; died Vienna, May 7, 1825. He attended the San Marco singing school in Venice. In 1766 he was taken to Vienna, where he continued his music study. Four years later he conducted the première of his first opera, *Le donne letterate*. Between 1770 and 1774 nine of his operas were produced at court. In 1774 he succeeded Florian Gassmann as conductor of Italian opera and chamber composer. Four years later he returned to Italy, writing operas for performance there. La Scala opened its doors (August 3, 1778) with one of his new operas: *Europa riconosciuta*. In 1784 Salieri was in Paris, where several of his new operas, strongly influenced by those of Gluck, were given successfully. Back in Vienna, he continued writing operas prolifically; on one occasion he had five new ones produced in a single year. From 1788 to 1824 he was court conductor, and after 1790 conductor of the court choir.

He was one of the most influential musicians in Vienna. At different periods he was a teacher of Beethoven, Schubert, and Liszt. He was a rival and enemy of Mozart, frequently obstructing the introduction of Mozart's operas, and just as often conniving for their failure. An unfounded rumor was circulated after Mozart's death that he had been poisoned by Salieri; this persistent legend was the basis of Rimsky-Korsakov's opera *Mozart and Salieri*.

After half a century of service at the Viennese court, Salieri retired in 1824 on a pension. Of his fifty or so operas, these were the most successful: *Armida* (1771); *Don Chisciotte alle nozze di Gamace* (1771); *La fiera di Venezia* (1772); *La secchia rapita* (1772); *Semiramide* (1782); *Les Danaïdes* (1784); *La grotta di Trofonio* (1785); *Tarare* (1787); *Il pastor fido* (1789); *Palmira* (1795); *Falstaff* (1799); *Angiolina* (1800); *Cesare di Farmacusa* (1800); *Die Neger* (1804); *Cyrus und Astyages* (1818).

Salignac, Eustase Thomas, tenor and teacher of singing. Born Generac, France, March 29, 1867. He studied singing at the Marseilles and Paris Conservatories. In 1893 he made his opera debut at the Opéra-Comique. After two seasons there he came to the United States, making his American debut at the Metropolitan Opera on December 11, 1896, singing the role of Don José. He remained there seven consecutive seasons, appearing in all the principal tenor parts of the French repertory. During this period he also appeared at Covent Garden. After returning to France in 1903 he joined the Opéra-Comique, remaining there ten seasons. In 1913 he was appointed director of the Nice Opera. A decade later he returned to the Opéra-Comique, and in 1926 he was director of an opera company that toured Canada and appeared in New York. In 1923 he became a professor of singing at the Fontainebleau Conservatory, and in 1924 professor of elocution at the Paris Conservatory. In 1933 and 1937 he organized competitions to discover new French singers. He created the role of Mârouf in Rabaud's opera of the same name (1918), and appeared in world premières of operas by Massenet, Laparra, and Milhaud, among others.

Salle Favart, see OPERA-COMIQUE.

Salome, one-act music drama by Richard Strauss. Libretto is the play by Oscar Wilde, in a German translation by Hedwig Lachmann. Première: Dresden Opera, December 9, 1905. American première: Metropolitan Opera House, New York, January 22, 1907.

Characters: Herod, Tetrarch of Judea (tenor); Herodias, his wife (mezzo-soprano); Salome, her daughter (soprano); Jokanaan, the prophet (baritone); Narraboth, a captain of the guards (tenor); Page of Herodias (mezzo-soprano); Jews; Nazarenes; soldiers. The setting is a terrace of Herod's palace in Galilee about A.D. 30.

During a banquet, Jokanaan (John the Baptist) proclaims the coming of the Messiah. Salome orders that the prophet be brought to her. When he arrives, he curses Herodias and entreats Salome not to imitate her mother's dissolute life. When Salome tries to get Jokanaan to kiss her, he pushes her aside and denounces her. Herod, seeking diversion, begs Salome to dance for him. She promises to do so but only if the Tetrarch will then reward her by granting any wish she may have. She performs the "Dance of the Seven Veils," then demands the head of Jokanaan. The horrified Tetrarch finally acquiesces and the severed head is brought to Salome. She addresses it passionately ("Ah! Du wolltest mich nicht deinen Mund küssen lassen!"), after which she lasciviously kisses the dead lips. Revolted by the spectacle, Herod orders his soldiers to crush Salome beneath their shields.

Wilde's lurid play was perfectly matched by Strauss's sensuous and erotic music. The result was an opera that shocked many an audience in the early 1900's. A première planned in Vienna had to be canceled when the censors stepped in. In Berlin, the Kaiser at first forbade its performance; when it was allowed, it inspired such adjectives as "repulsive" and "perverse." The London première was prevented by censors. In America, the première created a storm of protest that is described elsewhere (*see* METROPOLITAN OPERA HOUSE). But time has now cushioned the shock, and it is now conceded that *Salome* is a drama of unforgettable impact, as well as a remarkably successful example of fusion of text and music.

A French composer, Antoine Mariotte, also wrote an opera on Wilde's play. Though he completed it before Strauss wrote his opera, it was not introduced until 1908, in Lyons. Strauss had to give Mariotte written permission to have the latter's opera produced, waiving his rights to the exclusive use of the Wilde text. After the première of Mariotte's opera, Strauss had to issue a public denial that he had ever demanded that Mariotte destroy his manuscript, as rumor insisted that he did.

Salomé, Herodias' daughter (soprano) in Massenet's *Hérodiade.*

Salud, a gypsy girl (soprano) in Falla's *La vida breve.*

Salut à France, finale of Donizetti's *The Daughter of the Regiment.*

Salut à toi, soleil, see HYMN TO THE SUN.

Salut! demeure chaste et pure, Faust's cavatina in Act III of Gounod's *Faust.*

Salut! tombeau! sombre et silencieux, Roméo's aria in Act V of Gounod's *Roméo et Juliette.*

Salvezza alla Francia, see SALUT A FRANCE.

Salzburg Festival, one of the most significant of Europe's music festivals. It was created in the summer of 1920 through the combined efforts of Hugo von Hofmannsthal, Max Reinhardt, Franz Schalk, and Richard Strauss. The festival theater (Festspielhaus) was erected in 1926.

A pattern of activity was established which has continued through the years with only minor deviations. Since Salzburg was the birthplace of Mozart, his music predominates. The program—which continues approximately a month, beginning the last week in July—embraces operas and orchestral, chamber, and choral music. The opera performances, invariably of the highest order, enlist singers and the orchestra of the Vienna State Opera, together with guest singers and conductors from all parts of the world. Though Mozart's operas are the core of the repertory, operas of Richard Strauss, Gluck, Weber, Verdi, Wagner, Beethoven, and Rossini have also been given. Among the conductors who have led Salzburg performances are Arturo Toscanini, Richard Strauss, Franz Schalk, Clemens Krauss, Bruno Walter, Wilhelm Furtwaengler, Karl Böhm, Tullio Serafin, and Felix Weingartner. Singers have included Lotte Lehmann, Ezio Pinza, Elizabeth Schumann, Mariano Stabile, Cesare Siepi, Alexander Kipnis, Friedrich Schorr, Elisabeth Schwarzkopf, Irmgard Seefried, Richard Mayr, Salvatore Baccaloni, Erna Berger, and Tiana Lemnitz. The Festival suspended activity during World War II. It resumed operations on August 1, 1946, with the cooperation of the United States Army. Since World War II, the Festival has given first performances of a number of contemporary operas, including Boris Blacher's *Romeo and Juliet;* Werner Egk's *Irish Legend;* Gottfried von Einem's *Dantons Tod* and *Der Prozess;* Rolf Liebermann's *Penelope;* Frank Martin's *Le vin herbé;* Carl Orff's *Antigone;* Richard Strauss's *Die Liebe der Danae.*

Samoset, an Indian chief (bass) in Hanson's *Merry Mount.*

Samson, leader of the Israelites, the central character in Bernard Rogers' *The Warrior* and Saint-Saëns' *Samson* *et Dalila.* In both operas the role is for a tenor.

Samson et Dalila (Samson and Delilah), opera in three acts by Camille Saint-Saëns. Libretto by Ferdinand Lemaire, based on the Biblical story. Première: Weimar, December 2, 1877.

Characters: Samson, leader of the Israelites (tenor); Dalila, a Philistine, priestess of the Temple of Dagon (mezzo-soprano); High Priest of Dagon (baritone); Abimelech, Satrap of Gaza (bass); an old Hebrew (bass); Messenger of the Philistines (tenor); Hebrews; Philistines; people of Gaza; dancers. The setting is Gaza, Palestine, about 1150 B.C.

Act I. A public square. The Hebrews, in bondage to the Philistines, lament their lot. Samson tries to hearten them by urging them to praise God, not to complain ("Arrêtez ô mes frères"). Abimelech begins to harass Samson until the latter kills him with his own sword. The High Priest emerges from the Temple to call the Philistines to avenge the Satrap's death ("Maudite à jamais soit la race"). Rallying under Samson's leadership, the Hebrews attack the Philistines. They sing a hymn of victory ("Hymne de joie") as the Philistines disperse, carrying away Abimelech's body. From the temple come Dalila and her maidens bearing garlands for the Hebrews and singing a tribute ("Voici le printemps nous portant des fleurs"). Coquettishly, Dalila approaches Samson to lure him. She sings a song of spring ("Printemps qui commence") Bewitched by her beauty, Samson prays to heaven for the strength to resist her.

Act II. Before Dalila's house. Dalila invokes the magic of love to aid her in overpowering Samson ("Amour! viens aider ma faiblesse"). The High Priest urges Dalila to use her beauty to uncover the source of Samson's physical strength. When Samson arrives, a storm is brewing. Dalila lavishes her love on him ("Mon coeur s'ouvre à ta voix"), then begs him to disclose the secret of his strength. Samson denounces her as a temptress. Dalila goes inside her house. Samson follows, succumbing to her wiles. Within the house, Dalila cuts Samson's hair, then calls to the Philistine soldiers to enter and seize him.

Act III, Scene 1. The mill of Gaza. Samson is a prisoner of the Philistines. His eyes have been plucked out, and he is in chains. As he turns the Philistine's mill, he begs God to have mercy on him ("Vois ma misère, hélas!"). In the distance are heard the voices of the Hebrews denouncing Samson for having betrayed them. ("Samson, qu'as-tu fait du Dieu de tes pères").

Scene 2. The Temple of Dagon. The Philistines are celebrating their victory over the Hebrews with revelry ("Bacchanale"). Samson, in chains, is led into the temple by a child. He is mocked by the Philistines, particularly by Dalila. Samson asks to be led between the two great pillars supporting the temple roof. He then prays to God for a brief return of his former strength ("Souviens-toi de ton serviteur"). His prayer is answered. Samson sends the pillars over and the roof crashes down, burying Samson and his enemies.

Because of its Biblical subject and its wealth of choral music, *Samson et Dalila* is sometimes given as an oratorio. Its sound musical values—its highly effective mingling of French lyricism and Hebraic chants and Oriental dances—makes it a rewarding experience however it is performed. But it is unquestionably more effective in a stage production, since it is a consistently striking visual spectacle. In view of its dramatic and musical merits, it now seems strange that the opera had to wait so long for success. When one act was given in a concert version in Paris in 1875, critics complained about its "absence of melody" and "an instrumentation which nowhere rises above the level of the ordinary." There were no opera directors in Paris interested in the work, and it was finally introduced in Weimar (several years after completion) on December 2, 1877. It was not given in Paris for another thirteen years, and the Paris Opéra did not produce it until 1892. By then the opera had been acclaimed throughout Europe. Thus, paradoxically, France was one of the last nations in Europe to recognize one of the most French of French operas—an opera that was to become one of the best-loved works in the French repertory.

Samson, qu'as-tu fait du Dieu de tes pères, chorus of the Hebrews in Act III, Scene 1, of Saint-Saëns' *Samson et Dalila.*

San Carlo Opera, *see* TEATRO SAN CARLO.

San Carlo Opera Company, an American company named after the celebrated Neapolitan opera house. It was organized in 1909 by Fortune Gallo. Ever since, it has toured the United States in popular-priced opera performances. It is an unsubsidized, self-supporting company. While it has, for the most part, included young and unknown singers, it has also occasionally featured such famous performers as Martinelli, Jeritza, Schipa, Bori, Ruffo and Bonelli.

Sancho Panza, *see* PANZA, SANCHO.

Sand, Georges (born AMADINE AURORE DUPIN), novelist. Born Paris, July 1, 1804; died Nohant, France, June 8, 1876. Her novel *Consuelo* was made into operas by Giovanni Gordigiani, Vladimir Kashperov, Giacomo Orefice, and Alfonso Rendano. Other operas based on Sand's novels include Julius Benedict's *The Red Beard* and Augusto Machado's *Laureana.*

Sanderson, Sybil, soprano. Born Sacramento, California, December 7, 1865; died Paris, France, May 15, 1903. Jules Massenet wrote several operas for her,

including *Thaïs.* She went to Paris when she was nineteen and attended the Conservatory, where her teachers included Mathilde Marchesi and Giovanni Sbriglia. Her debut took place at The Hague in 1888 in the title role of *Manon.* She was acclaimed, and the Paris Exposition of 1889 engaged Massenet to write an opera for her. The work was *Esclarmonde,* and Sanderson made her Parisian debut in the title role at the Opéra-Comique on May 14, 1889. The opera was one of her triumphs; within a short period she appeared in it a hundred times in Paris, as well as in Brussels and St. Petersburg. Massenet now regarded her as the ideal interpreter of his female roles. She was next acclaimed as Thaïs, another role written for her. On January 16, 1895, she made her American debut at the Metropolitan Opera in *Manon.* However, she failed to duplicate her striking European successes in this country. In 1902 she returned to Paris.

Sandman, a character (soprano) in Humperdinck's *Hansel and Gretel.*

Sandman's Lullaby, *see* DER KLEINE SANDMANN BIN ICH.

Sandys, Lady Marigold, Sir Gower Lackland's fiancée (soprano) in Hanson's *Merry Mount.*

San Francisco Opera Company, the second leading opera company in the United States. It was organized by Gaetano Merola in 1923. Principal singers were recruited from the Metropolitan Opera, while minor roles and chorus parts were taken by local singers; the orchestra was the San Francisco Symphony. Performances of the initial season took place in the Civic Auditorium. The repertory embraced *La Bohème, Andrea Chénier, Il Trittico, Mefistofele, Tosca,* and *Roméo et Juliette.* The principal singers included Gigli, De Luca, Didur, and Muzio. In 1924 the company was incorporated as a nonprofit organization with Merola as general director and Robert I. Bentley as president. A campaign for funds yielded enough to assure a regular opera season. In 1932, on the completion of the War Memorial Opera House, the company was transferred to its permanent home, inaugurating it with a performance of *Tosca* on October 15.

In its first quarter of a century the company gave 548 performances of 74 operas. It has presented the following novelties: Mascagni's *L'amico Fritz;* Franco Vittadini's *Anima Allegra;* Giordano's *La cena delle beffe;* Boïto's *Mefistofele;* Cherubini's *The Portuguese Inn;* Massenet's *Werther;* and William Walton's *Troilus and Cressida.* In 1954 it presented the American première of Arthur Honneger's *Jeanne d'Arc au bûcher* in a staged version. Among the foreign singers who have made their American debuts with the San Francisco Opera Company are Elena Nikolaidi, Set Svanholm, and Renata Tebaldi. Kirsten Flagstad's return to the American opera stage after World War II was made with this company in 1949. Besides playing its yearly season in San Francisco, the company has regularly toured the Pacific Coast.

Though Gaetano Merola died only two weeks before the opening of the 1953 season, there was no postponement. Kurt Herbert Adler, formerly Merola's assistant, became artistic director.

Sanglot, French for "sob," a term implying the interpolation of an exclamation, or emotional accent, in a vocal part.

Sante, a servant (silent part) in Wolf-Ferrari's *The Secret of Suzanne.*

Santuzza, a village girl (soprano) in love with Turiddu in Mascagni's *Cavalleria rusticana.*

Sapho, (1) opera by Charles Gounod. Libretto by Émile Augier. Première: Paris Opéra, April 16, 1851. This was Gounod's first opera, written for the prima donna Pauline Viardot, who cre-

ated the title role. The opera was a failure and received only six performances. The central character is the Greek poetess of Lesbos. One of the arias, "O ma lyre immortelle" is familiar.

(2) Opera by Jules Massenet. Libretto by Henri Cain and Arthur Bernède, based on the novel of the same name by Alphonse Daudet. Première: Opéra-Comique, November 27, 1897. Jean Gaussin, a simple country fellow, falls in love with an artist's model, Fanny, who had posed for a statue of Sapho. Later, learning of her loose life, he leaves her but eventually returns because he cannot live without her. While he is asleep, Fanny disappears, convinced that she and Jean can never be happy together.

Sarabande, a sedate dance of Spanish origin, in triple time, with the phrase beginning on the first beat. Sarabandes appear in some seventeenth and eighteenth century operas. An aria found in Handel's *Rinaldo,* "Lascia ch' io pianga," is set to a sarabande from his earlier opera *Almira.*

Sarastro, priest of Isis (bass) in Mozart's *The Magic Flute.*

Sarastro's invocation, *see* O ISIS UND OSIRIS.

Sardou, Victorien, playwright. Born Paris, September 7, 1831; died there November 8, 1908. His dramas and comedies, among the most popular produced in France in the nineteenth century, were often used for operas. The most famous was Puccini's *Tosca.* Others derived from Sardou's plays, or for which Sardou wrote the text, were: Bizet's *Grisélidis;* Février's *Gismonda;* Giordano's *Fedora* and *Madame Sans-Gêne;* Leroux's *Cléopâtre* and *Théodora;* Emile Paladhile's *La Patrie;* Pierné's *La fille du Tabarin;* Saint-Saëns' *Les Barbares;* and Nikolai Soloviev's *Cordelia.*

Sarka, national opera by Zdeněk Fibich. Libretto by Anežka Schulzová. Première: National Theater, Prague, December 28, 1897. The Bohemian legend which inspired this opera was also the subject for one of Smetana's tone poems in the *My Country* cycle. Sarka is a Bohemian folk heroine who slew the knight Ctirad.

Sarti, Giuseppe, composer. Born Faenza, Italy, December 1, 1729; died Berlin, Germany, July 28, 1802. After studying with Padre Martini in Bologna he became organist of the Faenza Cathedral in 1748. His first opera, *Pompeo in Armenia,* was successfully given in his native city in 1752. *Il re pastore,* in 1753, made him famous. In the same year he was called to Copenhagen to direct an Italian opera troupe. During his prolonged stay in Denmark he wrote some twenty operas, many of them acclaimed in Italy. From 1775 to 1779 he was director of the Ospedaletto Conservatory in Venice. From 1779 to 1784 he was maestro di cappella of the Milan Cathedral, and from 1784 to 1802 court conductor to Catherine II in Russia. He died in Berlin on his way back to Italy from Russia. The most famous of his more than fifty operas were: *Il re pastore* (1753); *Ciro riconosciuto* (1756); *Armida* (1759); *Didone abbandonata* (1768); *Achille in Sciro* (1779); *Giulio Sabino* (1781); *I due litiganti* (1782); *Castore e Polluce* (1786). Mozart, in his *Don Giovanni,* has the Don's band entertain him with a melody from *I due litiganti.*

Satie, Erik, composer. Born Honfleur, France, May 17, 1866; died Paris, July 1, 1925. This extreme individualist— notorious for his tender and humorous music, his whimsical titles, and his peculiar instructions to performers— completed two operas. One was serious, the other characterized by his flair for wit. The serious opera was *Socrate* (1918), which Satie described as a "symphonic drama," was introduced in Paris on February 15, 1920. The other was *Le piège de Méduse* (*Me-*

dusa's Snare). Satie's libretto combines wit with nonsense and whimsy. Between scenes, a stuffed monkey performs dances. The action concerns the love affair of Frisette and Astalfo. Frisette's father consents to their marriage only after Astalfo is able to answer a preposterous question. The music is filled with popular idioms.

Sauguet, Henri, composer. Born Bordeaux, France, May 18, 1901. His music study took place with local teachers and with Charles Koechlin in Paris. He then allied himself with several young musicians who, as the "School of Arcueil," sought the inspiration of Erik Satie. Sauget's first major work was an opéra bouffe, *Le Plumet de Colonel,* given in Paris in 1924. After achieving success with several ballets, he wrote music criticisms for various Parisian journals and helped found the literary magazines *Candide* and *Revue Hebdomadaire.* He visited the United States in 1953. His operas: *La Contrabasse* (1930); *La Chartreuse de Parme* (1939); *La gageuse imprévue* (1944).

Sauvée! Christ est ressuscité! Closing chorus, Act V, Scene 2, of Gounod's *Faust.*

Savage, Henry Wilson, impresario. Born Alton, New Hampshire, March 21, 1859; died Boston, Massachusetts, November 29, 1927. Originally a real-estate promoter, when he was compelled to take over the Castle Theater in Boston, he decided to use it as a home for opera. Without previous musical experience, he organized a company in 1897, presenting grand and light operas. In 1900 he founded the English Grand Opera Company which, that year, gave opera in English at the Metropolitan Opera House in New York. In 1904 he organized a touring company to present *Parsifal* (in English) throughout the United States. On October 15, 1906, his company gave the American première of *Madama Butterfly,* in Washington,

D.C. His company toured the country again in 1911 with *The Girl of the Golden West* (in English). The company also gave the American premières of such celebrated operettas as Franz Lehár's *The Merry Widow* and Oscar Straus's *The Waltz Dream*.

Sayao, Bidu, soprano. Born Rio de Janeiro, Brazil, May 11, 1902. After preliminary study in her native city, she went to France and became a pupil of Jean de Reszke. Returning to Brazil, she made her concert debut in 1925, and her opera debut as Rosina a year after that. She was now engaged by the Opéra-Comique in Paris, where she appeared for several years. In 1935 she made her American debut in a recital in New York. On February 14, 1937, she made her bow at the Metropolitan Opera in *Manon*, and was hailed. She remained a permanent member of that company for the next decade and a half, particularly successful in such roles as those of Gilda, Violetta, Mimi, Juliette, Rosina, and Manon. Since leaving the Metropolitan, she has sung in South America and Europe.

Sbriglia, Giovanni, tenor and teacher of singing. Born Naples, Italy, June 23, 1832; died Paris, February 20, 1916. After attending the Naples Conservatory, he made his debut as tenor with the San Carlo Opera in his twenty-first year. He then appeared throughout Italy. Max Maretzek engaged him for the Academy of Music in New York. He appeared in opera in Havana, Mexico, and United States until 1875 when he settled in Paris as a teacher. He reformed the voice of Jean de Reszke from baritone to tenor, and also helped train Édouard and Josephine de Reszke. His many famous pupils included Lillian Nordica, Pol Plançon, and Sibyl Sanderson. In 1890 he was appointed a member of the Royal Academy in Florence. He was also an officer of the French Academy.

Scala, La, *see* TEATRO ALLA SCALA.

Scalchi, Sofia, contralto. Born Turin, Italy, November 29, 1850; died Rome, August 22, 1922. After studying with Augusta Boccabadati, she made her debut in Mantua in 1866 in *Un ballo in maschera*. She scored her first major success when she made her bow at Covent Garden, on November 5, 1868, as Azucena. She appeared at Covent Garden each season thereafter until 1890. Meanwhile, in 1882, she came to the United States with the Mapleson company. She made her debut at the Metropolitan Opera in the performance of *Faust* that opened that opera house on October 22, 1883. In 1884 she returned to the Mapleson company for four years; in 1891 she was back at the Metropolitan for five seasons. After appearing in the American premières of *Andrea Chénier, Falstaff, La Gioconda,* and *Otello,* she went into retirement in 1896.

Scaria, Emil, bass-baritone. Born Graz, Austria, September 18, 1838; died Blasewitz, Germany, July 22, 1886. His music study took place at the Vienna Conservatory. In 1860 he made his opera debut in Pest as St. Bris. He was such a failure that he abandoned the stage for further study, selecting Manuel García as his teacher. He returned to the opera stage in Dessau, but realized his first success at the Crystal Palace in London in 1862. In 1863 he appeared with the Leipzig Opera, and in 1864 with the Dresden Opera. In 1872 he was engaged by the Vienna Opera. He created the role of Gurnemanz when *Parsifal* was introduced in Bayreuth in 1882.

Scarlatti, Alessandro, composer. Born Palermo, Sicily, May 2, 1660; died Naples, Italy, October 24, 1725. He was the founder and leading figure of the Neapolitan school that developed the techniques and traditions of Italian opera. After studying with Giacomo Carissimi in Rome, he completed his first opera, *L'errore innocente,* which attracted considerable interest. In 1682 Scarlatti settled in Naples, his home for the next twenty years. There he became maestro di cappella to the Viceroy. He wrote prolifically for the stage, and his operas enjoyed immense favor both with royalty and the general public. Political disturbances in Naples sent Scarlatti to Rome in 1702. Through the influence of Cardinal Ottoboni, he was appointed assistant maestro di cappella of the church of Santa Maria Maggiore in 1703; four years later he became the maestro. In 1709 Scarlatti returned to Naples to resume his old post with the Viceroy. Except for occasional visits to Rome, he remained in Naples for the rest of his life. His pupils included Francesco Durante, Johann Adolph Hasse, and Nicola Logroscino —all later famous as composers.

The popularity of Scarlatti's operas, and the influence of his personality, shifted the center of Italian operatic activity from Venice to Naples. In Naples such characteristic forms as the Italian overture, the aria da capo, and the accompanied recitative were developed; a new importance was given to ensemble numbers, the chorus, and the orchestra. So widely were Scarlatti's operas imitated that Charles Burney, writing in the 1770's, declared: "I find part of his property among the stolen goods of all the best composers of the first forty or fifty years of the eighteenth century." The most celebrated of Scarlatti's hundred and some operas were: *Teodora* (1693); *Pirro e Demetrio* (1694); *L'Eraclea* (1700); *Mitridate Eupatore* (1707); *La principessa fedele* (1710); *Il Tigrane* (1715); *Il trionfo dell' onore* (1718); *Telemaco* (1718); *Griselda* (1721); *La virtù negli amori* (1721).

Alessandro's son, Domenico Scarlatti (1685–1757) is now remembered only for the five hundred and fifty sonatas he composed for the harpsichord. These fascinating works were all composed relatively late in Scarlatti's life. In his early years he wrote a dozen or so operas, all appreciated in their day. The most important were: *Ifigenia in Aulide* (1713); *Amleto* (1715); and *Narciso* (1720).

Scarlet Letter, The (1) novel by Nathaniel Hawthorne, the source of operas by Walter Damrosch and Vittorio Giannini. The setting is Boston in the seventeenth century. Hester Prynne is condemned to be an outcast and wear the scarlet letter "A" (for Adulteress) when she refuses to reveal the identity of her child's father. The father proves to be a young minister, Arthur Dimmesdale, who confesses, and receives his punishment at the scaffold. Hester commits suicide by drinking poison.

(2) Opera by Walter Damrosch. Libretto by George Parsons Lathrop, based on the Hawthorne novel. Première: Damrosch Opera Company, Boston, February 10, 1896.

(3) Opera by Vittorio Giannini. Libretto by the composer, based on the Hawthorne novel. Première: Hamburg State Opera, June 1, 1938.

Scarpia, chief of police (baritone) in Puccini's *Tosca.*

Scena, Italian for "scene," a term used in opera for a vocal number (generally, but not always, for one singer) more extended and dramatic character than an aria.

Scène des cheveux, *see* OH! QU'EST-CE QUE C'EST? . . . TES CHEVEUX.

Scène Infernale, orchestral excerpt from Prokofiev's *The Love for Three Oranges.*

Schalk, Franz, conductor. Born Vienna, Austria, May 27, 1863; died Edlach, Austria, September 2, 1931. His music study took place in Vienna with Anton Bruckner, among other teachers. In 1888 he conducted in Reichenbach. After appearances in Graz and Prague, he made his bow in Covent Garden in 1898. On December 14, 1898, he made his American debut at the Metropolitan

Opera, conducting *Die Walküre*. In 1900 he was engaged by the Vienna Opera, where he remained a principal conductor till the end of his life; from 1918 on, he was the institution's musical director (in collaboration with Richard Strauss between 1919 and 1924). He helped found the Salzburg Festival in 1920, and was one of its major conductors.

Schatz Walzer, waltz in Johann Strauss's *The Gypsy Baron*, known as *Treasure Waltz*.

Schau her, das ist ein Taler, Pedro's wolf narrative (Wolfserzählung) in Act I of D'Albert's *Tiefland*.

Schaunard, a musician (baritone) in Puccini's *La Bohème*.

Schauspieldirektor, Der (The Impresario), one-act opera by Mozart. Libretto by Gottlieb Stephanie. Première: Schönnbrunn, February 7, 1786. Originally, the text of this little comic opera was in two parts. The first, entirely dramatic, was filled with allusions to the theater of the day. In the musical part, an impresario is harassed by two temperamental prima donnas contending for the leading role in a new opera. The second part is the one performed today.

Scheff, Fritzi, soprano. Born Vienna, Austria, August 30, 1879; died New York City, April 8, 1954. While she achieved her greatest successes on the Broadway stage, particularly in Victor Herbert's operettas, she first received recognition in opera. Her mother, Hortense Scheff, was a member of the Vienna Opera. Fritzi Scheff attended the Hoch Conservatory in Frankfort-on-the-Main, then made her debut in that city in 1897 in *Roméo et Juliette*. She made her American debut at the Metropolitan Opera on December 28, 1900, as Marcellina in *Fidelio*. She was then heard in the Italian and French repertories, as well as in Wagner and Mozart operas. She abandoned opera in 1903 to appear in Victor Herbert's *Babette*, and in 1906 she achieved a

triumph in Herbert's *Mlle. Modiste*. She had a long and successful career in operettas, musical comedies, vaudeville, and the like.

Scherz, List und Rache (Jest, Cunning, and Revenge), *see* GOETHE, JOHANN WOLFGANG VON.

Schikaneder, Johann Emanuel, actor, singer, playwright, and impresario. Born Straubing, Austria, September 1, 1748; died Vienna, September 21, 1812. He was a member of a company of strolling players before settling in Vienna in 1784. Six years later he commissioned Mozart (whom he had previously befriended in Salzburg) to write an opera for his Theater-auf-der-Wieden. The opera, for which Schikaneder supplied the libretto, was *The Magic Flute*. Schikaneder created the role of Papageno. Later, opening another house, the Theater-an-der-Wien, he placed a statue of himself as Papageno on the roof. It was for this theater that the impresario commissioned Beethoven's *Fidelio*. Long successful, the talented Schikaneder ultimately died in poverty.

Schiller, Friedrich von, poet and dramatist. Born Marbach, Germany, November 10, 1759; died Weimar, Germany, May 9, 1805. An outstanding figure in German romantic literature, his many poetical dramas were a bountiful source of opera texts, among them: *Die Braut von Messina* (Fibich's *The Bride of Messina* and Nicola Vaccai's *La sposa di Messina*); *Die Bürgschaft* (operas by George Hellmesberger and Franz Lachner); *Don Carlos* (operas by Michael Costa and Verdi); *Fiesco* (Edouard Lalo's *Fiesque*); *Die Jungfrau von Orleans* (Balfe's *Joan of Arc*, Rezniček's *Die Jungfrau von Orleans*, Tchaikovsky's *Maid of Orleans*, Nicola Vaccai's *Giovanna d'Arco*, and Verdi's *Giovanna d'Arco*); *Kabale und Liebe* (Verdi's *Luisa Miller*); *Das Lied von der Glocke* (D'Indy's *Le chant de la* cloche); *Die Räuber* (Mercadante's *I briganti,* Verdi's *I masnadieri*); *Der Taucher* (opera by Reichardt); *Wallenstein* (operas by August von Adelburg and Jaromir Weinberger); *Wilhelm Tell* (Benjamin Carr's *The Archers,* Rossini's *William Tell*).

Schillings, Max von, conductor and composer. Born Düren, Germany, April 19, 1868; died Berlin, July 23, 1933. Trained for a career in science, he also studied music. While attending the Munich University, he was influenced by Richard Strauss and Ludwig Thuille to specialize in music. In 1892 he became an assistant conductor at the Bayreuth Festival, and a decade later chorus master. From 1908 to 1918 he was general music director of the city of Stuttgart; when, in 1912, a new opera house was opened, Schillings was given the honorary title of "von" by the king of Württemberg. In 1919 he succeeded Richard Strauss as musical director of the Berlin Opera. He resigned in 1925 after a dispute with the Prussian Ministry of Fine Arts. When the Nazis came into power in 1933, just before his death, he was appointed principal conductor of the Charlottenburg Opera in Berlin.

Schillings visited the United States for the first time in 1924, appearing as a conductor of German operas. He returned in 1931 as principal conductor of the German Grand Opera Company. His first opera, *Ingwelde,* was produced in Karlsruhe in 1894. It was more than two decades before he achieved a major success as a composer; it came with his opera *Mona Lisa,* given in Stuttgart on September 26, 1915, and soon thereafter performed extensively in Europe and the United States. Schillings' style was strongly influenced by Wagner. Besides the two operas already mentioned he wrote *Der Pfeifertag* (1899, revised 1931) and *Moloch* (1906).

In 1923 Schillings married the prima donna of the Berlin Opera, Barbara Kemp. She had created the title role in his *Mona Lisa.*

Schipa, Tito, tenor. Born Lecce, Italy, January 2, 1889. He attended the Lecce Conservatory, received vocal lessons in his fifteenth year, and when he was twenty-one studied voice with Emilio Piccoli in Milan. His debut took place in 1911 in Vercelli in *La traviata.* Success came in Rome, after which he appeared in leading opera houses of Europe and South America. In 1919 he was engaged by the Chicago Opera, where he remained until 1932. On November 23, 1932, he made his debut at the Metropolitan Opera in *L'elisir d'amore.* He remained at the Metropolitan until 1935, and returned during the 1940–1941 season. During this period he was often heard at La Scala and other major European opera houses. He made a concert tour of the United States in 1947. In 1955 he was a representative of the Italian government in an Italian opera festival held in the principal cities of Belgium.

Schirmer, G., Incorporated, one of the major music publishing houses in the United States, founded in New York City in 1861 by Gustav Schirmer and B. Beer. Schirmer acquired control five years later, and since then the company has been run first by Schirmer himself, then by members of his family. Its catalogue of American music is extensive and includes many American operas, among them: *Amahl and the Night Visitors; The Consul; Down in the Valley; The King's Henchman; Trouble in Tahiti.*

Schlafe wohl! und mag dich reuen, the good-night quartet of Lionel, Plunkett, "Martha," and "Julia" in Act II of Flotow's *Martha.*

Schlémil, Giulietta's lover (bass) in Offenbach's *The Tales of Hoffmann.*

Schlusnus, Heinrich, baritone. Born Braubach, Germany, August 6, 1888; died Frankfort-on-the-Main, June 19, 1952. He studied singing in Berlin with

Louis Bachner and in 1915 made his opera debut in Hamburg. From 1915 to 1917 he was a member of the Nuremberg Opera, and from 1917 on a principal baritone of the Berlin Opera. He also made many guest appearances in European opera houses and with the Chicago Opera. He was a noted concert artist.

Schmedes, Erik, dramatic tenor. Born Gjentofte, Denmark, August 27, 1866; died Vienna, Austria, March 23, 1931. He was trained as a pianist, but in 1888 Pauline Viardot-García heard him sing and advised him to specialize in singing. After studying with N. Rothmühl in Berlin and Désirée Artôt in Paris, he made his debut in Wiesbaden on January 11, 1891, in the baritone role of the Herald in *Lohengrin*. In 1894 he was engaged as first baritone of the Nuremberg Opera. After an additional period of study with Luise Ress in Vienna, he became a member of the Dresden Opera in 1896. He was advised by Bernhard Pollini to change the range of his voice to tenor. After training with A. Iffert in Dresden, he made his debut as tenor at the Vienna Opera on February 11, 1898, as Siegfried. He remained a principal tenor of the Vienna Opera until 1924. During this period he made many guest appearances in leading European opera houses and at the Bayreuth Festivals. On November 18, 1908, he made his American debut at the Metropolitan Opera as Siegmund; he remained there only a single season.

Schmelzlied, *see* NOTHUNG! NOTHUNG!
Schmiedelied, *see* HO-HO! SCHMIEDE, MEIN HAMMER.

Schnorr von Carolsfeld, Ludwig, tenor. Born Munich, July 2, 1836; died Dresden, July 21, 1865. He created the role of Tristan. After attending the Leipzig Conservatory, and studying privately with Eduard Devrient, he made his debut at the Karlsruhe Opera in Méhul's *Joseph.* After a brief period

there he was engaged by the Dresden Opera in 1860, where he remained five years. During this period, in 1862, he returned to the Karlsruhe Opera to appear in *Lohengrin,* making such an impression on Wagner that the latter decided to have him create the role of Tristan. Thus, Schnorr von Carolsfeld appeared in the world première of *Tristan und Isolde* in Munich on June 10, 1865, an occasion upon which his wife, Malwina Garrigues-Schnorr (1825–1904), created the part of Isolde. He died a little more than a month afterward of rheumatic fever.

Schoeck, Othmar, composer and conductor. Born Brünnen, Switzerland, September 1, 1886. He was the son of Alfred Schoeck, a famous painter, and for a while considered following in his father's footsteps. He decided on music in his seventeenth year. After studying at the Zurich Conservatory, and with Max Reger in Leipzig, he conducted various choral and orchestral groups in Switzerland. His first opera, *Don Ranudo,* was introduced in 1919. Subsequent compositions included several operas: *Das Wandbild* (1921); *Venus* (1922); *Penthesilea* (1927); *Massimilla Doni* (1937); *Das Schloss Durande* (1943). There was a Schoeck Festival in Berne, Switzerland, in April, 1934.

Schoenberg, Arnold, composer. Born Vienna, Austria, September 13, 1874; died Brentwood, California, July 13, 1951. The father of atonal music wrote several works for the stage. He began music study at the Realschule in Vienna, and continued it privately with Alexander Zemlinsky. His earliest works inspired a hostile reaction in Vienna in 1900. He began writing in an atonal style after 1908, and in this forbidding idiom he produced such iconoclastic works as *Erwartung* (1909), a monodrama, and the drama with music, *Die glückliche Hand* (1913). He began developing his twelve-tone technique in 1915, and by 1925 had codified a system which he espoused with the fervor of a prophet. The most important of his disciples in the field of opera composition were Alban Berg and Egon Wellesz.

Schoenberg left Europe immediately after the rise of Hitler and settled permanently in the United States. He became an American citizen in 1941. In this country he was a professor of composition at the University of Southern California and the University of California at Los Angeles. In America he wrote several major orchestral and choral works in which he tried combining his atonal writing with human values. In 1947 he received the Special Award of Merit from the National Institute of Arts and Letters. Besides the two stage works mentioned previously, Schoenberg wrote a one-act opera, *Von Heute auf Morgen* (1929), and an uncompleted two-act Biblical opera, *Moses and Aron* (1932).

Schoen-René, Anna, teacher of singing. Born Coblenz, Germany, January 12, 1864; died New York City, November 13, 1942. Her music study took place at the Berlin Royal Academy and privately with Pauline Viardot-García. In 1887 she made her debut at the Saxon-Altenburg Opera. She came to the United States in 1895 to appear at the Metropolitan Opera, but a serious illness prevented this, and she made no further stage appearances. After an additional period of study with Manuel García in England she concentrated on teaching singing, first in Minneapolis, and afterward for many years at the Juilliard School of Music. She was the author of *America's Musical Inheritance* (1941).

Schöffler, Paul, baritone. Born Dresden, Germany, September 15, 1897. After attending the Dresden Conservatory, he studied singing with various teachers, including Mario Sammarco in Milan. Fritz Busch heard him sing in 1925 and engaged him for the Dresden State Opera, where he appeared for the next twelve years. Since 1937 he has been the principal baritone of the Vienna State Opera. Since then he has been heard at the Bayreuth and Salzburg Festivals, at Covent Garden, and in the principal opera houses of Paris, Prague, Budapest, Brussels, Amsterdam, Milan, Rome, and Naples, particularly successful in the roles of Boris Godunov, Kurwenal, Iago, Figaro (in *The Marriage of Figaro*), Don Giovanni, and Amfortas. He has appeared in the world premières of *Cardillac, Schwanda, Capriccio, Dantons Tod,* and Wolf-Ferrari's *Sly.* He made his American debut at the Metropolitan Opera on January 26, 1949, as Jokanaan in Richard Strauss's *Salome,* and was acclaimed for his characterization as well as for his singing.

Schorr, Friedrich, dramatic baritone. Born Nagyvárad, Hungary, September 2, 1888; died Farmington, Connecticut, August 14, 1953. He combined the study of law with that of music, attending the University of Vienna, and studying singing privately with Adolph Robinson. During a brief visit to the United States in 1911 he appeared in minor roles with the Chicago Opera. His official debut in opera took place the same year in Graz, when he appeared as Wotan. After five years in Graz, he appeared with the National Opera in Prague and the Cologne Opera. In 1923 he was engaged as principal baritone of the Berlin Opera where, during the next decade, he established his reputation as one of the foremost Wagnerian baritones of his day. Meanwhile, in 1923, he returned to the United States as a member of the Wagnerian Opera Company, then touring America. On February 14, 1924, he made his bow at the Metropolitan Opera as Wolfram. He remained the principal German baritone of the Metropolitan for the next two decades.

His last appearance at the Metropolitan took place on March 2, 1943 as Wotan.

Schorr appeared at several Bayreuth Festivals, beginning in 1925, and at various times he made guest appearances with most of the major opera companies of Europe. In September 1938, he was appointed vocal advisor to the Wagner department of the Metropolitan Opera; in this capacity he guided many young Americans in Wagnerian traditions and style. After his retirement from the stage, he taught at the Hartt School of Music in Hartford, Connecticut, where he established an opera workshop. In 1950 he became advisor on German operas at the New York City Opera Company.

Besides his performances in all the principal baritone roles of the Wagnerian repertory, Schorr was heard at the Metropolitan Opera in the American premières of *Jonny spielt auf* and *Schwanda.*

Schott, Anton, dramatic tenor. Born Schloss Staufeneck, Bavaria, Germany, June 24, 1846; died Stuttgart, Germany, January 6, 1913. Before embracing music, he was an army officer. He began studying singing with Agnes Schebest-Strauss in 1871, and at the end of the same year made his debut at the Munich Opera. In 1872 he was engaged as leading lyric tenor of the Berlin Opera. In 1880 he made a successful London debut as Rienzi, and in 1882 he appeared in Italy with the Angelo Neumann troupe in the Wagner dramas. On November 17, 1884, he made his American debut at the Metropolitan Opera as Tannhäuser (the opening night of Leopold Damrosch's first season of German opera). He remained at the Metropolitan until 1887, and was subsequently heard in special guest performances in opera in Europe and South America, and on the concert stage.

Schotts Söhne, a music publishing house founded by Bernhard Schott in Mainz, Germany, in 1773. Long one of Germany's leading music houses, it was the first to issue the scores of Wagner's *Ring, Die Meistersinger,* and *Parsifal.*

Schreker, Franz, composer. Born Monaco, March 23, 1878; died Berlin, March 21, 1934. He attended the Vienna Conservatory and his first opera, *Flammen,* was written when he was only twenty. His failure to get it performed was so discouraging that for a while he abandoned composition for other musical activities. In 1911 he founded the Berlin Philharmonic Choir, with which he performed many new, provocative works; his espousal of the most advanced tendencies in music brought about his dismissal from the faculty of the Academy of Prussian Arts in 1913.

His first opera to be performed was *Der ferne Klang,* given in Frankfort-on-the-Main in 1912. It was a failure, denounced for its vigorous modern style and stark realism. Another opera, *Das Spielwerk und die Prinzessin,* created such a scandal when given in Vienna in 1913 that it had to be dropped. Recognition finally came with *Die Gezeichneten,* introduced in Frankfort in 1918. An even greater success followed in 1920, *Der Schatzgräber.* From this time on, Schreker's operas were given in leading European opera houses; there were over two hundred performances of his stage works in Austria and Germany before 1924. Schreker's musical style combined Wagnerian traits with a Debussyian impressionism. His texts, written by himself, ranged from naturalism to mysticism, and were strongly concerned with sexual psychology.

In 1920 Schreker became director of the Akademische Hochschule für Musik in Berlin, where his influence as a teacher was profound. He resigned this post in 1933 when the Nazis came to power. After *Der Schatzgräber,* Schreker wrote the following operas: *Irrelohe* (1924); *Der singende Teufel* (1928); *Christophorus* (1932); *Der Schmied von Gent* (1932).

Schröder-Devrient, Wilhelmine, soprano. Born Hamburg, Germany, December 6, 1804; died Coburg, Germany, January 26, 1860. She was guided to the stage from her childhood on, making public appearances until her seventeenth year, when she retired to concentrate on vocal study with J. Mazatti in Vienna. In 1821 she made a successful debut at the Burgtheater in Vienna as Pamina. After appearances in Prague and Dresden she scored a sensation as Leonore in *Fidelio* in Vienna (1822). In 1823 (the year she married her first husband, Karl Devrient, an actor) she was engaged as principal soprano of the Dresden Opera. She remained with this company for almost a quarter of a century. She appeared in the world premières of three Wagner operas, creating the roles of Adriano in *Rienzi,* Senta in *The Flying Dutchman,* and Venus in *Tannhäuser.* During this period she was also acclaimed in other European opera houses, making her Paris debut in 1830, her London debut in 1832. After leaving the opera stage, she made concert appearances throughout Germany.

Schubert, Franz Peter, composer. Born Vienna, Austria, January 31, 1797; died Vienna, November 19, 1828. The first of the great German romantics, and the father of the lied, was surprisingly ineffectual in opera. Most of his works designated as operas were actually singspiels, or operettas—plays with incidental songs. The few that can be characterized as operas were not performed in Schubert's lifetime and have rarely been heard since; they were burdened by ridiculous librettos and the composer's inability to bring to them any of the soaring inspiration found in other works. Schubert wrote the following complete operas: *Des Teufels Lustschloss* (1814); *Alfonso und Estrella* (1822); *Fierrabras* (1823). The following operas were not completed and exist only in fragments: *Adrast* (1819); *Die Bürgschaft* (1816); *Sakuntala* (1820). All Schubert's other stage works—*Die Zwillingsbrüder* (1820), *Die Zauberharfe* (1820), *Rosamunde* (1823), and a few more—are either singspiels, or plays with incidental music. Alfred Orel suggests the following explanation for Schubert's failure in opera: "The stage demands far and away coarser means of expression than the song, and these were unknown to Schubert. The dramatic accents of his works for the stage grow too much out of the lyrical; they do not breathe out the essential hot life of the dramatist. . . . Seen at the far greater distance of the theater audience, the delicate colors in which Schubert here works grow pale."

Schubert attended the school of the court chapel, after which for a while he earned his living as a schoolmaster. He produced his first masterpieces in the song form in his seventeenth year. He then abandoned teaching for composition, and for the remainder of his life was dependent upon the generosity of his friends for life's necessities. He lived in obscurity and poverty, in spite of which his musical production was prodigious. Recognition of his true stature as a composer did not come until many years after his death.

Schuch, Ernst von, conductor. Born Graz, Austria, November 23, 1846; died Dresden, Germany, May 10, 1914. He was a child prodigy who appeared on the concert stage both as violinist and pianist. After completing his music study in Vienna with Otto Dessoff, he was appointed in 1872 first conductor of the Dresden Opera, a post he held until the end of his life; in 1882 he was elevated to the position of musical director. He maintained the Dresden Opera on the highest artistic level throughout his regime, and was responsible for many significant premières, including those of most of the famous operas of Richard Strauss, Dohnányi's *Tante Simona*, Paderewski's *Manru*, and Wolf-Ferrari's *L'amore medico*. He was also responsible for introducing Puccini's operas to Germany. In 1897 he received a patent of nobility from the Emperor of Austria. In the spring of 1900 he visited the United States and led three concerts and a performance of *Lohengrin* at the Metropolitan Opera. His wife, Clementine Schuch-Proska, whom he married in 1875, was for many years a coloratura soprano of the Dresden Opera. Their daughter Lisa also appeared there for a while.

Schütz, Heinrich, composer. Born Köstritz, Saxony, October 8, 1585; died Dresden, November 6, 1672. His significance rests chiefly in his choral music; he was a towering figure in German music before the day of Johann Sebastian Bach. Schütz wrote one opera, *Dafne,* partly based on a translation of the Rinuccini libretto which had previously been used by the early Florentine opera composers. This opera has historic importance: it is the first with a German-language text, and is consequently the first German opera. Schütz wrote it in 1627 for the ceremonies attending the marriage of the daughter of the Saxon Elector to the Landgrave of Hesse-Darmstadt.

Schützendorf, Gustav, baritone. Born Cologne, Germany, 1883; died Berlin, April 27, 1937. He was born to a musical family; four of his brothers became opera singers. After music study in Cologne and Munich he made his debut in Düsseldorf in *Don Giovanni* in 1905. He then appeared extensively in German opera houses. On November 17, 1922, he made his American debut at the Metropolitan Opera in the role of Faninal. He remained with the Metropolitan until 1935, distinguishing himself in German roles, particularly as Beckmesser and Alberich.

Schuloper, German term for "school opera." This is a form of contemporary German opera which has an educational function. Hindemith's *Wir bauen eine Stadt* is such an opera.

Schumann, Elisabeth, soprano. Born Merseburg, Germany, June 13, 1885; died New York City, April 23, 1952. Her only teacher in singing was Alma Schadow in Hamburg. In 1910 she made her debut at the Hamburg Opera in a minor role in *Tannhäuser.* For the next seven years she appeared with that company in the German and Italian repertories. On November 20, 1914, she made her American debut at the Metropolitan Opera as Sophie in *Der Rosenkavalier,* one of her most celebrated roles. Schumann remained only a single season at the Metropolitan. At the recommendation of Richard Strauss she was engaged by the Vienna State Opera in 1919. She appeared there for the next two decades, a favorite of Viennese opera-goers. She appeared regularly at the Salzburg and Munich Festivals, as well as in leading European opera houses, achieving triumphs in operas by Mozart and Richard Strauss. She was decorated by the Danish government with the High Order for Art and Science, and was made honorary member of the Vienna Philharmonic Orchestra. She combined her career in the opera house with successes on the concert stage, where she was acclaimed as one of the foremost interpreters of lieder of her generation.

When Austria was annexed by Germany, she established her home permanently in the United States, becoming a citizen in 1944. In 1938 she became a faculty member of the Curtis Institute of Music, and at the time of her death was the head of the voice department.

Schumann, Robert, composer. Born

Zwickau, Germany, June 8, 1810; died Endenick, Germany, July 29, 1856. Like Schubert, Schumann was a giant figure in German romantic music whose contribution to opera was slight. He wrote only a single opera, *Genoveva*. Introduced in Leipzig on June 25, 1850, it received three performances. The opera has been rarely performed since then; a revival at the Florence May Music Festival of 1951 failed to stir new interest in it.

Schumann-Heink, Ernestine (born ERNESTINE ROSSLER), contralto. Born Lieben, near Prague, Bohemia, June 15, 1861; died Hollywood, California, November 16, 1936. She studied with Marietta von Leclair in Graz. Her concert debut took place when she was fifteen, in a performance of Beethoven's ninth symphony in Graz. On October 13, 1878, she made her opera debut at the Dresden Royal Opera as Azucena. Additional study now took place with Karl Krebs and Franz Wüllner. After marrying her first husband, Ernst Heink, in 1882, she was engaged by the Hamburg Opera, where she remained sixteen years, outstanding in the Wagnerian repertory. In 1892 she made her London debut as Erda. From 1896 to 1906 (except for 1904) she appeared regularly at the Bayreuth Festivals in the *Ring* cycle. From 1897 to 1900 she was heard at Covent Garden. In 1898 she signed a ten-year contract with the Berlin Opera. It gave her a leave of absence to appear in America, and on November 7, 1898, she made her American debut in Chicago as Ortrud. On January 9, 1899, she made her first appearance at the Metropolitan, once again as Ortrud. So great was her American success that when her leave of absence expired, she bought out her contract so that she might continue to appear at the Metropolitan. She remained at the Metropolitan until 1904. After leaving the Metropolitan, she appeared in leading German opera houses. In 1909 she created the role of *Klytemnestra* in Richard Strauss's *Elektra*. She now made periodic returns to the Metropolitan Opera: in 1909–1910, 1911–1913, 1915–1917, 1925–1926, 1928–1929, and 1931–1932. Her last Metropolitan appearance was as Erda on March 11, 1932. She had become an American citizen in 1908.

In 1926, the fiftieth anniversary of her Graz debut was celebrated at Carnegie Hall. This was followed by her last concert tour of the United States. In 1935 she appeared in a motion picture, *Here's to Romance*. Her opera repertory embraced some hundred and fifty roles, to which she brought not only exceptional vocal power and expressiveness, but a profound musicianship, pronounced dramatic feeling, and a striking temperament.

Schwanda der Dudelsackpfeifer (Schwanda the Bagpiper), folk opera by Jaromir Weinberger. Libretto by Miloš Kareš, based on a folk tale by Tyl. Première: Prague, April 27, 1927. Babinsky covets Dorota, wife of Schwanda. He induces Schwanda to try to win the heart of the wealthy Queen Ice-Heart with his magic pipings. Schwanda follows Babinsky's urging and wins the Queen; but when the latter discovers he is married, she orders his execution. Schwanda is able to elude death with his music, and with the aid of Babinsky's magic. In consequence of a rash oath, Schwanda is consigned to hell, from which he is rescued by Babinsky when the latter wins him in a card game with the devil. Babinsky finally recognizes that he can never win Dorota, and the married couple are happily reunited. The polka and fugue from this opera are well-known concert numbers.

Schwarzkopf, Elisabeth, soprano. Born Jarotschin, Poland, December 9, 1915. She attended the Berlin Hochschule für Musik, studying singing with Maria

Ivogün. Her debut took place on Easter Day, 1938, when she appeared as the first Flower Maiden in Parsifal at the Berlin Opera. After a period of additional study with Maria Ivogün, she established her reputation as a lieder singer, following a successful concert debut in Vienna in November, 1942. She then appeared in several guest performances in Mozart's operas at the Vienna State Opera. Acclaimed, she was engaged as a principal soprano. Her triumphs in Vienna were duplicated at Covent Garden and at the Bayreuth and Salzburg Festivals. In 1951 she appeared in the world première of Stravinsky's *The Rake's Progress* in Venice, and in 1953 in the première of Orff's *Trionfi* at La Scala. She has appeared extensively in America in song recitals and as soloist with orchestras.

schweigsame Frau, Die (The Silent Woman), comic opera by Richard Strauss. Libretto by Stefan Zweig, based on Ben Jonson's *Epicoene*. Première: Dresden Opera, June 29, 1935. The central character has such an aversion to noise that he decides to marry a silent woman. But immediately after the ceremony the silent woman suddenly finds her tongue. When the situation becoming intolerable, the victim tries to buy himself out of the marriage. He finally discovers that his "wife" is really a boy in disguise, and that a joke had been perpetrated on him.

Schweig' und tanze! Elektra's aria in the finale of Richard Strauss's *Elektra*.

Sciarrone, a gendarme (bass) in Puccini's *Tosca*.

Scindia, the king's minister (baritone) in Massenet's *Le Roi de Lahore*.

Scintille, diamant, Dapertutto's aria in Act II of Offenbach's *The Tales of Hoffmann*.

Scott, Sir Walter, poet and novelist. Born Edinburgh, Scotland, August 15, 1771; died Abbotsford, September 21, 1832. The originator of the English historical novel, Scott wrote many stories that have been used for operas, these being the most important: *The Bride of Lammermoor* (Michele Carafa's *La fiancée de Lammermoor,* Donizetti's *Lucia di Lammermoor,* Alberto Mazzucato's *La fidanzata di Lammermoor*); *The Eve of St. John* (opera by Mackenzie); *The Fair Maid of Perth* (Bizet's *La jolie fille de Perth*); *Guy Mannering* (opera by Henry Bishop, and Boieldieu's *La dame blanche*); *The Heart of Midlothian* (opera by Henry Bishop, Michele Carafa's *La prison d'Edimbourg,* Federico Ricci's *La prigione d'Edimburgo*); *Ivanhoe* (opera by Sir Arthur Sullivan, Nicolai's *Il Templario,* Marschner's *Der Templar und die Jüdin*); *Kenilworth* (Auber's *Leicester,* De Lara's *Amy Robsart,* Donizetti's *Il castello di Kenilworth*); *The Lady of the Lake* (Rossini's *La dame du lac*); *Maid Marian* (De Koven's *Robin Hood*); *Montrose* (opera by Henry Bishop); *Old Mortality* (Bellini's *I Puritani*); *Quentin Durward* (operas by Aleck Maclean and François Gevaert); *Rob Roy* (operas by Flotow and De Koven); *The Talisman* (opera by Balfe); *Waverly* (opera by Franz Holstein).

Scotti, Antonio, baritone. Born Naples, Italy, January 25, 1866; died there February 26, 1936. He was the principal baritone of the Metropolitan Opera for thirty-five years. After completing his studies with Mme. Trifari-Payanini and Vincenzo Lombardi, he made his debut in Malta in 1889 as Amonasro. Numerous appearances in Italy, Spain, Russia, Poland, and South America followed. He made his debut at La Scala in 1898 as Hans Sachs. In the fall of 1899 he made his American debut in Chicago, and on December 27 of the same year he made his first appearance at the Metropolitan Opera, as Don Giovanni. During the next thirty-five years Scotti was one of the principal members of the Metropolitan Opera company, outstandingly successful in such roles as Scarpia (which he created for America), Don Giovanni, Amonasro, Sharpless, Dr. Malatesta, Marcello, Falstaff, and Rigoletto. He appeared in the American premières of *Adriana Lecouvreur, Le donne curiose, Fedora, L'Oracolo, Tosca,* and De Lara's *Messaline.* He was featured in such significant Metropolitan premières and revivals as *Don Carlos, L'elisir d'amore, Falstaff, Iris, Lucrezia Borgia,* and *The Secret of Suzanne.* He was in the cast of *Rigoletto* when Caruso made his American debut; and he was the Scarpia for fifteen different Toscas.

Soon after World War I, Scotti formed his own company, the Scotti Opera Company, which he managed for four seasons in tours of the United States. On January 1, 1924, celebrating his twenty-fifth year at the Metropolitan, he appeared in a gala performance of *Tosca.* His last appearance at the Metropolitan took place on January 20, 1933, as Chim-Fen in *L'Oracolo,* a role he had created in 1905.

Scratch, the human form assumed by the devil (tenor) in Douglas Moore's *The Devil and Daniel Webster.*

Scribe, August-Eugène, dramatist and librettist. Born Paris, December 24, 1791; died there February 20, 1861. The most significant and at the same time most prolific of French librettists, Eugène Scribe wrote librettos that fill twenty-six volumes, while his more than three hundred plays require fifty volumes. The following are the most notable of the many operas for which he furnished librettos: Auber's *Fra Diavolo* and *La muette de Portici;* Bellini's *La sonnambula;* Boieldieu's *La dame blanche;* Cilèa's *Adriana Lecouvreur;* Donizetti's *La favorita;* Halévy's *La Juive;* Meyerbeer's *Les Huguenots, Le Prophète, Robert le Diable,* and *L'Africaine;* Rossini's *Le Comte d'Ory;* Verdi's *The Sicilian Vespers.* The librettos

of Donizetti's *L'elisir d'amore* and Verdi's *Un ballo in maschera* were based on stories by Scribe. In 1836 Scribe was made a member of the French Academy.

Scuoti quella fronda di ciliegio, Cio-Cio-San's and Suzuki's Flower Duet in Act II of Puccini's *Madama Butterfly.*

Sebastiano, rich landowner (baritone) in D'Albert's *Tiefland.*

Secondate, aurette amiche, duet of Ferrando and Guglielmo in Act II, Scene 2, of Mozart's *Così fan tutte.*

Second Hurricane, The, a children's opera by Aaron Copland. Libretto by Edward Denby. Première: Grand Street Theater, New York, April 21, 1937. The opera combines spoken scenes with songs, and was intended for performance by high-school children. The story concerns the efforts of four boys and two girls to bring flood victims food and aid. There is no curtain: a chorus drifts upon the stage, with it the principal characters. The head of the school appears to explain to the audience what is about to take place, and the opera begins.

Secret Marriage, The, *see* MATRIMONIO SEGRETO, IL.

Secret of Suzanne, The (Il segreto di Susanna), one-act opera (or intermezzo) by Ermanno Wolf-Ferrari. Libretto by Enrico Golisciani. Première: Munich Opera, December 4, 1909. American première: Philadelphia-Chicago Opera Company, playing in New York, March 14, 1911. The characters are the Countess Suzanne (soprano), Count Gil (baritone), and Sante, a servant (silent role). The setting is a living room in Piedmont around the turn of the century. Count Gil is disturbed, since every time he returns home he detects the odor of cigarette smoke. His suspicions that his wife is entertaining a lover mount as she evades answering his questions. In a fit of anger he smashes some furniture. His wife calms him by suggesting that he visit his club. The

Count goes, but decides to spy on his wife through a window. Thus, he learns the truth: that Suzanne, entirely faithful to him, is a secret smoker. Joyfully, he rushes into the room and joins her in a cigarette.

An astonishing contrast to the composer's tragic opera, *The Jewels of the Madonna*, *The Secret of Suzanne*, with its fresh and witty score, has remained one of the most successful of Wolf-Ferrari's several comic works for the stage.

Sedaine, Michel Jean, playwright and librettist. Born Paris, July 4, 1719; died there May 17, 1797. He published his first book in 1750, a volume of fables and songs; six years later he wrote his first opera libretto. Some of the best-known operas of his day were written to his librettos: Gluck's *Le Diable à quatre;* Gossec's *Les sabots et le cerisier;* Grétry's *Aucassin et Nicolette, Guillaume Tell, Raoul Barbe-Bleue,* and *Richard Coeur de lion;* Monsigny's *Aline, Le déserteur, Félix, Le roi et le fermier,* and *Rose et Colas;* Philidor's *Blaise le savetier* and *Le Diable à quatre.* At least one composer of the twentieth century has written an opera to a libretto by Sedaine: Henri Sauget, in his *La gageure imprévue.* Sedaine was elected to the French Academy in 1786.

Sedie, Enrico delle, see DELLE SEDIE, ENRICO.

Sedley, Mrs., a scandalmonger (contralto) in Britten's *Peter Grimes.*

Seefried, Irmgard, soprano. Born Köngetried, Swabia, Germany, October 9, 1919. She attended the Augsburg Conservatory and made her opera debut in her eleventh year as Gretel. Her first operatic engagement as a mature artist was with the Aachen Opera, where she was so successful that in 1943 she was engaged by the Vienna State Opera. She made her debut there as Eva in *Die Meistersinger* and became an immediate favorite. She now

appeared extensively in Europe, notably at the Salzburg and Edinburgh Festivals. She made her American debut as soloist with the Cincinnati Symphony in 1951, and her opera debut with the Metropolitan Opera on November 20, 1953, as Susanna in *The Marriage of Figaro.*

segreto di Susanna, Il, see SECRET OF SUZANNE, THE.

Seguidilla, a Spanish song and dance in triple time, usually in fast tempo, often accompanied with castanets. Carmen's aria "Près des remparts de Séville," in Act I of Bizet's *Carmen,* is a typical seguidilla.

Seidl, Anton, conductor. Born Budapest, Hungary, May 7, 1850; died New York City, March 28, 1898. One of the most noted of Wagnerian conductors, Seidl directed the American premières of several Wagner music dramas. He attended the Leipzig Conservatory, after which Hans Richter engaged him as chorus master of the Vienna Opera and introduced him to Wagner. In 1872 Seidl worked for Wagner in Bayreuth, copying the *Ring* cycle. He stayed with Wagner until 1876, assisting at the first Bayreuth festival. On Wagner's recommendation he became conductor of the Leipzig Opera in 1879, holding this post three years. He then became principal conductor of the Angelo Neumann company which presented Wagner's works throughout Europe until 1883. In that year he became principal conductor of the Bremen Opera.

Upon the sudden death of Leopold Damrosch in 1885, Seidl was invited to the United States to take over the German repertory at the Metropolitan Opera. He made his American debut on November 23, 1885, with *Lohengrin.* He remained a principal conductor of German operas at the Metropolitan until his death (except for the three-year period, beginning in 1892, when German opera took a secondary place in the repertory). During this time he led the American premières of *Die Meistersinger, Tristan und Isolde, Das Rheingold, Siegfried,* and *Die Götterdämmerung,* as well as the first cyclic performance of the *Ring.* From 1891 until his death he was also conductor of the New York Philharmonic Orchestra. In the spring of 1897 he directed at Covent Garden and in the summer of the same year he returned to Bayreuth after a prolonged absence. He died unexpectedly at the height of his career.

His wife, Augusta Seidl-Krauss, appeared with the Angelo Neumann company and with the Metropolitan Opera. She created for America the Wagnerian roles of Eva, Gutrune, and the Forest Bird.

Seien wir wieder gut, the composer's aria in Richard Strauss's *Ariadne auf Naxos.*

Se il mio nome saper, Count Almaviva's serenade in Act I, Scene 1, of Rossini's *The Barber of Seville.*

Seit er von dir geschieden, Waltraute's narrative in Act I, Scene 2, of Wagner's *Die Götterdämmerung.*

Selig, wie die Sonne, the quintet of Walther, Eva, Sachs, David, and Magdalena in Act III of *Die Meistersinger.*

Selika, African queen (soprano), the captive of Vasco da Gama, in Meyerbeer's *L'Africaine.*

Selim, the Pasha (speaking role) in Mozart's *The Abduction from the Seraglio.*

Sembrich, Marcella (born MARCELLINE KOCHANSKA), soprano. Born WISNIEWCZYK, Poland, February 18, 1858; died New York City, January 11, 1935. From her father, a concert violinist, she received her first music lessons when she was only four. Additional music study took place for four years at the Lemberg Conservatory. Upon Franz Liszt's advice, she decided to devote herself to singing, and studied with Viktor Rokitansky in Vienna and Francesco Lamperti in Milan. After marrying Wilhelm Stengel, one of her teachers at the Lemberg Conservatory, she went to Athens, where on June 3, 1877 she made her opera debut in *I Puritani.* Following more study with Richard Lewy in Vienna, she made her German debut in Dresden in October, 1878, in one of her most brilliant roles, Lucia. On June 12, 1880, she made her London debut at the Royal Italian Opera, once again as Lucia, and was so successful that she was invited back for four successive seasons. On October 24, 1883, she made her American debut at the Metropolitan Opera, singing the role of Lucia, and was acclaimed. From 1884 to 1898 she sang in the leading opera houses of Europe, and from 1898 to 1909 she was a principal soprano of the Metropolitan Opera. Her last appearance at the Metropolitan took place on February 6, 1909, when she appeared in acts from several operas. She now retired from the stage and devoted herself to teaching, first at the Curtis Institute of Music, later at the Juilliard School of Music. To her finest roles—Rosina, Violetta, Norina, Lucia, Gilda, Dinorah, Mimi, and Zerlina—she brought a surpassing beauty of tone and a brilliant technique. She was also a notable singer of lieder.

Semiramide, opera by Rossini. Libretto by Gaetano Rossi, based on Voltaire's drama. Première: Teatro la Fenice, Venice, February 3, 1823. Semiramis, Queen of Babylon, murders her husband, King Ninus, with the help of her lover, Assur. She subsequently falls in love with a handsome young warrior whom she believes to be a Scythian, but who is actually her son Arsaces. When she discovers his identity, she saves his life by receiving Assur's dagger blow intended for her son. Arsaces then kills Assur and ascends the throne. Rossini, who usually made his overtures independent of his operas, composed the overture of this opera from themes that are used in the body of the work. The overture is a familiar concert number.

The legendary Semiramis was a favorite heroine of the seventeenth and eighteenth centuries. Among the composers who made operas of her story were Caldara, Galuppi, Gluck, K. H. Graun, Hasse, Meyerbeer, Porpora, Sacchini, A. Scarlatti, and Vivaldi. Ottorino Respighi was also attracted to the story, composing an opera, *Semirama*, in 1910.

Semiseria, Italian for "semiserious," differentiating an opera with comic interludes from an opera seria.

Sempre libera, Violetta's aria in Act I of Verdi's *La traviata.*

Senesino (born FRANCESCO BERNARDI), male soprano. Born Siena, Italy, about 1680; died there about 1750. His stage name was derived from that of his native city. He studied with Antonio Bernacchi in Bologna, after which he appeared at the Dresden Opera. Handel heard him there and engaged him for London, where Senesino appeared for the first time in November, 1720, in Bononcini's *Astarto.* He was a sensation. For the next fifteen years he was the idol of the London opera public, creating in that time the principal roles in numerous Handel operas. He continued singing until 1733, when he quarreled with Handel and joined a rival company headed by Niccolò Porpora. He returned to Italy a wealthy man in 1735.

Senta, Daland's daughter (soprano) in Wagner's *Der fliegende Holländer.*

Senta's Ballad, *see* TRAFT IHR DAS SCHIFF.

Serafin, Tullio, conductor. Born Rottanova, Italy, December 8, 1878. He attended the Milan Conservatory. After playing the violin in the orchestra of La Scala he was engaged as a conductor of the Teatro Communale in Ferrara in 1900. In 1903 he conducted at the Teatro Regio in Turin, and in 1909 was engaged by La Scala, where he distinguished himself not only in the Italian repertory but also in operas by Wagner, Weber, and Gluck. His American debut took place at the Metropolitan Opera on November 3, 1924, with *Aïda.* He remained at the Metropolitan a decade, directing the Italian repertory, two Wagner operas, and the world premières of *The Emperor Jones, The King's Henchman, Merry Mount,* and *Peter Ibbetson.* He also conducted the American premières of *La cena delle beffe, The Fair at Sorochinsk, Giovanni Gallurese, Simon Boccanegra, Turandot,* and *La vida breve.* He left the Metropolitan in 1934 after a dispute with the management. He now became musical director of the Teatro Reale in Rome and later of La Scala. He also conducted in other European opera houses. He returned to America in the fall of 1952 to conduct the New York City Opera Company. His wife, the soprano Elena Rakowska, appeared in several European opera houses, and at the Metropolitan Opera between 1927 and 1930.

Seraglio, *see* ABDUCTION FROM THE SERAGLIO, THE.

Serena, (1) Robbins' wife (soprano) in Gershwin's *Porgy and Bess.*

(2) A Camorrist (soprano) in Wolf-Ferrari's *The Jewels of the Madonna.*

Serenade (Italian: SERENATA), most often, in operatic usage, a love song under a lady's window. Among the most famous operatic serenades are: "Im Mohrenland gefangen" (found in *The Abduction from the Seraglio*); "Ecco ridente in cielo" (*The Barber of Seville*); "Deh, vieni all finestra" (*Don Giovanni*); "Vous qui faites l'endormie" (*Faust*); "Aprila, o bella" (*The Jewels of the Madonna*); "Dans la cité lointaine" (*Louise*); "O Colombina" (*Pagliacci*); "Ah, lève-toi" (*Roméo et Juliette*); "Di rigori armato" (*Der Rosenkavalier*); "Deserto sulla terra" (*Il trovatore*).

Serge, Katerina's beloved (tenor) in Shostakovich's *Lady Macbeth of Mtsensk.*

Serov, Alexander, composer. Born St. Petersburg, Russia, January 23, 1820; died there February 1, 1871. He studied to be a lawyer but his friendship with Mikhail Glinka made him turn to music. After 1850 he wrote vigorous music criticisms in the Russian press. In 1857 he traveled to Germany and became an ardent Wagnerian. When he returned to Russia he posed himself as a vigorous opponent of musical nationalism. His first opera, *Judith,* produced in May, 1863, in St. Petersburg, was a major success, and did much to win support for Serov's musical position. His second opera, *Rogneda,* produced in 1865, was an even greater success, gaining the composer a handsome government pension.

While the style and esthetic approach of these operas were obviously Wagnerian, neither was, strictly speaking, a music drama in Wagner's sense of the term. The writing of such a work was now Serov's goal. In 1867 he composed *The Power of Evil,* but did not live to complete it; the last act was finished by N. T. Soloviev. The opera was produced on April 19, 1871, and for a while was very popular. Like Wagner, Serov wrote his own librettos.

Serpina, Uberto's maid (soprano) in Pergolesi's *La serva padrona.*

Serse (Xerxes), comic opera by Handel. Librettist unknown. Première: King's Theatre, London, April 15, 1738. Xerxes and Arsamenes, brothers, are in love with Romilda. Xerxes uses his royal power to win her, but fails. In the end he must satisfy himself with her sister, Atalanta, Romilda being united with Arsamenes. Handel's forty-third opera, this is the one in which the famous "Largo" appears. It is the opening aria, "Ombra mai fù," Xerxes' apostrophe to a shade tree. Though the opera is set in ancient Persia, Handel has included the tunes of London street cries in Elvira's song to the messenger.

serva padrona, La (The Servant Mistress), intermezzo, or opera buffa in two acts by Giovanni Pergolesi. Libretto by G. A. Federico. Première: Teatro San Bartolomeo, Naples, August 28, 1733.

Characters: Uberto, a bachelor (bass); Serpina, his maid (soprano); Vespone, Uberto's valet (silent role). The setting is Naples in the early eighteenth century.

Act I. A room in Uberto's house. Uberto is upset because Serpina has delayed bringing his chocolate. His anger grows when Serpina tells him she is in no hurry to obey his order ("Stizzoso, mio stizzoso"). Uberto now wants to take a walk, but Serpina announces firmly that if he does so she will lock him out of the house. His servant's effrontery convinces him that his house needs a mistress. Serpina agrees, insisting that she will become his wife. She enlists the aid of Uberto's valet.

Act II. The same scene. Serpina hides the valet in a closet. When Uberto appears, she tells him sadly that she has found another man and will no longer disturb Uberto with her marital designs. Uberto is curious about her lover, but her only reply is a lament that, surely, he will forget her completely when he himself gets married ("A Serpina penserete"). In an aside to the audience, Serpina expresses her belief that her intrigue is working. When Uberto is skeptical ("Son imbrogliato io già"), Serpina produces her supposed lover: Vespone disguised as a captain. Uberto is now convinced that all along he has wanted Serpina as a wife. Even after the identity of the captain is disclosed, Uberto is pleased at the turn of events.

La serva padrona is an intermezzo (which see), but it introduces so many of the stylistic elements afterward found in the Italian opera buffa that it may be considered the progenitor of this form. It is a work of exquisite perfection, the music catching every shade of comedy and burlesque, sentiment and poignancy, in its sparkling solos and duets. The work was tremendously popular in its day, and it exerted a wide influence on composers of comic operas. It is still occasionally performed, being an effective work for groups of limited resources.

Setti, Giulio, choral conductor. Born Traviglio, Italy, October 3, 1869; died Turin, Italy, October 2, 1938. After serving as chorus master in various opera houses in Italy, Cairo, Cologne, and Buenos Aires, he came to the United States in 1908 and was engaged as chorus master of the Metropolitan Opera. He remained in this post twenty-seven years, until his retirement in 1935, when he returned to Italy.

Se vuol ballare, Figaro's aria in Act I of Mozart's *The Marriage of Figaro.*

Sextet, *see* CHI MI FRENA.

Shadow Dance, *see* OMBRE LEGERE.

Shakespeare, William, poet and dramatist. Born Stratford-on-Avon, England, April, 1564; died there April 23, 1616. His comedies and tragedies have provided the material for a great number of operas, the following being the most significant. (Where opera titles are not given, they are the same as the play; composers identified by last names are treated elsewhere in this encyclopedia.)

Antony and Cleopatra: opera by Malipiero.

As You Like It: Francesco Veracini's *Rosalinda.*

The Comedy of Errors: Stephen Storace's *Gli equivoci;* Isa Krejci's *The Revolt at Ephesus.*

Coriolanus: opera by August Baeyens.

Cymbeline: opera by Arne Eggen; Kreutzer's *Imogène;* Edmond Missa's *Dinah;* Eduard Sobolevski's *Imogène.*

Hamlet: operas by Guiseppe Carcano, Luigi Caruso, Faccio, Francesco

Gasparini, Aristide Hignard, and Thomas.

Henry IV: Holst's *At the Boar's Head;* Giovanni Pacini's *La gioventù di Enrico V;* Verdi's *Falstaff* (with material from *The Merry Wives of Windsor*).

Henry VIII: opera by Saint-Saëns.

Julius Caesar: operas by Handel and Malipiero.

King Lear: opera by Frazzi.

Macbeth: operas by Bloch, Lawrence Collingwood, Nicholas Gatty, Karl Taubert, Verdi, and Lauro Rossi's *Biorn.*

Measure for Measure: Wagner's *Das Liebesverbot.*

The Merchant of Venice: operas by Fernand Brumagne, Hahn, Ciro Pinsuti, and Flor Alpaerts's *Shylock;* J. B. Foerster's *Jessica;* Otto Taubmann's *Porzia.*

The Merry Wives of Windsor: operas by Dittersdorf, Nicolai, Peter Ritter, and Adam's *Falstaff;* Balfe's *Falstaff;* Philidor's *Herne le chasseur;* Salieri's *Falstaff;* Vaughan Williams' *Sir John in Love;* Verdi's *Falstaff* (with material from Henry IV).

A Midsummer Night's Dream: operas by Dennis Arundell, Mancinelli, Thomas, Victor Vreuls, and Marcel Delannoy's *Puck;* Georges Huë's *Titania;* Purcell's *The Fair Queen;* John Christopher Smith's *The Fairies.*

Much Ado About Nothing: operas by Arpad Doppler, Stanford, and Berlioz' *Béatrice et Bénédict.*

Othello: operas by Rossini and Verdi.

Richard III: opera by Gaston Salvayre.

Romeo and Juliet: operas by Benda, Gounod, Pietro Carlo Guglielmi, Filippo Marchetti, Sutermeister, Nicola Vaccai, Zandonai, and Bellini's *I Capuletti e i Montecchi;* Conrado del Campo's *Los amantes de Verona;* Richard d'Ivry's *Les amants de Verone.*

The Taming of the Shrew: opera by Giannini, and Renzo Bossi's *Volpino il*

calderio; Goetz's *Der Widerspenstigen Zähmung;* Alick Maclean's *Petruccio;* Wolf-Ferrari's *Sly.*

The Tempest: operas by Atterberg, Fibich, Halévy, Lattuada, Reichardt, John Christopher Smith, Johann Zumsteeg.

Twelfth Night: Smetana's *Viola* (unfinished); Karl Taubert's *Cesario;* Karel Weis's *Viola.*

A Winter's Tale: opera by Goldmark, and Bruch's *Hermione;* Josef Nesvera's *Perditta.*

Shaklovity, a boyar (baritone) in Mussorgsky's *Khovantchina.*

Shaliapin, Feodor, *see* CHALIAPIN.

Shanewis (The Robin Woman), opera by Charles Wakefield Cadman. Libretto by Nelle Richmond Eberhart. Première: Metropolitan Opera, March 23, 1918. Mrs. Everton, a wealthy Californian, finances the musical career of the Indian girl Shanewis in New York. Shanewis meets and falls in love with Lionel Rhodes, fiancé of Mrs. Everton's daughter. Lionel wants to marry Shanewis. The girl is willing, if Lionel will first visit her on her reservation. After Shanewis returns to her tribe, Lionel follows and is fascinated by Indian customs. When Mrs. Everton and her daughter follow Lionel, Shanewis hears for the first time of his previous betrothal. She proudly rejects him. Finding Shanewis grief-stricken, her foster brother believes Lionel has deserted her. He kills Lionel with an arrow.

Filled with melodies and rhythms suggestive of Indian music, *Shanewis* is an early example of an American opera with an American setting. It was also the first American opera to survive more than a single season at the Metropolitan Opera.

Sharpless, the United States Consul (baritone) in Puccini's *Madama Butterfly.*

Shaw, George Bernard, dramatist and music critic. Born Dublin, Ireland, July

26, 1856; died Ayot-St.-Lawrence, England, November 2, 1950. The celebrated dramatist and wit, who won the Nobel Prize for literature in 1925, was a music critic before be began writing plays. In 1888 and 1889 he wrote criticisms for the London *Star* under the pen name of "Corno di Bassetto." From 1890 to 1894 his criticisms appeared in the London *World* over the initials G.B.S. An admirer of Wagner, he wrote a socialistic analysis of the *Ring* cycle entitled *The Perfect Wagnerite* (1898). His criticisms are collected in three volumes entitled *Music in London* (1932). His comedy *Arms and the Man* was the source of the popular operetta by Oscar Straus, *The Chocolate Soldier.* Ignace Lilien's *Die grosse Katharina* was derived from Shaw's *Great Catherine.*

Shepherds of the Delectable Mountains, The, a one-act opera by Ralph Vaughan Williams. Libretto by the composer, based on an episode in Bunyan's *The Pilgrim's Progress.* Première: London, July 11, 1922. Vaughan Williams subsequently incorporated this work into his opera *The Pilgrim's Progress.*

Sherasmin, Sir Huon de Bordeaux's squire (baritone) in Weber's *Oberon.*

Sheridan, Richard Brinsley, dramatist. Born Dublin, Ireland, October 30, 1751; died London, England, July 7, 1816. Three of his plays were the sources of a number of operas: *The Critic* (opera by Charles V. Stanford); *The Duenna* (operas by Ferdinando Bertoni, Robert Gerhard, Thomas Linley, and Serge Prokofiev's *The Duenna*); *The School for Scandal* (opera by Paul von Klenau).

She Stoops to Conquer, opera by George Alexander Macfarren. Libretto by E. Fitzball, based on Oliver Goldsmith's comedy of the same name. Première: Drury Lane, London, February 11, 1864. Marlow is a reticent young man who approaches his forthcoming meeting with Miss Hardcastle with foreboding; but he is an uninhibited lover when he mistakes her for a servant girl.

Shostakovich, Dmitri, composer. Born St. Petersburg, Russia, September 25, 1906. He attended the Glasser School of Music and the St. Petersburg Conservatory. Upon his graduation from the Conservatory he completed his first symphony, a work that made him famous throughout the world of music. Several failures followed this substantial success. One of these was an opera, *The Nose,* based on a Gogol story which, after its première in 1930, was officially denounced as "bourgeois" and "decadent." But several later compositions revived Shostakovich's popularity and added to his reputation; these included a ballet, *The Bolt,* and a concerto for piano and orchestra. On January 22, 1934, his opera *Lady Macbeth of Mtsensk* was introduced in Leningrad. It was a triumph, and for two years played to capacity houses. Then a violent attack was leveled against it from official quarters. For a time it almost seemed that this might be the end of the composer's career. But in 1937 Shostakovich succeeded in rehabilitating his position in Soviet music with his fifth symphony. In 1940 he won the Stalin Prize for his piano quintet. During World War II, as a defense worker in Leningrad, he was a public hero. He glorified the struggle of the Soviet people in several large works, among them his seventh symphony. But again Shostakovich became the object for official denunciation when, in 1948, the Central Committee of the Communist Party described some of his works as "formalistic" and "decadent." And once again he returned to grace, this time with an oratorio, *The Song of the Forests,* which won the Stalin Prize in 1949. In March, 1949, he briefly visited New York as a cultural emissary of the Soviet Union.

Shuisky, a prince (tenor) in Mussorgsky's *Boris Godunov.*

Siberia, opera by Umberto Giordano. Libretto by F. Civinni. Première: La Scala, December 19, 1903. Stephania, mistress of Prince Alexis, loves and is loved by Vassili. In a fit of jealousy, Alexis challenges Vassili to a duel and is wounded. For this, Vassili is exiled to Siberia. Stephania goes there to join him and share his fate. During an attempt at flight, Stephania is fatally wounded and Vassili is caught. Before she dies, Stephania is able to persuade the commandant of the post to free her beloved.

Siciliana, originally a Sicilian dance-song, in later times a vocal or instrumental piece in 12/8 or 6/8 time, generally in a minor key, and of moderate speed. A number of Handel's finest arias are basically sicilianas. A classic example of an instrumental siciliana is to be found in Gluck's *Armide.* The aria "O fortune, à ton caprice," in the finale of Act I of Meyerbeer's *Robert le Diable,* was designated a siciliana. Turiddu's "O Lola, bianca," in the opening of *Cavalleria rusticana* is another siciliana.

Sicilian Vespers, The (Les Vêpres Siciliennes; I Vespri Siciliani), opera in five acts by Verdi. Libretto by Eugène Scribe and Anne Honoré Duveyrier. Première: Paris Opera, June 13, 1855. The setting is thirteenth century Sicily, where the population rises in revolt against the occupying French. Against such a background, Elena, a Sicilian noblewoman and patriot, is in love with a commoner, Arrigo. But Arrigo turns out to be the son of Monforte, governor of Sicily, who is on the side of the French. When the governor consents to the marriage of Elena and Arrigo, she uses the wedding bells as the signal for the Sicilians to rise and massacre the French. The overture, made up of three principal themes from the opera, is the most popular part of the score.

Also popular are the bass aria "O tu Palermo" in Act II, the Ballet of the Seasons in Act III, and Elena's bolero, "Mercè, dilette amiche" in Act V.

Siebel, young man (mezzo-soprano) in love with Marguerite in Gounod's *Faust.*

Siegfried, (1) son (tenor) of Siegmund and Sieglinde in Wagner's *Siegfried* and *Die Götterdämmerung.*

(2) The third music drama in Wagner's *Der Ring des Nibelungen* (which see).

Siegfried's Death Music, the funeral music in Act III of Wagner's *Die Götterdämmerung.*

Siegfried's Rhine Journey, orchestral interlude between the prologue and Act I of Wagner's *Die Götterdämmerung.*

Sieglinde, Hunding's wife (soprano), sister of Siegmund, in Wagner's *Die Walküre.*

Sieglinde's Love Song, see DU BIST DER LENZ.

Siegmund, Sieglinde's brother (tenor) in Wagner's *Die Walküre.*

Siegmund's Love Song, see WINTERSTURME WICHEN.

Siepi, Cesare, bass. Born Milan, Italy, February 10, 1923. After studying with Chiesa he won first prize in a national singing contest in 1941. Three months later he made his opera debut as Sparafucile in Schio, near Venice. The war interrupted his career. His second debut took place in 1946 in *Nabucco,* the performance which reopened La Scala. He then sang throughout Italy, and was acclaimed in a performance of *Mefistofele* conducted by Arturo Toscanini. After appearances at the Edinburgh and Salzburg Festivals he made his American debut on the opening night of the 1950–1951 season of the Metropolitan Opera in *Don Carlos.* Siepi has made frequent television appearances in America. In Austria, he starred in a film version of *Don Giovanni.*

Si, fui soldato, Chénier's aria in Act III of Giordano's *Andrea Chénier.*

Sigismund, Polish king (bass) in Glinka's *A Life for the Czar.*

Signor Bruschino, Il, one-act comic opera by Rossini. Libretto by Giuseppe Foppa. Première: Teatro San Moïse, Venice, January 1813. In eighteenth century Italy, Sofia, ward of Gaudenzio, is about to be forced to marry Bruschino's son, even though she loves Florville. When Bruschino's son is imprisoned for debt, Florville impersonates him and is able to win Gaudenzio's consent to his marriage.

Sigurd, opera by Ernest Reyer. Libretto by Camille du Locle and Alfred Blau. Première: Théâtre de la Monnaie, Brussels, January 7, 1884. The text, like Wagner's *Ring* cycle, is based on the Nibelung sagas.

Si; io penso alla tortura, Madeleine's aria in Act I of Giordano's *Andrea Chénier.*

Si la rigueur et la vengeance, Cardinal Brogny's prayer in Act I of Halévy's *La Juive.*

Silva, Don Ruy Gomez de, Spanish grandee (bass) in Verdi's *Ernani.*

Silvio, a villager (baritone) in Leoncavallo's *Pagliacci.*

Si, me ne vo, Contessa, Gérard's aria in Act I of Giordano's *Andrea Chénier.*

Simon Boccanegra, opera by Verdi. Libretto by Francesco Maria Piave (later revised by Arrigo Boïto), based on a play by Antonio Garcia Gutiérrez. Première: Teatro La Fenice, Venice, March 12, 1857. The setting is fourteenth century Italy, when Genoa was a republic. Boccanegra, a pirate, becomes the tyrannical ruler of Genoa. He recovers his long-lost daughter, Amelia, now in love with Adorno, a nobleman. Boccanegra objects to their marriage. Not knowing that Boccanegra is Amelia's father, Adorno conspires to poison him. Before he dies, Boccanegra gives his belated consent to the marriage. The opera's most significant arias are Amelia's "Come in quest' ora bruna" in Act I, Fiesco's "Il lacerato spirito" (with chorus), also in Act I, and Adorno's "O Inferno, Amelia" in Act II.

Simone, Donati's cousin (bass) in Puccini's *Gianni Schicchi*.

Singher, Martial, baritone. Born Oloron-Sainte-Marie, France, August 14, 1904. He attended the Paris Conservatory. In 1930 he made his debut at the Paris Opéra, where he remained a principal baritone until 1941. During this period he was a guest of major European opera companies. He made a successful American debut at the Metropolitan Opera on December 10, 1943, in *The Tales of Hoffmann*. He has since then been acclaimed at the Metropolitan in the French repertory. He is the only singer to have sung the role of Pelléas as well as that of Golaud in *Pelléas et Mélisande*. A noted concert artist, he has also appeared in opera at the Teatro Colón in Buenos Aires, at Covent Garden, and at several Florence Festivals.

Singspiel, an early German form of comic opera, established in the middle of the eighteenth century by Johann Adam Hiller. Its chief characteristic is the use of spoken dialogue instead of recitatives. Hiller derived the form from the French comic theater, giving it a German personality by making German popular and folk songs an element of his music. Early singspiel composers were Johann Mattheson and Johann Friedrich Reichardt. The tradition was carried on by Mozart in his *The Abduction from the Seraglio* and *The Magic Flute*, and to a certain degree by Beethoven in his *Fidelio*.

Si, pel ciel marmoreo giuro! Otello's and Iago's vow of vengeance in Act II of Verdi's *Otello*.

Si può? Tonio's prologue in Leoncavallo's *Pagliacci*.

Sir John in Love, opera by Ralph Vaughan Williams. Libretto by the composer, based on Shakespeare's *The Merry Wives of Windsor*. Première:

London, March 21, 1929. The score makes notable use of English folk songs, including "Greensleeves."

Sirval, Marchioness de, Charles's mother (mezzo-soprano) in Donizetti's *Linda di Chamounix*.

Sister Angelica, see SUOR ANGELICA.

Sita, the woman (soprano) loved by the king of Lahore in Massenet's *Le Roi de Lahore*.

Si, vendetta, duet of Rigoletto and Gilda in the finale of Act II of Verdi's *Rigoletto*.

Six, Les (The Six), a group of six French composers who emerged to fame just after World War I. They were Georges Auric, Louis Durey, Germaine Tailleferre, Arthur Honegger, Darius Milhaud, and Francis Poulenc. The last three have made significant contributions to the art of opera. The name for the group was suggested by the critic Henri Collet in the periodical *Comoedia*, issue of January 16, 1920, when he reviewed the composers' album of piano pieces and likened the composers to the Russian school known as the "Five."

Slezak, Leo, tenor. Born Mährisch-Schönberg, Moravia, August 18, 1873; died Egern on Tegernsee, Bavaria, June 1, 1946. Described as "the second Tamagno," he was one of the most celebrated tenors of the early twentieth century. He planned to become an engineer, but while engaged in technical studies, decided to develop his voice. He took only a few lessons before joining the chorus of the Brünn Opera. He made his debut at the Brünn Opera in *Lohengrin* on March 17, 1896. After appearances at the Berlin Opera and at Covent Garden he received from the Vienna Opera a seven-year contract, making his debut there in *William Tell*. For a quarter of a century, Slezak was a principal tenor of the Vienna Opera, the idol of the Viennese. In 1908 he temporarily retired to study with Jean de Reszke. He returned to the opera stage triumphantly in London in May, 1909, as Otello. It was in this role that he made his American debut at the Metropolitan Opera on November 17, 1909. Slezak remained at the Metropolitan through the 1912–1913 season, outstandingly successful in the Wagner repertory. He also distinguished himself on the concert stage. He was appearing in recitals in Russia when World War I broke out. He escaped to Germany, where he joined the army and saw action. After the war, he continued his career at the Berlin Opera and the Vienna State Opera. After retiring from opera he continued to appear as a concert singer. He was the author of several volumes of reminiscences, one of which was published in English as *Song of Motley* (1938). His son, Walter Slezak, has been successful on the American musical-comedy stage and in motion pictures.

Smallens, Alexander, conductor. Born St. Petersburg, Russia, January 1, 1889. His musical training took place at the Institute of Musical Art in New York and at the Paris Conservatory. In 1911 he became assistant conductor of the Boston Opera Company, and soon after, first conductor of another company, the Boston National Opera. For two years he conducted the Anna Pavlova ballet troupe, touring the United States and South America. During this period he became the first North American conductor to direct operas at the Teatro Colón (Buenos Aires). In 1919 he was appointed first conductor of the Chicago Opera. He remained four seasons, conducting the premières of Reginald De Koven's *Rip Van Winkle* and Prokofiev's *The Love for Three Oranges*. During this period he also conducted opera in Europe. In 1924 he became musical director of the Philadelphia Civic Opera; in 1934 he began leading opera performances in Lewisohn Stadium in New York. He conducted the première of *Porgy and*

Bess (1935), a work that has ever since been identified with his direction; between 1952 and 1955 he led the American company that presented Gershwin's opera in a triumphant tour of Europe and the Near East. Other notable operas that Smallens has conducted have been Hindemith's *Hin und zurück,* Malipiero's *Sette canzoni,* Richard Strauss's *Feuersnot,* Stravinsky's *Mavra,* and Virgil Thomson's *Four Saints in Three Acts.* The last he led at its first performance.

Smanie implacabili, Dorabella's aria in Act I, Scene 3, of Mozart's *Così fan tutte.*

Smetana, Bedřich, composer. Born Litomischl, Bohemia, March 2, 1824; died Prague, May 12, 1884. The most significant composer of Bohemian national operas, he had little systematic musical training until his twenty-first year, though he interested himself in musical activities from childhood on. In his nineteenth year he fell in love with Katharina Kolař, who convinced him that he should turn to music seriously. He went to Prague in 1843 and became a pupil of Josef Proksch. A year later he was engaged as music teacher by Count Leopold Thun, holding this post four years. In 1848 he helped organize the first significant music school in Prague. In 1849 he married Katharina Kolař. A year later he was appointed pianist to the former Emperor of Austria, Ferdinand I, then residing in Prague. At the same time he began writing orchestral and chamber works. From 1856 to 1861 he lived in Gothenburg, Sweden, where he taught, played the piano, and conducted the city's orchestra. He interrupted his stay in 1859 with a return to Bohemia. His wife died during this trip. Remarrying in Prague, he returned to Gothenburg for an additional two years.

In 1861, again in his native land, Smetana assumed a dominating position in its musical life. He became director of the music school in Prague; led an important orchestra and chorus; wrote music criticisms in which he espoused the cause of Bohemian music; founded and directed a dramatic school for the Bohemian Theater in Prague; helped organize the Society of Artists. He did not neglect composition. In 1863 he completed his first opera, *The Brandenburgers in Bohemia,* which may be considered Bohemia's first major national opera. Produced in Prague on January 5, 1866, it was highly successful. His second opera, *The Bartered Bride,* was one of the great folk operas of all time. At its première (May 30, 1866) it was not well received, but after its third performance it was acclaimed. Smetana later wrote two other excellent folk operas: *Dalibor* (1868) and *Libussa* (1871).

After 1874 Smetana was afflicted with deafness. Despite this infirmity he continued producing important music, including his cycle of national tone poems entitled *My Country* (*Má Vlast*), one of which is *The Moldau.* He also wrote the following operas: *Two Widows* (1874); *The Kiss* (1876); *The Secret* (1878); *The Devil's Wall* (1882); and *Viola* (unfinished). His last complete opera, *The Devil's Wall,* was severely criticized at its première (though it later became popular), a disappointment which precipitated the composer's breakdown. Becoming insane in 1883, he was confined, dying a year later.

Smithers, Henry, a cockney trader (tenor) in Gruenberg's *The Emperor Jones.*

Smyth, Ethel, composer. Born London, England, April 23, 1858; died Woking, England, May 8, 1944. She attended the Leipzig Conservatory and soon after her graduation received recognition for her chamber music and a mass. A one-act opera, *Der Wald* (*The Forest*) was produced in Dresden in 1901. A three-act opera, *The Wreckers* (*Les Naufragés*) was introduced in Leipzig in 1906. Her most famous work, *The Boatswain's Mate,* was produced in London in 1916. These works placed her in the front rank of English opera composers of the early twentieth century. She also wrote the following operas: *Fantasio* (1898); *Fête galante* (1923); *Entente cordiale* (1925). An authoritative conductor of her own works, she also became known as a leader of the woman-suffrage movement. In 1920 she was made Dame of the British Empire. She wrote several autobiographical volumes, the best-known being *Impressions that Remained* (1919).

Snow Maiden, The (Snegurochka), opera by Rimsky-Korsakov. Libretto by the composer, based on a play by Alexander Ostrovsky, in turn derived from a folk tale. Première: Moscow, February 10, 1882. The composer subtitled this opera "a legend of springtime." The setting is the land of Berendeys in prehistoric times. The Snow Maiden is safe from death by the sun's rays only so long as she is innocent of love. She wants to live the life of a mortal, and is placed in the care of two villagers. There, her life is complicated by the fact that the merchant Mizguir falls in love with her, deserting his sweetheart. Ultimately the Snow Maiden falls in love with Mizguir. The sun touches her and she disappears. Grief-stricken, Mizguir commits suicide. The Dance of the Tumblers (or Buffoons) in Act III is a well-known excerpt.

So anch' io la virtù magica, Norina's aria in Act I, Scene 2, of Donizetti's *Don Pasquale.*

Sobinin, Antonida's lover (tenor) in Glinka's *A Life for the Czar.*

Socrate, *see* SATIE, ERIK.

Sodero, Cesare, conductor. Born Naples, Italy, August 2, 1886; died New York City, December 16, 1947.

He studied with Giuseppe Martucci, and graduated from the Naples Conservatory when he was only fourteen. After touring Europe as cellist he came to the United States in 1906, and for seven years directed various American opera companies, including the Chicago Opera. From 1913 to 1925 he was general music director of the Edison Phonograph Company. He turned to radio in 1925, and achieved significance as a pioneer in broadcasting operas. In 1926 he directed a series of fifty-three operas in tabloid form for NBC. From 1926 to 1934 he conducted several hundred symphony concerts over NBC. He then became musical director of the Mutual network. In 1942 he became a principal conductor of the Metropolitan Opera, making his debut there on November 28 with *Aïda*. He remained with the Metropolitan until his death. He wrote an opera, *Ombre Russe,* broadcast by NBC in 1929, and given its stage première in Venice in 1930.

So Do They All, *see* COSI FAN TUTTE.

Sofia, Gaudenzio's ward (soprano) in Rossini's *Il Signor Bruschino.*

Sogno soave e casto, Ernesto's aria in Act I, Scene 1, of Donizetti's *Don Pasquale.*

Sola, perduta, abbandonata, Manon's closing aria in Act IV of Puccini's *Manon Lescaut.*

Soldiers' Chorus, (1) *see* GLOIRE IMMORTELLE DE NOS AIEUX.

(2) See SQUILLI, ECHEGGI LA TROMBA.

Solenne in quest' ora, duet of Alvaro and Carlo in Act III, Scene 2, of Verdi's *La forza del destino.*

Solomon, the Hebrew king (baritone) in Goldmark's *The Queen of Sheba.*

Sommeil, French for "sleep"—in old French operas, a term signifying a quiet instrumental piece accompanying a scene of slumber or dreaming. An example occurs in Rameau's *Dardanus.*

Song of India, the Hindu's aria in Act II, Tableau 2, of Rimsky-Korsakov's *Sadko.*

Song of Roxane, Roxane's aria in Act II of Szymanowski's *King Roger.*

Song of the Flea, *see* CHANSON DE LA PUCE.

Song of the Gnat, the Nurse's song in Act II, Scene 2, of Mussorgsky's *Boris Godunov.*

Song of the Viking Guest, song in Act II, Tableau 2, of Rimsky-Korsakov's *Sadko.*

Song-speech, *see* SPRECHSTIMME.

Son imbrogliato io già, Uberto's aria in Act II of Pergolesi's *La serva padrona.*

sonnambula, La (The Sleepwalker), opera by Vincenzo Bellini. Libretto by Felice Romani. Première: Teatro Carcano, Milan, March 6, 1831. American première: Park Theater, New York, November 13, 1835. In a Swiss village, in the nineteenth century, Amina, fiancée of Elvino, is a sleepwalker. Their courtship becomes complicated when, in her sleep, Amina enters the window of a stranger, where she is discovered by her lover. Certain that Amina is unfaithful, Elvino denounces her and turns his affection to Lisa, whom he is now ready to marry. Count Rodolfo, whose room Amina entered, later proves to Elvino that Amina is innocent by pointing her out to him as she walks on a roof in her sleep. Amina is safely awakened by her lover, and all ends happily. The opera's most famous arias are Rodolfo's "Vi ravviso, o luoghi ameni" in Act I, Amina's "Ah! non credea mirarti" in Act III, and her "Ah! non giunge" at the end of the opera.

Sonnez, clairons, que vos chants de victoire, the chorus of homage in Act III of Halévy's *La Juive.*

Sontag, Henriette (born GERTRUD WALBURGA SONNTAG), soprano. Born Coblenz, Germany, January 3, 1806; died Mexico City, June 17, 1854. The daughter of actors, she made her stage debut at the age of six. After attending the Prague Conservatory, she made her opera debut in her fifteenth year as a last-minute replacement for an indisposed prima donna in Prague. In 1820 she appeared in German and Italian roles at the Vienna Opera. Weber was so impressed by her that he engaged her for *Euryanthe,* in which opera she created the title role in 1823 with outstanding success. She sang the soprano parts in the premières of Beethoven's ninth symphony and *Missa Solemnis.* On August 13, 1825, she made her debut in Berlin in *L'Italiana in Algeri.* Soon after, she made her bows in Paris and London in *The Barber of Seville.*

After a secret marriage to Count Rossi, a Sardinian diplomat, she retired from opera in 1830, but continued to appear in concerts. Her husband eventually resigned his post in order to acknowledge his marriage, and to follow his wife as she resumed her career. Sontag was once again acclaimed in the leading opera houses of London, Paris, and Germany. In 1852 she made a triumphal tour of the United States. She was singing in Mexico, when she was fatally stricken with cholera.

Sonzogno, Edoardo, publisher. Born Milan, Italy, April 21, 1836; died there March 14, 1920. He founded the music publishing house of Sonzogno in 1874, after inheriting his father's printing plant and bookstore. This house became celebrated in the field of opera by sponsoring contests for one-act operas, the first in 1883; *Cavalleria rusticana* won the prize in 1888. The house specialized in publishing cheap editions of old Italian music. From 1861 to 1909 Sonzogno was the owner and director of the newspaper *Il Secolo.* In 1894 he established a theater in Milan, the Lirico Internazionale.

Sophie, (1) Charlotte's sister (mezzo-soprano) in Massenet's *Werther.*

(2) Janusz' beloved (contralto) in Moniuszko's *Halka.*

(3) Herr von Faninal's daughter (soprano) in Richard Strauss's *Der Rosenkavalier.*

Sophocles, poet and dramatist. Born Colonus, Greece, about 496 B.C.; died, place unknown, 406 B.C. One of the great tragic poets and dramatists of ancient Greece, Sophocles wrote a number of dramas which were made into operas, notably: *Antigone* (operas by Arthur Honegger, Carl Orff, Menelaos Pallantios, Niccolò Zingarelli); *Elektra* (operas by Johann Haeffner, Jean Lemoyne, Richard Strauss); *Oedipus at Colonus* (operas by Charles Radoux-Rogier, Antonio Sacchini, Zingarelli); *Oedipus Tyrannus* (operas by Georges Enesco, Leoncavallo, Stravinsky).

Soprano, the highest female voice, normally ranging a little more than two octaves upward from the B-flat below middle C. Soprano voices are classified as dramatic, lyric, and coloratura; the latter, besides possessing an agility not required of the other types, having a compass of two octaves and a fourth above middle C. The term soprano was also given to the high male voices that used to sing women's parts in operas of the eighteenth century (*see* CASTRATO).

Soprano acuto, Italian for "high soprano."

Soprano Falcon, a type of dramatic soprano associated with such operatic roles as Rachel and Valentine. It was named after the singer Marie-Cornélie Falcon (1812–1897).

Soprano leggiero, a light, or agile, soprano.

Soprano sfogato, a high, thin soprano.

Sorgeva il dì del bosco in seno, *see* JOUR NAISSAIT DANS LE BOCAGE, LE.

Sorte amica, chorus of Sicilian knights in Act I of Meyerbeer's *Robert le Diable.*

Sortita, Italian for "coming out," an eighteenth century operatic term referring to the initial appearance and initial aria of the singer.

Sotto voce, Italian for "under the voice," in vocal music a direction to sing with a toneless quality, or in an undertone.

Soubrette, a French term designating, in opera, a young comedienne with a light soprano voice. Typical soubrette roles are those of Serpina, in *La serva padrona;* Despina, in *Così fan tutte;* Susanna, in *The Marriage of Figaro.*

Souviens-toi de ton serviteur, Samson's prayer in Act III, Scene 2, of Saint-Saëns' *Samson et Dalila.*

Spalanzani, scientist and inventor (tenor) in Offenbach's *The Tales of Hoffmann.*

Sparafucile, the assassin (bass) in Verdi's *Rigoletto.*

Spargi d'amaro pianto, Lucia's aria in Act III, Scene 1, of Donizetti's *Lucia di Lammermoor.*

Specht, Richard, critic and musicologist. Born Vienna, December 7, 1870; died there March 18, 1932. He studied architecture, but was advised by Brahms and Goldmark to turn to music criticism. In 1895 he became the critic of *Die Zeit,* and from 1908 to 1915 he was the critic of *Die Musik.* In 1909 he founded the musical journal *Der Merker.* Among his books are several on operatic subjects: *Gustav Mahler* (1906, revised 1913); *Johann Strauss* (1911); *Das Wiener Operntheater* (1919); *Richard Strauss* (1921); *Julius Bittner* (1921); *Wilhelm Furtwaengler* (1922); *E. von Rezniček* (1923); *Puccini* (1931).

Spinelloccio, the doctor (bass) in Puccini's *Gianni Schicchi.*

Spinning Chorus, *see* SUMM' UND BRUMM'.

Spira sul mare, Cio-Cio-San's first aria in Act I of Puccini's *Madama Butterfly.*

Spirito gentil, Fernando's aria in Act IV of Donizetti's *La favorita.*

Splendon più belle, aria of Baltasar with chorus of monks in Act IV of Donizetti's *La favorita.*

Spohr, Ludwig, violinist, composer, conductor. Born Brunswick, Germany, April 5, 1784; died Kassel, October 22, 1859. He received violin instruction from his seventh year, and in 1802 he began concertizing in Germany, achieving recognition as virtuoso two years later. In 1805 he became conductor of the ducal orchestra at Gotha, beginning an eventful career as conductor. In 1817 he went to Frankfort-on-the-Main to direct opera performances, and while there, led the premières of two of his operas: *Faust* and *Zemire und Azore.* In 1820, as a guest conductor of the Royal Philharmonic in London, he made conducting history by directing the orchestra with a baton; earlier performances of that orchestra had been led either by the concertmaster (while playing his violin) or by the piano player.

In 1822 Spohr became the director of the Kassel Court Theater, where he remained thirty-five years. A champion of Wagner, he led performances of *Der fliegende Holländer* in 1842 and *Tannhäuser* in 1853. He directed the première of his most important opera, *Jessonda,* in 1823. *Jessonda* has historical importance as one of the earliest German operas to use accompanied recitatives throughout instead of spoken dialogue. His operas: *Die Prüfung* (1806); *Alruna* (1808); *Die Eulenkönigin* (1808); *Der Zweikampf mit der Geliebten* (1811); *Faust* (1816, revised 1852); *Zemire und Azore* (1819); *Jessonda* (1823); *Der Berggeist* (1825); *Pietro von Abano* (1828); *Der Alchemist* (1830); *Die Kreuzfahrer* (1845).

Spoletta, a police agent (tenor) in Puccini's *Tosca.*

Spontini, Gasparo, composer. Born Majolati, Italy, November 14, 1774; died there January 24, 1851. His parents intended him for priesthood, but he preferred music. In 1793 he entered the Conservatorio de' Turchini in

Naples. He showed such promise as a student that the director of the Argentina Theater in Rome commissioned him to write an opera. Since conservatory students were forbidden to accept such commissions, he left and wrote *I puntigli delle donne*, which was such a success that the conservatory director allowed him to return. He continued to write operas, many of them comic. Among the most successful of these was *L'eroismo*, performed in many Italian theaters.

In 1803 Spontini went to Paris. His association with leading French composers led him to abandon his light style for a more serious one. On December 6, 1807, he was acclaimed for *La Vestale*, which had taken him three years to write, and which leading French musicians hailed as a masterwork; it won a prize for dramatic composition, the unanimous decision of the judges. The opera immediately became a fixture in the repertory of the Paris Opéra, receiving over two hundred performances by 1830; as early as 1828 it was given in the United States. To this day it remains the most frequently revived of Spontini's operas. Another substantial success followed on September 28, 1809, *Fernand Cortez*. A year later Spontini became conductor of Italian operas at the Théâtre de l'Impératrice, but was dismissed in 1812 because of differences with the director. In 1814 he became court composer for Louis XVIII, a post in which he wrote several operas glorifying the restoration of the Bourbons.

The failure of his *Olympie* in 1819 was such a blow to his pride that he left Paris for Berlin, where he was appointed general music director by Friedrich Wilhelm III. His operas in Berlin were not successful. Increasingly bitter, Spontini became involved in altercations with his patrons and co-workers. He was finally compelled to resign his post in 1841. For a while

he lived in Paris, but on an invitation from Wagner, he went to Dresden in 1844 to direct *La Vestale*. Toward the end of his life, Spontini lived in his native city, devoting himself to charity. His last years were marked by failing memory and hearing. He was the recipient of many honors, including the title of Conte de Sant' Andrea from the Pope, the knighthood of the Prussian Order of Merit, and membership in the Berlin Academy and the French Institute. His best operas, after those already mentioned, were: *La finta filosofa* (1799); *La fuga in maschera* (1800); *Milton* (1804); *Nurmahal* (1822); *Alcidor* (1825); *Agnes von Hohenstaufen* (1827, revised 1837).

Sportin' Life, dope peddler (tenor) in Gershwin's *Porgy and Bess*.

Sprechgesang, German for "speech song," a term used by Richard Wagner to characterize the musical style of his later works, in which the orchestra, sounding certain motifs intimately associated with the text, comments on the singer's words, or expresses their meaning in musical symbolism.

Sprechstimme, German for "speech voice," a kind of song-speech developed by Arnold Schoenberg, and used by composers of the atonal school. The words are half sung, half spoken, with their pitch not exactly notated. Sprechstimme is an important feature of Alban Berg's *Wozzeck* and *Lulu*.

Springer, manager of a theatrical troupe (bass) in Smetana's *The Bartered Bride*.

Squilli, echeggi la tromba guerriera (Or co' dadi), soldiers' chorus at the opening of Act III, Scene 1, of Verdi's *Il trovatore*.

Stabile, Mariano, baritone. Born Palermo, May 12, 1888. After attending the Santa Cecilia in Rome, he made his debut at the Teatro Biondo in Palermo in 1911 as Marcello in *La Bohème*. For the next decade he appeared in various Italian opera houses without attracting much attention. Fame came in 1921 when Arturo Toscanini selected him to appear as Falstaff in the season's opening performance at La Scala. He won an ovation in this role, and it has since been one of his outstanding characterizations. Remaining a principal baritone of La Scala, Stabile has appeared with outstanding success at Covent Garden and at the Glyndebourne and Edinburgh Festivals. His finest roles, besides that of Falstaff, include Don Pasquale, Don Giovanni, Scarpia, Gianni Schicchi, and Rigoletto.

Städtische Oper, see CHARLOTTENBURG OPERA.

Standestheater (or Stavoské Divadlo, or Nostitz theater), one of the most venerable theaters in Prague. It was built in 1781 by Count Anton von Nostitz-Rieneck for the presentation of plays and operas in German and Bohemian. It was here that Mozart directed the première of his *Don Giovanni* in 1787. When the repertory became exclusively German (Bohemian opera and drama acquiring a theater of its own), this house was known as the Deutsches Landestheater. Weber conducted here from 1813 to 1816. After the opening of the Neues Deutsches Theater in 1888 the Standestheater became the home for plays, though *Don Giovanni* returned to its stage in 1937 to honor the 150th anniversary of its première.

Standin' in de need of prayer, Brutus Jones's prayer in Act II of Gruenberg's *The Emperor Jones.*

Stanford, Charles Villiers, composer, conductor, and teacher. Born Dublin, Ireland, September 30, 1852; died London, England. March 29, 1924. He studied music privately in Dublin and London, after which he attended Queen's College, Cambridge, on an organ scholarship. In 1873 he became organist of Trinity College. After an additional two-year period of study in Germany with Carl Reinecke and Friedrich Kiel he made his debut as composer with incidental music to Tennyson's *Queen Mary,* written at the request of the poet, and performed in London in 1876. His first opera, *The Veiled Prophet of Khorassan,* was introduced in Hamburg in 1881. In 1883 he became a professor of composition at the Royal College of Music, and in 1887 professor of music at Cambridge; he held both posts all his life. His students included Ralph Vaughan Williams, Gustav Holst, Frank Bridge, and John Ireland. He was knighted in 1901, and in 1904 he became the first Englishman elected to the Berlin Academy of Arts. After *The Veiled Prophet of Khorassan* he wrote the following operas: *Savonarola* (1884); *The Canterbury Pilgrims* (1884); *Shamus O'Brien* (1889); *Much Ado About Nothing* (1901); *The Critic* (1916); *The Travelling Companion* (1917). He wrote several volumes of reminiscences, including *Pages from an Unwritten Diary* (1914) and *Interludes* (1922).

Starke Scheite schichtet mir dort, Brünnhilde's immolation scene at the close of Wagner's *Die Götterdämmerung.*

Steber, Eleanor, soprano. Born Wheeling, West Virginia, July 17, 1916. She attended the New England Conservatory of Music, and studied singing privately with William Whitney and Paul Althouse. In 1940 she won the Metropolitan Opera Auditions of the Air. On December 7, 1940, she made her debut at the Metropolitan Opera as Sophie in *Der Rosenkavalier.* She has since appeared at the Metropolitan in principal soprano roles of the French, Italian, and German repertories. In 1947 she appeared in the Glyndebourne and Edinburgh Festivals.

Steersman, The, a sailor (tenor) in Wagner's *Der fliegende Holländer.*

Steersman's Song, see MIT GEWITTER UND STURM

Stefan, Paul, writer on music. Born Brünn, Moravia, November 25, 1879; died New York City, November 12, 1943. His academic study took place at the Universities of Brünn and Vienna, his music study with Hermann Graedener and Arnold Schoenberg in Vienna. He wrote music criticisms for leading Viennese newspapers and magazines, and after 1921 edited the musical journal, *Anbruch.* He left Austria in 1938 and settled in New York. Among his books are several on operatic subjects: *Gustav Mahler* (1910); *Die Feindschaft gegen Wagner* (1918); *Das neue Haus,* a history of the Vienna Opera (1919); *Geschischte der Wiener Oper* (1932); *Toscanini* (1935); and *Dvořák* (1941). The last two books were published in the United States in English. He also edited the letters of Verdi and Wagner, translated into German Otakar Sourek's biography of Dvořák, and wrote a study of *Don Giovanni.*

Steffani, Agostino, composer. Born Castelfranco, Italy, July 25, 1654; died Frankfort-on-the-Main, Germany, February 12, 1728. His music study took place in Munich and Rome. In 1675 he was appointed court organist in Munich. Three years later he visited Paris, where he came under Lully's influence. In 1680 he decided to enter the church, and in 1682 he became Abbot of Leipzig. Meanwhile, in 1680, his first opera, *Marco Aurelio,* was produced in Munich. This was followed by five more operas given in the same city. He went to Hanover in 1688 to become court kapellmeister. His opera *Henrico Leone* opened a new opera house there in 1689. This work is particularly noteworthy for its advances in orchestration. In the next nine years Steffani completed nine more operas that were popular in Hanover.

Before the end of the century, he became involved in diplomacy, serving as special envoy to the German courts.

He participated in the complex negotiations resulting in the creation of a ninth Elector for Brunswick. His success brought him an appointment as Bishop of Spiga. Subsequently, he was privy councilor and Papal Protonotary at Düsseldorf. In 1711 he resigned his post as kapellmeister in Hanover (which he had retained even while engaged in diplomacy) and turned it over to Handel, whom he had met in Italy.

His most important operas were: *Marco Aurelio* (1681); *Solone* (1685); *Servio Tullio* (1686); *La lotta d'Ercole con Achelao* (1689); *La superbia d'Alessandro* (1690, revised 1691); *Orlando generoso* (1691); *I Baccanali* (1695); *Briseide* (1696); *Arminio* (1707); *Tassilone* (1709).

Stein, Gertrude, writer. Born Allegheny, Pennsylvania, February 3, 1874; died Neuilly, France, July 27, 1946. One of the leading experimental writers of her time, much of whose work struck many readers as being nonsensical, Miss Stein provided Virgil Thomson with the texts for two operas: *Four Saints in Three Acts* and *The Mother of Us All.*

Stella, (1) an opera singer (soprano) in Offenbach's *The Tales of Hoffmann.*

(2) Camorrist (soprano) in Wolf-Ferrari's *The Jewels of the Madonna.*

Stendhal (born MARIE HENRI BEYLE), novelist and writer on music. Born Grenoble, France, January 23, 1783; died Paris, March 23, 1842. One of his novels, *La Chartreuse de Parme,* was adapted as an opera of the same name by Henri Sauguet. Stendhal also wrote books on music, including the lives of Haydn, Mozart, Metastasio, and Rossini.

Stephana, Prince Alexis' mistress (soprano) in Giordano's *Siberia.*

Stephano, Roméo's page (soprano) in Gounod's *Roméo et Juliette.*

Steuermann! Lass die Wacht! Sailors' chorus in Act III of Wagner's *Der fliegende Holländer.*

Steuermannslied, *see* MIT GEWITTER UND STURM.

Stevens, Risë, contralto. Born New York City, June 11, 1913. She was a scholarship pupil at the Juilliard School of Music from 1932 to 1935, and her study was completed in Salzburg with Marie Gutheil-Schoder and Herbert Graf. Her debut took place in Prague in *Manon* in 1936, and her success brought her an engagement with the Vienna State Opera. Her American debut took place in Philadelphia on November 22, 1938, in *Der Rosenkavalier,* during a visit to that city of the Metropolitan Opera; a month later she appeared with the Metropolitan in New York in *Mignon.* She has since appeared in the principal contralto roles of the French and Italian repertories. In 1939 she became the first American singer to appear at the Glyndebourne Festival in England, in 1949 she appeared at the Paris Opéra, and in 1953 she was invited to La Scala to create the leading role in a new Italian opera, Virgilio Mortari's *La figlia del diavolo.* This was her first appearance in Italy. She has been seen in several motion pictures, including *The Chocolate Soldier* and *Going My Way,* and she has frequently sung on radio and television.

Stewa, Jenufa's stepbrother (tenor), father of her child, in Janáček's *Jenufa.*

Stiedry, Fritz, conductor. Born Vienna, October 11, 1883. He attended the Vienna University and the Vienna Conservatory. Gustav Mahler recommended him for the post of assistant conductor at the Dresden Opera in 1907. After one season there, and several seasons in other European opera houses, he was engaged in 1914 as a principal conductor of the Berlin Opera. Because of the outbreak of war, he was unable to assume this office until two years later. In 1924 he became principal conductor of the Volksoper in Vienna, and in 1929 he succeeded Bruno Walter as musical director of the Municipal Opera in Berlin. When the Nazis came to power, Stiedry went to the Soviet Union. For several years he was musical director of the Leningrad Philharmonic. He came to the United States in 1938, making it his permanent home. On November 15, 1946, he made his debut at the Metropolitan Opera, conducting *Siegfried.* He has remained there since, distinguishing himself particularly in the German repertory.

Stile rappresentativo, Italian for "theater style," that is, the style devised by the first composers of opera for their vocal melodies. The chief characteristic of the theater style was a faithful following of the inflections and rhythms of the words, leading to a sort of music that had little interest apart from text it accompanied.

Still, William Grant, composer. Born Woodville, Mississippi, May 11, 1895. He attended the Oberlin Conservatory and New England Conservatory, after which he studied privately with Edgar Varèse in New York. He first achieved recognition as a composer with orchestral works, including the *Afro-American Symphony,* written in 1931. Three years later he received a Guggenheim Fellowship and a Rosenwald Fellowship. He has written the following operas: *Blue Steel* (1935); *Troubled Island* (1938); *A Bayou Legend* (1940); *A Southern Interlude* (1942). *Troubled Island* was given by the New York City Opera in 1949.

Stizzoso mio stizzoso, Serpina's aria in Act I of Pergolesi's *La serva padrona.*

Stoltz, Rosine (born VICTORINE NOB), mezzo-soprano. Born Paris, France, February 13, 1815; died there July 28, 1903. After studying at Alexandre Choron's school in Paris she made her opera debut in Brussels in 1832. She was first acclaimed for her singing of Rachel in *La Juive,* as a result of which she was engaged by the Paris Opéra, where she made her debut on August 25, 1837, once again as Rachel. For a decade she was the idol of the Parisian opera public. Several operas were written for her, including Donizetti's *Don Sebastian* and *La favorita* and Halévy's *La Reine de Chypre*. Her last appearance in opera took place in 1860, after which she went into retirement.

Stolz, Teresa, soprano. Born Elbe Kosteletz, Bohemia, June 2, 1834; died Milan, Italy, August 23, 1902. After attending conservatories in Prague and Trieste, she made her opera debut in Tiflis. Between 1865 and 1879 she became one of the outstanding sopranos in Italy. She was a friend of Verdi, and she scored some of her greatest successes in his operas. She created the role of Leonora in *La forza del destino* and she appeared in the Italian première of *Aïda*. Her last public appearance was as soloist in Verdi's Requiem in 1879.

Stolzing, Walther von, Franconian knight (tenor) in Wagner's *Die Meistersinger*.

Stone, Jabez, a New England farmer (bass) in Douglas Moore's *The Devil and Daniel Webster*.

Stone, Mary, Jabez Stone's wife (mezzo-soprano) in Douglas Moore's *The Devil and Daniel Webster*.

Stone Guest, The, opera by Alexander Dargomizhsky (completed by César Cui and Rimsky-Korsakov). Libretto is Alexander Pushkin's play of the same name. Première: St. Petersburg, February 28, 1872. In this version of the Don Juan story, the stone guest is a statue of the Commandant, slain in a duel by the Don. Mockingly, the Don invites the statue to be his guest at dinner. The statue keeps the appointment and consigns the Don to the fires of hell. This was the composer's last opera and the one in which he brought his lifelong nationalist ambitions to fruition. Cui referred to it as "the very keystone of the new Russian opera."

It was written almost entirely in a dramatic recitative style.

Stradella, Alessandro, composer. Born Naples, Italy, about 1642; died Genoa, February 28, 1682. The romantic story of Stradella's life has been told in story and opera. As a youth he became famous as a singer and composer and was invited to Venice to write an opera for the carnival season. The Venetian senator Alvise Contarini engaged him to teach singing to his mistress, Hortensia. Stradella fell in love with her, and they fled from Venice. The senator engaged two assassins to pursue him. Legend would have us believe that the assassins caught up with Stradella in Rome, but were so moved by one of his oratorios that they warned him that his life was in danger. The pair now fled to Turin, where they acquired the protection of the Duchess of Savoy. One night, Stradella was waylaid on the street and stabbed, but not fatally. The Duchess arranged for Stradella and Hortensia to get married and live at her palace. But a year after that, on a visit to Genoa, Stradella was murdered. This largely unsubstantiated biography was the inspiration for Friedrich von Flotow's opera *Alessandro Stradella*. Stradella was a composer who brought to operatic lyricism a new expressiveness and dramatic feeling, entitling him to a place in musical history as a precursor of the Neapolitan school represented by Alessandro Scarlatti. His operas: *Corispera* (1665); *Orazio Cocle sul ponte* (1666); *Trespoulo tutore* (1667); *La forza del amore paterno* (1678); *La Doriclea*.

Strakosch, Maurice, impresario. Born Gross-Seelowitz, Moravia, 1825; died Paris, France, October 9, 1887. He attended the Vienna Conservatory after which he toured Europe as concert pianist. In 1848 he came to America, where for about a dozen years he was active as teacher and pianist. His first venture as an opera impresario took place in New York in 1857 when he managed a season of Italian operas; two years later he took his troupe to Chicago. In 1873 and 1874 he managed opera performances in Paris, and in 1884 and 1885 he collaborated with his brother Max in directing opera performances at the Teatro Apollo in Rome. He wrote two operas which were produced in New York: *Giovanni di Napoli* and *Sardanapalus.* His wife was the soprano Carlotta Patti, sister of Adelina Patti. He served as manager for Adelina Patti's concert tours in Europe. His autobiography, *Souvenirs d'un impresario,* appeared in 1887.

Strauss, Johann (II), composer. Born Vienna, October 25, 1825; died there June 3, 1899. His father, Johann Strauss I, was internationally famous as a composer of dance music and conductor of Viennese orchestras. Johann Strauss II made his debut as a composer and conductor of light music on October 15, 1844. With the composition of such waltzes as "The Beautiful Blue Danube," "Tales from the Vienna Woods," and "Wine, Women and Song," Strauss's popularity grew to prodigious proportions; he became the Waltz King, the idol of Vienna, the voice and symbol of Hapsburg Austria. His first operetta, *Indigo and the Forty Thieves, or A Thousand and One Nights,* was produced at the Theateran-der-Wien on February 10, 1871. *Die Fledermaus (The Bat),* was seen at the same theater on April 5, 1874. At first, it was a failure. But in Berlin, where it was produced soon after the première in Vienna, it was a sensation; its international popularity soon followed. *Der Zigeunerbaron (The Gypsy Baron)* was introduced eleven years after *Die Fledermaus,* on October 24, 1885, and was one of the triumphs of the composer's career. Meanwhile, in 1872, Strauss came to the United States to appear in concerts commemorating the centenary of American indepen-

dence. In 1894 the fiftieth anniversary of his debut as a conductor was celebrated for an entire week in Vienna. Though Strauss's stage works were operettas, intended primarily for the popular theater, they have often been produced in major opera houses. In addition to those mentioned, his most important stage works were: *Cagliostro* (1875), and *A Night in Venice* (1883).

Strauss, Richard, composer and conductor. Born Munich, Germany, June 11, 1864; died Garmisch-Partenkirchen, Germany, September 8, 1949. Strauss's father, the leading horn player of the Munich Opera, gave Richard Wagner practical help in the perfection of Siegfried's horn-call in his opera *Siegfried.* Strauss's mother was the daughter of a prosperous brewing family. Exceptionally precocious, Strauss was given piano lessons when he was four. At six, he began composition. While receiving musical instruction from August Tombo and Benno Walter, he gained his academic education at the University of Music. In 1880 three of his songs were performed in Munich; a year later his first symphony was introduced by Hermann Levi. In 1885 Strauss became the assistant of Hans von Bülow with the Meiningen Orchestra; the following year he succeeded von Bülow as principal conductor. His friendship with the poet-musician Alexander Ritter (who was married to Wagner's niece) brought about in Strauss a reevaluation of his music and the adoption of new principles. Ritter, a passionate Wagnerite, convinced Strauss that he should write music of a dramatic and programatic nature, within forms more flexible than the traditional symphony and suite. Forsaking his classic inclinations and freeing himself from the influence of Brahms, Strauss began the writing of the tone poems which were to make him one of the most provocative musical figures of his day: *Don Juan, Death and Transfigura-*

tion, Till Eulenspiegel's Merry Pranks, and their successors. He was also fertile in the field of the song, producing after 1883 some of the finest songs after those of Schumann and Brahms.

His first opera, *Guntram,* produced in Weimar in 1894, was a slavish imitation of Wagner and was a failure. His second, *Feuersnot,* produced in 1901, was also poorly received. But with *Salome,* first given in Dresden on December 9, 1905, he created a work that once again made him one of the most controversial and highly publicized figures in music. *Salome* was followed by another opera that excited enthusiasm and produced shock: *Elektra,* performed in Dresden on January 25, 1909. *Elektra* was the first opera in which Strauss collaborated with the Austrian poet and dramatist Hugo von Hofmannsthal, an arrangement that continued for the next quarter of a century, until Hofmannsthal's death. With *Der Rosenkavalier*—a comedy— given in Dresden on January 26, 1911, Strauss confirmed his position as the foremost German opera composer after Wagner.

After World War I, Strauss's artistic powers deteriorated, though he continued producing large works up to the end of his life, and operas up to 1941. While his technical mastery, facility, and charm prevail in many of these works, the old fire and passion, inspiration, and courageous independence are gone. In 1952, three years after his death, came the last première of one of his operas at the Salzburg Festival: *Die Liebe der Danae.*

Besides his eminence as a composer, Strauss had a worldwide reputation as a conductor. He was particularly noteworthy in his own works and in operas of Mozart and Wagner. The conductor of the Munich Opera in 1886, in 1889 and 1894 he was the first conductor at the Weimar Court. In 1898 he became musical director of the Berlin Opera,

remaining a dozen years. From 1919 to 1924 he was principal conductor and co-music director of the Vienna State Opera. Later, he conducted frequently at music festivals in Munich, Bayreuth, and Salzburg, as well as in major European opera houses. He visited the United States twice, but only as orchestral conductor, in 1904 and 1921.

With the rise of the Nazis in Germany, Strauss at first identified himself closely with the new regime, becoming President of the Third Reich Music Chamber. But he soon came into conflict with government officials, particularly after he collaborated with the Jewish writer Stefan Zweig on *Die schweigsame Frau.* From then on he lived in retirement, mostly at his home in Garmisch-Partenkirchen, where he died in 1949. During the war years he lived a great part of the time in Switzerland.

His operas: *Guntram* (1894); *Feuersnot* (1901); *Salome* (1905); *Elektra* (1909); *Der Rosenkavalier* (1911); *Ariadne auf Naxos* (1912, revised 1916); *Die Frau ohne Schatten* (1919); *Intermezzo* (1924); *Die aegyptische Helena* (1928); *Arabella* (1933); *Die schweigsame Frau* (1935); *Friedenstag* (1938); *Daphne* (1938); *Midas* (1939); *Die Liebe der Danae* (1940); *Capriccio* (1942).

Stravinsky, Igor, composer. Born Oranienbaum, Russia, June 17, 1882. The son of an opera singer at the Maryinsky Theater in St. Petersburg, Igor Stravinsky studied music privately while preparing for a legal career. In his twentieth year he met Rimsky-Korsakov, who encouraged him to undertake composition. After two years of study with Rimsky-Korsakov, Stravinsky completed several orchestral works that came to the notice of Serge Diaghilev, the impresario of the Ballet Russe. Diaghilev engaged Stravinsky to write music for his company. The resulting works, beginning with *The Firebird* in 1910, and including *Petrouchka, The Rites of Spring,* and *The Wedding* (*Les Noces*), made the composer one of the most celebrated figures in the world of music. The last-named work, *The Wedding,* while strictly speaking a cantata, is sometimes performed as an opera. Stravinsky also wrote two operas during this period: *The Nightingale* (*Le rossignol*) (1914), and *Renard* (1917), the last a chamber opera.

In 1910, Stravinsky settled in France, his home for the next decade and a half. Here he wrote the comic opera *Mavra* (1922) and an oratorio, *Oedipus Rex* (1927), a work that has sometimes been staged as an opera.

Stravinsky paid the first of several visits to the United States in 1925, appearing as a guest conductor in programs of his own works. He settled permanently in this country in 1939, becoming a citizen. Among his major works composed after coming to America was *The Rake's Progress,* an opera introduced at the Venice Festival on September 11, 1951, and soon after heard in most of the leading opera houses, including the Metropolitan Opera.

Streltsy, The, a band of Russian radicals conspiring to overthrow Peter the Great, in Mussorgsky's *Khovantchina.*

Strepponi, Giuseppina, soprano. Born Lodi, Italy, September 18, 1815; died Busseto, Italy, November 14, 1897. She was Giuseppe Verdi's second wife. After attending the Milan Conservatory she made her opera debut in Trieste in 1835. She became celebrated in tragic roles. On February 22, 1842, she made her debut at La Scala in Donizetti's *Belisario;* in the same year she created the role of Abigaile in Verdi's *Nabucco.* In 1849 she married Verdi, retiring from the stage.

Stretta, vocal music sung at a quickened tempo. "Di quella pira" in Verdi's *Il trovatore* is known as "stretta."

Stretti insiem tutti tre, *see* TOUS LES TROIS REUNIS.

Stride la vampa, Azucena's aria in Act II, Scene 1, of Verdi's *Il trovatore.*

Strindberg, August, novelist and dramatist. Born Stockholm, Sweden, January 22, 1849; died there May 14, 1912. Several of his plays have been made into operas, including: Erik Chisholm's *Simoon;* Ture Rangström's *Kronbruden;* Julius Röntgen's *Samum;* Edward Staempfli's *Ein Traumspiel;* Julius Weissmann's *Die Gespenstersonata, Schwanenweise,* and *Ein Traumspiel.*

Stromminger, Wally's father (bass) in Catalani's *La Wally.*

Stueckgold, Grete (born SCHMEIDT), soprano. Born London, England, June 6, 1895. She attended the Hochschule für Musik in Munich, and studied singing privately with Jacques Stueckgold, whom she married. Her opera debut took place in Nuremberg in 1913. She was then engaged by the Berlin Opera, where she appeared for several seasons. Her American debut took place at the Metropolitan Opera on November 2, 1927, in *Die Meistersinger.* She remained with the Metropolitan until 1931, returning for two additional periods: 1932–1934 and 1938–1939. Besides her appearances at the Metropolitan, she performed at Covent Garden, with the Chicago Civic Opera, and other major companies. She has distinguished herself primarily in the Wagnerian repertory, but has also been successful in the French and Italian repertories.

Styrienne, a slow melody in 2/4 time. Mignon's aria "Je connais un pauvre enfant" in Act II of Thomas's *Mignon* is a styrienne.

Sucher, Rosa (born HASSELBECK), dramatic soprano. Born Velburg, Germany, February 23, 1849; died Eschweiler, Germany, April 16, 1927. After attending the Munich Akademie she began her opera career in Treves. She then appeared in principal German opera houses. In 1877 she married the opera conductor Josef Sucher (1843–1908). In 1879 she became principal soprano of the Hamburg Opera, where she was acclaimed in Wagnerian roles. She appeared at the Bayreuth Festivals between 1886 and 1899. From 1888 to 1899 she was principal soprano at the Berlin Opera, where her husband was a principal conductor; she frequently sang in performances conducted by her husband. Her American debut took place at the Metropolitan Opera on February 25, 1895, as Isolde. Her farewell to the opera stage took place in Berlin on November 3, 1903, as Sieglinde. After 1908 she lived in Vienna, where she taught singing. Her autobiography, *Aus meinem Leben,* appeared in 1914.

Su! del Nilo al sacro lido! The King of Egypt's exhortation to battle in Act I, Scene 1, of Verdi's *Aïda.*

Suicidio! Aria of La Gioconda in Act IV of Ponchielli's *La Gioconda.*

Sukarev, Olga, a countess (soprano) in Giordano's *Fedora.*

Sulamith, the High Priest's daughter (soprano) in Goldmark's *The Queen of Sheba.*

Sulla vetta tu del monte, madrigal of the musicians in Act II of Puccini's *Manon Lescaut.*

Sullivan, Sir Arthur, composer. Born London, England, May 13, 1842; died there November 22, 1900. He is most famous as W. S. Gilbert's collaborator in writing comic operas. He wrote one grand opera, *Ivanhoe,* after his association with Gilbert ended. It was the initial production of the newly-founded Royal English Opera Company on January 31, 1891. A failure at its première, it has since been forgotten.

Sullivan received his musical education at the Royal Academy of Music in London and at the Leipzig Conservatory. After writing some serious instrumental and choral works he wrote his first comic opera, *Cox and Box,* in 1867. His fruitful collaboration with Gilbert began in 1875 with *Trial by Jury* and continued until 1896.

Sulpizio, Sergeant of the French 21st Regiment (bass) in Donizetti's *The Daughter of the Regiment.*

Summertime, Clara's lullaby in Act I of Gershwin's *Porgy and Bess.*

Summ' und brumm', the spinning chorus in Act II of Wagner's *Der fliegende Holländer.*

Sunken Bell, The, *see* CAMPANA SOMMERSA, LA.

Suor Angelica (Sister Angelica), one-act opera by Puccini. Libretto by Gioacchino Forzano. Première: Metropolitan Opera, December 14, 1918. This is the second of the three one-act operas that comprise *Il Trittico.* (The others are *Gianni Schicchi* and *Il Tabarro.*) The setting is a convent in the seventeenth century. Sister Angelica has sought refuge to expiate an old sin. When an aunt visits her, Angelica inquires about the fate of the child that she had abandoned before taking vows. The aunt replies that the child is dead. Angelica prays for forgiveness and commits suicide. An orchestral intermezzo and Angelica's arioso, "Senza mamma," are the best-known excerpts.

Suore, che riposate, Bertram's invocation in Act III of Meyerbeer's *Robert le Diable.*

Supervia, Conchita, mezzo-soprano. Born Barcelona, Spain, December 8, 1899; died London, England, March 30, 1936. At the age of fourteen she made her opera debut at the Teatro Colón in Buenos Aires. A year later she appeared in Italy, scoring a great success at La Scala, where she was heard in the Italian première of *L'heure espagnole.* The wide range, flexibility, and brilliance of her voice made her particularly effective in operas by Rossini, notably *The Barber of Seville, L'Italiana in Algeri,* and *La cenerentola.* She was also acclaimed at the

Paris Opéra, the Opéra-Comique, and at the Théâtre des Champs Elysées where, in 1929, she was heard in a season of Rossini operas. In 1932–1933 she was a member of the Chicago Civic Opera Company. In 1934 she made her first appearance at Covent Garden. She was also a noted concert artist.

Suppé, Franz von (born FRANCESCO SUPPE DEMELLI), composer. Born Spalato, Yugoslavia, April 18, 1819; died Vienna, Austria, May 21, 1895. A composer of operettas for the popular theater, Von Suppé wrote some works of sufficient stature to be performed in opera houses. In this respect his stage writings are comparable to those of Johann Strauss II. After attending the University of Padua, Von Suppé studied music at the Vienna Conservatory. He then conducted various theater orchestras until 1862, when he was appointed conductor of the Theater-an-der-Wien. Three years later he became conductor of the Leopoldstadt Theater, also in Vienna, holding this post till the end of his life. His first operetta, *Das Mädchen vom Lande*, was introduced in 1847 and was a huge success. He completed some one hundred and fifty similar pieces, the most famous being: *Die schöne Galatea* (*The Beautiful Galatea*) (1864); *Fatinitza* (1876); *Boccaccio* (1879); *Donna Juanita* (1880). *Donna Juanita* and *Boccaccio* were given by the Metropolitan Opera, the first in 1931, the second in 1932.

Suppliants, The, see AESCHYLUS.

Sur mes genoux, fils du soleil (In grembo a me), Selika's aria in Act II of Meyerbeer's *L'Africaine.*

Susanna, Figaro's betrothed (soprano) in Mozart's *The Marriage of Figaro.*

Susanin, Ivan, a peasant (bass) in Glinka's *A Life for the Czar.*

Su, su, marinar, see DEBOUT! MATELOTS.

Sutermeister, Heinrich, composer. Born Feuerthalen, Switzerland, August 12, 1910. He studied philology in Paris and Basel, and music with Carl Orff, Walter Courvoisier, and Hans Pfitzner. In 1934 he settled in Berne, subsequently becoming conductor of the Municipal Theater. After World War II he settled in Vaux-sur-Morges, Lake Geneva, to devote himself entirely to composition. His operas: *Romeo und Julia* (1940); *Die Zauberinsel* (1942); *Niobe* (1946); *Raskolnikoff* (1948); *Der rote Stiefel* (1951). He has also written a radio opera, *Die schwarze Spinne.*

Suzanne, Count Gil's wife (soprano) in Wolf-Ferrari's *The Secret of Suzanne.*

Suzel, a farmer's daughter (soprano) in Mascagni's *L'amico Fritz.*

Suzuki, Cio-Cio-San's servant (mezzo-soprano) in Puccini's *Madama Butterfly.*

Svanholm, Set, tenor. Born Västeras, Sweden, September 2, 1904. After attending the Stockholm Conservatory from 1927 to 1929, he studied singing with John Forsell. He made his opera debut as a baritone, in the role of Silvio, at the Stockholm Opera in 1930. He continued to appear in baritone roles for half a dozen years, then began appearing as a tenor at the Stockholm Opera, scoring major successes in the Wagnerian repertory. He appeared in other European houses, and at the Bayreuth and Salzburg Festivals, before making his American debut. This took place at the Metropolitan Opera on November 15, 1946, in the title role of *Siegfried.* He has been with the Metropolitan Opera since then, besides appearing at Covent Garden, and in Brussels and Copenhagen. He was appointed singer to the Swedish court in 1946.

Svietosar, Grand Duke of Kiev (bass) in Glinka's *Russlan and Ludmilla.*

Swallow, The, see RONDINE, LA.

Swarthout, Gladys, mezzo-soprano. Born Deepwater, Missouri, December 25, 1904. She attended the Bush Conservatory in Chicago and was prepared for opera by Leopoldo Mugnone. Her debut took place with the Chicago Civic Opera in 1924 in *Tosca*. During her initial season with this company she appeared in over half its performances. On November 15, 1929, she made her debut at the Metropolitan Opera as La Cieca. Two months later she was featured in the American première of *Sadko,* and in 1934 in the première of *Merry Mount.* She remained with the Metropolitan until 1938, returning for three additional periods: 1939–1941, 1942–1943, and 1944–1945. She has since appeared in concerts, motion pictures, and radio. Her husband is the baritone Frank Chapman; they were married in 1932.

Synge, John Millington, dramatist and poet. Born Rathfarnham, Ireland, April 16, 1871; died Dublin, March 24, 1909. The following of his plays dealing with Irish peasant life have been made into operas: *The Playboy of the Western World* (Leonid Polovinkin's *The Irish Hero*); *Riders to the Sea* (Rabaud's *L'appel de la mer;* Vaughan Williams' *Riders to the Sea*); and *The Shadow of the Glen* (Arrigo Pedrollo's *La Veglia*).

Szell, George, conductor. Born Budapest, June 7, 1897. His music study took place with Eusebius Mandyczewski, J. B. Forster, Richard Robert, and Max Reger. In 1917 he was recommended by Richard Strauss for a post as conductor of the Strassburg Municipal Opera. After conducting in Prague and Düsseldorf, Szell was engaged as principal conductor of the Berlin Opera, where he remained between 1924 and 1929. From 1930 to 1936 he was the principal conductor of German Opera in Prague. His American debut took place on August 16, 1940, when he led a Hollywood Bowl concert. In 1942 he was engaged by the Metropolitan Opera, making his debut there on December 9 with *Salome.* He remained with that company four years, specializing in the German repertory, but also

directing such operas as *Boris Godunov* and *Otello.* He left the Metropolitan in 1946 to become music director of the Cleveland Orchestra. He returned to the Metropolitan for some guest appearances in 1953, but a few months later he announced his decision to terminate his contract, due to differences with the management over artistic procedures.

Szenkar, Eugen, conductor. Born Budapest, April 9, 1891. He attended the Budapest Conservatory, and in 1912 was engaged as chorus master and assistant conductor of the Landestheater in Prague. From 1913 to 1915 he was a principal conductor of the Landestheater and the Budapest Volksoper. In 1923 he was engaged as musical director of the Berlin Volksoper, succeeding Otto Klemperer. A year later he became principal conductor of the Cologne Opera, remaining with this organization until 1933. Afterward, he conducted concerts in the United States, Europe, and Palestine. In 1939 he settled in Rio de Janeiro as conductor of the Brazilian Symphony Orchestra. After World War II he appeared as a guest conductor of major orchestras in England, Austria, Israel, and Egypt.

Szymanowski, Karol, composer. Born Timoshovka, Russia, October 6, 1882; died Lausanne, Switzerland, March 29, 1937. He studied with Sigismund Noskowski in Warsaw. Writing in an Oriental idiom, he composed his first opera, *Hagith,* in 1913. His second opera, one of his major works, came a decade later: *King Roger,* introduced at the Warsaw Opera on June 19, 1926. His writing in this opera was influenced by the folk songs and dances of Poland. This national style is found in other of his works, including the ballet *Harnasie.* In 1926 Szymanowski became director of the Warsaw Conservatory. His bad health, of which he had been a victim all his life, compelled him to resign in 1929. Later, he became president of the Warsaw Academy of Music. He left a large body of compositions in practically every form, most of them marked by a strong individuality and entitling him to his position as the outstanding Polish composer of the twentieth century.

T

tabarro, Il (The Cloak), one-act opera by Giacomo Puccini. Libretto by Giuseppe Adami, based on a play by Didier Gold, *La Houppelande.* Première: Metropolitan Opera, December 14, 1918. This is one of three one-act operas making up the trilogy *Il Trittico.* (The others are *Gianni Schicchi* and *Suor Angelica.*) The setting is a barge on the Seine River. Michele, a skipper, suspects his wife Giorgetta of being unfaithful, and tries to win back her love by reminding her how he used to protect her under his cloak. Giorgetta remains cold to him. She arranges a rendezvous with her lover, Luigi, using as a signal a lighted match. But when Michele lights his pipe, Luigi takes it for his signal. Michele kills him and covers his body with his cloak. When Giorgetta appears, Michele snatches the cloak from the body and hurls his wife on her dead lover.

Tacea la notte placida, Leonora's aria in Act I, Scene 2, of Verdi's *Il trovatore.*

Taddeo, Isabella's suitor (baritone) in Rossini's *L'Italiana in Algeri.*

Tagliavini, Ferruccio, tenor. Born Reggio, Italy, August 14, 1913. After attending the Parma Conservatory, he won first prize in a national singing contest conducted by the Florence May Music Festival in 1938. A year later he made his opera debut at the Teatro Communale in Florence as Rodolfo. Successful appearances in major Italian opera houses followed, including La Scala, the San Carlo, and the Teatro Reale in Rome. In 1946 he made an extensive tour of South America. On January 10, 1947, he was acclaimed as Rodolfo at his North American debut, at the Metropolitan Opera. He appeared in principal tenor roles at the Metropolitan Opera for the next few seasons, besides singing in recitals throughout the United States, and over the radio. Since leaving the Metropolitan he has appeared in most of the major European opera houses and has been starred in several motion pictures filmed in Italy. He has appeared in opera performances at the Metropolitan Opera and elsewhere with his wife, the soprano Pia Tassinari, whom he married in 1941.

Talbot, Lord Arthur, A Cavalier (tenor) in love with Elvira in Bellini's *I Puritani.*

Tale of a Real Man, The, opera by Serge Prokofiev. Libretto by Myra Mendelssohn (the composer's wife). Première: Leningrad, December 1948. Prokofiev wrote this opera to rehabilitate himself after the devastating attack on him by the General Committee of the Communist Party in 1948 (see PROKOFIEV). "In my new opera," the composer explained at the time, "I intend to use trios, duets, and contrapuntally developed choruses for which I will make use of some interesting northern folk songs." The text was also intended to appeal to Soviet officials, since the central character was a brave Soviet airplane pilot who loses both his legs, but insists on remaining in the service. But this opera failed to restore Prokofiev to the good graces of Soviet officialdom (this was to come later, with other works). *Sovietskaya Musica* wrote: "Prokofiev goes to all the negative and repulsive usages present in the music of the period of reckless infatuation with modernistic trickery."

Tale of Czar Saltan, *see* LEGEND OF CZAR SALTAN.

Tale of Two Cities, A, *see* DICKENS, CHARLES.

Tales of Hoffmann, The (Les contes d'Hoffmann), opera in three acts, with prologue and epilogue, by Jacques Offenbach. Libretto by Jules Barbier and Michel Carré, based on their play derived from stories by E. T. A. Hoffmann. Première: Opéra-Comique, February 10, 1881. American première: New York, Maurice Grau's French Opera Company, October 16, 1882.

Characters: Hoffmann, a poet (tenor); Nicklausse, his friend (contralto); Olympia, one of Hoffmann's loves (soprano); Giulietta, another love (soprano); Antonia, a third love (soprano); Coppélius, a magician, also appearing in the guises of Dr. Miracle and Dapertutto (baritone); Pittichinaccio, Giulietta's admirer (tenor); Lindorf, a Nuremberg councilor (bass); Stella, an opera singer (soprano); Andrès, her servant (tenor); Hermann, a student (baritone); Nathaniel, another student (tenor); Schlemil, Giulietta's lover (bass); Spalanzani, a scientist and inventor (tenor); Cochenille, his servant (tenor); Crespel, Antonia's father (bass); Frantz, his servant (tenor); Luther, a tavern keeper (bass); the Muse (soprano); Voice of Antonia's mother (soprano). The action takes place in Nuremberg, Venice, and Munich in the nineteenth century.

Prologue. The taproom of Luther's tavern in Nuremberg. Lindorf intercepts love note addressed to Hoffmann; it is an invitation for the poet to visit Stella after her performance in the opera house adjoining the tavern. Lindorf makes it plain that the poet will not keep the rendezvous. Hoffmann and a group of students enter the taproom during an intermission of Stella's opera. The students ask the poet for a song. Though dejected, he complies with a ballad about a hunchback jester at the Eisenach court (Légende de Kleinzach: "Il était une fois à la cour d'Eisenach"). But all at once Hoffmann abandons his ugly subject to speak of the beauty of a woman. The students twit him for being in love. The poet insists that he is through with love, having had three unfortunate experiences. Encouraged by a bowl of punch, he sets about describing them.

Act I. Spalanzani's drawing room. Spalanzani has collaborated with the magician Coppélius to create Olympia, a mechanical doll almost human in appearance. Hoffmann has seen Olympia from a distance and has fallen in love with her. When he confides to Nicklausse about his love, his friend tells a story about a mechanical doll that fell in love with a mechanical bird ("Une poupée aux yeux d'émail"). Hoffmann refuses to heed the story's warning. Spalanzani now entertains his guests by winding up Olympia so that she sings (Doll Song—Air de la poupée: "Les oiseaux dans la charmille"). More in love with her than ever, Hoffmann invites the doll to dance with him ("Waltz"). The dance becomes frenetic, and Hoffmann falls in a faint. Coppélius (an incarnation of Lindorf, Hoffmann's rival) now enters. He rages at Spalanzani because the latter has paid for Olympia with worthless currency. For revenge, he smashes the doll to pieces. It is only now that Hoffmann discovers he has been in love with clockwork.

Act II. The gallery of Giulietta's palace in Venice. The voices of Nicklausse and Giulietta are heard extolling the beauty of the night and the power of love (Barcarolle: "Belle nuit, o nuit d'amour"). Hoffmann finds the barcarolle melancholy, and he offers a happier tune ("Amis, l'amour tendre et rêveur"). Hoffmann loves Giulietta without realizing she is in the power of a magician, Dapertutto (another embodiment of Hoffmann's rival). Dapertutto points to a diamond on his finger and explains its powers ("Scintille diamant"). Schlemil, also in love with Giulietta, begins to quarrel with Hoffmann. In the ensuing duel, Schlemil is killed. But Hoffmann discovers that he cannot hope to win Giulietta's love—for the magician's slave throws herself into the arms of another admirer and disappears in a gondola.

Act III. A room in Crespel's house, Munich. Hoffmann is in love with Crespel's daughter, Antonia. Seated at her piano, she sings a lament about her lover who has gone away (Romance d'Antonia: "Elle a fui, la tourterelle"). Her song over, she faints. Dr. Miracle comes to her aid, reminding her she is a victim of consumption and that she must never again tax her health with singing. When Hoffmann arrives, he and Antonia confide their love for each other ("J'ai le bonheur dans l'âme"). But the power of Dr. Miracle (once again, Lindorf) is again triumphant. He evokes the ghost of Antonia's mother, who entreats the girl to sing. Unable to resist her dead mother's wish, Antonia sings, collapses, and dies in her father's arms.

Epilogue. Luther's Tavern. As Hoffmann finishes his remarkable story, Nicklausse suggests that his three women are in reality one—the singer, Stella. He proposes a toast to her. Hoffmann angrily shatters his glass and falls into a drunken stupor: the condition Lindorf had designed for him. When Stella appears, the triumphant Lindorf bears her off. But though the poet has failed her, Stella has thoughts for him. Just before she and Lindorf disappear, she throws a flower at Hoffmann's feet.

Genius of opéra-bouffe, Offenbach ended his triumphant career with a serious opera, the only one he ever wrote. The wonder is that with this single effort he was able to produce so successful a work. Responding sensitively to the book, he created a score in which E. T. A. Hoffmann's world of dreams and fantasies comes delightfully and movingly to life.

Seriously ill when he began *The Tales of Hoffmann*, Offenbach sensed that he had begun a race with death. His greatest hope was to complete what he felt would be his greatest composition. He did not live to see it performed. While making some minor revisions in the score he fainted; two days later he was dead. The opera, introduced a few months after his death, was such a huge success that it was given over a hundred performances during its first year.

Talley, Marion, soprano. Born Nevada, Missouri, December 20, 1907. She studied singing in New York and in Italy. Her debut at the Metropolitan Opera—on February 17, 1926 in *Rigoletto*—was accompanied by publicity attending few other debuts. For weeks her career was publicized in the newspapers. A delegation from her home town arrived on special trains. Her father (a telegraph operator) sent the world his impressions of the performance by means of a key installed backstage. Unfortunately, Talley did not rise to the occasion; her performance was disappointing. Nevertheless, she remained with the Metropolitan Opera four seasons, and in that time she appeared in the American première of Stravinsky's *Le rossignol*.

After leaving the Metropolitan, she appeared for a period in song recitals and in a motion picture, *Follow Your Heart.*

Tamagno, Francesco, tenor. Born Turin, Italy, December 28, 1850; died Varese, Italy, August 31, 1905. His music study took place at the Turin Conservatory. After a period of military service, and further study with Carlo Pedrotti, he made his opera debut in Palermo, in 1873, in *Un ballo in maschera.* Seven years later he was so successful at La Scala, particularly in the role of Ernani, that he was engaged to tour South America. Returning to Italy, he appeared in the major opera houses. Because of the exceptional power and brilliance of his voice, and his pronounced histrionic ability, Verdi selected him to create the title role in the première of his *Otello.* Tamagno's performance was a triumph, and contributed to the over-all success of the production; it became the standard by which later interpreters were measured. On March 24, 1891, he made his American debut at the Metropolitan Opera, again as Otello. He appeared at the Metropolitan through the 1894–1895 season.

Tambourin, a lively, drum-accompanied dance in 2/4 time, originating in Provençe. The characteristic drum beat is suggested in tambourins composed for other instruments. The famous Tambourin found in Rameau's harpsichord suite in E was drawn from the composer's opera *Les fêtes d'Hébé.* Other operatic instance of tambourins are those in Rameau's *Platée* and *Les Indes galantes,* Handel's *Alcina,* Gluck's *Iphigénie en Aulide,* and Paisiello's *Proserpina.*

Tamburini, Antonio, baritone. Born Faenza, Italy, March 28, 1800; died Nice, France, November 9, 1876. As a boy he received vocal lessons from Aldobrando Rossi, and sang in the opera chorus in his native city. He made his debut in Cento in Pietro Gen-

erali's *La contessa di colle erboso.* Appearances in other Italian cities followed, including two years in Rome, where he was heard in Rossini's *Mosè in Egitto.* For four years he sang for the impresario Domenico Barbaja in Naples, Milan, and Vienna, becoming one of the most highly acclaimed baritones of his time. In Vienna, he and the tenor Rubini were the first foreigners since the Duke of Wellington to receive the Order of the Savior. For almost a decade, beginning in 1832, he was an idol of the opera public in London and Paris. In 1841 he returned to Italy, a year later beginning a ten-year stay in Russia. After 1852 he sang in London, Holland, and Paris, even though his voice had greatly deteriorated. He made his last opera appearance in London in 1859, after which he went into retirement in Nice.

Taming of the Shrew, The, (1) a comedy by William Shakespeare, in which Petruchio, by an amusing stratagem, drives his wife Katherine to distraction and thus cures her of her terrible tempers and obstinacy.

(2) Comic opera by Hermann Goetz. Libretto by J. V. Widmann based on the Shakespeare comedy. Première: Mannheim, October 11, 1874.

(3) Opera by Vittorio Giannini. Libretto by Dorothea Fee and the composer, based on the Shakespeare comedy, with material from *Romeo and Juliet* and the *Sonnets.* Première: Cincinnati, Ohio, January 31, 1953.

Tamino, a prince (tenor), in love with Pamina, in Mozart's *The Magic Flute.*

T'amo, si, t'amo, e in lagrime, duet of Riccardo and Amelia in Act V of Verdi's *Un ballo in maschera.*

Tancredi, opera by Rossini. Libretto by Gaetano Rossi, based on Voltaire's *Tancrède,* derived from Tasso. Première: Teatro la Fenice, February 6, 1813. This was one of Rossini's greatest successes before *The Barber of Seville,* which it preceded by two years; it was also his first serious opera. The setting is Syracuse during the conflicts of the Christians and Moslems, and the story engages the hero and heroine in an assortment of trials and misunderstandings before they find true love. The opera contains one of Rossini's most beautiful love songs, "Di tanti palpiti." The overture was taken by the composer from one of his earlier operas, *La pietra del paragone*. The orchestration in this opera is so advanced that Stendhal remarked that it represented "an art of expressing by means of instruments that portion of their sentiments which the characters could not convey to us."

Tanglewood, *see* BERKSHIRE SYMPHONIC FESTIVAL.

Tannhäuser, opera in three acts by Richard Wagner. Libretto by the composer. Première: Dresden Opera, October 19, 1845. American première: Stadt Theater, New York, April 4, 1859.

Characters: Hermann, Landgrave of Thuringia (bass); Elisabeth, his niece (soprano); Tannhäuser, minstrel-knight (tenor); Wolfram von Eschenbach, his friend (baritone); Venus (soprano or contralto); minstrel knights, nobles, ladies, bacchantes, nymphs, pilgrims. The setting is Thuringia and the Wartburg at the beginning of the thirteenth century.

The overture begins with the Pilgrims' Chorus and contains the Venusberg music and Tannhäuser's hymn to Venus.

Act I, Scene 1. The Hill of Venus. Venus is reclining on a couch. Before her is the minstrel-knight, Tannhäuser, a fugitive from the world, now her partner in the enjoyment of sensual pleasures and revelry. Bacchantes are dancing ("Bacchanale"). When they finish, Tannhäuser sings a hymn to Venus ("Dir töne Lob!"). But he longs to return to his own world. Venus is

enraged, insisting that the world will never forgive him. Tannhäuser, however, puts his trust in the Virgin Mary. As he pronounces her name, Venus disappears, and darkness engulfs her realm.

Scene 2. A valley. Tannhäuser finds himself in a valley below the Castle Wartburg. A shepherd passes, singing a pastoral tune. Now is heard the chant of pilgrims on their way to Rome ("Pilgrims Chorus"). As they file past Tannhäuser, he falls on his knees in prayer. The sound of horns brings to the scene a group of minstrel-knights. Wolfram recognizes Tannhäuser and welcomes him warmly after his year's absence. Tannhäuser is reluctant to rejoin his old friends, but when Wolfram tells him how Elisabeth has been grieving over his absence, he decides to return with them to the Wartburg.

Act II. The Hall of Minstrels in Wartburg Castle. Elisabeth, overjoyed that Tannhäuser is returning, sings a hymn to the hall ("Dich, teure Halle"). When Tannhäuser appears, she questions him about his absence; his answers are evasive. Now the knights file in ("March"), followed by the nobles, ladies, and attendants ("Freudig begrüssen wir die edle Halle"). A song contest is about to take place, the prize to be Elisabeth's hand in marriage. The Landgrave announces that the subject of the songs will be Love. Wolfram sings a hymn to pure and unselfish love ("Blick' ich umher"). He is acclaimed. Tannhäuser sings a rhapsody to Venus ("Dir Göttin der Liebe"), glorifying sensual pleasures and carnal love. His audience is horrified. The ladies rush out of the hall, while some of the knights menace Tannhäuser with their swords. Elisabeth protects Tannhäuser, crying out that she will pray for his soul. Tannhäuser, contrite, promises to atone for his sins and begs for forgiveness. But the Landgrave banishes him, suggesting that he join the pilgrims and seek absolution from the Pope. Tannhäuser falls on his knees and kisses the hem of Elisabeth's garment. He then rushes out to join the pilgrims. Act III. The valley of Wartburg. Tannhäuser has been gone for several months. Elisabeth is waiting for his return. Pilgrims return from Rome, but Tannhäuser is not with them. Falling to her knees before a shrine, Elisabeth prays that Tannhäuser's sin be forgiven (Elisabeth's Prayer—Elisabeths Gebet: "Allmächt'ge Jungfrau"). After she leaves, the valley grows dark. Wolfram asks the evening star to guide Elisabeth and protect her (Ode to the Evening Star: "O du mein holder Abendstern"). Now he sees Tannhäuser stumbling toward him; the haggard knight is in rags. Tannhäuser tells him that the Pope has refused absolution, saying that his soul could never be reborn, just as the staff in the Pope's hand could never sprout leaves (Romerzählung—Rome Narrative). Doomed, Tannhäuser can only hope to return to Venus. But once again, when Wolfram mentions the name of Elisabeth Tannhäuser rejects the temptress. A funeral procession draws near. Minstrels and pilgrims are bearing the bier on which lies the dead Elisabeth. Sinking beside the bier, Tannhäuser dies. As morning dawns, more pilgrims arrive from Rome: they bear the Pope's staff, which has miraculously put forth leaves.

Tannhäuser, written between 1843 and 1845, belongs to Wagner's first creative period, in which he was still more or less subservient to tradition. The opera contains formal arias, ensemble numbers, choruses, scenes of pageantry, marches, and even a ballet. The musico-dramatic concept of his later dramas was not yet his ideal. Yet there is much in *Tannhäuser* to suggest the mature Wagner. There is a tentative, at times highly effective, use of leitmotifs, as in the recurrent use of themes designating the pilgrims, Venus-

berg, sensual love, and pure love. One also finds the first examples of the kind of narratives which would abound in his later works; for example, the Rome Narrative. *Tannhäuser* is also concerned with dramatic values. Though it is an opera and not a music drama, atmosphere, characterization, dramatic action, and climaxes are not undervalued.

The version of *Tannhäuser* most often heard today is not the one introduced in Dresden in 1845, but a revision prepared by Wagner for the Paris première in 1861. To meet Parisian partiality for ballet, Wagner interpolated an elaborate bacchanale in the opening scene, besides making other drastic alterations. The Paris première was a fiasco, largely brought about by Wagner's enemies, but the Paris version is the one most audiences now prefer, even though the original is dramatically more sound and artistically more valid.

Taras Bulba, *see* GOGOL, NIKOLAI.

Tartuffe, *see* MOLIERE.

Tasso, Torquato, poet. Born Sorrento, Italy, March 11, 1544; died Rome, April 25, 1595. Tasso's epic poems have been made into many operas. His magnum opus was *La Gerusalemme liberata,* which, as *Armida,* was made into operas by Dvořák, Haydn, Rossini, Sacchini, Salieri, and Tommaso Traetta; as *Armide,* by Gluck and Lully; under its original title, by Carlo Pallavacini and Vincenzo Righini; and as *Rinaldo,* by Handel. Other operas based on this epic were Johann Haeffner's *Renaud;* Sebastiano Moratelli's *Erminia ne' boschi* and *Erminia al campo;* Luis Persius' *Jérusalem délivrée;* Michel Angelo Rossi's *Erminia sul giordano;* and Niccolò Zingarelli's *La distruzione di Gerusalemme.*

Tasso's other epics were the source of such operas as: André Campra's *Tancrède;* Monteverdi's *Il combattimento di Tancredi e Clorinda;* Vin-

cenzo Righini's *La Selve incantata.*
Tasso appears as the central character
in Rossini's *Torquato Tasso.*

Tatiana, Mme. Larina's daughter (soprano), in love with Eugene Onegin,
in Tchaikovsky's *Eugene Onegin.*

Tauber, Richard (born ERNST SEIF-
FERT), lyric tenor. Born Linz, Austria,
May 16, 1892; died London, England,
January 8, 1948. He attended the Hoch
Conservatory, Frankfort-on-the-Main,
and studied singing with Carl Beines.
He made his opera debut in Chemnitz
as Tamino in 1913. He was immediately engaged by the Dresden Opera,
where for a decade he appeared in leading tenor roles of the German, Italian,
and French repertories. Meanwhile, in
1915, he also became a member of the
Berlin Opera. After World War I, he
appeared frequently as a guest in major
European opera houses, distinguishing
himself particularly in the operas of
Mozart.

Tauber became even more famous
in operetta, particularly in the operettas of Franz Lehár. He was the idol
of theater-goers in Germany, Austria,
England, and France. He was also distinguished on the concert stage. He
made his American debut on October
28, 1931, in a New York song recital,
and fifteen years later appeared on the
Broadway stage in Lehár's *The Land of
Smiles,* renamed *Yours Is My Heart.*
He also appeared in numerous motion
pictures. After the rise of Hitler,
Tauber settled permanently in England
where, in 1936, he married the British
actress Diana Napier.

Taylor, Deems, composer. Born New
York City, December 22, 1885. He began to study the piano in his eleventh
year. Later, he studied composition and
orchestration with Oscar Coon. While
attending college, Taylor wrote music
for student shows, one of which, *The
Echo,* was produced on Broadway.
After leaving college, Taylor acted in
vaudeville, then became an editor,

translator, and journalist. In 1919 he
achieved recognition as a composer
with his suite for orchestra, *Through
the Looking Glass.* Two years later he
became the music critic of the *New
York World,* but resigned after four
years to devote himself to composition.
In 1926, on a commission from the
Metropolitan Opera, he completed his
first opera, *The King's Henchman.* It
was introduced at the Metropolitan on
February 17, 1927. Its success led the
Metropolitan to commission a second
opera: *Peter Ibbetson,* seen on February 7, 1931, and so well liked that it
was given sixteen times in four seasons,
and opened the 1933–1934 season.
Taylor's third opera, *Ramuntcho,* was
produced by the Philadelphia Opera
Company in 1942. Taylor has also distinguished himself as an author, program annotator, and master-of-cere-
monies on radio programs.

Tchaikovsky, Peter Ilyich, composer.
Born Votinsk, Russia, May 7, 1840;
died St. Petersburg, November 6, 1893.
He attended the School for Jurispru-
dence in St. Petersburg, after which he
became clerk in the Ministry of Justice.
However, he had been fond of music
from childhood on, and in 1862 he resigned from his post with the Ministry
and enrolled in the newly founded St.
Petersburg Conservatory. After graduating, he became professor of harmony
at the new Moscow Conservatory. During this period he completed his first
symphony, performed in 1868, and an
opera, *The Voievoda,* performed in
Moscow on February 11, 1869, and
given five times. Tchaikovsky himself
was dissatisfied with the opera and subsequently he destroyed the score; in
recent times, however, the score has
been reconstructed from the orchestral
and vocal parts. Two more operas followed in the next few years: *Undine* in
1869, and *Oprichnik* in 1872. A third
opera, *Mandragora,* was never finished.
During this period Tchaikovsky also

produced his first orchestral masterwork, the symphonic poem *Romeo and Juliet*.

In 1877 Tchaikovsky embarked on a marriage which was unhappy from the first day. He did not love Antonina Miliukova either before or after he married her. It is possible that he used this alliance to conceal his homosexual tendencies. In any event, this unfortunate step upset his nervous system, and he tried to commit suicide. He then fled from his wife and traveled throughout Europe for a year. He never returned to her. Despite these emotional and physical upheavals, he completed several major works, one of which was his most important opera, *Eugene Onegin*, introduced in Moscow in 1879.

While traveling about Europe, Tchaikovsky learned that a wealthy patron, Nadezhda Filaretovna von Meck, stood ready to provide him with a handsome annual pension. This marked the beginning of a strange relationship between Madame von Meck and Tchaikovsky, carried on exclusively through correspondence. These letters were often passionate in their avowal of love, yet their authors never met. The reason for this condition, laid down by the patron, has never been satisfactorily explained. In any event, it is a curious fact that the great love affair of Tchaikovsky's life was carried on exclusively by letter. Financially independent, and stimulated by the outpouring admiration of his patron, he produced a succession of masterworks. Then, in 1890, after thirteen years, the strange friendship came to a sudden end. While in the Caucasus, he received word that Madame von Meck's financial reverses compelled her to terminate the pension. Since Tchaikovsky was now financially secure, he hastened to tell her he was no longer in need of her generosity, and to express the hope that their friendship might continue. This letter, and later ones, were not

answered. When he returned to Moscow, Tchaikovsky discovered that his patron had not suffered reverses, but had used this as an excuse to break off a relationship that had either begun to bore or embarrass her. The realization that she had thus discarded him was a blow from which the composer never completely recovered.

In 1891 Tchaikovsky visited the United States, conducting four concerts in New York, one in Baltimore, one in Philadelphia. Back in Russia he became a victim of emotional instability. In such a mood he completed his last symphony, appropriately named by his brother the *Pathétique*. The composer died less than two weeks after conducting the première of this work.

Though not heard as frequently as his other works, Tchaikovsky's best operas, while not consistent masterpieces, are filled with some of his finest melodic inspiration, and often have compelling dramatic power. His operas: *The Voyevoda* (1868); *Undine* (1869); *Oprichnik* (1872); *Vakula the Smith* (1885); *Eugene Onegin* (1878); *The Maid of Orleans (Joan of Arc)* (1879); *Mazeppa* (1883); *The Enchantress* (1887); *Pique Dame* (The Queen of Spades) (1890); *Iolanthe* (1891).

Tchekov, Anton, *see* CHEKHOV.

Tcherepnin, Alexander, composer. Born St. Petersburg, Russia, January 20, 1899. The son of Nicolai Tcherepnin (see below), Alexander studied with his father, then attended the Conservatories of St. Petersburg and Paris. He attracted attention with a piano concerto, but real success came with an opera, *Ol-Ol*, introduced in Weimar on January 31, 1928. In 1933 he toured as composer-pianist, making his first visit to the United States in 1934. From 1934 to 1937 he lived in the Orient. In 1948 he settled permanently in the United States. He has been a member of the music faculties of the San Francisco Music and Art Institute and the De Paul University Music School in Chicago. His wife, Lii Shiannmin, is a Chinese musician with whom he has often appeared in concerts of Chinese music. The Russian and Oriental influences in Tcherepnin's life are revealed in the exotic atmosphere and vivid colors of many of his works. A distinguishing technical trait is the use of a nine-tone scale, with which he is identified. His operas: *Ol-Ol* (1925); *Die Hochzeit der Sobeide* (1930); *The Farmer and the Fairy* (1952). He has also written a second act to, and revised the scoring of, Mussorgsky's unfinished opera *The Marriage;* this new version was introduced in Essen in 1937.

Tcherepnin, Nicolai, composer. Born St. Petersburg, Russia, May 14, 1873; died Issy-les-Moulineaux, France, June 26, 1945. He was a pupil of Rimsky-Korsakov. For five years, beginning in 1909, he was the conductor of Serge Diaghilev's Ballet Russe in Paris. Just before World War I he returned to Russia, but after the Revolution he settled permanently in Paris, where he devoted himself principally to writing music for the stage. His style was primarily influenced by the techniques and approaches of the Russian national school. He wrote three operas: *The Marriage Broker; Poverty Is Not a Crime;* and *Ivan the Chancellor.* He completed Mussorgsky's *The Fair at Sorochinsk,* a version introduced at the Monte Carlo Opera in 1923, and seven years later produced at the Metropolitan Opera.

Tcherevik, a peasant (bass) in Mussorgsky's *The Fair at Sorochinsk.*

Tchernomor, a wizard in Glinka's *Russlan and Ludmilla.*

Teatro alla Scala (La Scala), the leading opera house of Italy, and one of the most significant opera houses of the world. It was built in Milan in 1776 on the site of the church of Santa Maria alla Scala, by order of Empress Maria Theresa, replacing a theater which had burned. Designed by Piermarini of Fogliano, it was the finest and costliest theater of its day. It opened on August 3, 1778, with Antonio Salieri's *Europa riconosciuta.*

The theater was extensively remodeled in 1867. In 1872 it became the property of the municipality of Milan, administered by a commission elected by the city council and the theater's box owners. In 1897, in a wave of economy, the city administration stopped its funds, and the opera house closed. Such resentment was expressed by the Milanese that the city had to restore its financial support. The house closed again during the years of World War I. In 1920 a group of patrons provided the funds to reopen the theater. The building was completely modernized. It reopened with Arturo Toscanini as its artistic director. This period, ending with Toscanini's resignation in 1929, was one of the most brilliant in La Scala's history. The theater closed again during World War II, and was later severely bombed. After partial restoration, it reopened on May 11, 1946, with an orchestral concert conducted by Toscanini, the first of ten concerts he led to raise funds for the completion of the reconstruction. The first new opera performed by La Scala after the company resumed operations was Pizzetti's *L'Oro* on January 1, 1947.

The world premières at La Scala represent a sizable chapter in the history of Italian opera. This is a partial but representative list: *L'amore dei tre re; Andrea Chénier; La campana sommersa; La cena delle beffe; Debora e Jaele; Falstaff; Fra Gherardo; La Gioconda; Germania; I Lombardi; Lucrezia Borgia; Madama Butterfly; Mefistofele; Nabucco; Nerone; Norma; Otello; Turandot; La Wally; Zaza.*

Teatro Costanzi, the leading opera house in Rome, built in 1889 with funds provided by Domenico Costanzi, a wealthy hotel owner. It opened on November 27, 1880, with Rossini's *Semiramide*. It was then, and still is, one of the largest theaters in Italy, having been elaborately rebuilt in 1931, when it was renamed the Teatro Reale. The most important of its world premières have been those of *L'amico Fritz; Cavalleria rusticana; Lodoletta; Iris;* and *Tosca*.

Teatro della Pergola, the principal opera house of Florence, Italy, named after the street on which it is located. It was built in 1657 by the Medicis as a home for spoken drama. On the site, the first opera, Peri's *Dafne,* had been performed in 1597. Opera was presented at the Teatro della Pergola for the first time in 1738. For this house Meyerbeer wrote his *Il crociato in Egitto,* Verdi, his *Macbeth*.

Teatro di San Cassiano, the first public opera house, opened in Venice in 1637 (see OPERA PERFORMANCE).

Teatro la Fenice, the most important opera house of Venice, and one of Italy's greatest operatic institutions. Its construction began in 1790, but before the building was completed it was destroyed by fire. Living up to its name (in English, The Phoenix), it was rebuilt, opening two years later. Rossini's first serious opera, *Tancredi,* was introduced there in 1813. Other important premières included those of *Attila; Ernani; Rigoletto; Semiramide; Simon Boccanegra;* and *La traviata*.

Teatro Reale, *see* TEATRO COSTANZI.

Teatro San Carlo, the leading opera house in Naples, and one of Italy's most significant operatic organizations. It was built in 1737. From 1810 to 1839 it was under the artistic direction of Domenico Barbaja, becoming at that time one of the world's great opera houses. The present structure was built after a fire destroyed the original building in 1816. It was remodeled in 1844. After World War II it was remodeled once again. Under the artistic direction of Pasquale di Costanzo, it now embarked on a progressive program that included first performances in Italy of works by such composers as Handel, Haydn, Rimsky-Korsakov, and Roussel. Among the historic premières given here were those of *Lucia di Lammermoor;* Verdi's *Luisa Miller* and *Alzira;* Rossini's *Moïse* and *Otello;* D'Erlanger's *Tess*.

Tebaldi, Renata, soprano. Born Pesaro, Italy, February 1, 1922. For ten years she attended the Conservatories of Pesaro and Parma, where she at first specialized in the piano. When she decided to become a singer, she took private lessons from Carmen Mellis and dramatic coaching from Giuseppe Pais. She made her debut at the Rivigo Theater in *Mefistofele* in 1944. Success came in 1946, when Arturo Toscanini heard her and engaged her to appear under his direction at a concert at La Scala. Appearances in major Italian opera houses and at Covent Garden and the Edinburgh Festival brought her to the front rank of opera sopranos. She made her American debut at the San Francisco Opera on September 26, 1950. On January 31, 1955, she appeared for the first time at the Metropolitan Opera, singing the role of Desdemona. She appeared in a motion-picture version of *Lohengrin* filmed in Italy, and sang (but did not appear) in an Italian film version of *Aïda*.

Telephone, The, one-act opera by Gian-Carlo Menotti. Libretto by the composer. Première: New York, Heckscher Theater, February 18, 1947. Ben, trying to propose to Lucy, is continually interrupted by Lucy's passion for telephone conversations. In desperation he rushes out to a phone booth, telephones his proposal, and is accepted.

Television of opera, *see* OPERA PFRFORMANCE—TELEVISION.

Tell, *see* WILLIAM TELL.

Telramund, Frederick of, Count of

Brabant, Elsa's guardian (baritone) in Wagner's *Lohengrin*.

Telva, Marion, contralto. Born St. Louis, Missouri, September 26, 1897. She began studying singing seriously on the advice of Ernestine Schumann-Heink. On December 31, 1920, she made her debut at the Metropolitan Opera in *Manon Lescaut*. She remained at the Metropolitan for a decade, distinguishing herself both in Italian operas and in the Wagnerian repertory. She was seen in several important premières and revivals, including those of *Die aegyptische Helena, Norma, Peter Ibbetson,* and *The Snow Maiden.* She went into retirement after the 1930–1931 season. Re-engaged in 1935, she was unable to perform because of poor health.

Tempest, The, *see* SHAKESPEARE.

Temple Dancer, The, one-act opera by John Adam Hugo. Libretto by Bell-Ranske. Première: Metropolitan Opera, March 12, 1919. The temple dancer falls in love with a young man of her faith. Unable to gain a favorable sign from the gods for this love, she poisons the temple guard and curses the temple. A bolt of lightning kills her as she tries to steal the holy temple jewels.

Templer und die Jüdin, Der (The Templar and the Jewess), opera by Heinrich Marschner. Libretto by W. A. Wohlbrück and the composer, based on Sir Walter Scott's *Ivanhoe.* Première: Leipzig Opera, December 22, 1829. Robert Schumann quoted the aria "Wer ist der Ritter hoch geehrt" in the final variation of his *Etudes symphoniques,* for piano. *See also* IVANHOE.

Tender Land, The, opera by Aaron Copland. Libretto by Horace Everett. Première: New York City Opera, April 1, 1954. The setting is the Midwest, where a rural family engages two harvesters to aid with the crops. One of them, Martin, falls in love with the girl of the house, Laurie. They plan to elope, but Martin loses heart and one night disappears. Heartbroken, Laurie goes out into the world alone.

Tennyson, Alfred, Lord, poet. Born Somersby, England, August 6, 1809; died Haslemere, England, October 6, 1892. Tennyson's *Idylls of the King,* a poetic adaptation of the Arthurian legend, was made into operas by Herman Bemberg (*Elaine*), Walter Courvoisier (*Lanzelot und Elaine*), and Odon Michalovich (*Edin*). His *Enoch Arden* was adapted as an opera by Rezsö Raimann, also by Eduardo Sanchez de Fuentes (*Naufrago*).

Tenor, the highest range of the adult male voice, when produced naturally. Normally, it extends about two octaves upward from the C an octave below middle C.

Teresa, (1) a miller (mezzo-soprano) in Bellini's *La Sonnambula.*

(2) Balducci's daughter (soprano) in Berlioz' *Benvenuto Cellini.*

Ternina, Milka, dramatic soprano. Born Belgisč, Croatia, December 19, 1863; died Zagreb, May 18, 1941. She studied singing with Ida Winterberg in Zagreb and Joseph Gänsbacher in Vienna. Her debut took place while she was still studying in Zagreb, in the role of Amelia in *Un ballo in maschera.* On Anton Seidl's recommendation she was selected to succeed Katharina Klafsky as principal soprano of the Bremen Opera. In 1890 she was engaged by the Munich Royal Opera, where for a decade she distinguished herself as one of the outstanding Wagnerian sopranos of her time. Her American debut took place in Boston in 1896 with the Damrosch Opera Company; she sang the role of Brünnhilde in *Die Walküre.* She made her London debut, as Isolde, in 1898, and her first appearance at Bayreuth, as Kundry, in 1899. On January 27, 1900, she made her Metropolitan Opera debut as Elisabeth. During her association with the Metropolitan she created for America the role of Kundry; for participating in this performance, given against the wishes of the Wagner family, she was denounced and never again invited to Bayreuth. Ternina also created for America the role of Tosca. She went into retirement in 1906, at the height of her career, after an attack of paralysis. For a year she taught singing at the Institute of Musical Art in New York, after which she went into complete retirement in Zagreb. There she was credited with the discovery of Zinka Milanov.

Terzetto delle maschere, trio of Ping, Pang, and Pong, in Act II, Scene 1, of Puccini's *Turandot.*

Te souvient–il du lumineux voyage, duet of Thaïs and Athanaël in the closing scene of Massenet's *Thaïs.*

Tess, opera by Frederick d'Erlanger. Libretto by Luigi Illica, based on Thomas Hardy's *Tess of the d'Urbervilles.* Première: San Carlo Opera, April 10, 1906. Tess is seduced by an aristocrat. She confesses her sin to her husband in the bridal chamber. When he refuses to forgive her, she commits suicide.

Tetrazzini, Luisa, coloratura soprano. Born Florence, Italy, June 29, 1871; died Milan, April 28, 1940. One of the greatest coloratura sopranos of all time, she studied first with her sister Eva, a dramatic soprano, and then at the Musical Institute in Florence. Her debut took place in Florence in 1890 in the role of Inez. After many appearances in Italy and South America, she scored her first triumphs with a new company touring Mexico in 1905. The company later appeared in San Francisco, where she was a sensation. She made her Covent Garden debut on November 2, 1907, as Violetta. On January 15, 1908, she made her first appearance with the Manhattan Opera Company, again as Violetta. She became such a favorite in New York that in her first season with the Manhattan Opera she appeared twenty-two times, instead of the fifteen

originally scheduled. She remained with the company until its dissolution. On December 27, 1911, she made her bow at the Metropolitan Opera as Lucia. She stayed only a single season at the Metropolitan. For the next few years she toured the United States in recitals, and in 1913 she appeared with the Chicago Opera. After World War I she returned to America for several concert tours; her last appearance in this country was in 1931. She then devoted herself to teaching singing in Milan. Her older sister, Eva, made successful opera appearances in Europe and America, retiring when she married the opera conductor Cleofonte Campanini.

Teyte, Maggie (born TATE), soprano. Born Wolverhampton, England, April 17, 1888. After attending the Royal College of Music in London she studied privately for four years with Jean de Reszke. Her debut took place at the Opéra-Comique in 1908, as Mélisande. A year later in Munich she created the role of Suzanne in *The Secret of Suzanne.* On November 4, 1911, she made her American opera debut in Philadelphia, as Cherubino. For three seasons she was a permanent member of the Chicago Opera, and from 1915 to 1917 she appeared with the Boston Opera. After World War I she toured extensively in song recitals, becoming an outstanding interpreter of French song. She returned to opera on March 25, 1948, singing Mélisande with the New York City Opera Company. For her services to French music during World War II, she was decorated with the Croix de Lorraine in 1945.

Thaïs, opera in three acts by Jules Massenet. Libretto by Louis Gallet, based on the novel of the same name by Anatole France. Première: Paris Opéra, March 16, 1894. American première: Manhattan Opera House, New York, November 25, 1907.

Characters: Thaïs, a courtesan (soprano); Athanaël, a Cenobite monk (baritone); Nicias, a wealthy Alexandrian (tenor); Crobyle, his slave (soprano); Myrtale, another slave (mezzosoprano); Palemon, an old monk (bass); La Charmeuse, a dancer (soprano); Albine, an abbess (mezzosoprano); Cenobites, actors, dancers, nuns, citizens of Alexandria. The setting is Egypt in the fourth century.

Act I, Scene 1. A Cenobite community. Back from Alexandria, Athanaël tells his associates about the evil prevailing in that city, and the destructive influence of the courtesan Thaïs. In his sleep, Athanaël dreams that Thaïs is performing a sensual dance. He awakens with horror. Now determined to convert Thaïs, refusing to be dissuaded by Palemon's arguments, he sets out for Alexandria.

Scene 2. Nicias' house in Alexandria. Gazing at the city, Athanaël laments that it should have become so degenerate ("Voilà donc la terrible cité"). When Nicias appears, Athanaël welcomes his old friend warmly. Nicias is cynical when Athanaël tells him he has come to convert Thaïs, but for the sake of his friendship—and despite his own infatuation for the courtesan—he offers to help. He fits out the monk in handsome clothes. When Thaïs arrives, Athanaël is at first stunned by her beauty. She learns from Nicias that Athanaël is a philosopher who voluntarily lives in the desert and who has come to save her soul. Athanaël explains further that his teachings embrace the rejection of the flesh. Thaïs mockingly replies that her religion is that of love. Provocatively, she suggests to the monk that he try the delights she offers. Athanaël, horrified, rushes from the house.

Act II, Scene 1. Thaïs' house. Thaïs meditates on her world-weariness ("Ah, je suis seule"). She begs her mirror to tell her again that she is beautiful (Air du miroir: "Dis-moi que je suis belle").

Athanaël interrupts her reveries. While marveling at her beauty he remains inflexible in his resolve to save her. When Thaïs tries to lure him into making love to her, Athanaël reveals that he is really a monk, and will be her savior. A sudden fear seizes Thaïs; she falls on her knees and begs for mercy. Exultant, the monk promises her a new joy as the bride of Christ. From a distance comes Nicias' voice, calling to her. Recovered from her temporary fear, Thaïs exclaims she is, and always will be, a courtesan, and has no use for God.

Scene 2. Before Thaïs' house. Preceding the rise of the curtain, the orchestra plays the celebrated "Meditation," the beautiful melody symbolic of Thaïs' spiritual regeneration. Weary and spent from a night of revelry, Thaïs approaches Athanaël and confesses to him that her life has been wasted. She is ready to follow Athanaël, but begs that she may take with her a statue of Eros. When the monk learns that this statue was a gift from Nicias, he smashes it; then, entering the courtesan's house, he destroys all other symbols of physical pleasures. Meanwhile, Nicias and his friends appear, continuing their revelry. Voluptuous dances are performed, including one by La Charmeuse. When Athanaël and Thaïs emerge from the house, they announce that the old Thaïs is dead, and that a new spiritual woman has arisen in her place. Learning that the monk intends taking Thaïs away, the crowd rushes at Athanaël to kill him. Nicias saves the situation by throwing gold coins. In the mad scramble for them, the people forget Thaïs.

Act III, Scene 1. A desert oasis. Thaïs is in a state of exhaustion, but the monk urges her on, insisting she must mortify her flesh. But at the sight of her bleeding feet, Athanaël is filled with pity. He bathes her feet and brings her fruit and water. Thaïs now enters a state of exaltation. When nuns appear, Athanaël tells them he has brought them a sinner. The Abbess Albine and her sisters conduct Thaïs to a cell in their convent. At her departure, Athanaël is tortured by the thought that he will never again see Thaïs.

Scene 2. The Cenobite community. Athanaël confesses to Palemon that in saving Thaïs he has lost his soul. He cannot drive Thaïs from his mind. When Palemon leaves, Athanaël prays, but as he does so he sees Thaïs in a vision, sensuous and irresistible. Voices now proclaim that Thaïs must die. Horrified, Athanaël rushes away, determined to see her again.

Scene 3. The convent garden. Thaïs, dying, is surrounded by nuns. When Athanaël arrives, the nuns leave the monk and Thaïs alone. The monk pleads for Thaïs to return to Alexandria with him. Gently, Thaïs recalls their spiritual regeneration through the long journey in the desert; she has finally found peace (Death of Thaïs: "Te souvient-il du lumineux voyage"). Athanaël now tries to convince her that the only truth lies in physical pleasures. Thaïs raises herself, seeing a vision of Paradise. Athanaël begs her not to leave him. She falls back dead. The monk is inconsolable in his grief.

In adapting France's novel, Louis Gallet made a compromise between prose and poetry by using a free-flowing rhythmic prose. Massenet's music adapted itself to Gallet's style. As Ernest Newman has pointed out, the composer "cut his melodic periods to the size and shape of those of his librettist," making "his musical phrase-divisions, in the main, at the same points. The mood and feeling of play and music are also at one." The Gallet text, though it passes from sensual to spiritual love and back, from physical voluptuousness to religious exaltation, places greater emphasis on spiritual

than physical joy. And Massenet's music, in its over-all sweetness and radiance, is more soulful than passionate.

Massenet wrote his opera for the American soprano Sibyl Sanderson, who created the title role in 1894. The greatest of all Thaïses, Mary Garden, created the role for America in 1907.

Thanatos, god of death (bass) in Gluck's *Alceste.*

Theater-an-der-Wien, one of the most important theaters in Vienna, founded by Emmanuel Schikaneder with funds provided by the merchant Zitterbach. It was intended to rival the Burgtheater. It opened on June 13, 1801, with a play by Schikaneder, the music by the theater's kapellmeister, Franz Teyber. For a while, the productions were mainly spectacles, but before long operas were introduced. In 1803 Cherubini's *Lodoïska* and *Der Bernardsberg* were performed. It was for this theater that Schikaneder commissioned Beethoven to write his only opera, *Fidelio,* introduced in 1805. Schubert's *Die Zauberharfe* was produced in 1820. In 1821 Domenico Barbaja became director of the theater, which now distinguished itself with its performances of Rossini's operas. Jenny Lind appeared here in 1846. The Theater-an-der-Wien subsequently became a home for operettas, and it was here that such classics as *Die Fledermaus, The Gypsy Baron,* and *The Merry Widow* were introduced.

Théâtre de la Monnaie, the leading opera house in Belgium. Modeled after Italian opera houses, it opened in Brussels in 1700, presenting works by Lully and Rameau. The original theater was destroyed in 1820, and a larger one was built ten years later on the same site. Soon after the reopening, the first performance in Belgium of *La muette de Portici* had profound consequences, touching off the revolt against Dutch rule that resulted in Belgium's constitution as an independent state. It was at this opera house that *Carmen,* with Minnie Hauk singing the title role, achieved its first major success. In the last half century, the repertory has given prominence to operas by Belgian and French composers. In the thirty-year period beginning in 1918, it presented forty-one new operas by Belgians, twenty-four by Frenchmen and only fifteen by other composers. Among the theater's notable premières have been those of Ibert's and Honegger's *L'Aiglon;* Honegger's *Antigone;* D'Indy's *L'Etranger* and *Fervaal;* Massenet's *Hérodiade;* Milhaud's *Les malheurs d'Orphée;* Leroux's *La Reine Fiammette;* Chabrier's *Le roi malgré lui;* Reyer's *Salammbô* and *Sigurd.*

Théâtre de la Spectacle, the first opera house in the United States, built in New Orleans. See OPERA PERFORMANCE—UNITED STATES.

Théâtre Lyrique, an opera house inaugurated in Paris on September 21, 1851, under the direction of Edmond Souveste. It achieved its greatest significance as an operatic institution under the managership of Léon Carvalho between 1856 and 1860, and from 1862 to 1868. Among the celebrated French operas introduced there were Gounod's *Faust, Mireille, Philémon et Baucis,* and *Roméo et Juliette,* and Bizet's *Les pêcheurs de perles.*

Thélaïre, a character (soprano) in Rameau's *Castor et Pollux.*

Thill, Georges, tenor. Born Paris, December 14, 1897. After attending the Paris Conservatory, he studied privately with Fernando de Lucia in Naples and Ernest Dupré in Paris. In 1924 he made his debut at the Paris Opéra in *Thaïs.* He became an outstanding favorite at the Opéra, at Covent Garden, La Scala, and the Théâtre de la Monnaie. On March 20, 1931, he made his American debut at the Metropolitan Opera in *Roméo et Juliette.* He was unable to duplicate in America his

European successes, and he stayed at the Metropolitan only two seasons. After leaving, he returned to sing in France. He starred with Grace Moore in a motion-picture adaptation of *Louise* that was filmed in France.

Thoas, king of Scythia (bass) in Gluck's *Iphigénie en Tauride.*

Thomas, Ambroise, composer. Born Metz, France, August 5, 1811; died Paris, February 12, 1896. He attended the Paris Conservatory, where he won many prizes, including the Prix de Rome. In Rome he wrote several choral, orchestral and chamber works. After returning to Paris in 1836 he concentrated on music for the stage. He completed his first opera in 1837, *La double échelle,* produced at the Opéra-Comique. He wrote a number of other operas for the Opéra-Comique and some ballets for the Opéra, before achieving his first major success with *Mina,* given at the Opéra-Comique in 1843. A succession of operas continued to flow from his pen, all of them performed at the Opéra-Comique, culminating with *Mignon,* introduced on November 17, 1866. *Mignon* was Thomas's triumph. In less than thirty years it was given over a thousand performances; on the occasion of its thousandth performance (1894) Thomas was honored with the rank of Grand Cross of the Legion of Honor. After *Mignon,* his most important opera was *Hamlet* (1868), his first work since 1840 that was a grand opera rather than an opéra comique. Introduced by the Paris Opéra, it was a major success.

In 1851 Thomas became a member of the Institut de France. Five years later he was appointed a professor of composition at the Conservatory. In 1871 he succeeded Auber as the director of the Conservatory. His most important operas were: *Mina* (1843); *Le Caïd* (1849); *Le songe d'une nuit d'été* (1850); *Raymond* (1851); *Psyché*

(1857); *Le Carnaval de Venise* (1857); *Mignon* (1866); *Hamlet* (1868); *Françoise de Rimini* (1882).

Thomas, John Charles, baritone. Born Meyersdale, Pennsylvania, September 6, 1891. After completing music study at the Peabody Conservatory, he appeared extensively in operettas and musical comedies, achieving great success in Sigmund Romberg's *Maytime.* Thomas made his concert debut in 1918. Six years later he appeared in opera for the first time in a performance of *Aïda* in Washington, D.C. He went to Europe in 1925 and was acclaimed for his performance in *Hérodiade* at the Théâtre de la Monnaie. Thomas remained three years with that company, singing fifteen major roles; he appeared in the world première of Milhaud's *Les malheurs d'Orphée.* After other successful appearances in London, Berlin, and Vienna, he returned to the United States in 1930 and sang with the Chicago Civic Opera. On February 2, 1934, he made his debut at the Metropolitan Opera in *La traviata.* He remained with the Metropolitan through the 1942–1943 season. In 1940 he appeared in the motion picture *Kingdom Come.* Successful also in concerts and radio, Thomas was appointed executive director of the Santa Barbara (California) Music Academy in 1951.

Thomas, Theodore, conductor. Born Essen, Germany, October 11, 1835; died Chicago, Illinois, January 4, 1905. This pioneer in the development of musical culture in America made his conductorial debut in the opera house. As concertmaster of the orchestra at the Academy of Music in New York in 1858, he took over a performance of *La Juive,* when the regular conductor became indisposed. Though Thomas was known chiefly as conductor of concerts, in 1885 he became conductor of the newly formed American Opera Company (which see). Despite some

brilliant performances, the company was a failure and collapsed after a single season. When it was succeeded by the National Opera Company, Thomas continued as one of the conductors for a single season.

Thomson, Virgil, composer, critic, and writer on music. Born Kansas City, Missouri, November 25, 1896. He took courses in music at Harvard University. After his graduation (1922), he continued his music study in Paris with Nadia Boulanger. He remained in Paris until 1932, devoting himself to composition. His first major work was a provocative opera, *Four Saints in Three Acts,* the text by Gertrude Stein. The work was introduced in Hartford, Connecticut, in 1934. Thomson's second opera, *The Mother of Us All* (once again with text by Gertrude Stein) was commissioned by the Alice M. Ditson Fund and introduced in New York in 1947. In 1940 Thomson succeeded Lawrence Gilman as music critic of the New York *Herald Tribune.* He held this post until 1954, dropping it to concentrate on composition. He has written scores for several motion pictures; one of these, *Louisiana Story,* won the Pulitzer Prize for music in 1949. Thomson has also written a number of books on music.

Thorborg, Kerstin, contralto. Born Venjan, Kopparbergs Iän, Sweden, May 19, 1896. Her study took place at the opera school of the Stockholm Opera. She made her debut with this company in *Aïda.* Besides singing at the Stockholm Opera, she appeared successfully with the Berlin Opera, the Vienna State Opera, at Covent Garden, and at the Teatro Colón in Buenos Aires. Her success in the *Ring* dramas at Covent Garden in 1936 brought her a contract with the Metropolitan Opera, where she made her debut on December 21, 1936, in *Die Walküre.* During the next decade, she was acclaimed at the Metropolitan for her performances in the Wagner dramas, Strauss's *Elektra* and *Der Rosenkavalier,* and Gluck's *Orfeo ed Euridice.* She also was outstandingly successful at several Salzburg Festivals. In 1944 she was appointed singer to the Swedish court.

Thousand and One Nights, *see* ARABIAN NIGHTS.

Threepenny Opera, The (Die Dreigroschenoper), comic opera by Kurt Weill. Libretto by Bertolt Brecht. Première: Berlin, August 31, 1928. This cynical modern descendant of John Gay's *The Beggar's Opera,* set to German jazz, was founded on the popular eighteenth century work but bears little resemblance to it. It had a run of some four thousand performances in about one hundred and twenty German theaters, and was popular in other countries, including the United States. It was also made into a motion picture in Germany. In 1952 the American composer Marc Blitzstein brought Weill's opera up to date by writing new lyrics and adapting the text—he made no changes in the score. This new version was played successfully in New York.

Tibbett, Lawrence (born TIBBET), baritone. Born Bakersfield, California, November 16, 1896. He did not begin to study singing seriously until after he had made many church and light-opera appearances. His principal study took place with Frank La Forge and Basil Ruysdael. He made his debut at the Metropolitan Opera on November 23, 1923, as a monk in *Boris Godunov.* He continued appearing in minor roles until there took place an event that has been described as "without precedent in the annals of the Metropolitan." On January 2, 1925, at a revival of *Falstaff,* Tibbett had to serve as a last-minute replacement for the singer scheduled to appear as Ford. His exciting performance, both vocally and histrionically, elicited one of the most stirring ovations in the history of the opera house. That evening lifted Tibbett to stardom, and from then on he was seen in the principal baritone roles of many Italian and French operas; in the world premières of *The Emperor Jones, The King's Henchman,* and *Peter Ibbetson;* and in the Metropolitan premières of *Jonny spielt auf, Peter Grimes, Khovantchina,* and *Simon Boccanegra.* He was also successful on the concert stage, in radio appearances, and in motion pictures.

Tiefland (Lowland), opera by Eugène d'Albert. Libretto by Rudolph Lothar, adapted from *Terra Baixa,* a Spanish play by Angel Guimerá. Première: Prague, November 15, 1903. Pedro, a shepherd who lives atop a mountain in the Pyrenees, dreams that he will be sent a bride. The landowner Sebastiano gives him Martha, on condition that he live in the lowland. The reason for the condition becomes clear to Pedro when he learns that Sebastiano has betrayed Martha. Her shame revealed, Martha begs the shepherd to kill her, but his love is too great. When Sebastiano tries to detain the girl, Pedro strangles him, afterward returning to his mountain with Martha. The most celebrated aria is the Wolfserzählung of Pedro in Act I, "Schau her, das ist ein Taler," describing a battle with a wolf.

Tietjen, Heinz, conductor. Born Tangier, Morocco, June 24, 1881. He began conducting opera performances in Treves in 1904, where he remained for almost two decades, becoming the opera company's artistic director in 1907. After 1919 he held the post of artistic director with the Saarbrücken Opera and the Breslau Opera. In 1927 he became director of the Prussian State Theaters, and from 1925 to 1930 director of the Städtische Oper in Berlin. After 1931 he was for a number of years artistic director of the Bayreuth

Festivals, where he occasionally conducted some of the performances.

Timur, dethroned Tartar king (bass), in Puccini's *Turandot.*

Ti rincora, amata figlia, *see* AU SECOURS DE NOTRE FILLE.

Tisbe, Cinderella's stepsister (contralto) in Rossini's *La cenerentola.*

Titania, queen of the fairies (speaking role) in Weber's *Oberon.*

Titurel, retired king of the Knights of the Grail (bass), father of Amfortas, in Wagner's *Parsifal.*

Titus, Roman emperor (tenor) in Mozart's *La clemenza di Tito.*

Toby, Madama Flora's servant (mute) in Menotti's *The Medium.*

Toch, Ernst, composer. Born Vienna, Austria, December 7, 1887. The winning of the Mozart Prize in 1909 enabled him to attend the Hoch Conservatory in Frankfort-on-the-Main. Recognition of his talent came quickly. In 1910 he received the Mendelssohn Prize, and for four consecutive years the Austrian State Prize for chamber-music works. During World War I he served in the Austrian Army. He emerged as an important composer after the war with major orchestral and chamber works and a delightful chamber opera, *The Princess on the Pea.* In 1932 he visited the United States for the first time, appearing as soloist with the Boston Symphony in a program that included several of his works. After the rise of Hitler, Toch settled permanently in the United States and became a citizen. He wrote music for motion pictures and taught at the University of Southern California. His operas: *Wegwende* (1925); *The Princess on the Pea* (1927); *Egon und Emilie* (1928); *The Fan* (1930).

Todd, Miss, the old maid (contralto) in Menotti's *The Old Maid and the Thief.*

T'odio casa dorata, Gérard's aria in Act I of Giordano's *Andrea Chénier.*

Tokatyan, Armand, tenor. Born Plov-

div, Bulgaria, February 12, 1899. He appeared in operettas in Paris before studying singing seriously in Milan with Nino Cairone. In 1921 he made his opera debut in Milan in *Manon Lescaut.* In the same year he came to the United States and appeared with the Scotti Opera Company. On February 14, 1923, he made his Metropolitan Opera debut in the American première of Vittadini's *Anima Allegra.* He remained at the Metropolitan a decade, returning for three additional periods: 1935–1937, 1938–1942, and 1943–1946. He appeared in other American premières: those of *Le preziose ridicole, Il Signor Bruschino,* and *La vida breve.* He also made concert and radio appearances. After leaving the Metropolitan in 1946, he continued to sing in Europe.

Tolstoy, Leo, novelist. Born Tula, Russia, August 28, 1828; died Astapovo, Russia, November 20, 1910. Three of his novels have been made into operas: *Anna Karenina* by Jenö Hubay, *Resurrection* by Franco Alfano, *War and Peace* by Prokofiev.

Tomb Scene, *see* FRA POCO A ME RICOVERO.

Tommaso, a village patriarch (bass) in D'Albert's *Tiefland.*

Tonio, (1) Marie's lover (tenor) in Donizetti's *The Daughter of the Regiment.*

(2) A clown (baritone) in Leoncavallo's *Pagliacci.*

Toreador Song, Escamillo's aria in Act II of Bizet's *Carmen.*

Torquemada, a clockmaker (tenor) in Ravel's *L'heure espagnole.*

Tosca, opera in three acts by Giacomo Puccini. Libretto by Giuseppe Giacosa and Luigi Illica, based on Victorien Sardou's drama *La Tosca.* Première: Teatro Costanzi, Rome, January 14, 1900. United States première: New York, Metropolitan Opera House, February 4, 1901.

Characters: Floria Tosca, an opera

singer (soprano); Mario Cavaradossi, a painter (tenor); Baron Scarpia, chief of police (baritone); Cesare Angelotti, a political plotter (bass); Spoletta, a police agent (tenor); Sciarrone, a gendarme (bass); a Sacristan (baritone); a jailer, executioner, shepherd, judge; townspeople, guards. The setting is Rome in 1800.

Act I. The Church of Sant' Andrea della Valle. Angelotti, fleeing from the police, hides in the church, unseen by anybody. Mario Cavaradossi is here, too, painting a portrait of one of the worshipers who has caught his eye. He is unaware that his model is Angelotti's sister. In love with Tosca, Cavaradossi removes her miniature from his pocket and becomes rhapsodic over her beauty ("Recondita armonia"). As he starts painting again, he is accosted by Angelotti, his old friend, who asks for, and gets, his help. At the sound of Tosca's voice, Angelotti conceals himself. Tosca appears. She is enraged because, having heard whispered voices, she is suspicious that her lover has been having a secret meeting with a woman. Cavaradossi soothes her. The lovers exchange ardent sentiments ("Non la sospiri la nostra casetta"). When Tosca leaves, Cavaradossi guides Angelotti out of the church. Scarpia arrives, searching for Angelotti. The police chief comes upon a fan belonging to Angelotti's sister. When Tosca returns to spy on Cavaradossi, Scarpia shows her the fan and readily arouses her jealousy by suggesting it belongs to the woman of Cavaradossi's portrait. The church services now begin ("Te Deum"). As Scarpia kneels, he thinks of his forthcoming destruction of Cavaradossi and conquest of Tosca.

Act II. Scarpia's apartment. Angelotti cannot be found. Cavaradossi, who has been brought to Scarpia, hotly disclaims any knowledge of the refugee's whereabouts. When Tosca appears she rushes to her lover, but Cavaradossi is led to an adjoining room. Scarpia opens the door so that Tosca may hear her lover's anguished cries as he is being tortured, catch a glimpse of his suffering. Unable to stand the sight, Tosca reveals where Angelotti can be found (for Cavaradossi has told her his secret), and Cavaradossi is released. The news comes that Napoleon has won a great victory at Marengo. Cavaradossi shouts with joy. For this, Scarpia condemns him to be executed. After Cavaradossi is led away, Scarpia tries to win Tosca's love. With anguish, Tosca muses how cruelly fate has treated her, she who has devoted her life to art, love, and prayer ("Vissi d'arte"). Spoletta now brings the tidings that Angelotti has killed himself at the moment of his capture. Scarpia suggests that Cavaradossi will be the next to die, unless Tosca wishes to save him. Tosca promises to give herself to Scarpia if Cavaradossi's life be spared. Scarpia summons Spoletta and orders a mock execution for the prisoner, secretly adding a counterorder. He now approaches Tosca to claim his reward. Tosca, believing that she has preserved her lover's life, plunges a dagger into Scarpia's heart.

Act III. The terrace of the prison castle. Cavaradossi, in his cell, prepares for death by bidding his memory of Tosca farewell ("E lucevan le stelle"). Tosca arrives. She shows her lover the passport she obtained from Scarpia before she took his life, explaining that Cavaradossi must fall as if dead when blank cartridges are fired at him. Cavaradossi is led to the wall and shot —not with blanks. Tosca is stunned as she learns of Scarpia's final treachery. Spoletta and soldiers come to arrest her. She evades them by climbing the parapet and hurling herself into space.

Sardou's blood-and-thunder drama attracted two other composers before Puccini set it to music. One was Verdi, who decided he was too old to undertake the assignment. The other was Alberto Franchetti, who signed a contract with Sardou giving him exclusive rights to the play. Puccini became impressed with the operatic possibilities of the drama when he saw Sarah Bernhardt act it, but not until a decade later, when he read that Franchetti had acquired the opera rights, did he actively want to set the play. There followed a discreditable intrigue involving not only Puccini, but also Franchetti's librettist, Illica, and the publisher Ricordi. The conspirators were finally successful in convincing Franchetti not to write the opera. When Franchetti gave up his contract, Puccini made his own with Sardou. Understandably, Franchetti never forgave Puccini.

Tosca did not at first seem the kind of drama that suited Puccini's talent which, up to now, had been at its best in tender and sentimental plays. *Tosca* was lurid, filled with horror, sadism, murder, and suicide. However, with a true dramatist's instinct, Puccini changed his style to meet the demands of the play. His beautiful lyricism, however, was far from forgotten. The over-all effect of *Tosca* is one of compelling drama, but several of its arias are among the most memorable that Puccini ever wrote.

Toscanini, Arturo, conductor. Born Parma, Italy, March 25, 1867. For over half a century he was a giant figure in opera performances at La Scala, the Metropolitan Opera, Bayreuth, and Salzburg. He attended the Parma Conservatory from which he graduated in 1885 with the highest ratings. For a while he played the cello in various opera orchestras. The conductor of a touring company resigned just before a performance of *Aïda* in Rio de Janeiro, on June 26, 1886. Toscanini left his seat in the orchestra and took over the baton. Without opening the score, he directed with such authority

and brilliance that he was given an ovation. From that night on, Toscanini was the company's conductor; during the remainder of the tour he directed eighteen different operas, all of them from memory.

Back in Italy, he continued to conduct opera performances with such distinction that he was soon acclaimed as the most brilliant of the younger conductors. In 1896 he was given the most important operatic post in Italy: that of principal conductor and artistic director of La Scala. After his debut there on December 26, 1898, leading *Die Meistersinger,* Toscanini helped write one of the most brilliant chapters in the history of La Scala. He enriched the repertory through the introduction of many rarely heard or new German, French, and Russian operas. He instituted rigorous rehearsals, and made exacting demands on every department. Such a standard of performance was realized that La Scala, under Toscanini, became one of the greatest of the world's opera houses.

During this period, Toscanini remained only three seasons at La Scala. His regime came to an end with dramatic suddenness when, at a performance of *Un ballo in maschera,* the audience refused to comply with his rule against encores. When the audience persisted in its cries for a repetition of an aria, Toscanini left the opera house in the middle of the performance, and refused to conduct again. In 1906 he returned to La Scala, after promises had been made that every artistic demand would be adhered to. This time he stayed two seasons.

When Giulio Gatti-Casazza was engaged as one of the managers of the Metropolitan Opera in 1908, he induced Toscanini to come with him. Disturbed by his frequent clashes with La Scala officials, Toscanini welcomed a change of scene. On November 16, 1908, he made his first appearance at the Metropolitan Opera, conducting *Aïda,* "the finest performance of *Aïda* ever given in New York," as one of the critics wrote. A month later he directed his first Wagner performance in America, *Die Götterdämmerung.* Toscanini's performances at the Metropolitan set new criteria. He performed twenty-nine different operas, including the world premières of *The Girl of the Golden West* and *Madame Sans-Gêne,* and such novelties as *L'amore dei tre re, Ariane et Barbe-Bleue,* Gluck's *Armide,* and *Boris Godunov.*

Toscanini resigned from the Metropolitan after the 1914–1915 season. He had been involved in a Herculean struggle to achieve perfect performances, and he was weary of struggles with temperamental singers. When, in 1920, plans were made to reopen La Scala after its period of darkness during the war, a group of wealthy patrons offered to pay all the bills if Toscanini would return as artistic director. Toscanini consented on the condition that his word would be the law. He now had a free hand in matters of repertory, number of rehearsals, selection of singers. Limitless financial and artistic resources were placed at his command. On December 26, 1921, La Scala reopened with Toscanini directing *Falstaff.* Toscanini remained at his directorial post for eight years, a period that was another of La Scala's greatest. He resigned in 1929 because he felt he no longer had the physical strength to carry the crushing burdens of an opera house. His career was now devoted primarily to symphonic music. He became musical director of the New York Philharmonic-Symphony, and after that of the NBC Symphony Orchestra, which had been founded for him. On special occasions he returned to the theater: at the Bayreuth Festivals in 1930 and 1931, where he was the first Italian conductor, and at the Salzburg Festivals. He also led distinguished radio performances of operas with the NBC Symphony. He led his last concert with this orchestra on April 4, 1954, an all-Wagner program. After this, at the age of eighty-eight, he began his retirement.

tote Stadt, Die (The Dead City), opera by Erich Korngold. Libretto by Paul Schott, based on a play by Georges Rodenbach. Première: Hamburg Opera and Cologne Opera (simultaneously), December 4, 1920. The three acts are described as pictures, and are filled with dream sequences. Paul is a widower who lives in the past with haunting memories of his dead wife. He meets Marietta, a dancer who is the image of his dead wife, and he falls in love with her; but he loves in her only that which reminds him of his wife. Marietta is determined to have Paul love her for herself alone. When she desecrates the memory of the dead woman by putting on her hair, which Paul has saved and cherished, Paul kills her. But the murder turns out to be one of his dreams. He now realizes that he must forget his wife for good.

Tourel, Jennie, mezzo-soprano. Born Montreal, Canada, June 18, 1910. After studying with Anna El-Tour in Paris, she made her debut at the Opéra-Comique in 1933 as Carmen. The exceptional range of her voice enabled her to appear there in a great variety of roles. She made her American debut at the Metropolitan Opera on May 15, 1937 in *Mignon.* This was the sole appearance she made at the Metropolitan in this period. In 1942 she became famous when Toscanini selected her to sing in a performance of Berlioz' *Romeo and Juliet* by the New York Philharmonic Symphony. She returned to the Metropolitan Opera in 1943, and from then on appeared successfully in many roles, including that of Rosina in *The Barber of Seville,* which Metropolitan Opera audiences heard for the first time as Rossini wrote it, for coloratura mezzo-soprano. As a member of the Metropolitan, she made guest appearances with the New York City Opera during its initial season in 1944. In recent years she has sung in South America, at the Opéra-Comique, and at festival performances in Holland, Edinburgh, and Venice. She has also appeared extensively in the United States in recitals and as soloist with orchestras.

Tous les trois réunis (Stretti insiem tutti tre), trio of Marie, Tonio, and Sulpizio in Act II of Donizetti's *The Daughter of the Regiment.*

Traft ihr das Schiff, Senta's ballad in Act II of Wagner's *Der fliegende Holländer.*

Transatlantic, opera by George Antheil. Libretto by the composer. Première: Frankfurt Opera, May 25, 1930. A jazz opera, *Transatlantic* deals with an American presidential candidate and his hunt for a beautiful woman. The opera was thoroughly American in style and approach. One of its most provocative moments is an aria sung in a bathtub.

Transformation Scene, music accompanying the change of scene in Act I of Wagner's *Parsifal* from the forest to the castle of Monsalvat.

Traubel, Helen, soprano. Born St. Louis, Missouri, June 20, 1899. She began to study singing when she was thirteen with her first and only teacher, Vetta Kerst. Three years later she made her debut as soloist with the St. Louis Symphony. Though she was offered a contract by the Metropolitan Opera in 1926 she turned it down, feeling she was not yet ready. She continued studying for the next eight years, preparing roles, and singing in churches and synagogues. When she appeared as soloist with the St. Louis Symphony in an all-Wagner program conducted by Walter Damrosch, the conductor was so impressed that he asked her to appear in the leading female role of his new opera, *The Man Without a Country.*

When Traubel made her debut at the Metropolitan Opera on May 12, 1937, it was in the role of Mary Rutledge in Damrosch's opera.

After making many successful concert appearances, Traubel returned to the Metropolitan Opera on December 28, 1939, appearing in her first Wagner role, that of Sieglinde. She was such a success that for two years she divided the leading Wagnerian soprano roles with Kirsten Flagstad. When Flagstad left the Metropolitan Opera in 1941, Traubel became the principal Wagnerian soprano of the company. Within a few years her reputation as one of the great Wagnerian sopranos of our time was firmly established. Traubel resigned from the Metropolitan after the 1952–1953 season, following a disagreement over her right to appear in night clubs. Since then she has appeared in concerts, night clubs, on radio, and television, and in motion pictures.

Traurigkeit ward mir zum Lose, Constanza's aria in Act II of Mozart's *The Abduction from the Seraglio.*

Traveling Companions, The, *see* ANDERSEN, HANS CHRISTIAN.

traviata, La (The Lost One), opera in four acts by Giuseppe Verdi. Libretto by Francesco Maria Piave based on Alexandre Dumas's *La dame aux camélias.* Première: Teatro la Fenice, Venice, March 6, 1853. American première: New York, Academy of Music, December 3, 1856.

Characters: Violetta Valery, a courtesan (soprano); Annina, her maid (soprano or mezzo-soprano); Giuseppe, her servant (tenor); Alfredo Germont, her lover (tenor); Flora Bervoix, her friend (mezzo-soprano); Giorgio Germont, Alfredo's father (baritone); Baron Douphol, Alfredo's rival (baritone); Gastone, Viscount of Létorières (tenor); Marquis d'Obigny, a nobleman (bass); Dr. Grenvil, a physician (bass); servants, ladies, gentlemen. The setting is in and around Paris, about 1840 (though some productions make the year 1700).

Act I. Violetta's house. A brief prelude contains two themes from the opera, one connected with Violetta's illness, the other with her poignant farewell to Alfredo. At a party, Alfredo Germont is introduced to the hostess, Violetta, who invites him to sing a drinking song as the guests drink a toast. He complies, and she and the guests join in the refrain (Brindisi: "Libiamo, libiamo"). When the guests leave the room, Violetta is seized by a fainting spell. Alfredo offers his assistance; he grows solicitous over her delicate health. He then confesses that he has loved her for over a year ("Un di felice"). Violetta protests that she is not worthy of his love, but Alfredo grows more passionate. The guests now return to bid their hostess good night. When Violetta is alone, she muses about Alfredo's love and her own sympathetic reaction to it ("Ah, fors' è lui"). Then she proudly exclaims she lives only for pleasure and freedom ("Sempre libera").

Act II. A country house. Violetta and Alfredo are living together. He is overjoyed that she has renounced for his sake her former life, and he is grateful that she has taught him the meaning of love ("De' miei bollenti spiriti"). Annina confides that Violetta has been selling her jewels to support him. Enraged, he rushes off to Paris to raise some money. While he is gone, his father comes to denounce Violetta. He finds her to be a charming and generous woman, but even this does not keep him from trying to break off the liaison. He tells Violetta that his daughter, about to marry a nobleman, is threatened with desertion if the scandal surrounding the name of Germont is not terminated. Poignantly, Violetta realizes that her affair with Alfredo must ultimately destroy not only all those

related to him but Alfredo himself. She writes a letter of farewell to the man she loves. But before she can run away —and while the elder Germont is out of the house—Alfredo returns. She lies, telling him she is off to Paris to gain the consent of the elder Germont for their marriage. Only after she leaves does Alfredo come upon her farewell letter. Believing she deserted him because she does not love him any longer, and is lonesome for the gaiety of Paris, Alfredo is heartsick. His father reappears and tries to console him with reminders of their happy home in the Provençe ("Di Provenza il mar"). But Alfredo is inconsolable.

Act III. Flora Bervoix's house in Paris. At a party, Alfredo is gambling and winning. Violetta is also a guest, having come with Baron Douphol. When Violetta and Alfredo meet, he ignores her and continues his gambling. But when the guests drift into the dining salon, Violetta approaches him and implores him to leave the house before he gets into trouble. Alfredo promises to leave only if Violetta goes with him. She insists she is unable to do so since she is bound by a promise. Convinced that she will not go with him because of her tie to Baron Douphol, Alfredo calls loudly to the guests to return. Before their eyes he hurls his money at Violetta. The Baron challenges him to a duel. Alfredo's father appears, and denounces his son for his outrageous behavior.

Act IV. Violetta's bedroom. Dying of tuberculosis, Violetta reads a letter from the elder Germont in which he promises that Alfredo will be allowed to return to her. But Violetta knows it is too late. She bids the world farewell ("Addio del passato"). Alfredo arrives. He falls on his knees and begs Violetta to forgive him, for he has learned the truth about Violetta's renunciation of him. He promises her they will return to their idyllic home near Paris

("Parigi, o cara"). Violetta listens, then sinks back in her bed, exhausted. The elder Germont comes with a physician, but both are too late. Violetta emits a cry of anguish and dies.

Today it is difficult to believe that an opera as popular as *La traviata* should have been a failure when first produced. Yet this was the case. In Venice, in 1853, the opera was rejected vehemently by the first-night audience. "*La traviata* last night a fiasco," Verdi reported to one of his pupils. "Is the fault mine or the singers'? I don't know at all. Time will decide." Time has decided that the fault did not lie with the composer but in other directions. First, the opera was produced in contemporary costumes and scenery, and the novelty of seeing an opera in the dress of the day jarred the audience. Second, the principal tenor had a cold and was in poor voice. Third, the play was regarded as immoral. Fourth, the sight of a buxom soprano pretending to die of a wasting disease was ludicrous. The opera was withdrawn. A little more than a year later it was revived in Venice under more favorable auspices. The setting and costuming were set back a century. Greater care was taken in the casting and production. The audiences could now enjoy, and without disturbing diversions, the uninterrupted flow of some of the most wonderful melodies Verdi had written, together with the emotional impact of a play made up of self-sacrifice, a misunderstanding between lovers, and their reconciliation when it was too late. The opera was now a triumph, and it has since remained one of the best-loved operas in the repertory.

Tra voi, belle, Des Grieux's mocking serenade in Act I of Puccini's *Manon Lescaut.*

Treulich geführt, the bridal chorus in Act III, Scene 1, of Wagner's *Lohengrin.*

Trial, The (Der Prozess), opera by Gottfried von Einem. Libretto by Boris Blacher and Heinz von Cramer, based on the novel of the same name by Franz Kafka. Première: Salzburg Festival, August 17, 1953. The two acts are subdivided into nine "pictures." Joseph K., a respectable bank employee, is arrested for unexplained reasons. A chain of nightmarish incidents follows, until he is finally led to court and condemned, though still unable to uncover the crime of which he is accused. A unifying element is a series of rhythmic patterns that reappear. Some of the performers fill several different roles in order to emphasize the fact that "these Kafka characters are not so much separate beings as the embodiment of more general figures."

Trinke Liebchen, trinke schnell, the drinking duet of Alfred and Rosalinde in Act I Johann Strauss's *Die Fledermaus.*

Trionfi, a trilogy of operas or "scenic cantatas" by Carl Orff. Librettos by the composer. Première of complete trilogy: La Scala, February 13, 1953. The three operas are *Carmina Burana* (which see), Catulli Carmina (which see), and *Trionfo di Aphrodite.*

Trionfo di Afrodite (The Triumph of Aphrodite), opera or "scenic cantata" by Carl Orff. Libretto by the composer, based on a Latin poem of Catullus and Greek poems of Sapho and Euripides. This is the concluding opera of the trilogy *Trionfi* (see above). The opera is sung in Latin and Greek.

Triquet, a Frenchman (tenor) in Tchaikovsky's *Eugene Onegin.*

Tristan und Isolde, music drama in three acts by Richard Wagner. Libretto by the composer. Première: Munich Opera, June 10, 1865. American première: New York, Metropolitan Opera House, December 1, 1886.

Characters: Tristan, a Cornish knight (tenor); King Mark of Cornwall, his uncle (bass); Isolde, Princess of Ireland (soprano); Brangäne, her attendant (mezzo-soprano); Kurwenal, Tristan's servant (baritone); Melot, King Mark's courtier (tenor); a shepherd; a helmsman; sailors; knights. The action takes place aboard ship, in Cornwall, and in Brittany, in legendary times.

The prelude is made up of basic motives from the opera, principally the themes of the "love potion" and Tristan's "love glance." The prelude is a sustained crescendo, followed by a decrescendo, conveying the growing passions of the lovers and their tragic fate.

Act I. Tristan's ship. Isolde is upset, for, being taken to Cornwall as King Mark's bride-to-be, she has fallen in love with Tristan, the King's representative. When she sends for Tristan, he refuses to see her. Bitterly, Isolde recalls how she spared Tristan's life after he had slain her beloved Morold, brother of the King of Ireland. Aware that her love can never be satisfied, she directs Brangäne to prepare a death potion. Brangäne prepares a love potion instead. When Tristan finally appears, Isolde begs him to share with her a cup of peace. Though filled with foreboding, Tristan consents. They drink. As the potion works its magic, they look at each other with overwhelming love, then cling to one another as the ship arrives at Cornwall.

Act II. A garden before Isolde's chamber. Though Isolde is now married to King Mark, the passion consuming her and Tristan has not abated. King Mark, suspicious of them, leaves for a hunting trip, intending to return home unexpectedly and catch the lovers off guard. Brangäne guesses his intent but Isolde is deaf to advice and warnings: she can think only of her reunion with Tristan. When he appears, they embrace passionately, then speak of their limitless love ("O sink' hernieder, Nacht der Liebe!"). From a distance, Brangäne warns them to take heed ("Habet Acht"). But they are conscious only of their overwhelming emotions.

A scream of terror by Brangäne precedes the sudden arrival of Kurwenal, hurrying to warn Tristan and Isolde that King Mark is coming. The King appears, followed by Melot. He is so grief-stricken at this visible proof of Isolde's infidelity that he is incapable of anger. He asks Isolde if she is ready to follow Tristan wherever he chooses to go. When Isolde replies quickly in the affirmative, Melot draws his sword and challenges Tristan. Tristan makes no effort to defend himself and is seriously wounded.

Act III. The courtyard of Tristan's castle in Brittany. Kurwenal is tenderly nursing the stricken Tristan. A shepherd passes, playing a reed pipe. When Tristan regains consciousness, Kurwenal explains how he has been brought here, insisting that Isolde will surely follow. Feverishly, Tristan begs Kurwenal to scan the horizon for the sight of her ship. The pipings of the shepherd suddenly grow quick and gay. Kurwenal looks over the rampart and joyfully announces the approach of a ship. Tristan, wild with joy, struggles to his feet. The reunion is brief: as Isolde clasps her lover, he dies in her arms. A second ship brings King Mark, come to forgive Tristan and to allow the lovers to go their way together. Kurwenal, unaware of this, draws his sword, and is slain by Melot. Isolde now bids her dead lover farewell (Liebestod—Love Death: "Mild und leise wie er lächelt"), then falls dead on his body.

Wagner completed *Tristan und Isolde* in 1859, in what might be called a breathing spell from the harrowing labors of composing his monumental *Ring* cycle. *Tristan* itself is a work as vast in concept and design, as bold in execution, as revolutionary in its approach to operatic traditions, and as exacting in its demands on singers and orchestra, as any of the *Ring* dramas. Wagner's thinking about the esthetics of the musical drama, his leitmotive technique, and his new melodic and orchestral speech are as mature in *Tristan* as in the *Ring*, and are realized with equal mastery. Indeed, from certain points of view, *Tristan* is a finer work than any of the succeeding dramas, with the possible exception of *Die Meistersinger*. There is in *Tristan* greater integration, dramatic unity, and clarity of design. Music and drama are one, as the single theme of human passion unfolds with shattering effect in both play and score.

The opera was not immediately recognized as the masterpiece it is. It waited six years for its première in Munich, when it was received so coldly that it was discarded after three performances. Actually, it would have been dropped after one presentation had it not been for the influence of Wagner's patron, King Ludwig. Meanwhile, a projected première in Vienna had been abandoned after fifty-seven rehearsals; singers and orchestra insisted that Wagner's music was unsingable and unplayable. Yet time has placed *Tristan und Isolde* high on the list of enduring operas, largely because of all Wagner operas and music dramas it is one of the most appealing and affecting.

Tristes apprêts, Thélaïre's air in Act I of Rameau's *Castor et Pollux*.

Tristram, Sir, see MICKLEFORD.

trittico, Il (The Triptych), trilogy of one-act operas by Puccini. Première: Metropolitan Opera, December 14, 1918. The operas are: *Il tabarro, Suor Angelica,* and *Gianni Schicchi,* and they are discussed under their titles.

Triumph of Joan, The, opera by Norman Dello Joio. Libretto by Joseph Machlis. Première: New York City, May 9, 1950. Author and composer tried here to develop the social and psychological implications in the story of Joan of Arc, a woman who believed so intensely in an ideal she was willing to die for it.

Troilus and Cressida, opera by William Walton. Libretto by Christopher Hassail, based on Geoffrey Chaucer's *Troilus and Cressida.* Première: Covent Garden, December 3, 1954. With the aid of Pandarus, Troilus woos and wins Cressida, who ultimately deserts him for Diomedes.

Trompeter von Säkkingen, Der (The Trumpeter of Säkkingen), opera by Victor Nessler. Libretto by Rudolph Bunge, based on the poem of the same name by Joseph Victor von Scheffel. Première: Wiesbaden, May 4, 1884. Just after the Thirty Years' War, the trumpeter Werner is in love with the noble lady Maria, but her parents want her to marry Damian, a nobleman. During a peasants' attack, Werner proves himself a hero, while his rival betrays his cowardice; at the same time, Werner turns out to be of noble birth after all. The marriage of Werner and Maria now gets the blessings of her parents. Werner's aria in Act II, "Behüt dich Gott," is the best-known excerpt.

Troubled Island, The, opera by William Grant Still. Libretto by Langston Hughes, based on his play *The Drums of Haiti.* Première: New York City Opera, March 1949. The hero is Jean Jacques Dessalines, who helped establish Haitian independence.

Trouble in Tahiti, one-act opera by Leonard Bernstein. Libretto by the composer. Première: Waltham, Massachusetts, June 12, 1952. This comedy in seven scenes centers around the domestic bickerings of a typical American couple in a typical American suburb. The opera has had a number of presentations: during the Berkshire Festival at Tanglewood, Massachusetts, on television, and (1955) as part of a Broadway theatrical offering.

trovatore, Il (The Troubadour), opera in four acts by Giuseppe Verdi. Libretto by Salvatore Cammarano, based on a play by Antonio Garcia Gutierrez. Première: Teatro Apollo, Rome, January 19, 1853. American première: Academy of Music, New York, May 2, 1855.

Characters: Leonora, lady-in-waiting to the Queen (soprano); Inez, her attendant (soprano); Manrico, an officer serving the Prince of Biscay (tenor); Ruiz, a soldier in his service (tenor); Count di Luna, a nobleman (baritone); Ferrando, his captain of the guards (bass); Azucena, a gypsy (contralto); soldiers, nuns, gypsies, attendants, jailers. The action takes place in Biscay and Aragon in the middle of the fifteenth century.

Act I, Scene 1. The palace at Aliaferia. Ferrando regales the soldiers of the Queen with the story of Count di Luna and his long-lost brother ("Di due figli vivea"): A gypsy, burned by the Count's father as a witch, was avenged by her daughter, who kidnaped Count di Luna's younger brother. This happened many years ago, and it is believed that the stolen child was thrown into the flames that consumed the gypsy. As Ferrando finishes his gruesome story, a clock strikes midnight.

Scene 2. The palace gardens. Leonora reveals to Inez that a mysterious troubadour has been serenading her and that she has fallen in love with him ("Tacea la notte placida"). Inez fears that the unknown troubadour will bring her lady misfortune. When the ladies leave the garden, Count di Luna appears. In love with Leonora, he has come to tell her of his feelings. In the distance he hears the serenade of the troubadour ("Deserto sulla terra"). Leonora has also heard the song, and mistaking the Count for the troubadour she rushes to him ardently. When she recognizes her mistake, the Count becomes enraged and rushes into the shadows to challenge the troubadour

to a duel. He proves to be Manrico, leader of a rival army. As Count di Luna and Manrico cross swords, Leonora faints.

Act II, Scene 1. A gypsy camp in Biscay. Gypsies, working at a forge, sing as they swing their hammers (Anvil Chorus: "Vedi! le fosche notturne spoglie"). Near by, Azucena recalls an episode the time when her mother was burned as a witch ("Stride la vampa"). Some of the gypsies try to console her but Azucena is bitter and urges Manrico, who is with her, to avenge this cruel murder. She tells Manrico the rest of the story: how she stole the infant brother of Count di Luna in order to kill him, but instead, in her madness, threw her own babe into the fire. The story raises in Manrico's mind doubts about his own origin, but Azucena insists that he is rightfully hers, since she found him dying on a battlefield and brought him back to health. Manrico now reveals to Azucena that when he recently fought a duel with Count di Luna, for some inexplicable reason he spared the Count's life. The tale infuriates Azucena; she fiercely urges Manrico to strike down the Count ruthlessly. Ruiz arrives with word that Manrico must return to his troops. Manrico also learns that Leonora, believing that he has died in battle, is about to enter a convent. Azucena tries to restrain him from leaving, but Manrico pushes her aside.

Scene 2. The convent near Castellor. Count di Luna and his men have come to abduct Leonora. He speaks of his great love for her ("Il balen del suo sorriso"), then tells his followers he cannot live without her ("Per me ora fatale"). When Leonora and a group of nuns pass, the Count and his men seize her. Manrico and a band of henchmen arrive before the abduction; Manrico rescues his beloved and bears her off.

Act III, Scene 1. A military encampment. Manrico has taken Leonora to Castellor, which is now being attacked by Count di Luna. The Count's soldiers sing a hymn to war and victory ("Squilli, echeggi la tromba guerriera"). Bitter that the woman he loves is with another man, Count di Luna vows to avenge himself. Ferrando now brings him the news that the gypsy Azucena has been captured. When the Count subjects her to questioning, she tells him of her past ("Giorno poveri vivea"). She tells him she is Manrico's mother. When he suspects that she is guilty of his brother's death, he condemns her to death by burning.

Scene 2. The fortress of Castellor. Manrico and Leonora are about to be married. Manrico promises he will be true till death ("Ah si, ben mio"). Ruiz brings information about Azucena's imprisonment and forthcoming execution. Manrico vows to save her ("Di quella pira"). His soldiers shout their allegiance.

Act IV, Scene 1. The palace at Aliaferia. Captured, Manrico waits in a cell. Leonora comes in disguise, hoping to catch a glimpse of her beloved. She prays that her love for him will sustain him through his suffering ("D'amor sull' ali rosee"). A bell tolls. Within the castle voices intone a prayer for the doomed prisoners (Miserere: "Ah! che la morte ognora"); the solemn chanting impels Leonora to express her own concern. When Count di Luna appears, she hides, to emerge boldly when she overhears him giving orders concerning the executions. Leonora offers herself in return for Manrico's life. When the Count accepts the bargain, she secretly takes poison.

Scene 2. Manrico's cell. Azucena, tormenting herself with recollections of her mother's death, is soothed by Manrico. They console each other with the hope that someday they may return to their happy mountain land ("Ai nostri monti"). After Azucena falls asleep, Manrico welcomes Leonora who has come to tell him he is free to go. When she refuses to accompany him, Manrico learns the price of his freedom. While he denounces her for betraying their love, she sinks to the ground in agonizing pain. Manrico takes her in his arms. Learning that she has poisoned herself, he begs her to forgive him. Count di Luna enters the cell, takes in the situation, and orders Manrico's immediate execution. When he forces Azucena to watch the execution from the window, the gypsy, crazed with grief, cries out that Count di Luna has just killed his brother.

Though the Tiber River overflowed its banks the day that *Il trovatore* was first heard, and Roman opera-goers had to wade through water and mud to reach the Teatro Apollo, the première was a triumph. "The public listened to every number with religious silence and broke out with applause at every interval," reported the *Gazzetta Musicale*. "The end of the third act and the whole of the fourth arousing such enthusiasm that their repetition was demanded." To this day the score brings undiluted pleasure to the listener, and is a source of awe to the professional musician. Verdi's genius gave him the wings to soar above the confusion of an involved libretto, the inspiration for some of the most celebrated pages in all opera.

Troyens, Les (The Trojans), opera by Hector Berlioz. Libretto by the composer, based on Virgil. Originally conceived in six acts, it was subsequently divided into two operas bearing the titles *La prise de Troie* and *Les Troyens à Carthage*. Première of the two operas together: Karlsruhe Opera, December 6 and 7, 1890 (sung in German). *See* PRISE DE TROIE, LA.

Troyens à Carthage, Les (The Trojans at Carthage), opera by Hector Berlioz. Libretto by the composer, based on Virgil. Première: Paris, Théâtre Lyrique, November 4, 1863. This is the second of the two operas that make up *Les Troyens* (see above). Dido and Aeneas are in love, but Aeneas must sacrifice himself in order to fulfill his destiny of founding a state in Italy. Failing to delay his going, Dido kills herself with Aeneas' sword, prophesying that someone—Hannibal—will arise in time to come to avenge her.

Tu che la terra adora, *see* DIEU, QUE LE MONDE REVERE.

Tucker, Richard, tenor. Born Brooklyn, New York, August 28, 1916. He studied with Paul Althouse, then served as cantor at the Brooklyn Jewish Center and as leading tenor of the Chicago Theater of the Air. He made his debut at the Metropolitan Opera on January 25 1945, as Enzo in *La Gioconda*. Since then he has been a principal tenor of the Metropolitan Opera. In 1949 Toscanini selected him to sing the role of Radames in the concert performance of *Aïda* he conducted with the NBC Symphony. Tucker is the brother-in-law of Jan Peerce, another distinguished Metropolitan Opera tenor.

Tu m'as donné le plus doux rêve, duet of Lakmé and Gérald in Act III of Delibes's *Lakmé*.

Turandot, (1) a play by Carlo Gozzi which has been made into operas by numerous composers. *See* GOZZI, CARLO.

(2) Opera by Ferruccio Busoni. Libretto by the composer, based on the Gozzi play: Première: Zurich, May 11, 1917. Busoni originally wrote incidental music for Max Reinhardt's production of Karl Vollmöller's version of the Gozzi play, given at the Deutsches Theater on October 27, 1911. He then expanded his score into an opera.

(3) Opera in three acts by Giacomo Puccini. Libretto by Giuseppe Adami and Renato Simoni, based on Gozzi's play in an adaptation by Johann Friedrich Schiller. Première: La Scala, April 25, 1926. American première: New

York, Metropolitan Opera House, November 16, 1926.

Characters: The Emperor Altoum (tenor); Princess Turandot, his daughter (soprano); Timur, dethroned king of the Tartars (bass); Prince Calaf, the Unknown Prince, his son (tenor); Liù, a slave girl (soprano); Ping, Grand Chancellor (baritone); Pang, the General Purveyor (tenor); Pong, the Chief Cook (tenor); Pu-tin-Pao, executioner (baritone); a mandarin (baritone); Prince of Persia (baritone); priests, mandarins, slaves, people of Peking, imperial guards, eight wise men, slaves. The setting is Peking in legendary times.

Act I. Before the Imperial Palace in Peking. Princess Turandot has announced she will marry any man of noble blood who can answer three riddles; but he who tries and fails must die. The people listen to a mandarin pronounce the death sentence for a Persian prince who has failed his opportunity. In the crowd is an old man, Timur, who has been thrown to the ground. He is helped by Liù, and by his son, Calaf. When the Persian prince is led to his execution the crowd cries out for mercy ("O giovinetto"). Turandot's appearance silences the crowd, which now falls to its knees. But Calaf remains standing; he condemns the Princess for her cruelty. He is also so moved by her beauty that he offers to try answering her riddles. Liù, who loves him, begs him to desist from this folly ("Signore, ascolta"), reminding him that the only thing that sustained her during his exile was her memory of him. Calaf consoles her ("Non piangere, Liù"), but remains stubborn in his desire to win Turandot.

Act II, Scene 1. A pavilion of the palace. Ping, Pang, and Pong read the names of Turandot's victims and pray that she may desist from her game and find true love (Terzetto delle maschere). A trumpet fanfare announces that a new candidate is ready for the test.

Scene 2. A square outside the palace. Turandot explains to a large assemblage the reason for her strange decree. Her grandmother had met an unhappy fate at the hands of Tartars. Turandot, consequently, has sworn to avenge this cruelty on all who aspire to love her. She now warns the Unknown Prince, about to be a contestant for her love, that he will be destroyed if he fails the test. Calaf accepts the conditions. One by one the riddles are presented, and one by one they are answered. The crowd acclaims Calaf ("Gloria, o vincitore"). Now faced with the necessity of marrying Calaf, Turandot begs him to release her from her promise. He offers to do so, but only if, by the following dawn, she can uncover his true identity. The people sing the praises of the Emperor ("Ai tuoi piedi ci prostriamo").

Act III, Scene 1. The palace gardens. The Emperor's heralds are scouring the city, trying to learn the identity of the Unknown Prince. There has been a royal edict that no one in Peking will sleep until this is accomplished. Calaf comments on this ("Questa notte"), then muses on how the Princess is troubled and how he plans to resolve her problems with a kiss ("Nessun dorma"). Timur and Liù are brought to the palace and ordered to reveal the Unknown Prince's name. Liù insists that she alone knows it and that not even the threat of death will compel her to divulge it. When the executioner comes to seize her, Liù stabs herself and dies. Calaf now approaches Turandot and boldly takes her in his arms. Turandot realizes she loves him and is awed when, for the sake of his own love for her, he is willing to accept death and free her of her promise. When he reveals his name, Turandot discovers that he is the prince of the hated Tartars and insists upon his death.

Scene 2. The pavilion of the Palace. The Emperor is surrounded by his court. Turandot tells the Emperor that she has learned the prince's name. Turning to Calaf she jubilantly announces that it is—Love. Turandot and Calaf embrace and the people sing a hymn to love ("O sole! Vita! Eternità!").

Turandot was Puccini's last opera, and he died before he could finish it. Puccini's score ends shortly after Liù's suicide. The opera composer Franco Alfano completed the opera by adding a duet and the concluding scene, the version in which the opera is now performed. However, when Arturo Toscanini directed the world première at La Scala, two years after Puccini's death, he insisted on performing the opera as Puccini had left it. Toscanini was carrying out Puccini's wish, made just before his death: "If I do not succeed in finishing the opera, some one will come to the front of the stage and say, 'At this point the composer died.'" When the music stopped, Toscanini turned to the audience and said, "Here the Maestro put down his pen."

Besides being his last, *Turandot* was also Puccini's most original opera. Never had his musical thinking been so advanced. Unorthodox tonalities, scales, dissonances, timbres permeate the score, yet the tender and affecting Puccini lyricism is not sacrificed. His last opera proved that the composer was growing artistically all the time, that he had the courage to tap new veins in his indefatigable search for greater artistic truth.

Turgeniev, Ivan, author. Born Orel, Russia, November 9, 1818; died Bougival, France, September 3, 1883. Turgeniev's stories have been made into the following operas: Ippolitov-Ivanov's *Aca,* Alexander Kastalsky's *Clara Militch,* and Antoine Simon's

The Song of Love Triumphant. The prima donna Pauline Viardot-García wrote three operettas to Turgeniev's librettos.

Turiddu, a soldier (tenor) in Mascagni's *Cavalleria rusticana.*

Turiddu's Farewell, the aria "Mamma, quel vino e generoso" in *Cavalleria rusticana.*

Turn of the Screw, The, chamber opera by Benjamin Britten. Libretto by Myfawny Piper, based on the novel of the same name by Henry James. Première: Venice, September 14, 1954. The story concerns two orphan children who are haunted by the malevolent spirits of a dead groom and a dead governess.

Tutta su me ti sposa, Manon's aria in Act IV of Puccini's *Manon Lescaut.*

Tutte le feste, Gilda's aria in Act III, of Verdi's *Rigoletto.*

Tutti i fior, the Flower Duet of Cio-Cio-San and Suzuki in Act II of Puccini's *Madama Butterfly.*

Tutto nel mondo è burla, closing chorus of Verdi's *Falstaff.*

Tutto tace, Fritz' and Suzel's duet of the cherries in Act II of Mascagni's *L'Amico Fritz.*

Tu, tu piccolo Iddio! Cio-Cio-San's farewell to her son in Act III of Puccini's *Madama Butterfly.*

Twelfth Night, see SHAKESPEARE.

Twelve-tone technique, a system of musical composition devised by Arnold Schoenberg, and used by him, and other composers of the atonal school, for the composition of operas and other musical works. See ATONALITY.

Twilight of the Gods (Die Götterdämmerung), see RING DES NIBELUNGEN, DER.

Two Foscari, The, see BYRON, GEORGE NOEL GORDON, LORD.

Tybalt, Juliette's cousin (tenor) in Gounod's *Roméo et Juliette.*

Tyl, Father, a woodcutter (baritone) in Wolff's *L'oiseau bleu.*

Tyl, Grandfather, his father (bass).

Tyl, Grandmother, his mother (contralto).

Tyl, Mother, Father Tyl's wife (contralto).

U

Ubaldo, a knight (baritone) in Gluck's *Armide.*

Udite, udite, o rustici, Dr. Dulcamara's aria in Act I of Donizetti's *L'elisir d'amore.*

Ulrica, a fortuneteller (contralto) in Verdi's *Un ballo in maschera.*

Una cosa rara (A Rare Thing), opera by Martín y Soler. Libretto by Lorenzo da Ponte, based on the story of the same name by Luis Vélez de Guevara. Première: Burgtheater, Vienna, November 17, 1786. This charming little opera was applauded by Mozart's ene-

mies (including Antonio Salieri) to check the mounting success of Mozart's *The Marriage of Figaro. Una cosa rara* was so successful that the Viennese public forsook Mozart's opera, which closed after nine performances. Mozart made an amusing quotation from this opera in the finale of *Don Giovanni,* where an orchestra entertains the Don with music, and Leporello comments: "Hurrah! That's *Cosa rara."* This opera by Soler contains one of the earliest successful Viennese waltzes. *The Siege of Belgrade,* an opera by Stephen Storace, written in 1791, was an adaptation of *Una cosa rara.*

Una donna a quindici anni, Despina's aria in Act II, Scene 1, of Mozart's *Così fan tutte.*

Una furtiva lagrima, Nemorino's aria in Act II of Donizetti's *L'elisir d'amore.*

Una macchia è qui tuttora, Lady Macbeth's sleep-walking scene, Act IV, Scene 2, of Verdi's *Macbeth.*

Un' aura amorosa, Ferrando's aria in Act I, Scene 3, of Mozart's *Così fan tutte.*

Una vergine, un angiol di Dio, Fernando's romanza in Act I, Scene 1, of Donizetti's *La favorita.*

Una voce poco fa, Rosina's aria in Act I, Scene 2, of Rossini's *The Barber of Seville.*

Un bel dì vedremo, Cio-Cio-San's aria in Act II of Puccini's *Madama Butterfly.*

Un dì all' azzurro spazio, Chénier's aria in Act I of Giordano's *Andrea Chénier.*

Un dì felice, the love duet of Alfredo Germont and Violetta in Act I of Verdi's *La traviata.*

Undine, opera by Gustav Albert Lortzing. Libretto by the composer, based on the story of the same name by Friedrich de la Motte-Fouqué. Première: Magdeburg Opera, April 21, 1845. This was Lortzing's first serious opera and it was a major success. The knight Hugo von Ringstetten marries Undine, daughter of a fisherman. He then discovers that she is a water-fairy who can win herself a soul, and immortality, through the love of a faithful man. When Hugo is unfaithful with Berthalda, Undine returns to her watery realm, but on the day Hugo marries Berthalda, Undine appears and lures him to his death beneath the water.

Und ob die Wolke sie verhülle, Agathe's prayer in Act III, Scene 1, of Weber's *Der Freischütz.*

Une fièvre brûlante, an air from André

Gretry's *Richard Coeur de Lion* that Beethoven used as a theme for his *Eight Variations in C,* for piano (opus 184).

Une nuit de Cléopâtre, see GAUTIER, THÉOPHILE.

Une poupée aux yeux d'émail, Nicklausse's song about a mechanical doll in Act I of Offenbach's *The Tales of Hoffmann.*

Une voix mystérieuse, Irma's aria in Act II, Scene 2, of Charpentier's *Louise.*

Unger, Caroline, contralto. Born Vienna, Austria, October 28, 1803; died Florence, Italy, March 23, 1877. She studied in Italy with various teachers, including Domenico Ronconi. In 1821 she made her debut in Vienna in *Così fan tutte.* Three years later she sang the contralto parts in the first performance of Beethoven's ninth symphony and *Missa Solemnis.* She appeared extensively in opera performances in Italy, where many operas were written for her, including Bellini's *La straniera,* Donizetti's *Maria di Rudenz* and *Parisana,* Mercadante's *Le due illustre rivali,* and Pacini's *Niobe.* In 1833 she was an outstanding success at the Théâtre des Italiens in Paris. After marrying François Sabatier in 1841 she went into retirement.

Unis dès la plus tendre enfance, Pliades' aria in Act II of Gluck's *Iphigénie en Tauride.*

uragano, L', see HURRICANE, THE.

Urbain, Marguerite's page (soprano or contralto) in Meyerbeer's *Les Huguenots.*

Urban, Joseph, scene designer. Born Vienna, Austria, May 26, 1872; died New York City, July 10, 1933. He studied architecture in Vienna. In 1901 he came to the United States, where he subsequently designed sets for the Boston Opera Company, and, for many years, for the Metropolitan Opera. He also designed sets for Covent Garden and the Vienna State Opera.

Urlus, Jacques, tenor. Born Hegenrath,

Germany, January 9, 1867; died Noordwijk, Holland, June 6, 1935. He was trained to be an engineer, but in 1887 he began to study singing seriously in Utrecht. After attending the Amsterdam Conservatory, he made his debut in Amsterdam on September 20, 1894, in *Pagliacci*. He subsequently distinguished himself as a Wagnerian tenor, being acclaimed at Bayreuth, the Théâtre de la Monnaie, and Covent Garden. He made his American debut in Boston on February 12, 1912, as Tristan, and made his first appearance at the Metropolitan Opera, once again as Tristan, on February 8, 1913. He remained at the Metropolitan through the 1916–1917 season. His last impor-

tant appearances took place at Covent Garden in 1924 under the direction of Bruno Walter.

Ursuleac, Viorica, dramatic soprano. Born Czernowitz, Rumania, March 26, 1899. For many years she was the principal soprano of the Vienna State Opera. Her success there was combined with appearances in Germany and at the Salzburg Festivals. In 1933 she created the leading soprano role in Richard Strauss's *Arabella* in Dresden, and five years after that in *Friedenstag*, in Munich; Strauss dedicated the latter opera jointly to her and her husband Clemens Krauss, the conductor. Beginning in 1933, she was the principal soprano of the Berlin opera.

V

Va, crudele, Pollione's aria in Act I of Bellini's *Norma*.

Vainement, ma bien aimée, Mylio's aubade in Act III of Lalo's *Le Roi d'Ys*.

vaisseau fantôme, Le (The Phantom Vessel), opera by Pierre Louis Dietsch. Libretto by Benedict Henri Révoil and Paul Henri Foucher, based on Richard Wagner's scenario for *Der fliegende Holländer*. Première: Paris Opéra, November 9, 1842. Wagner wrote his scenario, intending to use it himself for an opera to be produced by the Paris Opéra. But the Opéra's interest did not extend as far as his music. It bought only the scenario, turning it over to Dietsch. Wagner consented to this arrangement because of his poverty, and because it did not preclude his writing his own opera later. Dietsch's opera was not successful.

Va! laisse-les couler, Charlotte's song

of tears in Act III of Massenet's *Werther*.

Valentine, (1) Marguerite's brother (baritone) in Gounod's *Faust*.

(2) St. Bris' daughter (soprano) in Meyerbeer's *Les Huguenots*.

Valentine's Invocation, *see* AVANT DE QUITTER CES LIEUX.

Valery, Violetta, a courtesan (soprano) in Verdi's *La traviata*.

Valhalla, the abode of the gods in Wagner's *Der Ring des Nibelungen*.

Valkyrie, The (Die Walküre), *see* RING DES NIBELUNGEN, DER.

Valleria, Alwina (born SCHOENING), dramatic soprano. Born Baltimore, Maryland, October 12, 1848; died Nice, France, February 17, 1925. She was the first American-born singer to appear in principal roles at the Metropolitan Opera. After attending the Royal Academy of Music in London, she made her opera debut in St. Petersburg in 1871. Appearances followed in Germany, at La Scala, and at Drury Lane in London. She was a favorite in London and was heard at her Majesty's Theatre in 1877–1878 and at Covent Garden from 1879 to 1882. She made her American debut on October 22, 1879, with the Mapleson company at the Academy of Music in *Faust*. On October 26, 1883, she made her debut at the Metropolitan Opera, singing the role of Leonora in *Il trovatore*. In 1878 she created the role of Micaëla for New York and London. In 1884 she joined the Carl Rosa company, being heard in the world premières of Alexander Mackenzie's *The Troubadour* and Arthur Goring Thomas's *Nadeshda*. She retired from the stage in 1886.

Valzacchi, an intriguing Italian (tenor) in Richard Strauss's *Der Rosenkavalier*.

Vampyr, Der (The Vampire), opera by Heinrich Marschner. Libretto by Wilhelm August Wohlbrück, based on a French melodrama by Nodier, Carmouche, and De Jouffroy, in turn derived from a story by John William Polidori. Première: Leipzig, March 29, 1828. Lord Ruthven escapes doom at the hands of the spirits for three years, on the condition that each year he bring a pure maiden for sacrifice. A vampire in disguise, he makes Ianthe and Emmy his victims before he is discovered, and destroyed by a bolt of lightning.

Van Dyck, Ernest, dramatic tenor. Born Antwerp, Belgium, April 2, 1861; died Berlaer-Lez-Lierre, Belgium, August 31, 1923. After studying both law and journalism, he decided to become a singer. He studied with Saint Yves-Bax in Paris, then made his opera debut at the Théâtre Eden on May 3, 1887, in the French première of *Lohengrin*. After an intensive period of study with Felix Mottl, he appeared as Parsifal at Bayreuth in 1888, with outstanding success. He was immediately engaged by the Vienna Opera, where he remained a decade. He also made frequent guest appearances in other leading European opera houses; he was featured in the world première of *Der Evangelimann* in London in 1897. On November 29, 1898, he made his American debut at the Metropolitan Opera as Tannhäuser. He stayed at the Metropolitan through the 1901–1902 season, acclaimed not only in the Wagner dramas but also in the French repertory. In 1907 he managed a season of German operas at Covent Garden, and in 1914 he appeared in the first performance in Paris of *Parsifal*.

Vanna, Marco, commander of the Florentine army (tenor) in Février's *Monna Vanna*.

Vanne, disse, Alice's romance in Act I of Meyerbeer's *Robert le Diable*.

Van Rooy, Anton, baritone. Born Rotterdam, Holland, January 1, 1870; died Munich, Germany, November 28, 1932. After studying with Julius Stockhausen in Germany, he made his opera debut at Bayreuth in 1897 as Wotan in the *Ring* cycle. In 1898 he made his debuts in Berlin and London, and on December 14, 1898, his American debut at the Metropolitan Opera as Wotan in *Die Walküre*. He remained at the Metropolitan until 1908, heard in all the leading baritone roles of the Wagner repertory, including that of Kurwenal in the American première of *Parsifal*. During these years he also sang regularly at Bayreuth and Covent Garden. After leaving the Metropolitan, he became the leading Wagnerian baritone of the Frankfurt Opera. He also distinguished himself as a concert singer and a soloist in oratorio performances.

Vanya, Ivan Susanin's adopted son (contralto) in Glinka's *A Life for the Czar*.

Van Zandt, Marie, soprano. Born New York City, October 8, 1861; died

Cannes, France, December 31, 1919. She created the role of Lakmé. Her mother, Jennie van Zandt, sang at La Scala and at the Academy of Music in New York. Marie studied with Francesco Lamperti in Milan and in 1879 made her debut in Turin as Zerlina. Successful appearances followed, particularly at Covent Garden in 1879, and at the Opéra-Comique in 1880. Delibes wrote *Lakmé* for her and she appeared in its first performance in 1883. An organized opposition at the Opéra-Comique contributed to discredit her at this time; on one occasion, the false rumor was circulated that she appeared on the stage while inebriated. On December 21, 1891, she made her debut at the Metropolitan Opera as Amina; she stayed at the Metropolitan only one season. In 1896 she returned to the Opéra-Comique and revived there her earlier successes. Soon after this, she married and went into retirement.

Va, pensiero sull' ali dorate, chorus in Act III of Verdi's *Nabucco*.

Varnay, Astrid, dramatic soprano. Born Stockholm, Sweden, April 25, 1918. Her father, a stage manager, founded the first opera company in Oslo; her mother was a coloratura soprano. She came to the United States when she was five and later became a citizen. She joined the Metropolitan Opera Company in 1941. When Lotte Lehmann was unable to appear as Sieglinde on December 6, 1941, Varnay stepped in as a last-minute replacement. Six days later she made another unscheduled appearance, this time substituting for Helen Traubel as Brünnhilde, again in *Die Walküre*. Since then she has appeared at the Metropolitan in leading soprano roles, not only in the Wagnerian repertory, in which she excels, but also in Richard Strauss's *Salome* and *Elektra*, and in Italian and French operas. She also appeared in the world première of Menotti's *The Island*

God. In 1947 she sang in the first performance of Wagner's entire *Ring* cycle given at the Teatro Colón in Buenos Aires. Her European debut took place in 1951 at the Florence May Music Festival, when she sang in Verdi's *Macbeth.* The same summer she sang at the Bayreuth Festival—the first of many such appearances. She was the first American artist to sing Brünnhilde at Bayreuth. She has sung in other opera houses of Europe and South America. In 1951 she was selected by the United States State Department to appear at the Berlin Opera in the Allied Festival of the Arts. In 1944 she married Hermann Weigert, a member of the musical staff of the Metropolitan Opera, and artistic adviser for the Bayreuth Festival, who died in 1955.

Vasco da Gama, officer in the Portuguese navy (tenor) in Meyerbeer's *L'Africaine.*

Vassilenko, Sergei, composer. Born Moscow, Russia, March 30, 1872. He attended the Moscow Conservatory, his teachers including Sergei Taneiev and Ippolitov-Ivanov. As his graduation exercise, in 1900, he wrote an opera-oratorio, *The Legend of the Great City of Kitezh,* a subject used by Rimsky-Korsakov two years later. In 1906 he became a professor at the Moscow Conservatory, where he remained over thirty years. His earlier works, most of them for orchestra, were national in feeling and style. He then became interested in the folk music of Oriental people, and this influence prevails in his later works. In this vein he wrote his opera, *Son of the Sun,* about the Boxer Rebellion; it was introduced in Moscow on May 23, 1929. His other operas: *Christopher Columbus* (1933); *The Snow Storm* (1938); *The Grand Canal* (1940); *Suvorov* (1942).

Vassili, Prince Alexis' rival (baritone) for Stefana's love in Giordano's *Siberia*.

Vaughan Williams, Ralph, composer. Born Down Ampney, England, October 12, 1872. His academic education took place at Charterhouse and Trinity College, Cambridge; his study of music at the Royal College of Music in London, with Max Bruch in Berlin, and in 1908 with Maurice Ravel in Paris. Meanwhile, in 1904, he became acquainted with English folk songs. This music exerted such a fascination on him that he became a member of the Folk-Song Society and devoted himself to research in the field of folk music. His artistic development was profoundly affected; henceforth, his compositions showed folk influences. An important work in this vein, the *Fantasia on a Theme by Tallis*, was composed in 1910. Other orchestral works followed, placing him in the front rank of contemporary English composers. In 1914 he completed his first opera, *Hugh the Drover*, in which the influence of English backgrounds and folk music is again in evidence.

During World War I, Vaughan Williams served in the Territorial Royal Army Military Corps, and also saw active service in the Artillery. After the war he became professor of composition at the Royal College of Music, a post he held with distinction for over three decades. He also completed many major works, among them *The Shepherds of the Delectable Mountains* (1922), an opera-oratorio, and *Sir John in Love* (1929), an opera dealing with Falstaff. Some years later Vaughan Williams incorporated the first of these works in his opera *The Pilgrim's Progress*.

In 1935 the composer received the Order of Merit, and in 1942 and 1952 his seventieth and eightieth birthdays were celebrated throughout England with concerts of his works. Vaughan Williams visited the United States for the first time in 1922 to direct a program of his music at the Norfolk Music Festival. He made two later visits, one in 1932, another in 1954.

His operas: *Hugh the Drover* (1914); *The Shepherds of the Delectable Mountains* (1922); *Sir John in Love* (1929); *Job*, a masque (1930); *The Poisoned Kiss* (1936); *Riders to the Sea* (1937); *The Pilgrim's Progress* (1951).

Veau d'or, see LE VEAU D'OR.

Vedi! le fosche notturne spoglie, the Anvil Chorus in Act II, Scene 1, of Verdi's *Il trovatore*.

Vedrai, carino, Zerlina's consolation of Masetto in Act II, Scene 1, of Mozart's *Don Giovanni*.

Vedrommi intorno, Idomeneo's aria in Act I, Scene 2, of Mozart's *Idomeneo*.

Veil, The, one-act opera by Bernard Rogers. Libretto by Robert Lawrence. Première: New York City, October 26, 1954. The setting is an English madhouse in 1825, where a physician goes mad and strangles one of his patients with a veil rather than let her go free.

Venus, Goddess of Love (soprano) in Wagner's *Tannhäuser*.

Venusberg Music, the bacchanale in Act I of Wagner's *Tannhäuser*.

Venus Descendeth, see LE JOUR NAISSAIT DANS LE BOCAGE.

Vénus d'Ille, La, see MERIMEE, PROSPER.

Vêpres Siciliennes, Les, see SICILIAN VESPERS, THE.

Verdi, Giuseppe, composer. Born Le Roncole, Italy, October 10, 1813; died Milan, January 27, 1901. He was given his first music lessons by a local organist, and by Ferdinando Provesi, town organist of Busseto. The Busseto townspeople recognized Verdi's talent and in 1832 raised a fund for him to go to the Milan Conservatory. Verdi, however, was denied admission there because he was too old, and too poorly trained. Instead, he studied privately with Vincenzo Lavigna. In 1833 he returned to Busseto, where he remained on and off for four years. He was appointed conductor of the Busseto Philharmonic Society, and in 1836 he married Margherita Barezzi. In Busseto, he also completed his first opera, *Oberto*. In 1837 he returned to Milan where, on November 17, *Oberto* was produced at La Scala and acclaimed. The young composer received from La Scala a commission to write three new operas, and the publishing house of Ricordi accepted *Oberto*.

The first of the new operas was a comedy, *Un giorno di regno*. Given in 1840, it was a distressing failure. But its successor, *Nabucco*, introduced on March 9, 1842, was such a triumph that overnight Verdi became an idol. Dishes and items of wearing apparel were named after him; he could demand, and get, the highest fee for future commissions. In the decade between 1842 and 1851 he completed a dozen operas, the most important being *I Lombardi, Ernani, Macbeth,* and *Luisa Miller*.

Though he was now the most popular opera composer in Italy, he had not yet hit his full stride. A new creative period began in 1851 with *Rigoletto*. Verdi proved that his earlier operas had merely been the apprenticeship of a master—who now produced a series that became the most extensively performed and the best loved Italian operas of all time. *Rigoletto* was followed by *Il trovatore, Simon Boccanegra, Un ballo in maschera, La forza del destino, Don Carlos,* and *Aïda*. What distinguished these operas was not only their exceptional lyricism (each had an apparently inexhaustible fund of unforgettable melodies and ensemble numbers) but their pronounced dramatic quality. Verdi was an artist who knew the theater: how to meet its demands through music, characterization, climax, and at times a profound humanity.

Becoming rich as well as world-famous, Verdi bought a large farm at Sant' Agata, and here he spent his summers almost to the end of his life. His first wife having died in 1840, he married again in 1859. His wife was Giuseppina Strepponi, a singer who had appeared in his *Nabucco*. She retired from the stage after her marriage to devote herself completely to her famous husband. Indicative of his tremendous popularity was the fact that when Cavour instituted the first Italian parliament, Verdi was elected a deputy. But Verdi hated politics, and never allowed himself to accept any kind of public office. When, in 1874, the King made him a senator, this was exclusively an honorary appointment without political demands.

The last opera in Verdi's rich second period, *Aïda,* was commissioned by the Khedive of Egypt to inaugurate a new opera house commemorating the opening of the Suez Canal. *Aïda* was introduced in Cairo, under magnificent auspices, in 1871. For the next fifteen years Verdi wrote no more operas, though he made several attempts to write a *King Lear*. He finally became convinced that he was through as a composer; that the new age which Richard Wagner had initiated made his kind of opera old-fashioned.

He was drawn out of his long, seemingly permanent, retirement by an eloquent libretto that Arrigo Boïto fashioned from Shakespeare's *Othello*. The première of *Otello* at La Scala in 1887 was an event attracting world attention, and the opera was one of the greatest triumphs of Verdi's long career. The audience, coming to pay homage to a master, did not fail to recognize that he had soared to new heights.

Verdi wrote one more opera, once again to a libretto by Boïto, adapted from Shakespeare. This time it was a comedy, *Falstaff*. Introduced at La Scala in 1893, it was no less a triumph than *Otello* had been. And it proved no less significant in revealing that the seventy-nine-year-old master was still growing, artistically.

With the death of his wife in 1897, Verdi lost his will to live. He could no longer bear staying at his beloved Sant' Agata, and he took rooms in a Milan hotel. His sight and hearing began to fail, and after that (as he once complained) "all my limbs no longer obey me." One day he suffered a paralytic stroke; six days later he was dead. All Italy mourned the death of her national hero. At his funeral, a quarter of a million of his admirers crowded the streets to pay him their last respects. During the procession, Arturo Toscanini led a chorus from *Nabucco*. The same night, Toscanini led it again at La Scala as a last tribute.

Verdi's operas: *Oberto* (1839); *Un giorno di regno* (1840); *Nabucco* (or *Nabucodonosor*) (1842); *I Lombardi* (1843); *Ernani* (1844); *I due Foscari* (1844); *Giovanna d'Arco* (1845); *Alzira* (1845); *Attila* (1846); *Macbeth* (1847); *I masnadieri* (1847); *Il corsaro* (1848); *La battaglia di Legnano* (1849); *Luisa Miller* (1849); *Stiffelio* (1850); *Rigoletto* (1851); *Il trovatore* (1853); *La traviata* (1853); *I Vespri Siciliani* (1855); *Simon Boccanegra* (1857); *Aroldo* (1857); *Un ballo in maschera* (1859); *La forza del destino* (1862); *Don Carlos* (1867); *Aïda* (1871); *Otello* (1887); *Falstaff* (1893).
Vere, captain of the British Navy (tenor) in Britten's *Billy Budd*.
Vergil, see VIRGIL.
Verismo, a naturalistic movement in Italian opera launched with *Cavalleria rusticana*. The emphasis was on librettos with everyday characters and situations, a complete departure from costume plays and episodes from history or legend. The Verismo movement was seen in the operas of Leoncavallo, Giordano, Puccini, and Zandonai. In France, its counterpart was found in the naturalistic operas of Alfred Bruneau and in Charpentier's *Louise*.
Verranno a te sull' aure, farewell duet of Lucia and Edgardo in Act I, Scene 2, of Donizetti's *Lucia di Lammermoor*.
Versiegelt (Sealed), one-act opera by Leo Blech. Libretto by Richard Barka. Première: Hamburg Opera, November 4, 1908. In Germany, in 1830, Else is in love with Bertel; but Else's father, the town burgomaster, objects to the match. While he is visiting Frau Schramm, he is forced to hide in a huge wardrobe when an intruder appears. The intruder happens to be the bailiff, come to seize the wardrobe for Frau Schramm's unpaid taxes. The bailiff puts an official seal on the wardrobe. The burgomaster is not released until he gives his consent to his daughter's marriage.
Vespone, Uberto's valet (silent role) in Pergolesi's *La serva padrona*.
Vespri Siciliani, I, see SICILIAN VESPERS, THE.
Vestale, La (The Vestal), opera by Gasparo Spontini. Libretto by Etienne de Jouy. Première: Paris Opéra, December 15, 1807. Spontini's most famous opera is a work of historical importance, a transition in opera development between the operas of Gluck and those of Meyerbeer. The setting is Rome during the Republic. Licino, back from a campaign in Gaul, discovers that his beloved Giulia has become a vestal virgin. He penetrates the temple and revives her love for him. In their passionate exchange, Giulia forgets that she must watch the holy fires; she allows it to be extinguished. For this sacrilege she is stripped of her veil and condemned to be buried alive. As she is being led to her death, a bolt of lightning relights the fires. The Romans take this as a sign that Giulia has been forgiven by the gods, and she is spared from death.
Vesti la giubba, Canio's aria closing Act I of Leoncavallo's *Pagliacci*.

Viardot-García, Pauline, mezzo-soprano and teacher of singing. Born Paris, July 18, 1821; died there May 18, 1910. Daughter of the tenor and singing teacher Manuel del Popolo García, she was the sister of the prima donna Malibran and the singing teacher Manuel García. She began to study music with her parents, later studying the piano with Franz Liszt, composition with Anton Reicha. In 1837 she made her debut as a singer in Brussels. She then appeared extensively throughout Europe. In 1839 she was a member of the Théâtre des Italiens in Paris, where, two years later, she married the theater's manager, Louis Viardot. As the principal soprano of the Paris Opéra, she appeared in the world première of *Le Prophète* (1849), creating the role of Fides. For a decade, beginning in 1848, she was a favorite with London opera-goers. From 1871 to 1875 she taught singing at the Paris Conservatory. Her daughter, Louise Héritte-Viardot (1841–1918) was a singing teacher at the St. Petersburg Conservatory, the Hoch Conservatory in Frankfort-on-the-Main, and in Berlin. Her son, Paul Viardot (1857–1941), a violinist, occasionally conducted at the Paris Opéra.

Via resti servita, duet of Marcellina and Susanna in Act I of Mozart's *The Marriage of Figaro.*

Vicar of Wakefield, The, *see* GOLD-SMITH, OLIVER.

Vicino a te s'acqueta, duet of Chénier and Madeleine in Act IV of Giordano's *Andrea Chénier.*

vida breve, La (Life Is Short), opera by Manuel de Falla. Libretto by Carlos Fernández-Shaw. Première: Nice, France, April 1, 1913. This was Falla's first major work, in which his pronounced leanings toward nationalist music first became evident. The gypsy girl Salud is in love with Paco; but though the latter keeps up a pretense that he is in love with her, he is actually about to marry somebody else. Salud discovers the truth at the wedding. She curses Paco, but the sight of him again transforms her anger to tenderness. Broken in heart, she falls dead at his feet. The opera's two Spanish dances are familiar excerpts, particularly the first one, which is heard in a variety of transcriptions.

Viene la sera, the love duet of Cio-Cio-San and Pinkerton in Act I of Puccini's *Madama Butterfly.*

Vieni, amor mio, Amneris' aria in Act II, Scene 1, of Verdi's *Aïda.*

Vienna State Opera (Staatsoper), the leading opera house of Austria, and one of the great operatic institutions of the world. Before World War I it was known as the Vienna Royal Opera. It was opened on the Ring in Vienna on May 25, 1869, under the artistic direction of J. F. von Herbeck, as the permanent home for opera, which up to then had been given at the Burgtheater. Under the artistic direction of Hans Richter (1880–1896) it became one of Europe's major opera houses; under the direction of Gustav Mahler (1896–1907), one of the world's foremost institutions. Later artistic and musical directors included Felix Weingartner, Hans Gregor, Franz Schalk, Clemens Krauss, Erwin Kerber and, after World War II, Egon Hilbert and Karl Böhm. When the Nazis invaded Austria in 1937, they smashed the bust of Mahler which stood in the vestibule of the opera house, and changed the artistic program of the company to meet their standards. The last performance at the Vienna State Opera before it closed was *Die Götterdämmerung,* given on June 30, 1944. On March 12, 1945, the theater was severely damaged by bombs. After the war, performances of the Vienna State Opera took place at the Theater-an-der-Wien. The original theater on the Ring was reconstructed, and it reopened on November 5, 1955, with *Fidelio.*

The Vienna State Opera's most frequently performed opera up to 1952 was *Aïda*, with 726 performances; *Lohengrin* ranked second with 719 performances. Between 1886 and 1952 the operas of Wagner were heard more often than those of any other composer; those of Verdi ranked next, followed by those of Mozart and Richard Strauss. World premières have been those of *The Cricket on the Hearth* and *The Queen of Sheba*, both by Karl Goldmark, and Massenet's *Werther*. *The Bartered Bride* first became a success after its brilliant revival here in 1892.

Viens ici, ne reste pas au bord du clair de lune, duet of Pelléas and Mélisande in Act IV, Scene 2, of Debussy's *Pelléas et Mélisande.*

vie Parisienne, La (Parisian Life), opéra bouffe by Jacques Offenbach. Libretto by Henri Meilhac and Ludovic Halévy. Première: Théâtre du Palais-Royal, Paris, October 31, 1866. Baron von Gardenfen finds the wife of Baron von Gondermark appealing. He seeks her out, disguised as an employee of the Grand Hotel, while a friend, assuming the name of Baron von Gardenfen, keeps the husband busy with an evening of Parisian gaiety. Thus, Gardenfen is able to make love to the Baroness von Gondermark, who is finally rescued by her aunt.

Vigny, Alfred de, author, poet, dramatist. Born Loches, France, March 27, 1799; died Paris, September 17, 1863. His historical novel *Cinq-Mars* was made into an opera of the same name by Charles Gounod, and his drama *Chatterton* was the source of Leoncavallo's first opera.

Village Romeo and Juliet, A, opera by Frederick Delius. Libretto by the composer, in collaboration with his wife, based on Gottfried Keller's story *Die Leute von Seldwyla*. Première: Komische Oper, Berlin, February 21, 1907 (as *Romeo und Julia auf dem Dorfe*). Described by the composer as a "lyric drama in six pictures," this opera is centered around a feud between two households, with the son of one falling in love with the daughter of another. The lovers run away, and end up committing suicide. An orchestral interlude, "The Walk to the Paradise Garden," has become independently popular. Delius wrote it five years after completing his opera, inserting it as an entr'acte between the fifth and sixth "pictures."

Villi, Le (The Witches), opera by Giacomo Puccini. Libretto by Ferdinando Fontana. Première: Teatro dal Verme, Milan, May 31, 1884 (one-act version); Teatro Regio, Turin, December 26, 1884 (two-act version). This was Puccini's first opera. He entered it in the Sonzogno competition but failed to gain even an honorable mention. It did, however, make an impression on Arrigo Boïto, who raised a fund to finance its production. That performance was so successful that La Scala accepted the opera for the following season and Ricordi published it and commissioned a new work from the composer. The setting is the Black Forest, where the betrothal of Robert and Anna is celebrated. Going to seek his fortune before marrying, Robert forgets his beloved in the pleasures of the city, and Anna dies of grief. Broken in spirit and fortune, Robert returns expecting to find Anna waiting for him, but confronts only an apparition who denounces him for his desertion. Witches dances around him until he falls dead at their feet.

Villon, François (born FRANÇOIS DE MONTCORBIER), poet. Born Paris, 1431; died there about 1463. The poet is the central character of Jean Nouguès' opera *Une aventure de Villon.*

Vinay, Ramón, tenor. Born Chile, 1912. He spent his boyhood in France, where he studied engineering. After a period of music study he joined a traveling opera company in 1934 with which he made his opera debut, as a baritone, in Mexico City in *Il trovatore.* He continued singing baritone roles until 1943. His second debut, this time as a tenor, took place with the National Opera Company of Havana, in June 1944, when he appeared in the title role of *Otello.* One year later, in the fall of 1945, he made his North American debut with the New York City Opera as Don José, and on February 22, 1946 made his Metropolitan Opera debut in the same role. He scored a great personal success at the Metropolitan on December 9, 1946 when, on ten hours' notice, he substituted for Torsten Ralf in the title role of *Otello.* Besides his appearances at the Metropolitan Opera, he has been heard at the Bayreuth, Salzburg, and Holland Festivals, and in leading opera houses of Europe and South America, including La Scala and Covent Garden.

Vincent, Mireille's suitor (tenor) in Gounod's *Mireille.*

Vin ou bière, chorus of the villagers in Act I, Scene 2, of Gounod's *Faust.*

Violanda, Arvino's wife (soprano) in Verdi's *I Lombardi.*

Violetta, *see* VALERY, VIOLETTA.

Vi ravviso, aria of the stranger (Count Rodolfo) in Act I of Bellini's *La sonnambula.*

Virgil (or Vergil), poet. Born Andes, Cisalpine Gaul, October 15, 70 B.C.; died Brundisium, Italy, September 21, 19 B.C. His Latin epic *The Aeneid* provided the material for several operas. The most famous is Henry Purcell's *Dido and Aeneas.* Others include: Thomas Arne's *Dido and Aeneas;* Hector Berlioz' *Les Troyens*; Pascal Colasse's *Enée et Lavinie;* Joseph Martin Kraus's *Aeneas i Carthago;* Nikolai Lissenko's *The Aeneid.*

Vision fugitive, Herod's aria in Act II of Massenet's *Hérodiade.*

Vissi d'arte, Tosca's aria of her dedication to art in Act II of Puccini's *Tosca.*

Vitellius, Roman proconsul (baritone) in Massenet's *Hérodiade.*

Viva il vino, Turiddu's drinking song in Mascagni's *Cavalleria rusticana.*

Vivaldi, Antonio, composer. Born Venice, Italy, about 1675; died Vienna, Austria, July 27, 1741. His music study took place with his father and with Giovanni Legrenzi in Venice, but he was prepared not for music but for the church. In 1703 he became a priest. His musical career began in 1709, when he became a teacher of the violin at the Ospedale della Pietà in Venice, rising to the post of music director seven years later. About 1720 he left Venice to become kapellmeister to Prince Philip of Hesse in Mantua. After a period of touring Europe as violin virtuoso, he returned to Italy, where he now became outstandingly popular for his operas. Though a dominant figure in the musical world of his time, he suffered poverty and neglect at the end of his life, and went to a pauper's grave when he died. Though he is known today chiefly for his instrumental music, of which he wrote an astonishing amount, he also composed some forty operas. The most successful were: *Nerone fatto Cesare* (1715); *L'Arsilda regina di Ponto* (1716); *L'incoronazione di Dario* (1716); *Armida al campo d'Egitto* (1718); *La Creola* (1723); *L'inganno trionfante in amore* (1725); *Farnace* (1726); *La Fida Ninfa* (1732); *Montezuma* (1733); *L'Olimpiade* (1734); *Griselda* (1735); *Rosmira* (1738); *Feraspe* (1739).

Viva Pagliaccio, chorus of the villagers in the opening scene of Leoncavallo's *Pagliacci.*

Vladimir, Prince Igor's son (tenor) in Borodin's *Prince Igor.*

Voce di donna, La Cieca's aria in Act I of Ponchielli's *La Gioconda.*

Voce di gola, Italian for "throat voice," or guttural voice.

Voce velata, Italian for "veiled voice," a muffled quality produced intentionally.

Vogl, Heinrich, tenor. Born Au, Germany, January 15, 1845; died Munich, April 21, 1900. After studying with Franz Lachner he made his debut at the Munich Opera in 1865. He became a permanent member of that company, and it was here that he scored his first major successes in the Wagnerian repertory. When Schnorr von Carolsfeld died in 1865, Vogl became the outstanding interpreter of Tristan. In 1869 he appeared in the world première of *Das Rheingold* as Loge, and a year later he sang Siegmund in the première of *Die Walküre.* He appeared at Bayreuth in 1876, and in 1882 toured Germany and Austria.

In 1868 he married the dramatic soprano Theresa Thoma, then the principal soprano of the Munich Opera, creator of the role of Sieglinde, and for many years a significant interpreter of Isolde.

Voi, che sapete, Cherubino's aria in Act II of Mozart's *The Marriage of Figaro.*

Voici ce qu'il écrit, the letter duet of Geneviève and Arkel in Act I, Scene 2, of Debussy's *Pelléas et Mélisande.*

Voici le printemps, nous portant des fleurs, the song of Dalila and her priestesses in Act I of Saint-Saëns' *Samson et Dalila.*

Voici ma misère, hélas! Samson's plea to God in Act III, Scene 1, of Saint-Saëns' *Samson et Dalila.*

Voilà donc la terrible cité, Athanaël's lament on the degeneration of Alexandria in Act I, Scene 2, of Massenet's *Thaïs.*

Voi lo sapete, Santuzza's aria in Mascagni's *Cavalleria rusticana.*

Volkhova, the King of the Ocean's daughter (soprano) in Rimsky-Korsakov's *Sadko.*

Volpone, satirical opera by George Antheil. Libretto by Alfred Perry, freely adapted from the Ben Jonson comedy of the same name. Première: Los Angeles, January 9, 1953. Volpone is a wily Venetian who pretends he is dying in order to gather gifts from his prospective legatees. Louis Gruenberg also wrote an opera on this play.

Voltaire (born FRANÇOIS MARIE AROUET), philosopher and author. Born Paris, November 21, 1694; died there May 30, 1778. His writings were the source of the following operas: *Alzira* (operas by Verdi and Niccolò Zingarelli); *La belle Arsène* (opera by Monsigny); *Candide* (Lev Knipper); *La fée urgèle* (Egidio Dune, Ignaz Pleyel); *Le Huron* (Grétry); *Isabelle et Gertrude* (Grétry); *Maometto II* (Rossini); *Olympie* (Spontini); *Semiramide* (Rossini); *Tancredi* (Rossini, using also material from Tasso); *Zaira* (Bellini). Voltaire wrote the librettos for the following operas by Jean Philippe Rameau: *Pandore; La Princesse de Navarre; Le temple de gloire; Les fêtes de Ramire; Samson.*

Vorspiel, German for "prelude." See PRELUDE.

Vous ne savez pas, the duet at the fountain of Pelléas and Mélisande, opening Act II of Debussy's *Pelléas et Mélisande.*

Vous qui du Dieu vivant outragez la puissance, the Cardinal's excommunication of Prince Leopold in Act III of Halévy's *La Juive.*

Vous qui faites l'endormie, Méphistophélès' serenade in Act IV, Scene 3, of Gounod's *Faust.*

Voyons, Manon, plus de chimères, Manon's revery in Act I of Massenet's *Manon.*

Vulcan, a god (bass) in Gounod's *Philémon et Baucis.*

W

Wach' auf, es nahet gen den Tag, chorus acclaiming Hans Sachs in Act III, Scene 2, of Wagner's *Die Meistersinger.*

Wagner, (1) a student (tenor) in Boïto's *Mefistofele.*

(2) A student (baritone) in Gounod's *Faust.*

Wagner, Cosima, second wife of Richard Wagner. Born Bellaggio, Italy, December 25, 1837; died Bayreuth, Germany, April 1, 1930. She was the daughter of Franz Liszt. Her first husband, whom she married in 1857, was the celebrated pianist conductor and Wagner enthusiast Hans von Bülow. Married to Wagner, she helped him prepare the first Bayreuth Festival, and after Wagner's death she maintained an autocratic rule over the Bayreuth performances. She wrote a memoir of her father, *Franz Liszt: Gedenkblatt von seiner Tochter* (1911).

Wagner, Johanna, soprano. Born Hanover, Germany, October 13, 1826; died Würzburg, Germany, October 16, 1894. The niece of Richard Wagner, she created the role of Elisabeth. She received her musical training from her father, a professional singer. After making various concert appearances she became, in 1844, a principal soprano of the Dresden Opera, where she appeared in the first performance of *Tannhäuser* (1845). For two years she studied in Paris with Pauline Viardot-García, then, between 1850 and 1852, became the principal soprano of the Berlin Opera. Losing her singing voice in 1861, she became an actress. From 1882 to 1884 she taught singing at the Royal School of Music in Munich.

Wagner, Richard, composer. Born Leipzig, Germany, May 22, 1813; died Venice, Italy, February 13, 1883.

It was long thought that Wagner was the son of Karl Friedrich Wagner, a Leipzig police official, but it is now generally believed that the actor Ludwig Geyer, a close friend of his mother, was his father. In any event, Karl Friedrich Wagner died a half year after Richard was born. About a year afterward Richard's mother married Geyer. A cultured man, Geyer instilled in Richard a love for the arts, particularly literature. As a boy, Richard nursed the ambition to become a writer; when he was eleven he wrote a four-act political drama in the style of Shakespeare.

Richard was enrolled in the Thomasschule in Leipzig, where he was so lax that he was expelled. Subsequently he entered the University of Leipzig, where once again he was indifferent to his studies. Only one serious interest absorbed him: music. He began his mastery of the art by studying a book on theory. In 1829 he wrote an overture that was performed in Leipzig. Two years later he studied theory for six months with Theodor Weinlig. He now completed a symphony (performed in Leipzig and Prague in 1833) and tried writing a first opera, *Die Hochzeit.* His first complete opera was *Die Feen,* written in 1834; it was not performed in the composer's lifetime. *Das Liebesverbot,* based on Shakespeare's *Measure for Measure,* followed in 1836.

Meanwhile, in 1834, he became the conductor of the Magdeburg Opera, where he made his debut leading *Don Giovanni.* On March 29, 1836, he introduced there his *Das Liebesverbot.* It was such a fiasco that the company (never too solvent) had to go into bankruptcy. He found a new post in Königsberg; it was there he met Minna Planer and married her on November 24, 1836. Between 1837 and 1839 he conducted operas in Riga. Heavily involved in debts, he was summarily dismissed and had his passport confiscated. To avoid imprisonment he had to flee from Riga by a smuggler's route.

He arrived in Paris on September 17, 1839, with bright hopes for the future. He had letters of introduction to Meyerbeer, then one of the most influential composers in the city; he also had parts of his new opera, *Rienzi.* But his three years in Paris proved a period of agonizing hardships and frustrations. Nevertheless, he completed *Rienzi* in 1840 and *Der fliegende Holländer* in 1841.

A change of fortune came in 1842 with an outstandingly successful performance of *Rienzi* at the Dresden Opera. While *Der fliegende Holländer,* given by the same company a year later, was a failure, Wagner's reputation had grown to such proportions that in 1843 he was appointed the kapellmeister of the Dresden Opera. During the next six years he elevated the artistic standards of the opera company to new heights. Wagner completed two new operas in Dresden. The first was *Tannhäuser,* given in Dresden in 1845, and a failure. The second, *Lohengrin,* had to wait for performance until Franz Liszt accepted it for Weimar, introducing it there on August 28, 1850. *Lohengrin* soon became popular throughout Germany. Wagner had not attended the première because

by then he was a political exile; having become involved in the revolutionary movement of 1848–1849 in Saxony, he had avoided arrest by fleeing from the country.

After a visit to Paris, he established his permanent home in Zurich. It was now that he began clarifying his new ideas about opera, and to expound them in essays and pamphlets. He had become impatient with the methods and patterns that for so long a time had constricted composers. He conceived opera as a synthesis of the theatrical arts (poetry, music, acting, scenery, drama). Old practices had to be discarded. The formal demarcations between recitative and aria had to make way for a continuous flow of melody. Such irrelevant elements as ballets had to be eliminated. Dramatic expressiveness was to be intensified by bringing symphonic breadth to orchestral and vocal writing. To realize his ideal of an inseparable text and score, he developed the technique of the leading motive (leitmotiv): a melodic idea or phrase associated with a character, situation, or idea. The ideal opera would be unified by a structure of these recurring motives. To put these theories into practice, Wagner outlined a vast musico-dramatic project based on the Nibelung legends. Originally, he planned a single music drama, as he called his new form, but he ended writing four: *Das Rheingold, Die Walküre, Siegfried,* and *Die Götterdämmerung.* This tetralogy was named *Der Ring des Nibelungen.* Wagner finished writing the texts and published them in 1852. The score of the last drama was completed in 1874. Thus, the creation of the *Ring* absorbed him for a quarter of a century—a period that also saw the composition and performance of two of his other music dramas: *Tristan und Isolde* (1859) and *Die Meistersinger von Nürnberg* (1867).

While these Herculean labors were

occupying him, his personal life was becoming complicated. His marriage to Minna had been unhappy and explosive, and he found solace in the love of other women, usually the wives of his benefactors. A ruthless egotist, he used people for his own needs and was unconcerned about the pain he caused. Thus, in 1853, he made love to Mathilde Wesendonck, even though her husband had provided him with a home in Zurich and had financed several of his concerts. Several years later he fell in love with the wife of another intimate friend, Hans von Bülow. She was Cosima, daughter of Franz Liszt. Though Von Bülow dedicated himself to the promotion of Wagner's music, Wagner did not hesitate to make Cosima his own. Far from attempting to conceal the relationship, he insisted upon calling their first child, born in 1865, Isolde. After a second daughter was born to them, Wagner and Cosima set up their own home on Lake Lucerne; a year after their third child— Siegfried—was born, Wagner and the now-divorced Cosima were married. Wagner remained devoted to her for the rest of his life.

The struggles of his artistic life were also to be resolved in victory. After being pardoned for his radical activities, Wagner returned to Saxony in 1862. Two years later he acquired a wealthy and powerful patron in Ludwig II, King of Bavaria. Under Ludwig's patronage, *Tristan und Isolde* was introduced in Munich in 1865, *Die Meistersinger* in 1868, *Das Rheingold* in 1869, and *Die Walküre* in 1870. But getting his dramas performed did not completely satisfy Wagner. He nursed an ambition to have a special theater built where they could be performed according to his own ideas of staging. He overcame seemingly insurmountable obstacles to make his dream a reality. On August 13, 1876, his vision became a fact. In a theater built

according to his specifications in Bayreuth (see BAYREUTH), his *Ring* tetralogy was given its first complete performance, with the great of the world attending.

Wagner completed one more drama, the consecrational play *Parsifal,* introduced in Bayreuth on July 26, 1882. After the harrowing task of bringing this work to performance, Wagner went with Cosima on a vacation to Venice. He suffered a heart attack there and died. His body was brought back to Bayreuth, to be buried in the garden of his home, Villa Wahnfried.

His operas: *Die Feen* (1834); *Das Liebesverbot* (1836); *Rienzi* (1840); *Der fliegende Holländer* (1841); *Tannhäuser* (1845); *Lohengrin* (1848); *Das Rheingold* (1854); *Die Walküre* (1856); *Tristan und Isolde* (1859); *Die Meistersinger von Nürnberg* (1867); *Siegfried* (1871); *Die Götterdämmerung* (1874); *Parsifal* (1882).

Wagner, Siegfried, conductor and composer. Born Triebschen, Lucerne, Switzerland, June 6, 1869; died Bayreuth, Bavaria, August 4, 1930. The only son of Richard and Cosima Wagner, he was educated as an architect, but turned to music, studying with Engelbert Humperdinck and Julius Kniese. In 1894 he became an assistant conductor at Bayreuth, and from 1896 on conducted there regularly. From 1909 until his death he supervised all the productions at Bayreuth. He visited the United States in 1923–1924, conducting several concerts of his father's music to raise funds for the reopening of the Bayreuth Theater after World War I. He married Winifred Williams in 1915; after his death, she became the guiding hand at the Bayreuth Festivals until the beginning of World War II.

Siegfried Wagner wrote fourteen operas, all to his own texts. The following were the most successful: *Der Kobold* (1904); *Sternengebot* (1908);

Schwarzschwanenreich (1918); *Sonnenflammen* (1918); *Der Heidenkönig* (1915); *Der Friedensengel* (1915); *Der Schmied von Marienburg* (1920).

Wahn! Wahn! Überall Wahn! Hans Sachs's monologue in Act III, Scene 1, of Wagner's *Die Meistersinger*.

Waldner, Count, Arabella's father (bass) in Richard Strauss's *Arabella*.

Waldweben (Forest Murmurs), a scene in Act II of Wagner's *Siegfried*.

Walk to the Paradise Garden, The, orchestral entr'acte between the fifth and sixth "pictures" of Delius' *A Village Romeo and Juliet*.

Walküre, Die, *see* RING DES NIBELUNGEN, DER.

Walkürenritt, *see* RIDE OF THE VALKYRIES, THE.

Wallace, Jake, traveling camp minstrel (baritone) in Puccini's *The Girl of the Golden West*.

Wallenstein, *see* SCHILLER, FRIEDRICH.

Wallerstein, Lothar, stage director. Born Prague, Czechoslovakia, November 6, 1882; died New Orleans, Louisiana, November 13, 1949. He studied to be a doctor, but abandoned medicine for music. From 1910 to 1914 he was conductor and stage director of the Posen Opera. From 1918 to 1922 he held a similar position with the Breslau Opera, and from 1924 to 1927 with the Frankfort Opera. In 1927 he was appointed stage director of the Vienna State Opera, where he remained eleven years and staged over seventy-five new works. During this period he also staged operas at La Scala and the Salzburg Festivals. He came to the United States just before World War II, and in 1949 was appointed resident stage director of the New Orleans Opera Association.

Wally, La, opera by Alfredo Catalani. Libretto by Luigi Illica, based on *Die Geyer-Wally*, a novel by Wilhelmine von Hillern. Première: La Scala, January 20, 1892. In nineteenth century Switzerland, Wally, daughter of Strom-

minger, refuses to marry Gellner; she is in love with Hagenbach. Gellner tries to murder Hagenbach, which only brings the lovers closer together. But the lovers finally meet their doom in an avalanche. The preludes to Acts III and IV, the "Walzer del bacio" in Act III, and Wally's aria "Ebben? ne andrò lontana" in Act I are the best-known excerpts.

Walpurgis Night, ballet music in Act IV, Scene 3, of Gounod's *Faust*.

Walter, Bruno (born SCHLESINGER), conductor. Born Berlin, September 15, 1876. After receiving his musical education at the Stern Conservatory in Berlin, he served his apprenticeship in various smaller German opera houses. Gustav Mahler engaged him as assistant conductor of the Vienna Opera in 1901. Walter worked under Mahler for eleven years and developed into a mature artist. In 1914 Walter was engaged as general music director in Munich, in succession to Felix Mottl. His performances there of Mozart and Wagner gained him an international reputation. In 1923 Walter made his American debut as guest conductor of the New York Symphony Society. Since then he has appeared extensively in the United States as a conductor of its major orchestras. Many years passed, however, before Walter appeared in America as a conductor of operas. The first instance was at the Metropolitan Opera on February 14, 1941, when he led *Fidelio*. Walter continued to conduct special performances at the Metropolitan Opera through the 1945–1946 season.

Walter's long association with the Salzburg Festival began in 1922. He appeared at Covent Garden for the first time in 1924; for the next seven years he was one of this institution's principal conductors of Wagner and Mozart. In 1925 he was appointed principal conductor of the Charlottenburg Opera in Berlin, and in 1930 principal conductor of the Leipzig Gewand-haus Orchestra. He had to resign his posts in Germany and leave the country when the Nazis came to power. When Felix Weingartner resigned as general music director of the Vienna State Opera in 1936, Walter replaced him as principal conductor and "artistic advisor." When Hitler took over Austria, Walter made his home in Paris. During World War II, he continued conducting opera and symphony performances throughout the free world. After the war, he was given a hero's welcome when he returned for guest performances at the Vienna State Opera. He has written an autobiography, *Theme and Variations* (1946), and a biography of Gustav Mahler (1936).

Walther, see STOLZING, WALTHER VON.

Walton, Lord Gautier, Elvira's father (bass) in Bellini's *I Puritani*.

Walton, Sir George, Elvira's uncle (bass) in Bellini's *I Puritani*.

Walton, William, composer. Born Oldham, England, March 29, 1902. He received his musical education at the Christ Church Cathedral Choir School, Oxford, and first attracted attention as a composer in 1922 with *Façade,* a provocative setting of some spirited poems by Edith Sitwell. He did not write an opera until late in his career, and after he had established himself as one of the outstanding English composers of his day. The opera was *Troilus and Cressida,* introduced in London in 1954. Besides writing outstanding works for chorus, orchestra, ballet, and chamber combinations, Walton has written music for several important films, including *Henry V, Hamlet,* and *Major Barbara.*

Waltraute, a valkyrie (mezzo-soprano) in Wagner's *Die Walküre* and *Die Götterdämmerung.*

Waltz, an Austrian dance of peasant origin, it acquired its polish and its fame in Vienna in the late eighteenth century, then spread throughout the world. One of the earliest examples of a typical Viennese waltz is found in Martín y Soler's opera *Una cosa rara,* produced in Vienna in 1785. The following operas are notable for containing waltzes: *La Bohème* ("Musetta's Waltz," Act II); *Eugene Onegin* (Act II); *Faust* ("Ainsi que la brise," Act II); *Hansel and Gretel* ("Gingerbread Waltz," Act III); *Mefistofele* ("Peasants' Waltz," Act I); *Roméo et Juliette* ("Je veux vivre dans ce rêve," Act I). Johann Strauss's *Die Fledermaus* and Richard Strauss's *Der Rosenkavalier* are works that might be called waltz-inspired.

Wanderer, The, the god Wotan in mortal guise in Wagner's *Siegfried.*

War and Peace, opera in five acts (twelve scenes) by Serge Prokofiev. Libretto by Myra Mendelssohn (the composer's wife), based on the novel by Leo Tolstoy. Première: Leningrad, 1946. The opera is devoted to that portion of Tolstoy's novel that portrays Napoleon's invasion of Russia. Prokofiev wrote it during World War II, after the Nazi invasion of the Soviet Union, then his homeland. The opera is on a grand scale, calling for sixty characters, and requiring two evenings for performance. Acts I and II are described as "Scenes of Peace," and portray the peaceful lives and relationships of the main characters: Natasha Rostova, Andrei Volkonsky, Anatoli Kurgin, and Pierre Bezukhov. The remaining three acts are "Scenes of War." Here we have, in the composer's explanation, "the Russian people's struggle, their sufferings, wrath, courage, and victory over the invaders. In this part, the people themselves constitute the hero of the opera in the person of the peasants, of the popular militia, the regular Russian army, the Cossacks, the guerrillas. . . . The destinies of the main characters introduced in the first six scenes are closely linked with the war events." Coursing throughout the opera is the thread of the love of Natasha and Andrei. Some of the most effective pages are the mass scenes, such as the chorus of the Smolensk refugees, and the battle of Borodino; but the opera contains other effective numbers, notably the spring nocturne in the first scene. Throughout the opera are found tone poems describing such scenes as that of ruined Moscow (opening of the ninth scene) and the battle of the partisans and the French (eleventh scene). Prokofiev revised his score extensively just before his death. The new version was heard at the Maly Theater, Leningrad, March 31, 1955.

War es so schmählich? Brünnhilde's plea to Wotan in Act III of Wagner's *Die Walküre*.

Warren, Leonard, baritone. Born New York City, April 21, 1911. He engaged in business activities until 1933, when he decided to become a professional singer. His vocal studies took place with Sidney Dietsch, and subsequently with Giuseppe de Luca. After singing with the chorus of the Radio City Music Hall, he appeared on the Metropolitan Auditions of the Air and won first prize. As a result he made his opera debut at the Metropolitan Opera on January 13, 1939 in *Simon Boccanegra*. He has since been a principal baritone of the company, besides making successful appearances in Mexico, South America, Puerto Rico, Italy, and Canada. He has also appeared over radio, television, and in the motion picture *When Irish Eyes Are Smiling*.

Warrior, The, one-act opera by Bernard Rogers. Libretto by Norman Corwin. Première: Metropolitan Opera, January 11, 1947. This opera is based on the Biblical story of Samson and Delilah.

Wartburg, the castle of the minstrelknights in Thuringia in Wagner's *Tannhäuser*.

Was bluten muss? Elektra's ecstatic threatening of her mother in Richard Strauss's *Elektra*.

Was gleicht wohl auf Erden, the Huntsmen's Chorus in Act III, Scene 3, of Weber's *Der Freischütz*.

Wasps, The, see ARISTOPHANES.

Water Carrier, The (Der Wasserträger; Les Deux Journées), opera by Luigi Cherubini. Libretto by Jean-Nicolas Bouilly. Première: Théâtre Feydeau, Paris, January 16, 1800. When Count Armand falls into disfavor with Mazarin, Michele, the water carrier, arranges for his escape from Paris in a water barrel. The Count and his wife are seized by soldiers, but before they are taken back to Paris, Michele comes with the news that Mazarin has forgiven the Count and restored to him his former high station.

Weber, Carl Maria von, composer. Born Eutin, Oldenburg, Germany, November 18, 1786; died London, England, June 5, 1826. He was born a sickly child, with a disease of the hip that gave him a life-long limp. Despite this infirmity, he was made to travel continually with his parents, since his father played the violin in various small orchestras. The father compelled the boy to study music industriously, bent on developing a prodigy. When Carl was eleven, he studied for six months with Michael Haydn in Salzburg. His later study took place in Munich. It was there that he completed his first opera, *Die Macht der Liebe und des Weins*. His second, *Das Waldmädchen*, was performed in 1800 and was a failure.

In 1803, Weber went to Vienna and studied for two years with Abbé Vogler. Upon Vogler's recommendation he received a post as conductor with the Breslau Opera in 1805. His three years in Breslau were unhappy, since he was in perpetual conflict with the management and members of the company, while his dissolute and irresponsible behavior aroused the hostility of the public. He left Breslau and assumed two other musical posts. The second, in Stuttgart, came to a sudden end when he was accused of having stolen some funds. A period of travel followed, during which Weber appeared as a concert pianist and composed several large works, including a comic opera, *Abu Hassan,* successful when given in Munich in 1811. Finally, in 1813, he settled in Prague and became director of the Opera. Three years later he received his most important conductorial assignment when he was engaged as musical director of the Dresden Opera. His success was so substantial that his post was confirmed for life. His future assured, Weber married the singer Caroline Brandt. Weber devoted himself to conducting German operas, and this inflamed him with an ideal: He would write a national opera. The task took him three years. The opera, *Der Freischütz,* was introduced in Berlin on August 18, 1821 (his wife appearing as Agathe), and was such a sensation that Weber became the man of the hour. His opera received fifty performances in a year and a half, then duplicated its successes the following year in Dresden and Vienna. German audiences went wild over *Der Freischütz,* in which the romantic tendencies of the times were crystallized, and in which German traditions, backgrounds, and culture were glorified.

Weber was now commissioned by the impresario Domenico Barbaja to write a new opera for Vienna. His *Euryanthe* was introduced in that city in 1823 and was acclaimed. His last opera, *Oberon,* was written for Covent Garden on commission. Though ill at the time, Weber made the arduous journey to London to write *Oberon* and supervise its production. The première in 1826—Weber himself conducted—was such a triumph that Weber described it as "the greatest success of my life." But the supreme effort of completing and producing his new opera undermined his health completely. He died in his sleep, just before he was to make his journey home. He was buried in London. Eighteen years later his body was transferred to Dresden. For this second burial, Wagner wrote special music and delivered the eulogy.

While Weber's three operatic masterworks—*Der Freischütz, Euryanthe,* and *Oberon*—are no longer in the permanent repertory, and are heard only when an opera company undertakes an adventurous revival, their significance cannot be overestimated. Through their exploitation of German backgrounds and culture, and their indebtedness to German folk song and dance, they helped to establish a national operatic movement that could rival the then ascendant Italian school. The road from Weber leads directly to Wagner, as Wagner himself conceded. Before Wagner, Weber made tentative use of the leitmotiv method; he gave greater dramatic significance to the recitative and greater symphonic importance to the orchestra than any composer before his time; and he integrated his plays and music more successfully than any composer since Gluck.

His operas: *Die Macht der Liebe und des Weins* (1798); *Das Waldmädchen* (1800); *Peter Schmoll und seine Nachbarn* (1803); *Rübezahl* (1805); *Silvana* (1810); *Abu Hassan* (1811); *Der Freischütz* (1821); *Die drei Pintos* (unfinished); *Euryanthe* (1823); *Oberon* (1826).

Webster, Daniel, the celebrated American statesman and orator, a principal character (baritone) in Douglas Moore's *The Devil and Daniel Webster.*

Wedekind, Frank, dramatist. Born Hanover, Germany, July 24, 1864; died Munich, March 9, 1918. Two of his plays—*Earth Spirit (Der Erdgeist)* and *Pandora's Box* (*Die Büchse der Pandora*)—were made into a single libretto for Alban Berg's opera *Lulu*. Max Ettinger's *Frühlingserwachen* was based on Wedekind's drama of the same name.

Weh! Nun ist all unser Glück dahin! Lohengrin's revelation that his happiness with Elsa has ended, in Act III, Scene 1, of Wagner's *Lohengrin*.

Weiche, Wotan, weiche! Erda's Warning, in the final scene of Wagner's *Das Rheingold*.

Weill, Kurt, composer. Born Dessau, Germany, March 2, 1900; died New York City, April 3, 1950. He attended the Berlin Hochschule für Musik, where his teachers included Engelbert Humperdinck; he subsequently studied privately with Ferruccio Busoni. His first opera, *The Protagonist*—introduced in Dresden in 1927—made extensive use of popular-music idioms. This element was pronounced in his succeeding operas, making him an outstanding exponent of a German cultural movement, known as *Zeitkunst,* which glorified contemporary subjects treated in a racy, modern style. In 1928 he wrote his greatest success, a modern adaptation of *The Beggar's Opera: The Threepenny Opera* (*Die Dreigroschenoper*). Introduced in Berlin on August 31, it enjoyed a sensational success, being given a total of over four thousand performances in some hundred and twenty German theaters. In 1930 came another provocative opera, *The Rise and Fall of Mahagonny.* Here Weill perfected his popular-song form and used it to replace the traditional opera aria; one of these numbers, "The Alabamy Song," became a great hit.

Weill's last German opera, *The Silver Lake,* opened simultaneously in 1933 in eleven different German cities. The following morning the Reichstag was burned. Weill's opera closed and the composer fled to Paris. In 1935 he came to the United States, later becom-

ing a citizen. He soon assumed a leading position in the Broadway musical world, producing a succession of stage triumphs that included *Lady in the Dark* and *One Touch of Venus.* He also completed a one-act American folk opera, *Down in the Valley,* introduced at Indiana University on July 15, 1948.

His operas: *The Protagonist* (1926); *The Royal Palace* (1927); *The Czar Has Himself Photographed* (1928); *The Threepenny Opera* (1928); *Happy End* (1929); *The Rise and Fall of Mahagonny* (1930); *Der Jasager* (1930); *Die Bürgschaft* (1932); *The Silver Lake* (1933); *Down in the Valley* (1947).

Weinberger, Jaromir, composer. Born Prague, Czechoslovakia, January 8, 1896. He was a pupil of Vítězslav Novák and Jaroslav Krička at the Prague Conservatory, after which he studied privately with Max Reger in Berlin. In 1922 he came to the United States and taught composition at the Ithaca Conservatory. He returned to Europe four years later, becoming director of opera at the National Theater in Bratislava, and head of the Eger School of Music. Success as a composer came with a folk opera, *Schwanda der Dudelsackpfeifer,* first given in Prague on April 27, 1927. Between 1927 and 1931 it was heard over two thousand times in Europe; on November 7, 1931, it was given at the Metropolitan Opera.

With the rise of the Nazi threat to Czechoslovakia, Weinberger escaped to Paris. In 1939 he came to the United States, where he became a citizen. He has since written works for orchestra that draw on American historical themes and at times use American idioms.

His operas: *Schwanda der Dudelsackpfeifer* (1927); *The Beloved Voice* (1931); *The Outcasts of Poker Flat* (1932); *A Bed of Roses* (1934);

Wallenstein (1937); *A Bird's Opera* (1941).

Weingartner, Felix, conductor and composer. Born Zara, Dalmatia, June 2, 1863; died Winterthur, Switzerland, May 7, 1942. His music study began in Graz. In his eighteenth year he entered the Leipzig Conservatory, where he won the Mozart Award. After leaving Leipzig, his first opera, *Sakuntala,* was performed in Weimar in 1884. There he met and became a friend of Liszt, who convinced him that he ought to become a conductor. Through Liszt's recommendation, Weingartner became Hans von Bülow's assistant with the Meiningen Orchestra. In 1891 Weingartner was appointed principal conductor of the Berlin Opera. After a brief period as conductor of the Kaim Concerts in Munich, he was summoned to Vienna in 1908 to succeed Gustav Mahler as artistic director of the Opera and principal conductor of the Vienna Philharmonic. Weingartner became famous in Vienna for his outstanding performances of the Beethoven symphonies and the Wagner music dramas. He left Vienna in 1911, and from 1912 to 1914 was principal conductor of the Hamburg Stadttheater, and from 1914 to 1919 music director in Darmstadt.

He first came to the United States in 1905 as a guest conductor of the New York Philharmonic Orchestra. In 1912 he made his American debut as opera conductor by directing *Tristan und Isolde* with the Boston Opera.

In 1927 he settled in Basel, Switzerland, to become director of the Conservatory and conductor of symphony concerts. In 1935 he was recalled to Vienna to replace Clemens Krauss as artistic director. He remained in this post only until the fall of 1936. He also appeared as guest conductor in most of the major opera houses of the world, and at the Salzburg Festivals. He wrote several books, including a valuable treatise on conducting (1895) and a

history of Bayreuth (1896). He also prepared new editions of several operas, including Weber's *Oberon* and Méhul's *Joseph.*

His operas were: *Sakuntala* (1884); *Malawika* (1886); *Genesius* (1892); *Orestes,* a trilogy including *Agamemnon, Das Totenopfer,* and *Die Errinnyen* (1902); *Kain und Abel* (1914); *Dame Kobold* (1916); *Die Dorfschule* (1920); *Meister Andrea* (1920); *Der Apostat* (1938).

Weis, Karel, composer. Born Prague, Czechoslovakia, February 13, 1862; died there April 4, 1944. After attending the Prague Conservatory and the Organ School, he became a church organist and music teacher. In 1886 he was appointed conductor at the Brünn National Theater. He wrote his first opera, *Viola,* in 1892. Nine years later his outstanding success, *The Polish Jew,* was introduced in Prague. Later operas: *The Attack on the Mill* (1912); *The Blacksmith of Lešetin* (1920).

Welche Wonne, welche Lust, Blonde's aria in Act II of Mozart's *The Abduction from the Seraglio.*

Welitch (or Welitsch), Ljuba, soprano. Born Borissova, Bulgaria, July 10, 1913. She studied singing in Sofia and in Vienna, then joined the Graz Opera Company, where she made her debut as Nedda. After appearances with several provincial opera companies, during which period she was heard in over forty roles, she became a member of the Hamburg Opera. In 1943 she was engaged by the Vienna State Opera, where she scored her first major successes. Her London debut took place at Covent Garden in 1947 in a performance of Richard Strauss's *Salome.* She enjoyed a triumphant success. Her American debut, also in *Salome,* was no less a sensation at the Metropolitan Opera on February 4, 1949. She has since been a principal soprano of both Covent Garden and the Metropolitan Opera, besides giving guest performances with leading European opera companies.

Wellesz, Egon, composer and musicologist. Born Vienna, Austria, October 21, 1885. He attended the University of Vienna, and studied music privately with Arnold Schoenberg. From 1911 to 1915 he taught music history in Vienna. In 1919 he joined the faculty of the Vienna University where, from 1928 on, he was a professor of music history. After the annexation of Austria by Germany, he left his native land and settled in England, where he received a research fellowship at Oxford. He visited the United States in 1947 and delivered lectures at Princeton University and Columbia University. He is an authority on Byzantine music, having written a definitive study of it in 1922, and an allied work, *Eastern Elements in Western Chant,* in 1947. In 1948 he was appointed University Reader in Byzantine Music at Oxford. He has also written a study on the early history of opera, *Cavalli und der Stil der venetianischen Oper* (1913), and a biography of Arnold Schoenberg (1921). As a composer, he is represented by a number of operas: *Die Prinzessin Girnara* (1921); *Alkestis* (1924); *Die Opferung des Gefangenen* (1926); *Scherz, List und Rache* (1928); *Die Bacchantinen* (1931); *Incognita* (1951).

Wellgunde, a Rhine maiden (soprano) in Wagner's *Das Rheingold* and *Die Götterdämmerung.*

Wenn der Freude Thränen fliessen, love duet of Belmonte and Constanza in Act II of Mozart's *The Abduction from the Seraglio.*

Wenzel, Micha's bumpkin son (tenor) in Smetana's *The Bartered Bride.*

Werdenberg, see PRINCESS VON WERDENBERG.

Wer ein holdes Weib errungen, the closing chorus of Beethoven's *Fidelio.*

Wer ein Liebchen hat gefunden, Osmin's cynical opinion of women in Act I of Mozart's *The Abduction from the Seraglio.*

Werfel, Franz, novelist and dramatist. Born Prague, Bohemia, September 10, 1890; died Beverly Hills, California, August 26, 1945. His novel *The Forty Days of Musa Dagh* was the source of Lodvico Rocca's opera *Monte Ivnor.* Milhaud's *Maximilien* was derived from Werfel's play *Juarez and Maximilian.* A discriminating music lover, Werfel wrote a novel entitled *Verdi: A Novel of the Opera* (1925), and collaborated with Paul Stefan in editing *Verdi: The Man in His Letters* (1942). He also translated the librettos of several of Verdi's operas into German. He married the widow of Gustav Mahler.

Werner Kirchhofer, the trumpeter (baritone) in Nessler's *Der Trompeter von Säkkingen.*

Werther, lyric drama by Jules Massenet. Libretto by Edouard Blau, Paul Milliet, and Georges Hartmann, based on Goethe's novel *Die Leiden des jungen Werther.* Première: Vienna Opera, February 16, 1892. This is one of the few operas by Massenet in which the heroine is not a courtesan but a virtuous woman. In Germany, in 1772, Werther falls in love with Charlotte, who is betrothed to his friend, Albert. Charlotte, who returns Werther's love but feels bound to marry Albert, urges Werther to leave her forever. When she discovers that he has asked Albert for his pistols, she becomes apprehensive, and rushes to him late one night in a blinding snowstorm. Werther has shot himself, and he dies in her arms. Werther's aria "Pourquoi me réveiller" and Charlotte's Song of Tears, "Va! laisse-les couler," both in Act III, are among the more familiar vocal excerpts.

When I am laid in earth, Dido's lament in Act III of Purcell's *Dido and Aeneas.*

Whitehill, Clarence, baritone. Born

Marengo, Iowa, November 5, 1871; died New York City, December 19, 1932. He was employed as a clerk when Nellie Melba heard him and advised him to study singing. He went to Paris and studied with Giovanni Sbriglia and Alfred Auguste Giraudet. His debut took place at the Théâtre de la Monnaie in 1899 in *Roméo et Juliette*. He made such a good impression that he was engaged by the Opéra-Comique. In 1900 he returned to the United States and sang with the Henry Savage Opera Company. After an additional period of study in Germany, and appearances in minor German opera houses, he made his Metropolitan Opera debut on November 25, 1909, as Amfortas, a role for which he became noted. During this period he remained at the Metropolitan only two seasons. After an engagement with the Chicago Opera, he returned to the Metropolitan in 1918, now remaining eighteen years. He was acclaimed for his interpretations of Wagnerian roles, particularly those of Amfortas, Hans Sachs, and Wotan. He was also heard in French operas, including the Metropolitan premières of *Louise* and *Pelléas et Mélisande*. He resigned from the Metropolitan after the 1931–1932 season because of differences with the management.

White Wings, opera by Douglas Moore. Libretto is Philip Barry's play of the same name. Première: Hartford, Connecticut, February 9, 1949. "White Wings" is an organization of street cleaners in an unnamed American city in 1895. Archie Inch and Mary Todd are in love, but are kept apart by their differing views on horses. Archie comes from a generation of White Wingers who revere the horse while Mary has faith in the future of the automobile. She grows wealthy when her father becomes a successful automobile manufacturer. After the last horse in town is shot, the now impoverished and humble Archie is able to marry the girl he loves.

Widerspänstigen Zähmung, Der, *see* TAMING OF THE SHREW, THE.

Wieland, Christoph Martin, poet. Born Biberich, Germany, September 5, 1733; died Weimar, Germany, January 20, 1813. Weber's *Oberon* was based on Wieland's verse epic of the same name. Paul Wranitzky's *Oberon* and Friedrich Kuntzen's *Holger Danske* were drawn from the same poem. Friedrich Benda's *Alceste* was based on Wieland's poem of this name, and Anton Schweitzer wrote operas to his *Alceste, Aurora, Rosamund,* and *Die Wahl des Herkules.*

Wie oft in Meeres tiefsten Schlund, the Dutchman's aria in Act I of Wagner's *Der fliegende Holländer.*

Wilde, Oscar, poet, dramatist, novelist. Born Dublin, Ireland, October 16, 1856; died Paris, France, November 30, 1900. His poetic drama *Salome* (which he wrote in French) was the libretto of Richard Strauss's opera of the same name. Another composer, Antoine Mariotte, made an operatic version of Wilde's play at about the same time, causing a conflict (*see* SALOME). Other operas made from Wilde's writings include: Renzo Bossi's *L'usignuolo e la rosa;* Jaroslav Křička's *The Gentleman in White;* Alexander Zemlinsky's *Eine florentinische Tragödie* and *Der Zwerg,* William Orchard's *The Picture of Dorian Gray;* Hans Schaeuble's *Dorian Gray.*

Wilhelm Meister, *see* GOETHE, JOHANN WOLFGANG VON.

Wilhelm Tell, a poetic drama by Friedrich Schiller, the source of operas by Benjamin Carr (*The Archers*), André Grétry, and Rossini.

William Ratcliff, a drama by Heinrich Heine, the source of operas by César Cui and Pietro Mascagni, among others. Mascagni's, a student effort, was later revised.

William Tell (Guillaume Tell), opera

in four acts by Gioacchino Rossini. Libretto by Etienne de Jouy and Hippolyte Bis, based on the drama by Friedrich Schiller. Première: Paris Opéra, August 3, 1829. American première: New York, Park Theater, September 19, 1831.

Characters: William Tell, a Swiss patriot (bass); Hedwig, his wife (soprano); Jemmy, his son (soprano); Arnold, another Swiss patriot (tenor); Melcthal, his father (bass); Walter Fürst, another Swiss patriot (bass); Gessler, governor of Schwitz and Uri (bass); Mathilde, his daughter (soprano); Rudolph, captain of Gessler's guards (tenor); Ruodi, a fisherman (tenor); Leuthold, a shepherd (bass); knights, peasants, pages, ladies, hunters, soldiers. The setting is Switzerland in the fourteenth century.

The famous overture is a veritable tone poem, beginning with a description of a Swiss dawn. A storm erupts, followed by a pastoral section. The overture ends with a vigorous march, introduced by a fanfare.

Act I. Tell's chalet on Lake Lucerne. Swiss patriots are conspiring to overthrow the tyrant Gessler. Arnold, one of them, is in love with Gessler's daughter. A marriage celebration is taking place. Shepherds participate in a folk dance (Passo a sei). The festivities are disturbed when the shepherd Leuthold appears and asks for help: one of Gessler's soldiers has tried to abduct his daughter and he has killed the man. William Tell starts ferrying Leuthold across the lake. Gessler's soldiers arrive. Since they cannot find Leuthold, they seize Melcthal instead.

Act II, Scene 1. A forest. Mathilde, in love with Arnold, muses on how she prefers a simple life with her beloved to the luxury of her father's palace ("Sombre forêt"). When Arnold appears, the lovers greet each other passionately and curse the destiny that keeps them apart. Tell now comes to inform Arnold that his father, Melcthal, has been killed by Gessler.

Scene 2. A secret meeting place in a wood. The patriots, banded to plan Gessler's overthrow, are inspired by a rousing speech by William Tell.

Act III, Scene 1. A ruined chapel near Gessler's palace. Arnold comes to bid his beloved Mathilde farewell ("Pour notre amour") because it has become his mission to destroy her father. Mathilde promises to remain true to Arnold.

Scene 2. The market place of Altdorf. Gessler addresses his people on the occasion of the centenary of Austrian rule. The people, in festive mood, celebrate the observance with songs and dances. William Tell and his son Jemmy are present. Noticing that Tell is not paying proper homage to Gessler, the captain of the guards arrests him and brings him to the governor. Hoping to humiliate Tell, Gessler orders him to place an apple on his son's head and split the apple with an arrow. Tell begs his son to remain immobile and put his trust in God ("Sois immobile"). He then takes aim and shoots the apple squarely. Bitterly, Tell informs the governor that had he missed his target and hurt his son he would have sent a second arrow into Gessler's heart. The governor orders Tell's arrest.

Act IV, Scene 1. Before Melcthal's house. Arnold recalls the happy days of his youth ("Asile héréditaire"). The patriots appear with the news of Tell's arrest.

Scene 2. The shore of Lake Lucerne. Mathilde tells Hedwig Tell that her husband has escaped. As a storm is brewing, the patriots appear with Tell at their head. Gessler arrives, hunting for Tell. Tell kills him with an arrow. The patriots give voice to their rejoicing, which is further intensified with the news that Gessler's palace has fallen. Switzerland is now free. The patriots sing a hymn of joy.

In writing his last opera, Rossini was consciously—perhaps *too* consciously—creating his crowning masterpiece. He built the work on monumental lines. It requires six hours for a complete performance; it is filled with big scenes and pageantry. All this was new to Rossini, who heretofore had been at his best in light, spontaneous music for comparatively trivial episodes. Also new was the dramatic power, sublimity of expression, psychological insight into character, and symphonic breadth and harmonic richness of the musical writing found in *William Tell.* These elements compensate for the dull stretches and the lapses of inspiration. Audiences have never been wholeheartedly fond of *William Tell.* Students of opera, however, consider it a surpassing creation. Bellini said it reduced all operas of his day, including his own, to pygmies. Wagner considered that it anticipated his own revolutions in dramatic thought and stylistic approaches.

Willow Song, *see* SALCE, SALCE.

Willst jenes Tag's du nicht dich mehr entsinnen, Erik's plea to Senta in Act III of Wagner's *Der fliegende Holländer.*

Win-San-Luy, Win-Shee's son (tenor) in Leoni's *L'Oracolo.*

Win-Shee, a doctor (bass) in Leoni's *L'Oracolo.*

Winter's Tale, A, *see* SHAKESPEARE.

Witch, The, a character (mezzo-soprano) in Humperdinck's *Hansel and Gretel.*

Witch of Salem, A, opera by Charles Wakefield Cadman. Libretto by Nelle Richmond Eberhart. Première: Chicago Civic Opera, December 8, 1926. The background is the witch trials in Salem, Massachusetts, in 1692. Sheila loves Arnold, who prefers Claris. When she cannot win Arnold, she accuses Claris of being a witch. Just before Claris is executed, Sheila confesses to Arnold that she lied and offers herself as a substitute for Claris at the scaffold—but only if he will kiss her once. Arnold does so, and Sheila dies in Claris' place.

Witch's House, The (Das Knusperhäuschen), the prelude to Act III of Humperdinck's *Hansel and Gretel.*

Witch's Ride, *see* HEXENRITT.

Witch's Song, *see* HURR, HOPP, HOPP, HOPP.

With drooping wings, concluding chorus of Purcell's *Dido and Aeneas.*

Witherspoon, Herbert, bass. Born Buffalo, New York, July 21, 1873; died New York City, May 10, 1935. After graduating from Yale University in 1895, he studied music with Gustav Stoeckel, Horatio Parker, and Edward MacDowell. He received instruction in singing from Walter Henry Hall and Max Treumann in New York, Jean Baptiste Faure and Jacques Bouhy in Paris, and Francesco Lamperti in Milan. He made his debut in 1898 with a small opera company in New York. After extensive appearances in concerts and oratorio performances, he made his debut at the Metropolitan Opera on November 26, 1908, as Titurel. He remained at the Metropolitan Opera until 1914 when he retired from the stage to concentrate on teaching. In 1925 he became president of the Chicago Musical College. In 1930 he was engaged as artistic director of the Chicago Civic Opera and in 1931 as president of the Cincinnati Conservatory. When Giulio Gatti-Casazza retired as general manager of the Metropolitan Opera in 1935, Witherspoon was selected as his successor. But before his first season began he died of a heart attack in his office at the Metropolitan.

Woglinde, a Rhine maiden (soprano) in Wagner's *Das Rheingold* and *Die Götterdämmerung.*

Wolf, Hugo, composer. Born Windischgräz, Austria, March 13, 1860; died

Vienna, February 22, 1903. He attended the Vienna Conservatory for a brief period, but for the most part was self-taught. He became a music critic of the Vienna *Salonblatt* in 1884. Three years later he published his first volumes of the songs for which he is today remembered and admired. His opera *Der Corregidor*, introduced in Mannheim in 1896, was a failure. He was working on a second opera, *Manuel Venegas*, when he lost his mind, and was confined to a private hospital for the rest of his life.

Wolff, Albert Louis, conductor and composer. Born Paris, France, January 19, 1884. He graduated from the Paris Conservatory. After serving as church organist for four years he became, in 1908, chorus master of the Opéra-Comique. His debut as a conductor took place there in 1911 in the world première of Laparra's *La Jota*. For several seasons he conducted at the Opéra-Comique. He made his American bow on November 21, 1919, at the Metropolitan Opera, leading *Faust*. A month later, on December 27, he conducted there the world première of his opera *L'oiseau bleu* (*The Bluebird*). He remained at the Metropolitan Opera until 1921, specializing in French operas. In 1922 he returned to the Opéra-Comique, and in 1924 he became artistic director of a new opera company at the Théâtre des Champs Elysées. He subsequently distinguished himself as conductor of symphonic music.

Wolf-Ferrari, Ermanno, composer. Born Venice, Italy, January 12, 1876; died Venice, January 21, 1948. He was trained to be an artist but a visit to Bayreuth turned him to music. He completed his music study in Munich with Josef Rheinberger. In 1899 he made his debut as composer with a Biblical cantata, *La Sulamite*, a success when performed in Venice. A year later *La cenerentola*, his first opera, was

introduced at the Teatro la Fenice. Success came with an opera buffa, *Le donne curiose*, given in Munich in 1903. His masterwork in the comic style, *The Secret of Suzanne*, was introduced in Munich in 1909. His tragic opera, *The Jewels of the Madonna*, first heard in Berlin in 1911, was also acclaimed. From 1902 to 1909 he was the director of the Liceo Benedetto Marcello in Venice. In 1912 he visited the United States to supervise the American première of *The Jewels of the Madonna* at the Chicago Opera.

His operas: *La cenerentola* (1900); *Le donne curiose* (1903); *I quattro rusteghi* (1906); *Il segreto di Susanna* (1909); *I gioielli della Madonna* (*The Jewels of the Madonna*) (1911); *L'amore medico* (1913); *Gli amanti sposi* (1925); *Veste di Cielo* (1927); *Sly* (1927); *La vedova scaltra* (1931); *Il Campiello* (1936); *La Dama Boba* (1938); *Gli dei a Tebe* (1943).

Wolfram von Eschenbach, a minstrel-knight (baritone), Tannhäuser's friend, in Wagner's *Tannhäuser.*

Wolfserzählung, *see* SCHAU HER, DAS IST EIN TALER.

Worms, Carlo, Ricki's lover (baritone) in Franchetti's *Germania.*

Wotan, ruler of the gods (bass-baritone) in Wagner's *Das Rheingold, Die Walküre,* and *Siegfried.*

Wotan's Farewell, *see* LEB' WOHL.

Wowkle, a squaw (mezzo-soprano) in Puccini's *The Girl of the Golden West.*

Wozzeck, opera in three acts by Alban Berg. Libretto by the composer, based on the drama of the same name by Georg Büchner. Première: Berlin State Opera, December 14, 1925. American première: Philadelphia, Philadelphia Grand Opera Association, March 19, 1931.

Characters: Wozzeck, a soldier (tenor); Marie, his mistress (soprano); a drum major (baritone); a captain (bass); a doctor (bass); Margaret (contralto). The setting is a small town in Germany, the time about 1820.

Act I. In the captain's room, the captain quizzes Wozzeck about his illegitimate child and lectures him about morality. The scene shifts to the open country, where Wozzeck and a fellow soldier are chopping wood. Wozzeck is filled with dread at the sound of strange noises. We now enter Marie's room. Looking out the window, Marie flirts with a passing drum major. The frightened Wozzeck enters and talks incoherently. And now Wozzeck is in a study, where he is the subject for experiments by a fanatical doctor. When Wozzeck coughs, the doctor scolds him. Partaking of a special diet prepared for him, Wozzeck begins to see hallucinations. The doctor is delighted, for his experiments are successful. In front of Marie's house, meanwhile, Marie and the drum major get acquainted. They embrace and enter Marie's room.

Act II. Marie, in her room, admires the earrings the drum major has given her. When Wozzeck sees the earrings he grows suspicious. Later, in a street, the captain taunts Wozzeck with hints at Marie's infidelity. Wozzeck now openly accuses Marie of having been unfaithful. Marie defies him. The scene changes to a garden, where Wozzeck finds Marie dancing with the drum major. Later, in the barracks, the drum major boasts before Wozzeck of his affair with Marie. He invites Wozzeck to drink with him and when Wozzeck refuses, he beats him soundly.

Act III. Marie is repentant, and seeks solace in the Bible. She takes a walk with Wozzeck, who is distraught and incoherent. At the edge of a pond, Wozzeck kills her with a knife. Later on, in a tavern, Wozzeck is drinking with Margaret who notices blood on his hands. Wozzeck leaves impetuously. Returning to the pond, he searches for the knife he had thrown into the water;

seeing it, he wades out to recover it and drowns. The captain and doctor hear Wozzeck's cries for help. The captain wants to investigate these cries, the doctor—superstitious and afraid—prevents him from doing so. In the concluding scene, Marie's son is playing with a hobby horse. His little friends come to tell him that his mother is dead, but he does not understand what they are saying.

There are few works in the entire history of opera so remarkably original and, at the same time, so effective as *Wozzeck*. Berg uses a vocal style that is freely declamatory, at some points approaching the sound of speech through the use of sprechstimme (which see). This lyricism is stark, at times gruesome, in its dressing of atonal harmonies, and it reflects the morbidness of the text with striking success. Involved in the structure of the score are Berg's versions of such traditional musical forms as a passacaglia, a rhapsody, a suite, a sonata, a fantasia and fuge, and a set of inventions. These forms are not necessarily evident to the person hearing *Wozzeck,* and the composer did not intend them to be; they are present for their symbolic appropriateness to various moments of the story. Berg's orchestration is rich and unusual: at various points he calls for a chamber orchestra, a military band, a restaurant orchestra of high-pitched violins, an accordion, and an out-of-tune upright piano.

The music world first became acquainted with *Wozzeck* through three excerpts introduced at the Frankfurt Music Festival in 1924. These pieces created a sensation. In 1925, after 137 rehearsals, the work was produced in its entirety by the Berlin Opera. There were widely divergent opinions. A critic in the *Deutsche Zeitung,* "had the sensation of having been not in a public theater but in an insane asylum."

Others considered it the greatest opera since *Pelléas et Mélisande*. A booklet was published in Vienna quoting and analyzing these strange differences of critical thought. Regardless of controversy, *Wozzeck* soon established itself as one of the most compelling of modern operas. Within a decade of its première it was given over twelve hundred times in twenty-eight European cities. In the period following World War II the opera has had a number of significant revivals, in the United States as well as in Europe.

Manfred Gurlitt's *Wozzeck*, another opera based on Georg Büchner's drama, has been entirely eclipsed by Berg's masterpiece.

Wreckers, The (Les Naufrageurs), opera by Ethel Smyth. Libretto (in French) by H. B. Laforestier (pseudonym of H. B. Brewster). Première: Leipzig, November 11, 1896. The setting is the Cornish coast in the eighteenth century. The inhabitants regard shipwrecks as gifts from heaven, and conspire to bring on many such wrecks. When Pasco's wife Thirzen (who loves Mark) lights a bonfire, she is falsely suspected of trying to warn ships away from the coast. For this, she and Mark are condemned to death by the wreckers.

Wrestling Bradford, the Puritan clergyman (baritone) in Hanson's *Merry Mount*.

X

Xenia, Boris' daughter (soprano) in Mussorgsky's *Boris Godunov*.

Xerxes, *see* SERSE.

Y

Yamadori, a Japanese prince (baritone) in love with Cio-Cio-San in Puccini's *Madama Butterfly*.

Yeats, William Butler, poet and dramatist. Born Dublin, Ireland, June 13, 1865; died France, January 28, 1939. Operas based on Yeats's writings include: Werner Egk's *Irish Legend;* Lou Harrison's *The Only Jealousy of Emer;* Fritz Hart's *The Land of Heart's Desire;* and Manolis Kalomiris' *The Shadowy Waters*.

Yniold, Golaud's young son (soprano) in Debussy's *Pelléas et Mélisande*.

Z

Zacharias, an Anabaptist preacher (bass) in Meyerbeer's *Le Prophète*.

Zamiel, the Black Huntsman (speaking role) in Weber's *Der Freischütz*.

Zampa, opéra-comique by Louis Hérold. Libretto by Mélesville. Première: Opéra-Comique, May 3, 1831. Zampa is the leader of a pirate band invading the island of Castel Lugano. He compels Camille to abandon her betrothed and marry him instead. During the pirates' celebration of this event, Zampa derisively places a ring on the finger of a statue of Alice, a girl he has betrayed. The statue refuses to release the ring. Camille now escapes. As Zampa attempts to pursue her, the statue drags him to his death in the sea. The overture is a well-known excerpt.

Zandonai, Riccardo, composer. Born Sacco, Italy, May 28, 1883; died Pesaro, Italy, June 5, 1944. His music study took place in Roveredo with V. Gianferrari and at the Pesaro Liceo with Mascagni. Boïto introduced him to the publisher Ricordi, who commissioned him to write his first opera: *The Cricket on the Hearth,* introduced in Turin in 1908 with moderate success. He was acclaimed in 1914 for his best opera, *Francesca da Rimini,* first performed that year in Turin and afterward given by many major opera houses. In 1921 he completed another important opera, *Giulietta e Romeo,* introduced at the Teatro Costanzi on February 14, 1922. From 1939 to the time of his death he was director of the Liceo Rossini in Pesaro. Besides operas already mentioned, he wrote: *Conchita* (1911); *Melenis* (1912); *La via della finestra* (1919); *I cavalieri di Ekebù* (1925); *Giuliano* (1928); *La farsa amorosa* (1933); *Una partita* (1933).

Zandt, Marie Van, see VAN ZANDT, MARIE.

Zaretski, Lensky's friend (baritone) in Tchaikovsky's *Eugene Onegin.*

Zar und Zimmermann (The Czar and the Carpenter), comic opera by Albert Lortzing. Libretto by the composer. Première: Leipzig Stadttheater, December 22, 1837. Peter I of Russia assumes the identity of a carpenter, Peter Michailov, in Sardam. There, a carpenter, Peter Ivanov, is mistaken for the disguised Czar by foreign envoys seeking to negotiate treaties. The ensuing complications are finally straightened out, the Czar returns to Russia, and Ivanov marries the girl he loves, Marie.

Zarzuela, a form of Spanish opera that arose in the seventeenth century and took its name from the Palace of Zarzuela, near Madrid, where these entertainments first became popular. A few zarzuelas were sung throughout, but a usual feature of this form is the utilization of spoken dialogue between the songs. The works were not long, allowing three or four to be given in an evening. Some zarzuelas were tragic or melodramatic; most were of a humorous cast, ranging from satire to burlesque. The form retained its popularity, leading to the founding of the Teatro de la Zarzuela in Madrid in 1856. Many leading Spanish composers have composed zarzuelas, and the popularity of the form and style continues today.

Zauberflöte, Die, see MAGIC FLUTE, THE.

Zauberoper, see MAGIC OPERA.

Zaza, opera by Ruggiero Leoncavallo. Libretto by the composer, based on the play of the same name by Pierre Berton and Charles Simon. Première; Teatro Lirico, Milan, November 10, 1900. Zaza is a café singer who loves Milio Dufresne and becomes his mistress. Learning that he is married, she frightens him by threatening to disclose their affair to his wife. Milio suddenly realizes how much his family means to him, and Zaza, going to Milio's home, leaves without making her scene. Later, she renounces her beloved and sends him back to his family.

Zeffiretti lusinghieri, Ilia's aria in Act III, Scene 1, of Mozart's *Idomeneo.*

Zeitkunst, German for "art of the times," a term used in Germany in the 1920's to describe works dealing with current themes, treated in a modern style. The German operas of Kurt Weill were representative examples.

Zemlinsky, Alexander, composer and conductor. Born Vienna, Austria, October 4, 1872; died New York City, March 16, 1942. He attended the Vienna Conservatory. His first opera, *Sarema,* won the Lenpold Award in 1897. It was a conductor of operas that he first became known. In 1906 he became first conductor of the Vienna Volksoper. Two years later he was engaged as first conductor of the Vienna Opera. In 1911 he was appointed musical director of the German Opera in Prague, where he remained sixteen years. From 1927 to 1933 he was principal conductor of the Berlin Opera. When the Nazis came to power, Zemlinsky returned to Vienna. In 1938 he came to the United States, where he died four years later. He was a distinguished teacher, his pupils including Arnold Schoenberg, Erich Korngold, and Artur Bodanzky. His operas: *Sarema* (1897); *Es war einmal* (1900); *Kleider machen Leute* (1910); *Eine florentinische Tragödie* (1917); *Der Zwerg* (1921); *Der Kreidekreis* (1933).

Zenatello, Giovanni, dramatic tenor. Born Verona, Italy, February 22, 1876; died New York City, February 11, 1949. He attended the Scuola di Canto, after which he studied privately with Giovanni Moretti in Milan. In 1901 he made his debut at the San Carlo in Naples as Canio. From 1903 to 1907 he appeared at La Scala, where he created the role of Pinkerton, and that of Vassili in *Germania.* After successful appearances at Covent Garden he made his American debut at the Manhattan Opera on November 4, 1907, in *La Gioconda.* He stayed with the Manhattan Opera two seasons, then appeared for five more with the Boston Opera, and for a single season with the Chicago Opera. He retired from the stage in 1930, devoting himself to teaching singing in New York.

Zeno, Apostolo, poet and librettist. Born Venice, Italy, December 11,

1668; died there November 11, 1750. He was the most significant opera librettist before Metastasio. In 1710 he founded in Venice the *Giornale dei letterati d'Italia.* Eight years later he settled in Vienna, where he served as court poet. Returning to Venice, he lived there the rest of his life. He wrote over seventy librettos which were set to music by most of the famous composers of his generation, including: Bononcini (*Astarto*); Caldara (*Ifigenia in Aulide*); Francesco Gasparini (*Merope*); Handel (*Faramondo* and *Scipio*); Hasse (*Lucio Papirio*); Jommelli; Pergolesi (*Salustia*); Porpora (*Temistocle*); Antonio Lotti; Sacchini; Domenico Scarlatti, Vivaldi, Niccolò Zingarelli.

Zerbinetta, a character (soprano) in Richard Strauss's *Ariadne auf Naxos.*

Zerlina, (1) the innkeeper's daughter (soprano) in Auber's *Fra Diavolo.*

(2) Masetto's betrothed (soprano) in Mozart's *Don Giovanni.*

Ziegler, Edward, opera manager. Born Baltimore, Maryland, March 25, 1870; died New York City, October 25, 1947. His professional career in music began when he became assistant music critic to James Gibbons Huneker on the *New York Sun.* After holding various other posts as music critic in New York, he became administrative secretary of the Metropolitan Opera in 1916. In 1920 he was engaged as assistant general manager to Gatti-Casazza. He remained at the Metropolitan until the end of his life, supervising most of its administrative and financial operations, and scouting Europe for new singers. He arranged the broadcasts of Metropolitan Opera from its stage, and in 1940 was one of the leaders of the successful public drive to raise a million dollars for the Metropolitan.

Zigeuenerbaron, Der, *see* GYPSY BARON, THE.

Zita, Donati's cousin (mezzo-soprano) in Puccini's *Gianni Schicchi.*

Zitti, zitti, (1) Figaro's plea for quiet and haste in Act III of Rossini's *The Barber of Seville*.

(2) Chorus of the courtiers in the finale of Act II of Verdi's *Rigoletto*.

Zola, Émile, author. Born Paris, France, April 2, 1840; died there September 29, 1902. The foremost exponent of French literary naturalism profoundly affected the development of the French opera composer Alfred Bruneau, who adapted many of his novels for operas, while writing other works to Zola's librettos (*see* BRUNEAU). Manfred Gurlitt wrote an opera on Zola's *Nana*. Karel Weis used Zola's *Soirées de Medaic* for his opera *The Attack on the Mill*. The same story was used by Bruneau for his *L'attaque du moulin*.

Zuane, a gondolier (bass) in Ponchielli's *La Gioconda*.

Zukunftsart, and Zukunftsmusik, German for "art of the future" and "music of the future." The terms were used by Richard Wagner to describe his music, and for the next half century they were current in the literary battles fought over Wagner's esthetics.

Zum Leiden bin ich auserkoren, the aria of the Queen of the Night in Act I, Scene 1, of Mozart's *The Magic Flute*.

Zum letzten Liebesmahle, chorus of the Knights of the Holy Grail in Act I, Scene 2, of Wagner's *Parsifal*.

Zuniga, a captain of the guards (bass) in Bizet's *Carmen*.

Zurga, tribal chieftain (baritone) in Bizet's *Les pêcheurs de perles*.

Pronunciation Guide

Pronunciation Guide

Foreign names and titles are pronounced in whatever form is known or believed to be current in American usage; Anglicizations and other conventional forms are recorded if known; otherwise, the word is pronounced with whatever sounds best imitate the original foreign pronunciation. The pronunciation symbols employed are those used in *The American College Dictionary;* most of these symbols are either common and unambiguous English letters (such as b, ch, and d), or are so widely used in other texts and reference books that they are familiar to most Americans (ā, ă, ē, ĕ, and so on). The sounds represented are those of American English, plus a few foreign sounds fairly familiar to many Americans: French a in *patte*, French eu and German ö, French u and German ü, the consonant of German *ach*, and nasal vowels as in French *un bon vin blanc*.

This pronunciation section was prepared by Reason A. Goodwin, pronunciation editor of the *New Century Cyclopedia of Names*, and formerly of the Department of Foreign Languages of the University of Louisville and of the Department of Linguistics of the University of Chicago.

PRONUNCIATION KEY *

b, d, f, h, j, k, l, m, n, p, r, sh, t, v, w, y, z—usual English values.

ă	act, bat
ā	able, cape
â	air, dare
ä	art, calm
å	Fr. *ami* (intermediate between the ă of *cat* and the ä of *calm*)
ch	chief, beach
ĕ	ebb, set
ē	equal, bee
g	give, beg
ĭ	if, big
ī	ice, bite
KH	Ger. *ach*; Scottish *loch* (made by bringing the tongue toward the position for *k* in *key, coo*, while pronouncing a strong *h*)
N	indicates that the vowel before it is a nasal vowel. There are four such vowels in Fr. *un bon vin blanc* (œN bôN văN bläN)
ng	sing, singer
ŏ	odd, not
ō	over, no
ô	order, ball
œ	Fr. *feu*; Ger. *schön* (lips rounded in position for ō as in *over*, while trying to say ā as in *able*)
oi	oil, joy
o͝o	book, put
o͞o	ooze, rule
ou	out, loud
s	see, miss
th	thin, path
TH	that, other
ŭ	up, love
ū	use, cute
û	urge, burn
Y	Fr. *tu*; Ger. *über* (lips rounded in position for o͞o as in *ooze*, while trying to say ē as in *easy*)
zh	vision, measure
ə	occurs in unaccented syllables and indicates the sound of *a* in *alone*, *e* in *system*, *i* in *easily*, *o* in *gallop*, *u* in *circus*

Accents

′   primary (the loudest syllable), as in the first syllable of *mother* (mŭTH′ər)

‵   secondary, as in the second syllable of *grandmother* (grănd′mŭTH′ər); a syllable so marked is pronounced with less prominence than the one marked (′) but with more prominence than syllables in the same word bearing no accent mark

---

* The system of symbols used in the pronunciation guide is from *The American College Dictionary*, copyright 1947 by Random House, Inc., and is used here with the permission of Random House.

A

Abdallah (ăb dăl′ə)

Abduction from the Seraglio (sĭ răl′yō, -räl′-), The

Abends, will ich schlafen geh'n (ä′bənts, vĭl ĭкн shlä′fən gän)

Abigaille (ä bē gĭl′lä)

Abimelech (ə bĭm′ə lek)

Abscheulicher! Wo eilst du hin? (äp shoi′ lĭкн ər! vō ĭlst′ dōō hĭn?)

Abstrakte Oper (äp sträk′tə ō′pər)

Abul Hassan (ä′bōōl häs′än)

Accompagnato (äk kôm′pä nyä′tō)

Ach, das Lied hab' ich getragen (äкн′, däs lēt′ häp ĭкн gə trä′gən)

Ach, ich fühl's (äкн′, ĭкн fȳls′)

Ach, ich liebte (äкн′, ĭкн lēp′tə)

Achilles (ə kĭl′ēz)

Acis (ā′sĭs) and Galatea (găl ə tē′ə)

Ackté (äk tā′), Aïno (ī′nô)

Adalgisa (ä däl jē′zä)

Adam (à dän′), Adolphe Charles (Fr. à dôlf′ shàrl)

Adamastor, roi des vagues profondes (à dà màs tôr′, rwà dā vàg prô fônd′)

Addio alla madre (äd dē′ō äl lä mä′drä)

Addio del passato (äd dē′ō dĕl päs sä′tō)

Addio dolce svegliare (äd dē′ō dōl′chä zvä lyä′rä)

Addio, fiorito asil (äd dē′ō, fyô rē′tō ä zēl′)

Addio, senza rancor (äd dē′ō, sĕn′tsä räng kōr′)

Adele (ä dā′lə)

Adieu donc, vains objets (à dyœ′ dôn, văn zôb zhĕ′)

Adieu, Mignon, courage! (à dyœ′, mē nyôn′, kōō ràzh′!)

Adieu, mon doux rivage (à dyœ′, môn dōō rē vàzh′)

Adieu, notre petite table (à dyœ′, nô′trə pə tēt′ tá′blə)

Adina (ä dē′nä)

Adina, credimi (ä dē′nä, krä′dē mē)

Admetos (ăd mē′təs)

Adolar (ä dō lär′)

Adorno (ä dōr′nō), Gabriele (gä brē ĕ′lä)

Adriana Lecouvreur (ä drē ä′nä lə kōō vrœr′)

Adriano (ä drē ä′nō)

Aegisthus (ē jĭs′thəs)

Aegyptische Helena, Die (dē ä gĭp′tĭsh ə hä′lä nä)

Aelfrida (ăl frē′də)

Aeneas (ĭ nē′əs)

Aeneid (ĭ nē′ĭd)

Aeschylus (ĕs′kə ləs; ēs′—)

Aethelwold (ăth′əl wōld)

Africaine, L' (là frē kĕn′)

Afron or Aphron (ä frôn′)

Agamemnon (ăg ə mĕn′nŏn, —nən)

Agathe (ä gä′tə)

Ah! bello, a me ritorna (ä′! bĕl′lō, ä mä rē tōr′nä)

Ah, chacun le sait (ä′, shà kœn′ lə sĕ′)

Ah! che la morte ognora (ä′! kä lä môr′tä ōn nyō′rä)

Ah! che tutta in un momento (ä′! kä tōōt′tä ēn ōōm mô män′tō)

Ah, ciascun lo dice (ä′, chä skōōn′ lō dē′chä)

Ah! dite alla giovine (ä′! dē′tä äl lä jô′vē nä)

Ah! du wolltest mich nicht deinen Mund küssen lassen! (ä′! dōō vôl′təst mĭкн nĭкнt dī′nən mōōnt kȳs′ən läs′ən!)

Ah! fuyez, douce image (ä′! fȳē yä′, dōōs ē mázh′)

Ah guarda, sorella (ä′ gwär′dä, sô rĕl′lä)

Ah! io veggio quell' anima bella (ä′! ē′ō väd′jō kwĕl lä′nē mä bĕl′lä)

Ah! je respire enfin! (ä′! zhə rĕs pēr′ än fän′!)

Ah! je suis seule (ä′! zhə sȳē sœl′)

Ah! lève-toi, soleil (ä′! lĕ′və twä, sô lĕ′y)

Ah! Louise, si tu m'aimes (ä'! lōō ēz', sē ty mĕm')

Ah! Manon, mi tradisce (ä'! mȧ nôN', mē trä dēsh'shä)

Ah, mon fils! (ä', môN fēs'!)

Ah! non credea mirarti (ä'! nōn krā dā'ä ɪnē rär'tē)

Ah! non giunge (ä'! nōn jōōn'jä)

Ah, qual colpo (ä', kwäl kôl'pō)

Ah! ritrovarla nella sua capanna (ä'! rē trō vär'lä nĕl lä sōō'ä kä pän'nä)

Ah! se intorno a quest' urna funesta (ä'! sä ēn tôr'nō ä quȧst ōōr'nä fōō nĕ'stä)

Ah! sì, ben mio (ä'! sē', bĕn mē'ō)

Ah! si, fa core e abbraccia (ä'! sē', fä kô'rä ā äb brät'chä)

Ah, un foco insolito (ä', ōōn fô'kō en sô'lē tō)

Aïda (ä ē'dä)

Aiglon, L' (lĕ glôN')

Ai nostri monti (ī nô'strē mōn'tē)

Ainsi que la brise légère (ăN sē' kə lä brēz lä zhĕr')

Air de la poupée (ĕr də lȧ pōō pā')

Air du miroir (ĕr dy mē rwȧr')

Aithra (ī'trä)

Alain (ȧ lăN')

Albanese (äl bä nä'zä), Licia (lē'chä)

Albani (äl bä'nē), Emma

Alberich (äl'bə rĭKH)

Albine (ȧl bēn')

Alboni (äl bō'nē), Marietta

Alceste (ȧl sĕst'; It. äl chĕ'stä)

Alcindoro (äl chēn dō'rō)

Alda (äl'də), Frances

Aleko (ə lĕk'ō; Rus. ä lyô'kə)

Alerte! Alerte! (ȧ lĕr'tə! ȧ lĕr'tə!)

Alessandro Stradella (ä läs sän'drō strä dĕl'lä)

Alessio (ä lĕs'syō)

Alexis (ə lĕk'sĭs), Prince

Alfano (äl fä'nō), Franco (fräng'kō)

Al fato dan legge (äl fä'tō dän lĕd'jä)

Alfio (äl'fyō)

Alfonso (äl fôn'sŏ; It. äl fôn'sō; Sp. äl fôn'sō)

Alfonso, Don (dŏn; It., Sp. dôn)

Alfred (Ger. äl'frĕt)

Alfredo (äl frä'dō)

Alice (Fr. ȧ lēs')

Alidoro (ä lē dō'rō)

Alim (ȧ lēm')

Alisa (ä lē'sä)

Alkestis (äl kĕs'tĭs)

Alla Câ d'Oro (äl lä kä' dô'rō)

All' idea di quel metallo (äl lē dĕ'ä dē kwĕl mä täl'lō)

Allmächt'ge Jungfrau (äl mĕKHt'gə yōōng'frou)

Allmächt'ger Vater, blick' herab (äl mĕKHt'gər fä'tər,blĭk hə räp')

Almanzor (äl män'zôr)

Almaviva (äl mä vē'vä)

Alphonse (ȧl fôNs')

Altair (äl tīr')

Althouse (ôlt'hous), Paul

Altoum (äl tōōm')

Alvar, Don (dôn äl vär')

Alvarez (ȧl vȧ rä'), Albert Raymond (Fr. äl bĕr' rä môN')

Alvaro, Don (dôn äl'vä rō)

Alvary (äl vä'rē), Max (Ger. mäks)

Alvise (äl vē'zä)

Amahl (ä mäl') and the Night Visitors

Amantio di Nicolao (ä män'tyō dē nē kô lä'ō)

Amato (ä mä'tō), Pasquale (pä skwä'lä)

Amelia (ə mēl'yə; It. ä mä'lyä)

Amelia Grimaldi (ä mä'lyä grē mäl'dē)

Amfiparnaso, L' (läm'fē pär nä'zō)

Amfortas (äm fôr'täs)

Amico Fritz, L' (lä mē'kō frĭts)

Amina (ä mē'nä)

Amis, l'amour tendre et rêveur (ȧ mē', lȧ mōōr' tän'drä rĕ vœr')

Am Jordan Sankt Johannes stand (äm yôr'dän zängkt yō hän'əs shtänt)

Amleto (äm lä'tō)

Amneris (äm nĕr'ĭs)

Amonasro (ȧ mô näz'rō)

Amor (ä'môr)

Amore dei tre re, L' (lä mō'rä dĕ ē trā' rā')

Amore medico, L' (lä mō'rä mĕ'dē kō)

Amore o grillo (ä mō'rä ō grēl'lō)

Amor ti vieta (ä mōr' tē vyĕ'tä)

Amour des Trois Oranges, L' (lȧ mōōr' dā trwä zô räNzh')

Amour est un oiseau rebelle, L' (lȧ mōōr' ĕ tœn nwä zō' rə bĕl')

Amour médecin, L' (lȧ mōōr' mäd săN')

Amour! viens aider ma faiblesse! (ȧ mōōr'! vyän zĕ dä' mȧ fĕ blĕs'!)

Am stillen Herd (äm shtĭl'ən hĕrt)

Andrea Chénier (än drĕ'ä shä nyä')

Andrei (än drā')

Andrès (äN drĕs')

Andreyev (än drā'ĭf), Leonid (lē'ə nĭd; Rus. lĭ ŏ nyēt')

Ange adorable (äNzh ȧ dô rȧ'blə)

Angelica (It. än jĕ'lē kä)

Angelotti (än jĕ lŏt′tē), Cesare (chĕ′zä rä)
Anges du paradis (änzh dy på rå dē′)
Anges purs, anges radieux (än′zhǝ pyr, än′zhǝ rå dē œ′)
Anita (ä nē′tä)
An jenem Tag (än yä′nǝm täk′)
Anna, Donna (dôn′nä än′nä)
Ännchen (ĕn′shǝn)
Annina (än nē′nä)
Antheil (än′tĭl), George
Antigone (än tĭg′ǝ nē)
Antigono (än tē′gô nō)
Antonia (Fr. än tô nē å′)
Antonida (än tŏ nē′dǝ)
Antonio (än tō′nĭ ō; It. än tô′nyō)
Antonio e Cleopatra (än tô′nyō ā klä ō pä′trä)
Aphrodité (Fr. å frô dē tä′)
Aphron (ä frôn′)
Apostrophe, L' (là pôs trôf′)
Aprila, o bella (ä′prē lä, ō bĕl′lä)
Arabella (ä rä bĕl′ä)
Araquil (ä rä kēl′)
Archibaldo (är kē bäl′dō)
Ardon gl'incensi (är′dôn lyēn chĕn′sē)
Ariadne auf Naxos (ä rē äd′nǝ ouf näk′sôs)
Ariane et Barbe-Bleue (å ryán′ ā bår′bǝ-blœ′); Ariadne (är ĭ ăd′nĭ) and Blue-beard
Arianna (ä rē än′nä)
Aricie (å rē sē′)
Aristophanes (ăr ǝs tŏf′ǝ nēz)
Arkas (är′kǝs)
Arkel (är kĕl′)
Arlecchino (är läk kē′nō)
Arlesiana, L' (lär lä zyä′nä)
Armida (är mē′dä)
Armide (år mēd′)
Arnolfo (är nôl′fō)
Arnould (år nōō′), Sophie (Fr. sô fē′)
Arrêtez, ô mes frères (å rĕ tä′, ō mä frĕr′)
Arrigo (är rē′gō)
Artemidor (år tä mē dôr′)
Artemis (är′tǝ mĭs); Diana (dī ăn′ǝ)
Artôt (år tō′), Désirée (dä zē rä′)
Arvino (är vē′nō)
Ase (ä′sǝ)
A Serpina penserete (ä sĕr pē′nä pĕn sĕ rä′tä)
Asrael (ăz′rĭ ǝl)
Assad (ăs′ăd)

Assoluta (äs sô lōō′tä)
Assur (äs sōōr′)
Astaroth (ăs′tǝ rŏth)
A terra! si nel livido (ä tĕr′rä! sē nĕl lē′vē dō)
Athanaël (å tá ná ĕl′)
A travers le désert (å trá vĕr lǝ dä zĕr′)
Atterberg (ät′tǝr bĕr′y), Kurt (kōōrt)
Aubade (ō bád′)
Auber (ō bĕr′), Daniel François (Fr. då nyĕl′ frän swá′)
Au bruit de la guerre (ō bryē dǝ là gĕr′)
Au bruit des lourds marteaux (ō bryē dä lōōr már tō′)
Aucassin et Nicolette (ō kå săn′ ā nē kô lĕt′)
Au fond du temple (ō fôn dy tän′plǝ)
A un dottor della mia sorte (ä ōōn dōt tōr′ dĕl lä mē′ä sôr′tä)
Au secours de notre fille (ō sǝ kōōr′ dǝ nô′trǝ fē′y)
Aus einem Totenhaus (ous ī nǝm tō′tǝn-hous)
Austral (ôs′trǝl), Florence
Avant de quitter ces lieux (å vän′ dǝ kē tä′ sä lyœ′)
Ave Maria (ä′vä mä rē′ä)
Avis de clochettes (å vē′ dǝ klô shĕt′)
Avito (ä vē′tō)
Azora (ǝ zō′rǝ)
Azucena (ä dzōō chä′nä)

B

Baba Mustapha (bä′bä mōō′stä fä)
Babekan (bä bĕ kän′)
Babinsky (bä′bĭn skĭ)
Baccaloni (bäk kä lō′nē), Salvatore (säl vä tō′rä)
Bacchanale (bå kå nál′)
Bacchus (bäk′ǝs)
Bach (bäкн), Johann Christian (Ger. yō′hän krĭs′tē än)
Balducci (bäl dōōt′chē)
Ballata del fischio (bäl lä′tä dĕl fē′skyō)
Ballatella (bäl lä tĕl′lä)
Ballo in maschera, Un (ōōm bäl′lō ēm mä′skĕ rä)
Balstrode (băl′strōd)
Baltasar (băl tä′zǝr)
Balthazar (băl thä′zǝr)

ăct, āble, dâre, ärt; å, Fr. ami; ĕbb, ēqual; ĭf, īce; кн, Ger. ach; n indicates nasal vowel; ŏdd, ōver, ôrder; œ, Fr. feu, Ger. schön; bŏŏk, ōōze; ŭp, ūse, ûrge; y, Fr. tu, Ger. über; zh, vision; ǝ, a in alone. (Full key on page 558)

Balzac (bôl'zăk; *Fr.* bál zák'), Honoré de (ô nô rā' də)

Barbaja (bär bä'yä), Domenico dô mā'nē kō)

Barbarina (bär bä rē'nä)

Barbarino (bär bä rē'nō)

Barbe-Bleue (bàr'bə blœ')

Barber of Bagdad, The: Der Barbier von Bagdad (dər bär bēr' fôn bäk'dät)

Barber of Seville, The: Il barbiere di Siviglia (ēl bär byě'rä dē sē vē'lyä)

Barbier (bàr byā'), Jules (zhyl)

Barbieri (bär byě'rē), Fedora (fä dō'rä)

Bardi (bär'dē), Giovanni (jō vän'nē)

Bardolph (bär'dŏlf)

Barnaba (bär'nä bä)

Barrientos (bär ryěn'tōs), Maria (mä rē'ä)

Bartolo (bär'tô lō)

Basilio, Don (dŏn bə zĭl'ĭ ō; *It.* dôm bä zē'lyō)

Bassi (bäs'sē)

Bastien und Bastienne (bás tyăN' ŏŏnt bás tyěn')

Batti, batti, o bel Masetto (bät'tē, bät'tē, ō běl mä zät'tō)

Battistini (bät tēstē'nē), Mattia (mättē'ä)

Baucis (bô'sĭs; *Fr.* bō sēs')

Bayreuth (bī roit')

Beatrice (*It.* bä ä trē'chä)

Béatrice et Bénédict (bā á trēs' ā bä nā-dēkt')

Beaumarchais (bō mär shä'; *Fr.* bō már-shě'), Pierre Augustin Caron de (pyěr ō gYs tăN' ká rôN' də)

Beckmesser (běk'měs'ər)

Beethoven (bā'tō'vən), Ludwig van (lōōt'vĭKH fän)

Behüt' dich Gott (bə hYt' dĭKH gôt)

Bei Männern, welche Liebe fühlen (bī měn'ərn, věl'KHə lē'bə fY'lən)

Bekker (běk'ər), Paul

Bel Canto (běl kän'tō)

Belcore (běl kô'rä)

Belinda (bə lĭn'də)

Bella figlia dell' amore (běl'lä fēl'lyä děl lä mō'rä)

Bella siccome un angelo (běl'lä sēk kō'mä ōōn än'jě lō)

Belle, ayez pitié de nous (běl, ě yä' pē tyä' də nōō')

Belle nuit, ô nuit d'amour (bě'lə nYē, ō nYē dá mōōr')

Bellincioni (běl lēn chō'nē), Gemma (jěm'mä)

Bellini (běl lē'nē), Vincenzo (vēn chěn'tsō)

Belmonte (běl mōn'tä)

Benda (běn'dä), Georg (gä ôrk', gä'ôrk); Jiři (*Czech* yē'rzhē)

Benelli (bě něl'lē), Sem (säm)

Benoît (běn'wä; *Fr.* bə nwá')

Benoît (bə nwá'), Pierre Léonard (pyěr lä ô nàr')

Benvenuto Cellini (běn vě nōō'tō chěl lē' nē)

Benvolio (běn vō'lĭ ō)

Beppe (běp'pä)

Berceuse (běr sœz')

Berg (bûrg; *Ger.* běrk), Alban (ôl'bən, ăl'-; *Ger.* äl'bän)

Berger (běr'gər), Erna (ěr'nä)

Bergmann (bûrg'mən; *Ger.* běrk'män), Carl

Berkenfeld (bûr'kən fěld), Countess of

Berlin State Opera: Staatsoper (shtäts'-ō'pər)

Berlioz (běr'lĭ ōz; *Fr.* běr lyôz'), Hector (*Fr.* ěk tôr')

Bernauerin, Die (dē běr'nou'ə rĭn)

Bersi (běr'zē)

Berta (běr'tä)

Bertha (bûr'thə; *Ger.* běr'tä)

Bertram (bûr'trəm)

Bervoix (běr vwä'), Flora

Betto di Signa (bät'tō dē sēn'nyä)

Betz (běts), Franz (fränts)

Biaiso (bē ī 'zō)

Bianca al par hermine (byäng'kä äl pär' ěr mē'nä)

Bildnis (bĭlt'nĭs) Aria

Bis (bēs)

Bispham (bĭs'fəm), David

Biterolf (bē'tə rôlf)

Bittner (bĭt'nər), Julius (*Ger.* ū'lě ōōs)

Bizet (bē zā'; *Fr.* bē zě'), Georges (zhôrzh)

Bjoerling (byœr'lĭng), Jussi (*Sw.* yōōs'sē)

Blacher (bläKH'ər), Boris

Blanche Dourge (bläNsh dōōrzh)

Blaze (bláz), François-Henri-Joseph (fräN swá' äN rē' zhō zěf')

Blech (blěKH), Leo (*Ger.* lä'ō)

Blick' ich umher (blĭk ĭKH ōōm här')

Blitzstein (blĭts'stīn), Marc

Blockx (blôks), Jan (yän)

Blonde (blôn'də); Blonda (blôn'də)

Boccaccio (bō kä'chĭ ō; *It.* bōk kät'chō), Giovanni (jō vän'nē)

Boccanegra (bōk kä nä'grä)

Bodanzky (bō dänts'kĭ), Artur (*Ger.* är'tŏŏr)

Böhm (bœm), Karl

Bohème, La (là bô ěm')

Boieldieu (bwȧl dyœ′), François Adrien (fräN swȧ′ ȧ drē ăN′)
Boisfleury (bwä flœ rē′), Marquis de (də)
Boïto (bô′ē tō), Arrigo (är rē′gō)
Bonci (bōn′chē), Alessandro (älässän′drō)
Boniface (bŏn′ĭ fäs; Fr. bô nē fȧs′)
Bonocini (bô nôn chē′nē), Giovanni Battista (jō vän′nē bät tē′stä)
Bonze (bŏnz), The; Bonzo (bōn′dzō)
Bori (bō′rē), Lucrezia (lōō krä′tsyä)
Boris Godunov (bōr′ĭs gō′də nôf; Rus. bŏ rēs′ gə dōō nôf′)
Borodin (bōr′ə dēn), Alexander
Borov (bō′rôf)
Borromeo, Carlo (bôr rô mě′ō), Carlo
Borsa (bōr′sä)
Bostana (bō stä′nä)
Boughton (bou′tən), Rutland
Bouillon (bōō yôN′), Princesse de (präN sěs′ də)
Bourgeois Gentilhomme, La (lə bōōr zhwä′ zhäN tē yôm′)
Brander (brän′dər)
Brangäne (bräng gä′nə)
Braslau (brăs′lou), Sophie
Braut von Messina, Die (dē brout fôn mě sē′nä)
Bravour (brä vōōr′) Aria
Bravura (brä vōō′rä)
Brecht (brěкнt), Bertolt (běr′tŏlt)
Breisach (brī′zäкн), Paul
Breitkopf und Härtel (brīt′kŏpf ōōnt hěr′təl)
Bréval (brä vȧl′), Lucienne (lʏ syěn′)
Bride of Abydos (ə bī′dŏs), The
Bride of Lammermoor (lăm′ər mōōr), The
Brindisi (brēn′dē zē)
Brogny (brô nyē′), Cardinal
Brothers Karamazov (kä rä mä′zəf)
Bruch (brōōкн), Max (Ger. mäks)
Brüderchen, komm tanz' mit mir (brʏ′dər shən, kôm tänts mĭt mēr)
Brüll (brʏl), Ignaz (ĭg′näts)
Brünnhilde (brʏn hĭl′də)
Brünnhilde! Heilige Braut! (brʏn hĭl′də! hī′lĭ gə brout!)
Bruneau (brʏ no′), Alfred (Fr. ȧl frěd′)
Büchner (bʏкн′nər), Georg (gä ôrk′, ga′-ôrk)
Buffa (bōōf′fä); buffo (bōōf′fō)
Bülow (bʏ′lō), Hans von (häns fôn)
Buona Figliuola, La (lä bwô′nä fēl lū ô′lä)

Buononcini (bwô nôn chē′nē)
Burgschaft, Die (dē bōōrk′shäft)
Burgtheater (bōōrk′tä ä′tər)
Busch (bōōsh), Fritz
Busch, Hans (häns)
Busoni (bōō zō′nē), Ferruccio (fěr rōōt′-chō)
Büsser (bʏ sěr′), Paul Henri (Fr. pôl äN rē′)
Buzzy (bŭz′ĭ)

C

Cabaletta (It. kä bä lät′tä)
Caccini (kät chē′nē), Giulio (jōō′lyō)
Cachés dans cet asile (kȧ shä′ däN sět ȧ zēl′)
Caffarelli (käf fä rěl′lē)
Caius (kä′əs), Dr.
Calaf (kä lȧf′)
Calatrava (kä lä trä′vä), Marquis of
Calchas (kăl′kəs)
Caldara (käl dä′rä), Antonio (än tô′nyō)
Calderón de la Barca (käl dä rōn′ dä lä bär′kä), Pedro (pä′drō)
Callas (kăl′əs), Maria Meneghini (mä rē′-ä mě ně gē′nē)
Calvé (kȧl vä′), Emma
Calzabigi (käl tsä bē′jē), Ranieri da (rä nyě′rē dä)
Cambert (käN běr′), Robert (Fr. rô běr′)
Camerata (kä mě rä′tä)
Camille (kə mĭl′)
Cammarano (käm mä rä′nō), Salvatore (säl vä tō′rä)
Campanari (käm pä nä′rē), Giuseppe (jōō-zěp′pä)
Campana sommersa, La (lä käm pä′nä sôm měr′sä)
Campanini (käm pä nē′nē), Cleofonte (klä ō fōn′tä)
Campanini, Italo (ē′tä lō)
Campiello, Il (ēl käm pyěl′lō)
Campra (käN prä′), Andre (äN drä′)
Canio (kä′nyō)
Caponsacchi (kăp ən săk′ĭ)
Capriccio (kə prē′chĭ ō; It. kä prēt′chō)
Capulet (kăp′yə lət)
Cardillac (kȧr dē yȧk′)
Carestini (kä rä stē′nē), Giovanni (jō-vän′nē)
Carlo, Don (It. dôn kär′lō)

Carlos, Don (dŏn kär'ləs, dôn kär'lōs)
Carmela (kär mā'lä)
Carmen (kär'mən)
Carmina Burana (kär'mĭ nä bōō rä'nä)
Carmosine (kár mô zēn')
Carolina (kä rô lē'nä)
Caro nome (kä'rō nô'mā)
Carré (ká rā'), Albert (*Fr.* ál bĕr')
Carrosse du Saint-Sacrement, La (lä ká-rôs' dy sǎn sá krə män')
Caruso (kə rōō'sō; *It.* kä rōō'zō), Enrico (ĕn rē'kō)
Carvalho (kár vá yō'), Léon (lā ôɴ')
Casella (kä sĕl'lä), Alfredo (äl frä'dō)
Caspar (*Ger.* käs'pär)
Cassandra (kə sǎn'drə; *It.* käs sän'drä)
Cassio (käslı'ĭ ō, kǎs'-; *It.* käs'syō)
Casta Diva (kä'stä dē'vä)
Castagna (kä stän'nyä), Bruna (brōō'nä)
Castil-Blaze (kás tēl' blàz')
Castor et Pollux (kás tôr' ā pô lyks')
Castrato (kä strä'tō)
Catalani (kä tä lä'nē), Alfredo (äl frä'dō)
Catalani, Angelica (än jĕ'lē kä)
Catherine (*Fr.* ká trēn')
Catulli Carmina (kä tōōl'ē kär'mĭ nä)
Cavalieri (kä vä lyĕ'rē), Emilio de' (ĕ mē'-lyō dä)
Cavalieri, Lina (lē'nä)
Cavalleria rusticana (kä'väl lĕ rē'ä rōō stē-kä'nä)
Cavalli (kä väl'lē), Francesco (frän chä'-skō)
Cavaradossi (kä'vä rä dôs'sē), Mario (mä'-rē ō)
Cavatina (kǎv ə tē'nə; *It.* kä vä tē'nä)
Cavatine du page (ká vá tēn' dy pázh')
Cebotari (chä bō tä'rē), Maria (mä rē'ä)
Ce bruit de l'or . . . ce rire (sə bryē də lôr' . . . sə rēr')
Cecchina, La (lä chäk kē'nä)
Celeste Aïda (chĕ lĕ'stä ä ē'dä)
Cellini (chĕl lē'nē), Benvenuto (bĕn vĕ-nōō'tō)
Cena delle beffe, La (lä chä'nä dĕl lä bĕf'fä)
Cendrillon (säɴ drē yôɴ')
Cenerentola, La (lä chĕ nĕ rĕn'tô lä)
C'en est donc fait et mon coeur va changer (säɴ nĕ dôɴ fĕ' ā môɴ kœr vá shäɴ zhä')
Ceprano (chä prä'nō)
Cervantes (sər vǎn'tēz; *Sp.* thĕr vän'täs), Miguel de (mē gĕl' dä)
Cesare Angelotti (chĕ'zä rä än jĕ lôt'tē)
Cesti (chä'stē), Marc' Antonio (märk än-tô'nyō)

C'est ici le séjour (sĕ tē sē' lə sä zhōōr')
C'est l'histoire amoureuse (sĕ lēs twâr' á mōō rœz')
Chabrier (shá brē ā'), Emmanuel (*Fr.* ĕ má ny ĕl')
Chaconne (shá kôn')
Chaliapin (shä lyä'pĭn), Feodor (fyô'dər)
Champs paternels (shäɴ pá tĕr nĕl')
Chanson bachique(shäɴ sôɴ' bá shēk')
Chanson de la puce (shäɴ sôɴ' də lá pys')
Chanson hindoue (shäɴ sôɴ' näɴ dōō')
Chanson huguenote (shäɴ sôɴ' ɣ gə nôt')
Charfreitagszauber (kär frī'täks tsou'bər)
Charlotte (*Fr.* shár lôt')
Charlottenburg (shär lŏt'ən bûrg; *Ger.* shär lôt'ən bōōrk) Opera
Charmant oiseau (shár mäɴ' twá zō')
Charpentier (shár päɴ tyä'), Gustave (*Fr.* gys táv')
Charton-Demeur (shár tôɴ' də mœr'), Anne
Chartreuse de Parme, La (lä shár trœz' də párm')
Che farò senza Euridice? (kä fä rō' sĕn'-tsä ā ōō'rē dē'chä?)
Che gelida manina (kä jĕ'lē dä mä nē'nä)
Chekhov (chĕ'кнəf), Anton
Ch'ella mi creda libero (kĕl'lä mē krä'dä lē'bĕ rō)
Chénier (shä nyä')
Che puro ciel! (kä pōō'rō chĕl!)
Cherubini (kĕ rōō bē'nē), Maria Luigi (mä-rē'ä lōō ē'jē)
Cherubino (kĕ rōō bē'nō)
Che soave zeffiretto (kä sō ä'vä dzĕf fē-rāt'tō)
Che vita maledetta (kä vē'tä mä lä dät'tä)
Che volo d'augelli (kä vō'lō dou jĕl'lē)
Chi del gitano i giorni abella? (kē dĕl jē-tä'nō ē jōr'nē äb bĕl'lä?)
Chi mai fra gli inni e i plausi (kē mĭ frä lyē ēn'nē ā ē plou'ze)
Chimène (shē mĕn')
Chi mi frena? (kē mē frä'nä?)
Chi vide mai a bimbo (kē vē'dä mĭ ä bēm'-bō)
Choéphores, Les (lä kô ä fôr')
Chorus of the Levites (lē'vīts)
Christophe Colomb (krēs tôf' kô lôɴ')
Chrysis (krī'sĭs)
Chrysothemis (krĭ sŏth'ə mĭs)
Cicillo (chē chēl'lō)
Cid, Le (lə sēd')
Cieco (chĕ'kō)
Cielo e mar! (chĕ'lo ä mär!)
Cilèa (chē lĕ'ä), Francesco (frän chä'skō)

Cimarosa (chē mä rō'zä), Domenico (dô-
(mä'nē kō)
Cio-Cio-San (chō chō sän')
Claudel (klō dĕl'), Paul
Claudius (klô'dĭ əs)
Claussen (klou'sən), Julia (*Sw.* ū'lē ä)
Clément (klä män'), Edmond (*Fr.* ĕd-
môn')
Clemenza di Tito, La (lä klĕ mĕn'tsä dē
tē'tō)
Cléophas (klä ô fäs')
Clitandro (klē tän'drō)
Clotilda (klō tēl'dä)
Cobblers Song: Schusterlied (shoo'stər-
lēt')
Cochenille (kô shə nē'y)
Cocteau (kôk tō'), Jean (zhän)
Colas Breugnon (kô lä' brœ nyôn')
Colbran (kōl'brän), Isabella
Colline (kô lēn')
Colomba (kô lôn bä')
Coloratura (kŭl'ə rə tū'rə, -tōō'-)
Combien tu m'es chère (kôn byän' tȳ mĕ
shĕr')
Com' è gentil (kōm ĕ gĕn tēl')
Come in quest' ora bruna (kō'mä ēng
kwäst ō'rä broo'nä)
Come scoglio (kō'mä skôl'lyō)
Come un bel dì di maggio (kō'mä ōōm bĕl
dē' dē mäd'jō)
Comme autrefois dans la nuit sombre
(kôm ō trə fwä' dän lä nȳē sôn'brə)
Commedia per musica (kôm mĕ'dyä pĕr
mōō'zē kä)
Comme une pâle fleur (kôm ȳn päl flœr')
Comte Ory, Le (lə kônt ô rē')
Comus (kō'məs)
Concetta (kôn chät'tä)
Concitato (kôn chē tä'tō)
Connais-tu le pays? (kô nĕ tȳ' lə pä ē'?)
Conried (kŏn'rēd), Heinrich (hīn'rĭkн)
Constanza (kôn stän'tsä)
Constanza! dich wiederzusehen! (kôn-
stän'tsä! dĭкн vē'dər tsōō zā'ən!)
Consuelo (kŏn sōō ā'lō; *Fr.* kôn sȳä lō')
Contes d'Hoffmann, Les (lä kônt dôf-
män')
Contratador dos Diamantes (kōn trä'tə-
dōr' dōōz dyä mŭn'tēs)
Convien partir, o miei compagni d'arme
(kôn vyän' pär tər', ō myĕ'ē kôm pän'-
nyē där'mä)

Copland (kōp'lənd), Aaron
Coppélius (kô pä lē ȳs')
Coq d'or, Le (lə kôk dôr')
Corneille (kôr nĕ'y), Pierre (pyĕr)
Cornelius (kôr nä'lē ōōs), Peter (*Ger.* pä'-
tər)
Coro delle campane (kô'rō dĕl lä käm pä'-
nä)
Coronation of Poppea (pŏ pē'ə)
Corps de Ballet (kôr də bà lĕ')
Corregidor, Der (dər kôr rä'gē dôr')
Corsi (kôr'sē), Jacopo (yä'kō pō)
Così fan tutte (kō sē' fän tōōt'tä)
Costa (kŏs'tə), Michael
Costanzi (kō stän'tsē)
Couplets Bacchiques (kōō plĕ' bà shēk'
Covent (kŭv'ənt, kŏv'-) Garden
Credo a una possanza arcana (krä'dō ä
ōō'nä pōs sän'tsä är kä'nä)
Credo in uno Dio crudel (krä'do ēn ōō'nō
dē'ō krōō dĕl')
Crespel (krĕs pĕl')
Cristoforo Colombo (krē stô'fô rō kô lōm'-
bō)
Crobyle (krô bēl')
Crudel! perchè finora (krōō dĕl'! pĕr kä'
fē nō'rä)
Csárdás (chär'däsh)
Cui (kȳ ē'), César (sä zàr')
Curra (kōōr'rä)
Cyrano de Bergerac (sĭr'ə nō də bĕr'zhə-
räk; *Fr.* sē rà nō' də bĕr zhə ràk')

D

Da capo (dä kä'pō) aria
Dafne (däf'nĭ; *It.* däf'nä)
Da Gama (də gä'mə)
Dai campi, dai prati (dä ē käm'pē, dä ē
prä'tē)
Daland (dä'länt)
D'Albert (däl'bərt, däl'-), Eugène (œ-
zhĕn')
Dalibor (dä'lē bôr)
Dalila (dä lē lä')
Dal labbro il canto estasiato (däl läb'brō
ēl kän'tō ä stä zyä'tō)
Dalla sua pace (däl lä sōō'ä pä'chä)
Dalle stanze, ove Lucia (däl lä stän'tsä,
o'vä lōō chē'ä)
Dalmorès (dàl mô rĕs'), Charles (*Fr.* shàrl)
D'Alvarez (dàl vär'əz), Marguerite

ăct, āble, dâre, ärt; à, *Fr.* ami; ĕbb, ēqual; ĭf, īce; кн, *Ger.* ach; n indicates nasal vowel;
ŏdd, ōver, ôrder; œ, *Fr.* feu, *Ger.* schön; bŏŏk, ōōze; ŭp, ūse, ûrge; ȳ, *Fr.* tu, *Ger.* über;
zh, vision; ə, a in alone. (Full key on page 558)

Dame aux camélias, La (là dàm ō kà mä-lyà')

Damian (*Ger.* dä mē än')

Damnation of Faust (foust)

D'amor sull' ali rosee (dà mōr' sōōl lä'lē rô'zä ä)

D'amour l'ardente flamme (dà mōōr' làr-dänt' flàm')

Damrosch (dăm'rŏsh), Leopold

Dancaïre, Le (lə dän kà ēr')

Dance of the Camorristi (kä môr rē'stē)

Dandini (dän dē'nē)

Daniello (dä nyĕl'lō)

D'Annunzio (dän nōōn'tsyō), Gabriele (gä brē ĕ'lä)

Dans la cité lointaine (dän là sē tä' lwăn-tĕn')

Dante Alighieri (dän'tä ä lē gyĕ'rē)

Dantons Tod (dän tôns' tōt); Danton's (dän tônz') Death

Dapertutto (dä pĕr tōōt'tō)

Daphne (dăf'nĭ)

Da Ponte (dä pōn'tä), Lorenzo (lô rĕn'tsō)

Dardanus (där'də nəs)

Dargomizhsky (där gə mĭsh'skĭ), Alexander

Das schöne Fest (däs shœ'nə fĕst)

Das süsse Lied verhallt (däs zys'ə lēt fər-hält')

Da-ud (dä ōōt')

Daudet (dō dĕ'), Alphonse (àl fôns')

David (dà vēd'), Félicien (fä lē syän')

Da zu Dir der Heiland kam (dä tsōō dēr dər hī'länt käm); Kirchenchor (kĭr'khən kōr')

Debora e Jaele (dĕ'bô rä ā yä ä'lä)

Debout! matelots (də bōō'! màt lō')

De Brétigny (də brä tē nyē')

Debussy (də by sē'), Claude Achille (klōd à shēl')

Decameron (dĭ kăm'ər ən)

De Falla (dä fä'lyä), Manuel (mä nwĕl')

Deh! con te li prendi (dĕ! kōn tä lē prĕn'-dē)

Deh non parlare al misero (dĕ nōn pär-lä'rä äl mē'zĕ rō)

Deh! proteggimi o Dio! (dĕ! prō tĕd'jē-mē ō dē'ō!)

Deh, vieni alla finestra (dĕ, vyĕ'nē äl lä fē nĕ'strä)

Deh vieni, non tardar (dĕ vyĕ'nē, nōn tär där')

Déjà les hirondelles (dā zhà' lā zē rôn-dĕl')

De Koven (də kō'vən), Reginald

De Lara (də là rá'), Isidore (*Fr.* ē zē dôr')

De l'enfer qui vient émousser (də län fĕr' kē vyăn tä mōō sä')

Delibes (də lēb'), Léo (lä ō')

Delilah (dĭ lī'lə)

Delius (dē'lĭ əs, dēl'yəs), Frederick

Dell' aura tua profetica (dĕl lou'rä tōō'ä prō fĕ'tĕ kä)

Della città all' occaso (dĕl lä chēt tä' äl lōk kä'zō)

Della vittoria agli arbitri (dĕl lä vēt tô'ryä äl lyē är'bē trē)

Delle Sedie (dĕl lā sĕ'dyä), Enrico (ĕn-rē'kō)

Delmas (dĕl màs'), Jean-François (zhän frän swä')

De los Angeles (dä lōs än'khä läs), Victoria

Del Puente (dĕl pwĕn'tä), Giuseppe (jōō-zĕp'pä)

Del Tago sponde addio (dĕl tä'gō spōn'-dä äd dē'ō)

DeLuca (dä lōō'kä), Giuseppe (jōō zĕp'-pä)

Demetrios (dĭ mē'trĭ əs)

Demeur (də mœr'), Anne Arsène (àn àr-sĕn')

De' miei bollenti spiriti (dä myĕ'ē bôl-lĕn'tē spē'rē tē)

Demoni, fatale (dĕ'mô nē, fä tá'lä)

De Musset (də my sĕ'), Alfred (*Fr.* àl-frĕd')

De Nangis (də nän zhē'), Raoul (rà ōōl')

De Paris tout en fête (də pa rē' tōō tän fĕt')

Depuis le jour (də pyē' lə zhōōr')

Depuis l'instant où dans mes bras (də-pyē' läns tän' ōō dän mä brä')

Depuis longtemps j'habitais cette chambre (də pyē' lôn tän' zhà bē tĕ' sĕt shän'-brə)

De Reszke (də rĕsh'kĕ), Edouard (ā-dwár')

De Reszke, Jean (zhän)

Der kleine Sandmann bin ich (dər klī'nə zänt'män bĭn ĭkh)

Der kleine Taumann heiss' ich (dər klī-nə tou'män hīs ĭkh)

Der Vogelfänger bin ich (dər fō'gəl fĕng'ər bĭn ĭkh)

De Saxe (də sàks'), Maurice (mô rēs')

Desdemona (dĕz də mō'nə; *It.* dä zdĕ'-mô nä)

Deserto sulla terra (dä zĕr'tō sōōl lä tĕr'rä)

Des Grieux, Chevalier (shə vá lyä´ dä grē œ´)
Des Grieux, Comte (kôɴt)
Désiré (dä zē rä´)
De Siriex (də sē rē ĕks´)
De Sirval (də sēr vȧl´), Arthur
De son coeur j'ai calmé la fièvre (də sôɴ kœr zhä kȧl mä´ lȧ fyĕ´ vrə)
Despina (dä spē´nä)
Destinn (dĕs´tĭn), Emmy
Deutsches Opernhaus (doi´chəs ō´pərnhous)
Deux Journées, Les (lä dœ zhōōr nä´)
De Valois (də vȧl´wä; Fr. də vá lwȧ´), Marguerite
Devin du village, Le (lə də văɴ´ dy vē lázh´)
Dich, teure Halle (dĭкн, toi´rə häl´ə)
D'ici je vois la mer immense (dē sē´ zhə vwȧ lȧ mĕr ēm mäɴs´)
Dido (dī´dō) and Aeneas (ĭ nē´əs)
Di due figli vivea (dē dōō´ä fēl´lyē vē vä´ä)
Didur (dē´dōōr), Adamo (ä dä´mō)
Diego, Don (dŏn dĭ ā´gō; dôn dyä´ gō)
Die Majestät wird anerkannt (dē mä yĕs tĕt´ vĭrt än ĕr känt´)
Dies Bildnis ist bezaubernd schön (dēs bĭlt´nĭs ĭst bə tsou´bərnt shœn)
Dietsch (dēch), Pierre-Louis (pyĕr lwē´)
Dieu, que le monde révère (dyœ, kə lə môɴd rä vĕr´)
Dieu, que ma voix tremblante (dyœ, kə mȧ vwä träɴ bläɴt´)
Die Zukunft soll mein Herz bewahren (dē tsōō´kōōnft zôl mīn hĕrts bə vä´rən
Di Luna (dē lōō´nä), Count
D'immenso giubilo (dēm mĕn´sō jōō´bēlō)
Dimmi, Fiora, perchè ti veggo ancora (dēm´mē, fyō´rä, pĕr kä´ tē väg´gō äng kō´rä)
Din, don, suona vespero (dēn, dôn, swō´nä vĕ´spĕ rō)
D'Indy (dăɴ dē´), Vincent (Fr. văɴ säɴ´)
Dinorah (dĭ nō´rə)
Dio! mi potevi scagliar (dē´ō! mē pō tä´vē skäl lyär´)
Dio ti giocondi (dē´ō tē jō kōn´dē)
Di pescatore ignobile (dē pä skä tō´rä ēn nyō´bē lä)
Dippel (dĭp´əl), Andreas (än drä´äs)

Di Provenza il mar (dē prō vĕn´tsä ēl mär)
Di quella pira (dē kwĕl´lä pē´rä)
Di qui io vedo il mar (dē kwē ē´ō vä´dō ēl mär)
Di rigori armato (dē rē gō´rē är mä´tō)
Dir töne Lob (dēr tœ´nə lōp)
Di Signa (dē sēn´nyä), Betto (bät´tō)
Dis-moi que je suis belle (dē´mwȧ kə zhə syē bĕl)
Di tanti palpiti (dē tän´tē päl´pē tē)
Dittersdorf (dĭt´ərs dôrf), Karl Ditters von (kärl dĭt´ərs fôn)
Di' tu se fedele (dē tōō sä fä dā´lä)
Divinités du Styx (dē vē nē tä´ dy stēks´)
Dmitri (də mē´trĭ)
D'Obigny (dô bē nyē´), Marquis
Dobrowen (dô´brô vĕn), Issai (ē sī´)
Doctor and Apothecary: Der Doktor und der Apotheker (dər dôk´tôr ōōnt dər ä pō tā´kər)
Dodon (dô dôn´)
Doktor Faust (Ger. dôk´tôr foust)
Dolores, La (lä dō lō´rĕs)
Don Carlos (dŏn kär´ləs, dôn kär´lōs)
Don Giovanni (dŏn jə vän´ĭ; It. dôn jō vän´nē)
Donizetti (dŏn ə zĕt´ĭ; It. dô nē dzät´tē), Gaetano (gä ä tä´nō)
Don José (dŏn hō zā´; Fr. dôɴ zhō zā´)
Don Juan (dŏn wän´, dŏn jōō´ən; Sp. dôn кнwän´)
Donna Diana (dŏn´ə dĭ ȧn´ə)
Donna non vidi mai (dôn´nä nōn vē´dē mī)
Donna serpente, La (lä dôn´nä sĕr pĕn´tä)
Donne curiose, Le (lä dôn´nä kōō ryō´sä)
Donner (dôn´ər)
Don Pasquale (dôn pä skwä´lä)
Don Pedro (dŏn pē´drō; dôn pä´drō)
Don Quichotte (dôɴ kē shôt´)
Don Quixote (dŏn kē hō´tĭ, kwĭk´sət; Sp. dôn kē кнō´tä)
Dorabella (dô rä bĕl´lä)
Dormirò sol nel manto (dôr mē rō´ sōl nĕl män´tō)
Dorota (dō´rō tä)
Dositheus (də sĭth´ĭ əs)
Dostoyevsky (dŏs tô yĕf´skĭ), Feodor (fyô´dər)
Douphol (dōō fôl´), Baron

ăct, āble, dȧre, ärt; à, Fr. ami; ĕbb, ēqual; ĭf, īce; кн, Ger. ach; ɴ indicates nasal vowel; ŏdd, ōver, ôrder; œ, Fr. feu, Ger. schön; bŏŏk, ōōze; ŭp, ūse, ûrge; y, Fr. tu, Ger. über; zh, vision; ə, a in alone. (Full key on page 558)

Dove son? O qual gioia (dō'vä sōn? ō kwäl jô'yä)

Dove sono (dō'vä sō'nō)

Dramma giocoso (dräm'mä jō kō'sō)

Dram.na per musica (dräm'mä pĕr mōō'zē kä)

Dreigroschenoper, Die (dē drī grôsh'ən-ō'pər)

Dryad (drī'əd, -ăd)

Du bist der Lenz (dōō bĭst dər lĕnts)

Du und Du (dōō ōŏnt dōō) Waltzes

Due Foscari, I (ē dōō'ā fō'skä rē)

Dufresne (dy frĕn'), Milio (mē'lē ō)

Dukas, (dy kȧ'), Paul

Duke of Mantua (măn'chōō ə)

Dulcamara (dōōl kä mä'rä)

Dulcinea (dŭl sĭn'ĭ ə; Sp. dōōl thē nä'ä, -sē-)

Du Locle (dy lô'klə), Camille (kȧ mē'y)

Dumas (dōō'mä; Fr. dy mȧ'), Alexandre (père) (ȧ lĕk säɴ'drə, pĕr)

Dumas, Alexandre (fils) (fēs)

Duo de la fontaine (dy ō' də lȧ fôɴ tĕn')

Duo de la lettre (dy ō' də lȧ lĕ'trə)

Durand et Compagnie (dy räɴ' ā kôɴ-pȧ nyē')

Durante (dōō rän'tä), Francesco (frän-chā'skō)

Durch die Wälder, durch die Auen (dōŏʀкʜ dē vĕl'dər, dōŏʀкʜ dē ou'ən)

Du trugest zu ihm meine Klage (dōō trōō'gəst tsōō ēm mī'nə klä'gə)

Dvořák (dvôr'zhäk), Antonín (än'tô nēn)

Dybbuk (dĭb'ək), The

Dzerzhinsky (dzĕr zhĭn'skĭ), Ivan (ē vän')

E

Eadgar (ĕd'gər, ăd'gär) of Wessex

Eames (āmz), Emma

E Amore un ladroncello (ĕ ä mō'rä ōōn lä drôn chĕl'lō)

Ebben? ne andrò lontana (āb bĕn'? nä än drō' lôn tä'nä)

Ebert (ā'bərt), Carl

Eboli (ĕ'bô lē), Princess

E casta a par di neve! (ĕ kä'stä äl pär dē nĕ'vä!)

Echo (ĕk'ō)

Ecoute, écoute, compagnon (ā kōōt', ā kōōt', kôɴ pȧ nyôɴ')

Ecoute, mon ami (ā kōōt', môɴ nȧ mē')

Edgardo (ĕd gär'dō) of Ravenswood

Edipo Re (ā dē'pō or ä'dē pō rä)

Edmondo (ĕd mōn'dō)

Egk (ĕk), Werner (vĕr'nər)

Eglantine de Puiset (ā gläɴ tēn' də pʀē zĕ')

Egli è salvo (āl'lyē ĕ säl'vō)

E il sol dell' anima (ĕ ēl sōl dĕl lä'nē mä)

Einem (ī'nəm), Gottfried von (gôt'frēt fôn)

Ein' feste Burg ist unser Gott (īn fĕs'tə bŏŏrk ĭst ōŏn'zər gôt)

Ein Mädchen oder Weibchen (īn māt'-shən ō'dər vīp'shən)

Ein Männlein steht im Walde (īn mĕn'līn shtāt ĭm väl'də)

Einsam in trüben Tagen (īn'zäm ĭn trʏ'bən tä'gən)

Ein Schwert verhiess mir der Vater (īn shvärt fər hēs' mēr dər fä'tər)

Einstein (īn'stīn), Alfred

Eisenstein (ī'zən shtīn), Baron von (fôn)

Eleazar (ĕl ĭ ä'zər)

Elektra (ĭ lĕk'trə)

Elena (ĕ'lĕ nä)

Eleonora (ĕ'lä ô nō'rä)

Elisa (ĕ lē'zä), Princess

Elisabeth (ĭ lĭz'ə bəth; Ger. ā lē'zä bĕt)

Elisetta (ĕ lē zät'tä)

Elisir d'amore, L' (lĕ lē zēr' dä mō'rä)

Elle a fui, la tourterelle (ĕl ȧ fʏē, lȧ tōōr tə rĕl')

Elle ne croyait pas (ĕl nə krwȧ yĕ' pä')

Elmendorff (ĕl'mən dôrf), Karl

Elsa (ĕl'zä)

Elvira, Donna (dôn'nä ĕl vē'rä)

Emilia (ĕ mē'lyä)

Enfant et les sortilèges, L' (läɴ fäɴ' ā lä sôr tē lĕzh')

En fermant les yeux (äɴ fĕr mäɴ' lä zyœ')

Enrico (ĕn rē'kō) Ashton

En silence, pourquoi souffrir? (äɴ sē-läɴs', pōōr kwȧ' sōō frēr'?)

Entführung aus dem Serail, Die (dē ĕnt fʏ'rōŏng ous dəm zä rīl')

Entr'acte (äɴ träkt')

Entrada (ĕn trä'тнä)

Entrée (äɴ trä')

En vain pour éviter (äɴ văɴ pōōr ā vē tä')

Enzo Grimaldo (ĕn'tsō grē mäl'dō)

E quest' asilo ameno e grato (ĕ kwäst' ä zē'lō ä mä'nō ā grä'tō)

Eravate possente (ĕ rä vä'tä pōs sĕn'tä)

Erckmann-Chatrian (ĕrk män' shȧ trē äɴ')

Erda (ĕr'dä)

Erede (ĕ rĕ'dä), Alberto (äl bĕr'tō)

Erik (ĕr'ĭk; Ger. ā'rĭk)

Eri tu che macchiavi (ĕ'rē tōō kä mäk kyä'vē)

Erkel (ĕr'kĕl), Franz (fränts); Ferenc (fĕ'rĕnts)
Erlanger (ĕr läN zhā'), Camille (kȧ mē'y)
Ernani (ĕr nä'nē)
Ernani, involami (ĕr nä'nē, ēn vō'lä mē)
Ernesto (ĕr nĕ'stō)
Ero e Leandro (ā'rō ā lā än'drō)
Erwartung (ĕr vär'tŏŏng)
Escamillo (ĕs kä mē'lyō)
Eschenbach (ĕsh'ən bäKH), Wolfram von (vôlf'räm fôn)
E scherzo od e follia (ĕ skĕr'tsō ōd ĕ fôl lē'ä)
Esmeralda (ĕz mə rӑl'də, ĕs-)
E sogno? o realtà? (ĕ sōn'nyō? ō rӑ äl tä'?)
Esultate! (ā zōōl tä'tä!)
Etoile du nord, L' (lä twȧl' dY nôr')
Euch Lüften, die mein Klagen (oiKH lyf'tən, dē mīn klä'gən)
Eudoxie (œ dôk sē'), Princess
Eugene Onegin (ŏ nyā'gĭn)
Eumenides (ū mĕn'Ĭ dēz)
Euridice (ā ōō'rē dē'chä)
Euripides (ū rĭp'Ĭ dēz)
Euryanthe (ū rĬ än'thĬ; Ger. oi rē än'tə)
Eva (ē'və; Ger. ä'vä)
Evander (Ĭ văn'dər)
Evangelimann, Der (dər ä väng gä'lē män)
Evangeline (Ĭ văn'jə lĭn)
Evenings on a Farm Near Dikanka (dē kän'kə)

F

Faccio (fät'chō), Franco (fräng'kō)
Fafner (fäf'nər)
Fair at Sorochinsk (sŏ rô'chĬnsk), The
Faites-lui mes aveux (fĕt'lyē mä zȧ vœ')
Falcon (fȧl kôN'), Marie-Cornélie (mȧ rē' kôr nä lē')
Falla (fä'lyä), Manuel de (mä nwĕl' dä)
Fanciulla del West, La (lä fän chōōl'lä dĕl wĕst)
Fandango (fӑn dӑng'gō)
Fanget an! So rief der Lenz in den Wald (fäng'ət än'! zō rēf dər lĕnts Ĭn dən vält)
Faninal (fä nē näl')
Farandole (fӑr'ən dōl; Fr. fȧ räN dôl')
Farfarello (fär fä rĕl'lō)
Farinelli (fä rē nĕl'lē)
Farlaf (fär läf')
Farrar (fə rär'), Geraldine

Fasolt (fä'zôlt)
Fata Morgana (fä'tä môr gä'nä)
Fatima (fä'tē mä, fӑt'Ĭ mə, fə tē'mə)
Fauré (fō rä'), Gabriel (Fr. gȧ brē ĕl')
Fauré, Jean Baptiste (zhäN bȧ tēst')
Faust (foust)
Favart (fȧ vȧr'), Charles Simon (Fr. shȧrl sē môN')
Favola del figlio cambiato, La (lä fä'vô lä dĕl fēl'lyō käm byä'tō)
Favola per musica (fä'vô lä pĕr mōō'zē kä)
Favorita, La (lä fä vô rē'tä)
Federica (fā dĕ rē'kä)
Federico (fä dĕ rē'kō)
Fedora (fä dō'rä)
Feen, Die (dē fä'ən)
Feldlager in Schlesien, Ein (Ĭn fĕlt'lä'gər Ĭn shlä'zē ən)
Feldmarschallin (fĕlt'mär'shä lĬn), The
Fenena (fĕ nä'nä)
Fenice, La (lä fĕ nē'chä)
Fennimore (fĕn'Ĭ mōr) and Gerda (gĕr'dä)
Feodor (fyŏ'dər)
Fernando (fĕr nän'dō, fər nӑn'dō)
Fernando, Don (dôn, dŏn)
Ferrando (fĕr rän'dō)
Feste! Pane! (fĕ'stä! pä'nä!)
Festa Teatrale (fĕ'stä tä ä trä'lä)
Festspiel (fĕst'shpēl')
Feuersnot (foi'ərs nōt')
Feuerzauber (foi'ər tsou'bər)
Février (fä vrē ā'), Henri (äN rē')
Fibich (fē'bēKH), Zdĕnek (zdyĕ'nĕk)
Fidalma (fē däl'mä)
Fidelio (fĭ dā'lĬ ō)
Fidès (fē dĕs')
Fieramosca (fyĕ rä mō'skä)
Fiesco (fyĕ'skō), Jacopo (yä'kō pō)
Figaro (fĬg'ə rō, fē'gä rō)
Figlia che reggi il tremulo piè (fēl'lyä kä rĕd'jē ēl trĕ'mōō lō pyĕ')
Figlia del reggimento, La (lä fēl'lyä dĕl räd jē män'tō)
Figlia di re, a te l'omaggio (fēl'lyä dē rä, ä tä lô mäd'jō)
Filipievna (fē lēp'yĬv nə)
Fille des rois, à toi l'hommage (fē'y dä rwȧ, ȧ twȧ lô mäzh')
Fille du régiment, La (lȧ fē'y dY rä zhē mäN')
Finita è per frati (fē nē'tä ĕ pĕr frä'tē)

Fiora (fyō'rä)
Fior di giaggiolo (fyōr dē jäd jô'lō)
Fiordiligi (fyōr dē lē'jē)
Fioriture (fyō rē tōō'rä)
Firenze è come un albero fiorito (fē rĕn'-
  tsä ĕ kō'mä ōōn äl'bĕ rō fyô rē'tō)
Fischer-Dieskau (fĭsh'ər dē'skou), Die-
  trich (dē'trĭKH)
Flagstad (fläg'städ), Kirsten (kĭr'stən)
Flamand (flä män')
Flaminio (flä mē'nyō)
Flammen (flä mĕn')
Flammen, perdonami (flä mĕn', pĕr dō'-
  nä mē)
Flaubert (flō bĕr'), Gustave (*Fr.* gʏs tàv')
Flavio (flä'vyō)
Fledermaus, Die (dē flä'dər mous)
Fleg (flĕg), Edmond (*Fr.* ĕd môN')
Fleur que tu m'avais jetée, La (là flœr
  kə tʏ mà vĕ· zhə tä')
Fleurissait une sauge (flœ rē sĕ' ʏn sōzh')
Fliegende Holländer, Der (dər flē'gən də
  hôl'ĕn'dər)
Florestan (flôr'əs tän)
Floriana (flô rē ä'nä)
Floria Tosca (flô'rē ä tō'skä)
Florindo (flô rēn'dō)
Florville (flôr vēl')
Flosshilde (flôs'hĭl'də)
Flotow (flō'tō), Friedrich, Freiherr von
  (frē'drĭKH, frī'hĕr fôn)
Fluth (flōōt)
Fontana, Signor (sēn nyōr' fôn tä'nä)
Forsell (fôr sĕl'), John
Forty Days of Musa Dagh (mōō'sä däKH'),
  The
Forza del destino, La (lä fôr'tsä dĕl
  dä stē'nō)
Foss (fŏs; *Ger.* fôs), Lukas
Fouché (fōō shä')
Fra Diavolo (frä dĭ ä'və lō; *It.* frä dyä'-
  vô lō)
Fra Gherardo (frä gĕ rär'dō)
Françaix (frän sä'), Jean (zhänN)
France (fräns), Anatole (à nà tôl')
Francesca da Rimini (frän chä'skä dä
  rē'mē nē)
Franchetti (fräng kät'tē), Alberto
  (äl bĕr'tō)
Franck (fränK), César (sä zàr')
Frantz (fränts)
Fra poco a me ricovero (frä pô'kō ä mä
  rē kō vĕ'rō)
Frasquita (fräs kē'tä)
Frau ohne Schatten, Die (dē frou ō'nə
  shät'ən)

Frazzi (frät'tsē), Vito (vē'tō)
Frédéric (frä dä rēk')
Freia (frī'ä)
Freischütz, Der (dər frī'shʏts)
Fremstad (frĕm'städ), Olive
Freudig begrüssen wir die edle Halle
  (froi'dĭKH bə grʏs'ən vēr dē äd'lə häl'ə)
Fricka (frĭk'ä)
Friedenstag, Der (dər frē'dəns täk')
Froh (frō)
Fugitif et tremblant (fʏ zhē tēf' ä trän-
  bläN')
Fuoco di gioia (fwô'kō dē jô'yä)
Furiant (fōō rē änt')
Furtwaengler (fōōrt'vĕng'lər), Wilhelm
  (vĭl'hĕlm)

G

Gadski (gät'skĭ), Johanna (yō hän'ä)
Gailhard (gà yàr'), Pierre (pyĕr)
Galatea (gäl ə tē'ə)
Galitsky (gä lyēts'kĭ), Prince
Galitzin (gä lyē'tsĭn), Prince Vasily
  (vä sē'lyē)
Galli-Curci (gäl ĭ kûr'chĭ; *It.* gäl'lē kōōr'-
  chē), Amelita (ä mĕ lē'tä)
Galli-Marié (gà lē' mà ryä'), Marie
  Célestine (mà rē' sä lĕs tēn')
Gallo (gäl'ō, gäl'lō), Fortune
Galuppi (gä lōōp'pē), Baldassare (bäl-
  däs sä'rä)
García (gär thē'ä, -sē'ä), Manuel del
  Popolo Vicente (mä nwĕl' dĕl pō'pō lō
  bē thĕn'tä, -sĕn'-)
Garcias (gär sē'äs)
Garrido (gär rē'dō)
Gasparo, Don (dôn gä'spä rō)
Gastone (gä stō'nä)
Gatti (gät'tē), Guido Maria (gwē'dō
  mä rē'ä)
Gatti-Casazza (gät'tē kä zät'tsä), Giulio
  (jōō'lyō)
Gaubert (gō bĕr'), Philippe (fē lēp')
Gaudenzio (gou dĕn'tsyō)
Gaudio son al cuore queste pene dell'
  amor (gou'dyō sōn äl kwô'rä kwä'stä
  pā'nä dĕl lä mōr')
Gaussin (gō säN')
Gautier (gō tyä'), Théophile (tä ô fēl')
Gazza ladra, La (lä gäd'dzä lä'drä)
Gebrauchsmusik (gə brouKHs'mōō zēk')
Gellner (gĕl'nər)
Geneviève (*Fr.* zhĕn vyĕv')
Gennaro (jĕn nä'rō)
Genoveva (gä nō fä'fä)

Gérald (jĕr'əld; *Fr.* zhā ráld')
Gérard (zhā rár'), Charles (*Fr.* shárl)
Germania (jĕr mä'nyä)
Germont (zhĕr môN'), Alfredo (äl frä'dō)
Germont, Giorgio (jôr'jō)
Geronimo (jĕ rô'nē mō)
Gerster (gĕr'stər), Etelka (ĕ'tĕl kô)
Gerusalemme liberata (jĕ rōō'zä lĕm'mä lē bĕ rä'tä)
Gerville-Réache (zhĕr vēl' rā ásh'), Jeanne (zhän)
Gessler (gĕs'lər)
Gezeichneten, Die (dē gə tsīKH'nə tən)
Ghedini (gā dē'nē), Giorgio Federico (jôr'jō fä dĕ rē'kō)
Gherardino (gĕ rär dē'nō)
Gherardo (gĕ rär'dō)
Ghione (gē ō'nä), Francesco (frän chä'skō)
Ghislanzoni (gē zlän tsō'nē), Antonio (äntô'nyō)
Già i sacerdoti adunansi (jä ē sä chĕrdô'tē ä dōō'nän sē)
Già nella notte densa (jä nĕl lä nôt'tä dĕn'sä)
Gianetta (jä nät'tä)
Gianetto (jä nät'tō)
Giannini (jə nē'nĭ; *It.* jän nē'nē), Dusolina (dōō zə lē'nə)
Giannini, Vittorio (vĭ tōr'ĭ ō; *It.* vēt tô'ryō)
Gianni Schicchi (jän'nē skēk'kē)
Giarno (jär'nō)
Gibichungs (gē'bĭKH ŏŏngz)
Gigli (jēl'lyē), Beniamino (bĕ nyä mē'nō)
Gil (jēl), Count
Gil Blas (zhēl blás)
Gilda (jēl'dä)
Gilgamesj (gĭl'gä mĕsh)
Gillot (gē yō')
Gioconda, La (lä jō kōn'dä)
Gioielli della Madonna, I (ē jō yĕl'lē dĕl lä mä dôn'nä)
Giordano (jôr dä'nō), Umberto (ōōmbĕr'tō)
Giorgetta (jôr jät'tä)
Giorgio Germont (jôr'jō zhĕr môN')
Giovanna (jō vän'nä)
Giovanni (jō vän'nē)
Giulia (jōō'lyä)
Giulietta (jōō lyät'tä)
Giulio Cesare (jōō'lyc̄ chĕ'zä rä)
Giuseppe (jōō zĕp'pä)

Glaives pieux, saintes épées (glĕv pē œ', sǎn'tə zā pā')
Glaz (gläts), Herta (hĕr'tä)
Glinka (glĭng'kə), Michael
Gloire immortelle de nos aïeux (glwár ēm môr tĕl' də nō zá yœ')
Gloria all' Egitto (glô'ryä äl lä jēt'tō)
Gluck (glŏŏk), Alma
Gluck, Christoph Willibald (krĭs'tôf vĭl'ē bält)
Glyndebourne (glĭn'də bərn, glĭn'bôrn) Opera
Godard (gô dár'), Benjamin (*Fr.* bǎNzhá mǎN')
Godunov (gō'də nôf; *Rus.* gə dōō nôf')
Goethe (gœ'tə), Johann Wolfgang von (yō'hän vôlf'gäng fôn)
Götterdämmerung, Die (dē gœt'ər dĕm'ə rōōng)
Goetz (gœts), Hermann (*Ger.* hĕr'män)
Götz von Berlichingen (gœts fôn bĕr'lĭKH ĭng ən)
Gogol (gô'gəl), Nikolai (nē kŏ lī')
Golaud (gô lō')
Goldene Kreuz, Das (däs gôl'də nə kroits)
Goldmark (gōld'märk; *Ger.* gôlt'märk), Karl
Goldoni (gôl dō'nē), Carlo
Goldovsky (gôl dôf'skĭ), Boris
Gonzalve (gôn säl'vä)
Goossens (gōō'sənz), Eugene
Gopak (*Rus.* gŏ päk')
Gorislava (gôr i slä'və)
Gormas (gôr'mäs), Count de (dä)
Goro (gō'rō)
Gossec (gŏ sĕk'), François-Joseph (frÄNswá' zhō zĕf')
Gottfried (gôt'frēt)
Gounod (gōō'nō; *Fr.* gōō nō'), Charlev François (*Fr.* shárl frÄN swá')
Goyescas (gō yĕs'käs)
Gozzi (gôt'tsē), Carlo
Graener (grä'nər), Paul (*Ger.* poul)
Graf (gräf), Herbert
Gralserzählung (gräls'ĕr tsä'lŏŏng)
Granados (grä nä'dōs; *Sp.* grä nä'thōs), Enrique (ĕn rē'kä)
Grane (grä'nə)
Grassini (gräs sē'nē), Josephina (jō zə fē'nə)
Grau (grou), Maurice

ǎct, āble, dâre, ärt; á, *Fr.* ami; ĕbb, ēqual; ĭf, īce; KH, *Ger.* ach; N indicates nasal vowel; ŏdd, ōver, ôrder; œ, *Fr.* feu, *Ger.* schön; bŏŏk, ōōze; ŭp, ūse, ûrge; Y, *Fr.* tu, *Ger.* über; zh, vision; ȇ, a in alone. (Full key on page 558)

Graun (groun), Karl Heinrich (kärl hīn'-rĭкн)

Graupner (group'nər), Christoph (krĭs'-tôf)

Grech (grĕch)

Gregorio (grĭ gōr'ĭ ō, grä gô'ryō)

Gremin (grä'mĭn), Prince

Grenvil (grĕn vēl'), Dr.

Gretchaninov (grĕ chä nyē'nəf), Alexander

Gretel (grä'təl, grĕt'əl)

Grétry (grä trē'), André Ernest (än drä' ĕr nĕst')

Grimaldo ⟨grē mäl'dō), Enzo (ĕn'tsō)

Grimm, Friedrich Melchior (frē'drĭкн mĕl'кнē ôr), Baron von (fôn)

Grimm, Jakob Ludwig (yä'kôp lōōt'-vĭкн)

Grisélidis (grē zä lē dēs')

Grisi (grē'zē), Giulia (jōō'lyä)

Gritzko (grēts'kō)

Grossmächtigste Prinzessin (grōs'mĕкн'-tĭкн stə prĭn tsĕs'ĭn)

Gruenberg (grōō'ən bûrg), Louis

Guadagni (gwä dän'nyē), Gaetano (gä ä-tä'nō)

Guarany, Il (ēl gwä rä nē')

Guardiano, Padre (pä'drä gwär dyä'nō)

Gueden (gy'dən), Hilde (hĭl'də)

Guerre des Bouffons (gĕr dā bōō fôN')

Guglielmo (gōōl lyĕl'mō)

Gui (gwē), Vittorio (vēt tô'ryō)

Guido (gwē'dō)

Guidon (gvē dôn'), Prince

Guillaume Tell (gē yōm' tĕl)

Guiraud (gē rō'), Ernest (Fr. ĕr nĕst')

Gunther (gōōn'tər)

Guntram (gōōn'träm)

Gura (gōō'rä), Eugen (oi'gän, oi gän')

Gurnemanz (gōōr'nə mänts)

Gustav Hinrichs (gōōs'täf hĭn'rĭks) Opera Company

Gutheil-Schoder (gōōt'hĭl shō'dər), Marie (Ger. mä rē'ə, mä rē')

Gut'n Abend, Meister (gōō'tən ä'bənt, mī'stər)

Gutrune (gōōt'rōō'nə)

Gwendoline (gwĕn'də lĭn, -lēn)

H

Hába (hä'bä), Alois (ə lois'; Czech ä'lô-ēs, ä'lois)

Habanera (ä bä nä'rä)

Habet Acht! (hä'bət aкнt!)

Hab' mir's gelobt (häp mērs gə lōpt')

Hadji (hä'jē)

Hageman (hä'gə män), Richard

Hagen (hä'gən)

Hagenbach (hä'gən bäкн)

Hagith (hä'gēt, hä'gĭth)

Hahn (hän), Reynaldo (rä näl'dō)

Halász (hä'läs), Lászlo (läs'lō)

Halévy (hăl'ə vĭ; Fr. à lä vē'), Jacques-François (zhäk fräN swà')

Halka (häl'kä)

Hallström (häl'strœm), Ivar (ī'vər; Sw. ē'vär)

Haltière (àl tyĕr'), Madame de la (mà-dàm' də là)

Hammerstein (hăm'ər stīn), Oscar

Handel (hăn'dəl), George Frideric

Handlung (händ'lŏŏng)

Hans (häns)

Hansel (hăn'səl) and Gretel (grä'təl, grĕt'əl); Hänsel und Gretel (hĕn'səl ŏŏnt grä'təl)

Hans Heiling (häns hī'lĭng)

Hanslick (häns'lĭk), Eduard (ā'dōō ärt)

Hans Sachs (häns zäks)

Harlequin's (här'lə kwĭnz, -kĭnz) Serenade

Harmonie der Welt, Die (dē här mō nē' dər vĕlt)

Harun-al-Rashid (hä rōōn' äl rä shēd', äl räsh'ĭd)

Háry János (hä'rē yä'nōsh)

Haug (houk), Hans (häns)

Hauk (houk), Minnie

Hauptmann (houpt'män), Gerhart (gĕr'-härt)

Ha! wie will ich triumphieren! (hä! vē vĭl ĭкн trē ŏŏm fē'rən!)

Haydn (hī'dən), Franz Joseph (Ger. fränts yō'zef)

Hebbel (hĕb'əl), Friedrich (frē'drĭкн)

Hedwig (hĕd'wĭg; Ger. hät'vĭкн)

Heger (hä'gər), Robert

Heil dir, Sonne! (hĭl dēr, zôn'ə!)

Heil Sachs! (hĭl zäks!)

Heimchen am Herd, Das (däs hīm'shən äm hĕrt)

Heine (hī'nə), Heinrich (hīn'rĭкн)

Heldentenor (hĕl'dən tä'nôr)

Hempel (hĕm'pəl), Frieda

Henze (hĕn'tsə), Hans Werner (häns vĕr'nər)

Hercules (hûr'kyə lēz)

Hermann (hĕr'män)

Hero (hĭr'ō) and Leander (lĭ ăn'dər)

Herod (hĕr'əd); Hérode (ā rôd')

Hérodiade (ā rô dyàd')

Herodias (hĭ rō'dĭ əs)

Hérold (ā rōld'), Louis Joseph Ferdinand (*Fr.* lwē zhō zĕf' fĕr dē näN')

Herr Kavalier! (hĕr kä vä lēr'!)

Hertz (hĕrts), Alfred

Herzeleide (hĕr'tsə lī'də) Scene

Heure Espagnole, L' (lœr ĕs på nyôl')

Hexenlied (hĕk'sən lēt')

Hexenritt (hĕk'sən rĭt')

Hidroat (ē drô å')

Hier soll ich dich denn sehen (hēr zôl Ĭкн dĭкн dĕn zā'ən)

Hiller (hĭl'ər), Johann Adam (yō'hän ä'-däm)

Hindemith (hĭn'də mĭt), Paul

Hinrichs (hĭn'rĭks), Gustav (gŏŏs'täf)

Hin und zurück (hĭn ŏŏnt tsŏŏ rʏk')

Hippodameia (hĭ pŏd'ə mī'ə)

Hippolyte et Aricie (ē pô lēt' ā å rē sē'); Hippolytus (hĭ pŏl'ĭ təs) and Aricia (ə-rĭsh'ə)

Hoël (hō ĕl')

Hölle Rache kocht in meinem Herzen, Der (dər hœl'ə rä'кнə kôкнt ĭn mī'nəm hĕr'tsən)

Hoffmann (hôf'mən)

Hoffmann (hôf'män), Ernst Theodor Amadeus (*Ger.* ĕrnst tā'ō dôr ä mä dā'ŏŏs)

Hofmannsthal (hōf'mäns täl, hôf'-), Hugo von (*Ger.* hŏŏ'gō fôn)

Hof und National (hōf' ŏŏnt nä tsyō näl') Theater

Ho! He! Ho! (hō! hā! hō!)

Ho-ho! Schmiede, mein Hammer (hō hō! shmē'də, mīn häm'ər)

Ho-Jo-To-Ho! (hō yō tō hō!)

Holberg (hôl'bĕrg), Ludvig (lŏŏd'vĭg; *Dän.* lŏŏth'vē)

Honegger (hŭn'ə gər, hŏn'-; *Fr.* ô nĕ gĕr'), Arthur

Hoo-Chee (hŏŏ chē')

Hoo-Tsin (hŏŏ tsĭn')

Hopak (hō'pãk)

Horch, die Lerche (hôrкн, dē lĕr'кнə

Ho sete! Ho sete! (ô sā'tä! ô sā'tä!)

Hua-Quee (hwä kwē')

Hubay (hŏŏ'boi), Jenö (yĕ'nœ)

Hübsch (hʏpsh), Gerhard (gĕr'härt)

Hugo (hū'gō; *Fr* ʏ gō'), Victor (*Fr.* vēk-tôr')

Huguenots, Les (lä ʏ gə nō')

Humperdinck (hŏŏm'pər dĭngk), Engelbert (ĕng'əl bĕrt)

Hunchback of Notre Dame (nô'trə dåm', nō'tər dām'), The

Hunding (hŏŏn'dĭng)

Huon de Bordeaux (ʏ ôN' də bôr dō'; hū'ən —), Sir

Hurr, hopp, hopp, hopp (hŏŏr, hôp, hôp, hôp)

Hymne de joie (ēmn də zhwä')

I

Iago (Ĭ ä'gō; *It.* yä'gō)

Ibert (ē bĕr'), Jacques (zhäk)

Ibsen (ĭb'sən), Henrik (hĕn'rĭk)

Ich baue ganz auf deine Stärke (Ĭкн bou'ə gänts ouf dī'nə shtĕr'kə)

Ich gehe, doch rathe ich dir (Ĭкн gā'ə, dôкн rä'tə Ĭкн dēr)

Ich sah das Kind an seiner Mutter Brust (Ĭкн zä däs kĭnt än zī'nər mŏŏt'ər brŏŏst)

Idamante (ē dä män'tā)

Idomeneo, rè di Creta (ē dô'mĕ nĕ'ō, rä dē krä'tä)

Ifigenia in Aulide (ē fē jĕ'nyä ēn ou'lē dä)

Ifigenia in Tauride (ē fē jĕ'nyä ēn tou*̱-rē dä)

Igor (ē'gər), Prince

Il balen del suo sorriso (ēl bä län' dĕl sŏŏ'ō sôr rē'sō)

Il cavallo scalpita (ēl kä väl'lō skäl'pē tä)

Il core vi dono (ēl kô'rä vē dō'nō)

Il est beau de mourir en s'aimant (ēl ĕ bō də mŏŏ rēr' äN sĕ män')

Il est des Musulmans (ēl ĕ dä mʏ zʏl-män')

Il était une fois à la cour d'Eisenach (ēl ā tĕ' tʏn fwä å lä kŏŏr dī zĕ nåk')

Il était un roi de Thulé (ēl ā tĕ' tœn rwå də tʏ lä')

Il faut partir, mes bons compagnons (ēl fō pår tēr', mä bôN kôN på nyôN'}

Ilia (ē'lyä)

Il lacerato spirito (ēl lä chĕ rä'tō spē'rē tō)

Iliad (ĭl'ĭ əd), The

Illica (ēl'lē kä), Luigi (lŏŏ ē'jē)

Il mio tesoro (ēl mē'ō tä zô'rō)

Il partit au printemps (ēl pår tĕ' ō prãN-tãN')

Il segreto per essere felice (ēl sä grä'tō pĕr ĕs'sĕ rä fĕ lē'chä)

Imbroglio (ĭm brōl'yo; *It.* ēm brōl'lyō)

Im Mohrenland gefangen war (ĭm mō'-rən länt' gə fäng'ən vär)

Inaffia l'ugola! (ē näf'fyä lōō'gô lä!)

Incoronazione di Poppea, L' (lēng kô'rô-nä tsyō'nä dē pōp pĕ'ä)

Indes galantes, Les (lä zăɴd gà länt')

In des Lebens Frühlingstagen (ĭn dĕs lä'bəns frȳ'lĭngs tä'gən)

In dieser feierlichen Stunde (ĭn dē'zər fī'ər lĭĸʜ ən shtŏŏn'də)

In diesen heil'gen Hallen (ĭn dē'zən hīl'-gən häl'ən)

Inez (ī'nĕz, ē'nĕz, ī nĕz'; Sp. ē nāth', ē nās')

In fernem Land (ĭn fĕr'nəm länt)

Ingelbrecht (ăɴ gĕl brĕsht'), Désiré (dā-zē rā')

In grembo a me (ēng grĕm'bō ä mā)

In mia man alfin tu sei (ēm mē'ä män äl fēn' tōō sĕ'ē

Inigo, Don (dôn ē'nē gō)

In preda al duol (ēm prĕ'dä äl dwŏl)

In quelle trine morbide (ēng kwēl'lä trē'nä môr'bē dä)

Inquisitore, L' (lēng kwē'zē tō'rä)

Intendant (ĭn tĕn dänt')

Intermède (ăɴ tĕr mĕd')

Intermezzo (ĭn tər mĕt'sō, -mĕd'zō; It. ēn tĕr mĕd'dzō)

In the Town of Kazan (kä zän')

Introduzione (ēn'trō dōō tsyō'nä)

In uomini, in soldati (ēn wô'mē nē, en sôl dä'tē)

Invano Alvaro (ēn vä'nō äl'vä rō)

Invisible City of Kitezh (kē tĕsh'), The

Invocation à la nature (ăɴ vô kà syôɴ' à lä nà tyr')

Io sono l'umile ancella (ē'ō sō'nō lōō'mē-lä än chĕl'lä)

Io vidi la luce nel camp guerrier (ē'ō vē'dē lä lōō'chä nĕl kämp gwĕr ryĕr')

Io vidi miei signori (ē'ō vē'dē myĕ'ē sēn-nyō'rē)

Ipanov (ē pä'nəf), Count Loris (lŏ rēs')

Iphigenia in Aulis (ĭf'ĭ jĭ nī'ə) in Aulis (ô'lĭs)

Iphigenia in Tauris (tôr'ĭs)

Iphigénie en Aulide (ē fē zhä nē' äɴ nō-lēd')

Iphigénie en Tauride (ē fē zhä nē' äɴ tō-rēd')

Irene (Fr. ē rĕn'; Ger. ē rā'nə)

Iris (It. ē'rēs)

Isabella (It. ē zä bĕl'lä)

Isepo (ē zĕ'pō)

Isolde (ē zôl'də)

Ist ein Traum (ĭst ĭn troum)

Italia! Italia! e tutto il mio ricordo! (ē tä'-lyä! ē ta'lyä! ĕ tōōt'tō ēl mē'ō rē-kôr'dō!)

Italiana in Algeri, L' (lē tä lyä'nä ēn äl-jĕ'rē)

Ivan Susanin (ē vän' sōō sä'nyĭn)

Ivogün (ē'vō gyn), Maria (mä rē'ä)

Ivrogne corrigé, L' (lē vrôn'y kô rē zhā')

J

Ja, das alles auf Ehr (yä, däs äl'əs ouf ār)

Jadlowker (yäd lôf'kər), Herman

Jagel (yä'gəl), Frederick

J'ai le bonheur dans l'âme (zhä lə bô nœr' däɴ läm')

J'ai vu, nobles seigneurs (zhä vy', nô'blə sĕ nyœr')

Ja! Lasst uns zum Himmel die Blicke erheben (yä! läst ŏŏns tsŏŏm hĭm'əl dē blĭk'ə ĕr hā'bən)

Janáček (yä'nä chĕk), Leoš (lĕ'ôsh)

Janssen (yän'sən), Herbert (Ger. hĕr'bərt)

Janusz (yä'nōōsh)

Jaquino (yä kē'nō)

Jaroslavna (yä rə släv'nə)

Jasager, Der (dər yä'zä'gər)

Jean (zhäɴ)

Jeanne d'Arc au bûcher (zhäɴ dárk ö bȳ shä')

Je connais un pauvre enfant (zhə kô nĕ zœn pô'vräɴ fäɴ')

J'écris à mon père (zhä krē' à môɴ pĕr')

Je crois entendre (zhə krwä zän tän'drə)

Je crois entendre encore (zhə krwä zän-tän'dräɴ kôr')

Je dis que rien ne m'épouvante (zhə dē kə ryăɴ nə mä pōō vänt')

Je marche sur tous les chemins (zhə mársh sȳr tōō lä shə măɴ')

Je me souviens, sans voix, inanimée (zhə mə sōō vyäɴ, säɴ vwä, ē nà nē mä')

Jemmy (jĕm'ĭ)

Jenufa (yĕ'nōō fä)

Je ris de me voir (zhə rē də mə vwár')

Jeritza (yĕ'rē tsä), Maria (mä rē'ä)

Jerum! Jerum! (yĕr'ŏŏm! yĕr'ŏŏm!)

Jessner (yĕs'nər), Irene

Jessonda (jĕ sŏn'də, yĕ sôn'dä)

Je suis encore tout étourdie (zhə sȳ zäɴ kôr' tōō tä tōōr dē')

Je suis l'oiseau (zhə sȳĕ lwá zō')

Je suis Titania (zhə sȳē tē tà nē á')

Je veux vivre dans ce rêve (zhə vœ vē'vrə däɴ sə rĕv')

Jochum (yō′кнȱm), Eugen (oi′gän, oi-gän′)
Jokanaan (jō kăn′ən)
Jommelli (yôm měl′lē), Niccolò (nēk kô-lô′)
Jonas (zhô näs′)
Jongleur de Notre Dame, Le (lə zhôn-glœr′ də nô′trə dåm′)
Jonny spielt auf! (jŏn′ĭ shpēlt ouf′!)
Jontek's (yôn′těks) Revenge
Joseph (Fr. zhō zěf′)
Jour naissait dans le bocage, Le (lə zhȱōr ně sě′ dän lə bô kàzh′)
Journet (zhȱōr ně′), Marcel (màr sěl′)
Jouy (zhwē), Victor Joseph Etienne de (vēk tôr′ zhō zěf′ ā tyěn′ də)
Juarez (кнwä′rěs) and Maximilian
Juch (yȱōкн), Emma
Juchhei, nun ist die Hexe todt! (yȱōкн-hī′, nōōn ĭst dē hěk′sə tōt!)
Juif Polonais, Le (lə zhyēf pô lô ně′)
Juive, La (là zhyēv′)
Julien (zhy lyăN′)
Jumping Frog of Calaveras (kăl ə věr′əs) County
Jungfrau von Orleans, Die (dē yȱōng′frou fôn ȯr lā äN′)
Jurinac (ū′rē näts), Sena (sě′nä)

K

Kabale und Liebe (kä bä′lə ȯ̆ōnt lē′bə)
Kabalevsky (kä bä lyěf′skĭ), Dmitri (də-mē′trĭ)
Kafka (käf′kä), Franz (fränts)
Kaiser (kī′zər), Georg (gä ôrk′, gä′ôrk)
Kalidasa (kä lĭ dä′sä)
Kammeroper (käm′ər ō′pər)
Kammersänger (käm′ər zěng′ər); Kammersängerin (käm′ər zěng′ə rĭn)
Kann ich mich auch an ein Mädel erinnern (kän ĭкн mĭкн ouкн än ĭn mä′dəl ěr-ĭn′ərn)
Kapellmeister (kä pel′mī′stər)
Kapp (käp), Julius (Ger. ū′lē ȱōs)
Kappel (käp′əl), Gertrude
Karajan (kä′rä yän), Herbert von (Ger. hěr′bərt fôn)
Kärntnerthortheater (kěrnt′nər tōr′tä ä′-tər)
Katerina (kăt ə rē′nə; Rus. kä tyĭ rē′nə)
Kathinka (kä′tĭng kä)

Keilberth (kĭl′běrt), Joseph (Ger. yō′zěf)
Keiser (kī′zər), Reinhard (rīn′härt)
Keller (kěl′ər), Gottfried (gôt′frēt)
Kezal (kě′tsäl)
Khan Kontchak (кнän kŏn chäk′)
Khivria (кнēv′rĭ ə)
Khovantchina (кнŏ vän′shchĭ nə)
Khovantsky (кнŏ vän′tskĭ), Prince Ivan (Rus. ē vän′)
Khrennikov (кнrě′nyĭ kəf), Tikhon (tē′-кнən)
Kienzl (kēn′tsəl), Wilhelm (vĭl′hělm)
Kiepura (kyě pōō′rä), Jan (yän)
Kilian (kě′lē än)
Kipnis (kĭp′nĭs, kēp′-; Rus. kēp nyēs′), Alexander
Kirchhofer (kĭrкн′hō′fər), Werner (věr′-nər)
Kirsten (kĭr′stən), Dorothy
Klänge der Heimat (klěng′ə dər hī′mät)
Klafsky (kläf′skĭ), Katharina (kä tä rē′nä)
Kleiber (klī′bər), Erich (ā′rĭкн)
Klein (klīn), Herman
Kleist (klīst), Heinrich Wilhelm von (hīn′-rĭкн vĭl′hělm fôn)
Klemperer (klěm′pə rər), Otto
Klenau (klä′nou), Paul von (Ger. poul fôn)
Klingsor (klĭng′zôr)
Kluge, Die (dē klȱō′gə)
Klytemnestra (klĭ təm něs′trə, klĭt əm-)
Knappertsbusch (knäp′ərts bȱōsh), Hans (häns)
Knote (knō′tə), Heinrich (hīn′rĭкн)
Knusperhäuschen, Das (däs knȯ̆ōs′pər hois′shən)
Knusperwalzer (knȯ̆ōs′pər val′tsər)
Koanga (kō ăng′gə)
Kobbé (kō′bä), Gustav (gŭs′təv)
Kobus (kō′bȱōs), Fritz
Kodály (kō′dĭ), Zoltan (zōl′tän)
Kolodin (kə lō′dən), Irving
Königin von Saba, Die (dē kœ′nē gĭn fôn zä′bä)
Königskinder, Die (dē kœ′nĭкнs kĭn′dər)
Konrad (Ger. kôn′rät)
Konstanze (kôn stän′tsə)
Kontchak, Khan (кнän kŏn chäk′)
Kontchakovna (kŏn chä′kəv nə)
Körner (kœr′nər), Karl Theodor (kärl tä′ō dôr)
Korngold (kôrn′gōld; Ger. kôrn′gôlt), Erich Wolfgang (ěr′ĭk; Ger. ā′rĭкн; vôlf′gäng)

ăct, āble, dâre, ärt; à, Fr. ami; ĕbb, ēqual; ĭf, īce; кн, Ger. ach; N indicates nasal vowel; ŏdd, ōver, ôrder; œ, Fr. feu, Ger. schön; bȱōk, ōōze; ŭp, ūse, ûrge; Y, Fr. tu, Ger. über; zh, vision; ə, a in alone. (Full key on page 558)

Kotzebue (kŏt′sə bōō), August Friedrich von (ou′gōōst frē′drĭĸн fôn)
Koupava (kōō pä′və)
Kraus (krous), Ernst (ĕrnst)
Kraus, Felix von (fā′lĭks fôn)
Krauss (krous), Clemens (klā′məns)
Krauss, Gabrielle (gä brĭ ĕl′, găb rĭ ĕl′)
Krehbiel (krā′bĕl), Henry Edward
Křenek (krĕn′ĕk), Ernest
Kreutzer (kroit′sər), Conradin (kôn′rädēn)
Kreutzer (Fr. krœ tsĕr′), Rodolphe (rôdôlf′)
Kronold (krō′nôlt), Selma
Kruschina (krōō′shĭ nä)
Kullman (kŭl′mən), Charles
Kundry (kŏŏn′drĭ)
Kuno (kōō′nō)
Kunz (kōŏnts), Erich (ā′rĭĸн)
Kurt (kŏŏrt), Melanie (mā lä nē′)
Kurwenal (kōōr′və näl)
Kurz (kŏŏrts), Selma (Ger. zĕl′mä)
Kyoto (kyō′tō)

L

Lablache (là blàsh′), Luigi (lōō ē′jē)
Laca (lä′tsä)
La calunnia è un venticello (lä kä lōōn′ nyä ĕ ōōn vĕn tē chĕl′lō)
La Charmeuse (lä shàr mœz′)
Là ci darem la mano (lä chē dä räm′ lä mä′nō)
La Cieca (lä chĕ′kä)
La Ciesca (lä chĕ′skä)
La donna è mobile (lä dôn′nä ĕ mô′bē lä)
Lady Macbeth of Mtsensk (əm tsĕnsk′)
Laërtes (lā ûr′tēz)
Laetitia (lē tĭsh′ə)
La fatal pietra (lä fä täl′ pyĕ′trä)
La Fontaine (là fôɴ tĕn′), Jean de (zhäɴ də)
Lagerlöf (lä′yər lœv), Selma
Lakmé (läk′mä)
Lalla Roukh (läl′ə rŏŏk)
Lalo (là lō′), Edouard (ā dwàr′)
L'altra notte (läl′trä nôt′tä)
La mamma morta (lä mäm′mä môr′tä)
Lamartine (là màr tēn′), Alphonse de (àl fôɴs′ də)
Lamoureux (là mōō rœ′), Charles (Fr. shàrl)
Lamperti (läm pĕr′tē), Francesco (frän chä′skō)
L'anima ho stanca (lä′nē mä ô stäng′kä)
La notte il giorno (lä nôt′tä ĕl jōr′nō)

Laparra (là pà rá′), Raoul (rà ōōl′)
Lara (là rá′), Isidore de (Fr. ē zē dôr′ də)
Largo al factotum (lär′gō äl fäk tō′tōōm)
Larina (lä′rĭ nə), Madame
La rivedrà nell′ estasi (lä rē vä drä′ nĕl lĕ′stä zē)
Larsen–Todsen (lär′sən tōd′sən), Nanny
La Scala (lä skä′lä)
Lascia ch′ io pianga (läsh′shä kē′ō pyäng′gä)
Lasciatemi morire (läsh shä′tä mē mô rē′rä)
Lassalle (là sàl′), Jean (zhäɴ)
Lasst mich euch fragen (läst mĭĸн oiĸн frä′gən)
Lattuada (lät twä′dä), Felice (fĕ lē′chä)
Laubenthal (lou′bən täl), Rudolf
Laura (It. lou′rä)
Lauretta (It. lou rät′ta)
Lauretta mia (lou rät′tä mē′ä)
Lauri–Volpi (lou′rē vōl′pē), Giacomo (jä kô mō)
Lausanne (lō zăn′) International Competition
La Vergine degli angeli (lä vĕr′jē nä dĕl lyē än′jĕ lē)
Lazzari (läd′dzä rē), Virgilio (vēr jē′lyō)
Leander (lĭ ăn′dər)
Leben des Orest, Das (däs lā′bən dĕs ō rĕst′)
Leb′ wohl (läp vōl′)
Lefebvre (lə fĕ′vrə)
Le Flem (lə flĕm′), Paul
Légende de la sauge (lā zhäɴd′ də là sōzh′)
Legend of Czar Saltan (tsär säl tän′)
Legend of Kleinzach (klīn zàk′)
Lehmann (lā′män), Lilli (lĭl′ĭ)
Lehmann, Lotte (lŏt′ə; Ger. lôt′ə)
Leiden des jungen Werthers, Die (dē lī′ dən dĕs yŏŏng′ən vĕr′tərs)
Leider (lī′dər), Frida (frē′dä)
Leila (lā′lä)
Leinsdorf (līnz′dôrf), Erich (ĕr′ĭk; Ger. ā′rĭĸн)
Leise, leise, fromme Weise (lī′zə, lī′zə, frôm′ə vī′zə)
Leitmotiv (līt′mō tēf′)
Lel (lĕl)
Lelio (lĕ′lyō)
Lemoyne (lə mwàn′), Jean-Baptiste (zhäɴ bà tēst′)
Lensky (lĕn′skĭ)
Leo (lĕ′ō), Leonardo (lä ô när′dō)
Leoncavallo (lä ōn′kä väl′lō), Ruggiero (rōōd jĕ′rō)

Leoni (lā ō'nē), Franco (fräng'kō)
Leonora (lā ô nō'rä)
Leonora de Guzman (lā ô nō'rä dä gōōzmän', gōōth män')
Leonore (lā ō nō'rə)
Leporello (lĕp ə rĕl'ō; *It.* lä pô rĕl'lō)
Leroux (lə rōō'), Xavier (*Fr.* gzà vyä')
Lert (lĕrt), Ernst (ĕrnst)
Le Sage (lə sàzh'), Alain René (à lăN' rə nä')
Lescaut (lĕs kō')
Lesueur (lə sy œr'), Jean-François (zhäN fräN swà')
Leuchtende Liebe! (loiKH'tən də lē'bə!)
Levasseur (lə và sœr'), Nicolas-Prosper (nē kô lä' prôs pĕr')
Le veau d'or (lə vō dôr')
Levi (lā'vē), Hermann (*Ger.*) hēr'män)
Libiamo, libiamo (lē byä'mō, lē byä'mō)
Libuše (lē'bōō shĕ)
Licino (lē chē'nō)
Liebe der Danae, Die (dē lē'bə dər dä'nä ä)
Liebermann (lē'bər män), Rolf (*Ger.* rôlf)
Liebesnacht (lē'bəs näKHt')
Liebestod (lē'bəs tōt')
Liebesverbot Das (däs lē'bəs fər bōt')
Liederspiel (lē'dər shpēl')
Lieti Signori, salute! (lyĕ'tē sēn nyō'rē, sä lōō'tä!)
Linda di Chamounix (lēn'dä dē shä mōōnē')
Lindorf (lǐn'dôrf)
Lindoro (lēn dō'rō)
Linette (lǐ nĕt')
Lionel (lǐ'ə nəl, —nĕl)
Lisa (lē'zä)
Litvinne (lēt vēn'), Félia (fä'lǐ ə)
Liù (lū)
Ljungberg (yōōng'bĕr'y), Göta (yœ'tä)
Lobetanz (lō'bə tänts)
L'oca del Cairo (lô'kä dĕl kǐ'rō)
Lodoletta (lō dō lät'tä)
Lodovico (lō dō vē'kō)
Loge (lō'gə)
Logroscino (lō grō shē'nō), Nicola nēkô'lä)
Lohengrin (lō'ən grǐn)
Lohse (lō'zə), Otto
Lola (lō'lä)
Lombardi, I (ē lôm bär'dē)
Lorek (lô'rᵘk)

Loreley, Die (dē lō'rə lǐ)
Lorenz (lō'rĕnts), Max (*Ger.* mäks)
Lortzing (lôr'tsǐng), Gustav Albert (*Ger.* gōōs'täf äl'bərt)
Lo sposo deluso (lō spô'zō dä lōō'zō)
Lothario (lō thâr'ǐ ō)
Loti (lô tē'), Pierre (pyĕr)
Louÿs (lwē), Pierre (pyĕr)
Lualdi (lōō äl'dē), Adriano (ä drē ä'nō)
Lubava (lū bä'və)
Lubin (ly băN'), Germaine (zhĕr mĕn')
Luca (lōō'kä), Giuseppe de (jōō zĕp'pä dä)
Lucca (lōōk'kä), Pauline (*Ger.* pou lē'nə)
Lucia (lōō chē'ä)
Lucia di Lammermoor (lōō chē'ä dē lăm'ər mōōr)
Lucinda (*It.* lōō chēn'dä)
Lucrezia Borgia (lōō krä'tsyä bôr'jä)
Ludikar (lōō'dē kär), Pavel (pä'vĕl)
Ludmilla (lūd mē'lə)
Luigi (lōō ē'jē)
Luisa Miller (lōō ē'zä mǐl'ər)
Lully (ly lē'), Jean-Baptiste (zhàN bàtēst')
Lulu (lōō'lōō)
Luna (lōō'nä), Count di (dē)
Lustigen Weiber von Windsor, Die (dē lōōs'tǐ gən vī'bərn fôn vǐnt'zôr
Luther *Fr.* (ly tér')
Lycidas (lǐs'ǐ dəs)
Lysiart (lē zē àr')
Lysistrata (lǐ sǐs'trə tə, lǐs ǐ strä'tə)

M

Maccus (măk'əs)
Machiavelli (măk'ǐ ə vĕl'ǐ; *It.* mä kyävĕl'lē), Niccolò (nēk kô lô')
Ma dall' arido (mä däl lä'rē dō)
Madama (mä dä'mä) Butterfly
Madame Sans-Gêne (mà dàm' säN zhĕn')
Madamina, il catalogo è questo (mä dämē'nä, ēl kä tä'lō gô ĕ kwä'stō)
Maddalena (mäd dä lä'nä)
Madeleine (măd'ə lǐn, —lān; *Fr.* màdlĕn')
Mademoiselle Fifi (màd mwà zĕl' fē fē')
Madre, pietosa Vergine (mä'drä, pyä tō'sä vĕr'jē nä)
Maestro di cappella (mä ĕ'strō dē käppĕl'lä)

Maeterlinck (mä'tər lĭngk, mĕt'ər—; Fr. mä tĕr länk'; Flemish mä'tər lĭngk), Maurice (mô rēs')

Magda (mäg'dä)

Magdalena (mäg dä lä'nä)

Maggio Musicale Fiorentino (mäd'jō mōō zē kä'lä fyô rĕn tē'nō)

Maggiorivoglio (mäd jō'rē vôl'lyō), Marchioness of

Magnifico, Don (dôn män nyē'fē kō)

Mahabharata (mə hä'bä'rə tə), The

Mahler (mä lər), Gustav (Ger. gōōs'täf)

Maid of Orleans (ôr lēnz', ôr'lĭ ənz), The

Maid of Pskov (pskôf), The

Main de gloire, La (lä mǎn də glwȧr')

Maintenant que le père de Pelléas est sauvé (mǎn tə nǎn' kə lə pěr da pĕlä äs' ĕ sō vä')

Maison (mĕ zŏn'), René (rə nä')

Maître de chapelle (mĕ'trə də shȧ pĕl')

Malatesta (mä lä tĕs'tä), Dr.

Malazarte (mä lə zär'tĭ)

Malheurs d'Orphée, Les (lä mȧ lœr' dôr fä')

Malibran (mȧ lē brän'), Maria Felicita (mä rē'ä fĕ lē' chē tä')

Maliella (mä lyĕl'lä)

Malipiero (mä lē pyĕ'rō), Gian Francesco (jän frän chä'skō)

Mallika (mä'lĭ kə)

Mallinger (mäl'ĭng ər), Mathilde (mätĭl'də)

Mal reggendo all' aspro assalto (mäl rädjĕn'dō äl lä'sprō äs säl'tō)

Malvolio (mäl vō'lyō)

Mamelles de Tirésias, Les (lä mȧ mĕl' də tē rä zyȧs')

Ma mère, je la vois (mȧ mĕr', zhə lä vwä')

Mamma Lucia (mäm'mä lōō chē'ä)

Mamma, quel vino è generoso (mäm'mä, kwĕl vē'nō ĕ jĕ nĕ rō'sō)

Mancinelli (män chē nĕl'lē), Luigi (lōōē'jē)

Manfred (mǎn'frəd)

Manfredo (män frä'dō)

Mann (män), Thomas

Manon (mȧ nôn')

Manon Lescaut (mȧ nôn' lĕs kō')

Manrico (män rē'kō)

Manru (män'rōō)

Manzoni (män dzō'nē), Alessandro (äläs sän'drō)

M'appari (mäp pä rē')

Marcel (mȧr sĕl')

Marcellina (mär sə lē'nə; It. mär chĕllē'nä)

Marcello (mär chĕl'lō)

Marche du couronnement (märsh dɤ kōō rôn män')

Marchesi (märkä'zē), Mathilde de Castrone (mä tĭl'də dä kä strō'nä)

Marchesi, Salvatore (säl vä tō'rä)

Maretzek (mä'rĕ tsĕk), Max

Marfa (mär'fə)

Magared (mär gä rĕd')

Margaretha (mär gä rä'tä)

Margiana (mär jä'nä)

Maria (mä rē'ä)

Marianne (mä rē än'ə)

Marie (mə rē'; Fr. mȧ rē'; Ger. mä rē'ə, mä rē'

Marietta (mär ĭ ĕt'ə; It. mä rē ät'tä)

Marina (mä rē'nə)

Marinaresca (mä'rē nä rä'skä)

Marino Faliero (mə rē'nō fäl yȧr'ō)

Marinuzzi (mä rē nōōt'tsē), Gino (jĕ'nō); Giuseppe (jōō zĕp'pä)

Mario (mä'rē ō)

Mario (mȧr'ĭ ō), Queena (kwē'nə)

Mariola (mä rē ō'lä)

Marmontel (mȧr môn tĕl'), Jean François (zhän frän swä')

Mârouf (mä rōōf')

Marriage of Figaro (fĭg'ə rō, fē'gä rō), The

Marschallin (mär'shä lĭn), The

Marschner (märsh'nər), Heinrich (hīn'rĭкн)

Martern aller Arten (mär'tərn äl'ər är'tən)

Martha (mär'thə; Ger. mär'tä)

Martin, Riccardo (rĭ kär'dō)

Martinelli (mär tĭ nĕl'ĭ; It. mär tē nĕl'lē), Giovanni (jō vän'nē)

Martini (mär tē'nē), Nino (nē'nō)

Martinu (mär'tē nōō), Bohuslav (bô'hōōsläf)

Martín y Soler (mär tēn' ē sō lĕr'), Vicente (vē sĕn'tä; Sp. bē thĕn'tä)

Marullo (mä rōōl'lō)

Masaniello (mä zä nyĕl'lō)

Mascagni (mä skän'nyē), Pietro (pyĕ'trō)

Ma se m'è forza perderti (mä sä mĕ fôr'tsä pĕr'dĕr tē)

Masetto (mäɬ zä'tō)

Massé (mȧ sā'), Victor (Fr. vēk tôr')

Massenet (mǎs ə nä'; Fr. mȧs nĕ'), Jules Emile Frédéric (zhɤl ā mēl' frä därēk')

Mastersingers of Nuremberg (nōō'rəm-bûrg, nū'—), The
Materna (mä tĕr'nä), Amalie (ä mä'lē ə)
Mathilde (mä tĭl'də)
Mathis der Maler (mä'tĭs dər mä'lər)
Mathisen (mä'tē sĕn)
Matho (má tō')
Matrimonio segreto, Il (ēl mä trē mô'-nyō sä grä'tō)
Ma tu, o Re, tu possente (mä tōō, ō rä, tōō pōs sĕn'tä)
Matzenauer (mät'sə nou'ər), Margarete (mär gä rä'tə)
Maudite à jamais soit la race (mō dēt' á zhá mĕ' swá lá rás')
Maupassant (mō pá sän'), Guy de (gē də)
Maurel (mō rĕl'), Victor
Maurice de Saxe (mô rēs' də sáks')
Mavra (mäv'rə)
Max (Ger. mäks)
Maximilien (mák sē mē lyăn')
Mayr (mīr), Richard (Ger. rĭкн'ärt)
Mazeppa (mə zĕp'ə)
Mazurka (mə zûr'kə, -zōōr'—)
Meco all' altar di Venere (mä'kō äl läl-tär dē vĕ'nĕ rä)
Medea (mē dē'ə)
Médecin malgré lui, Le (lə mād săn' mál grä' lɥē)
Médée (mä dä')
Mefistofele (mä fē stô'fĕ lä)
Méhul (mä ɥl'), Etienne Nicholas (ä-tyĕn' nē kô lä'); Henri (än rē')
Mein Herr Marquis (mīn hĕr' mär kē')
Mein Herr und Gott, nun ruf' ich Dich (mīn hĕr ōōnt gôt, nōōn rōōf ĭкн dĭкн)
Meistersinger (mī'stər zĭng'ər)
Meistersinger von Nürnberg, Die (dē mī'stər zĭng'ər fôn nɥrn'bĕrk)
Melchior (mĕl'kĭ ôr), Lauritz (lou'rĭts)
Melchthal (mĕlкн'täl)
Mélesville (mä les vēl')
Mélisande (mä lē zänd')
Melitone, Fra (frä mĕ lē tō'nä)
Melot (mä'lôt)
Mendelssohn (mĕn'dəl sən,—sōn), Felix (Ger. fä'lĭks)
Mendès (män dĕs'), Catulle (ká tɥl')
Menelas (mä nə läs')

Menotti (mə nŏt'ĭ; It. mĕ nôt'tē), Gian-Carlo (jän kär'lō)
Me pellegrina ed orfana (mä pĕl lä grē'-nä äd ôr'fä nä)
Mephisto's (mē fĭs'tōz) Serenade
Mephistopheles (mĕf i stŏf'ə lēz)
Mercadante (mĕr kä dän'tä), Saverio (sä vĕ'ryō)
Mercédès (mĕr sä dĕs')
Mercè, dilette amiche (mĕr chä', dē lät'-tä ä mē'kä)
Mercutio (mər kū'shĭ ō)
Mérimée (mä rē mä'), Prosper (Fr. prôs pĕr')
Merlin (mûr'lĭn; Ger. mĕr'lĭn)
Merola (mä'rô lä), Gaetano (gä ä tä'nō)
Mes amis, écoutez l'histoire (mä zá mē', ä kōō tä' lēs twár')
Me sedur han creduto (mä sä dōōr' än krä dōō'tō)
Mesrour (mĕs rōōr')
Messa di voce (mäs'sä dē vō'chä)
Messager (mĕ sá zhä'), André Charles Prosper (än drä' shárl prôs pĕr')
Metastasio (mä tä stä'zyō), Pietro (pyĕ'-trō)
Me voici dans son boudoir (mə vwá sē' dän sôn bōō dwár')
Meyerbeer (mī'ər bār, -bĭr), Giacomo (jä'kô mō)
Mezza aria (mĕd'dzä ä'ryä)
Mezza voce (mĕd'dzä vō'chä)
Mezzo-soprano (It. mĕd'dzō sō prä'nō)
Micaëla (mē kä ä'lä)
Micha (mē'кнä)
Michael Kohlhaas (mĭкн'ä ĕl kōl'häs)
Michele (mē kĕ'lä)
Mi chiamano Mimi (mē kyä'mä nō mē mē')
Mignon (mē nyôn')
Milanov (mē'lä nôv), Zinka (zēng'kä)
Mildenburg (mĭl'dən bōōrk), Anna von (Ger. än'ä fôn)
Milder-Hauptmann (mĭl'dər houpt'män), Pauline Anna (Ger. pou lē'nə än'ä)
Mild und leise wie er lächelt (mĭlt ōōnt li'zə vē ĕr lĕкн'əlt)
Milhaud (mē yō'), Darius (də rī'əs; Fr. dá ryɥs')
Mime (mē'mə)
Mimi (mē mē')
Mimodrama (mĭ'mə drä'mə, -dräm'ə)

ăct, āble, dâre, ärt; á, Fr. ami; ĕbb, ēqual; ĭf, īce; кн, Ger. ach; n indicates nasal vowel; ŏdd, ōver, ôrder; œ, Fr. feu, Ger. schön; bŏŏk, ōōze; ŭp, ūse, ûrge; ɥ, Fr. tu, Ger. über; zh, vision; ə, a in alone. (Full key on page 558)

Minnesingers (mĭn'ə sĭng'ərz)
Minnie della mia casa (mĭn'ĭ dĕl lä mē'ä kä'sä)
Miracle (mē rä'klə), Dr.
Mira, o Norma (mē'rä, ō nôr'mä)
Mireille (mē rĕ'y)
Mir ist so wunderbar! (mēr ĭst zō vŏŏn'dər bär!)
Miserere (mĭz ə rär'i, -rĭr'i, mē sĕ rĕ'rĕ)
Missail (mē sä ēl')
Mistral (mēs trȧl'), Frédéric (frä dä rēk')
Mit Gewitter und Sturm (mĭt gə vĭt'ər ŏŏnt shtŏŏrm')
Mit mir, mit mir (mĭt mēr', mĭt mēr')
Mitropoulos (mĭ trŏp'ə ləs), Dimitri (dĭ mē'trĭ)
Mizguir (mēz gēr')
Moïse (mô ēz')
Molière (mô lyĕr')
Mona Lisa (mō'nə lē'zə)
Mon coeur s'ouvre à ta voix (môN kœr sōō vrȧ tȧ vwä')
Monforte (môn fôr'tä)
Monica (môn'ĭ kə)
Moniuszko (mô nū'shkô), Stanislaus (stän'ĭs lôs)
Monna Vanna (môn'nä vän'nä)
Monostatos (mō nō stä'tôs)
Monsalvat (môn säl'vät)
Monsigny (môN sē nyē'), Pierre-Alexandre (pyĕr ȧ lĕk sän'drə)
Montano (môn tä'nō)
Montemezzi (mōn tĕ mĕd'dzē), Italo (ē'tä lō)
Monterone (mōn tĕ rō'nä), Count
Monteux (môN tœ'), Pierre (pyĕr)
Monteverdi (mōn tä vĕr'dē), Claudio (klou'dyō)
Montfleury (môN flœ rē')
Morales (mō rä'läs)
Morfontaine (môr fôN tĕn'), Guillot de (gē yō' də)
Morgenlich leuchtend im rosigen Schein (môr'gən lĭкн loiкн'tənt ĭm rō'zĭ gən shīn')
Morrò, ma prima in grazia (môr rō', mä prē'mä ēng grä'tsyä)
Mort de Don Quichotte (môr də dôN kē shôt')
Moscona (mō skō'nä), Nicola (nē kô'lä)
Mosè in Egitto (mō zä' ēn ä jēt'tō)
Mosenthal (mō'zən täl), Salomon Hermann von (zä'lō môn hĕr'män fôn)
Mottl (môt'əl), Felix (Ger. fā'lĭks)
Mozart (mō'tsärt), Wolfgang Amadeus (vôlf'gäng ä mä dä'ŏŏs)

Muck (mŏŏk), Karl
Muette di Portici, La (lä myĕt dē pôr'tē chē)
Muff (mŭf)
Mugnone (mōōn nyō'nä), Leopoldo (lä ō pôl'dō)
Müller (myl'ər), Maria
Muratore (my rä tôr'), Lucien (Fr. lȳsyăn')
Musetta (mōō zät'tä)
Musette (mū zĕt')
Musica parlante (mōō'zē kä pär län'tä)
Musset (my sĕ'), Alfred de (ȧl frĕd' də)
Mussorgsky (mōō sŏrk'skĭ), Modest (mŏdĕst', -dyĕst')
Mustafa (mōō stä'fä)
Mutter, Die (dē mŏŏt'ər)
Muzio (mōō'tsē ō), Claudia (It. klou'dyä)
Mylio (mē lyō')
Myrtale (mēr täl')
Mytyl (mē tēl')

N

Nabucco (nä'bōōk kō); Nabucodonosor (nä'bōō kō dō'nō sōr')
Nachbaur (näкн'bour), Franz (fränts)
Nachtlager von Granada, Das (däs näкнt'lä'gər fôn grä nä'dä)
Nadir (nä dēr')
Naiad (nä'ăd, nī'-)
Naina (nä ē'nə)
Nannetta (nä nĕt'ə; It. nän nät'tä)
Nápravník (nä'präv nēk), Eduard (ĕ'dōō ärt)
Narraboth (när'ä bōt)
Natoma (nə tō'mə)
Nature immense (nȧ tyr' ēm mäns')
Navarraise, La (lä nȧ vȧ rĕz')
Nave, La (lä nä'vä)
Nedda (nĕd'dä)
Neipering (nī'pə rĭng), Count
Nella (nĕl'lä)
Nelusko (nĕ lōōs'kō)
Nemico della patria (nä'mē kō dĕl lä pä'trē ä)
Nemorino (nĕ mô rē'nō)
Nerone (nĕ rō'nä)
Nerto (nĕr'tō)
Nerval (nĕr vȧl'), Gérard de (zhä rär' də)
Nessler (nĕs'lər), Victor (Ger. vĭk'tôr)
Nessun dorma (näs sōōn' dôr'mä)
Neuendorff (noi'ən dôrf), Adolf (Ger. ä'dôlf)
Neues vom Tage (noi'əs fôm tä'gə)

Neumann (noi′män), Angelo (ăn′jə lō; *Ger.* än′jä lō)

Nevers (nə vĕr′), Count de (də)

Nibelhelm (nē′bəl hĕlm)

Nibelung (nē′bə lŏong) Ring, The

Nibelung Saga, The: Das Nibelungenlied (däs nē′bə lŏong ən lēt′)

Nicias (nĭsh′ī əs)

Nicklausse (nē klous′)

Nicolai (nē′kō lī), Carl Otto

Nicolette (nĭk ə lĕt′)

Nicolini (nē kô lē′nē)

Niemann (nē′män), Albert (*Ger.* äl′bərt)

Nie werd' ich deine Huld verkennen (nē vĕrt ĭĸн dī′nə hŏolt fər kĕn′ən)

Nikisch (nē′kĭsh), Arthur (*Ger.* är′tŏor)

Nikolaidi (nē kô lī′dē), Elena (ĕ′lĕ nä)

Nilakantha (nē lə kän′thə)

Nilsson (nĭls′sôn), Christine (*Sw.* krĭs tēn′)

Ninette (nĭ nĕt′)

Niun mi tema (nūn mē tä′mä)

Nobili acciar, nobili e santi (nô′bē lē ät chär′, nô′bē lē ä sän′tē)

Nobles Seigneurs, salut! (nô′blə sĕ nyœr′, sȧ lĭ′!)

Noëmi (nō′ə mĭ)

Non ho colpa (nōn ô kôl′pä)

Non imprecare, umiliati (nōn ēm prä-kä′rä, ōō mē lyä′tē)

Non la sospiri la nostra casetta (nōn lä sō spē′rē lä nô′strä kä sät′tä)

Non mi dir (nōn mē dēr′)

Non mi resta che il pianto (nōn mē rĕ′stä kä ēl pyän′tō)

Non più andrai (nōn pū än drī′)

Non siate ritrosi (nōn syä′tä rē trō′sē)

Non so più cosa son (nōn sô pū kô′sä sōn)

No! pazzo son! guardate (nô! pät′tsō sōn! gwär dä′tä)

Nordica (nôr′dĭ kə), Lillian

Noréna (nō rā′nä), Eidé (ā′dä)

Norina (nô rē′nä)

Norma (nôr′mä)

Normando (nôr män′dō)

Nothung! Nothung! Neidliches Schwert! (nō′tŏong! nō′tŏong! nīt′lĭĸн əs shvärt!)

Notre Dame de Paris (nô′trə dȧm də pȧ rē′)

Nouguès (nōō gĕs′), Jean (zhäɴ)

Nourrit (nōō rē′), Adolphe (*Fr.* ȧ dôlf′)

Nous vivrons à Paris (nōō vē vrôɴ′ zȧ pȧ rē′)

Novák (nô′väk), Vítězslav (vē′tyĕs släf)

Novotna (nô′vôt nä), Jarmila (yär′mē lä)

Nozze di Figaro, Le (lä nôt′tsä dē fē′gä rō)

Nozze di Teti e Peleo (nôt′tsä dē tä′tē ä pĕ lĕ′ō): The Wedding of Thetis (thē′tĭs) and Peleus (pē′lūs, pē′lĭ əs)

Nuit d'hyménéé (nɣē dē mä nä′)

Nuitter (nɣē tä′), Charles Louis (*Fr.* shȧrl lwē)

Nume, custode e vindice (nōō′mä, kōō-stô′dä ä vēn′dē chä)

Nun eilt herbei (nōōn īlt hĕr bī′)

Nun sei bedankt, mein lieber Schwan (nōōn zī bə dängkt′, mīn lē′bər shvän)

Nureddin (nōō rĕd dēn′)

Nuri (nōō′rē)

O

O Amore, o bella luce (ō ä mō′rä, ō bĕl′lä lōō′chä)

O beau pays de la Touraine (ō bō pä ē′ də lä tōō rĕn′)

Obéissons, quand leur voix appelle (ô bā ē sôɴ′, käɴ lœr vwä zȧ pĕl′)

O belle enfant! Je t'aime (ō bĕl äɴ fäɴ′! zhə tĕm′)

Oberon ō′bə rŏn)

Oberthal (ō′bər täl), Count

O Carlo, ascolta (ō kär′lō, ä skōl′tä)

Ochs (ôks), Baron

O ciel! dove vai tu? (ō chĕl! dō′vä vī′ tōō?)

O cieli azzurri (ō chĕ′lē äd dzoor′rē)

O Colombina (ō kô lôm bē′nä)

Octavian (ŏk tä′vĭ ən)

O del mio dolce ardor (ō dĕl mē′ō dōl′-chä är dōr′)

O Dieu, Dieu de nos pères (ō dyœ, dyœ də nō pĕr′)

O! di qual sei tu (ō! dē kwäl sĕ′ē tōō)

O du mein holder Abendstern (ō dōō mīn hôl′dər ä′bənt shtĕrn′)

Odyssey (ŏd′ə sĭ), The

Oedipus (ĕd′ĭ pəs, ē′dĭ-)

Offenbach (ôf′ən bäk; *Fr.* ô fĕn bȧk′), Jacques (zhäk)

O grandi occhi (ō grän′dē ôk′kē)

O grido di quest' anima (ō grē′dō dē kwäst ä′nē mä)

Oh ciel, où courez-vous (ō syĕl, ōō kōō-rä′vōō)

O holdes Bild (ō hôl′dəs bĭlt)

Oh! qu'est-ce que c'est? . . . tes cheveux (ō! kĕs kə sĕ′? . . . tä shə vœ′)

O inferno, Amelia qui! (ō ēn fĕr′nō, ä mä′lyä kwē!)

Oiseau Bleu, L' (lwà zō′ blœ′)

Oiseaux dans la charmille, Les (lä zwà-zō′ däɴ là shàr mē′y)

O Isis und Osiris (ō ē′zĭs ŏŏnt ō zē′rĭs)

O légère hirondelle (ō lä zhēr′ ē rôɴ dĕl′)

Olga (ŏl′gə)

O Liberté ma mie (ō lē bĕr tä′ mà mē′)

Olivier (ô lē vyä′)

O Lola bianca (ō lō′lä byäng′kä)

O luce di quest' anima (ō lōō′chä dē kwäst ä′nē mä)

Olympia (Fr. ô läɴ pē à′)

O ma lyre immortelle (ō mà lēr ēm-môr tĕl′)

Ombra mai fu (ōm′brä mī fōō)

Ombra leggiera (ōm′brä läd jĕ′rä)

Ombre légère (ôɴ′brə lä zhēr′)

O Mimi, tu più non torni (ō mē mē′, tōō pū nōn tōr′nē)

O mio babbino caro (ō mē′ō bäb bē′nō kä′rō)

O mio Fernando (ō mē′ō fĕr nän′dō)

O mon enfant (ō môɴ näɴ fäɴ′)

O monumento (ō mô nōō män′tō)

O namenlose Freude (ō nä′mən lō′zə froi′də)

On ne badine pas avec l'amour (ôɴ nə bà dēn′ pä zà vĕk′ là mōōr′)

Onegin (ŏ nyä′gĭn)

Onégin (ŏ nyä′gĭn), Sigrid (sĭg′rĭd, sē′grĭd)

O nuit, étends sur eux ton ombre (ō nɥē, ä täɴ syr œ tôɴ nôɴ′brə)

O padre mio (ō pä′drä mē′ō)

O Paradis (ō pà rà dē′)

O Paradiso (ō pä rä dē′zō)

O patria mia (ō pä′trē ä mē′ä)

Opéra, L' (lô pä rà′); Académie de Musique (à kà dä mē′ də my zēk′)

Opéra bouffe (ô pä rà′ bōōf′)

Opera buffa (ô′pĕ rä bōōf′fä)

Opéra comique (ô pä rà′ kô mēk′)

Opéra-Comique, L' (lô pä rà′ kô mēk′)

Opera seria (ô′pĕ rä sĕ′ryä)

Ophelia (ō fē′lĭ ə, ō fēl′yə)

O prêtres de Baal (ō prĕ′trə də bäl′)

O pur bonheur (ō pyr bô nœr′)

O qual soave brivido (ȝ kwäl sō ä′vä brē′vē dō)

Oracolo, L' (lô rä′kô lō)

Ora e per sempre addio sante memorie (ō′rä ā pĕr sĕm′prä äd dē′ō sän′tä mĕ mô′ryä)

Ora soave, sublime ora d'amore (ō′rä sō ä′vä, sōō blē′mä ō′rä dä mō′rä)

Or co' dadi (ōr kō dä′dē)

Ordgar (ôrd′gär)

Oreste (ô rĕst′)

Orestes (ō rĕs′tēz)

Orest! Es rührt sich niemand! (ō rĕst′! ĕs ryrt′ zĭкн nē′mänt!)

Orfeo (ôr fĕ′ō)

Orfeo ed Euridice (ôr fĕ′ō äd ā ōō′rē-dē′chä; Orpheus and Eurydice (ū rĭd′-i sē)

O riante nature (ō ryäɴ′tə nà tyr′)

O Richard, O mon Roi! (ō rē shàr′, ō môɴ rwà′!)

Orlovsky (ŏr lôf′skĭ), Prince

Oroveso (ō rō vä′zō)

Orpheus (ôr′fĭ əs, ôr′fūs)

Orpheus in the Underworld: Orphée aux Enfers (ôr fä′ ō zäɴ fĕr′)

Orsini (ôr sē′nē), Maffio (mäf′fyō)

Orsini, Paolo (pä′ô lō)

Ortensio (ôr tĕn′syō)

Ortrud (ôr′trōōt)

Orzse (ôr′zhĕ)

Osaka (ō sä′kä)

O Signore, dal tetto natio (ō sēn nyō′rä, däl tät′tō nä tē′ō)

O sink hernieder, Nacht der Liebe (ō zĭngk hĕr nē′dər, näкнт dər lē′bə)

Osmin (ŏs′mĭn)

O soave fanciulla (ō sō ä′vä fän chōōl′lä)

O sommo Carlo (ō sōm′mō kär′lō)

O Souverain! O Juge! O Père (ō sōō-vräɴ′! ō zhyzh′! ō pĕr′!)

Otello (ō tĕl′lō)

Otello, il Moro di Venezia (ō tĕl′lō, ēl mô′rō dē vĕ nĕ′tsyä)

O tempio sontuoso (ō tĕm′pyō sôn tōō-ō′sō)

O temple magnifique (ō täɴ′plə mà nyē-fēk′)

O terra, addio (ō tĕr′rä, äd dē′ō)

Othello (ō thĕl′ō)

O transport, o douce extase (ō träɴs pôr′, ō dōōs ĕks täz′)

Ottavio (ōt tä′vyō)

Ottokar (ōt′tō kär)

O tu che in seno agli angeli (ō tōō kä ēn sä′nō ä lyē än′jĕ lē)

Oui, je veux par le monde (wē, zhə vœ pàr lə môɴd′)

Ourrias (ōō rē äs′)

Où va la jeune hindoue? (ōō và là zhœn ăn dōō'?)

O vago suol della Turrena (ō vä'gō swôl' dĕl lä tōōr rä'nä)

O vin dissipe la tristesse (ō văN dē sēp' là trēs tĕs')

O welche Lust! (ō vĕl'KHə lōōst!)

Ozean, du Ungeheuer! (ō'tsä än, dōō ŏŏn gə hoi'ər!)

P

Pace e gioia (pä'chä ā jô'yä)

Pace, pace, mio Dio! (pä'chä, pä'chä, mē'ō dē'ō!)

Pace t'imploro (pä'chä tēm plô'rō)

Paco (pä'kō)

Paër (pä'ĕr), Ferdinando (fĕr dē nän'dō)

Pagano (pä gä'nō)

Pagliacci (päl lyät'chē)

Paillasse (pà yàs')

Paisiello (pä ē zyĕl'lō), Giovanni (jō vän'-nē)

Palemon (pà lä môN')

Palestrina (păl əs trē'nə; *It*. pä lä strē'nä)

Pamela (*It*. pä mä'lä)

Pamina (pä mē'nä)

Pandolphe (päN dôlf')

Pang (päng)

Panizza (pä nēt'tsä), Ettore (ĕt'tô rä)

Pantolone (pän tä lō'nä)

Panza (păn'zə; *Sp*. pän'tha), Sancho (săn'chō; *Sp*. sän'chō)

Paolino (pä ô lē'nō)

Paolo e Francesca (pä'ô lō ā frän chä'skä)

Papagena (pä pä gā'nä)

Papageno (pä pä gā'nō)

Papi (pä'pē), Genarro (jĕ när'rō)

Paquiro (pä kē'rō)

Parassia (pä räs'yə)

Pardon de Ploërmel, Le (lə pàr dôN' də plô ĕr mĕl')

Paride ed Elena (pä'rē dä ād ĕ'lĕ nä)

Parigi, o cara (pä rē'jē, ō kä'rä)

Paris (păr'ĭs; *Fr*. pà rē')

Parlando (pär län'dō); parlante (pär län'-tä)

Parmi les pleurs (pàr mē' lä plœr')

Parmi veder le lagrime (pär'mē vä där' lä lä'grē mä)

Parpignol (pàr pē nyôl')

Parsifal (pär'sĭ fäl, -fəl)

Partagez-vous mes fleurs! (pàr tà zhā'vōō mä flœr'!)

Pasquale, Don (dôn pä skwa'lä)

Passacaglia (päs sä käl'lyä)

Passepied (pä sə pyä')

Passo a sei (päs'sō ä sĕ'ē)

Pasta (pä'stä), Giuditta (jōō dēt'tä)

Pasticcio (pä stēt'chō)

Pastorale (pàs tô ràl')

Patti (pàt'ĭ; *It*. pät'tē), Adelina (ăd ə lĭ'-nə; *It*. ä dĕ lē'nä)

Patzak (pät'säk; *Ger*. pät'säk), Julius (*Ger*. ū'lē ŏŏs)

Paul (*Ger*. poul)

Paul et Virginie (pôl ā vēr zhē nē')

Pauly (pou'lē), Rosa

Paur (pour), Emil (ā'mĭl)

Pauvre Matelot, Le (lə pō'vrə mà tlō')

Pêcheurs de perles, Les (lä pĕ shœr' də pĕrl')

Pedrillo (pä drĭl'ō)

Pedro (pä'drō)

Peerce (pĭrs), Jan (jăn)

Peer Gynt (pĭr gĭnt; *Norw*. pār gynt)

Pelléas et Mélisande (pĕ lä äs' ā mä lē-zäNd')

Penelope (pə nĕl'ə pĭ)

Penthesilea (pĕn'thĕ sĭ lē'ə)

Pepa (pä'pä)

Pepusch (pä'pŏŏsh), Johann Christoph (yō'hän krĭs'tôf)

Perchè ciò volle, il mio voler possente (pĕr kä' chô vôl'lä, ēl mē'ō vô lär' pōs sĕn'tä)

Perchè v'amo (pĕr kä 'vä'mō)

Pergolesi (pĕr gô lā'zē), Giovanni Battista (jō vän'nē bät tē'stä)

Peri (pä'rē), Jacopo (yä'kō pō)

Perlea (pĕr lä ä'), Jonel (zhō nĕl')

Per me ora fatale (pĕr mä ō'rä fä tä'lä)

Per pietà, ben mio perdona (pĕr pyä tä', bĕn mē'ō pĕr dō'nä)

Perrault (pĕ rō'), Charles (*Fr*. shàrl)

Perrin (pĕ răN'), Emile Césare (ā mēl' sä zàr')

Perrin, Pierre (pyĕr)

Persephone (pər sĕf'ə nĭ)

Persiani (pĕr sē ä'nē), Fanny

Pertile (pĕr tē'lä), Aureliano (ou rä lyä'-nō)

Pescator, affonda l'esca (pä skä tōr', äf-fōn'dä lä'skä)

ăct, āble, dâre, ärt; à, *Fr*. ami; ĕbb, ēqual; ĭf, īce; KH, *Ger*. ach; N indicates nasal vowel; ŏdd, ōver, ôrder; œ, *Fr*. feu, *Ger*. schön; bŏŏk, ōōze; ŭp, ūse, ûrge; Y, *Fr*. tu, *Ger*. über; zh, vision; ə, a in alone. (Full key on page 558)

Petrarch (pē'trärk)
Petroff (pĕt'rôf; *Rus.* pĭ trôf'), **Ossip** (ô'sĭp)
Petrovich (pĭ trô'vĭch)
Petruchio (pĭ trōō'chĭ ō, -kĭ ō)
Pfitzner (pfĭts'nər), **Hans** (häns)
Phanuel (făn'ū əl)
Philémon et Baucis (fē lā môɴ' ā bō sēs')
Philidor (fē lē dôr'), **François André Danican** (frän swȧ' äɴ drä' dȧ nē käɴ')
Philine (fē lēn')
Philomela (fĭl ə mē'lə)
Piangendo (pyän jĕn'dō)
Piave (pyä'vā), **Francesco Maria** (frän-chä'skō mä rē'ä)
Piccinni (pēt chēn'nē), **Nicola** (nē kô'la); **Niccolò** (nēk kô lô')
Piège de Méduse, Le (lə pyĕzh də mā-dȳz')
Pierné (pyĕr nā'), **Gabriel** (*Fr.* gȧ brē ĕl')
Pierroto (pyĕr rō'tō)
Pilar (pē lär')
Pimen (pē'mĕn)
Pinellino (pē nĕl lē'nō)
Ping (pĭng)
Pini-Corsi (pē'nē kôr'sē), **Antonio** (än-tô'nyō)
Pinza (pēn'zə; *It.* pēn'tsä), **Ezio** (ĕt'sĭ ō; *It.* ĕ'tsyō)
Pique Dame (pēk dȧm')
Pistol (pĭs'təl)
Pittichinaccio (pēt'tē kē nät'chō)
Pizarro (pē thär'rō, -sär')
Pizzetti (pēt tsät'tē), **Ildebrando** (ēl dä-brän'dō)
Plançon (pläɴ sôɴ'), **Pol-Henri** (pôl äɴ-rē')
Planquette (pläɴ kĕt'), **Jean-Robert** (zhän rô bĕr')
Pleurez, pleurez mes yeux (plœ rā', plœ rā' mā zyœ')
Plus blanche que la blanche hermine (plȳ bläɴsh kə là bläɴsh ĕr mēn')
Pogner (pōg'nər)
Polacco (pô läk'kō), **Giorgio** (jôr'jō)
Polkan (pôl kän'), **General**
Pollack (pôl'ək), **Egon** (ē'gən; *Ger.* ā'gôn)
Pollione (pôl lyō'nä)
Pollux (pôl'əks)
Polonaise (pôl ə nāz', pō lə-)
Polonius (pə lō'nĭ əs)
Polovtsian (pô'ləf tsĭ ən, pə lôf'-) **Dances**
Polyeucte (pô lyœkt')
Polyphemus (pôl ĭ fē'məs)
Pompeo (pôm pĕ'o)

Ponchielli (pông kyĕl'lē), **Amilcare** (ä-mēl'kä rä)
Pong (pông)
Pons (pŏnz; *Fr.* pôɴs), **Lily**
Ponselle (pŏn sĕl'), **Carmella** (kär mĕl'ə)
Porgi amor (pôr'jē ä mōr')
Porgy (pôr'gĭ) **and Bess**
Porpora (pōr'pô rä), **Niccolò** (nēk kô lô')
Poulenc (pōō läɴk'), **Francis** (*Fr.* fräɴ-sēs')
Pouplinière (pōō plē nyĕr'), **Alexandre Jean-Joseph le Riche de la** (ȧ lĕk-säɴ'drə zhäɴ zhō zĕf' lə rēsh' də lȧ)
Pour Bertha moi je soupire (pōōr bĕr tȧ' mwȧ zhə sōō pēr')
Pour les couvents c'est fini (pōōr lā kōō väɴ' sĕ fē nē')
Pour la Vierge (pōōr lȧ vyĕrzh')
Pourquoi me réveiller? (pōōr kwȧ' mə rā vĕ yā'?
Poveri fiori (pô'vĕ rē fyō'rē)
Pratella (prä tĕl'lä), **Francesco** (frän-chä'skō)
Précieuses ridicules, Les (lā prä syœz' rē dē kȳl')
Prelude: Vorspiel (fōr'shpēl')
Prenderò quel brunettino (prĕn dĕ rō' kwĕl brōō nät tē'nō)
Prendre le dessin d'un bijou (präɴ'drə lə də säɴ' dœn bē zhōō')
Presago il core della tua condanna (prä-sä'gō ēl kô'rä dĕl lä tōō'ä kôn dän'nä)
Près des remparts de Séville (prĕ dä räɴ pȧr' də sä vē'y)
Preussisches Märchen (prois'ish əs mĕr'-shən)
Prévost (prā vō'), **Abbé** (ȧ bā'); **Antoine François Prévost d'Exiles** (äɴ twȧn' fräɴ swȧ' prä vō' dĕg zēl')
Preziose Ridicole, Le (lā prä tsyō'sä rē-dĕ'kô lä)
Preziosilla (prä tsyō sēl'lä)
Prigioniero, Il (ēl prē jô nyĕ'rō)
Prima donna (prē'mə dŏn'ə; *It.* prē'mä dôn'nä)
Primo uomo (prē'mō wô'mō)
Prince Igor (ē'gər)
Princesse d'Auberge, La (lȧ präɴ sĕs' dō-bĕrzh')
Princess von Werdenberg (fôn vĕr'dən-bĕrk)
Printemps qui commence (präɴ täɴ' kē kô mäɴs')
Prinzivalle (prēn tsē väl'lä)
Prinzregenten (prĭnts'rä gĕn'tən) **Theater**
Prise de Troie, La (lā prēz də trwä')

Pritzko (prēts'kō)
Prodaná Nevěsta (prŏ'dä nä ně'vyěs tä)
Prokofiev (prŏ kôf'yĭf), Serge (Sěrzh)
Promesse de mon avenir (prŏ měs' də
mȏN nàv nēr')
Promessi sposi, I (ē prŏ mās'sē spŏ'zē)
Prometheus (prə mē'thĭ əs, -thōōs)
Bound
Prophète, Le (lə prŏ fět')
Protagonist, Der (dər prō tä'gō nĭst')
Prozess, Der (dər prō tsěs')
Prunières (prʏ nyěr'), Henri (äN rē')
Puccini (pōōt chē'nē), Giacomo (jä'kô-
mō)
Purcell (pûr'səl), Henry
Puritani, I (ē pōō rē tä'nē)
Pushkin (pōōsh'kĭn), Alexander
Pu-Tin-Pao (pōō tĭn pä'ō)
Pylades (pĭl'ə dēz)

Q

Quadrille (kwə drĭl', kə-); Quadrille de
contredanses (Fr. kà drē'y də kȏN trə-
däNs)
Qual cor tradisti (kwäl kȏr trä dē'stē)
Quand apparaissent les étoiles (käN tà-
pà rěs' lā zä twàl')
Quand du Seigneur le jour luira (käN
dʏ sě nyœr' lə zhōōr lʏē rà')
Quand le destin au milieu (käN lə děs-
tăN' ō mē lyœ'); Quando fanciulla
ancor l'avverso (kwän'dō fän chōōl'
lä äng kȏr' läv věr'sō)
Quand'ero paggio del Duca di Norfolck
(kwändě rō päd'jō děl dōō'kä dē
nȏr'fōk)
Quando le sere al placido (kwän'dō lā
sā'rä äl plä'chē dō)
Quando me'n vo' soletta (kwän'dō män
vô sô lät'tä)
Quando rapita in estasi (kwän'dō rä-
pē'tä ēn ě'stä zē)
Quanto è bella! (kwän'tō ě běl'lä!)
Quattro rusteghi, I (ē kwät'trō rōō'stä gē)
Queen of Shemakha (shě mä кнä')
Que fais-tu, blanche tourterelle (kə
fě'tʏ, bläNsh tōōr tə rěl')
Quel plaisir! quelle joie! (kěl plě zēr⁴!
kěl zhwä'!)
Questa notte (kwä'stä nôt'tä)
Questa o quella (kwä'stä ō kwěl'lä)

Questo e quel pezzo (kwä'stō ā kwěl
pět'tsō)
Questo sol è il soggiorno (kwä'stō sȏl ě
ēl sōd jȏr'nō)
Qui la voce sua soave (kwē lä vō'chä
sōō'ä sō ä'vä)
Quinault (kē nō'), Philippe (fē lēp')
Qui posa il fianco (kwē pŏ'sä ēl fyäng'-
kō)
Qui sola, vergin rosa (kwē sō'lä, věr'jēn
rŏ'zä)
Quo Vadis (kwō vä'dĭs)

R

Rabaud (rà bō'), Henri (äN rē')
Rabelais (răb'ə lä; Fr. rà blě'), François
(fräN swà')
Rachel (rä'chəl; Fr. rà shěl')
Rachel, quand du Seigneur (rà shěl',
käN dʏ sě nyœr')
Rachmaninoff (räкн mä'nĭ nôf), Sergei
(sěr gä')
Racine (rà sēn'), Jean Baptiste (zhäN
bà tēst')
Radames (rä'də mēz; It. rä'dä mās)
Radames, è deciso il tuo fato (rä'dä-
mās, ě dä chē'zō ēl tōō'ō fä'tō)
Rafaele (rä fä ě'lä)
Raimondo (rī mōn'dō)
Raisa (rä ē'sä), Rosa
Rákóczy (rä'kō tsē) March
Rambaldo (räm bäl'dō)
Rameau (rà mō'), Jean-Philippe (zhäN
fē lēp')
Ramerrez (rə měr'əz)
Ramfis (räm'fĭs)
Ramiro (rä mē'rō)
Ramon (rä mōn')
Ramuntcho (rə mŭn'chō)
Rangstrom (räng'strœm'), Ture (tōō'rě)
Ranz des vaches (räNts dä vàsh')
Rape of Lucretia (lōō krē'shə), The
Raskolnikoff (räs kôl'nĭ kôf; Rus. räs-
kôl'nyĭ kəf)
Rataplan (rà tà pläN')
Ratmir (rät mēr')
Räuber, Die (dē roi'bər)
Ravel (rà věl'), Maurice (mô rēs')
Rebikov (rä'byĭ kəf), Vladimir (vlä dyē'-
mĭr)
Recitative: Recitativo (rä'chē tä tē'vō)

ăct, āble, dâre, ärt; à, Fr. ami; ěbb, ēqual; ĭf, īce; кн, Ger. ach; N indicates nasal vowel;
ŏdd, ōver, ôrder; œ, Fr. feu, Ger. schön; bŏŏk, ōōze; ŭp, ūse, ûrgə; ʏ, Fr. ⋯, Ger. über;
zh, vision; ə, a in alone. (Full key on page 558)

Recondita armonia (rä kôn'dē tä är mô-nē'ä)

Re dell' abisso affrettati (rä děl lä bēs'sō äf frāt'tä tē)

Regina (rǐ jē'nə, -jǐ'-)

Regnava nel silenzio (rän nyä'vä něl sē lěn'tsyō)

Reich (rīкн)

Reichardt (rī'кнärt), Johann Friedrich (yō'hän frē'drǐкн)

Reichmann (rīкн'män), Theodor (Ger. tä'ō dôr)

Reine de Saba, La (lå rěn də så bá')

Reine Fiammette, La (lå rěn fyå mět')

Reiner (rī'nər), Fritz

Reiss (rīs), Albert

Re Lear (rä lēr')

Rembrandt van Rijn (rěm'bränt, Dutch rěm'bränt, vän rīn')

Remendado, Le (lə rä měn dä'dō)

Remigio (rä mē'кнyō)

Renard (rěn'ərd; Fr. rə når')

Renato (rě nä'tō)

Renaud (rə nō'), Maurice (mô rēs')

Répétition (rä pä tē syôn')

Residenztheater (rä zē děnts'tä ä'tər)

Respighi (rä spē'gē), Ottorino (ōt tô-rē'nō)

Reste, repose-toi (rěs'tə, rə pō'zə twá)

Retablo de Maese Pedro, El (ěl rä tä'-blō dä mä ä'sä pä'drō)

Rethberg (rěth'bûrg; Ger. rět'běrk), Elizabeth

Rêve, Le (lə rěv')

Reyer (rä yěr'), Ernest (Fr. ěr něst')

Rezia (rē'zǐ ə)

Rezniček (rěz'nē chěk), Emil (ā'mǐl)

Rheingold, Das (däs rīn'gôlt)

Riccardo (rēk kär'dō)

Richard Coeur de Lion (rē shår' kœr də lyôn')

Richepin (rēsh pǎn'), Jean (zhän)

Richter (rīкн'tər), Hans (häns)

Ricki (rǐk'ē)

Ricordi (rē kôr'dē) & Company

Rienzi (rǐ ěn'zǐ)

Rigaudon (rē gō dôn')

Rigoletto (rǐg ə lět'ō; It. rē gô lät'tō)

Rimsky-Korsakov (rǐm'skǐ kôr'sə kôf), Nicholai (nē kô lī')

Rinaldo (rǐ näl'dō)

Ring des Nibelungen, Der (dər rǐng' děs nē'bə lōong ən)

Rinuccini (rē nōot chē'nē), Ottavio (ōt-tä'vyō)

Rinuccio (rē nōot'chō)

Rise and Fall of Mahagonny (mə hǎg'-ə nǐ), The

Risurrezione, La (lä rē'sōōr rä tsyō'nä)

Ritorna vincitor! (rē tôr'nä vēn chē tôr'!)

Ritornello (rē tôr něl'lō)

Robert le Diable (rô běr' lə dyä'blə)

Rocca (rôk'kä), Lodovico (lō dō vē'kō)

Rocco rôk'kō)

Rodelinda (rǒd ə lǐn'də)

Roderigo (rō dě rē'gō)

Rodolfo (rō dôl'fō)

Rodrigo (rō drē'gō)

Rodrigue (rô drēg')

Roi de Lahore, Le (lə rwá də lá ôr'); The King of Lahore (lə hōr')

Roi d'Ys, Le (lə rwá dēs'); The King of Ys (ēs)

Roi d'Yvetot, Le (lə rwá dēv tō'); The King of Yvetot (ēv tō')

Roi l'a dit, Le (iə rwá lá dē')

Rolland (rô län'), Romain (rô mǎn')

Roller (rôl'ər), Alfred (Ger. äl'frět)

Romance d'Antonia (Fr. rô mäns' dän-tô nē á')

Romani (rô mä'nē), Felice (fě lē'chä)

Romeo (rō'mǐ ō) and Juliet (jōō'lǐ ət, jōō lǐ et'); Roméo et Juliette (rô mä ō' ä zhʏ lyět')

Romeo und Julia auf dem Dorfe (rō'mä ō ōōnt ū'lē ä ouf dəm dôr'fə)

Romerzählung (rōm'ěr tsä'lōong)

Ronde du postillon (rônd dʏ pôs tē yôn')

Rondine, La (lä rōn'dē nä)

Rosalinde (rō zä lǐn'də)

Rosario (rō sä'ryō)

Rosaura (rō zou'rä)

Roschana (rō shä'nə)

Rosenberg (rō'zən bûrg; Sw. rō'sən běr'y), Hilding (hǐl'dǐng)

Rosenkavalier, Der (dər rō'zən kä vä lēr')

Rosina (rō zē'nä)

Rosing (rō'zǐng), Vladimir (vlǎd'ǐ mǐr)

Rossi (rōs'sē), Gaetano (gä ä tä'nō)

Rossignol, Le (lə rô sē nyôl')

Rossi-Lemeni (rōs'sē lā'mě nē), Nicola (nē kô'lä)

Rossini (rô sē'nǐ; It. rōs sē'nē), Gioacchino Antonio (jō äk kē'nō än tô'nyō)

Rostand (rôs tän'), Edmond (ěd môn')

Rothier (rô tyä'), Léon (lä ôn')

Rothmüller (rōt'mʏl'ər), Marko

Rôtisserie de la Reine Pédauque, La (lä rô tē srē' də lá rěn pä dōk')

Rousseau (rōō sō'), Jean Jacques (zhän zhäk)

Roussel (roo sĕl'), Albert Charles (*Fr.* àl bĕr' shárl)

Rouvel (roo vĕl'), Baron

Roxane (rŏk săn')

Roze (rōz), Marie

Rozenn (rō zĕn')

Rubini (roo bē'nē), Giovanni Battista (jō vän'nē bät tē'stä)

Rubinstein (roo'bĭn stīn), Anton

Rudolf (roo'dŏlf), Max

Rühlmann (rɣl mȧn'), François (fräɴ swȧ')

Ruffo (roof'fō), Titta (tēt'tä)

Ruggero (rood jĕ'rō)

Ruggiero (rood jĕ'rō)

Ruiz (roo ēs')

Rusalka (roo säl'kə; *Czech* roo'säl kä)

Russalka (roo säl'kə)

Russlan (roos län') and Ludmilla (lood-mē'lə)

S

Saamschedine (säm shĕ dēn'), Princess

Sabata (sä'bä tä), Victor de (dä)

Sacchini (säk kē'nē), Antonio (än tô'nyō)

Sachs (zäks), Hans (häns)

Sachse (zäk'sə), Leopold (lē'ə pōld; *Ger.* lä'ō pôlt)

Sadko (sät kô')

St. Bris (săɴ brē'), Count de (də)

Saint Daniel et Saint Michel (săɴ dȧ nyĕl' ä săɴ mē shĕl')

Saint-Georges (săɴ zhôrzh'), Jules Henri Vernoy de (zhɣl äɴ rē' vĕr nwȧ' də)

Saint Julien l'Hospitalier (săɴ zhɣ lyăɴ' lôs pē tȧ lyä')

Saint-Pierre (săɴ pyĕr'), Jacques Henri Bernardin de (zhäk äɴ rē' bĕr nȧr dăɴ' də)

Saint-Saëns (săɴ säɴs'), Camille (kȧ mē'y)

Sakuntala (shə koon'tə lä)

Salambô (sä läm bō')

Salce! Salce! (säl'chä! säl'chä!)

Salieri (sä lyĕ'rē), Antonio (än tô'nyō)

Salignac (sȧ lē nyȧk'), Eustase Thomas (œs tȧz' tô mä')

Salle Favart (sȧl fȧ vȧr')

Salome (săl'ə mä, sə lō'mĭ)

Salomé (sȧ lô mä')

Salud (sä looтн')

Salut à France (sȧ lɣ'tȧ fräɴs')

Salut à toi, soleil (sȧ lɣ' tȧ twȧ', sô lĕ'y)

Salut! demeure chaste et pure (sȧ lɣ'! də mœr' shȧst ä pɣr')

Salut! tombeau! sombre et silencieux (sȧ lɣ'! tôɴ bō'! sôɴ'brä sē läɴ syœ')

Salvezza alla Francia (säl vät'tsä äl lä frän'chä)

Salzburg (sôlz'bûrg; *Ger.* zälts'boŏrk) Festival

Samoset (săm'ə sĕt)

Samson et Dalila (säɴ sôɴ' ä dȧ lē lȧ')

Samson, qu'as-tu fait du Dieu de tes pères (säɴ sôɴ', kȧ'tɣ fĕ dɣ dyœ də tä pĕr')

San Carlo (säng kär'lō) Opera

San Carlo (săn kär'lō) Opera Company

Sancho Panza (săn'chō păn'zə; *Sp.* säɴ'-chō pän'thä)

Sand (sănd; *Fr.* säɴd), George (jôrj; *Fr.* zhôrzh)

Sandys (săndz), Lady Marigold

Sanglot (säɴ glō')

Sante (sän'tä)

Santuzza (sän toot'tsä)

Sapho (săf'ō; *Fr.* sȧ fō')

Sarabande (săr'ə bănd)

Sarastro (zä räs'trō)

Sardou (sȧr doo'), Victorien (vēk tô ryăɴ')

Sarka (sär'kä)

Sarti (sär'tē), Giuseppe (joo zĕp'pä)

Satie (sȧ tē'), Erik (*Fr.* ä rēk')

Sauguet (sō gĕ'), Henri (äɴ rē')

Sauvée! Christ est ressuscité! (sō vä' krēst ĕ rĕ sɣ sē tä'!)

Sayao (sä youɴ'), Bidu (bē doo')

Sbriglia (zbrēl'lyä), Giovanni (jō vän'nē)

Scala, La (lä skä'lä)

Scalchi (skäl'kē), Sofia (*It.* sō fē'ä)

Scaria (skä'rē ä), Emil (*Ger.* ä'mĭl)

Scarlatti (skär lät'tē), Alessandro (ä läs-sän'drō)

Scarpia (skär'pyä)

Scena (shĕ'nä)

Scène des cheveux (sĕn dä shə vœ')

Scène infernale (sĕn ăɴ fĕr nȧl')

Schalk (shälk), Franz (fränts)

Schatz Walzer (shäts väl'tsər)

Schau her, das ist ein Taler (shou här'ɩ däs ĭst ĭn tä'lər)

Schaunard (shō nȧr')

Schauspieldirektor, Der (dər shou'shpēl'-dē rĕk'tôr)

Scheff (shĕf), Fritzi (frĭt'sĭ)

ăct, āble, dâre, ärt; à, *Fr.* ami; ĕbb, ēqual; ĭf, īce; кн, *Ger.* ach; ɴ indicates nasal vowel; ŏdd, ōver, ôrder; œ, *Fr.* feu, *Ger.* schön; boŏk, ooze; ŭp, ūse, ûrge; ɣ, *Fr.* tu, *Ger.* über; zh, vision; ə, a in alone. (Full key on page 558)

Scherz, List und Rache (shĕrts, lĭst ŏŏnt rä′кнə)

Schikaneder (shē kä nä′dər), **Johann Emanuel** (yō′hän ā mä′nōō ĕl)

Schiller (shĭl′ər), **Friedrich von** (frē′ drĭкн fôn)

Schillings (shĭl′ĭngs), **Max von** (*Ger.* mäks fôn)

Schipa (skē′pä), **Tito** (tē′tō)

Schlafe wohl! und mag Dich reuen (shlä′fə vōl′! ŏŏnt mäk dĭкн roi′ən)

Schlémil (shlä mēl′)

Schlusnus (shlōōs′nōōs), **Heinrich** (hīn′- rĭкн)

Schmedes (shmä′dəs), **Erik** (*Dan.*, *Ger.* ä′rĭk)

Schmelzlied (shmĕlts′lēt′)

Schmiedelied (shmē′də lēt′)

Schnorr von Carolsfeld (shnôr fôn kä′- rôls fĕlt), **Ludwig** (lōōt′vĭкн)

Schoeck (shœk), **Othmar** (ôt′mär)

Schoenberg (shœn′bûrg; *Ger.* shœn′bĕrk) **Arnold**

Schoen-René (shœn′ rə nä′), **Anna**

Schöffler (shœf′lər), **Paul** (*Ger.* poul)

Schorr (shôr), **Friedrich** (frē′drĭкн)

Schott (shôt), **Anton** (*Ger.* än′tōn)

Schotts Söhne (shôts′ zœ′nə)

Schubert (shōō′bərt), **Franz Peter** (fränts pä′tər)

Schröder-Devrient (shrœ′dər də vrēnt′), **Wilhelmine** (vĭl hĕl mē′nə)

Schubert (shōō′bərt), **Franz** (fränts)

Schuch (shōōкн), **Ernst von** (ĕrnst fôn)

Schuloper (shōōl′ō′pər)

Schumann (shōō′män), **Elisabeth**

Schumann-Heink (shōō′män hīngk′), **Ernestine** (*Ger.* ĕr nĕs tē′nə)

Schütz (shуts), **Heinrich** (hīn′rĭкн)

Schützendorf (shуt′sən dôrf), **Gustav** (gōōs′täf)

Schwanda der Dudelsackpfeifer (shvän′- dä dər dōō′dəl zäk′ pfī′fər)

Schwarzkopf (shvärts′kôpf), **Elisabeth**

Schweigsame Frau, Die (dē shvīk′zäm ə frou)

Schweig' und tanze! (shvīk′ ŏŏnt tän′ tsə!)

Sciarrone (shär rō′nä)

Scindia (sĭn′dĭ ə)

Scintille, diamant (sǎn tē′y, dyȧ män′)

Scotti (skôt′tē), **Antonio** (än tô′nyō)

Scribe (skrēb), **August-Eugène** (ō gуst′- œ zhĕn′)

Scuoti quella fronda di ciliegio (skwȯ′tē kwĕl lä frōn′dä dē chē lyĕ′jō)

Sebastiano (sä bäs tyä′nō)

Secondate, aurette amiche (sä kôn dä′tä ou rät′tä ä mē′kä)

Sedaine (sə dĕn′), **Michel Jean** (mē shĕl′ zhäɴ)

Seefried (zä′frēt), **Irmgard** (ĭrm′gärt)

Segreto di Susanna, Il (ēl sä grä′tō dē sōō zän′nä)

Seguidilla (sä gē dē′lyä)

Seidl (zī′dəl), **Anton** (*Ger.* än′tōn)

Seien wir wieder gut (zī′ən vēr vē′dər gōōt)

Se il mio nome saper (sä ēl mē′ō nȯ′mä sä pär′)

Seit er von dir geschieden (zīt ĕr fôn dēr gə shē′dən)

Selig, wie die Sonne (zä′lĭкн, vē dē zôn′ə)

Selika (sĕ lē′kä)

Selim (zä′lĭm)

Sembrich (sĕm′brĭk; *Ger.* zĕm′brĭкн), **Marcella**

Semiramide (sĕ mē rä′mē dä)

Semiseria (sĕ mē sĕ′ryä)

Sempre libera (sĕm′prä lē′bĕ rä)

Senesino (sĕ nä sē′nō)

Senta (sĕn′tə); *Ger.* zĕn′tä)

Serafin (sĕ rä fēn′), **Tullio** (tōōl′lyō)

Serena (sə rē′nə; *It.* sĕ rä′nä)

Serenade: Serenata (sĕ rĕ nä′tä)

Serge (sûrj, sĕrzh)

Serov (sĕ′rəf), **Alexander**

Serpina (sĕr pē′nä)

Serse (sĕr′sä); **Xerxes** (zúrk′sēs)

Serva padrona, La (lä sĕr′vä pä drō′nä)

Setti (sĕt′tē), **Giulio** (jōō′lyō)

Se vuol ballare (sä vwôl bäl lä′rä)

Shaklovity (shäk lȯ vē′tĭ)

Shanewis (shǎn′ə wĭs)

Sherasmin (shĕ räz′mĭn)

Shostakovich (shȯs tə kȯ′vĭch), **Dmitri** (də mē′trĭ)

Shuisky (shōō′ĭ skĭ)

Siberia (sĭ bĭr′ĭ ə)

Siciliana (sĭ sĭl′ĭ ä′nə; *It.* sē chē lyä′nä)

Sicilian (sĭ sĭl′yən) **Vespers, The**

Siebel (sē′bəl)

Siegfried (sēg′frēd; *Ger.* zēk′frēt)

Sieglinde (zēk′lĭn′də)

Siegmund (zēk′mŏŏnt)

Siepi (syĕ pē), **Cesare** (chĕ′zä rä)

Si, fui soldato (sē, fōō′ē sôl dä′tō)

Sigismund (sĭj′ĭs mənd, sĭg′—)

Signor Bruschino, Il (ēl sēn nyōr′ brōō- skē′nō)

Sigurd (sĭg′ərd)

Si; io penso alla tortura (sē; ē'ō pĕn'sō äl lä tôr tōo'rä)

Si la rigueur et la vengeance (sē là rē gər' ā là vän zhäns')

Silva (sēl'vä), Don Ruy Gomez de (dôn rōo'ĭ gō'mĕth dä)

Silvio (sēl'vyō)

Si, me ne vo, Contessa (sē, mā nā vô', kôn tās'sä)

Simon Boccanegra (sē mōn' bōk kä nä'- grä)

Simone (sē mō'nä)

Singher (sän gĕr'), Martial (Fr. màr syàl')

Singspiel (zĭng'shpēl')

Si, pel ciel marmoreo giuro! (sē, pāl chĕl' mär mô'rä ō jōo'rō!)

Si può? (sē pwô'?)

Sirval (sēr vàl'), Marchioness de (də)

Sister Angelica (än jĕl'ĭ kə)

Sita (sē'tə)

Si, vendetta (sē, vĕn dāt'tä)

Six, Les (lä sēs')

Slezak (slä'zäk), Leo

Smallens (smôl'ənz), Alexander

Smanie implacabili (zmä'nyä ēm plä kä'- bē lē)

Smetana (smĕ'tä nä), Bedřich (bĕd'rĭk; Czech bĕ'drzhēκн)

Snow Maiden, The: Snegurochka (snyĭ- gōo'rəch kə)

So anch' io la virtù magica (sô' ängk ē'ō lä vēr tōo' mä'jē kä)

Sobinin (sŏ bē'nyĭn)

Socrate (sô kràt')

Sodero (sō dä'rō), Cesare (chĕ'zä rä)

Sofia (sō fē'ä)

Sogno soave e casto (sōn'nyō sō ä'vä ā kä'stō)

Sola, perduta, abbandonata (sō'lä, pĕr- dōo'tä, äb bän'dô nä'tä)

Solenne in quest' ora (sô lĕn'nä ēng kwäst ō'rä)

Sommeil (sô mĕ'y)

Son imbrogliato io già (sōn ēm brō lyä'tō ē'ō jä)

Sonnambula, La (lä sôn näm'bōo lä)

Sonnez, clairons, que vos chants de vic- toire (sô nä', klĕ rôn', kə vō shän də vēk twàr')

Sontag (zôn'täk), Henriette (hĕn rĭ ĕt'ə)

Sonzogno (sôn tsōn'nyō), Edoardo (ā dō- är'dō)

Sophie (sō'fĭ; Fr. sô fē'; Ger. zō fē'ə, zō fē')

Sophocles (sŏf'ə klēz)

Soprano acuto (sō prä'nō ä kɔ̄o'tō)

Soprano Falcon (fàl kôn')

Soprano leggiero (läd jĕ'rō)

Soprano sfogato (sfō gä'tō)

Sorgeva il dì del bosco in seno (sôr jä'vä ēl dē' dĕl bô'skō ēn sä'nō)

Sorte amica (sôr'tä ä mē'kä)

Sortita (sôr tē'tä)

Sotto voce (sōt'tō vō'chä)

Soubrette (sōo brĕt')

Souviens-toi de ton serviteur (sōo vyän'- twà də tôn sĕr vē tœr')

Spalanzani (spä län dzä'nē)

Sparafucile (spä'rä fōo chē'lä)

Spargi d'amaro pianto (spär'jē dä mä'rō pyän'tō)

Specht (shpĕκнt), Richard (Ger. rĭκн'ärt)

Spinelloccio (spē nĕl lōt'chō)

Spira sul mare (spē'rä sōol mä'rä)

Spirito gentil (spē'rē tō jĕn tēl')

Splendon più belle (splĕn'dôn pū bĕl'lä)

Spohr (shpōr), Ludwig (lōōt'vĭκн)

Spoletta (spô lät'tä)

Spontini (spôn tē'nē), Gasparo (gä'spä rō)

Sprechgesang (shprĕκн'gə zäng')

Sprechstimme (shprĕκн'shtĭm'ə)

Springer (sprĭng'ər)

Squilli, echeggi la tromba guerriera (skwēl' lē, ā kàd'jē lä trōm'bä gwĕr ryĕ'rä)

Stabile (stä'bē lä), Mariano (mä ryä'nō)

Städtische Oper (shtä'tĭsh ə ō'pər)

Standestheater (shtän'dəs tä ä'tər); Sta- voské Divadlo (stä'vô skĕ dē'vä dlô); Nostitz (nôs'tĭts) Theater

Starke Scheite schichtet mir dort (shtär'kə shī'tə shīκн'tət mēr dôrt)

Stefan (stĕ'fän), Paul

Steffani (stĕf'fä nē), Agostino (ä gō- stĕ'nō)

Stein (stīn), Gertrude

Stella (stĕl'ə) It. stĕl'lä)

Stendhal (stĕn'däl; Fr. stän dàl')

Stephana (stĕ fä'nə)

Stephano (stĕf'ə nō)

Steuermann! Lass die Wacht! (shtoi'ər- män! läs dē väκнt!)

Steuermannslied (shtoi'ər mäns lēt')

Stevens, Risë (rē'sə)

Stewa (stä'vä)

Stiedry (stĕ'drĭ), Fritz

ăct, āble, dâre, ärt; à, Fr. ami; ĕbb. ēqual; ĭf, īce; κн, Ger. ach; ν indicates nasal vowel; ŏdd, ōver, ôrder; œ, Fr. feu, Ger. schön; bŏŏk, ōōze; ŭp, ūse, ûrge; ɣ, Fr. tu, Ger. über; zh, vision; ə, a in alone. (Full key on page 558)

Stile rappresentativo (stē'lä räp'prä zĕn'-tä tē'vō)

Stizzoso mio stizzoso (stĕt tsō'sō mē'ō stĕt tso'sō)

Stoltz (stôlts), Rosine (rō zēn')

Stolz (shtôlts), Teresa (tä rä'zä)

Stolzing (shtôl'tsĭng), Walther von (väl'tər fôn)

Stradella (strä dĕl'lä), Alessandro (ä lässän'drō)

Strakosch (strä'kôsh), Maurice (mô rēs')

Strauss (strous; Ger. shtrous), Johann (yō'hän)

Strauss, Richard (Ger. rĭкн'ärt)

Stravinsky (strə vĭn'skĭ), Igor (ē'gər)

Streltsy (strĕl'tsĭ; Rus. strĕl tsĭ'), The

Strepponi (sträp pō'nē), Giuseppina (jōōzäp pē'nä)

Stretta (strāt'tä)

Stretti insiem tutti tre (strāt'tē ēn syĕm' tōōt'tē trä)

Stride la vampa (strē'dä lä väm'pä)

Strindberg (strĭnd'bûrg; Sw. strĭn'bĕr'y), August (Sw. ou'gōōst)

Stromminger (shtrôm'ĭng ər)

Stueckgold (shtʏk'gôlt'), Grete (grā'tə)

Styrienne (stē rē ĕn')

Sucher (zōō'кнər) Rosa

Su! del Nilo al sacro lido! (sōō! dĕl nē'lō äl sä'krō lē'do!)

Suicidio! (sōō ē chē'dyō!)

Sukarev (sōō kä rĕf'), Olga

Sulamith (sōō'lə mĭth)

Sulla vetta tu del monte (sōōl lä vät'tä tōō dĕl mōn'tä)

Sulpizio (sōōl pē'tsyō)

Summ' und brumm' (zōŏm' ŏŏnt brŏŏm')

Suor Angelica (swôr än jĕ'lē kä)

Suore, che riposate (swô'rä, kä rē pōsä'tä)

Supervia (sōō pĕr'vyä), Conchita (kônchē'tä)

Suppé (zŏŏp'ä), Frantz von (fränts fôn)

Sur mes genoux, fils du soleil (sʏr mä zhə nōō', fēs dʏ sô lĕ'y)

Susanna (sōō zän'nä)

Susanin (sōō sä'nyĭn), Ivan (ē vän')

Su, su, marinar (sōō, sōō, mä rē när')

Sutermeister (zōō'tər mĭ'stər), Heinrich (hīn'rĭкн)

Suzel (sōō zĕl')

Suzuki (sōō zōō'kē)

Svanholm (svän'hôlm), Set (sĕt)

Sviestosar (svyĕ'tə zär)

Swarthout (swôr'thout), Gladys

Synge (sĭng), John Millington

Szell (sĕl), George

Szenkar (sĕng'kär), Eugen (oi'gän, oigän')

Szymanowski (shĭ mä nôf'skē), Karol (kä'rôl)

T

Tabarro, Il (ēl tä bär'rō)

Tacea la notte placida (tä chä'ä lä nôt'tä plä'chē dä)

Taddeo (täd dĕ'ō)

Tagliavini (täl lyä vō'nē), Ferruccio (fĕrrōōt'chō)

Tamagno (tä män'nyō), Francesco (fränchä'skō)

Tambourin (täɴ bōō răɴ')

Tamburini (täm bōō rē'nē), Antonio (äntô'nyō)

Tamino (tä mē'nō)

T'amo, si, t'amo, e in lagrime (tä'mō, sē, tä'mō, ā ēn lä'grē mä)

Tancredi (täng krä'dē)

Tannhäuser (tän'hoi'zər)

Taras Bulba (tä räs' bōōl'bə)

Tartuffe (tár tʏf')

Tasso (täs'sō), Torquato (tôr kwä'tō)

Tatiana (tä tyä'nə)

Tauber (tou'bər), Richard (Ger. rĭкн'ärt)

Tchaikovsky (chĭ kôf'skĭ), Peter Ilyich (ēl yēch')

Tcherepnin (chĕ rĕp nēn'), Alexander

Tcherepnin, Nicolai (nē kŏ lī')

Tcherevik (chĕ rĕ vēk')

Tchernomor (chĕr nŏ môr')

Teatro alla Scala (tä ä'trō äl lä skä'lä)

Teatro Costanzi (tä ä'trō kō stän'tsē)

Teatro della Pergola (tä ä'trō dĕl lä pĕr'gô lä)

Teatro di San Cassiano (tä ä'trō dē säng käs syä'nö)

Teatro la Fenice (tä ä'trō lä fĕ nē'chä)

Teatro Reale (tä ä'trō rä ä'lä)

Teatro San Carlo (tä ä'trō säng kär'lō)

Tebaldi (tä bäl'dē), Renata (rĕ nä'tä)

Telramund (tĕl'rä mŏŏnt), Frederick of

Telva (tĕl'və) Marion

Templer und die Jüdin, Der (dər tĕm'plər ŏŏnt dē yʏ'dĭn)

Teresa (tä rä'zä)

Ternina (tĕr nē'nä), Milka (mēl'kä)

Terzetto delle maschere (tĕr tsät'tō dĕl lä mä'skĕ tä)

Te souvient-il du lumineux voyage (tə sōō vyăn'tēl dʏ lʏ mē nœ'vyá yàzh')

Tetrazzini (tā trät tsē'nē), Luisa (lo͞o ē'zä)
Teyte (tāt), Maggie
Thaïs (tä ēs', tä'ēs)
Thanatos (thăn'ə tŏs)
Theater-an-der-Wien (tä ä'tər än dər vēn')
Théâtre de la Monnaie (tä ä'trə də lä mô nĕ')
Théâtre de la Spectacle (tä ä'trə də lä spĕk tá'klə)
Théâtre Lyrique (tä ä'trə lē rēk')
Thélaïre (tä lä ēr')
Thill (tēl), Georges (zhôrzh)
Thoas (thō'əs)
Thomas (tô mä'), Ambroise (äɴ brwàz')
Thomas (tŏm'əs), Theodore
Thorborg (thôr'bôrg; Sw. to͞or'bôr'y), Kerstin (kûr'stĭn; Sw. chĕr'stĭn)
Tiefland (tēf'länt')
Tietjen (tēt'yən), Heinz (hīnts)
Timur (tē mo͞or')
Ti rincora, amæta figlia (tē rēng kô'rä, ä mä'tä fēl'lyä)
Tisbe (tē'zbä)
Titania (tĭ tā'nĭ ə)
Titurel (tē'to͞o rĕl)
Titus (tī'təs)
Toch (tŏk), Ernst (ûrnst, ĕrnst)
T'odio casa dorata (tô'dyō kä'sä dô rä'tä)
Tokatyan (tô kät'yän), Armand (àr mäɴ')
Tolstoy (tŏl stoi'), Leo
Tommaso (tō mä'sō)
Tonio (tô'nyō)
Torquemada (tôr kä mä'ᴛʜä)
Tosca (tŏs'kə; It. tô'skä)
Toscanini (tŏs kə nē'nĭ; It. tō skä nē'nē), Arturo (àr to͞o'rō)
Tote Stadt, Die (dē tō'tə shtät)
Tourel (to͞o rĕl'), Jennie
Tous les trois réunis (to͞o lä trwä rä ʏ nē')
Traft ihr das Schiff (träft' ēr däs shĭf')
Traubel (trou'bəl), Helen
Traurigkeit ward mir zum Lose (trou'-rĭᴋʜ kīt värt mēr tso͞om lō'zə)
Traviata, La (lä trä vyä'tä)
Tra voi, belle (trä vō'ē, bĕl'lä)
Treulich geführt (troi'lĭᴋʜ gə fʏrt')
Trinke Liebchen, trinke schnell (trĭng'kə lēp'shən, trĭng'kə shnĕl')
Trionfi (trē ōn'fē)

Trionfo di Afrodite (trē ōn'fō dē ä frō-dē'tä); The Triumph of Aphrodite (äf rə dī'tĭ)
Triquet (trē kĕ')
Tristan und Isolde (trĭs'tän ŏŏnt ē zôl'də)
Tristes apprêts (trēs'tə zà prĕ')
Tristram (trĭs'trəm), Sir
Trittico, Il (ēl trēt'tē kō)
Troilus (troi'ləs, trō'ĭ ləs) and Cressida (krĕs'ĭ də)
Trompeter von Säkkingen, Der (dər trôm-pā'tər fôn zĕk'ĭng ən); The Trumpeter of Säkkingen (zĕk'ĭng ən)
Trouble in Tahiti (tä hē'tē, tī'tē)
Trovatore, Il (ēl trō vä tō'rä)
Troyens, Les (lä trwá yäɴ')
Troyens à Carthage, Les (lä trwá yäɴ' zà kár tázh')
Tu che la terra adora (to͞o kä lä tĕr'rä ä dō'rä)
Tu m'as donné le plus doux rêve (tʏ má dô nā' lə plʏ do͞o rĕv')
Turandot (to͞o rän dôt')
Turgeniev (to͞or gä'nyĭf), Ivan (ē vän')
Turiddu (to͞o rēd'do͞o)
Tutta su me ti sposa (to͞ot'tä so͞o mä tē spô'zä)
Tutte le feste (to͞ot'tä lä fĕ'stä)
Tutti i fior (to͞ot'tē ē fyôr')
Tutto nel mondo è burla (to͞ot'tō nĕl mōn'dō ĕ bo͞or'lä)
Tutto tace (to͞ot'tō tä'chä)
Tu, tu piccolo Iddio! (to͞o, to͞o pēk'kô lō ēd dē'ō!)
Two Foscari (fŏs'kə rē), The
Tybalt (tĭb'əlt)
Tyl (tēl), Father

U

Ubaldo (o͞o bäl'dō)
Udite, udite, o rustici (o͞o dē'tä, o͞o dē'tä, ō ro͞o'stē chē)
Ulrica (o͞ol rē'kä)
Una cosa rara (o͞o'nä kô'sä rä'rä)
Una donna a quindici anni (o͞o'nä dôn'nä ä kwēn'dē chē än'nē)
Una furtiva lagrima (o͞o'nä fo͞or'tē vä lä'-grē mä)
Una macchia è qui tuttora (o͞o'nä mäk'-kyä ĕ kwē to͞ot tō'rä)

ăct, āble, dâre, ärt; à, Fr. ami; ĕbb, ēqual; ĭf, īce; ᴋʜ. Ger. ach; ɴ indicates nasal vowel; ŏdd, ōver, ôrder; œ, Fr. feu, Ger. schön; bo͝ok, o͞oze; ŭp, ūse, ûrgə; ʏ, Fr. tu, Ger. über; zh, vision; ə, a in alone. (Full key on page 558)

Un' aura amorosa (ōon ou'rä ä mô rō'sä)

Una vergine, un angiol di Dio (ōō'nä věr'jě nä, ōon än'jōl dē dē'ō)

Una voce poco fa (ōō'nä vō'chä pô'kō fä')

Un bel dì vedremo (ōōm běl dē' vä drä'mō)

Un dì all' azzurro spazio (ōon dē' äl läd-dzōōr'rō spä'tsyō)

Un dì felice (ōon dē' fĕ lē'chä)

Undine (ŏŏn dē'nə)

Und ob die Wolke sie verhülle (ŏŏnt ôp dē vôl'kə zē fər hYl'ə)

Une fièvre brûlante (Yn fyĕ'vrə brY länt')

Une nuit de Cléopâtre (Yn nYē də klä ô-pä'trə)

Une poupée aux yeux d'émail (Yn pōō pä' ō zyœ dä má'y)

Une voix mystérieuse (Yn vwä mēs tä-ryœz')

Unger (ŏŏng'ər), Caroline

Unis dès la plus tendre enfance (Y nē' dĕ lá plY tän'dräN fäNs')

Uragano, L' (lōō rä gä'nō)

Urbain (Yr bǎN')

Urban (ûr'bən), Joseph

Urlus (ōōr'lōōs), Jacques (zhäk)

Ursuleac (ōōr sōō'lyäts), Viorica (vē ō-rē'kä)

V

Va, crudele (vä', krōō dĕ'lä)

Vainement, ma bien aimée (věn mäN', má byǎN ně mä')

Vaisseau fantôme, Le (lə vě sō' fäN tōm')

Va! laisse-les couler (vå! lĕ'sə lä kōō lä')

Valentine (väl'ən tīn; Fr., woman's name, vå läN tēn')

Valery (väl'ə rǐ), Violetta (vī ə lĕt'ə; It. vyô lät'tä)

Valhalla (väl hǎl'ə)

Valkyrie (väl kǐr'ǐ,—kī'rǐ, väl'kǐr ǐ), The

Valleria (və lǐr'ǐ ə), Alwina (äl wī'nə)

Valzacchi (väl tsäk'kē)

Vampyr, Der (dər väm'pēr)

Van Dyck (văn dīk'; Flemish vän dīk'), Ernest (Flemish ěr něst')

Vanna (vän'nä), Marco

Vanne, disse (vän'nä, dēs'sä)

Van Rooy (vän rō'ǐ), Anton (Dutch än'tôn)

Vanya (vä'nyə)

Van Zandt (văn zănt'), Marie

Va, pensiero sull' ali dorate (vä', pěn-syĕ'rō sōōl lä'lē dô rä'tä)

Varnay (vär'nī), Astrid (äs'trǐd)

Vasco da Gama (väs'kō də gä'mə)

Vassilenko (vä sǐ lěng'kō), Sergei (sĕr gä')

Vassili (vä sē'lē)

Vedi! le fosche notturne spoglie (vä'dē! lä fō'skä nōt tōōr'nä spôl'lyä)

Vedrai, carino (vä drī', kä rē'nō)

Vedrommi intorno (vä drōm'mē ēn tōr'nō)

Venusberg (vē'nəs bûrg) Music

Vénus d'Ille, La (là vä nYs' dēl')

Vêpres Siciliennes, Les (lä vě'prə sē sē-lyěn')

Verdi (věr'dē), Giuseppe (jōō zěp'pä)

Verismo (vě rēz'mō)

Verranno a te sull' aure (věr rän'nō ä tä sōōl lou'rä)

Versiegelt (fər zē'gəlt)

Vespone (vä spō'nä)

Vespri Siciliani, I (ē vě'sprē sē chē lyä'nē)

Vestale, La (lä vä stä'lä)

Vesti la giubba (vě'stē lä jōōb'bä)

Viardot-García (vyár dō' gär thē'ä), Pauline

Via resti servita (vē'ä rě'stē sěr vē'tä)

Vicino a te s'acqueta (vē chē'nō ä tä säk-kwē'tä)

Vida breve, La (lä vē'thä brä'vä)

Viene la sera (vyě'nä lä sä'rä)

Vieni, amor mio (vyě'nē, ä mōr' mē'ō)

Vienna State Opera: Staatsoper (shtäts'-ō'pər)

Viens ici, ne reste pas au bord du clair de lune (vyǎN zē sē', nə rěs'tə pä zō bôr dY klěr də lYn')

Vie Parisienne, La (là vē på rē zyěn')

Vigny (vē nyē'), Alfred de (Fr. àl frěd' də)

Villi, Le (lä vēl'lē)

Villon (vē yôN'), François (fräN swà')

Vinay (vē nī'), Ramón (rä mōn')

Vincent (Fr. văN säN')

Vin ou bière (văN ōō byěr')

Violanda (vyô län'dä)

Violetta (vī ə lĕt'ə; It. vyô lät'tä)

Vi ravviso (vē räv vē'zō)

Vision fugitive (vē zyôN' fY zhē tēv')

Vissi d'arte (vēs'sē där'tä)

Vitellius (vǐ těl'ǐ əs)

Viva il vino (vē'vä ēl vē'nō)

Vivaldi (vē väl'dē), Antonio (än tô'nyō)

Viva Pagliaccio (vē'vä päl lyät'chō)

Vladimir (vlä dyē'mǐr)

Voce di donna (vō'chä dē dôn'nä)

Voce di gola (vō'chä dē gō'lä)

Voce velata (vō'chä vě lä'tä)

Vogl (fō'gəl), Heinrich (hīn'rǐKH)

Voi, che sapete (voi, kä sä pä'tä)

Voici ce qu'il écrit (vwä sē' sə kěl ā krē')

Voici le printemps, nous portant des fleurs (vwä sē' lə prǎN täN', nōō pôr täN' dä flœr')

Voici ma misère, hélas! (vwá sē' má mē-zĕr', ā läs'!)

Voilà donc la terrible cité (vwá lá' dôn lá tĕ rē'blə sē tā')

Voi lo sapete (voi lō sä pā'tā)

Volkhova (vŏl'кнə və)

Volpone (vŏl pō'nĭ)

Voltaire (vŏl târ')

Vorspiel (fōr'shpēl')

Vous ne savez pas (vo͞o nə sá vä' pä')

Vous qui du Dieu vivant outragez la puissance (vo͞o kē dɣ dyœ vē vän' o͞o trá-zha' lá pȳē säns')

Vous qui faites l'endormie (vo͞o kē fĕt län dôr mē')

Voyons, Manon, plus de chimères (vwá-yôn', má nôn', plɣ də shē mĕr')

Vulcan (vŭl'kən)

W

Wach' auf, es nahet gen den Tag (väкн ouf', ĕs nä'ət gĕn dən täk')

Wagner (väg'nər)

Wagner, Cosima (Ger. kō'zē mä)

Wagner, Johanna (yō hän'ä)

Wagner, Richard (Ger. rĭкн'ärt)

Wagner, Siegfried (Ger. zēk'frēt)

Wahn! Wahn! Überall Wahn! (vän! vän! ɣ bər äl' vän!)

Waldner (vält'nər), Count

Waldweben (vält'vä'bən)

Walküre, Die (dē väl kɣ'rə, väl'kɣ' rə)

Walkürenritt (väl kɣ' rən rit', väl'kɣ'—)

Wallenstcin (wŏl'ən stīn; Ger. väl'ən shtīn)

Wallerstein (väl'ər shtīn), Lothar (lō'tär)

Wally, La (lä väl lē')

Walpurgis (väl po͞or'gĭs) Night

Walter (väl'tər), Bruno

Walther (väl'tər)

Walton, Lord Gautier (gō tyä')

Waltraute (väl'tou'tə)

War es so schmählich? (vär ĕs zō shmä'-lĭкн?)

Wartburg (värt'bo͞ork)

Was bluten muss? (väs blo͞o'tən mo͝os?)

Was gleicht wohl auf Erden (väs glīкнt' vŏl ouf ĕr'dən)

Water Carrier, The: Der Wasserträger (dər väs'ər trä'gər)

Weber (vä'bər), Carl Maria von (kärl ɾɑä rē'ä fôn)

Wedekind (vä'də kĭnt), Frank (Ger. frängk)

Weh! Nun ist all unser Glück dahin! (vä! no͞on ĭst äl' o͞on zər glɣk' dä hĭn'!)

Weiche, Wotan, weiche! (vī'кнə, vō'tän, vī'кнə!)

Weill (wīl, Ger. vīl), Kurt (kûrt, Ger. ko͞ort)

Weinberger (vīn'bĕr'gər), Jaromir (yä'-rô mēr)

Weingartner (vīn'gärt'nər), Felix (Ger. fä'lĭks)

Weis (vīs), Karel (kä'rĕl)

Welche Wonne, welche Lust (vĕl'кнə vôn'ə, vĕl'кнə lo͞ost')

Welitch, Welitsch (vĕ'lēch), Ljuba (lū'bä)

Wellesz (vĕl'ĕs), Egon (Ger. ā'gôn)

Wellgunde (vĕl'go͞on'də)

Wenn der Freude Thränen fliessen (vĕn dər froi'də trä'nən flē'sən)

Wenzel (vĕn'tsəl)

Werdenberg (vĕr'dən bĕrk)

Wer ein holdes Weib errungen (vĕr ɪn hôl'dəs vīp' ĕr ro͝ong'ən)

Wer ein Liebchen hat gefunden (vĕr ɪn lēp'shən hät gə fo͞on'dən)

Werfel (vĕr'fəl), Franz (fränts)

Werner Kirchhofer (vĕr'nər kĭrкн'hō'fər)

Werther (vĕr'tər; Fr. vĕr tĕr')

Widerspänstigen Zähmung, Der (dər vē'-dər shpĕn'stɪ gən tsä'mo͞ong)

Wieland (vē'länt), Christoph Martin

Wie oft in Meeres tiefsten Schlund (vē ôft ɪn mā'rəs tēf'stən shlo͞ont')

Wilhelm Meister (vĭl'hĕlm mī'stər)

Wilhelm Tell (vĭl'hĕlm tel')

William Tell: Guillaume Tell (gē yōm' tel')

Willst jenes Tag's du nicht dich mehr entsinnen (vĭlst yä'nəs täks do͞o nĭкнt dĭкн mär ĕnt zĭn'ən)

Win-San-Luy (wĭn săn lo͞o'ĭ)

Win-Shee (wĭn shē')

Woglinde (vōk'lĭn'də)

Wolf (vôlf), Hugo (Ger. ho͞o'gō)

Wolff (vôlf), Albert Louis (Fr. ál bĕr' lwē')

Wolf-Ferrari (vôlf' fĕr rä'rē), Ermanno (ĕr män'nō)

Wolfram von Eschenbach (vôlf'räm fôn ĕsh'ən bäкн)

Wolfserzählung (vôlfs'ĕr tsä'lo͞ong)

Worms (vôrms), Carlo

Wotan (wō'tən; Ger. vō'tän)

ăct, āble, dâre, ärt; á, Fr. ami; ĕbb, ēqual; ĭf, īce; кн, Ger. ach; ɴ indicates nasal vowel; ŏdd, ōver, ôrder; œ, Fr. feu, Ger. schön; bo͝ok, o͞oze; ŭp, ūse, ûrge; ɣ, Fr. tu, Ger. über; ɀh, vision; ə, a in alone. (Full key on page 558)

Wowkle (wou'kəl)
Wozzeck (vôt'sĕk)
Wreckers, The: Les Naufrageurs (lä nō-
fră zhœr')

X

Xenia (zēn'yə, zē'nĭ ə)
Xerxes (zûrk'sēz)

Y

Yamadori (yä mä dō'rē)
Yeats (yāts), William Butler
Yniold (ē nyôl')

Z

Zacharias (zăk ə rī'əs)
Zamiel (zā'mĭ əl)
Zampa (zäm'pə; *Fr.* zän pà')
Zandonai (dzän dô nä'ē), Riccardo (*It.*
rēk kär'dō)
Zaretski (zä rĕts'kĭ)
Zar und Zimmermann (tsär' ŏŏnt tsĭm'-
ər män)
Zarzuela (thär thwä'lä, sär swä'lä)

Zauberflöte, Die (dē tsou'bər flœ'tə)
Zauberoper (tsou'bər ō'pər)
Zaza (zä zä')
Zeffiretti lusinghieri (dzĕf fē rät'tē lōō-
zēng gyĕ'rē)
Zeitkunst (tsīt'kŏŏnst')
Zemlinsky (zĕm lĭn'skĭ), Alexander
Zenatello (tsĕ nä tĕl'lō), Giovanni
vän'nē)
Zeno (dzä'nō), Apostolo (ä pôs'tô lō)
Zerbinetta (tsèr bē nĕt'ä)
Zerlina (zĕr lē'nä, dzĕr-)
Ziegler (zē'glər), Edward
Zigeunerbaron, Der (dər tsē goi'nər bä-
rōn')
Zita (tsē'tä)
Zitti, zitti (tsēt'tē, tsēt'tē)
Zola (zō'lə, zō lä'; *Fr.* zô lä'), Emile (ā mēl')
Zuane (tsōō ä'nä)
Zukunftsart (tsōō'kŏŏnfts ärt): Zukunfts-
musik (tsōō'kŏŏnfts mōō zēk')
Zum Leiden bin ich auserkoren (tsōōm
lī'dən bĭn ĬKH ous ər kō'rən)
Zum letzten Liebesmahle (tsōōm lĕts'tən
lē'bəs mä'lə)
Zuniga (zōō nē'gə)
Zurga (zŏŏr'gə)